Plant Regeneration from Seeds
A Global Warming Perspective

Plant Regeneration from Seeds
A Global Warming Perspective

Edited by

Carol C. Baskin
Department of Biology, University of Kentucky,
Lexington, KY, United States
Department of Plant and Soil Sciences, University of Kentucky,
Lexington, KY, United States

Jerry M. Baskin
Department of Biology, University of Kentucky,
Lexington, KY, United States

Academic Press is an imprint of Elsevier
125 London Wall, London EC2Y 5AS, United Kingdom
525 B Street, Suite 1650, San Diego, CA 92101, United States
50 Hampshire Street, 5th Floor, Cambridge, MA 02139, United States
The Boulevard, Langford Lane, Kidlington, Oxford OX5 1GB, United Kingdom

Copyright © 2022 Elsevier Inc. All rights reserved.

No part of this publication may be reproduced or transmitted in any form or by any means, electronic or mechanical, including photocopying, recording, or any information storage and retrieval system, without permission in writing from the publisher. Details on how to seek permission, further information about the Publisher's permissions policies and our arrangements with organizations such as the Copyright Clearance Center and the Copyright Licensing Agency, can be found at our website: www.elsevier.com/permissions.

This book and the individual contributions contained in it are protected under copyright by the Publisher (other than as may be noted herein).

Notices

Knowledge and best practice in this field are constantly changing. As new research and experience broaden our understanding, changes in research methods, professional practices, or medical treatment may become necessary.

Practitioners and researchers must always rely on their own experience and knowledge in evaluating and using any information, methods, compounds, or experiments described herein. In using such information or methods they should be mindful of their own safety and the safety of others, including parties for whom they have a professional responsibility.

To the fullest extent of the law, neither the Publisher nor the authors, contributors, or editors, assume any liability for any injury and/or damage to persons or property as a matter of products liability, negligence or otherwise, or from any use or operation of any methods, products, instructions, or ideas contained in the material herein.

ISBN: 978-0-12-823731-1

For Information on all Academic Press publications
visit our website at https://www.elsevier.com/books-and-journals

Publisher: Nikki P. Levy
Acquisitions Editor: Nancy J. Maragioglio
Editorial Project Manager: Jai Marie Jose
Production Project Manager: Swapna Srinivasan
Cover Designer: Christian J. Bilbow

Typeset by MPS Limited, Chennai, India

Contents

About the authors   xi
Introduction   xix

Section I
Biogeography

1. Effect of climate change on plant regeneration from seeds in the arctic and alpine biome   3

Andrea Mondoni, Borja Jiménez-Alfaro and Lohengrin A. Cavieres

Introduction   3
Climate warming in the arctic–alpine life zones   4
Effects of climate warming on arctic–alpine plants   4
Climate warming and plant regeneration from seeds in alpine–arctic environments   5
    Seed production and seed mass   6
    Seed dispersal in time and space   7
    Seed dormancy and germination   8
    Seedling emergence and establishment   9
Conclusions and future research   11
References   12
Further reading   18

2. Effects of climate change on regeneration of plants from seeds in boreal, subarctic, and subalpine regions   19

Bente J. Graae, Kristin O. Nystuen, Vigdis Vandvik and Amy E. Eycott

The boreal ecosystem   19
Recruitment from seeds in the boreal zone   21
Predicted climate change and its impact on the boreal zone   21
Sampling and characterizing the knowledge base   22
Seed production   23
Seed dormancy and germination   24
Seedling survival and growth   26
Long-term consequences of climate change on seedling recruitment   27
Future research   27
References   28

3. Effects of climate change on plant regeneration from seeds in the cold deserts of Central Asia   33

Juanjuan Lu, Dunyan Tan, Carol C. Baskin and Jerry M. Baskin

Introduction   33
Climate change in cold deserts of Central Asia   34
Plant life history traits   34
    Seed dormancy/germination   34
    Seedling survival and growth   36
    Plant growth, development, biomass accumulation/allocation, and seed production   37
    Phenology   37
    Plant morphological characters   40
    Dry mass accumulation and allocation   40
    Number of seeds produced   41
    Importance of heterodiaspory   41
Vegetation and community dynamics   42
Future research needs   43
Acknowledgments   43
References   43

4. Plant regeneration by seeds in hot deserts   47

Marina L. LaForgia, D. Lawrence Venable and Jennifer R. Gremer

Introduction   47
Climate change in deserts   48
Effects of increasing temperature   49
    Increasing temperature effects on seed survival   49

    Increasing temperature effects on dormancy and seed banks   50
    Increasing temperature effects on germination   51
  Effects of decreasing precipitation   51
    Decreasing precipitation effects on dormancy and seed banks   54
    Decreasing precipitation effects on germination   54
  Increasing variability and interactions between temperature and precipitation   54
    Effects of increasing variability and temperature-precipitation interactions on germination   55
  Other climatic changes   55
  Summary and conclusions   56
  Future research   56
  References   56

5. Effect of climate change on plant regeneration from seeds in steppes and semideserts of North America   61

Susan E. Meyer

Introduction   61
Current and future climates   61
Seed regeneration strategies and climate change   62
  The genus *Penstemon*   63
  Regionally important shrub species   65
  Native perennial grasses   68
  Native herbaceous dicots   68
  An invasive annual grass   69
Conclusions and recommendations for future research   70
References   70

6. Effect of climate change on plant regeneration from seeds in steppes and semideserts of northern China   75

Xuejun Yang, Gaohua Fan and Zhenying Huang

Introduction   75
Projected climate changes and effects on natural ecosystems in northern China   75
Effects of projected climate changes in northern China on regeneration of plants from seeds   76
Future directions   83
References   83

7. Effect of climate change on regeneration of plants from seeds in grasslands   87

Eric Rae and Yuguang Bai

Introduction   87
Climate change in grasslands   87
Grassland species composition and population dynamics   88
Reproductive phenology and seed production   89
Seed germination   92
Seed physiology   93
Conclusions and future research   94
References   95

8. Climate change and plant regeneration from seeds in Mediterranean regions of the Northern Hemisphere   101

Efisio Mattana, Angelino Carta, Eduardo Fernández-Pascual, Jon E. Keeley and Hugh W. Pritchard

Introduction   101
Plant regeneration from seeds under a changing climate   103
  Seed production   103
  Mild winters, decreased cold stratification, and seed dormancy break   103
  Increased germination temperatures   104
  Altered precipitation regimes and germination   105
  Sea level rise and salinity stress   105
  Altered fire regimes   105
  Seedling survival   106
  Facilitation and drought stress   107
  Local adaptation and phenotypic plasticity in the face of climate change   108
Future directions   108
Concluding remarks   109
Acknowledgments   110
References   110

9. Plant regeneration from seeds in the southern Mediterranean regions under a changing climate   115

Jennifer A. Cochrane and Sarah Barrett

Southern Hemisphere Mediterranean-type ecosystems   115
Current and predicted environmental changes due to global warming   116
  Southwestern and southern Australia   116
  Central Chile   117

Cape Region of South Africa	117
Current knowledge on plant regeneration from seeds	117
Southwestern and southern Australia	117
Central Chile	120
South Africa Cape Region	121
Future research needs	123
References	124

10. Plant regeneration from seeds in the temperate deciduous forest zone under a changing climate — 131

Jeffrey L. Walck and Siti N. Hidayati

Introduction	131
Changes in temperature and precipitation	132
Changes in temperature	132
Interaction between warming and changes in precipitation	134
Interactions of warming and/or changes in moisture with biotic and nonclimatic abiotic factors	135
Snow cover reduction	135
Elevated CO_2	136
Seed production	136
Soil seed banks	137
Geographical range shifts	137
Future considerations	138
References	139

11. Plant regeneration from seeds: Tibet Plateau in China — 145

Kun Liu, Miaojun Ma, Carol C. Baskin and Jerry M. Baskin

Introduction	145
Climate change	146
Seed production	147
Soil seed bank	148
Seed dormancy and germination	148
Seedling emergence and establishment	149
Change in community structure and composition	150
Research needs	151
References	151

12. Effect of climate change on regeneration of plants from seeds in tropical wet forests — 157

James Dalling, Lucas A. Cernusak, Yu-Yun Chen, Martijn Slot, Carolina Sarmiento and Paul-Camilo Zalamea

Introduction	157
Observed and predicted climate change in tropical forests	158
Effects of climate change on reproductive phenology	159
Seed responses to elevated temperature and drought	160
Seedling responses to increased temperatures	161
Drought tolerance and shade tolerance trade-offs for seedlings	162
Effects of climate change on nutrient availability	163
Interactions between elevated CO_2 and nutrient limitation	163
Concluding remarks and future research challenges	163
References	164

13. Climate change and plant regeneration from seeds in tropical dry forests — 169

Guillermo Ibarra-Manríquez, Jorge Cortés-Flores, María Esther Sánchez-Coronado, Diana Soriano, Ivonne Reyes-Ortega, Alma Orozco-Segovia, Carol C. Baskin and Jerry M. Baskin

Introduction	169
Predicted climate changes	170
Flowering, seed production, and dispersal	171
Seed dormancy and germination	171
Physical and physiological dormancy	172
Effect of temperature on seed germination	173
Seedling growth and survival	174
Soil seed bank	174
Community composition and shifts in species distribution	176
Future research needs	176
Acknowledgments	177
References	177

14. Regeneration from seeds in South American savannas, in particular the Brazilian Cerrado — 183

L. Felipe Daibes, Carlos A. Ordóñez-Parra, Roberta L.C. Dayrell and Fernando A.O. Silveira

Introduction	183
The Cerrado vegetation	183
Climate change effects in the Cerrado	184
Germination ecology in the Cerrado: how much do we know?	185
Environmental drivers of Cerrado regeneration from seeds under a changing climate	187
Increased fire frequency	187
Increase in mean temperature	188

Decrease in mean annual rainfall	188
Predicted consequences of climate change on plant populations and communities	189
Changes in species distribution	189
Changes in species abundance and community composition	190
Knowledge gaps and research needs	190
Conclusions	192
Acknowledgments	193
References	193

15. Plant regeneration from seeds in savanna woodlands of Southern Africa 199

Emmanuel N. Chidumayo and Gudeta W. Sileshi

Introduction	199
Data acquisition and analytical approach	200
Regional climate trends in southern Africa	201
Local climate trends in southern African woodlands	201
Field observational studies on plant regeneration from seeds	201
Fruit production and climate	203
Seedling emergence and climate	206
Seedling mortality and climate	208
Conclusions and recommendations for future research	209
References	209

Section II
Special topics

16. Effects of climate change on annual crops: the case of maize production in Africa 213

Carol C. Baskin

Introduction	213
Effect of climate change on life cycle of maize	215
Seed dormancy/germination	215
Plant growth from germination to tasseling	216
Gamete formation, pollination, and fertilization	216
Grain (seed) formation	217
Grain (seed) filling	218
Predicted effect of climate change on maize production in Africa	219
Mitigation	221
Broad implications for wild plant species	221
References	222

17. Fire and regeneration from seeds in a warming world, with emphasis on Australia 229

Mark K.J. Ooi, Ryan Tangney and Tony D. Auld

Introduction	229
Climate-related shifts in the fire regime and plant regeneration	231
Fire intensity and severity—what can happen when fires burn hotter?	231
Survival of seeds and changes in fire severity	231
Fire season—what happens when fire season shifts from historic norms?	232
Germination timing	232
Survival of seeds and changes to fire season	233
Postfire flowering species and fire seasonality	233
Postfire seed predation and mortality	233
Fire frequency—what happens when fires occur at shorter intervals?	234
Seed bank depletion	234
Seed dispersal and persistence of postfire obligate colonizers	235
Interactions of multiple factors on plant regeneration from seeds under climate change	235
Fire-prone regions and the potential for "winners" and "losers" under climate change	235
Adaptive capacity to ameliorate impacts of climate change in fire-prone regions	236
Conclusions and future research needs	237
References	237

18. Effects of global climate change on regeneration of invasive plant species from seeds 243

Cynthia D. Huebner

Introduction and background	243
Mating systems and phenology	245
Sexual reproductive capacity and seed dispersal	245
Seed dormancy	246
Seed germination and viability	247
Soil seed banks	248
Biotic interactions of invasive plant species	249
Linking regeneration by seeds with climate change mitigation	250
Future research needs	250
References	251

19. Regeneration in recalcitrant-seeded species and risks from climate change   259

Hugh W. Pritchard, Sershen, Fui Ying Tsan, Bin Wen, Ganesh K. Jaganathan, Geângelo Calvi, Valerie C. Pence, Efisio Mattana, Isolde D.K. Ferraz and Charlotte E. Seal

Introduction	259
Case study 1: Regeneration and light	260
Seed development	261
Case study 2: Regeneration in *Coffea* species	262
Is dormancy present in recalcitrant seeds?	262
Germination in relation to precipitation and temperature	263
Case study 3: Regeneration timing in relation to precipitation in seasonally dry tropical vegetation	264
Stopping germination during seed storage	266
Seed and seedling banks	267
Future research needs	267
Acknowledgments	268
References	268

20. Effect of climate change on regeneration of seagrasses from seeds   275

Gary A. Kendrick, Robert J. Orth, Elizabeth A. Sinclair and John Statton

Introduction	275
Evolution of seagrasses	276
Seagrass mating systems	277
Flowering, seed production, and seed germination	278
Seed dispersal	279
Seed settlement and early seedling survival	279
Climate change and seagrass regeneration from seeds—the future	280
Acknowledgments	280
References	281

21. Soil seed banks under a warming climate   285

Margherita Gioria, Bruce A. Osborne and Petr Pyšek

Introduction	285
Effects of a warming climate on seed bank persistence and density	286
Changes in the composition and structure of seed banks under a warming climate	288
Mountain ecosystems and elevation gradients	288
Seed banks confer increased resilience in extreme environments	289
Buffering the effects of climate warming: temporary resilience?	290
Seed banks and plant migration potential under a warming climate	290
Challenges and future research directions	291
Concluding remarks	292
References	292

Section III
Conclusion

22. Summary and general conclusions   301

Carol C. Baskin and Jerry M. Baskin

Seed production	301
Seed dormancy and germination	301
Seedling survival and growth	302
Shifts in species composition of plant communities	303
Future research needs	303
Conclusions	305

Index   307

About the authors

Tony D. Auld is the president of the Australian Network for Plant Conservation, and he recently retired as a senior principal research scientist in New South Wales. He is a professorial fellow at the University of Wollongong and the University of New South Wales, Australia. He worked in conservation for more than 30 years, and his research focus is on understanding how different threats, including, for example, changed fire regimes, grazing, and invasive species, impact the persistence of plant species across their native range, with a particular interest in soil seed bank dynamics. He has been involved in developing policy and on-ground management for how best to mitigate threats to biodiversity.

Yuguang Bai is a professor and head of Department of Plant Sciences, University of Saskatchewan (Canada). He holds a BSc in Physical Geography, an MSc in Ecology, and a PhD in Plant Ecology. His research interests include ecology and management of grassland and forest ecosystems, impact of climate change on these ecosystems, seed ecology, modeling of seed germination, and effects of fire and smoke on seed germination and seedling recruitment. He served as an associate editor for *Rangeland Ecology and Management*, the *Western North American Naturalist*, and *Journal of Plant Ecology*.

Sarah Barrett works with threatened flora conservation for the Western Australian Department of Biodiversity, Conservation, and Attractions, South Coast Region (Australia). Her research and management efforts are focused on plant disease, fire ecology, long-term monitoring, reproductive biology, seedling ecology, seed conservation, and threatened flora translocation. She has a keen interest in the impact of climate change on ecosystem health.

Carol C. Baskin is a professor of Biology and of Plant and Soil Sciences at the University of Kentucky (United States). Her research interests are ecology, biogeography, and evolution of seed dormancy and germination.

Jerry M. Baskin is a retired professor of Biology at the University of Kentucky (United States). His research interests include ecology, biogeography, and evolution of seed dormancy and germination.

Geângelo Petene Calvi is a forest engineer with a doctorate in Tropical Forest Science from the National Institute of Amazonian Research in Manaus (Brazil), where he works as a technician in the Seed Laboratory. His research interests are the classification of seed storage behavior, germination and storage of recalcitrant seeds, and recovery of degraded areas.

Angelino Carta is an associate professor at the Department of Biology of the University of Pisa, Italy. His research deals with macroevolution, biogeography, and reproductive biology of vascular plants, with a special focus on regeneration strategies from seeds and plant systematics in the Mediterranean Basin. Angelino is the author of several scientific articles in plant evolutionary biology, reproductive ecology, and conservation.

Lohengrin A. Cavieres is a professor in the Botany Department, University of Concepción, Chile, and associate researcher in the Institute of Ecology and Biodiversity—IEB, Chile. Most of his work is devoted to the ecology and ecophysiology of alpine plant species in the Andes, with a special interest in the consequences of global change drivers (climate and biological invasions) on the diversity and functioning of fragile Andean ecosystems.

Lucas A. Cernusak is an associate professor in the Ecology and Zoology group at James Cook University—Cairns, Australia. His main research interest is the environmental and biological controls on carbon dioxide and water vapor exchange between plants and the atmosphere and how these processes are responding to climate change.

Yu-Yun Chen is an associate professor at the National Dong Hwa University, Taiwan. She has been involved in the monitoring of flower/seed rain and seedling dynamics at the Pasoh Forest Reserve in Malaysia and the Fushan Forest Dynamics Plot in Taiwan for more than a decade. Her research interests include the evolution of reproductive phenology in tropical and subtropical forests and the impact of climate change on flowering phenology, seed production, and seedling establishment.

Emmanuel N. Chidumayo is a plant ecologist and author of *Miombo ecology and management* (Intermediate Technology Publications 1997) and coeditor of *The dry forests and woodlands of Africa* (Earthscan, 2010) and *Climate change and African forests and wildlife resources* (African Forest Forum, 2011). He is currently involved in long-term research in Zambia on responses of African tropical dry forest plants to disturbance.

Jennifer Anne Cochrane is a research associate with the Western Australian Department of Biodiversity, Conservation and Attractions and Campus Visitor at the Australian National University. Before retiring in 2018, she was employed for 25 years as a research scientist managing the department's seed conservation facility for rare and threatened species of Western Australia. Aside from the collection and storage of native seeds, Anne's research has focused on seed biology and ecology, including the production of seedlings for threatened plant translocation.

Jorge Cortés-Flores is a research associate in the Jardín Botánico, Instituto de Biología, Universidad Nacional Autónoma de México. His research focuses on restoration ecology, including plant reproduction, plant biotic interactions, and seed germination in seasonal Neotropical forests in Mexico.

L. Felipe Daibes obtained his PhD in Plant Biology from Universidade Estadual Paulista, Rio Claro, Brazil. His research interests include regeneration ecology in environmental gradients with a focus on seed and germination traits.

James Dalling is a professor of Tropical Forest Ecology at the University of Illinois at Urbana-Champaign (United States). His research explores the determinants of tropical tree distribution patterns along soil fertility gradients in lowland and montane forests and the role of plant–fungal interactions in shaping seed survival and seedling recruitment.

Roberta L.C. Dayrell is a postdoctoral researcher at the University of Regensburg (Germany). Her research focuses on seed traits relevant to seed identification and new technologies. She has completed a joint PhD on ecological and evolutionary drivers of regeneration from seeds in harsh megadiverse ecosystems at the Universidade Federal de Minas Gerais (Brazil) and the University of Western Australia.

Amy E. Eycott is an associate professor of Terrestrial Ecology at Nord University, in Steinkjer, Norway. She studies the effects of human activity on biodiversity and ecosystem function. Her work is largely observational and based on understanding the consequences of management decisions. She works in boreal, temperate, and tropical forests, as well as in cultural landscapes such as heathlands. She has particular interests in endozoochorous seed dispersal and in systematic reviews and knowledge mapping.

Gaohua Fan is a PhD student in the Institute of Botany, Chinese Academy of Sciences in Beijing. His research focuses on the impact of global change on plant regeneration from seeds.

Diana Soriano Fernández is a professor at the Universidad Nacional Autónoma de México. She is a specialist in plant ecophysiology in the area of germination, seedling establishment, and plant hydraulics. She has worked in the Chamela Dry Forest in Mexico for more than 15 years.

Eduardo Fernández-Pascual is the scientific curator of the Atlantic Botanic Garden (Spain), with responsibilities on the analysis of taxonomical and functional data. He is a plant ecologist with a special focus on seed ecophysiology and the study of germination traits in response to climatic drivers, especially in mires, alpine habitats, and temperate forests.

Isolde D.K. Ferraz is a senior scientist of the National Institute for Amazonian Research in Manaus (Brazil), where she is a teacher and adviser in the postgraduate programs in Tropical Forest Science and in Botany. Her research interests include basic information on Amazonian tree seeds and seedlings, including germination, storage, and morphology.

Margherita Gioria is a researcher in the Department of Invasion Ecology at the Institute of Botany, Czech Academy of Sciences. Her main research interests are the ecology and biology of invasive alien plants and their long-term impacts on native plant communities and of all aspects of the ecology of seeds and seed banks. She is especially interested in understanding the role of soil seed banks in determining plant community dynamics and how global environmental changes might affect regeneration from seeds, from local to global scales.

Bente J. Graae is a professor in Plant Ecology at the Norwegian University of Science and Technology in Trondheim, Norway. Her research focuses on how plant recruitment and vegetation dynamics respond to climate and land-use changes, mostly in temperate and boreal forest and arctic and alpine ecosystems in Northern Europe, but she also has worked on African savannas. She has a special interest in understanding the early phases of plant recruitment related to dispersal, dormancy break, germination, and seedling establishment in relation to the microclimatic landscape and therefore has worked with seeds and seedlings in growth chambers, greenhouse, and field experiments and often in combination.

Jennifer R. Gremer is an associate professor of Ecology at the University of California, Davis (United States). She is fascinated by the diversity of strategies that plants exhibit in variable and changing environments. Her research seeks to identify traits and mechanisms underlying those strategies, as well as the implications for individual fitness, population dynamics, and patterns of biodiversity in communities. A prominent focus of her work is understanding these strategies through bet-hedging theory, including research on seed germination and dormancy in variable environments.

Siti N. Hidayati is a lecturer at Middle Tennessee State University in Tennessee, United States. She has been a Fulbright US Senior Scholar in Indonesia. Her research interests include seed germination ecology and tropical ecology, especially the biology of *Rafflesia*.

Zhenying Huang is a professor and principal investigator of the Seed Ecology Research Group of the Institute of Botany, Chinese Academy of Sciences in Beijing. His research includes seed ecology, plant ecophysiology, and ecological restoration of the sandland ecosystem. He is the coeditor-in-chief of the journal *Plant Physiology and Biochemistry* (Elsevier).

Cynthia D. Huebner is a research botanist with the Northern Research Station, USDA Forest Service in Morgantown, West Virginia (United States) and an adjunct associate professor at West Virginia University. Her research is on invasive plant species biology and ecology and how invasive and native plant species interact in response to changing environmental conditions and disturbances.

Guillermo Ibarra-Manríquez is a researcher in the Instituto de Investigaciones en Ecosistemas y Sustentabilidad, Universidad Nacional Autónoma de México. His research is focused on community ecology of tropical and temperate forests. His work on tropical dry forests includes reproductive phenology, seed dispersal syndromes, seed germination, and aspects of composition, structure, diversity, and conservation of these forests in several regions of Mexico.

Ganesh K. Jaganathan is a recent Postdoc associated with the Institute of Biothermal Technology, University of Shanghai for Science and Technology, Shanghai, China. His research interests include germination ecology of recalcitrant seeds, cryopreservation of difficult-to-store seeds, seed dormancy, seed germination in a changing climate, and soil seed banks.

Borja Jiménez-Alfaro is a distinguished researcher at the Biodiversity Research Institute (University of Oviedo, CSIC, Principality of Asturias, Spain) and the director of Science of the Atlantic Botanic Garden in Gijón, Asturias. He is mainly interested in the biogeography and regeneration of plant communities, with a special focus on high-mountain vegetation in southern Europe.

Jon E. Keeley is a senior research scientist with the US Geological Survey and an adjunct professor at the University of California, Los Angeles (United States). His research includes ecological life history strategies of plants from fire-prone ecosystems, fire-stimulated seed germination, invasive species, taxonomy of *Arctostaphylos*, and biochemical pathways of photosynthesis in aquatic plants. He is the senior author of *Fire in Mediterranean Climate Ecosystems: Ecology, Evolution and Management* (Cambridge University Press, 2012).

Gary A. Kendrick is a professor in the Oceans Institute and School of Biological Sciences, the University of Western Australia. He has focused his research on marine benthic biodiversity, resource mapping, and seagrass ecology. He has worked with industry and government to determine needs and to develop solutions for climate change threats and environmental and conservation issues in Australia and globally. He is presently working on developing science for management solutions in the areas of seagrass conservation and restoration.

Marina L. LaForgia is a postdoctoral scholar at the University of California, Davis and the University of Oregon. She is broadly interested in the strategies plants use to cope with climate variability and how plants with those strategies will fare under future climate variability. Her current research is focused on understanding seed persistence and dispersal in dryland annuals through seed traits and long-term demographic studies. She obtained her PhD from UC Davis investigating the interactive effects of climate change and invasive species on native species in semiarid grasslands.

Kun Liu is an assistant professor at Lanzhou University, Lanzhou, China. His research is on plant community ecology in alpine/subalpine meadows on the Tibetan Plateau, and it focuses on ecology of seed dormancy and germination, seedling emergence phenology, and plant functional traits in relation to plant community ecology.

Juanjuan Lu is a professor in the College of Grassland and Environment Sciences, Xinjiang Agricultural University, China. Her research is focused on the ecology and evolution of seed dispersal, seed dormancy and germination, seed banks in Central Asian cold deserts plants in relation to climate change, and the impact of climate change on life history traits of desert species.

Miaojun Ma is a professor at Lanzhou University, Lanzhou, China. His research is focused on the effects of climate change on soil seed bank ecology in alpine ecosystems on the Tibetan Plateau, and he studies the role of the soil seed bank in community regeneration, resilience, and restoration.

Efisio Mattana is a research fellow at the Royal Botanic Gardens, Kew (United Kingdom). His research interests include seed biology and ecology and the impact of climate change on the regeneration by seeds of wild, threatened and neglected, and underutilized species in Mediterranean ecosystems.

Susan E. Meyer is a senior research plant ecologist with the US Forest Service Rocky Mountain Research Station (RMRS), Provo, Utah (United States), where she has worked at the RMRS Shrub Sciences Laboratory for over 35 years. Susan's research has focused on many aspects of seed ecology, including hydrothermal time modeling, genetic variation in seed germination syndromes, and seed–pathogen interactions

Andrea Mondoni is a senior research scientist and professor of Applied Botany at the University of Pavia (Italy). He is a plant ecologist with a particular interest in alpine plants, germination and dormancy traits of native species, risks of natural regeneration from climate change, and ex situ conservation in gene banks.

Sershen Naidoo is the executive director of the Institute of Natural Resources NPO (South Africa). He is a multidisciplinarian working across the Natural and Social Sciences on projects focused on plant ecophysiology, water and sanitation, waste management, urban ecology, tourism, climate change, sustainability science, and public health. He is an applied researcher and is focused on integrated environmental management.

Kristin Odden Nystuen has a PhD in Plant Ecology from the Norwegian University of Science and Technology in Trondheim, Norway. Her specialty is seedling recruitment patterns and drivers of seedling emergence and survival of various species and functional groups in Scandinavian alpine and subalpine vegetation. Her main research focus is on the impacts of microclimate and herbivory.

Mark K.J. Ooi is a senior research fellow at the University of New South Wales, Sydney, Australia. His research interests include fire and plant ecology (particularly seed ecology), and he heads a lab focused on understanding shifting fire regimes and conservation.

Carlos A. Ordóñez-Parra is an MSc student at the Plant Biology Program at Universidade Federal de Minas Gerais (Brazil). He is interested in natural regeneration of tropical ecosystems, and his current research aims to describe the germination strategies of plants from Brazilian rock outcrop vegetation by assessing potential trade-offs across seed and germination functional traits.

Alma Orozco-Segovia is a researcher in the Instituto de Ecología, Universidad Nacional Autónoma de México. Her professional expertise is ecophysiology of tropical species, and she has focused mainly on seed and fern spore germination and early seedling establishment in relation to restoration strategies and effects of deforestation and climatic change. She has been studying the effects of hydric priming and experimental burial of propagules in the soil as pretreatments to invigorate seed germination of native tropical forests tree species and enhance seedling establishment in the field.

Robert J. Orth is a Professor Emeritus at the Virginia Institute of Marine Science, College of William and Mary (United States). His research has focused on faunal relationships, seed ecology, monitoring, and restoration strategies of seagrasses. His work has figured prominently in the management and conservation of seagrasses in the mid-Atlantic USA.

Bruce A. Osborne is a professor of Plant Ecophysiology in University College Dublin (UCD) in Ireland and a principal investigator in the UCD Earth Institute, with wide interests in functional biology and the ecology of plants, which encompasses research from the individual to ecosystem scales of inquiry. He has a particular interest in the ecophysiology of the *Gunnera-Nostoc* symbiosis and the impacts of plant invasions on ecosystems and ecosystem processes.

Valerie C. Pence is the director of Plant Research at the Center for Conservation and Research of Endangered Wildlife at the Cincinnati Zoo & Botanical Garden. Her work focuses on applying in vitro methods and cryobiotechnologies to the conservation of plant species that cannot be conserved in traditional seed banks, known as *exceptional species*, many of which are critically endangered from across the United States.

Hugh W. Pritchard is a senior research leader, Royal Botanic Gardens, Kew, Wakehurst (United Kingdom) and full professor, Kunming Institute of Botany, Chinese Academy of Sciences. His research interests include the mechanistic basis of seed desiccation sensitivity and the ecology, germination, and storage of recalcitrant seeds.

Petr Pyšek is a senior research scientist at the Institute of Botany, Czech Academy of Sciences and professor of Ecology at the Charles University in Prague. His research focuses on biological invasions, including the role of macroecological drivers and their context dependence, global biogeography of alien floras, mechanisms of species invasiveness at various organizational and spatial scales, habitat invisibility, and impact of invasive species.

Eric Rae received his MSc degree in Plant Sciences from the University of Saskatchewan (Canada) in 2021. His master's thesis was conducted on seed germination, and it introduced a novel way of comparing the germination of samples. Previously, he has published a paper and textbook chapter on how to match mouse models of Alzheimer's disease with human clinical studies. Eric is particularly passionate about applying computational power to scientific problems, whether through analysis of images or statistics.

Ivonne Reyes-Ortega teaches Plant Biology in the Faculty of Science of the Universidad Nacional Autónoma de México. She specializes in germination ecophysiology, and recently her research has focused on the effect of temperature on germination. She is interested in the effect of priming on the invigoration of seeds from species growing in the tropical dry forest that has been stored for many years under suboptimal conditions.

María Esther Sánchez-Coronado is an academic technician in the Laboratory of Physiological Ecology, Instituto de Ecología, Universidad Nacional Autónoma de México (UNAM). Her research is focused on the ecophysiology of germination and plant establishment and growth. She also teaches biophysics, biostatistics, and plant physiology at UNAM.

Carolina Sarmiento is a research associate at the Department of Integrative Biology at the University of South Florida (United States) and at the Smithsonian Tropical Research Institute (Panama). She is a tropical ecologist with a general interest in plant–fungal interactions, seed ecology, and microbial ecology. Carolina holds a Masters' Degree from Universidad de Los Andes (Colombia) and enjoys scientific illustration.

Charlotte E. Seal is a research leader at the Millennium Seed Bank, Royal Botanic Gardens, Kew (United Kingdom). She is a seed biologist with a focus on seed functional traits. Of particular interest are species from environments considered stressful, such as those affected by salinity or drought. Her current research takes a comparative approach across a global scale to quantify seed germination traits, assess germination resilience to climate change, and elucidate mechanisms of stress tolerance.

Gudeta W. Sileshi is a fellow of the African Academy of Sciences, an adjunct professor of Biology and Biodiversity Management at Addis Ababa University in Ethiopia, an honorary fellow at the University of KwaZulu-Natal in South Africa, and formerly the regional representative of the World Agroforestry Centre in Southern Africa. His research interest is production ecology. He has coedited a book entitled *Indigenous Fruit Trees in the Tropics* (CABI, 2008) and coauthored two books entitled *Termite Management in Agroforestry* (University of Zambia Press, 2016) and *Bamboos: Climate change adaptation and mitigation* (Apple Academic Press, 2020).

Fernando A.O. Silveira is a professor of Ecology at the Department of Genetics, Ecology and Evolution, Universidade Federal de Minas Gerais (Brazil). His main research interests are the evolutionary ecology of seed traits and the ecology, conservation, and restoration of open ecosystems such as grasslands, shrublands, and savannas.

Elizabeth A. Sinclair is a senior research fellow at the University of Western Australia in Perth and a research scientist at Kings Park Science in the Western Australian Department of Biodiversity, Conservation and Attractions. Her current research is developing an understanding of genomic diversity, gene expression, and adaptation in changing environmental conditions for large temperate seagrasses and integrating this information into marine restoration practices.

Martijn Slot is a staff scientist at the Smithsonian Tropical Research Institute (Panama). His research focuses on the effects of climate change on the physiology, growth, and reproduction of tropical forest vegetation.

John Statton is a research associate in the Oceans Institute and School of Biological Sciences at the University of Western Australia. His seagrass restoration research has focused on how to successfully utilize seeds to increase the scale and effectiveness of seagrass restoration. He has experienced first-hand the challenges associated with seagrass seed ecology and restoration ecology within a changing climate both in simulated laboratory settings and the natural environment.

Dunyan Tan is a professor in the College of Grassland and Environment Sciences, Xinjiang Agricultural University, China. His research focuses on plant systematics, pollination ecology, breeding systems, seed dispersal, seed dormancy, seed banks, and the impact of climate change on life history traits of cold desert species.

Ryan Tangney is a postdoctoral research fellow at the University of New South Wales (Australia) and a research associate at the Western Australian Department of Biodiversity, Conservation and Attractions. His research is focused on understanding how changes in fire season affect recruitment from seeds, including how recruitment from seeds following fire may change under climate change.

Fui Ying Tsan is an associate professor in the Faculty of Plantation and Agrotechnology of the Universiti Teknologi MARA, Malacca, Malaysia. Her research interests are seed science and plant physiology. Her PhD dissertation was on the physiology of dipterocarp seedlings stored as slow-growing planting materials at high densities in the forest. Subsequent to this work, she did modeling of moisture loss from recalcitrant seeds and desiccation research on embryos. Recently, she has been involved in in vivo regeneration from cut seeds of *Syzygium* (Myrtaceae).

Vigdis Vandvik is a professor of Plant Ecology at the University of Bergen in Norway. Her research is focused on how major global change drivers—especially climate and land-use change—affect plants from their physiological responses via population and community dynamics to ecosystem functioning. Her research focuses on alpine and heathland ecosystems and on understanding the roles and responses of seeds and seed recruitment in the community and ecosystem dynamics of these systems. She is an elected member of the Norwegian Academy of Science and Letters and is actively involved in science communication and the science–policy interface, both within Norway and internationally.

D. Lawrence Venable is a professor of Ecology and Evolutionary Biology at the University of Arizona (United States). He studies plant population and community dynamics and plant reproductive ecology. His interest in adaptation to variable environments lies in the interplay between bet-hedging, phenotypic plasticity, and adaptive genetic change. His theoretical work on plant reproductive ecology deals with seed dormancy and dispersal, sex allocation, sexual system evolution, pollen evolution, seed size, seed heteromorphism, hierarchical packaging of reproduction, and the evolution of inflorescence design.

Jeffrey L. Walck is a professor in the Department of Biology at the Middle Tennessee State University. His research focuses on seed germination ecology, conservation biology, and climate change.

Bin Wen is head of the Seedbank of Xishuangbanna Tropical Botanical Garden, Chinese Academy of Sciences. His research interests are tropical seeds, including seed germination, storage, development, and ecology.

Xuejun Yang is an associate professor in the Institute of Botany, Chinese Academy of Sciences, in Beijing. His research includes seed ecology and plant geography. He is an editorial board member of the journal *Flora*.

Paul-Camilo Zalamea is an assistant professor at the Department of Integrative Biology at the University of South Florida (United States). He is a tropical plant community ecologist working at the interface of plant–microbial dynamics. His research integrates large-scale field, greenhouse, and laboratory experiments with molecular techniques to address questions fundamental to understanding the distribution and maintenance of biodiversity in tropical forests.

Introduction

Carol C. Baskin[1,2] and Jerry M. Baskin[1]
[1]*Department of Biology, University of Kentucky, Lexington, KY, United States,* [2]*Department of Plant and Soil Sciences, University of Kentucky, Lexington, KY, United States*

Background

By the end of the 21st century, the annual average global temperature is projected to be 1.5°C−2.0°C higher than it was between 1850 and 1900, but this projection is based on a significant reduction in emission of greenhouse gases (Wuebbles et al., 2017; IPCC, 2018). Without a significant reduction in gaseous emissions, the projected temperature increase is 5°C. Scientists working to understand the consequences of global warming have identified various possible changes in the earth's environment. These changes include increased temperature of the earth's surface, atmosphere, and oceans; melting of glaciers and sea ice; rising sea levels; increased water vapor in the atmosphere; and changes in patterns of precipitation that could increase the risk of flooding or drought, depending on location (Wuebbles et al., 2017; IPCC Intergovernment Panel on Climate Change, 2018; Manabe, 2019).

The changes in temperature and precipitation are not occurring uniformly over the earth. Temperatures are increasing faster/more at high than at low elevations, in winter than in summer, and at night than during the day (Wuebbles et al., 2017; IPCC Intergovernment Panel on Climate Change, 2018). The extent of winter snow cover at high latitudes is decreasing, and with increased temperatures, snowmelt is occurring earlier in the spring than in the past, resulting in an increase in the length of the growing season. Greve et al. (2014) reached the conclusion that for about 11% of the earth's land area, wet regions will become wetter and dry regions will become drier in response to global warming. For another 10% of the land surface, they predicted the opposite trend, that is, wet regions become drier and dry regions become wetter. Trenberth (2011) predicted an increase in rainfall, a decrease in snowfall, and an increase in summer drought in continental areas. In a review of climate change metrics commonly used to assess changes in biodiversity, Garcia et al. (2014) concluded that the area with hot, arid conditions will expand in Africa and Australia, whereas the area of cool conditions will shrink in polar regions and on mountain tops. Further, they predicted that novel environments (e.g., hotter than at present in tropical and subtropical regions) would develop by the end of the 21st century.

With increases in air temperature, the amount of water vapor in the atmosphere increases, thereby increasing the probability of heavy rainfall events (Neelin et al., 2017; Feng et al., 2019). Feng et al. (2019) predicted moderate decreases in total precipitation from the tropics to mid-latitudes and increases at high latitudes. Further, global warming is one of the contributing factors to droughts and food shortages (Connolly-Boutin and Smit, 2016; Kamali et al., 2018; Haile et al., 2019), as well as to the increased frequency of fires (Gillett et al., 2004; Flannigan et al., 2005; Vilà-Vilardell et al., 2020).

The regeneration of plants from seeds requires suitable environmental factors, primarily favorable temperature and soil moisture conditions, for dormancy break as well as for germination, seedling establishment, and plant growth to reproductive maturity. Depending on the species and the kind (class) of dormancy, high (at moist and/or dry conditions) temperatures of summer and/or low (at moist conditions) temperatures of winter are required for dormancy to be broken (Baskin and Baskin, 2014). After seeds germinate, soil moisture and temperature in the habitat must be favorable long enough for an annual plant to complete its life cycle or for a perennial/biennial to reach a perennating size; otherwise, the plant will die without producing offspring ($R_o = 0$). Thus, the timing of germination so that conditions are favorable for growth after the seedling emerges is an important aspect of the adaptation of a plant species to its habitat.

Seed dormancy/germination and seedling survival

Research on the effects of global warming on plant species has included various aspects of the seed and seedling phases of the life cycle, including seed production, dormancy break, temperature requirements for germination, timing of germination in nature, and seedling survival. These topics will be briefly surveyed to provide an overview of the various kinds of changes that might be expected to occur in response to global warming.

Seed production and quality

Heat stress can reduce pollen viability and pollen tube growth (Chiluwal et al., 2019), and increased night-time temperatures can reduce seed set (Wang et al., 2020). Also, elevated temperatures during development can affect the stored reserves in seeds, for example, their fatty acid composition (Pokharel et al., 2020). The production of *Zostera marina* seeds was higher in relatively cool than relatively warm water (Qin et al., 2020), and warming decreased the number of infructescences produced by plants of *Potamogenton crispus* (Xu et al., 2020). High temperatures have been shown to decrease the seed yield of various crops, including *Brassica napus* (Wu et al., 2020), *Glycine max* (Lippmann et al., 2019), *Oryza sativa* (Hsuan et al., 2019), *Triticum aestivum* (Lippmann et al., 2019) *Vicia sativa* (Li et al., 2017), and *Zea mays* (Iqbal et al., 2020) and of noncrop species such as *Acacia koa* and *Metrosiderous polymorpha* (Pau et al., 2020). Seeds of *Arabidopsis thaliana* produced at a high temperature had a lower level of dormancy than those produced at a low temperature (maternal environmental effect) (Huang et al., 2018).

Dormancy break

Depending on the species and its geographical range, seeds may require exposure to dry and/or wet conditions at relatively high ($\geq 15°C$) temperatures for dormancy break to occur (Baskin and Baskin, 2014). Seeds of some species with physiological dormancy (PD), that is, they have a physiological inhibiting mechanism in the embryo, may undergo dormancy break during exposure to dry or wet/dry conditions at high temperatures. Also, seeds of some species with water-impermeable seed or fruit coats may become sensitive to dormancy-breaking conditions, while they are exposed to hot, dry summer-like conditions (Baskin and Baskin, 2014). For seeds that require exposure to dry, hot conditions for dormancy break to proceed, it seems reasonable that an increase in temperatures due to global warming would not have a significant effect on dormancy break per se, except to possibly increase its rate (speed).

On the other hand, seeds of many species require several weeks (number varies with the species) of exposure to moist cold (c.$0°C - 10°C$) conditions for dormancy break to occur. The concern about species whose seeds require moist cold conditions for dormancy break is that global warming might reduce the length of the period during which conditions are favorable for dormancy break. In various species, simulated global warming that reduced the length of the moist cold period resulted in PD of seeds not being fully broken (Mondoni et al., 2012; Orrù et al., 2012; Sommerville et al., 2013; Hoyle et al., 2014; Solarik et al., 2016; Footitt et al., 2018; Cuena-Lombaraña et al., 2020). That is, the germination percentage was low, and/or seeds did not germinate over the full range of environmental conditions possible for nondormant seeds of the study species to do so; seeds were still in conditional dormancy.

Temperature requirements for germination

Germination tests have been conducted in which seeds were incubated at elevated temperatures, that is, at those predicted to occur in the habitat of the species due to global warming. As might be expected, depending on the species elevated temperatures may decrease (Kim and Han, 2018), have little or no effect (Cochrane, 2019), or promote (Kim and Han, 2018; Cochrane, 2019; Tietze et al., 2019) germination. For 38 of 41 species of the Neotropical family Bromeliaceae, a $3°C$ increase above the mean annual temperature in the habitat of each species during 1950–2000 did not exceed the thermal optima for germination (Müller et al., 2017). Seeds of some species incubated at elevated temperatures germinate to high percentages at very high rates (speed) (Frenne et al., 2012; Cochrane, 2017a; Monllor et al., 2018). However, will seedlings from seeds that germinate at elevated temperatures survive? For example, many seedlings of Bromeliaceae from seeds germinating at high temperatures died (Müller et al., 2017). Seedlings of the Neotropical legume *Dipteryx alata* from seeds germinated at $40°C$ were significantly smaller than those from seeds germinated at 32°C and 36°C (Ribeiro et al., 2019). If simulated global warming temperatures exceed the optimal temperatures for seed germination of a species, researchers usually predict that the range of the species will be reduced or extinction will occur (e.g., Catelotti et al., 2020).

Shifts in timing of germination

The germination of nondormant seeds occurs in nature when soil moisture and temperatures are favorable. If temperatures are favorable but the onset of the wet season is delayed, the timing of germination will be delayed (Kimball et al., 2010). Although soil moisture is favorable, germination will be delayed until temperatures (in temperate regions) increase or decrease enough to overlap with those required for germination (Baskin and Baskin, 1985). Thus, changes in the time when soil moisture and temperature are favorable for the germination in the habit can shift the time of germination.

In the case of temperate-zone, spring-germinating species, global warming could result in a shift in the timing of seedling emergence to the relatively short days of early spring. Germination in early spring leads to concerns about the possibility of damage to seedlings due to unexpected late frosts (Bianchi et al., 2019) or of death due to drought damage in the case of seedlings not protected by snow especially in alpine regions (García-Fernández et al., 2015; Fernández-Pascual et al., 2017). In the alpine zone of the southeastern Tibetan Plateau in China, however, global warming is predicted to increase snowfall, resulting in delayed snowmelt in spring (Deng et al., 2017). Experimentally prolonged spring snow cover in the alpine habitat did not adversely affect seed germination of four alpine species on the Tibet Plateau, but it had a negative effect on seedling survival of three of the four species (Wang et al., 2018).

For temperate-zone, autumn-germinating species that normally germinate as temperatures decrease in autumn, relatively high temperatures at this time potentially could delay germination (Baskin and Baskin, 1971). Thus, a delay in the time of germination results in seedlings becoming established under relatively short days of mid- to late autumn. On the other hand, seeds of some species are nondormant at maturity/dispersal in autumn, but their germination is inhibited by low autumn temperatures (Baskin and Baskin, 1973). Increased temperatures in autumn may promote germination of seeds that normally do not germinate until temperatures increase in spring (Mondoni et al., 2012; Bandara et al., 2019), which leads to questions regarding winter survival of seedlings and plant fecundity (Lu et al., 2014; Gremer et al., 2020).

If the timing of germination is changed, the resulting seedlings will grow in a novel postgermination environment (Donohue, 2002; Donohue et al., 2010), which may have effects on seedling survival, plant size, phenology, seed production, and dry mass accumulation and allocation (Lu et al., 2014). Further, the seeds produced by these plants will develop under environmental conditions that differ from those that were present before the climate changed; thus, new variations of traits may be subjected to selection by the environment (Donohue 2002; Donohue et al., 2010; Zhang et al., 2014; Burghardt et al., 2016). Donohue (2002) found that the timing of germination of *Arabidopsis thaliana* seeds could alter the mode and strength of natural selection on rosette diameter, number of plant branches, bolting date, and flowering interval.

Seedling survival and drought stress

In view of the predicted increase in drought stress in some regions due to global warming, much research is underway to identify the effects of reduced precipitation on regeneration of plant species from seeds. Depending on the species, habitat, and how individuals respond to drought, seed production may be decreased (Pérez-Ramos et al., 2010), not affected (Bogdziewicz et al., 2020), or increased (Wright and Calderón, 2006). Further, decreased precipitation is likely to reduce germination and seedling growth/survival of crop species; thus, research is being conducted to identify genotypes of annual crops that can tolerate drought (Carvalho et al., 2019; Sauca and Anton, 2019).

For some wild plant species, the increased water stress associated with climate change may decrease natural regeneration and consequently lead to local extinction and thus reduction in the size of the geographical range (Zhang et al., 2017). For example, elevated temperatures did not inhibit germination of *Picea jezoensis* from the subalpine zone on the Korean Peninsula when seeds were placed on woody debris (the normal microsite for seedling establishment in the forest), litter, sand, or soil (Han et al., 2018). However, after 60 days at evaluated temperatures, seedling mortality on woody debris, litter, sand, and soil was 53%, 77%, 53%, and 41%, respectively. Although *P. jezoensis* normally regenerates from seeds germinating on nurse logs in the forests, woody debris was not a favorable substrate for seedling survival under elevated temperatures. The authors suggested that drying of woody debris and litter at elevated temperatures was the cause of seedling death.

There is a particular concern not only about the effects of elevated temperatures on regeneration in high mountains but also about the effects of drought on seedling survival in these habitats (Giménez-Benavides et al., 2018). How much resilience do mountain species have, and can they evolve fast enough to keep up with climate change (Giménez-Benavides et al., 2018; Dickman et al., 2019)? Increased drought stress on seedlings of the endemic cushion plant *Heterotheca brandegeei* that grows on rock outcrops in the alpine zone of the Sierra de San Pedro Mártin in Baja California (Mexico) decreased growth and biomass accumulation (Winkler et al., 2019). Further, drought stress in combination with elevated temperatures significantly increased seedling mortality.

Consequences of global warming on species, populations, and communities

Global warming may lead to the extinction and/or migration of species and perhaps to phenotypic adjustment and/or adaptation to the changed environment. Thus, depending on what happens at the species level the size of populations could decrease, stay the same, or increase. The species composition of communities may shift, and the relative abundance of individual species in the community may change. There could even be invasion of alien plant species.

Extinction

The worst possible effect of global warming on a plant species that cannot reproduce asexually would be failure to regenerate via seeds, eventually resulting in the species becoming extinct in parts, or all, of its range. Using a modeling approach that included species distribution area, predicted temperature increases, and ability of an organism to disperse, Thomas et al. (2004) predicted different percentages of species extinction in different regions of the earth. For example, with maximum warming ($>2°C$) extinction with and without dispersal in the Amazon was 69% and 87%, respectively, and in Europe 8% and 29%, respectively. For the European Alps, Dullinger et al. (2012) predicted that due to global warming, the range size of the high-mountain species will be reduced by 44%–50% by the end of the 21st century.

In a survey of the literature on climate change, local extinctions, and range shifts of organisms, Wiens (2016) found data for 260 plant species, of which 99 had one or more local extinctions. However, it is difficult to know if local extinctions of plant populations are due to effects of global warming or to other factors, but long-term warming studies in natural habitats can provide some insight. At the Rocky Mountain Biology Laboratory in Colorado (USA), a study on the effects of warming on species composition of a subalpine meadow was initiated in 1990 (Harte and Shaw, 1995; Harte et al., 2015) in which soil temperatures were increased by $1.4°C–3.8°C$ (Panetta et al., 2018). Increased soil temperatures lead to a 15%–20% decrease in soil moisture and over time (25 years) advanced snowmelt in spring by 1–4 weeks (Harte et al., 2015; Panetta et al., 2018). After 25 years, *Androsace septentrionalis* was extinct in one of the five heated plots, and the number of individuals of this species in the other heated plots ranged from 6 to 14. In the non-heated plots, the number of *A. septentrionalis* individuals ranged from ca. 35 to 200 (Panetta et al., 2018). These authors found that warming promoted germination, decreased seedling survival, and decreased the probably that an established plant would live to reproduce, all of which contributed to reduced seed production and depletion of the soil seed bank.

Migration

Migration to a new site of a plant species that reproduces only by seeds involves seed dispersal, dormancy break and germination, seedling establishment, and growth of plants to reproductive maturity. In a meta-analysis of changes in range boundaries and phenology of various kinds of organisms, including woody and herbaceous species of plants, Parmesan and Yohe (2003) found that 80% of the changes in range and abundance agreed with the predictions of climate change and that species ranges have moved an average of 6.1 km poleward per decade. Further, 62% of the 677 species in their phenology analysis showed an advancement of spring by 2.3 days per decade. A comparison of the optimum elevation of 171 forest species in western Europe between 1905–1985 and 1986–2005 revealed a significant upward shift of about 20 m per decade (Lenoir et al., 2008).

A meta-analysis revealed that increasing temperatures were correlated with range shifts in latitude and elevation (Chen et al., 2011). For example, the movement of species away from the equator was 16.9 km per decade, and movement up mountains was 11.0 m per decade. Corlett and Westcott (2013) estimated that the predicted increase in mean annual temperature from 2000 to 2100 would require a movement of 0.42 km per year for a species to keep up with climate change. Using a pseudo-spatial ecosystem model and a vegetation migration submodel, Flanaga et al. (2019) concluded that the potential for vegetation to shift in response to climate change was most strongly influenced by dispersal rate and least influenced by dispersal mode.

When specific habitats and species are considered, much variation in predicted migration direction and distance is found. For the Tibetan Plateau, Yan and Tang (2019) predicted that 15%–59% of the endemic plant species will lose 30% or more of their current habitat by 2070. Thus, they predict major shifts in the range of species toward the west and north, with the greatest habitat loss and possible extinction, especially for herbaceous species, on the interior part of the Plateau. Models of the effects of climate change on the distribution of *Haloxylon* species in the cold deserts of Central Asia predict range increases, decreases, and shifts (Li et al., 2019). The range/habitat of *H. persicum* may increase 44%–62%, whereas that of *H. ammodendron* may decrease 22%–34% by the end of the 21st century. Further, habitat suitable for *Haloxyon* is predicted to decrease in Xinjiang (northwest China) and increase in southern Kazakhstan.

The movement of trees into alpine and arctic regions has been the focus of much research (Kueppers et al., 2017; Lubetkin et al., 2017; Kambo and Danvy, 2018; Lett and Dorrepaal, 2018; Shen et al., 2018; Ensing and Eckert, 2019; Davis et al., 2020; Rees et al., 2020). However, dispersal to these relatively cool (new) habitats does not necessarily ensure the successful establishment of the species. Although seeds may germinate, water stress (Kueppers et al., 2017; Lubetkin et al., 2017; Lett and Dorrepaal, 2018), late frosts, strong light (Shen et al, 2018), low depth of winter snowpack (Lubetkin et al., 2017), and a combination of high soil surface temperatures in the growing season, unfavorable soil microsites and low winter snowpack (Davis and Gedalof, 2018) may limit seedling establishment. Another issue with migration to high elevations and latitudes is that length of the growing season decreases poleward/upward. Thus, the plant growth of some species is limited, resulting in insufficient seed production for the population to be sustainable (Ensing and Eckert, 2019).

Transgenerational and phenotypic plasticity and adaptation

Many plant species exhibit considerable population-to-population differences in various life history characteristics, for example, the temperature requirements for seed germination (Baskin and Baskin, 2014; Blok et al., 2018; Gray et al., 2019). However, the source of this variation often is not known, and it may be due to maternal environmental effects (Roach and Wulff, 1987; Wulff, 1995; Fernández-Pascual et al., 2019), genetics (Adkins et al., 1986; Lampei and Tielbőrger, 2010), or genetic × environment interactions (Awan et al., 2018). The pollen parent has much less influence on the progeny via environmental paternal effects than the mother plant does through environmental maternal effects (Baskin and Baskin, 2019).

In terms of response to the effects of global warming, offspring may benefit from transgenerational plasticity (Wadgymar et al., 2018a,b). That is, the environment experienced by the mother (or grandmother) plant may help mediate the effects of climate change on the offspring. For example, seeds of *Brachypodium hybridum* produced at 30°C/20°C were more dormant and more tolerant of osmotic stress during germination than those produced at 25°C/15°C (Elgabra et al., 2019). Fernández-Pascual et al. (2019) used the term "thermal memory" to refer to the remembrance of "past thermal environments experienced by the seed and its recent ancestors." In reference to this phenomenon for forest trees, Kijowska-Oberc et al. (2020) called it thermal memory or epigenetic memory.

Phenotypic plasticity is the ability of a single genotype to have variable phenotypes in different environments. The plastic response to environmental factors could be harmful, neutral, or highly beneficial to the survival of the individual (Hendry, 2016). Phenotypic plasticity allows a species to respond to both short- and long-term variation in the environment (Gratani, 2014), and much attention is being given to its role in the survival of species in local habitats as climate change occurs (Anderson et al., 2012; Franks et al., 2014; Gratani, 2014; Becklin et al., 2016; Ahrens et al., 2020). If the environment is highly variable, it may promote the evolution of trait plasticity (Springate et al., 2011; Grenier et al., 2016; Hendry, 2016; Chevin and Hoffmann, 2017; Oostra et al., 2018; King and Hadfield, 2019; Rago et al., 2019).

Local (genetic) adaptation is another way in which plant species can be fine-tuned/adjusted to the conditions in their habitat (Solarik et al., 2018; Gárate-Escamilla et al., 2019; Sang et al., 2019; Anderson and Wadgymar, 2020; Blumstein et al., 2020; Chen et al., 2020; Etterson et al., 2020). Also, plant species may have both phenotypic plasticity and local adaptation (Franks et al., 2014; Gárate-Escamilla et al., 2019). Much research attention has been given to the evolvability of local adaptation in relation to climate change (Lampei and Tielbőrger, 2010; Anderson et al., 2012; Franks et al., 2014; Rubio de Casas et al., 2015; Mitchell and Whitney, 2018; Ensing and Eckert, 2019; Ahrens et al., 2020; Blumstein et al., 2020), and some evidence for genetic adaptation to climate change has been found (Lampei and Tielbőrger, 2010; Merilä and Hendry, 2014; Franks et al., 2014; Mitchell and Whitney, 2018).

There are both negative and positive aspects of local adaptation when considered from a response to global warming perspective. Local adaptation may slow the rate of migration of some species in response to climate change (Solarik et al., 2018) and thus increase the risk of extinction (Anderson and Wadgymar, 2020). However, in the establishment of new populations of a species for conservation purposes, using seeds from warmer (Etterson et al, 2020), lower elevation (Anderson and Wadgymar, 2020; Peterson et al., 2020) populations, or from drier habitats (Axelsson et al., 2020) than the restoration site, potentially can result in a new population that has "migrated" with climate change.

Both plastic and evolutionary responses to climate change may occur in the same species (Franks et al., 2014). Plasticity may increase fitness of individuals, as well as influence genetic evolution (Grenier et al., 2016). According to Hendry (2016), plasticity could either "protect" a genotype from selection or increase selection pressure, resulting in slowed or accelerated evolution, respectively.

Changes in community composition

Knowing that plant species may become extinct at a particular site or migrate to other regions in response to global warming leads to the conclusion that species composition and abundance of individual species in plant communities could change. Evidence that plant community structure has changed (is changing) comes from studies involving use of (1) supplemental heat in natural plant communities (Sternberg et al., 1999; Grime et al., 2000; Lloret et al., 2009; Yang et al., 2011; Shi et al, 2015), (2) open-top chambers in the field to increase temperatures during the growing season (Klanderud and Totland, 2005; Walker et al., 2006), (3) increase/decrease of natural amount of precipitation in the field (Sternberg et al., 1999 Grime et al., 2000; Lloret et al, 2009; Yang et al., 2011; Castillioni et al., 2020), (4) varying temperatures, watering regimes, and CO_2 on monoliths [i.e., samples (0.5 m × 0.5 m × 0.4 m depth of soil and vegetation) removed from grassland (Cantarel et al., 2013)], (5) long-term monitoring of vegetation (Bertrand et al., 2011; Munson et al., 2012; Aguirre-Gutiérrez et al., 2019), (6) establishment of transects/plots on altitudinal gradients (Bässler et al., 2010; Sheldon et al., 2011), (7) resampling existing vegetation relevés (Van Der Veken et al., 2004; Lenoir et al., 2010; Petriccione and Bricca, 2019), and (8) models (Frenette-Dussault et al., 2013; Mokany et al., 2015).

The responses of a plant community to the effects of increased temperature vary with the plant species and geographical location. A plant community may exhibit (1) resistance to change, at least for several years (Grime et al., 2000; Klanderud and Totland, 2005; Bertrand et al., 2011), (2) a reduction in species richness (Walker et al., 2006; Yang et al., 2011), (3) community disassembly (Sheldon et al., 2011), (4) increase in drought-tolerant species (Petriccione and Bricca, 2019), (5) increase in deciduous species (Aguirre-Gutiérrez et al., 2019), and (6) an influx for species from lower elevations (Lenoir et al., 2010) or lower latitudes (Van Der Veken et al., 2004). Also, Seidl et al. (2017) predicted that global warming will increase the possibility of increased fire frequency and invasion of insects and pathogens that will greatly impact plant communities.

Changes in precipitation also can have effects on plant communities. With increased summer precipitation, species richness increased in calcareous grasslands in the United Kingdom (Sternberg et al., 1999), temperate steppes in China (Yang et al., 2011), and prairies in Oklahoma (USA) (Castillioni et al., 2020). However, decreased precipitation can significantly decrease seedling survival, which ultimately results in changes in the species composition of the community. For example, persistent drought in Mediterranean shrublands decreased seedling establishment of all species, but there was a differential decrease in the number of established seedlings. Fewer seedlings of *Dorycnium pentaphyllum* and *Erica multiflora* than of *Coris monspeliensis* and *Fumana thymifolia* became established (Lloret et al., 2009). With decreased annual precipitation in the Sonoran Desert (USA), perennial grasses declined and cacti increased (Munson et al., 2012).

Interactions between fire, seed banks, and invasive species

Studies have shown that global warming may be related to changes in timing, intensify, and frequency of fire (Miller et al., 2019; Fernández-García et al., 2020), size of seed banks in the habitat (Ooi et al., 2009; Del Cacho et al., 2012; Cochrane, 2017b; Aragón-Gastélum et al., 2018), and increased numbers of invasive species (Huebner et al., 2018; Albuquerque et al., 2019; Kerns et al., 2020), all of which can result in shifts in species composition of plant communities. Although fire, seed banks, and invasive species can change in response to increased temperatures and changes in precipitation, they also can interact with each other. Fire can promote (Cóbar-Carranza et al., 2015; Kerns et al., 2020) or impede (Moreschi et al., 2019) the establishment of invasive species, and the presence of invasive grasses in some ecosystems may increase the occurrence of fire (Fusco et al., 2019).

Depending on the species, invasives may form a large (Fumanal et al., 2008; Gioria and Osborne, 2009), relatively small (Blossey et al., 2017; Hopenfensperger et al., 2019; Skálová et al., 2019), or no (Gioria and Osborne, 2010) persistent soil seed bank. One way in which fire can prevent, or at least reduce, the number of seedlings emerging at a site is that it destroys seeds (Tesfaye et al., 2004; Vermeire and Rinella, 2009; Daibes et al., 2017). Further, fire may kill seeds in the soil seed bank of both native (Cao et al., 2000) and invasive (Beckstead et al., 2011) species. However, the heat from fires can break physical dormancy (water-impermeable seed or fruit coat), thereby promoting germination (Daibes et al., 2017) and cause the opening of serotinous (closed) fruits/cones of species with aerial seed banks, thereby releasing the seeds into a favorable environment for germination and seedling establishment (Lamont et al., 1991; Alexander and Cruz, 2012).

Special concerns about seagrasses, species with desiccation-sensitive seeds, and annual food crops

Seagrasses

Seagrass meadows are found in coastal waters in subtropical and tropical regions at depths of 1−90 m, depending on the species and light attenuation (Duarte, 1991). Although the plants grow while submerged in seawater, their habitat is not immune to the effects of global climate change. The kinds of changes that could occur (are occurring) in the seagrass habitat that can be correlated with climate change include increased temperatures and water turbidity; changes in water levels, salinity, CO_2 concentrations, pH, and tidal depth; and increased water motion (Short and Neckles, 1999; Koch et al., 2007; Statton et al., 2017; Sunny, 2017; Brodie and N'Yeurt, 2018; Ontoria et al., 2019; Aoki et al., 2020). It seems that any/all of these changes in the environment could affect seed germination and seedling survival and establishment and thus ultimately have a major impact on population stability.

Species with recalcitrant seeds

The concern about species with recalcitrant (desiccation-sensitive) seeds is that they lose viability if their moisture content decreases below a certain level, generally approximately 20%−40% depending on the species (Roberts, 1973. In contrast, orthodox (desiccation-tolerant) seeds can be dried to a moisture content of 2%−5% without loss of viability. Species with desiccation-sensitive seeds occur in a range of vegetation zones, but they are the most common in warm, wet evergreen rain forests (Tweddle et al., 2003). The majority of desiccation-sensitive seeds are nondormant, metabolically active, and thus germinate shortly after dispersal (Berjak and Pammenter, 2008; Marques et al., 2018); however, some desiccation-sensitive seeds are dormant at maturity and require a period of incubation at dormancy-breaking conditions before they germinate (Pritchard et al., 1999; Yu et al., 2008). Regardless of whether seeds are nondormant or dormant at maturity, their moisture content must be kept at a species-specific high level, or they will die. Given that climate change may affect temperature and water relations in the habitat, we need to know the impact of increased temperatures (Sershen et al., 2014) and drought stress (Joët et al., 2016) on seed germination and seedling establishment and ultimately the survival of the species with desiccation-sensitive seeds.

Annual crops

The effects of climate change on the growth and yield of annual crops are related to the food security of human populations (Jones and Thornton, 2003; Serdeczny et al., 2017; Ray et al., 2019; WMO, 2019). The most common annual food crops in terms of hectares of production are sugar cane (*Saccharum* spp.), maize (*Zea mays*), wheat (*Triticum aestivum*), and rice (*Oryza sativa*) (White et al., 2011; Lobell and Gourdji, 2012). Increased concentrations of CO_2 in the atmosphere and relatively small temperature increases (e.g., 0.5°C) may increase crop yields if soil moisture is not limiting, but this varies with the crop (Devkota et al., 2013; Chen et al., 2014; Bao et al., 2015; Rose et al., 2016; Van Oort and Zwart, 2017). However, large temperature increases (e.g., 2°C) can decrease the length of the growing season by causing early maturation of seeds (Rose et al., 2016) and cause low spikelet (pollen) fertility (Devkota et al., 2013; Van Oort and Zwart, 2017). With global warming, there may be changes in the geographical regions that are the most suitable for the production of various crops (Ortiz et al., 2008; Ray et al., 2019). Thus, agronomic research is underway to develop strategies, such as changes in planting date (Bao et al., 2015; Rose et al., 2016; Paymard et al., 2018), development of heat-tolerant (Ortiz et al., 2008; Devkota et al., 2013) drought-tolerant (Raza et al., 2019) varieties, and construction of irrigation systems (Lobell and Gourdji, 2012; Paymard et al., 2018), to deal with the effects of climate change and thus maintain high yields of annual crops.

Aims of this book

The paper by Walck et al. (2011) entitled "Climate change and plant regeneration from seed" clearly articulated the need for a broad understanding of this topic. These authors emphasized that we lack a good knowledge of the effects of climate change on the regeneration of species from seeds in many vegetation regions on earth and of various key issues related to seed germination ecology. One aim of this book is to provide a review of the effects of climate change on regeneration from seeds in a wide range of vegetation types from tropical wet forests to Arctic/alpine tundra (Fig. 1). Each of the chapters on a type of vegetation provides background on the expected changes in environmental conditions due to global warming, what is known about regeneration from seeds under the modified environmental conditions, consequences on species dynamics, and community composition and research questions that need to be addressed.

For each type of vegetation, will increases in atmospheric CO_2 concentration (ppmv) and temperature and increases/decreases in precipitation have an effect on seed production, dormancy break, timing of germination, and other life history traits? Also, will there be loss and/or migration of species and thus modifications in plant community structure and composition? When the reviews for the various types of vegetation are assembled, we will have information with which to address two important questions. (1) In what types of vegetation are the most (or highest proportion of) species extinctions and/or migrations likely to occur? (2) In what types of vegetation, might we expect large shifts in community structure and composition?

As a consequence of global warming, changes are predicted in frequency and intensity of fire, composition of soil seed banks, and number and abundance of invasive plant species in many vegetation regions (see above). A second aim of this book is to evaluate how fire, soil seed banks, and invasive species could be changed by global warming and what effect these changes might have on species and community dynamics.

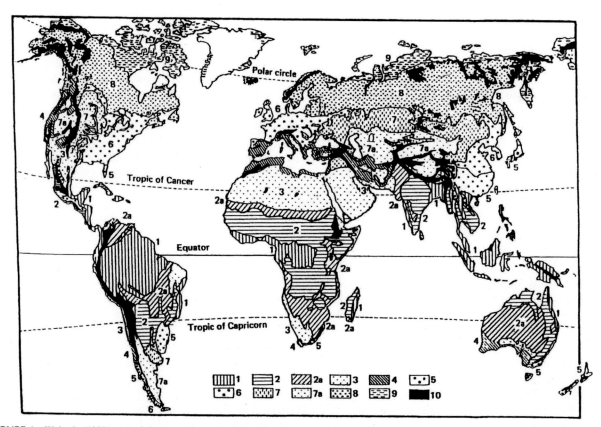

FIGURE 1 Walter's (1979) map of the vegetation zones of the world. (I) Tropical and subtropical zones: (1) evergreen rain forests of the lowlands and mountainsides (cloud forests); (2) semievergreen and deciduous forests; (2a) dry woodlands, natural savannas or grasslands; (3) hot semideserts and deserts, poleward up to a latitude of 35°C (see also 7a). (II) Temperate and arctic zones: (4) sclerophyllous woodlands with winter rain; (5) moist warm temperature woodlands; (6) deciduous (nemoral) forests; (7) steppes; (7a) semideserts and deserts with cold winters; (8) boreal coniferous zone; (9) tundra; (10) mountains.

A third aim of the book is to evaluate the effects of climate change on what we perceive to be highly vulnerable kinds of plants, such as seagrasses, species with recalcitrant seeds, and annual crops. How sensitive are these special kinds of plants to increased CO_2 levels, temperatures, and/or changes in precipitation? Are seagrasses and species with recalcitrant seeds threatened with extinction, and is migration a possibility for them? Can the effects of climate change on annual crops, for example, maize, be mitigated, thus ensuring adequate food production for the health and wellbeing of humans? Can we gain insight from results of the detailed studies conducted on the life cycle of maize that will generate exciting new research questions about the effects of heat and water stress on the early life history stages of wild species?

To accomplish the aims outlined for this book, we have secured the help of a large group of plant scientists from 20 countries. The research interests of these scientists cover many aspects of plant ecology, resulting in a team whose combined knowledge, expertise, and experience will provide new insight into the issues related to the regeneration of plant species as we contemplate the future in terms of global warming.

References

Adkins, S., Loewen, M., Symons, S.J., 1986. Variation with pure lines of wild oats (*Avena fatua*) in relation to degree of primary dormancy. Weed Sci. 84, 859–864.

Aguirre-Gutiérrez, J., Oliveras, I., Rifai, S., Fauset, S., Adu-Bredu, S., Affum-Baffoe, K., et al., 2019. Drier tropical forests are susceptible to functional changes in response to a long-term drought. Ecol. Lett. 22, 855–865.

Ahrens, C.W., Andrew, M.E., Mazanec, R.A., Ruthrof, K.X., Challis, A., Hardy, G., et al., 2020. Plant functional traits differ in adaptability and are predicted to be differentially affected by climate change. Ecol. Evol. 10, 232–248.

Albuquerque, F.A., Macías-Rodríguez, M.A., Búrquez, Astudillo-Scalia, Y., 2019. Climate change and the potential expansion of buffelgrass (*Cenchrus ciliaris* L., Poaceae) in biotic communities of southwest United States and northern Mexico. Biol. Inv. 21, 3335–3347.

Alexander, M.E., Cruz, M.G., 2012. Modelling the effects of surface and crown fires behavior on serotinous cone opening in jack pine and lodgepole pine forests. Int. J. Wildl. Fire 21, 709–721.

Anderson, J.T., Panetta, A.M., Mitchell-Olds, T., 2012. Evolutionary and ecological responses to anthropogenic climate change. Plant Physiol. 160, 1728–1740.

Anderson, J.T., Wadgymar, S.M., 2020. Climate change disrupts local adaptation and favours upslope migration. Ecol. Lett. 23, 181–192.

Aoki, L.R., McGlatherym, K.J., Wiberg, P.L., Al-Haj, A., 2020. Depth affects seagrass restoration success and resilience to marine heat wave disturbance. Estuaries Coasts 43, 316–328.

Aragón-Gastélum, J.L., Flores, J., Jurado, E., Ramírez-Tobías, H.M., Robles-Díaz, E., Rodas-Ortiz, J.P., et al., 2018. Potential impact of global warming on seed bank, dormancy and germination of three succulent species from the Chihuahuan Desert. Seed Sci. Res. 28, 312–318.

Awan, S., Footitt, S., Finch-Savage, W.E., 2018. Interaction of maternal environment and allelic differences in seed vigour genes determines seed performance in *Brassica oleracea*. Plant J. 94, 1098–1108.

Axelsson, E.P., Grady, K.C., Lardizabal, M.L.T., Nair, I.B.S., Rinus, D., Ilstedt, U., 2020. A pre-adaptive approach for tropical forest restoration during climate change using naturally occurring genetic variation in response to water limitation. Restor. Ecol. 28, 156–165.

Bandara, R.G., Finch, J., Walck, J.L., Hidayati, S.N., Havens, K., 2019. Germination niche breadth and potential response to climate change differ among three North American perennials. Folia Geobot. 54, 5–17.

Bao, Y., Hoogenboom, G., McClendon, R.W., Paz, J.O., 2015. Potential adaptation strategies for rainfed soybean production in the south-eastern USA under climate change based on the CSM-CROPGRO-Soybean model. J. Agric. Sci. 153, 798–824.

Baskin, C.C., Baskin, J.M., 2014. Seeds: Ecology, Biogeography and Evolution of Dormancy and Germination, 2nd ed. Academic Press/Elsevier, San Diego.

Baskin, J.M., Baskin, C.C., 1971. Germination ecology and adaptation to habitat in *Leavenworthia* spp. (Cruciferae). Amer. Midl. Nat. 85, 22–35.

Baskin, J.M., Baskin, C.C., 1973. Ecological life cycle of *Helenium amarum* in central Tennessee. Bull. Torrey Bot. Club 100, 117–124.

Baskin, J.M., Baskin, C.C., 1985. The annual dormancy cycle in buried weed seeds: a continuum. BioScience 35, 492–498.

Baskin, J.M., Baskin, C.C., 2019. How much influence does the paternal parent have on seed germination? Seed Sci. Res. 29, 1–11.

Bässler, C., Müller, J., Dziock, F., 2010. Detection of climate-sensitive zones and identification of climate change indicators: a case study from the Bavarian Forest National Park. Folia Geobot. 45, 163–182.

Becklin, K.M., Anderson, J.T., Gerhart, L.M., Wadgymar, S.M., Wessinger, C.A., Ward, J.K., 2016. Examining plant physiological responses to climate change through an evolutionary lens. Plant Physiol. 172, 635–649.

Beckstead, J., Street, L.E., Meyer, S.E., Allen, P.S., 2011. Fire effects on the cheatgrass seed bank pathogen *Pyrenophora semeniperda*. Rangel. Ecol. Manage 64, 148–157.

Berjak, P., Pammenter, N.W., 2008. From *Avicennia* to *Zizania*: seed recalcitrance in perspective. Ann. Bot. 101, 213–228.

Bertrand, R., Lenoir, J., Piedallu, C., Riofrio-Dillon, G., de Raffray, P., Vidal, C., et al., 2011. Changes in plant community composition lag behind climate warming in lowland forests. Nature 479, 517–520.

Bianchi, E., Bugmann, H., Bigler, C., 2019. Early emergence increases survival of tree seedlings in central European temperate forests despite severe late frost. Ecol. Evol. 9, 8238–8252.

Blok, S.E., Olesen, B., Krause-Jensen, D., 2018. Life history events of eelgrass *Zostera marina* L. populations across gradients of latitude and temperature. Mar. Ecol. Prog. Ser. 590, 79–93.

Blossey, B., Nuzzo, V., Dávalos, A., 2017. Climate and rapid local adaptation as drivers of germination and seed bank dynamics of *Alliaria petiolata* (garlic mustard) in North America. J. Ecol. 105, 1485–1495.

Blumstein, M., Richardson, A., Weston, D., Zhang, J., Muchero, W., Hopkins, R., 2020. A new perspective on ecological prediction reveals limits to climate adaptation in a temperate tree species. Curr. Biol. 30, 1447–1453.

Bogdziewicz, M., Fernández-Martínez, M., Espelta, J.M., Ogaya, R., Penuelas, J., 2020. Is forest fecundity resistant to drought? Results from an 18-yr rainfall-reduction experiment. New Phytol. 227, 1073–1080.

Brodie, G., N'Yeurt, A.D.R., 2018. Effects of climate change on seagrasses and seagrass habitats relevant to the Pacific Islands. Pac. Mar. Clim. Change Rept. Card. Sci. Rev. 2018, 112–131.

Burghardt, L.T., Edwards, B.R., Donohue, K., 2016. Multiple paths to similar germination behavior in *Arabidopsis thaliana*. New Phytol. 209, 1301–1312.

Cantarel, A.A.M., Bloor, J.M.G., Soussana, J.-F., 2013. Four years of simulated climate change reduces above-ground productivity and alters functional diversity in a grassland ecosystem. J. Veg. Sci. 24, 113–126.

Cao, M., Tang, Y., Sheng, C., Zhang, J., 2000. Viable seeds buried in the tropical forest soils of Xishuangbanna, SW China. Seed Sci. Res. 10, 255–264.

Carvalho, M., Matos, M., Castro, I., Monteiro, E., Rosa, E., Lino-Neto, T.V., et al., 2019. Screening of worldwide cowpea collection to drought tolerant at a germination stage. Scient. Hortic. 247, 107–115.

Castillioni, K., Wilcox, K., Jiang, L., Luo, Y., Jung, C.G., Souza, L., 2020. Drought mildly reduces plant dominance in a temperature prairie ecosystem across years. Ecol. Evol. 10, 6702–6713.

Catelotti, K., Bino, G., Offord, C.A., 2020. Thermal germination niches of *Persoonia* species and projected spatiotemporal shifts under a changing climate. Divers. Distrib. 26, 589–609.

Chen, C., Zhou, G.-S., Zhou, L., 2014. Impacts of climate change on rice yield in China from 1961 to 2010 based on provincial data. J. Integr. Agric. 13, 1555–1564.

Chen, I.-C., Hill, J.K., Ohlemüller, R., Roy, D.B., Thomas, C.D., 2011. Rapid range shifts of species associated with high levels of climate warming. Science 333, 1024–1026.

Chen, Q., Yin, Y., Zhao, R., Yang, Y., Teixeira da Silva, J.A., Yu, X., 2020. Incorporating local adaptation into species distribution modeling of *Paeonia mairei*, an endemic plant to China. Front. Plant. Sci. 10, e1717. Available from: https://doi.org/10.3389/fpls.2019.01717.

Chevin, L.-M., Hoffmann, A.A., 2017. Evolution of phenotypic plasticity in extreme environments. Phil. Trans. R. Soc. B 372, 20160138. Available from: https://doi.org/10.1098/rstb.2016.0138.

Chiluwal, A., Bheemanahalli, R., Kanaganahalli, V., Boyle, D., Perumal, R., Pokharel, M., et al., 2019. Deterioration of ovary plays a key role in heat stress-induced spikelet sterility in sorghum. Plant. Cell Environ. 43, 448–462.

Cóbar-Carranza, A.J., García, R.A., Pauchard, A., Peñña, E., 2015. Efecto de la alta temperatura en la germinacion y supervivencia de semillas del la especie invasora *Pinus contorta* y dos especies nativas del sur de Chile. Bosque 36, 53–60.

Cochrane, A., 2017a. Modelling seed germination response to temperature in *Eucalyptus* L'Her. (Myrtaceae) species in the context of global warming. Seed Sci. Res. 27, 99–109.

Cochrane, A., 2017b. Are we underestimating the impact of rising summer temperatures on dormancy loss in hard-seeded species? Aust. J. Bot. 65, 248–256.

Cochrane, A., 2019. Effects of temperature on germination of eight Western Australian herbaceous species. Folia Geobot. 54, 29–42.

Connolly-Boutin, L., Smit, B., 2016. Climate change, food security, and livelihoods in sub-Saharan Africa. Reg. Environ. Change 16, 385–399.

Corlett, R.T., Westcott, D.A., 2013. Will plant movements keep up with climate change? Trends Ecol. Evol. 28, 482–488.

Cuena-Lombaraña, A., Porceddu, M., Dettori, C.A., Bacchetta, G., 2020. Predicting the consequences of global warming on *Gentiana lutea* germination at the edge of its distributional and ecological range. PeerJ 8, e8894. Available from: https://doi.org/10.7717/2Fpeerj.8894.

Daibes, L.F., Zupo, T., Silveira, F.A.O., Fidelis, A., 2017. A field perspective on effects of fire and temperature fluctuation on cerrado legume seeds. Seed Sci. Res. 27, 74–83.

Davis, E.L., Brown, R., Daniels, L., Kavanagh, T., Gedalof, Z., 2020. Regional variability in the response of alpine treelines to climate change. Clim. Change 162, 1365–1384.

Davis, E.L., Gedalof, Z., 2018. Limited prospects for future alpine treeline advance in the Canadian Rocky Mountains. Glob. Change Biol. 24, 4489–4504.

Del Cacho, M., Saura-Mas, S., Estiarte, M., Peñuelas, J., Lloret, F., 2012. Effect of experimentally induced climate change on the seed bank of a Mediterranean shrubland. J. Veg. Sci. 23, 280–291.

Deng, H., Pepin, N.C., Chen, Y., 2017. Changes of snowfall under warming in the Tibetan Plateau. J. Geophy. Res. Atmos. 122, 7323–7341.

Devkota, K.P., Manschadi, A.M., Devkota, M., Lamers, J.P.A., Ruzibaev, E., Egamberdiev, O., et al., 2013. Simulating the impact of climate change on rice phenology and grain yield in irrigated drylands of Central Asia. J. Appl. Meteorol. Climatol. 52, 2033–2050.

Dickman, E.E., Pennington, L.K., Franks, S.J., Sexton, J.P., 2019. Evidence for adaptive responses to historic drought across a native plant species range. Evol. Appl. 12, 1569–1582.

Donohue, K., 2002. Germination timing influences natural selection of life-history characters in *Arabidopsis thaliana*. Ecology 83, 1006–1016.

Donohue, R., Rubio de Cases, R., Burghardt, L., Kovach, K., Willis, C.G., 2010. Germination, postgermination adaptation, and species ecological ranges. Annu. Rev. Ecol., Evol. Syst. 41, 293–319.

Duarte, C.M., 1991. Seagrass depth limits. Aquat. Bot. 40, 363–377.

Dullinger, S., Gattringer, A., Thuiller, W., Moser, D., Zimmermann, N.E., Guisan, A., et al., 2012. Extinction debt of high-mountain plants under twenty-first-century climate change. Nat. Clim. Change 2, 619–622.

Elgabra, M., El-Keblaway, A., Mosa, K.A., Soliman, S., 2019. Factors controlling seed dormancy and germination response of *Brachypodium hybridum* growing in the hot arid mountains of the Arabian Desert. Botany 97, 371–379.

Ensing, D.J., Eckert, C.G., 2019. Interannual variation in season length is linked to strong co-gradient plasticity of phenology in a montane annual plant. New Phytol. 224, 1184–1200.

Etterson, J.R., Cornet, M.W., White, M.A., Kavajecz, L.C., 2020. Assisted migration across fixed seed zones detects adaptation lags in two major North American tree species. Ecol. Appl. 30, e02092. Available from: https://doi.org/10.1002/eap.2092.

Feng, X., Liu, C., Xie, F., Lu, J., Chiu, L.S., Tintera, G., et al., 2019. Precipitation characteristic changes due to global warming in a high-resolution (16 km) ECMWF simulation. Quart. J. R. Meteorol. Soc. 145, 303–317.

Fernández-García, V., Marcos, E., Fulé, P.Z., Reyes, O., Santana, V.M., Calvo, L., 2020. Fire regimes shape diversity and traits of vegetation under different climatic conditions. Sci. Total Environ. 716, 137137. Available from: https://doi.org/10.1002/eap.2092.

Fernández-Pascual, E., Jiménez-Alfaro, Bueno, Á., 2017. Comparative seed germination traits in alpine and subalpine grasslands: higher elevations are associated with warmer germination temperatures. Plant Biol. 19, 32–40.

Fernández-Pascual, E., Mattana, E., Pritchard, H.W., 2019. Seeds of future past: climate change and the thermal memory of plant reproductive traits. Biol. Rev. 94, 439–456.

Flannigan, M.D., Amiro, B.D., Logan, K.A., Stocks, B.J., Wotton, B.M., 2005. Forest fires and climate change in the 21st century. Mitig. Adapt. Strateg. Glob. Change 11, 847–859.

Flanaga, S.A., Hurtt, G.C., Fisk, J.P., Sahajpal, R., Zhao, M., Dubayah, R., et al., 2019. Potential transient response of terrestrial vegetation and carbon in northern North America from climate change. Climate 7, 113. Available from: https://doi.org/10.3390/cli7090113.

Footitt, S., Huang, Z., Ölcer-Footitt, H., Clay, H., Finch-Savage, W.E., 2018. The impact of global warming on germination and seedling emergence in *Alliaria petiolata*, a woodland species with dormancy loss dependent on low temperature. Plant. Biol. 20, 682–690.

Franks, S.J., Weber, J.J., Aitken, S.N., 2014. Evolutionary and plastic responses to climate change in terrestrial plant populations. Evol. Appl. 7, 123–139.

Frenette-Dussault, C., Shipley, B., Meziane, D., Hingrat, Y., 2013. Trait-based climate change predictions of plant community structure in arid steppes. J. Ecol. 101, 484–492.

Frenne, P.D., Graae, B.J., Brunet, J., Shevtsova, A., Schrijver, A.D., Chabrerie, O., et al., 2012. The response of forest plant regeneration to temperature variation along a latitudinal gradient. Ann. Bot. 109, 1037–1046.

Fumanal, B., Gaudot, I., Bretagnolle, F., 2008. Seed-bank dynamics in the invasive plant, *Ambrosia artemisiifolia* L. Seed Sci. Res. 18, 101–114.

Fusco, E.J., Finn, J.T., Balch, J.K., Nagy, R.C., Bradley, B.A., 2019. Invasive grasses increase fire occurrence and frequency across US ecoregions. Proc. Nat. Acad. Sci. U.S.A. 116, 23594–23599.

Gárate-Escamilla, H., Hampe, A., Vizaíno-Palomar, N., Robson, T.M., Garzón, M.B., 2019. Range-wide variation in local adaptation and phenotypic plasticity of fitness-related traits in *Fagus sylvatica* and their implications under climate change. Glob. Ecol. Biogeogr. 28, 1336–1350.

Garcia, R.A., Cabeza, M., Rahbek, C., Araújo, M.B., 2014. Multiple dimensions of climate change and their implications for biodiversity. Science 344 (6183), 1247579. Available from: https://doi.org/10.1126/science.1247579.

García-Fernández, A., Escudero, A., Lara-Romero, C., Iriondo, J.M., 2015. Effects of the duration of cold stratification on early life stages of the Mediterranean alpine plant *Silene ciliata*. Plant Biol. 17, 344–350.

Gillett, N.P., Weaver, A.J., Zwiers, F.W., Flannigan, M.D., 2004. Detecting the effect of climate change on Canadian forest fires. Geophy. Res. Lett. 31, L18211. Available from: https://doi.org/10.1029/2004GL020876.

Giménez-Benavides, L., Escudero, A., García-Camacho, R., García-Fernández, A., Iriondo, J.M., Lara-Romero, C., et al., 2018. How does climate change affect regeneration of Mediterranean high-mountain plants? An integration and synthesis of current knowledge. Plant. Biol. 20, 50–62.

Gioria, M., Osborne, B., 2009. The impact of *Gunnera tinctoria* (Molina) Mirbel invasions on soil seed bank communities. J. Plant. Ecol. 2, 153–167.

Gioria, M., Osborne, B., 2010. Similarities in the impact of three large invasive plant species on soil seed bank communities. Biol. Inv. 12, 1671–1683.

Gratani, L., 2014. Plant phenotypic plasticity in response to environmental factors. Adv. Bot. 2014, 208747. Available from: https://doi.org/10.1155/2014/208747.

Gray, F., Cochrane, A., Poot, P., 2019. Provenance modulates sensitivity of stored seeds of the Australian native grass *Neurachne alopecuoides* to temperature and moisture availability. Aust. J. Bot. 67, 106–115.

Gremer, J.R., Wilcox, C.J., Chiono, A., Suglia, E., Schmitt, J., 2020. Germination timing and chilling exposure create contingency in life history and influence fitness in the native wildflower *Streptanthus tortuosus*. J. Ecol. 108, 239–255.

Grenier, S., Barre, P., Litrico, I., 2016. Phenotypic plasticity and selection: nonexclusive mechanisms of adaptation. Scientifica 2016, 7021701. Available from: https://dx.doi.org/10.1155/2016/7021701.

Greve, P., Orlowsky, B., Mueller, B., Sheffield, J., Reichstein, M., Seneviratne, S.I., 2014. Global assessment of trends in wetting and drying over land. Nat. Geosci. 7, 716–721.

Grime, J.P., Brown, V.K., Thompson, K., Masters, G.J., Hillier, S.H., Clarke, I.P., et al., 2000. The response of two contrasting limestone grasslands to simulated climate change. Science 289, 762−765.

Haile, G.G., Tang, Q., Sun, S., Huang, Z., Zhang, X., Liu, X., 2019. Droughts in East Africa: causes, impact and resilience. Earth-Sci. Rev. 193, 146−161.

Han, A.R., Kim, H.J., Jung, J.B., Park, P.S., 2018. Seed germination and initial seedling survival of the subalpine tree species, *Picea jezoensis*, on different forest floor substrates under elevated temperature. For. Ecol. Manage. 429, 579−588.

Harte, J., Saleska, S.R., Levy, C., 2015. Convergent ecosystem responses to 23-year ambient and manipulated warming link advancing snowmelt and shrub encroachment to transient and long-term climate-soil carbon feedback. Glob. Change Biol. 21, 2349−2356.

Harte, J., Shaw, R., 1995. Shifting dominance within a montane vegetation community: results of a climate-warming experiment. Science 267, 876−880.

Hendry, A.P., 2016. Key questions on the role of phenotypic plasticity in eco-evolutionary dynamics. J. Heredity 107, 25−41.

Hopenfensperger, K.N., Boyce, R.L., Schenk, D., 2019. Potential reinvasion of *Lonicera maackii* after urban riparian forest restoration. Ecol. Restor. 37, 25−33.

Hoyle, G.L., Cordiner, H., Good, R.B., Nicotra, A.B., 2014. Effects of reduced winter duration on seed dormancy and germination in six populations of the alpine herb *Aciphyllya glacialis* (Apiaceae). Conserv. Physiol. 2, cou015. Available from: https://doi.org/10.1093/conphys/cou015.

Hsuan, T.-P., Jhuang, P.-R., Wu, W.-C., Lur, H.-S., 2019. Thermotolerance evaluation of Taiwan Japonica type rice cultivars at the seedling stage. Bot. Stud. 60, 29. Available from: https://doi.org/10.1186/s40529-019-0277-7.

Huang, Z., Footitt, S., Tang, A., Finch-Savage, W.E., 2018. Predicted global warming scenarios impact on the mother plant to alter seed dormancy and germination behavior in *Arabidopsis*. Plant. Cell Environ. 41, 187−197.

Huebner, C.D., Regula, A.E., McGill, D.W., 2018. Germination, survival, and early growth of three invasive plants in response to five forest management regimes common to US northeastern deciduous forests. For. Ecol. Manage. 425, 100−118.

IPCC (Intergovernment Panel on Climate Change), 2018. Summary for policymakers. In: Masson-Delmotte, V., Zhai, P., Pőrtner, H.-O., Skea, J., Shukla, P.R., Pirani, A., et al.,Global Warming of 1.5°C. Special Report on the Impacts of Global Warming of 1.5°C Above Pre-industrial Levels and Related Global Greenhouse Gas Emission Pathways, in the Context of Strengthening the Global Response to the Threat of Climate Change, Sustainable Development, and Efforts to Eradicate Poverty. United Nations Environment Programme.

Iqbal, H., Yaning, C., Rehman, H., Waqas, M., Ahmed, Z., Rasa, S.T., et al., 2020. Improving heat stress tolerance in late planted spring maize by using different exogenous elicitors. Chil. J. Agric. Res. 80, 30−40.

Joët, T., Ourcival, J.-M., Capelli, M., Dussert, S., Morin, X., 2016. Explanatory ecological factors for the persistence of desiccation-sensitive seeds in transient soil seed banks: *Quercus ilex* as a case study. Ann. Bot. 117, 165−176.

Jones, P.G., Thornton, P.K., 2003. The potential impacts of climate change on maize production in Africa and Latin American in 2055. Glob. Environ. Change 13, 51−59.

Kamali, B., Abbaspour, K.C., Lehmann, A., Wehrli, B., Yang, H., 2018. Spatial assessment of maize physical drought vulnerability in sub-Saharan Africa: linking drought exposure with crop failure. Environ. Res. Lett. 13, 074010. Available from: https://doi.org/10.1088/1748-9326/aacb37.

Kambo, D., Danvy, R.K., 2018. Constraints on treeline advance in a warming climate: a test of the reproduction limitation hypothesis. J. Plant. Ecol. 11, 411−422.

Kerns, B.K., Tortorelli, C., Day, M.A., Nietupski, T., Barros, A.M.G., Kim, J.B., et al., 2020. Invasive grasses: a new perfect storm for forested ecosystems? For. Ecol. Manage. 463, 117985. Available from: https://doi.org/10.1016/j.foreco.2020.117985.

Kijowska-Oberc, J., Staszak, A.M., Kamiński, J., Ratajczak, E., 2020. Adaptation of forest trees to rapidly changing climate. Forests 11, 123. Available from: https://doi.org/10.3390/f11020123.

Kim, D.H., Han, S.H., 2018. Direct effects on seed germination of 17 tree species under elevated temperature and CO_2 conditions. Open Life Sci. 13, 137−148.

Kimball, S., Angert, A.L., Huxman, T.E., Venable, D.L., 2010. Contemporary climate change in the Sonoran Desert favors cold-adapted species. Glob. Change Biol. 16, 1555−1565.

King, J.G., Hadfield, J.D., 2019. The evolution of phenotypic plasticity when environments fluctuate in time and space. Evol. Lett. 3, 15−27.

Klanderud, K., Totland, Ø., 2005. Simulated climate change altered dominance hierarchies and diversity of an alpine biodiversity hotspot. Ecology 86, 2047−2054.

Kueppers, L.M., Faist, A., Ferrenberg, S., Castanha, C., Conlisk, E., Wolf, J., 2017. Lab and field warming similarly advance germination date and limited germination rate for high and low elevation provenances of two widespread subalpine conifers. Forests 8, 433. Available from: https://doi.org/10.3390/f8110433.

Lamont, B.B., Le Maitre, D.C., Cowling, R.M., Enright, N.J., 1991. Canopy seed storage in woody plants. Bot. Rev. 57, 277−317.

Lampei, C., Tielbőrger, K., 2010. Evolvability of between-year seed dormancy in populations along an aridity gradient. Biol. J. Linn. Soc. 100, 924−934.

Lenoir, J., Gégout, J.C., Dupouey, J.L., Bert, D., Svenning, J.-C., 2010. Forest plant community change during 1989-2007 in response to climate warming in the Jura Mountains (France and Switzerland). J. Veg. Sci. 21, 949−964.

Lenoir, J., Gégout, J.C., Marquet, P.A., de Ruffray, P., Brisse, H., 2008. A significant upward shift in plant species optimum elevation during the 20th century. Science 320, 1768−1771.

Lett, S., Dorrepaal, E., 2018. Global drivers of tree seedling establishment at alpine treelines in a changing climate. Funct. Ecol. 32, 1666−1680.

Li, J., Chang, H., Liu, T., Zhang, C., 2019. The potential geographical distribution of *Haloxylon* across Central Asia under climate change in the 21st century. Agric. For. Meteorol. 275, 243−254.

Li, R., Chen, L., Wu, Y., Zhang, R., Baskin, C.C., Baskin, J.M., et al., 2017. Effects of cultivar and maternal environment on seed quality in *Vicia sativa*. Front. Plant. Sci. 8, 1411. Available from: https://doi.org/10.3389/fpls.2017.01411.

Lippmann, R., Babben, S., Menger, A., Delker, C., Quint, M., 2019. Development of wild and cultivated plants under global warming conditions. Curr. Biol. 29, R1326−R1338.

Lloret, F., Peñuelas, J., Prieto, P., Llorens, L., Estiarte, M., 2009. Plant community changes induced by experimental climate changes: seedling and adult species composition. Persp. Plant. Ecol. Evol. Syst. 11, 53−63.

Lobell, D.B., Gourdji, S.M., 2012. The influence of climate change in global crop productivity. Plant Physiol. 160, 1687−1697.

Lu, J.J., Tan, D.Y., Baskin, J.M., Baskin, C.C., 2014. Germination season and watering regime, but not seed morph, affect life history traits in a cold desert diaspore-heteromorphic annual. PLoS ONE 9, e102018. Available from: https://doi.org/10.1371/journal.pone.0102018.

Lubetkin, K.C., Westerling, A.L.-R., Kueppers, L.M., 2017. Climate and landscape drive the pace and pattern of conifer encroachment into subalpine meadows. Ecol. Applic 27, 1876−1887.

Manabe, S., 2019. Role of greenhouse gas in climate change. Tellus 71, 1620078. Available from: https://doi.org/10.1080/16000870.2019.1620078.

Marques, A., Buijs, G., Ligterink, W., Hilhorst, H., 2018. Evolutionary ecophysiology of seed desiccation sensitivity. Funct. Plant. Biol. 45, 1083−1095.

Merilä, J., Hendry, A.P., 2014. Climate change, adaptation, and phenotypic plasticity: the problem and the evidence. Evol. Applic. 7, 1−14.

Miller, R.G., Tangney, R., Enright, N.J., Fontaine, J.B., Merritt, D.J., Ooi, M.K., et al., 2019. Mechanisms of fire seasonality effects on plant populations. Trends Ecol. Evol. 34, 1104−1117.

Mitchell, N., Whitney, K.D., 2018. Can plants evolve to meet a changing climate? The potential of field experimental evolution studies. Am. J. Bot. 105, 1613−1616.

Mokany, K., Thomson, J.J., Lynch, A.J.J., Jordan, G.J., Ferrier, S., 2015. Linking changes in community composition and function under climate change. Ecol. Monogr. 25, 2132−2141.

Mondoni, A., Rossi, G., Orsenigo, S., Probert, R.J., 2012. Climate warming could shift the timing of seed germination in alpine plants. Ann. Bot. 110, 155−164.

Monllor, M., Soriano, P., Llinares, J.V., Boscaiu, M., Estrelles, E., 2018. Assessing effects of temperature change on four *Limonium* species from threatened Mediterranean salt-affected habitats. Not. Bot. Horti. Agrobo 46, 286−291.

Moreschi, E.G., Funes, G., Zeballos, S.R., Tecco, P.A., 2019. Post-burning germination responses of woody invaders in a fire-prone ecosystem. Aust. Ecol. 44, 1163−1173.

Müller, L.-L.B., Albach, D.C., Zotz, G., 2017. 'Are 3 °C too much?': thermal niche breadth in Bromeliaceae and global warming. J. Ecol. 105, 507−516.

Munson, S.M., Webb, R.J., Belnap, J., Hubbard, A., Swann, D.E., Rutman, S., 2012. Forecasting climate change impacts to plant community composition in the Sonoran Desert region. Glob. Change Biol. 18, 1083−1095.

Neelin, J.D., Sahany, S., Stechmann, S.N., Bernstein, D.N., 2017. Global warming precipitation accumulation increases above the current-climate cut-off scale. Proc. Nat. Acad. Sci. USA 114, 1258−1263.

Ontoria, Y., Cuesta-Garcia, A., Ruiz, J.M., Romero, J., Pérez, M., 2019. The negative effects of short-term extreme thermal events on the seagrass *Posidonia oceanica* are exacerbated by ammonium additions. PLoS ONE 14, e0222798. Available from: https://doi.org/10.1371/journal.pone.0222798.

Ooi, M.J., Auld, T.D., Denham, A.J., 2009. Climate change and bet-hedging: interactions between increased soil temperatures and seed bank persistence. Glob. Change Biol. 15, 2375−2386.

Oostra, V., Saastamoinen, M., Zwaan, B.J., Wheat, C.W., 2018. Strong phenotypic plasticity limits potential for evolutionary responses to climate change. Nat. Comm. 9, 1005. Available from: https://doi.org/10.1038/s41467-018-03384-9.

Orrù, M., Mattana, E., Pritchard, H.W., Bacchetta, G., 2012. Thermal thresholds as predictors of seed dormancy release and germination timing: altitude-related risks from climate warming from the wild grapevine *Vitis vinifera* subsp. *sylvestris*. Ann. Bot. 110, 1651−1660.

Ortiz, R., Sayre, K.D., Govaerts, B., Gupta, R., Subbarao, G.V., Ban, T., et al., 2008. Climate change: can wheat beat the heat? Agric. Ecosyst. Environ. 126, 46−58.

Panetta, A.M., Stanton, M.L., Harte, J., 2018. Climate warming drives local extinction: evidence from observation and experimentation. Sci. Adv. 4, eaaq1819. Available from: https://doi.10.1126/sciadv.aag1819.

Parmesan, C., Yohe, G., 2003. A globally coherent fingerprint of climate change impacts across natural systems. Nature 421, 37−42.

Pau, S., Cordell, S., Ostertag, R., Inman, F., Sack, L., 2020. Climatic sensitivity of species' vegetative and reproductive phenology in a Hawaiian montane wet forest. BioTropica 52, 825−835.

Paymard, P., Bannayan, M., Haghighi, R.S., 2018. Analysis of the climate change effect on wheat production systems and investigate the potential of management strategies. Nat. Hazards 91, 1237−1255.

Pérez-Ramos, I.M., Ourcival, J.M., Limousin, J.M., Rambal, S., 2010. Mast seeding under increasing drought: results from a long-term data set and from a rainfall exclusion experiment. Ecology 91, 3057−3068.

Peterson, M.L., Angert, A.L., Kay, K.M., 2020. Experimental migration upward in elevation is associated with strong selection on life history traits. Ecol. Evol. 10, 612−625.

Petriccione, B., Bricca, A., 2019. Thirty years of ecological research at the Gran Sasso d'Italia LTER site: climate change in action. Nat. Conserv. 34, 9–39.

Pokharel, M., Chiluwal, A., Stamm, M., Min, D., Rhodes, D., Jagadish, S.V.K., 2020. High night-time temperature during flowering and pod filling affects flower opening, yield and seed fatty acid composition in canola. J. Agron. Crop. Sci. 206, 579–596.

Pritchard, H.W., Steadman, K.J., Nash, J.V., Jones, C., 1999. Kinetics of dormancy release and the high temperature germination response in *Aesculus hippocastanum* seeds. J. Exp. Bot. 50, 1507–1514.

Qin, L.-Z., Kim, S.H., Song, H.-J., 2020. Influence of regional water temperature variability on the flowering phenology and sexual reproduction of the seagrass *Zostera marina* in Korean coastal water. Estuaries Coasts 43, 449–462.

Rago, A., Kouvaris, K., Uller, T., Watson, R., 2019. How adaptive plasticity evolves when selected against. PLoS ONE 15, e1006260. Available from: https://doi.org/10.1371/journal.pcbi.1006260.

Ray, D.K., West, P.C., Clark, M., Gerber, J.S., Prishchepov, A.V., Chatterjee, S., 2019. Climate change has likely already affected global food production. PLoS ONE 14, e0217148. Available from: https://doi.org/10.1371/journal.pone.0217148.

Raza, A., Razzaq, A., Mehmood, S.S., Zou, X., Zhang, X., Lv, Y., et al., 2019. Impact of climate change on crops adaptation and strategies to tackle its outcome: a review. Plants 8, 34. Available from: https://doi.org/10.3390/plants8020034.

Rees, W.G., Hofgaard, A., Boudreau, S., Cairns, D.M., Harper, K., Mamet, S., et al., 2020. Is subarctic forest advance able to keep pace with climate change? Glob. Change Biol. 26, 3965–3977.

Ribeiro, R.M., Tessarolo, G., Soares, T.N., Teixeira, I.R., Nabout, J.C., 2019. Global warming decreases the morphological traits of germination and environmental suitability of *Dipteryx alata* (Fabaceae) in Brazilian cerrado. Acta Bot. Bras. 33, 446–453.

Roach, D.A., Wulff, R.D., 1987. Maternal effects in plants. Annu. Rev. Ecol. Syst. 18, 209–235.

Roberts, E.H., 1973. Predicting the storage life of seeds. Seed. Sci. Technol. 1, 499–514.

Rose, G., Osborne, T., Greatrex, H., Wheeler, T., 2016. Impact of progressive global warming on the global-scale yield of maize and soybean. Clim. Change 134, 417–428.

Rubio de Casas, R., Donohue, K., Venable, D.L., Cheptou, P.-O., 2015. Gene-flow through space and time: dispersal, dormancy and adaptation to changing environments. Evol. Ecol. 29, 813–831.

Sang, Z., Sebastian-Azona, J., Hamann, A., Menzel, A., Hackle, U., 2019. Adaptive limitations of white spruce populations to drought imply vulnerability to climate change in its western range. Evol. Applic 12, 1850–1860.

Sauca, F., Anton, F.G., 2019. EW sources for genetic variability with resistance at drought obtained by interspecific hybridization between cultivated sunflower and the annual wild species *Helianthus argophyllus*. Sci. Pap. Ser. A—Agron. 62, 422–427.

Seidl, R., Thom, D., Kautz, M., Martin-Benito, D., Peltoniemi, M., Vacchiano, G., et al., 2017. Forest disturbances under climate change. Nat. Clim. Change 7, 395–402.

Serdeczny, O., Adams, S., Baarsch, F., Coumou, D., Robinson, A., Hare, W., et al., 2017. Climate change impacts in Sub-Saharan Africa: from physical changes to their social repercussions. Reg. Environ. Change 17, 1585–1600.

Sheldon, K.S., Yang, S., Tewksbury, J.J., 2011. Climate change and community disassembly: impacts of warming on tropical and temperate montane community structure. Ecol. Lett. 14, 1191–1200.

Shen, W., Zhang, L., Guo, Y., Luo, T., 2018. Causes for treeline stability under climate warming: evidence from seeds and seedling transplant experiments in southeast Tibet. For. Ecol. Manage. 408, 45–53.

Sershen, Perumal, A., Varghese, B., Govender, P., Ramdhani, S., Berjak, P., 2014. Effects of elevated temperatures on germination and subsequent seedling vigour in recalcitrant *Trichilia emetica* seeds. S. Afr. J. Bot. 90, 153–162.

Shi, Z., Sherry, R., Xu, X., Hararuk, O., Sousa, L., Jiang, L., et al., 2015. Evidence for long-term shift in plant community composition under decadal experimental warming. J. Ecol. 103, 1131–1140.

Short, F.T., Neckles, H.A., 1999. The effects of global climate change on seagrasses. Aquat. Bot. 63, 169–196.

Skálová, H., Moravcová, L., Čuda, J., Pyšek, P., 2019. Seed-bank dynamics of native and invasive *Impatiens* species during a five-year field experiment under various environmental conditions. NeoBiota 50, 75–95.

Solarik, K.A., Gravel, D., Ameztequi, A., Bergeron, Y., Messier, C., 2016. Assessing tree germination resilience to global warming: a manipulative experiment using sugar maple (*Acer saccharum*). Seed Sci. Res. 26, 153–164.

Solarik, K.A., Messier, C., Ouimet, R., Bergeron, Y., Gravel, D., 2018. Local adaptation of trees to the range margins impacts range shifts in the face of climate change. Glob. Ecol. Biogeogr. 27, 1507–1519.

Sommerville, K.D., Martyn, A.J., Offord, C.A., 2013. Can seed characteristics or species distribution be used to predict the stratification requirements of herbs in the Australian Alps? Bot. J. Linn. Soc. 172, 187–204.

Springate, D.A., Scarcelli, N., Rowntree, J., Kover, P.X., 2011. Correlated response in plasticity to selection for early flowering in *Arabidopsis thaliana*. J. Evol. Biol. 24, 2280–2288.

Statton, J., Sellers, R., Dixon, K.W., Kilminster, K., Merritt, D.J., Kendrick, G.A., 2017. Seed dormancy and germination of *Halophila ovalis* mediated by simulated seasonal temperature changes. Estuar. Coast. Shelf Sci. 198, 156–162.

Sternberg, M., Brown, V.K., Masters, G.J., Clarke, I.P., 1999. Plant community dynamics in a calcareous grassland under climate change manipulations. Plant. Ecol. 143, 29–37.

Sunny, A.R., 2017. A review on effect of global climate change on seaweed and seagrass. Int. J. Fish. Aquat. Stud. 5, 19–22.

Tesfaye, G., Assefa, Y., Teketay, D., Fetene, M., 2004. The impact of fire on the soil seed bank and regeneration of Harenna Forest, southeastern Ethiopia. Mount. Res. Develop. 24, 354–361.

Thomas, C.D., Cameron, A., Green, R.E., Bakkenes, M., Beaumont, L.J., Collingham, Y.C., et al., 2004. Extinction risk from climate change. Nature 427, 145–148.

Tietze, H.S.E., Joshi, J., Pugnaire, F.I., Dechoum, M.D.A., 2019. Seed germination and seedling establishment of an invasive tropical tree species under different climate change scenarios. Austral Ecol. 44, 1351–1358.
Trenberth, K.E., 2011. Changes in precipitation with climate change. Clim. Res. 47, 123–138.
Tweddle, J.C., Dickie, J.B., Baskin, C.C., Baskin, J.M., 2003. Ecological aspects of seed desiccation sensitivity. J. Ecol. 91, 294–304.
Van Der Veken, S., Bossuyt, B., Hermy, M., 2004. Climate gradients explain changes in plant community composition of the forest understorey: an extrapolation after climate warming. Belg. J. Bot. 137, 55–69.
Van Oort, P.A.J., Zwart, S.J., 2017. Impacts of climate change on rice production in Africa and causes of simulated yield changes. Glob. Change Biol. 24, 1029–1045.
Vermeire, L.T., Rinella, M.J., 2009. Fire alters emergence of invasive plant species from soil surface-deposited seeds. Weed Sci. 57, 304–310.
Vilà-Vilardell, L., Keeton, W.S., Thom, D., Gyeltshen, C., Tshering, K., Gratzxer, G., 2020. Climate change effects on wildfires in the wildland-urban-interface-Blue pine forests of Bhutan. For. Ecol. Manage. 461, 117927. Available from: https://doi.org/10.1016/j.foreco.2020.117927.
Wadgymar, S.M., Mactavish, R.M., Anderson, J.T., 2018a. Transgenerational and within-generation plasticity in response to climate change: insights from a manipulative field experiment across an elevational gradient. Am. Nat. 192, 698–714.
Wadgymar, S.M., Ogilvie, J.E., Inouye, D.W., Weiss, A.E., Anderson, J.T., 2018b. Phenological responses to multiple environmental drivers under climate change: insights from a long-term observational study and a manipulative field experiment. New Phytol. 218, 517–529.
Walck, J.L., Hidayati, S.N., Dixon, K.W., Thompson, K., Poschlod, P., 2011. Climate change and plant regeneration from seed. Glob. Change Biol. 17, 2145–2161.
Walker, M.D., Wahren, C.H., Hollister, R.D., Henry, G.H.R., Ahlquist, L.E., Alatalo, J.M., et al., 2006. Plant community responses to experimental warming across the tundra biome. Proc. Nat. Acad. Sci. U.S.A. 103, 1342–1346.
Wang, G., Baskin, C.C., Baskin, J.M., Yang, X., Liu, G., Ye, X., et al., 2018. Effects of climate warming and prolonged snow cover on phenology of the early life history stages of four alpine herbs on the southeastern Tibetan Plateau. Am. J. Bot. 105, 967–976.
Wang, Y., Tao, H., Zhang, P., Hou, X., Sheng, D., Tian, B., et al., 2020. Reduction in seed set upon exposure to high night temperature during flowering in maize. Physiol. Plant. 169, 73–82.
White, J.W., Hoogenboom, G., Kimball, B.A., Wall, G.W., 2011. Methodologies for simulating impacts of climate change on crop production. Field Crops Res. 124, 357–368.
Wiens, J.J., 2016. Climate-related local extinctions are already widespread among plant and animal species. PLoS ONE 14, e2001104. Available from: https://doi.org/10.1371/journal.pbio.2001104.
Winkler, D.E., Lin, M.Y.-C., Delgadillo, J., Chapin, K.J., Huxman, T.E., 2019. Early life history responses and phenotypic shifts in a rare endemic plant responding to climate change. Conserv. Physiol. 7, coz076. Available from: https://doi.org/10.1093/conphys/coz076.
WMO (World Meteorological Organization), 2019. WMO provisional statement on the state of the global climate in 2019. https://public.wmo.int/en/resources/library/wmo-provisional-statement-state-of-global-climate-2019.
Wright, S.J., Calderón, O., 2006. Seasonal, El Niño and longer term changes in flower and seed production in a moist tropical forest. Ecol. Lett. 9, 35–44.
Wu, W., Shah, F., Duncan, R.W., Ma, B.L., 2020. Grain yield, root growth habit and lodging of eight oilseed rape genotypes in response to a short period of heat stress during flowering. Agric. For. Meteorol. 287, 107954. Available from: https://doi.org/10.1016/j.agrformet.2020.107954.
Executive summary. In: Wuebbles, D.J., Fahey, D.W., Hibbard, K.A., DeAngelo, D., Doherty, S., Hayhoe, K., et al.,Climate Science Special Report: Fourth National Climate Assessment, Volume 1. U.S. Global Change Research Program, Washington, DC, pp. 12–34.
Wulff, R.D., 1995. Environmental maternal effects on seed quality and germination. In: Kigel, J., Galili, F. (Eds.), Seed Development and Germination. Marcel Dekker, New York, pp. 491–505.
Xu, J., Molinos, J.G., Li, C., Hu, B., Pan, M., 2020. Effects of warming, climate extremes and phosphorus enrichment on the growth, sexual reproduction and propagule carbon and nitrogen stoichiometry of *Potamogeton crispus* L. Environ. Int. 137, 105502. Available from: https://doi.org/10.1016/j.envint.2020.105502.
Yan, Y., Tang, Z., 2019. Protecting endemic seed plants on the Tibetan Plateau under future climate change: migration matters. J. Plant. Ecol. 12, 962–971.
Yang, H., Wu, M., Liu, W., Zhang, Z., Zhang, N., Wan, S., 2011. Community structure and composition in response to climate change in a temperate steppe. Glob. Change Biol. 17, 452–465.
Yu, Y., Baskin, J.M., Baskin, C.C., Tang, Y., Cao, M., 2008. Ecology of seed germination of eight non-pioneer tree species from a a tropical seasonal rain forest in southwest China. Plant. Ecol. 197, 1–16.
Zhang, C., Willis, C.G., Burghardt, L.T., Qi, W., Souza-Filho, P.R.M., Ma, Z., et al., 2014. The community-level effect of light on germination timing in relation to seed mass: a source of regeneration niche differentiation. New Phytol. 204, 496–506.
Zhang, S., Kang, H., Yang, W., 2017. Climate change-induced water stress suppresses the regeneration of the critically endangered forest tree *Nyssa yunnanensis*. PLoS One 12, e0182012. Available from: https://doi.org/10.1371/journal.pone.0182012.

Further reading

Koch, M.S., Schopmeyer, S.A., Kyhn-Hansen, C., Madden, C.J., Peters, J.S., 2007. Tropical seagrass species tolerance to hypersalinity stress. Aquat. Bot. 86, 14–24.

Section I

Biogeography

Chapter 1

Effect of climate change on plant regeneration from seeds in the arctic and alpine biome

Andrea Mondoni[1], Borja Jiménez-Alfaro[2] and Lohengrin A. Cavieres[3,4]

[1]Department of Earth and Environmental Science, University of Pavia, Pavia, Italy, [2]Research Unit of Biodiversity, University of Oviedo, Mieres, Spain, [3]Botany Department, Faculty of Natural and Oceanographic Sciences, University of Concepción, CP, Concepcion, Chile, [4]Institute of Ecology and Biodiversity (IEB), Chile

Introduction

The alpine life zone encompasses all the habitats found above the elevation of natural treeline (Körner, 2003). As such, this area is the only biogeographic unit with a global distribution (Nagy and Grabherr, 2009), accounting for approximately 3% of the vegetated lands on Earth (Körner et al., 2011; Testolin et al., 2020). Alpine habitats exhibit a high degree of taxonomic richness (Winkler et al., 2016), hosting approximately 4% of the Earth's flora (8000–10,000 species of vascular plants) with a diversity peak at middle latitudes and in the Andean páramo (Testolin et al., 2021). Arctic and Antarctic life zones are found beyond the high-latitude limit for trees to grow and have approximately 2000–2500 vascular species worldwide (Peterson 2014). Thus the Arctic and Antarctic life zones are related to those of alpine areas due to the absence of trees, which is largely the result of the low temperatures that characterize these zones. As a whole, arctic–alpine ecosystems have similar challenges for plant life that largely determine their biotic composition, including low mean annual temperatures, snow cover, strong winds, nutrient-poor substrates, short growing seasons, and large microclimate variation at a short geographic distance (Bliss, 1962; Greenland and Losleben, 2001).

Arctic-alpine habitats are highly threatened by climate change, since they are experiencing the highest rates of warming (Diaz and Bradley, 1997; Beniston, 2003; Pepin et al., 2015). Indeed, the rate of warming is amplified with both elevation and latitude, such that high-mountain and arctic environments have experienced more rapid changes in temperature than those at lower elevations and latitudes (Mountain Research Initiative EDW Working Group, 2015; Pepin et al., 2015). In many alpine and arctic ecosystems, surface air temperature has more than doubled that of the global average (e.g., Beniston et al., 1997; Jones and Moberg, 2003; Notz and Stroeve, 2016; Richter-Menge et al., 2017), with approximately 1.2 times faster increases in annual mean temperatures from 1961 to 2010 (Wang et al., 2016). This temperature increase has been particularly marked in the last few decades, especially in spring and summer (e.g., Ceppi et al., 2012; Marty and Meister, 2012; Overland et al., 2018). Since arctic and alpine habitats are characterized by species adapted to low temperatures, a small temperature increment experienced by plants in this environment may have a greater biological impact than the same increment experienced in less extreme environments. As a result, the alpine and arctic life zones provide opportunities to detect climate change and assess climate-related impacts.

In this chapter, we review the research related to the effects of global warming on regeneration of arctic and alpine plant species from seeds. First, we introduce the general trends on climate change known/predicted to occur in alpine-arctic regions and the general effects of these changes on alpine and arctic plants. Second, we evaluate the response of key seed traits to climate change based on empirical evidence obtained from a review of research conducted in arctic–alpine regions. Finally, we use this information to identify the major effects of climate change on regeneration by seeds and to suggest future research on the topic.

Climate warming in the arctic–alpine life zones

Alpine climates have undergone significant change over the past century, although there are contrasting patterns depending on the region and season (see Hock et al., 2019 for a review). For example, mountain surface air temperature observations in Western North America, the European Alps, and High Mountain Asia have shown contemporary warming at an average rate of 0.3°C per decade, thereby outpacing the global warming rate of 0.2°C ± 0.1°C per decade (IPCC, 2018). In contrast, warming has been less intense in other alpine zones such as in Australia, with increases in air temperature of approximately 0.2°C per decade over the last 35 years (Hennessy et al., 2003; Green and Pickering, 2009), the Tropical Andes (0.13°C per decade; Vuille, 2013), and mountain chains of South and East Africa (0.14°C per decade; Pepin and Seidel, 2005). Climate warming also has shown contrasting trends between Arctic and Antarctic regions (see Meredith et al., 2019 for a review). Unlike the Arctic, which has warmed by 2°C–3°C over the last century (Overland et al., 2014; Notz and Stroeve, 2016; Richter-Menge et al., 2017), regions of Antarctica have experienced more pronounced variation, with warming over parts of West Antarctica and no significant changes over East Antarctica (Nicolas and Bromwich, 2014; Jones et al., 2016). Indeed, Turner et al. (2016) reported that warming in the Antarctic Peninsula has stopped in the last decade, which seems to be a consequence of a short-term natural climate variability, and new warming phases are expected to occur across the Antarctic Peninsula (Lee et al., 2017).

Unlike temperature, changes in precipitation are less well quantified and show high regional and seasonal variability, even within mountain regions (Giorgi et al., 2016; Napoli et al., 2019). Despite this, warming has consistently decreased the proportion of snow precipitation falling in both alpine (Beniston and Stoffel, 2016; Frei et al., 2018; Marty, et al., 2017) and arctic (Brown et al., 2017; Luomaranta et al., 2019) zones, thereby influencing the depth and duration of snow cover (Cramer et al., 2014; Gottfried et al., 2011). This trend has been recorded especially in spring (Gobiet et al., 2014), leading to longer growing seasons with lower soil water contents (Calanca 2007; Harte et al., 1995; Taylor and Seastedt, 1994) and higher frequency of frost events (Gerdol et al., 2013).

Warming in alpine, Arctic, and Antarctic regions is projected to continue. Assessments of the impacts of climate change on global mountain systems predict increases in temperature between 0.25°C and 0.48°C per decade, depending on the CO_2 emissions scenario (Noguès-Bravo et al., 2007). Again, the warming trend varies depending on the mountain region and season (Hock et al., 2019). For example, by the end of 21st century the average mean annual temperature increase per decade is estimated to be 0.49°C for the mountains of North America (<55°N; Noguès-Bravo et al., 2007), 0.36°C for the European Alps (Gobiet et al., 2014), 0.45°C in Caucasus and Middle East (Babaeian et al., 2015), and 0.34°C for the Southern Andes (Noguès-Bravo et al., 2007). Other predictions for temperature increase report +0.6°C to +2.9°C for the Australian alpine areas (Hennessy et al., 2003) by 2050 relative to 1990, and +1.8°C to +4°C in the European Alps for the period 2051–80 (Zimmermann et al., 2013). As the Earth approaches a warming of 2°C, the Arctic and Antarctic may reach 4°C and 2°C mean annual warmings, respectively, relative to 1981–2005 (Post et al., 2019).

Summer warming is also expected to be coupled to more frequent extreme climatic events, such as heat waves (Schär et al., 2004), high temperature extremes, and drought episodes (Cubasch et al., 2001). Future projections of annual precipitation indicate increases of 5%–20% over the 21st century in many mountain regions, including the Hindu Kush and Himalaya, East Asia, eastern Africa, the European Alps, and the Carpathian region, and decreases in the Mediterranean and Southern Andes (Hock et al., 2019). In addition, the annual cycle of precipitation is expected to change considerably, with increases in precipitation during winter and decreases during summer (Gobiet et al., 2014). In Mediterranean alpine environments, a decrease in precipitation has been predicted to occur mainly in spring (Noguès-Bravo et al., 2007), with a magnitude of 5% per century (Brunetti et al., 2006). Nevertheless, mountain snow cover is projected to continue to decline in many mountain ranges, likely between 10% and 40% by 2031–50, regardless of the greenhouse gas emission scenario (Hock et al., 2019).

Effects of climate warming on arctic–alpine plants

There is a large body of literature dealing with the effects of climate change on arctic–alpine plants, for which reference is made to specific reviews (e.g., Winkler et al., 2019; Bjorkman et al., 2020). Theurillat and Guisan (2001) theorized three essential ways in which arctic–alpine plants may respond to climatic changes: (1) persistence in the modified climate, (2) migration to more suitable climates, and (3) extinction. Persistence may occur due to gradual ecological buffering, phenotypic plasticity, or genetic adaptation, and strong evidence has been provided for changes in growth, metabolism, and timing of plant life cycle events (Kudernatsch et al., 2008; Hulber et al., 2010; Anderson et al., 2012). Typical responses of arctic–alpine plants to climate warming include increased growth rate and accelerated/advanced flowering and seed development. The extent to which these changes will lead to an overall increase in plant fitness is still a matter of debate. For example, increased temperatures enhanced growth in spring, but summer heat reduced survival (Nicole et al., 2011). Summer heat waves reduced the number of flowers in two orophytic species (Abeli et al., 2012), while advanced snowmelt increased the probability of frost damage (Gerdol et al., 2013). On the other hand, heat waves and early snowmelt may favor plant performance by increasing photosynthetic rate (Bokhorst et al., 2011) and providing longer periods for growth.

In addition to physiological and phenological changes, many studies indicate that arctic−alpine species respond to climate warming by migrating. Upward shifts in elevation have been reported for treeline ecotones (Du et al., 2018) and alpine species above tree line (e.g., Parolo and Rossi, 2008; Pauli et al., 2012), while only a few poleward range shifts have been reported (see Lenoir and Svenning, 2015 for a review). Extension of species ranges in elevation has resulted in a net increase in species richness in alpine areas (e.g., Pauli et al., 2012; Rixen and Wipf, 2017; Rogora et al., 2018) and a progressive homogenization of summit vegetation (Gottfried et al., 2009), while species disappearance has been observed only recently (Lamprecht et al., 2018). Recent observations also revealed that species range shift has accelerated in both temperate (Steinbauer et al., 2018) and Mediterranean (Jiménez-Alfaro et al., 2014) alpine regions, especially during the last 20−30 years.

Model projections using climate change scenarios predict a dramatic reduction of suitable habitats for alpine plants (Engler et al., 2011), especially for those that already live near the upper margins of low elevation mountains (Theurillat and Guisan, 2001; Colwell et al., 2008) and/or for range-restricted species (Parmesan and Martens, 2006) that often are habitat specialists, weak dispersers, and/or weak competitors against incoming species (Theurillat and Guisan, 2001; Casazza et al., 2014). In some regions, projections using species distribution models predict up to 100% species turnover in alpine plant communities by 2100 (Engler et al., 2011). Nonetheless, more recent models suggest that species turnover and/or decline may be delayed for several decades because of the long-lived nature of many alpine plants, which allows them to persist under unsuitable climates (Dullinger et al., 2012; Cotto et al., 2017). Also, dispersal and establishment limitations of new species from lower elevations will reduce competition (Alexander et al., 2018). Loss of climatically suitable habitat may be further buffered by the large topographic diversity of alpine habitats, which may offer thermally suitable "escape" habitats within short distances (Scherrer and Körner, 2011; Jiménez-Alfaro et al., 2016).

Climate warming and plant regeneration from seeds in alpine−arctic environments

Although clonal reproduction is common in arctic−alpine plants, the colorful flowers of alpine communities in summer indicate that plants allocate much biomass to sexual reproduction (Körner, 2003). Indeed, studies have found high percentages of seedling establishment (e.g., Niederfriniger and Erschbamer, 2000; Forbis, 2003), gene flow (Pluess and Stöcklin, 2004), and a diversity of genotypes (Jonsson et al., 1996; Gabrielsen, 1998), arguing for the importance of sexual reproduction in the maintenance of arctic−alpine plant diversity.

Here, we outline the seed ecological spectrum of arctic−alpine plants, which is a set of seed-trait ecological responses that determines the ability of plants to disperse, persist, germinate, and establish in their environment (Saatkamp et al., 2019). In particular, we describe the ecological framework of four groups of traits and functions with expected responses to climate change: (1) seed production and mass, (2) seed dispersal in time and space, (3) seed germination and dormancy, and (4) seedling emergence and establishment. Then, we highlight the potential impacts of climate change, particularly increased temperatures (hereafter warming), water stress, and reduced winter duration on all these aspects of seeds for species typical of alpine−arctic environments.

To illustrate these effects, we conducted a literature search in the Web of Science using a combination of the terms "seeds," "alpine," "arctic," "plants," and "climate change" that reported a total of 434 studies. We reviewed each study individually and kept only those with original results on climate-change effects on seed production and seed mass, seed dispersal in time or space, seed germination and dormancy, and seedling emergence and establishment on species we identified as "alpine" or "arctic." After excluding the studies conducted on the Tibetan Plateau (discussed in a separate chapter of this book), we evaluated 45 studies carried out mostly in the last decade that provided data on 242 species from alpine temperate mountains (57%) in Europe (35%), Australia (20%), and North America (2%) and from arctic areas of Europe (43%). Note that species from tropical mountains are not represented in this review, given the general lack of information about the effects of climate change on plant regeneration from seeds in these systems.

Most of the species evaluated (80%) have been used to investigate the effect of warming, while a minority of species were used to infer related effects like reduced snow cover (16%), increased drought, reduced duration of winter (i.e., cold stratification, 8%−12%), and increased ozone concentration (2%). All traits connected to main seed functions have been covered in the current literature, although to a different extent. In particular, seed germination (i.e., radicle emergence) is the most studied trait, being investigated in 67% of the species, followed by seedling emergence (28%), seed production (16%), seed dispersal in time (14%), seedling establishment (11%), seed mass (7%), and seed dispersal in space (0.4%). Eighty-nine species have been included in more than one study for either the same or different seed traits, thus we have analyzed a total of 489 case studies (i.e., individual treatments for one response trait and one arctic—alpine species).

Seed production and seed mass

In arctic—alpine environments, the production of seeds is seriously constrained by the time of plant reproduction, that is, between snowmelt (or thaw) and the end of the growing period. Flowering phenology is a major determinant for pollination and seed development, both of which are regulated by temperature and photoperiod (Körner, 2003). However, plant species may exhibit a different phenology even if they occur in the same community, ranging from early flowering to late-flowering species (Molau, 1993; Straka and Starzomski, 2015). Assuming that pollen limitation is not a limiting factor (García-Camacho and Totland, 2009) and flowers are fertilized (i.e., perhaps by delayed self-fertilization mechanisms), seed development requires between 16 and 69 days to be completed, depending on the species and its flowering phenology (Körner, 2003). During this period, the production of seeds is influenced by frost risk (especially in early flowering species) as well as by the temperature and water stress experienced by the mother plant. These conditions also regulate the seed:ovule ratio (i.e., number of ovules formed/number of seeds matured), which is expected to be lower on early flowering than late-flowering species (Molau, 1993), thus influencing the plant's investment in the quantity and quality of seeds. As occurs in most seed plants (Moles et al., 2005), the size of seeds (and fruits) of arctic—alpine species is strongly regulated by phylogeny, although related traits such as relative mass of endosperm may be linked to germination responses (Fernández-Pascual et al., 2021). Also, in alpine and arctic plants there is a seed size-number trade-off, by which plants invest more resources in larger, but a smaller number, of seeds and vice versa, depending on environmental conditions (Moles and Westoby, 2006). Thus, given the phylogenetic constraints of each species and its flowering strategy, the extent of the growing season and the specific climatic conditions will determine the number and the size of the seeds that finally will be dispersed.

The effects of climate change on seed production and seed mass were assessed in 51 and 25 case studies, respectively. In 22 cases, the two traits were evaluated together, reflecting the general interest in assessing both aspects of sexual reproduction. Interestingly, the studies showed consistent responses for seed production and seed mass to different climate-change factors (Fig. 1.1). Anticipated snowmelt (ASM) was the factor evaluated in the highest number of cases. A decrease in seed

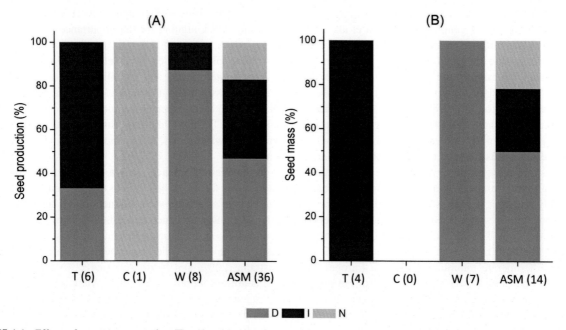

FIGURE 1.1 Effects of temperature warming (T), reduced length of winter period (C), reduced water availability (W), and anticipated snow melt (ASM) on seed production (A) and seed mass (B) in Arctic and alpine areas. Decrease (D), increase (I), none (N). Values on y-axis are percentages of total number of cases (i.e., numbers in parentheses on x-axis) for each environmental factor.

production with ASM was consistently found in the European Alps (Hüelber et al., 2011; Rosbakh et al., 2017; Tonin et al., 2019), the Rocky Mountains (Gezon et al., 2016), and the Arctic (Cooper et al., 2011; Mallik et al., 2011), with a few species from the European Alps and the Arctic showing no effects. A single study conducted in the Pyrenees (Lluent et al., 2013) found an increase of seed production with ASM, and the high number of species analyzed equaled the number of case studies with decreasing trends. ASM was correlated with decreases in seed mass in the Bavarian Alps (Rosbakh et al., 2017), positive or neutral effects in the southern Alps (Tonin et al., 2019), and positive effects in the USA Rocky Mountains (Galen and Stanton, 1993). Since divergent results are strongly linked to specific study systems, it is difficult to evaluate whether the differences are based on regional drivers or other properties of the study sites, including habitat types and/or experimental designs.

The effect of water stress derived from warmer temperatures and/or decreases in precipitation provided consistent results across species, although most examples are restricted to a single region. In response to water stress, most case studies showed a decrease in seed mass and seed production in the Bavarian Alps (Rosbakh et al., 2017), with only one species increasing seed production in the Arctic (Wookey et al., 1995).

In response to warming, both seed production and seed mass increased in the European Alps (Wagner and Reichegger, 1997) and Scandinavia (Wookey et al., 1995; Totland, 1997). Seed production also decreased with warming in one species in the New Zealand Alps (Cranston et al., 2015) and one species in alpine communities of the USA Rocky Mountains (Kettenbach et al., 2017). The response to cold stratification was only evaluated in one species, on Mount Teide in the Canary Islands (Segui et al., 2017), without a clear effect on seedling production.

Overall, these results suggest that climate warming and water stress have contrasting effects on seed production and seed mass. While increasing temperature may promote an increase in number and size of seeds, this effect is regulated by water availability during seed development. Indeed, the effect of climate change on alpine communities is expected to be modulated by soil moisture mainly on graminoids and forbs (Winkler et al., 2019), which are the functional groups of most of the species analyzed in the reviewed studies. Since arctic—alpine ecosystems generally are limited by low temperatures, it is reasonable to assume that increasing temperature (a few degrees Celsius) will favor plant and seed development (Körner, 2003). However, alpine plants are also adapted to specific soil moisture regimes that may change drastically at the regional and landscape scales, suggesting that the net effect of warming (or earlier snowmelt) on seed production and seed mass will ultimately depend on changes in soil water availability.

Seed dispersal in time and space

Seed dispersal is a key process in the life cycle of plants since it is one of the ways by which genetic exchange occurs within and between populations. Seed dispersal can occur in space by the transport of propagules away from the mother plants or in time through the formation of persistent soil seed banks (Fenner and Thompson, 2005).

Seed dormancy and the formation of a persistent seed bank are advantageous in temporally and spatially unpredictable habitats (Venable and Brown, 1988). Hence, in Arctic, Antarctic, and alpine habitats, where conditions unfavorable for seed production, germination, and/or seedling establishment may persist for two or more consecutive years (Körner, 2003), formation of soil seed banks is a key survival strategy (Jaganathan et al., 2015). Indeed, long-term persistent seed banks have been found for several alpine species from all the world's mountains, including the European Alps (Schwienbacher et al., 2010), the Andes (Cavieres, 1999; Cavieres and Arroyo, 2001), and the Australian Alps (Venn and Morgan, 2010), as well as in Antarctic (McGraw and Day, 1997) and several Arctic plant communities (e.g., McGraw and Vavrek, 1989; Cooper et al., 2004; Graae et al., 2008 and references therein). As mentioned above, climate change may exacerbate the interannual variation in length of the growing season due to yearly differences in amount of winter snow, suggesting that seed dormancy and formation of persistent seed banks may continue to play a relevant role in species persistence by allowing species to cue their germination to more favorable years.

Seed dispersal allows spatial exploration of environmental heterogeneity and the arrival of new propagules to suitable sites for seedling survival, facilitating population persistence, and promoting colonization of new areas. Studies conducted in alpine habitats (e.g., Willson et al., 1990; Cavieres et al., 1999; Bu et al., 2008) generally agree that most alpine species do not have specialized morphological structures (e.g., wings, pappus, hooks) for either active (via animals) or passive (e.g., wind) long-distance dispersal. Indeed, experimental studies have shown that alpine plants have a limited capacity for long-distance seed dispersal (Cichini et al., 2011; Morgan and Venn, 2017). Nevertheless, differences in diaspore traits such as mass and presence of structures enhancing long-distance dispersal may help distinguish weak from strong colonizers in alpine plants (Vittoz et al., 2009), thereby playing a crucial role in the response of alpine plants to climate warming. Furthermore, the dogma that seed dispersal agents are rare in alpine environments (McGraw and Vavrek, 1989) has been refuted by several studies during the last decades, in which zoochory by lizards, ants, and birds has been demonstrated for different alpine plant species (e.g., Munoz and Cavieres, 2006; Celedon-Neghme et al., 2008; Young et al., 2012). Since climate change may affect the phenology of alpine plants (see below), there is a potential to generate a mismatch between plant and animal phenological cycles, thus precluding seed dispersal.

Very few studies have evaluated the effects of warming on seed dispersal in time (mainly in the soil seed bank) and in space ($n = 2$ and 1, respectively). Regarding dispersal in time, Hoyle et al. (2013) evaluated seedling emergence from the seed bank under warming conditions in the laboratory and reported that while warming increased the number of species emerging from the seed bank, the number of germinants decreased. Nonetheless, responses to warming were highly species specific, with Poaceae species showing decreased emergence from the soil seed bank and Cyperaceae species showing an increase. Seed longevity, which is an important trait for persistence in the soil, also has been shown to be affected by warming. Bernareggi et al. (2015) reported that seed longevity of four arctic−alpine species growing under warming conditions in the field increased, suggesting a potential to remain viable for long periods of time in the soil (see Mondoni et al., 2014 for a similar result in *Silene vulgaris*).

Regarding dispersal in space, although much work has been done on flowering phenology of arctic−alpine plants, which is largely determined by snowmelt regimes and temperature sum (e.g., Molau et al., 2005; Quaglia et al., 2020), only one study specifically addressed the effect of temperature on the start of the seed dispersal period. Carbognani et al. (2018) showed that warming advanced the onset of seed dispersal of three alpine species in the Italian Alps by 1 week. The consequences of such changes will depend on the dispersal mode of the seeds, for which there is a risk for animal-dispersed species of a mismatch between the plant and animal phenological cycles.

Seed dormancy and germination

In arctic−alpine environments, high species and habitat diversity have resulted in highly variable germination patterns, making it difficult to define a common "arctic−alpine" germination syndrome. One of the most controversial aspects is the germination response to temperature, with arctic and arctic−alpine species showing either relatively high (Bliss, 1962; Billings and Mooney, 1968) or relatively low temperature requirements for seed germination that may even occur after various periods of cold stratification (Schwienbacher et al., 2011; Hoyle et al., 2015; Fernández-Pascual et al., 2017; Cavieres and Sierra-Almeida, 2018). Thus arctic−alpine seeds have different kinds and degrees of dormancy, although physiological dormancy has been found repeatedly to be the most frequent (Schwienbacher et al., 2011; Tudela-Isanta, Ladouceur et al., 2018). There are several factors involved in the kind and degree of dormancy, including species' elevation range (Fernández-Pascual et al., 2017; Walder and Erschbamer, 2015), successional niche (Schwienbacher, et al., 2012), and habitat preference (Tudela-Isanta et al., 2018a,b; Veselá et al., 2020). In a recent global *meta*-analysis, Fernández-Pascual et al. (2021) highlighted the frequent requirement for cold stratification to break dormancy and warm temperatures for germination of truly alpine species (i.e., those occurring predominantly above treeline), while generalists (i.e., those also occurring at lower elevations) show more heterogeneous responses. Consequently, physiological dormancy and overwintering are key climate-driven traits of germination in arctic−alpine species.

Seed germination (i.e., radicle emergence) has been the most investigated regenerative trait in response to climate change ($n = 239$ case studies), especially for the effects of warming (Wookey et al., 1995; Graae et al., 2008; Milbau et al., 2009, Shevtsova et al., 2009; Wagner and Simons, 2009; Muller et al., 2011; Mondoni et al., 2012; Alsos et al., 2013; Orsenigo et al., 2015; Sanhueza et al., 2017; Segui et al., 2017; Satyanti et al., 2019; Shi et al., 2019), followed by the reduced duration of winter (Wagner and Simons, 2009; Mondoni et al., 2012; Sommerville et al., 2013; Hoyle et al., 2014), increased drought (Orsenigo et al., 2015), anticipated snowmelt (Cooper et al., 2011; Hüelber et al., 2011; Mallik et al., 2011), and acute ozone exposure (Abeli et al., 2017). Most studies showed that warming after dispersal either increased or had no effect on seed germination in arctic−alpine species, while a minority reported a decrease in germination and a shift in time (i.e., Td, Fig. 1.2). Thus seeds of arctic−alpine plants germinate better at increased temperatures, suggesting that seed germination might benefit from a warming climate.

Nevertheless, current studies used constant water supplies in the germination tests, thereby excluding possible effects of drought driven by climate change. Arctic−alpine plants have low germination percentages under water stress compared with species from lower elevations (Walder and Erschbamer, 2015), and accordingly water stress consistently reduces seed germination in these species (Orsenigo et al., 2015). Therefore the extent to which warming might increase seed germination in arctic−alpine species likely will depend on how climate change affects water availability. Considering that summer precipitation is expected to decrease by the end of the 21st century in the alpine life zone (e.g., Beniston and Stoffel, 2016; Gobiet et al., 2014) and hence the number of drought events during the growing season might increase, recruitment success of alpine species may be disadvantaged in a warmer climate.

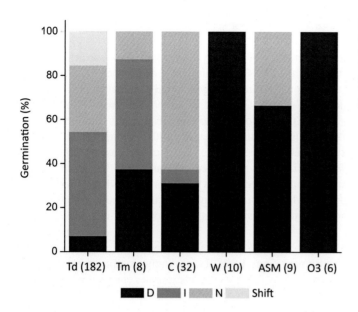

FIGURE 1.2 Effects of temperature warming [after seed dispersal (Td) and during seed maturation (Tm)], reduced length of winter cold (C), water availability (W), anticipated snow melt (ASM), and increased O_3 concentration (O3) on seed germination in Arctic and alpine areas. Decrease (D), increase (I), none (N), and shift in time. Values on y-axis are percentages of total number of cases (i.e., numbers in parentheses on x-axis) for each environmental factor.

The reviewed studies also show that seed germination of arctic—alpine species can be reduced by early snowmelt (Fig. 1.2), likely because the requirements for seed dormancy break are not met during a shortened winter (Dong et al., 2010). Nevertheless, shorter cold stratification periods reduced seed germination in one-third of the species tested so far, while the remaining ones were unaffected (Fig. 1.2). These results indicate that requirements for dormancy break and thereby the response to climate warming may vary significantly across arctic—alpine plants. This concurs with the findings of the global analysis of Fernández-Pascual et al. (2021), which found that physiological dormancy and overwintering are especially important for strict alpine species (i.e., those that occur exclusively above the treeline). Therefore changes in snow cover duration will unevenly affect the recruitment success among species inhabiting alpine habitats, indirectly favoring those with a wide elevational range and/or no stratification requirement. This possibility might be partly alleviated by the plastic physiological response of seeds to maturation temperature, a mechanism of "seed memory" that facilitates acclimation to changing environments (Fernández-Pascual et al., 2019). However, the few studies available on this issue provide contrasting results, showing either a reduction (Bernareggi et al., 2016) or an increase (Mondoni et al., 2018) in the degree of dormancy for seeds developed and matured under a warmer environment (see Tm on Fig. 1.2).

Seedling emergence and establishment

A major barrier for plant recruitment in arctic—alpine environments is the short growing season, as it constrains seedling growth and the period favorable for seedling establishment (Chambers et al., 1990; Forbis, 2003). Freezing temperatures during the growing season, summer drought, and heat waves are other major causes of seedling mortality in arctic—alpine ecosystems (Forbis, 2003; Marcante et al., 2012, 2014; Neuner, 2012). Therefore seedlings have to develop a deep root system to survive during topsoil desiccation in summer (Kammer and Mohl, 2016) and a critical biomass and perennating bud by the end of growing season to withstand the harsh and long winter conditions (Billings and Mooney, 1968). Despite these constraints, there is evidence of high percentages of seedling emergence and survival in alpine habitats (Niederfriniger and Erschbamer, 2000; Erschbamer et al., 2001; Forbis, 2003). Recruitment success in these harsh environments is determined by the occurrence of sites that facilitate seed trapping, improve moisture retention around seeds, and protect seedlings from desiccation (i.e., safe sites, *sensu* Harper et al., 1961). These sites may be provided either by biotic or abiotic factors, such as the presence of neighbor plants (Niederfriniger and Erschbamer, 2000; Callaway et al., 2002; Cavieres et al., 2005, 2006, 2008; Erschbamer et al., 2008) and/or coarse substrate and large rocks (Jumpponen et al., 1999). Nevertheless, facilitation by resident vegetation is especially important in the pioneer stage (Niederfriniger and Erschbamer, 2000; Erschbamer et al., 2008) or in sites with high environmental harshness (Callaway et al., 2002; Cavieres et al., 2014). As succession proceeds, nutrient availability and productivity increase, thereby increasing negative interactions among neighbor plants due to competition (e.g., Chapin et al., 1994; Jumpponen et al., 1999; Forbis, 2003; Klanderud et al., 2017). When plant cover is high, the "safe sites" for successful recruitment are vegetation gaps with reduced competition for space, light, and nutrients (e.g., Silvertown, 1981; Eriksson and Froborg, 1996; Klanderud and

Totland 2005). Some evidence suggests that vegetation gaps also are relevant for seedling recruitment in primary succession in some alpine environments (Cichini et al., 2011).

The effects of climate change on seedling emergence (i.e., emergence of the cotyledons above soil surface, $n = 87$) and establishment (i.e., survival after emergence, $n = 36$) have received less attention than radicle emergence (see section on "Seed dormancy and germination" above). Of the studies on seedling emergence and establishment, effects of warming have been the most investigated (Klanderud et al., 2007, 2017; Graae et al., 2009; Shevtsova et al., 2009; Hoyle et al., 2013; Kim & Donohue, 2013; Meineri et al., 2013; Angers-Blondin and Boudreau, 2017 (Mondoni et al., 2015, 2020), followed by the effects of ASM and increased drought (Klanderud et al., 2007; Meineri et al., 2013). Unlike radicle emergence, warming mostly decreased cotyledon emergence (approx. 50% of the cases), although in some cases it was increased (c. 34%) or unaffected (c. 13%) (Fig. 1.3A). A plausible explanation for this contradictory response is that radicle emergence is often monitored under constant water supply in the laboratory, while observations of cotyledon emergence are conducted in the field or greenhouse, where warming may indirectly reduce water availability in the soil. Accordingly, in the few studies available increased drought significantly reduced both radicle (Fig. 1.2) and cotyledon (Fig. 1.3A) emergence.

On the other hand, the effects of warming on seedling establishment did not show a clear trend, with a similar number of cases for reduced, increased, or unaffected seedling establishment (Fig. 1.3B). Interestingly, responses to warming showed stronger trends after studies were sorted into similar methodology categories. For example, emergence and establishment were reduced when warming was applied during a specific season (in spring/summer, Graae et al., 2009; Shevtsova et al., 2009; Hoyle et al., 2013), while repeated field observations during a natural warming trend (Angers-Blondin and Boudreau, 2017) or simulated all-season warming (Klanderud et al., 2007; Kim & Donohue, 2013; Klanderud et al., 2017; Meineri et al., 2013; Mondoni et al., 2015) had an overall positive effect on both seedling traits (Fig. 1.4). Assuming that the latter two methods include both direct and indirect effects of warming, it seems likely that warming per se may obscure the true responses to climate change, at least in early spring, when most seedlings emerge. In support of this view, Körner (1992) suggested that climate warming might operate on alpine plants primarily via indirect effects, for example, via responses attributable to extension of the snow-free period. A reduction of snow cover duration in early spring increased recruitment success (see ASM, Fig. 1.3A and B) but not the soil surface temperature in this season (e.g., Mondoni et al., 2012; 2015), indicating that alpine seeds might not necessarily experience higher temperatures during the most favorable period for their emergence. Conversely, direct increases in temperature may act on seedling recruitment in following seasons, for example, by shifting timing of emergence from spring to autumn (e.g., in seeds with low dormancy, Mondoni et al., 2015) or by reducing survival in summer due to water shortage. Surprisingly, little is known about heat and drought stress on seedling recruitment in arctic–alpine areas (see Marcante et al., 2014).

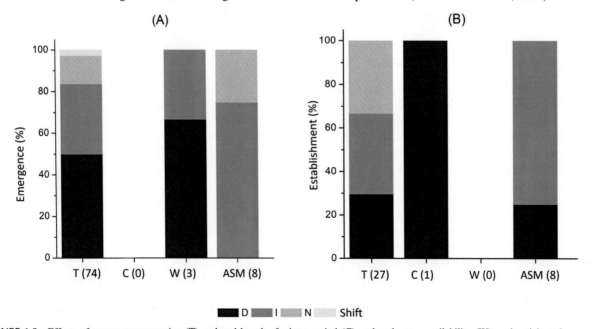

FIGURE 1.3 Effects of temperature warming (T), reduced length of winter period (C), reduced water availability (W), and anticipated snow melt (ASM) on (A) seedling emergence and (B) establishment in Arctic and alpine areas. Decrease (D), increase (I), none (N), and shift in time. Values on y-axis are percentages of total number of cases (i.e., numbers in parentheses on x-axis) for each environmental factor.

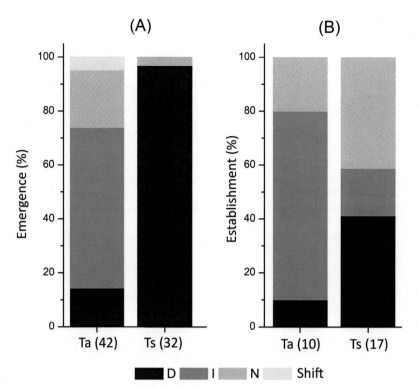

FIGURE 1.4 Effects of temperature warming on all seasons (Ta) or a specific season (Ts) on (A) seedling emergence (EME) and (B) seedling establishment in Arctic and alpine areas. Decrease (D), increase (I), none (N), and shift in time. Values on y-axis are percentages of total number of cases (i.e., numbers in parentheses on x-axis) for each environmental factor.

Conclusions and future research

Despite the crucial role of seeds for plant persistence and migration in arctic–alpine ecosystems, our understanding of the effects of climate change on seedling recruitment relies on a relatively small number of studies ($N = 45$) carried out exclusively at Arctic and temperate-alpine latitudes, while species from tropical mountains remain neglected. Most studies focused on the effects of warming on seed germination, while other traits received less attention, especially seed dispersal in space and time, as well as seedling establishment. Overall, there is compelling evidence that warming and water stress have contrasting effects on several traits including seed production, seed mass, and germination. While increasing temperatures favor the number, size, and germinability of seeds, water stress has a consistent negative effect on these traits, suggesting that the net effect of warming will depend mostly on changes in water availability. Considering that summer precipitation is expected to decrease in several arctic–alpine zones, there is an urgent need to better understand the effects of the new precipitation regimes and water stress on seedlings recruitment.

Field observations during a natural or simulated all-season warming trend showed that arctic–alpine plants should not experience water stress during the most favorable season for seedling emergence (i.e., early spring/summer), because in this period snowmelt provides ample water. However, changes in snow cover and precipitation due to climate warming may affect water availability and other relevant factors for seedling recruitment. For example, with ASM it is unlikely that arctic–alpine seeds will experience increased temperatures in early spring but rather higher probabilities of early season frost and longer (but warmer) growing seasons. Therefore desiccation risk may be higher in a later than in an earlier stage of seedling development (e.g., in summer). ASM also consistently decreased seed mass and seed production, while it unevenly affected the germination of arctic–alpine species, indirectly constraining those with physiologically dormant seeds. Hence, germination of species with physiological dormancy might be reduced in a warming climate, unless plastic physiological response of seeds to maturation temperature facilitates acclimation to changing environments, for example, by reducing dormancy or increasing longevity in the soil. Despite these latter possibilities that have been poorly investigated (and show contrasting results), the number of species emerging from the soil seed bank has been shown to increase with warming, suggesting an overall increase in germination of arctic–alpine species. In this regard, a low degree of dormancy should not be considered a successful trait, since in this case anomalous climate conditions may induce some species to germinate in seasons unsuitable for seedling survival.

In conclusion, our review highlights the need for a better understanding of the effects of warming on seedling recruitment in arctic–alpine plants, especially in response to frost and desiccation risks. Most importantly, assuming climate warming is not a seasonal selective event more comprehensive approaches are necessary to account for its effects during seed maturation and after dispersal periods. We still need to develop studies covering more seed traits and to assess climate-driven responses in different regions of the arctic—alpine climatic spectrum, with a special effort on less studied ecosystems like alpine-tropical mountains. Such studies also need to consider topographical variation in arctic—alpine habitats under new climatic scenarios, which ultimately will determine whether populations adapt, persist, or go extinct.

References

Abeli, T., Rossi, G., Gentili, R., Gandini, M., Mondoni, A., Cristofanelli, P., 2012. Effect of the extreme summer heat waves on isolated populations of two orophitic plants in the north Apennines (Italy). Nordic J. Bot. 30, 109—115.

Abeli, T., Guasconi, D.B., Mondoni, A., Dondi, A., Bentivoglio, A., Buttafava, et al., 2017. Acute and chronic ozone exposure temporarily affects seed germination in alpine plants. Plant. Biosyst. 151, 304—315.

Alexander, J.M., Chalmandrier, L., Lenoir, J., Burgess, T.I., Essl, F., Haider, S., et al., 2018. Lags in the response of mountain plant communities to climate change. Glob. Change Biol. 24, 563—579.

Alsos, I.G., Muller, E., Eidesen, P.B., 2013. Germinating seeds or bulbils in 87 of 113 tested Arctic species indicate potential for ex situ seed bank storage. Polar Biol. 36, 819—830.

Anderson, J.T., Inouye, D.W., McKinney, A.M., Colautti, R.I., Mitchell-Olds, T., 2012. Phenotypic plasticity and adaptive evolution contribute to advancing flowering phenology in response to climate change. Proc. R. Soc. B Biol. Sci. 279, 3843—3852.

Angers-Blondin, S., Boudreau, S., 2017. Expansion dynamics and performance of the dwarf shrub *Empetrum hermaphroditum* (Ericaceae) on a subarctic sand dune system, Nunavik (Canada). Arct. Antarct. Alp. Res. 49, 201—211.

Babaeian, M.R., Modirian, R., Karimian, M., Zarghami, M., 2015. Simulation of climate change in Iran during 2071—2100 using PRECIS regional climate modelling system. Desert 20 (2), 123—134.

Beniston, M., 2003. Climatic change in mountain regions: a review of possible impacts. Clim. Change 59, 5—31.

Beniston, M., Diaz, H.F., Bradley, R.S., 1997. Climatic change at high elevation sites: an overview. Clim. Change 36, 233—251.

Beniston, M., Stoffel, M., 2016. Rain-on-snow events, floods and climate change in the Alps: events may increase with warming up to 4°C and decrease thereafter. Sci. Total Environ. 571, 228—236.

Bernareggi, G., Carbognani, M., Mondoni, A., Petraglia, A., 2016. Seed dormancy and germination changes of snowbed species under climate warming: the role of pre- and post- dispersal temperatures. Ann. Bot. 118, 529—539.

Bernareggi, G., Carbognani, M., Petraglia, A., Mondoni, A., 2015. Climate warming could increase seed longevity of alpine snowbed plants. Alp. Bot. 125, 69—78.

Billings, W.D., Mooney, H.A., 1968. The ecology of arctic and alpine plants. Biol. Rev. 43, 481—529.

Bjorkman, A.D., Mariana García, C., Myers-Smith, I.H., Ravolainen, V., Jónsdóttir, I.S., Westergaard, K.B., et al., 2020. Status and trends in Arctic vegetation: evidence from experimental warming and long-term monitoring. Ambio 49, 678—692.

Bliss, L.C., 1962. Adaptations of arctic and alpine plants to environmental conditions. Arctic 15, 117—144.

Bokhorst, S., Bjerke, J.W., Street, L.E., Callaghan, T.V., Phoenix, G.K., 2011. Impacts of multiple extreme winter warming events on sub-arctic heathland: phenology, reproduction, growth and CO_2 flux responses. Glob. Change Biol. 17, 2817—2830.

Brown, R., Schuler, D.V., Bulygina, O., Derksen, C., Luojus, K., Mudryk, L., et al., 2017: Arctic terrestrial snow cover. In: Snow, Water, Ice and Permafrost in the Arctic (SWIPA) 2017 Assessment. Arctic Monitoring and Assessment Programme, Oslo, Norway.

Brunetti, M., Maugeri, M., Monti, F., Nanni, T., 2006. Temperature and precipitation variability in Italy in the last two centuries from homogenised instrumental time series. Int. J. Climatol. 26, 345—381.

Bu, H., Du, G., Chen, X., Xu, X., Liu, K., Wen, S., 2008. Community-wide germination strategies in an alpine meadow on the eastern Qinghai-Tibet plateau: phylogenetic and life-history correlates. Plant. Ecol. 195, 87—98.

Calanca, P., 2007. Climate change and drought occurrence in the Alpine region: how severe are becoming the extremes? Glob. Planet. Change 57, 151—160.

Callaway, R.M., Brooker, R.W., Choler, P., Kikvidze, Z., Lortie, C.J., Michalet, R., et al., 2002. Positive interactions among alpine plants increase with stress. Nature 417, 844—848.

Carbognani, M., Tomaselli, M., Petraglia, A., 2018. Different temperature perception in high-elevation plants: new insight into phenological development and implications for climate change in the alpine tundra. Oikos 127, 1014—1023.

Casazza, G., Giordani, P., Benesperi, R., Foggi, B., Viciani, D., Filigheddu, R., et al., 2014. Climate change hastens the urgency of conservation for range-restricted plant species in the central-northern Mediterranean region. Biol. Conserv. 179, 129—138.

Cavieres, L.A., 1999. Persistent seed banks: delayed seed germination models and their application to alpine environments. Rev. Chil. Hist. Nat. 72, 457—466.

Cavieres, L.A., Arroyo, M.T.K., 2001. Persistent soil seed banks in *Phacelia secunda* (Hydrophyllaceae): experimental detection of variation along an altitudinal gradient in the Andes of central Chile (33 degrees S). J. Ecol. 89, 31—39.

Cavieres, L.A., Badano, E.I., Sierra-Almeida, A., Gómez-González, S., Molina-Montenegro, M.A., 2006. Positive interactions between alpine plant species and the nurse cushion plant *Laretia acaulis* do not increase with elevation in the Andes of central Chile. New Phytol. 169, 59—69.

Cavieres, L.A., Brooker, R.W., Butterfield, B.J., Cook, B.J., Kikvidze, Z., Lortie, C.J., et al., 2014. Facilitative plant interactions and climate simultaneously drive alpine plant diversity. Ecol. Lett. 17, 193–202.

Cavieres, L.A., Papic, C., Castor, C., 1999. Variación altitudinal en los síndromes de dispersión de semillas de la vegetación andina de la cuenca del Rio Molina, Chile central (33°S). Gayana Bot. 56, 115–123.

Cavieres, L.A., Quiroz, C.L., Molina-Montenegro, M.A., Muñoz, A.A., Pauchardb, A., 2005. Nurse effect of the native cushion plant *Azorella monantha* on the invasive non-native *Taraxacum officinale* in the high Andes of central Chile. Persp. Plant. Ecol. Evol. Syst. 7, 217–226.

Cavieres, L.A., Quiroz, C.L., Molina-Montenegro, M.A., 2008. Facilitation of the non-native *Taraxacum officinale* by native nurse cushion species in the high Andes of central Chile: are there differences between nurses? Funct. Ecol. 22, 148–156.

Cavieres, L.A., Sierra-Almeida, A., 2018. Assessing the importance of cold-stratification for seed germination in alpine plant species of the High-Andes of central Chile. Persp. Plant Ecol. Evol. Syst. 30, 125–131.

Celedon-Neghme, C., San Martin, L.A., Victoriano, P.F., Cavieresa, L.A., 2008. Legitimate seed dispersal by lizards in an alpine habitat: the case of *Berberis empetrifolia* (Berberidaceae) dispersed by *Liolaemus belii* (Tropiduridae). Acta Oecol. 33, 265–271.

Ceppi, P., Scherrer, S.C., Fischer, A.M., Appenzeller, C., 2012. Revisiting Swiss temperature trends 1959–2008. Int. J. Climatol. 32, 203–213.

Chambers, J.C., MacMahon, J.A., Brown, R.W., 1990. Alpine seedling establishment: the influence of disturbance type. Ecology 71, 1323–1341.

Chapin, F.S., Walker, L.R., Fastie, C.L., Sharman, L.C., 1994. Mechanisms of primary succession following deglaciation at Glacier Bay, Alaska. Ecol. Monogr. 64, 149–175.

Cichini, K., Schwienbacher, E., Marcante, S., Seeber, G.U.G., Erschbamer, B., 2011. Colonization of experimentally created gaps along an alpine successional gradient. Plant. Ecol. 212, 1613–1627.

Colwell, R.K., Brehm, G., Cardelús, C.L., Gilman, A.C., Longino, J.T., 2008. Global warming, elevational range shifts, and lowland biotic attrition in the wet tropics. Science 322, 258–261.

Cooper, E.J., Alsos, I.G., Hagen, D., Smith, F.M., Coulson, S.J., Hodkinson, I.D., 2004. Plant recruitment in the High Arctic: seed bank and seedling emergence on Svalbard. J. Veg. Sci. 15, 115–124.

Cooper, E.J., Dullinger, S., Semenchuk, P., 2011. Late snowmelt delays plant development and results in lower reproductive success in the High Arctic. Plant. Sci. 180, 157–167.

Cotto, O., Wessely, J., Georges, D., Klonner, G., Schmid, G.M., Dullinger, S., et al., 2017. A dynamic eco-evolutionary model predicts slow response of alpine plants to climate warming. Nat. Commun. 8, 15399. Available from: https://doi.org/10.1038/ncomms15399.

Cramer, W., Yohe, G.W., Auffhammer, M., Huggel, C., Molau, U., da Silva Dias, M.A.F., et al., 2014. Detection and attribution of observed impacts. In: Field, C.B., Barros, V.R., Dokken, D.J., Mach, K.J., Mastrandrea, M.D., Bilir, T.E., et al.,Climate Change 2014: Impacts, Adaptation, and Vulnerability. Part A: global and sectoral aspects. Contribution of Working Group II to the Fifth Assessment Report of the Intergovernmental Panel on Climate Change. Cambridge University Press, Cambridge, pp. 979–1037.

Cranston, B.H., Monks, A., Whigham, P.A., Dickinson, K.J.M., 2015. Variation and response to experimental warming in a New Zealand cushion plant species. Austral Ecol. 40, 642–650.

Cubasch, U., Meehl, G. A., Boer, G. J., Stouffer, R. J., Dix, M., Noda, A. et al., 2001. Projections of future climate change. In: J.T., Houghton, Y., Ding, D.J., Griggs, D.J., Noguer, M., van der Linden, P.J. Maskell K., et al., (Eds.), Climate Change 2001: The Scientific Basis. Contribution of Working Group I to the Third Assessment Report of the Intergovernmental Panel. Cambridge University Press, Cambridge, pp. 526–582.

Diaz, H.F., Bradley, R.S., 1997. Temperature variations during the last century at high elevation sites. In: Diaz, H.F., Beniston, M., Bradley, R.S. (Eds.), Climatic Change at High Elevation Sites. Springer, Dordrecht, pp. 21–47.

Dong, W., Jiang, Y., Yang, S., 2010. Response of the starting dates and the lengths of seasons in mainland China to global warming. Clim. Change 99, 81–91.

Dullinger, S., Willner, W., Plutzar, C., Englisch, T., Schratt-Ehrendorfer, L., Moser, D., et al., 2012. Post-glacial migration lag restricts range filling of plants in the European Alps. Glob. Ecol. Biogeogr. 21, 829–840.

Du, H., Liu, J., Li, M.H., Büntgen, U., Yang, Y., Wang, L., et al., 2018. Warming-induced upward migration of the alpine treeline in the Changbai Mountains, northeast China. Glob. Change Biol. 24, 1256–1266.

Engler, R., Randin, C.F., Thuiller, W., Dullinger, S., Zimmermann, N.E., Araùjo, M.B., et al., 2011. 21st century climate change threatens mountain flora unequally across Europe. Glob. Change Biol. 17, 2330–2341.

Eriksson, O., Froborg, H., 1996. "Windows of opportunity" for recruitment in long-lived clonal plants: experimental studies of seedling establishment in *Vaccinium* shrubs. Can. J. Bot. 74, 1369–1374.

Erschbamer, B., Kneringer, E., Niederfriniger, S.R., 2001. Seed rain, soil seed bank, seedling recruitment, and survival of seedlings on a glacier foreland in the Central Alps. Flora 196, 304–312.

Erschbamer, B., Niederfriniger, S.R., Winkler, E., 2008. Colonization processes on a central Alpine glacier foreland. J. Veg. Sci. 19, 855–862.

Fenner, M., Thompson, K., 2005. The Ecology of Seeds. Cambridge University Press, Cambridge.

Fernández-Pascual, E., Carta, A., Mondoni, A., Cavieres, L.A., Rosbakh, S., Venn, S., et al., 2021. The seed germination spectrum of alpine plants: a global *meta*-analysis. New Phytol. 229, 3573–3586.

Fernández-Pascual, E., Jiménez-Alfaro, B., Bueno, Á., 2017. Comparative seed germination traits in alpine and subalpine grasslands: higher elevations are associated with warmer germination temperatures. Plant. Biol. 19, 1–9.

Fernández-Pascual, E., Mattana, E., Pritchard, H.W., 2019. Seeds of future past: climate change and the thermal memory of plant reproductive traits. Biol. Rev. 94, 439–456.

Forbis, T.A., 2003. Seedling demography in an alpine ecosystem. Am. J. Bot. 90, 1197–1206.

Frei, P., Kotlarski, S., Liniger, M.A., Schär, C., 2018. Future snowfall in the Alps: projections based on the EURO-CORDEX regional climate models. Cryosphere 12, 1−24.

Gabrielsen, B., 1998. Sex after all: high levels of diversity detected in the arctic clonal plant *Saxifraga cernua* using RAPD markers. Mol. Ecol. 7, 1701−1708.

Galen, C., Stanton, M.L., 1993. Short-term responses of alpine buttercups to experimental manipulations of growing-season length. Ecology 74, 1052−1058.

García-Camacho, R., Totland, Ø., 2009. Pollen limitation in the Alpine: a *meta*-analysis. Arct. Antarct. Alp. Res. 41, 103−111.

Gerdol, R., Siffi, C., Iacumin, P., Gualmini, M., Gualmini, M., Tomaselli, M., 2013. Advanced snowmelt affects vegetative growth and sexual reproduction of *Vaccinium myrtillus* in a sub-alpine heath. J. Veg. Sci. 24, 569−579.

Gezon, Z.J., Inouye, D.W., Irwin, R.E., 2016. Phenological change in a spring ephemeral: implications for pollination and plant reproduction. Glob. Change Biol. 22, 1779−1793.

Giorgi, F., Torma, C., Coppola, E., Ban, N., Schär, C., Somot, S., 2016. Enhanced summer convective rainfall at alpine high elevations in response to climate warming. Nat. Geosci. 9, 584−589.

Gobiet, A., Kotlarski, S., Beniston, M., Heinrich, G., Rajczak, J., Stoffel, M., 2014. 21st century CC in the European Alps—a review. Sci. Total Environ. 493, 1138−1151.

Gottfried, M., Hantel, M., Maurer, C., Toechterle, R., Pauli, H., Grabherr, G., 2011. Coincidence of the alpine-nival ecotone with the summer snowline. Environ. Res. Lett. 6, 014013. Available from: https://doi.org/10.1088/1748-9326/6/1/014013.

Graae, B.J., Ejrnæs, R., Marchand, F.L., Milbau, A., Shevtsova, A., Beyens, L., et al., 2009. The effect of an early-season short-term heat pulse on plant recruitment in the Arctic. Polar Biol. 32, 1117−1126.

Graae, B.J., Alsos, I.G., Ejrnaes, R., 2008. The impact of temperature regimes on development, dormancy breaking and germination of dwarf shrub seeds from arctic, alpine and boreal sites. Plant Ecol. 198, 275−284.

Greenland, D., Losleben, M., 2001. Climate. In: Bowman, W.D., Seastedt, T.R. (Eds.), Structure and Function of an Alpine Ecosystem: Niwot Ridge, Colorado. Oxford University Press, New York.

Green, K., Pickering, C., 2009. The decline of snowpatches in the Snowy Mountains of Australia: importance of climate warming, variable snow and wind. Arct. Antarct. Alp. Res. 41, 212−218.

Harper, J.L., Clatworthy, J.N., Mc Naughton, I.H., Sagar, G.R., 1961. The evolution of closely related species living in the same area. Evolution 15, 209−227.

Harte, J., Torn, M.S., Chang, F.-R., Feifarek, B., Kinzig, A.P., Shaw, R., et al., 1995. Global warming and soil microclimate: results from a meadow-warming experiment. Ecol. Appl. 5, 132−150.

Hennessy K., Whetton P., Smith I., Bathols J., Hutchinson, M., Sharples, J., 2003. The Impact of Climate Change on Snow Conditions in Mainland Australia. CSIRO Atmospheric Research, Aspendale, Victoria.

Hock, R., G. Rasul, C. Adler, Cáceres, B., Gruber, S., Hirabayashi, Y., et al. 2019. High mountain areas. In: Pörtner, H.O., Roberts, D.C., Masson-Delmotte, V., Zhai, P. Tignor, M., Poloczanska, E., et al. (Eds.), IPCC Special Report on the Ocean and Cryosphere in a Changing Climate. <https://www.ipcc.ch/site/assets/uploads/sites/3/2019/12/02_SROCC_FM_FINAL.pdf>

Hoyle, G.L., Cordiner, H., Good, R.B., Nicotra, A.B., 2014. Effects of reduced winter duration on seed dormancy and germination in six populations of the alpine herb *Aciphyllya glacialis* (Apiaceae). Conserv. Physiol. 2, cou015.

Hoyle, G.L., Steadman, K.J., Good, R.B., McIntosh, E.J., Galea, L.M.E., Nicotra, A.B., 2015. Seed germination strategies: an evolutionary trajectory independent of vegetative functional traits. Front. Plant Sci. 6, 731. Available from: https://doi.org/10.3389/fpls.2015.00731.

Hoyle, G.L., Venn, S.E., Steadman, K.J., Good, R.B., McAuliffe, E.J., Williams, E.R., et al., 2013. Soil warming increases plant species richness but decreases germination from the alpine soil seed bank. Glob. Change Biol. 19, 1549−1561.

Hüelber, K., Bardy, K., Dulinger, S., 2011. Effects of snowmelt timing and competition on the performance of alpine snowbed plants. Persp. Plant Ecol. Evol. Syst. 13, 15−26.

Hulber, K., Winkler, M., Grabherr, G., 2010. Intraseasonal climate and habitat-specific variability controls the flowering phenology of high alpine plant species. Funct. Ecol. 24, 245−252.

IPCC, 2018. Global Warming of 1.5°C. Special Report on the Impacts of Global Warming of 1.5°C Above Pre-industrial Levels and Related Global Greenhouse Gas Emission Pathways, in the Context of Strengthening the Global Response to the Threat of Climate Change, Sustainable Development, and Efforts to Eradicate Poverty. An IPCC. https://www.ipcc.ch/sr15/download/.

Jaganathan, G.K., Dalrymple, S.E., Liu, B., 2015. Towards an understanding of factors controlling seed bank composition and longevity in the alpine environment. Bot. Rev. 81, 70−103.

Jiménez-Alfaro, B., Garcia-Calvo, L., Garcia, P., Acebes, J.L., 2016. Anticipating extinctions of glacial relict populations in mountain refugia. Biol. Conserv. 201, 243−251.

Jiménez-Alfaro, B., Gavilán, R.G., Escudero, A., Iriondo, J.M., Fernández-González, F., 2014. Decline of dry grassland specialists in Mediterranean high-mountain communities influenced by recent climate warming. J. Veg. Sci. 25, 1394−1404.

Jones, J.M., Gille, S., Goosse, H., Abram, N.J., Canziani, P.O., Charman, D.J., et al., 2016. Assessing recent trends in high-latitude Southern Hemisphere surface climate. Nat. Clim. Change 6, 917−926.

Jones, P.D., Moberg, A., 2003. Hemispheric and large-scale surface air temperature variations: an extensive revision and an update to 2001. J. Clim. 16, 206−223.

Jonsson, B.O., Jonsdottir, I.S., Cronberg, N., 1996. Clonal diversity and allozyme variation in populations of the arctic sedge *Carex bigelowii* (Cyperaceae). J. Ecol. 84, 449−459.

Jumpponen, A., Mattson, K., Trappe, J., 1999. Effects of established willows on primary succession on Lyman Glacier Forefront, North Cascade Range, Washington, U.S.A.: evidence for simultaneous canopy inhibition and soil facilitation. Arct. Alp. Res. 30, 21–39.

Jumpponen, A., Väre, H., Mattson, K.G., Ohtonen, R., Trappe, J.M., 1999. Characterization of 'safe sites' for pioneers in primary succession on recently deglaciated terrain. J. Ecol. 87, 98–105.

Kammer, P.M., Mohl, A., 2016. Factors controlling species richness in alpine plant communities: an assessment of the importance of stress and disturbance. Arct. Antarct. Alp. Res. 34, 398–407.

Kettenbach, J.A., Miller-Struttmann, N., Moffett, Z., Galen, C., 2017. How shrub encroachment under climate change could threaten pollination services for alpine wildflowers: a case study using the alpine skypilot, *Polemonium viscosum*. Ecol. Evol. 7, 6963–6971.

Kim, E., Donohue, K., 2013. Local adaptation and plasticity of *Erysimum capitatum* to altitude: its implications for responses to climate change. J. Ecol. 101, 96–805.

Klanderud, K., Meineri, E., Topper, J., Michel, P., Vandvik, V., 2017. Biotic interaction effects on seedling recruitment along bioclimatic gradients: testing the stress-gradient hypothesis. J. Veg. Sci. 28, 347–356.

Klanderud, K., Totland, Ø., 2005. The relative importance of neighbours and abiotic environmental conditions for population dynamic parameters of two alpine plant species. J. Ecol. 93, 493–501.

Klanderud, K., Totland, Ø., 2007. The relative role of dispersal and local interactions for alpine plant community diversity under simulated climate warming. Oikos 116, 1279–1288.

Körner, C., 1992. Response of alpine vegetation to global climate change. In: Proceedings of the International Conference on Landscape Ecological Impact of Climate Change. Reiskirchen: Catena Verlag, 85–96.

Körner, C., 2003. Alpine Plant Life: Functional Plant Ecology of High Mountain Ecosystems, Second ed. Springer, Basel.

Körner, C., Paulsen, J., Spehn, E.M., 2011. A definition of mountains and their bioclimatic belts for global comparisons of biodiversity data. Alp. Bot. 121, 73,. Available from: https://doi.org/10.1007/s00035-011-0094-4.

Kudernatsch, T., Fischer, A., Bernhardt-Roemermann, M., Abs, C., 2008. Short-term effects of temperature enhancement on growth and reproduction of alpine grassland species. Basic Appl. Ecol. 9, 263–274.

Lamprecht, A., Semenchuk, P.R., Steinbauer, K., Winkler, M., Pauli, H., 2018. Climate change leads to accelerated transformation of high-elevation vegetation in the central Alps. New Phytol. 220, 447–459.

Lee, J.R., Raymond, B., Bracegirdle, T.J., Chadès, L., Fuller, R.A., Shaw, J.D., et al., 2017. Climate change drives expansion of Antarctic ice-free habitat. Nature 547, 49–54.

Lenoir, J., Svenning, J.C., 2015. Climate-related range shifts—a global multidimensional synthesis and new research directions. Ecography 38, 15–28.

Lluent, A., Anadon-Rosell, A., Ninot, J.M., Grau, O., Carrillo, E., 2013. Phenology and seed setting success of snowbed plant species in contrasting snowmelt regimes in the Central Pyrenees. Flora 208, 220–231.

Luomaranta, A., Aalto, J., Jylhä, K., 2019. Snow cover trends in Finland over 1961–2014 based on gridded snow depth observations. Int. J. Climatol. 39, 3147–3159.

Mallik, A.U., Wdowiak, J.V., Cooper, E.J., 2011. Growth and reproductive responses of *Cassiope tetragona*, a circumpolar evergreen shrub, to experimentally delayed snowmelt. Arct. Antarct. Alp. Res. 43, 404–409.

Marcante, S., Erschbamer, B., Buchner, O., Neuner, G., 2014. Heat tolerance of early developmental stages of glacier foreland species in the growth chamber and in the field. Plant Ecol. 215, 747–758.

Marcante, S., Sierra-Almeida, A., Spindelböck, J.P., Erschbamer, B., Neuner, G., 2012. Frost as a limiting factor for recruitment and establishment of early development stages in an alpine glacier foreland? J. Veg. Sci. 23, 858–868.

Marty, C., Meister, R., 2012. Long-term snow and weather observations at Weissfluhjoch and its relation to other high-altitude observatories in the Alps. Theor. Appl. Climatol. 110, 573–583.

Marty, C., Tilg, A.M., Jonas, T., 2017. Recent evidence of large-scale receding snow water equivalents in the European Alps. J. Hydrometeorol. 18, 1021–1031.

McGraw, J.B., Day, T.A., 1997. Size and characteristics of a natural seed bank in Antarctica. Arct. Alp. Res. 29, 213–216.

McGraw, J.B., Vavrek, M.C., 1989. The role of buried viable seeds in arctic and alpine plant communities. In: Leck, M.A., Parker, V.T., Simpson, R.L. (Eds.), Ecology of Soil Seed Banks. Academic Press, San Diego, pp. 91–106.

Meineri, E., Spindelbock, J., Vandvik, V., 2013. Seedling emergence responds to both seed source and recruitment site climates: a climate change experiment combining transplant and gradient approaches. Plant Ecol. 214, 607–619.

Meredith, M., Sommerkorn, M., Cassotta, S., Derksen C., Ekaykin, A., Hollowed, A., et al., 2019. Polar regions. In: Pörtner, H.O., Roberts, D.C., Masson-Delmotte, V., Zhai, P. Tignor, M., Poloczanska, E., et al. (Eds.), IPCC Special Report on the Ocean and Cryosphere in a Changing Climate. <https://www.ipcc.ch/site/assets/uploads/sites/3/2019/12/02_SROCC_FM_FINAL.pdf>.

Milbau, A., Graae, B.J., Shevtsova, A., Nijs, I., 2009. Effects of a warmer climate on seed germination in the subarctic. Ann. Bot. 104, 155–164.

Molau, U., 1993. Relationships between flowering phenology and life-history strategies in tundra plants. Arct. Antarct. Alp. Res. 25, 391–402.

Molau, U., Nordenhäll, U., Eriksen, B., 2005. Onset of flowering and climate variability in an alpine landscape: a 10-year study from Swedish Lapland. Am. J. Bot. 92, 422–431.

Moles, A.T., Ackerly, D.D., Webb, C.O., Tweddle, J.C., Dickie, J.B., Westoby, M., 2005. A brief history of seed size. Science 307, 576–580.

Moles, A.T., Westoby, M., 2006. Seed size and plant strategy across the whole life cycle. Oikos 113, 91–105.

Mondoni, A., Orsenigo, S., Donà, M., Balestrazzi, A., Probert, R.J., Hay, F.R., et al., 2014. Environmental-induced transgenerational changes in seed longevity: maternal and genetic influence. Ann. Bot. 113, 1257–1263.

Mondoni, A., Orsenigo, S., Muller, J.V., Carlsson-Graner, U., Jiménez-Alfaro, B., Abeli, T., 2018. Seed dormancy and longevity in subarctic and alpine populations of *Silene suecica*. Alp. Bot. 128, 71–81.

Mondoni, A., Pedrini, S., Bernareggi, G., Rossi, G., Abeli, T., Probert, R.J., et al., 2015. Climate warming could increase recruitment success in glacier foreland plants. Ann. Bot. 116, 907–916.

Mondoni, A., Rossi, G., Orsenigo, S., Probert, R.J., 2012. Climate warming could shift the timing of seed germination in alpine plants. Ann. Bot. 110, 155–164.

Morgan, J.W., Venn, S.E., 2017. Alpine plant species have limited capacity for long-distance seed dispersal. Plant Ecol. 218, 813–819.

Mountain Research Initiative EDW Working Group, 2015. Elevation-dependent warming in mountain regions of the world. Nat. Clim. Change 5, 424–430.

Muller, E., Cooper, E.J., Alsos, I.G., 2011. Germinability of arctic plants is high in perceived optimal conditions but low in the field. Botany 89, 337–348.

Munoz, A.A., Cavieres, L.A., 2006. A multi-species assessment of post-dispersal seed predation in the central Chilean Andes. Ann. Bot. 98, 193–201.

Nagy, L., Grabherr, G., 2009. The Biology of Alpine Habitats. Oxford University Press, New York.

Napoli, A., Crespi, A., Ragone, F., Maugeri, M., Pasquero, C., 2019. Variability of orographic enhancement of precipitation in the Alpine region. Sci. Rep. 9, 13352. Available from: https://doi.org/10.1038/s41598-019-49974-5.

Neuner, G., 2012. Frost as a limiting factor for recruitment and establishment of early development stages in an alpine glacier foreland? J. Veg. Sci. 23, 858–868.

Nicolas, J.P., Bromwich, D.H., 2014. New reconstruction of antarctic near-surface temperatures: multidecadal trends and reliability of global reanalyses. J. Clim. 27, 8070–8093.

Nicole, F., Dahlgren, J.P., Vivat, A., Till-Bottraud, I., Ehrlén, J., 2011. Interdependent effects of habitat quality and climate on population growth of an endangered plant. J. Ecol. 99, 1211–1218.

Niederfriniger, S., Erschbamer, B., 2000. Germination and establishment of seedlings on a glacier foreland in the central Alps, Austria. Arct. Antarct. Alp. Res. 32, 270–277.

Noguès-Bravo, D., Araùjo, M.B., Errea, M.P., Martínez-Rica, J.P., 2007. Exposure of global mountain systems to climate warming during the 21st Century. Glob. Environ. Change 17, 420–428.

Notz, D., Stroeve, J., 2016. Observed Arctic sea-ice loss directly follows anthropogenic CO_2 emission. Science 354, 747–750.

Orsenigo, S., Abeli, T., Rossi, G., Bonasoni, P., Pasquaretta, C., Gandini, M., et al., 2015. Effects of autumn and spring heat waves on seed germination of High Mountain plants. PLoS One 10, e0133626. Available from: https://doi.org/10.1371/journal.pone.0133626.

Overland, J.E., Hanna, E., Hanssen-Bauer, I., Kim, S.J., Walsh, J.E., Wang, M., et al., 2018. Surface air temperature. In: Arctic Report Card 2018. <https://www.arctic.noaa.gov/Report-Card>.

Overland, J.E., Wang, M., Walsh, J.E., Stroeve, J.C., 2014. Future arctic climate changes: adaptation and mitigation time scales. Earth's Future 2, 68–74.

Parmesan, C., Martens, P., 2006. Climate change. In: Sala, O., Meyerson, L., Parmesan, C. (Eds.), Biodiversity, Health and the Environment: SCOPE/Diversitas Rapid Assessment Project. Island Press, Washington.

Parolo, G., Rossi, G., 2008. Upward migration of vascular plants following a climate warming trend in the Alps. Basic Appl. Ecol. 9, 100–107.

Pauli, H., Gottfried, M., Dullinger, S., Abdaladze, O., Akhalkatsi, M., Alonso, J.L.B., et al., 2012. Recent plant diversity changes on Europe's mountain summits. Science 336, 353–355.

Pepin, N., Bradley, R.S., Diaz, H.F., Baraer, M., Caceres, E.B., Forsythe, N., et al., 2015. Elevation-dependent warming in mountain regions of the world. Nat. Clim. Change 5, 424–430.

Pepin, N., Seidel, D.J., 2005. A global comparison of surface and free-air temperatures at high elevations. J. Geophys. Res. 110, D03104. Available from: https://doi:10.1029/2004JD005047.

Peterson, K.M., 2014. Plants in Arctic environments. In: Monson, R. (Ed.), Ecology and the Environment. The Plant Sciences, vol. 8. Springer, New York.

Pluess, A.R., Stöcklin, J., 2004. Population genetic diversity of the clonal plant *Geum reptans* (Rosaceae) in the Swiss Alps. Am. J. Bot. 91, 2013–2021.

Post, E., Alley, R.B., Christensen, T.R., Macias-Fauria, M., Forbes, B.C., Gooseff, M.N., et al., 2019. The polar regions in a 2°C warmer world. Sci. Adv. 5, https://doi.org/10.1126/sciadv.aaw9883 eaaw9883.

Quaglia, E., Ravetto Enri, S., Perotti, E., Probo, M., Lombardi, G., Lonati, M., 2020. Alpine tundra species phenology is mostly driven by climate-related variables rather than by photoperiod. J. Mtn. Sci. 17, 2081–2096.

Richter-Menge, J., Overland J.E., Mathis J.T., 2017. Arctic Report Card 2017. <http://www.arctic.noaa.gov/Report-Card>.

Rixen, C., Wipf, S., 2017. Non-equilibrium in Alpine plant assemblages: shifts in Europe's summit floras. In: Catalan, J., Ninot, J., Aniz, M. (Eds.), High Mountain Conservation in a Changing World. Advances in Global Change Research, 62. Springer, Cham.

Rogora, M., Frate, L., Carranza, M., Freppaz, M., Stanisci, A., Bertani, et al., 2018. Assessment of climate change effects on mountain ecosystems through a cross-site analysis in the Alps and Apennines. Sci. Total Environ. 624, 1429–1442.

Rosbakh, S., Leingaertner, A., Hoiss, B., 2017. Contrasting effects of extreme drought and snowmelt patterns on mountain plants along an elevation gradient. Front. Plant. Sci. 8, 1478. Available from: https://doi.org/10.3389/fpls.2017.01478.

Saatkamp, A., Cochrane, A., Commander, L., Guja, L.K., Jimenez-Alfaro, B., Larson, J., et al., 2019. A research agenda for seed-trait functional ecology. New Phytol. 221, 1764–1775.

Sanhueza, C., Vallejos, V., Cavieres, L.A., Saez, P., Bravo, L.A., Corcuera, L.J., 2017. Growing temperature affects seed germination of the Antarctic plant *Colobanthus quitensis* (Kunth) Bartl (Caryophyllaceae). Polar Biol. 40, 449–455.

Satyanti, A., Guja, L.K., Nicotra, A.B., 2019. Temperature variability drives within-species variation in germination strategy and establishment characteristics of an alpine herb. Oecologia 189, 407–419.

Schär, C., Vidale, P., Lüthi, D., Frei, C., Häberli, C., Liniger, M.A., et al., 2004. The role of increasing temperature variability in European summer heatwaves. Nature 427, 332–336.

Scherrer, D., Körner, C., 2011. Topographically controlled thermal-habitat differentiation buffers alpine plant diversity against climate warming. J. Biogeogr. 38, 406–416.

Schwienbacher, E., Marcante, S., Erschbamer, B., 2010. Alpine species seed longevity in the soil in relation to seed size and shape—A 5-year burial experiment in the Central Alps. Flora 205, 19–25.

Schwienbacher, E., Navarro-Cano, J.A., Neuner, G., Erschbamer, B., 2011. Seed dormancy in alpine species. Flora 206, 845–856.

Schwienbacher, E., Navarro-Cano, J.A., Neuner, G., Erschbamer, B., 2012. Correspondence of seed traits with niche position in glacier foreland succession. Plant Ecol. 213, 371–382.

Segui, J., Lopez-Darias, M., Perez, A.J., Nogales, M., Traveset, A., 2017. Species-environment interactions changed by introduced herbivores in an oceanic high-mountain ecosystem. AoB Plants 9, plw091. Available from: https://doi.org/10.1093/aobpla/plw091.

Shevtsova, A., Graae, B.J., Jochum, T., Milbau, A., Kockelbergh, F., Beyens, L., et al., 2009. Critical periods for impact of climate warming on early seedling establishment in subarctic tundra. Glob. Change Biol. 15, 2662–2680.

Shi, P., Quinn, B.K., Zhang, Y., Bao, X., Lin, S., 2019. Comparison of the intrinsic optimum temperatures for seed germination between two bamboo species based on a thermodynamic model. Glob. Ecol. Conserv. 17, e00568.

Silvertown, J., 1981. Micro-spatial heterogeneity and seedling demography in species rich grassland. N. Phytol. 88, 117–125.

Sommerville, K.D., Martyn, A.J., Offord, C.A., 2013. Can seed characteristics or species distribution be used to predict the stratification requirements of herbs in the Australian Alps? Bot. J. Linn. Soc. 172, 187–204.

Steinbauer, M.J., Grytnes, J.A., Jurasinsk, G., Kulonen, A., Lenoir, J., Pauli, H., et al., 2018. Accelerated increase in plant species richness on mountain summits is linked to warming. Nature 556, 231–234.

Straka, J.R., Starzomski, B.M., 2015. Fruitful factors: what limits seed production of flowering plants in the alpine? Oecologia 178, 249–260.

Taylor, R.V., Seastedt, T.R., 1994. Short- and long-term patterns of soil moisture in alpine tundra. Arct. Alp. Res. 26, 14–20.

Testolin, R., Attore, F., Borchardt, P., Brand, R.F., Bruelheide, H., Chytrý, M., et al., 2021. Global hospots of alpine diversity. Glob. Ecol. Biogeogr.

Testolin, R., Attorre, F., Jiménez-Alfaro, B., 2020. Global distribution and bioclimatic characterization of alpine biomes. Ecography 43, 779–788.

Theurillat, J.P., Guisan, A., 2001. Potential impact of climate change on vegetation in the European Alps: a review. Clim. Change 50, 77–109.

Tonin, R., Gerdol, R., Tomaselli, Petraglia, A., Carbognani, M., Wellstein, C., 2019. Intraspecific functional trait response to advanced snowmelt suggests increase of growth potential but decrease of seed production in snowbed plant species. Front. Plant Sci. 10, 289.

Totland, O., 1997. Effects of flowering time and temperature on growth and reproduction in *Leontodon autumnalis* var. *taraxaci* a late-flowering alpine plant. Arct. Alp. Res. 29, 285–290.

Tudela-Isanta, M., Fernández-Pascual, E., Wijayasinghe, M., Orsenigo, S., Rossi, G., Pritchard, H.W., et al., 2018b. Habitat-related seed germination traits in alpine habitats. Ecol. Evol. 8, 150–161.

Tudela-Isanta, M., Ladouceur, E., Wijayasinghe, M., Pritchard, H.W., Mondoni, A., 2018a. The seed germination niche limits the distribution of some plant species in calcareous or siliceous alpine bedrocks. Alp. Bot. 128, 83–95.

Turner, J., Lu, H., White, I., King, J.C., Phillips, T., Hosking, J.S., et al., 2016. Absence of 21st century warming on Antarctic Peninsula consistent with natural variability. Nature 535, 411–423.

Venable, D.L., Brown, J.S., 1988. The selective interactions of dispersal, dormancy, and seed size as adaptations for reducing risk in variable environments. Am. Nat. 131, 360–384.

Venn, S.E., Morgan, J.W., 2010. Soil seedbank composition and dynamics across alpine summits in south-eastern Australia. Aust. J. Bot. 58, 349–362.

Veselá, A., Dostálek, T., Rokaya, M.B., Münzbergová, Z., 2020. Seed mass and plant home site environment interact to determine alpine species germination patterns along an elevation gradient. Alp. Bot. 130, 101–113.

Vittoz, P., Dussex, N., Wassef, J., Guisan, A., 2009. Diaspore traits discriminate good from weak colonisers on high-elevation summits. Basic Appl. Ecol. 10, 508–515.

Vuille, M., 2013. Climate change and water resources in the tropical Andes. Inter-American Development Bank Technical Note 515. <https://publications.iadb.org/handle/11319/5827>.

Wagner, J., Reichegger, B., 1997. Phenology and seed development of the alpine sedges *Carex curvula* and *Carex firma* in response to contrasting topoclimates. Arct. Alp. Res. 29, 291–299.

Wagner, I., Simons, A.M., 2009. Divergent norms of reaction to temperature in germination characteristics among populations of the arctic-alpine annual, *Koenigia islandica*. Arct. Antarct. Alp. Res. 41, 388–395.

Walder, T., Erschbamer, B., 2015. Temperature and drought drive differences in germination responses between congeneric species along altitudinal gradients. Plant Ecol. 216, 1297–1309.

Wang, Q., Fan, X., Wang, M., 2016. Evidence of high-elevation amplification vs Arctic amplification. Sci. Rep. 6, 19219.

Willson, M.F., Rice, B.L., Westoby, M., 1990. Seed dispersal spectra: a comparison of temperate plant communities. J. Veg. Sci. 1, 547–562.

Winkler, D.E., Lubetkin, K.C., Carrell, A.C., 2019. Responses of alpine plant communities to climate warming. In: Mohan, J.E. (Ed.), Ecosystem Consequences of Soil Warming Microbes, Vegetation, Fauna and Soil Biogeochemistry. Academic Press/Elsevier, London, London, pp. 297–346.

Winkler, M., Lamprecht, A., Steinbauer, K., Hülber, K., Theurillat, J.P., Breineret, F., et al., 2016. The rich sides of mountain summits — a pan-European view on aspect preferences of alpine plants. J. Biogeogr. 43, 2261–2273.

Wookey, P.A., Robinson, C.H., Parsons, A.N., Welker, J.M., Press, M.C., Callaghanet, T.V., et al., 1995. Environmental constraints on the growth, photosynthesis and reproductive development of *Dryas octopetala* at a high arctic polar semi-desert, Svalbard. Oecologia 102, 478–489.

Young, L.M., Kelly, D., Nelson, X.J., 2012. Alpine flora may depend on declining frugivorous parrot for seed dispersal. Biol. Conserv. 147, 133–142.

Zimmermann, N.E., Gebetsroither, E., Zuger, J., Schmatz, D., Psomas, A., 2013. Future climate of the European Alps. In: Cerbu, G., Hanewinkel, M., Gerosa, G., Jandl, R. (Eds.), Management Strategies to Adapt Alpine Space Forests to Climate Change Risks. InTechOpen, pp. 27–36.

Further reading

Mondoni, A., Orsenigo, S., Abeli, T., Rossi, G., Brancaleoni, L., Corli, A., et al., 2020. Plant regeneration above the species elevational leading edge: trade-off between seedling recruitment and plant production. Front. Ecol. Evol. 8, 572878. Available from: https://doi.org/10.3389/fevo.2020.572878.

Chapter 2

Effects of climate change on regeneration of plants from seeds in boreal, subarctic, and subalpine regions

Bente J. Graae[1], Kristin O. Nystuen[1,2], Vigdis Vandvik[3] and Amy E. Eycott[4]
[1]*Department of Biology, NTNU, Norwegian University of Science and Technology, Trondheim, Norway,* [2]*Norwegian Biodiversity Information Centre, Trondheim, Norway,* [3]*Department of Biological Sciences and Bjerknes Centre for Climate Research, University of Bergen, Norway,* [4]*Faculty of Biosciences and Aquaculture, Nord University, Steinkjer, Norway*

The boreal ecosystem

The boreal forest belt, adjacent subarctic transition regions, and subalpine areas on mountains, hereafter the boreal zone, is the earth's largest vegetation zone. The boreal zone covers about 18.9 million km^2— about 11% of the world's terrestrial surface—and it largely occurs between 45°N and 70°N with northern and southern boundaries determined by the July 13°C and July 18°C isotherms (Stocks et al., 2001; Brandt et al., 2013). It extends from Kamchatka in East Asia over vast regions of northern China, Siberia, northern Russia, and Fennoscandia, across Iceland and over major portions of northern North America, covering much of Canada and large parts of interior and southern Alaska (Fig. 1.1). Since the boreal zone occurs at high latitudes, it is subjected to a marked annual cycle in day length, large annual ranges in temperature (Fig. 2.1), and receives modest amounts of precipitation that mostly are concentrated in summer (Brandt et al., 2013). Thus, the seasonal climatic variation in the boreal zone is characterized by short, cool summers with long to very long day lengths and long, cold winters with short day lengths (Brandt et al., 2013). The large landmasses in North America and across Eurasia mean that much of the boreal zone has a continental climate with relatively low precipitation and a very large annual amplitude in temperature (Price et al., 2013). In contrast, boreal areas in oceanic regions, such as those along the northwestern coast of the Eurasian continent, experience significantly milder climate and thus smaller annual temperature amplitudes than boreal areas in continental regions.

The boreal zone is dominated by evergreen coniferous forests. The successively colder climate toward higher latitudes and altitudes results in decreased tree density, with the trees often being increasingly dwarfed in stature because of growth limitations imposed by short and cool summers, low soil nutrient availability, and harsh winter conditions related to snowpack effects, ground frost, and permafrost active layer depth and dynamics (Körner, 2012; Helbig et al., 2016). Limiting the boreal zone toward the south is the temperate zone with deciduous forests in oceanic and subcontinental regions, grading into parklands with sparse woody vegetation and steppe or prairie grasslands in low-precipitation continental regions such as in western Canada and central Asia (Kurz et al., 2013).

The most dominant trees of the boreal forest are conifers including *Abies, Larix, Picea*, and *Pinus* and deciduous hardwoods such as *Alnus, Betula, Populus*, and *Salix*. Due to the combination of cool temperatures, a general precipitation surplus, and dominance of evergreen species with decay-resistant litter produced in both the canopy (conifers) and understory (*Ericaceae*), boreal forest soils are mostly podzolized, with the more productive brown soils only on more climatically and edaphically favorable sites (Fischer et al., 2008). A well-developed ground layer of mosses and lichens also is characteristic of boreal vegetation (Nilsson and Wardle, 2005). The boreal zone also contains extensive areas of peatland and wetlands, in particular mires, which are not covered in this chapter.

While precipitation generally is concentrated in summer (Brandt et al., 2013), winter precipitation generally falls as snow, which has a substantial impact on several aspects of ecosystem functioning. Snowmelt delays onset of the growing season, thus the growing season is shorter than that of more southern vegetation regions. The snow layer also affects

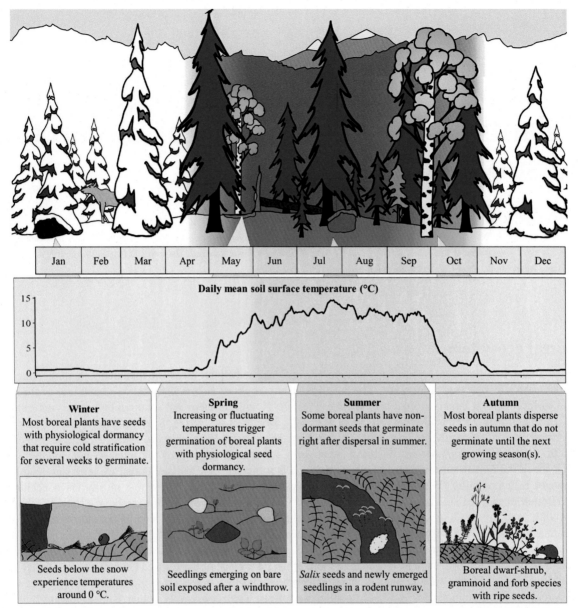

FIGURE 2.1 A year in the boreal zone. The recruitment phases seed formation, dispersal, dormancy-break, germination, and seedling establishment depend on seasonal differences in temperature, moisture, and/or daylength. Biotic interactions also affect seedling recruitment. Disturbances create gaps in the otherwise dense understory of mosses, lichens, and extant seed plant vegetation often promoting seedling emergence and survival. The soil surface temperature data are from a boreal spruce-dominated forest in mid-Norway (site 10, Kolstad et al., 2019). Kolstad and colleagues logged temperatures during 1 year from May 2016 and onward; thus, the January to May data in this figure are from 2017, whereas the May to December data are from 2016 (hence the temperature break in May).

the winter microclimate, insulating the ground from cold air and keeping soil surface temperatures just above freezing for long periods of the winter and spring (Zhang, 2005; Kolstad et al., 2019). These winter conditions also indirectly impact plant community assembly, trophic interactions, and nutrient cycling.

Boreal ecosystems are characterized by recurrent large- and small-scale disturbances that are important and integral to boreal ecosystem functioning, dynamics, and biodiversity. The most important large-scale disturbances prior to the appearance of humans were forest fires, large storm windthrows, insect outbreaks, and diseases. The natural boreal fire regime (fire frequency, severity, and size) is driven by climatic, landscape, and biotic features (Wardle et al., 2003). Autumn and winter storms often cause extensive windthrows. In particular, windthrows occur in stands dominated by shallow-rooted trees such as those on waterlogged sites (Stokes, 2002), evergreen trees such as spruce (Peterson, 2007), and early succession, shade-intolerant species (Rich et al., 2007). Other important large-scale, natural disturbances in the system are caused by insect outbreaks and diseases. For example, the autumnal moth (*Epirrita autumnata*) in northern boreal birch forests creates large areas of dead forest (Jepsen et al., 2008; Netherer and Schopf, 2010).

Small-scale natural disturbances are part of the dynamics of boreal ecosystems, and they also are important for recruitment of new individuals. Death of single trees occurs regularly in natural forests and creates a light gap and exposes soil (Fig. 2.1, spring). Furthermore, the boreal biome is home to a number of large ungulate herbivores, including moose, deer, reindeer, muskoxen, and in modern times domestic sheep and cattle, in addition to rodents (Austrheim and Eriksson, 2001). These animals are important in shaping vegetation and landscape structure by selectively grazing particular functional groups (Lorentzen et al., 2018; Sitters et al., 2019), creating gaps for seedling recruitment (Ericson et al., 1992; Dufour-Tremblay and Boudreau, 2011; Fig. 2.1, summer), aiding seed dispersal (Jaroszewicz, 2013; Steyaert et al., 2018) and redistributing nutrients within the landscape (Barthelemy et al., 2015).

Recruitment from seeds in the boreal zone

Most plant species in the boreal zone are long-lived, and many of them have extensive clonal growth. Still, seed production is abundant, and it is necessary for responding to climate change, whether via migration of species or through microevolution under new environmental conditions. Boreal plant species are adapted to the characteristic annual climatic cycle and to the longer-term spatiotemporal high climatic variability of the boreal zone and the associated characteristic ecosystem functioning. For instance, boreal populations typically have fast phenological development and seed germination traits, which ensure that shoots, leaves, and seedlings will emerge early in spring, enabling plants to take advantage of the short growing season (Fig. 2.1). The selection pressures for withstanding the highly varied climate diurnally, seasonally, interannually, and historically may mean that boreal species have wide climatic niches and thus resistance to climate change.

The dense tree, field, and ground vegetation layers of many boreal ecosystems imply strong biotic filtering on seed germination and seedling establishment. Thus, recruitment of new individuals via seeds usually depends on disturbances or other safe sites for recruitment (Eriksson and Ehrlén, 1992; Eriksson and Fröborg, 1996). In the tree layer, light gaps and nurse logs resulting from individual tree deaths can provide such safe sites for seedlings (Hofgaard, 1993; Kuuluvainen, 1994; Eriksson and Fröborg, 1996; Simard et al., 1998). Long-lived and clonally propagating dwarf shrub, forb, and graminoid species that dominate the field layer and dense bryophyte mats in the ground layer can monopolize nutrients and space. Also, small-scale disturbances by animals are important in creating vegetation gaps for seedling recruitment in boreal systems (Fig. 2.1). Soil seed banks can be important reservoirs of seeds for gap recruitment (Vandvik et al., 2016). The importance of gaps for seedling establishment in boreal, subarctic, and subalpine vegetation has been demonstrated repeatedly (Eriksson and Fröborg, 1996; Vandvik, 2004; Vikane et al., 2013; Tingstad et al., 2015; Klanderud et al., 2017). In contrast, lichen and moss mats may facilitate seed regeneration (Steijlen et al., 1995; Zackrisson et al., 1995; Soudzilovskaia et al., 2011), depending on the ground layer species and/or the recruiting plant species in question (Soudzilovskaia et al., 2011; Nystuen et al., 2019). Furthermore, many boreal species, especially ericads, lichens, and bryophytes, produce allelopathic compounds that may restrain seed germination and early growth of new individuals (Zackrisson and Nilsson, 1992; Vikane et al., 2013; Pilsbacher et al., 2021).

Fires can remove tree layer and/or ground layer vegetation in boreal systems, and boreal plant recruitment can benefit from fire both directly and indirectly. Burns through the moss and litter layers generally promote germination of seeds in the persistent buried seed pool in boreal systems (Schimmel and Granström, 1996; Måren and Vandvik, 2009), and high temperatures can function as a germination cue for certain species (Granström and Schimmel, 1993). Plants also may evolve germination responses to chemical fire cues, as demonstrated for the boreal dwarf shrub *Calluna vulgaris* (Vandvik et al., 2014) and other plant functional types, including perennial forbs and grasses (Bargmann et al., 2014). Finally, ash from fire can neutralize phytotoxic substances released as part of allelopathic interactions (Vikane et al., 2013).

Predicted climate change and its impact on the boreal zone

The boreal zone is among the biomes expected to experience the most severe climate change, with increases in land surface temperatures as high as 2.5°C from 1901 to 2012 (IPCC, 2014). Both increased and decreased amounts of annual precipitation were recorded in boreal regions from 1951 to 2010 (IPCC, 2014). RCP2.8, IPCC's most dramatic scenario, predicts temperatures for the boreal biome to be 4°C−11°C warmer by 2100 compared to the 1986−2005 period, with the strongest warming at higher latitudes (IPCC, 2014). Furthermore, warming is expected to be pronounced in spring, resulting in earlier snowmelt and earlier onset of the growing season (Derksen and Brown, 2012; IPCC, 2014; Jungqvist et al., 2014). Without a protective snow layer, the vegetation will be more prone to early season frost events, which have increased in frequency in many boreal regions during the last three decades (Liu et al., 2018). Increase in precipitation by 0%—30% is also expected under RCP2.8, and while there are regional differences higher latitudes generally will experience the greatest precipitation increases (Price et al., 2013; IPCC, 2014; Gauthier et al., 2015). Across the boreal zone, change in precipitation is modest compared to changes in temperature, thus the increase in precipitation may not compensate for the increased evapotranspiration demand caused by increased temperatures (Gauthier et al., 2015). These effects may be enhanced by regional variation in the timing of precipitation. In North American boreal forests, large increases in winter precipitation are expected, whereas enhanced summer precipitation is expected in boreal northern Asia (IPCC, 2013). Thus, periods with summer drought are expected to increase, and boreal regions with low soil moisture are especially vulnerable to these droughts (Dulamsuren et al., 2010; IPCC, 2013; Price et al., 2013). Wildfires are expected to increase in frequency and severity with warmer climate and longer periods of drought (Weber and Flannigan, 1997; Stevens-Rumann et al., 2018; Parisien et al., 2020). Climate change also is associated with increased frequency of extreme weather events. For example, in western Canada the increase in extreme precipitation events may be as much as 300% (Li et al., 2019).

According to some studies, climate change in the boreal zone is expected to increase storm frequency and intensity, potentially resulting in an increase in windthrows (Blennow and Olofsson, 2008). A warmer climate with more fires and more productive vegetation also may increase the density of herbivores, which may further speed up succession towards more nutrient-rich and dynamic systems. Climate change also is predicted to increase outbreaks of insects and pathogens, which would create large and more frequent disturbances of the tree layer (Netherer and Schopf, 2010). Thus dramatic changes in fire and insect and pathogen outbreak frequencies, as well as composition shifts in tree-layer species, may have great effects on the ground layer vegetation. Typically, an increase in soil nutrients, fire frequency, and productivity under the forest canopy will be less favorable for the cryptogam species in competition with herbaceous vegetation. A drier climate also will limit paludification (peat-land formation) in the boreal zone (Laamrani et al., 2020).

Sampling and characterizing the knowledge base

We sampled the literature published in English on the effects of climate change on seeds in the boreal zone, using a keyword string and predefined inclusion criteria (see Box 2.1). We use "sampled," rather than "systematically reviewed" because of challenges in obtaining a complete coverage of the literature since few studies use "boreal" in their titles or keywords. Also, we would have needed comprehensive geographic or taxonomic locators for a complete coverage. Our search string yielded 1977 hits, and a check of each abstract identified 232 relevant articles. We coded each article according to plant functional group, climatic variables, and seed responses. However, some articles could not be placed/classified into one of the three coding variables, and we did not have access to the full text of all articles.

More than half of the articles had germination as one of the response variables, and about a quarter of them evaluated seed production. Seed rain, seed dispersal, and seed bank/seed pool studies accounted for only around 15% of the

---

BOX 2.1 Literature search.

Search strategy

Search string: [(seed* OR propagule* OR germin*) AND (boreal OR circumboreal OR taiga OR subalpine OR "subalpine" OR subarctic OR "subarctic") AND (climat* OR weather OR rain* OR temperature OR light OR season* OR daylength OR daylight OR frost* OR drought* OR storm* OR wind* OR gale* OR snow*)]

Note that the asterisk indicates records with any word ending, for example, climat* returned hits for climate, climatic, and climates. The search was performed in ISI Web of Science Expanded Collection on 24th November 2020.

Inclusion criteria

Response: Direct measures of seed germination, seed dormancy or recruitment in the sense of seedling establishment, seed mortality, seed production, or seed dispersal (including temporal).

Intervention: Direct measures of an effect related to climate, including sowing experiments at different elevations and testing soil moisture.

Scope: Must include species of boreal forests, boreal subalpine zones (in the sense of the zone between forest line and treeline), or boreal—arctic transitions. Food-crop species such as barley were not included.

Climate tests must be more or less within the potential bounds of a boreal/subalpine zone.

articles. Two-thirds of the papers addressed trees, while one-fifth adressed herbs. Ten papers adressed shrubs (primarily ericaceous), four graminoids, and only 11 papers addressed more than one functional group.

Articles mostly addressed temperature effects and only a few investigated precipitation (rain or snow). Temperature was mostly investigated using temperature-controlled growth chambers, but some used "natural experiments" over elevational gradients or common garden experiments. Distribution of the different climate and response variables was approximately equal across the different functional groups, although seedbank studies of trees were underrepresented. The following topic summaries are based on a combination of the publications retrieved during the literature-searching process and our own collective knowledge of relevant studies. Due to the many studies on single tree species, we focused on review papers for trees.

Seed production

Many studies document how seed production in the boreal zone can be favored by warming. Seed production has been extensively studied in trees, in which interannual climate variation drives fecundity and seed quality. Low temperatures are usually a factor limiting seed production (including flower bud differentiation, flowering, pollination, and seed maturation) at the northern or upslope limits of the range of a species (Stinziano and Way, 2014; Boucher et al., 2020), while at the southern or downslope edge of the range, temperature extremes eventually can limit viable seed output through damage to flowers and fruits (Ågren, 1988). A warmer and longer growing season increases overall productivity and thus the amount of resources available for reproduction. Boucher et al. (2020) reviewed the effects of high temperature and drought on postfire recruitment of four North American boreal tree species (*Betula papyrifera*, *Picea mariana*, *Pinus banksiana*, and *Populus tremuloides*), and they reported species-specific variation in responses to climate. The effects of precipitation on seed production often are minor compared to those of temperature (Zamorano et al., 2018). However, seed production can be reduced by drought, and desiccation has been implicated as one of the key factors involved in seed mortality that occurs during the first summer following fire (Boucher et al., 2020). Seed production usually peaked when warm temperature was combined with high precipitation (Boucher et al., 2020).

For forest floor and other short-stature plants, increased temperatures during seed development increased seed size (Graae et al., 2009) and number of germinable seeds (De Frenne et al., 2010, 2011). For the species in these studies, seeds were collected along a latitudinal gradient from temperate to boreal, and those from boreal populations showed a positive response to warming (De Frenne et al., 2009, 2010). Molau and Larsson (2000) studied seed set along an elevational gradient from subalpine to alpine in different years and found that number of diaspores produced decreased with elevation and that the cool and rainy summer of 1995 reduced seed set substantially in all plant functional types and in herbs and graminoids in particular. However, results of different studies reveal different patterns, depending on temperature range tested and species. Seed production in the important boreal shrub species *Empetrum nigrum* ssp. *hermaphroditum* did not increase in response to warming (Wookey et al., 1993). In *Vaccinium myrtillus*, reproduction was affected by several habitat-related factors, but it was relatively unaffected by climate. Throughout its elevation range in the boreal and alpine zone, seed production of *V. myrtillus* was high, whereas photosynthesis and growth decreased with elevation after a peak at 950 m a.s.l., which is about the timberline (Pato and Obeso, 2012). For the subalpine herb *Primula farinosa*, McKee and Richards (1998) suggested that an increase in temperature at flowering might adversely affect the quantity and quality of seeds set because plants from two upland populations kept at 26°C during flowering produced fewer seeds with lower germinability than those exposed to lower temperatures during flowering.

Timing of warming also has an effect on seed production. In a snow-removal treatment for subalpine herbaceous species in Colorado (United States), two early flowering species had decreased fruit and seed set due to frost damage of plants. However, two later-flowering species escaped frost damage and had increased pollinator visitation and reproduction (Pardee et al., 2019). Similar results for seed production were found for the herb *Ipomopsis tenuituba* in the same region (Campbell, 2019). Late snowmelt increases water availability in the soil and thereby increases the ability of plants to produce viable seeds after pollination (Campbell and Powers, 2015; Gallagher and Campbell, 2017). Lack of water due to either decreased precipitation or soil warming may be a problem for flowering even for species in the boreal zone (Gauthier et al., 2015). Adding water equivalent to a 50% increase in natural summer precipitation increased seed set in *Empetrum nigrum* ssp. *hermaphroditum* (Wookey et al., 1993). In years with low snow accumulation, *Delphinium nelsonii* produced fewer flowers, probably because the plants were exposed to colder temperatures and frost damage during floral development (Inouye and McGuire, 1991). Many studies have shown that spring frosts may damage flowers and decrease seed set (Usui et al., 2005; Pardee et al., 2018).

Seed dormancy and germination

The climate of the boreal zone favors species with seed dormancy (or timing of seed germination) that prevents seedling emergence during the cold autumn or winter. Seeds of 72%, 66%, and 57% of boreal trees, shrubs, and herbaceous species have physiological dormancy (PD), respectively, that require cold stratification to germinate (Baskin and Baskin, 2014). In general, seeds require several weeks of (moist) cold stratification (0°C–10°C) for dormancy-break. When the depth of the snow layer reaches about 40 cm, the temperature of the soil under the snow is stabilized at around 0°C–1°C (Zhang, 2005), thereby promoting dormancy-break by cold stratification. Cold stratification not only breaks seed dormancy but also lowers the temperature requirements for germination. Thus, seeds can germinate at relatively low temperatures in spring, thereby giving seedlings the full growing season to become established. The temperature window for germination following cold stratification usually is broad (Densmore, 1997; Baskin and Baskin, 2014), and the increase in temperature that follows snowmelt triggers germination (Fig. 2.1). Requirements for dormancy-break can vary between seeds of close relatives and between different regions. For example, the physiologically dormant seeds of *Rhinanthus minor* collected from plants growing at sea level in The Netherlands or in the subarctic or subalpine in Switzerland require a long period of cold stratification for dormancy-break, whereas seeds of *R. alectorolophus* collected from plants growing at sea level zone in The Netherlands and in the montane in Switzerland required a short period of cold stratification for dormancy-break (Ter Borg, 2005). Thus, warming may have a greater negative impact on dormancy-break and germination of *R. minor* than *R. alectorolophus*.

Seeds of the dominant boreal dwarf shrub *Empetrum nigrum* have PD. About 40% of the seeds require only cold stratification for dormancy-break, but the remainder requires warm stratification followed by cold stratification to germinate. A period of warm stratification can decrease the length of the cold stratification period required to break PD in seeds of this species (Baskin et al., 2002). Even after three cycles of 3 months freezing at −4°C followed by incubation at 20°C some seeds of *E. nigrum* had not germinated (Molau and Larsson, 2000). Thus, seeds of *E. nigrum* are difficult to germinate to high percentages (Baskin et al., 2002; Graae et al., 2008, 2011), and the germination requirements may result in a persistent seedbank with an average turnover of >200 years (Molau and Larsson, 2000). Since a reduction of the depth of winter snow layer by one-half in Sweden did not significantly reduce the percentage of seeds that came out of dormancy (Baskin et al., 2002) and since the species has a persistent soil seed bank (Molau and Larsson, 2000), it is predicted that climate warming will not have an immediate negative effect on germination of *E. nigrum*.

Other ecologically important and abundant dwarf shrubs in boreal regions that belong to the genus *Vaccinium* require cold, but not warm, stratification to break PD (Baskin et al., 2000; Graae et al., 2008, 2011). Warm spells during winter may interrupt the required cold stratification period of seeds (e.g., Sommerville et al., 2013). If precipitation falls as rain, it decreases the depth of or even removes the snow layer. Thinning or removal of the insulating snow layer lowers the temperatures to which seeds are exposed during subsequent cold periods (Milbau et al., 2009), resulting in temperatures too low for effective cold stratification, thereby preventing or delaying dormancy-break and thus germination in spring.

According to Baskin and Baskin (2014), seeds of 17% of boreal shrub species and 25% of boreal herbaceous species have morphophysiological dormancy (MPD). *Anemone nemorosa* is a typical herb with MPD that emerges in early spring in deciduous forests throughout most of Europe but later in boreal and subalpine forests due to the long-lying snow cover. Following dispersal, seeds require a warm period for embryo development, after which the radicle emerges as temperatures decrease in autumn. In a study in Italy, the radicle did not emerge until the temperature was <15°C for lowland elevation (74 m a.s.l.) and <10°C for high elevation (1350 m a.s.l.) plants, and the shoot did not emerge until the temperature was <10°C and <6°C (in late autumn/early winter), respectively (Mondoni et al., 2008). Thus, seed dormancy/germination of *A. nemorosa* is adapted to the local climate, and this has been confirmed for *A. nemorosa* and other species such as the European understory plant *Milium effusum* (De Frenne et al., 2012). The response of seeds of locally-adapted species to a warming climate needs to be determined.

Throughout its wide geographical range, seeds of the winter-deciduous shrub *Viburnum opulus* have deep simple epicotyl MPD. Embryo growth and radicle (root) emergence require a period of warm stratification (moist, ≥15°C), while emergence of the epicotyl (shoot) requires that seeds with an emerged root be cold stratified. Thus, the seeds, which mature in late autumn, are not exposed to warm stratification until the following summer, which results in embryo growth and root emergence in late summer–early autumn of the first year following seed maturity. Seeds with an emerged root are cold stratified during winter, and the shoot emerges the second spring following seed maturity (Walck et al., 2012). Will a warming climate shorten the length of time from dormancy-break (root and shoot) in this species? For example, in a warmed climate will the root emerge the first autumn and the shoot the first spring, thereby decreasing the length of the dormancy-breaking process in *V. opulus*?

Seeds of many boreal zone willows (*Salix* spp.) are nondormant and thus germinate rapidly over a wide range of temperatures immediately after dispersal in summer or early autumn (Zasada and Viereck, 1975; Graae et al., 2011;

summer panel in Fig. 2.1). Seeds of several boreal shrubs and some boreal trees and herbaceous species can germinate at high, but not at low, temperatures right after maturity and dispersal. Thus, germination is prevented in autumn because temperatures generally are low, and therefore it does not occur until the following spring, after cold stratification has decreased the minimum temperature for germination (Densmore & Zasada, 1983; Densmore, 1997; Baskin and Baskin, 2014). A warmer climate could result in these species germinating in autumn, with potentially high mortality risks for young seedlings during winter. The effect of autumn germination on seedling survival and species persistence in the community needs to be determined.

There is considerable variation in seed germination strategies along bioclimatic gradients within the boreal zone, which reflects germination-cueing mechanisms that enable risk-avoidance strategies and increase seedling survival. Two contrasting patterns have been documented for germination temperature requirements. One is increased germination at low temperatures (cold tolerance) in seeds from colder climates. For example, Bevington (1986) and Benowicz et al. (2001) found generally higher and faster germination at cold temperatures for seeds of paper birch from colder-climates populations. Similarly, some studies found faster germination for species growing at higher than at lower elevations or latitudes (Benowicz et al., 2001; De Frenne et al., 2010). Such germination strategies may confer a selective advantage by enabling earlier germination and thus more effective exploitation of the relatively short growing season. The other pattern is decreased germination at low temperatures (cold avoidance) in seeds from colder climates.

At the interspecific level, higher temperature requirements for germination have been documented in cold-climate relative to warm-climate boreal species (Chabot and Billings, 1972; Baskin and Baskin, 2014). At the intraspecific level, seeds from colder-climate populations have higher temperature requirements and germinate more slowly than seeds from warmer-climate populations (Meyer et al., 1989; Meyer and Monsen, 1991; Cavieres and Arroyo, 2000; Vandvik and Vange, 2003). Seeds of the dominant boreal field layer dwarf shrub *Calluna vulgaris* were collected from 11 sites across broad bioclimatic gradients in Norway and tested at temperatures ranging from 10°C to 25°C. Seeds from all populations germinated relatively fast and to high percentages at high (25°C) temperatures, but those from cold-climate populations germinated to low percentages and rates at low temperatures (10°C) (Spindelböck et al., 2013). The authors suggest that these temperature responses reflect a shift in the major drivers of seedling mortality risk from late-spring frosts in colder climates, towards competitive effects from the vegetation in warmer climates. This would imply a shift in selective pressure for delayed germination and hence a cold-avoidance strategy in cold climates, and advanced germination and hence a cold-tolerance germination strategy in warm climates. On the other hand, in warmer climates competition from established vegetation and/or mid-summer drought risk would be a selective pressure for early germination and hence cold tolerance. For species with such cold-avoidance strategies, climate warming could promote earlier germination in colder-climate populations, potentially increasing exposure of seedlings to late-spring frost events since climate variability and extremes are predicted to increase in the future.

Studies on seed germination of boreal species, including many trees, report that germination percentages and rates are temperature-dependent and increase with increasing temperatures, with optima generally being relatively high (Graae et al., 2008, 2011; Milbau et al., 2009; Vandvik et al., 2017; Boucher et al., 2020). In general, temperatures resulting in the best germination are higher for boreal trees (mean ± SE: 23.0 ± 0.6°C) than for shrubs (16.1 ± 0.5°C) and herbs (13.4 ± 0.2°C) (Baskin and Baskin, 2014). Many studies that found positive effects of warm temperatures on seed germination were done in growth chambers, and the seed-germination substrate was kept continuously moist. However, field experiments on the effects of high temperatures on plants often are associated with risks of desiccation or even drought (Stinziano and Way, 2014), which can limit germination. In a review on recruitment of four North American boreal tree species at high temperatures and drought on postfire burned areas, Boucher et al. (2020) reported that the ability of seeds to germinate under moisture deficient conditions differed between species.

In a study of 10 species in subarctic and subalpine Scandinavia, Shevtsova et al. (2009) heated plots in the field 3°C above ambient temperature during different seasons. They also had a treatment in which normal precipitation was increased by 30% in accordance with the highest expected summer precipitation increase for Scandinavia in the IPCC's climate scenario (Christensen et al., 2007). In contrast to results from growth chamber experiments, almost all species had lower recruitment at the higher temperatures, and watering did not mitigate the effects of warming (Shevtsova et al., 2009). The late-germinating species did best since the seedlings were not affected by early and peak season warming effects. Warming may have heat-stressed the cold-adapted species, but most likely the detrimental effects were due to soil drying. Therefore, earlier germination in a warmer climate may be detrimental to seedling establishment, unless germination is associated with increased moisture, for instance, more rainfall. Earlier snowmelt, as a consequence of a warmer spring or winter, also may lead to decreased soil moisture during the subsequent growing season, which likewise may restrict germination.

In nature, a warming macroclimate may be strongly buffered by vegetation, rocks and logs, and topography in local sites, resulting in a microclimate with altered temperature and moisture extremes. This is particularly true for the boreal zone during winter due to the snow cover (Graae et al., 2012; Zhang et al., 2018) and also for forests in summer. Forest cover has a strong buffering effect on the microclimate (De Frenne et al., 2013; Holtmeier and Broll, 2020). For seeds and seedlings, intact understory of seed plants (Klanderud et al., 2021), mosses (Soudzilovskaia et al., 2011), lichens (Nystuen et al., 2019; van Zuijlen et al., 2020), and litter (Sayer, 2006) may further limit temperature extremes and decrease the risk of droughts that otherwise would damage the newly emerged radicle or shoot.

Fire reduces the density of vegetation, creates a flush of available nutrients, and produces ash that can neutralize phytotoxic substances, and thus germination cues that promote postfire seedling recruitment are common and widespread in fire-adapted floras worldwide (Keeley et al., 2011). For example, smoke-derived chemicals such as karrikinolide and glyceronitrile play key roles in smoke-stimulated germination (Flematti et al., 2004, 2011). Natural fire regimes are prominent in boreal systems (Stocks et al., 2001), and fire has been used as a management tool by human populations in boreal zone for millennia, creating seminatural ecosystems with characteristic postfire successional dynamics such as those in the coastal heathlands of northwest Europe (Gimingham, 1972; Prøsch-Danielsen and Simonsen, 2000; Vandvik et al., 2005). Smoke-stimulated germination has been documented in several boreal forbs, graminoids, and dwarf shrubs (Bargmann et al., 2014), contributing to fast recruitment from soil seed banks after fire (Måren and Vandvik, 2009; Måren et al., 2010). A study comparing heathland habitats differing in historic fire regimes suggests that smoke-stimulated germination can evolve relatively rapidly in response to changes in fire regimes (Vandvik et al., 2014).

Climate change may result in an increase in widespread and extreme fires in the boreal zone, which could destroy the temperature-buffering and moisture-protecting understory layers, thereby exposing seeds to conditions too warm and too dry for successful germination (Stevens-Rumann and Morgan, 2019; Boucher et al., 2020). Furthermore, extreme fires are threats to seed banks, especially when they burn deep into the soil (Schimmel and Granström, 1996). Thus, an increase in fire frequency as a consequence of climate change may have unpredictable effects on regeneration of plants from seeds, depending on species, ecological context, and fire conditions.

There are many interacting factors moderating the effects of climatic impacts on seed germination in nature (Milbau et al., 2013), in addition to the ones discussed above. For example, interannual climatic variations in combination with seed bank dynamics may further challenge our capacity to disentangle the effects of a warmer climate on germination and seedling emergence in nature (Molau and Larsson, 2000; Stevens-Rumann and Morgan, 2019).

Seedling survival and growth

Drought and desiccation are the biggest threats to seedling survival in all biomes (Moles and Westoby, 2004b). In general, seedling growth in the boreal zones benefits from warmer temperatures up to a certain species-specific threshold (Dang and Cheng, 2004; Danby and Hik, 2007; Boucher et al., 2020). Seedlings of deciduous tree species potentially have a particularly big advantage in a warmer climate (Landhäusser et al., 1996; Peng and Dang, 2003). The projected global warming increase in length of the growing season in cold regions may benefit seedling growth if summer droughts do not increase (Clark et al., 2016; Stevens-Rumann et al., 2018; Boucher et al., 2020).

In their review paper, Boucher et al. (2020) describe how mild drought may affect tree seedlings through effects on growth and photosynthesis. Severe drought can kill trees via hydraulic failure or carbon starvation. They also describe how seedlings in warmer climates tend to allocate less biomass to roots and thereby become more drought sensitive. Seed size in trees plays a role in the ability of seedlings to withstand drought, with seedlings from large seeds tending to have higher survival than those from small seeds (Moles and Westoby, 2004a; Boucher et al., 2020). Seed provenance may confer an advantage. Seedlings derived from seeds produced in warm provenances outperformed those from locally sourced seeds in warmed common gardens in situ for limber pine (*Pinus flexilis*), though not Engelmann spruce (*Picea engelmannii*) (Kueppers et al., 2017). In a study along an altitudinal gradient in northern Sweden, seedling emergence and establishment were very weakly related to summer soil temperature but strongly related to pH and soil moisture (Milbau et al., 2013). The established vegetation can have a strong competitive effect on seedling emergence and seedling survival, an effect that decreases as climates become colder (Milbau et al., 2013; Klanderud et al., 2017). Eriksson and Fröborg (1996) described the importance of gaps or safe sites in vegetation as "windows of opportunity" essential for recruitment in four boreal *Vaccinium* species.

Thus, while warmer temperatures generally favor seedling growth seedling survival may be compromised as climate changes, especially with increased climatic variability and increased risks of drought, fire, and other disturbances. Seedlings often are more vulnerable to disturbances such as drought, herbivores, and fire than mature individuals (Anderson-Teixeira et al., 2013). However, another important factor for seedling survival and establishment relates to their interactions with the extant vegetation, which also will be affected by climate change (e.g., Klanderud et al., 2017,

2021). In particular, gaps are needed for seedling establishment, and such gaps are consequences of disturbances from herbivores and other animals, fire, storms, fallen logs, and land use by humans (Eriksson and Fröborg, 1996; Vandvik, 2004).

Long-term consequences of climate change on seedling recruitment

Predicting how the seed ecology of boreal species will respond to climate change is complicated by the great variation in climate across the biome and the differences in responsiveness and temperature requirements among species. The projected increase in climatic extremes and changes in seasonal dynamics are expected to have greater impacts on plant recruitment than the projected average changes in climate. For example, late frosts in spring or early summer or extremes in precipitation can severely damage flower buds, flowers, and seedlings (Brandt et al., 2013; Moreau et al., 2020). Warm spells during winter may cause rain-on-snow events and snowmelt during winter that can expose seeds and seedlings to much lower temperatures than they would experience if covered by an insulating layer of snow (Groffman et al., 2001).

In the literature surveyed, we could not distinguish clear differences in responses associated with different functional groups in terms of temperature requirements for plant regeneration from seeds. In very general terms, seed production and germination tend to increase with a warmer climate, but they could be compromised if maternal plant fitness is decreased due to climatic extremes or drought. The effects of climate change on dormancy-break are even more complicated than those on seedling establishment and often will depend on snow conditions during winter. Thus, snowmelt with winter warming, especially in combination with rain, may decrease the depth of the snow layer, which in turn may result in lower soil temperatures due to decreased insulation from the thinner snow layer. Seedling establishment and survival will depend on interactions between temperature, moisture, and biotic interactions. That is, there will be increased germination but also increased risk of drought-related mortality and competitive effects from the established vegetation due to climate change. Since soil moisture is closely related to snow abundance and timing of snowmelt, there often are effects of winter temperatures on the subsequent seedling recruitment during spring and summer. While laboratory studies often show a positive effect of warming on seed germination, at least some field studies show subsequent disadvantages of warming, likely because of risks of drought. The surrounding vegetation may buffer both temperature and moisture extremes and thereby be important for recruitment success. However, the surrounding vegetation also will increase competition for light and space in a warmer climate.

Variation in phenology further limits our capacity for prediction. In cold years, or in years with deep snow, seed set and germination will be delayed. However, in years with deep snow, germination and seedling establishment occur when there is high soil moisture due to snowmelt and plenty of light due to long days. Length of the season for seedling establishment will be shorter in deep-snow years, but conditions for seedling establishment may be optimal, especially for high latitude species adapted to rapid growth during short seasons.

Climate change also may have many indirect effects on recruitment through its effects on the standing vegetation. Reduction in dominance by cryptogams potentially will increase opportunities for establishment of plants from seeds. However, cryptogams have a buffering effect on soil temperatures and thus facilitate establishment of newly emerged seedlings in harsh climates (Eckstein et al., 2011; Lett et al., 2017). An increase in temperature generally will increase productivity in many temperature-limited boreal systems, which in turn will increase the standing fuel load. Due to low moisture content of the vegetation and increased frequency of lightning events in warmer weather, an increase in fire frequency is expected (Price et al., 2013). Increased fire frequency will limit recruitment of some species (Rayfield et al., 2021), whereas it may create opportunities for seedling recruitment for others such as fire-adapted and ruderal species and newcomers that can become established quickly in sites where fire has destroyed forest or heaths (Gärtner et al., 2011). The outcome of changes in fire regimes will depend on fire severity and availability of seeds in the soil seed bank and on moisture availability during postfire seedling recruitment (Greene et al., 2007; Stevens-Rumann et al., 2018).

Future research

To fully understand the impact of climate change on recruitment of plants from seeds in the boreal zone, there is a need for combined studies in growth chambers, common gardens, and the field. While growth chamber studies may pinpoint the important physiological requirements, they need to be seen in the context of ecological complexity. Ecological/field studies are needed to identify the most important factors limiting seedling recruitment of a diversity of boreal species in the "real world". There is a clear need for experiments that are realistic in terms of the projected temperature increases due to climate change. Perhaps "gold standard" studies are warming experiments in which seeds are sown in soil in open-top warming chambers in the field and in an array of "common gardens" over a climatic gradient so that seed dormancy/germination and seedling establishment can be evaluated under a range of field conditions. For trees, the long-term responses to climate change often need to be extracted by use of dendrochronological data or from data collected along large-scale or steep climatic gradients.

In closing, then, the physiological responses to temperature and moisture in laboratory and controlled common garden studies need to be evaluated in the context of highly complex interactions with winter conditions, buffering effects of vegetation, other biotic interactions, and phenological patterns. All these naturally varying conditions also will be dependent on the weather and climate and indirectly affect the reproduction of plants from seeds in ways hard to predict.

References

Ågren, J., 1988. Between-year variation in flowering and fruit set in frost-prone and frost-sheltered populations of dioecious *Rubus chamaemorus*. Oecologia 76, 175–183.

Anderson-Teixeira, K.J., Miller, A.D., Mohan, J.E., Hudiburg, T.W., Duval, B.D., DeLucia, E.H., 2013. Altered dynamics of forest recovery under a changing climate. Glob. Change Biol. 19, 2001–2021.

Austrheim, G., Eriksson, O., 2001. Plant species diversity and grazing in the Scandinavian mountains - patterns and processes at different spatial scales. Ecography 24, 683–695.

Bargmann, T., Måren, I.E., Vandvik, V., 2014. Life after fire: smoke and ash as germination cues in ericads, herbs and graminoids of northern heathlands. Appl. Veg. Sci. 17, 670–679.

Barthelemy, H., Stark, S., Olofsson, J., 2015. Strong responses of subarctic plant communities to long-term reindeer feces manipulation. Ecosystems 18, 740–751.

Baskin, C.C., Baskin, J.M., 2014. Seeds: Ecology, Biogeography and Evolution of Dormancy and Germination, Second ed. Academic Press/Elsevier, San Diego.

Baskin, C.C., Milberg, P., Andersson, L., Baskin, J.M., 2000. Germination studies of three dwarf shrubs (*Vaccinium*, Ericaceae) of Northern Hemisphere coniferous forests. Can. J. Bot. 78, 1552–1560.

Baskin, C.C., Zackrisson, O., Baskin, J.M., 2002. Role of warm stratification in promoting germination of seeds of *Empetrum hermaphroditum* (Empetraceae), a circumboreal species with a stony endocarp. Am. J. Bot. 89, 486–493.

Benowicz, A., Guy, R., Carlson, M., El-Kassaby, Y., 2001. Genetic variation among paper birch (*Betula papyrifera* MARSH.) populations in germination, frost hardiness, gas exchange and growth. Silvae Genet. 50, 7–12.

Bevington, J., 1986. Geographic differences in the seed germination of paper birch (*Betula papyrifera*). Am. J. Bot. 73, 564–573.

Blennow, K., Olofsson, E., 2008. The probability of wind damage in forestry under a changed wind climate. Clim. Change 87, 347–360.

Boucher, D., Gauthier, S., Thiffault, N., Marchand, W., Girardin, M., Urli, M., 2020. How climate change might affect tree regeneration following fire at northern latitudes: a review. New For. 51, 543–571.

Brandt, J.P., Flannigan, M.D., Maynard, D.G., Thompson, I.D., Volney, W.J.A., 2013. An introduction to Canada's boreal zone: ecosystem processes, health, sustainability, and environmental issues. Environ. Rev. 21, 207–226.

Campbell, D.R., 2019. Early snowmelt projected to cause population decline in a subalpine plant. Proc. Natl. Acad. Sci. USA 116, 12901–12906.

Campbell, D.R., Powers, J.M., 2015. Natural selection on floral morphology can be influenced by climate. Proc. R. Soc. B 282, 20150178. Available from: https://doi.org/10.1098/rspb.2015.0178.

Cavieres, L.A., Arroyo, M.T., 2000. Seed germination response to cold stratification period and thermal regime in *Phacelia secunda* (Hydrophyllaceae) − altitudinal variation in the mediterranean Andes of central Chile. Plant Ecol. 149, 1–8.

Chabot, B.F., Billings, W.D., 1972. Origins and ecology of the Sierran alpine flora and vegetation. Ecol. Monogr. 42, 163–199.

Christensen, J.H., Hewitson, B., Busuioc, A., Chen, A., Gao, X., Held, I., et al., 2007. Regional Climate Projections. Chapter 11. Climate Change 2007: The Physical Science Basis. Contribution of Working Group I to the Fourth Assessment Report of the Intergovernmental Panel on Climate Change. Cambridge University Press, Cambridge, pp. 847–940.

Clark, J.S., Iverson, L., Woodall, C.W., Allen, C.D., Bell, D.M., Bragg, D.C., et al., 2016. The impacts of increasing drought on forest dynamics, structure, and biodiversity in the United States. Glob. Change Biol. 22, 2329–2352.

Danby, R.K., Hik, D.S., 2007. Responses of white spruce (*Picea glauca*) to experimental warming at a subarctic alpine treeline. Glob. Change Biol. 13, 437–451.

Dang, Q.-L., Cheng, S., 2004. Effects of soil temperature on ecophysiological traits in seedlings of four boreal tree species. For. Ecol. Manage. 194, 379–387.

De Frenne, P., Kolb, A., Verheyen, K., Brunet, J., Chabrerie, O., Decocq, G., et al., 2009. Unravelling the effects of temperature, latitude and local environment on the reproduction of forest herbs. Glob. Ecol. Biogeogr. 18, 641–651.

De Frenne, P., Graae, B.J., Kolb, A., Brunet, J., Chabrerie, O., Cousins, S.A., et al., 2010. Significant effects of temperature on the reproductive output of the forest herb *Anemone nemorosa* L. For. Ecol. Manage. 259, 809–817.

De Frenne, P., Brunet, J., Shevtsova, A., Kolb, A., Graae, B.J., Chabrerie, O., et al., 2011. Temperature effects on forest herbs assessed by warming and transplant experiments along a latitudinal gradient. Glob. Change Biol. 17, 3240–3253.

De Frenne, P., Graae, B.J., Brunet, J., Shevtsova, A., De Schrijver, A., Chabrerie, O., et al., 2012. The response of forest plant regeneration to temperature variation along a latitudinal gradient. Ann. Bot. 109, 1037–1046.

De Frenne, P., Rodríguez-Sánchez, F., Coomes, D.A., Baeten, L., Verstraeten, G., Vellend, M., et al., 2013. Microclimate moderates plant responses to macroclimate warming. Proc. Natl. Acad. Sci. USA 110, 18561–18565.

Densmore, R.V., 1997. Effect of day length on germination of seeds collected in Alaska. Am. J. Bot. 84, 274–278.

Densmore, R.V., Zasada, J., 1983. Seed dispersal and dormancy patterns in northern willows: ecological and evolutionary significance. Can. J. Bot. 61, 3207–3216.

Derksen, C., Brown, R., 2012. Spring snow cover extent reductions in the 2008–2012 period exceeding climate model projections. Geophys. Res. Lett. 39, 19504. Available from: https://doi.org/10.1029/2012GL053387.

Dufour-Tremblay, G., Boudreau, S., 2011. Black spruce regeneration at the treeline ecotone: synergistic impacts of climate change and caribou activity. Can. J. For. Res. 41, 460–468.

Dulamsuren, C., Hauck, M., Leuschner, C., 2010. Recent drought stress leads to growth reductions in *Larix sibirica* in the western Khentey, Mongolia. Glob. Change Biol. 16, 3024–3035.

Eckstein, R.L., Pereira, E., Milbau, A., Graae, B.J., 2011. Predicted changes in vegetation structure affect the susceptibility to invasion of bryophyte-dominated subarctic heath. Ann. Bot. 108, 177–183.

Ericson, L., Elmqvist, T., Jakobsson, K., Danell, K., Salomonson, A., 1992. Age structure of boreal willows and fluctuations in herbivore populations. Proc. R. Soc. B 98, 75–89.

Eriksson, O., Ehrlén, J., 1992. Seed and microsite limitation of recruitment in plant populations. Oecologia 91, 360–364.

Eriksson, O., Fröborg, H., 1996. "Windows of opportunity" for recruitment in long-lived clonal plants: experimental studies of seedling establishment in *Vaccinium* shrubs. Can. J. Bot. 74, 1369–1374.

Fischer, G., Nachtergaele, F., Prieler, S., van Velthuizen, H. T., Verelst, L., Wiberg, D., 2008. Global Agro-ecological Zones Assessment for Agriculture (GAEZ 2008). Laxenburg, Austria and FAO, Rome, Italy, IIASA.

Flematti, G.R., Ghisalberti, E.L., Dixon, K.W., Trengove, R.D., 2004. A compound from smoke that promotes seed germination. Science 305, 977.

Flematti, G.R., Merritt, D.J., Piggott, M.J., Trengove, R.D., Smith, S.M., Dixon, K.W., et al., 2011. Burning vegetation produces cyanohydrins that liberate cyanide and stimulate seed germination. Nat. Commun. 2, 360. Available from: https://doi.org/10.1038/ncomms1356.

Gallagher, M.K., Campbell, D.R., 2017. Shifts in water availability mediate plant–pollinator interactions. New Phytol. 215, 792–802.

Gärtner, S.M., Lieffers, V.J., MacDonald, E.S., 2011. Ecology and management of natural regeneration of white spruce in the boreal forest. Environ. Rev. 19, 461–478.

Gauthier, S., Bernier, P., Kuuluvainen, T., Shvidenko, A.Z., Schepaschenko, D.G., 2015. Boreal forest health and global change. Science 349, 819–822.

Gimingham, C.H., 1972. Ecology of Heathlands, First ed. Chapman and Hall, London.

Graae, B.J., Alsos, I.G., Ejrnaes, R., 2008. The impact of temperature regimes on development, dormancy breaking and germination of dwarf shrub seeds from arctic, alpine and boreal sites. Plant Ecol. 198, 275–284.

Graae, B.J., Verheyen, K., Kolb, A., Van Der Veken, S., Heinken, T., Chabrerie, O., et al., 2009. Germination requirements and seed mass of slow- and fast-colonizing temperate forest herbs along a latitudinal gradient. Ecoscience 16, 248–257.

Graae, B.J., Ejrnæs, R., Lang, S.I., Meineri, E., Ibarra, P.T., Bruun, H.H., 2011. Strong microsite control of seedling recruitment in tundra. Oecologia 166, 565–576.

Graae, B.J., De Frenne, P., Kolb, A., Brunet, J., Chabrerie, O., Verheyen, K., et al., 2012. On the use of weather data in ecological studies along altitudinal and latitudinal gradients. Oikos 121, 3–19.

Granström, A., Schimmel, J., 1993. Heat effects on seeds and rhizomes of a selection of boreal forest plants and potential reaction to fire. Oecologia 94, 307–313.

Greene, D.F., Macdonald, S.E., Haeussler, S., Domenicano, S., Noël, J., Jayen, K., et al., 2007. The reduction of organic-layer depth by wildfire in the North American boreal forest and its effect on tree recruitment by seed. Can. J. For. Res. 37, 1012–1023.

Groffman, P.M., Driscoll, C.T., Fahey, T.J., Hardy, J.P., Fitzhugh, R.D., Tierney, G.L., 2001. Colder soils in a warmer world: a snow manipulation study in a northern hardwood forest ecosystem. Biogeochemistry 56, 135–150.

Helbig, M., Pappas, C., Sonnentag, O., 2016. Permafrost thaw and wildfire: equally important drivers of boreal tree cover changes in the Taiga Plains, Canada. Geophys. Res. Lett. 43, 1598–1606.

Hofgaard, A., 1993. Structure and regeneration patterns in a virgin *Picea abies* forest in northern Sweden. J. Veg. Sci. 4, 601–608.

Holtmeier, F.-K., Broll, G., 2020. Treeline research - from the roots of the past to present time. A review. Forests 11, 38. Available from: https://doi.org/10.3390/f11010038.

Inouye, D.W., McGuire, A.D., 1991. Effects of snowpack on timing and abundance of flowering in *Delphinium nelsonii* (Ranunculaceae): implications for climate change. Am. J. Bot. 78, 997–1001.

IPCC, 2013. In: Stocker, T.F., Qin, D., Plattner, G.-K., Tignor, M., Allen, S.K., Boschung, J., et al.,Climate Change 2013: The Physical Science Basis. Contribution of Working Group I to the Fifth Assessment Report of the Intergovernmental Panel on Climate Change. Cambridge University Press, Cambridge.

IPCC, 2014. In: Pachauri, R.K., Meyer, L.A. (Eds.), Climate Change 2014: Synthesis Report. Contribution of Working Groups I, II and III to the Fifth Assessment Report of the Intergovernmental Panel on Climate Change. IPCC, Geneva.

Jaroszewicz, B., 2013. Endozoochory by European bison influences the build-up of the soil seed bank in subcontinental coniferous forest. Eur. J. For. Res. 132, 445–452.

Jepsen, J.U., Hagen, S.B., Ims, R.A., Yoccoz, N.G., 2008. Climate change and outbreaks of the geometrids *Operophtera brumata* and *Epirrita autumnata* in subarctic birch forest: evidence of a recent outbreak range expansion. J. Anim. Ecol. 77, 257–264.

Jungqvist, G., Oni, S.K., Teutschbein, C., Futter, M.N., 2014. Effect of climate change on soil temperature in Swedish boreal forests. PLoS One 9, e93957. Available from: https://doi.org/10.1371/journal.pone.0093957.
Keeley, J.E., Pausas, J.G., Rundel, P.W., Bond, W.J., Bradstock, R.A., 2011. Fire as an evolutionary pressure shaping plant traits. Trends Plant Sci. 16, 406−411.
Klanderud, K., Meineri, E., Goldberg, D.E., Michel, P., Berge, A., Guittar, J.L., et al., 2021. Vital rates in early life history underlie shifts in biotic interactions along bioclimatic gradients: an experimental test of the stress gradient hypothesis. J. Veg. Sci. 32, e13006. Available from: https://doi.org/10.1111/jvs.13006.
Klanderud, K., Meineri, E., Töpper, J., Michel, P., Vandvik, V., 2017. Biotic interaction effects on seedling recruitment along bioclimatic gradients: testing the stress-gradient hypothesis. J. Veg. Sci. 28, 347−356.
Kolstad, A.L., Austrheim, G., Graae, B.J., Solberg, E.J., Strimbeck, G.R., Speed, J.D., 2019. Moose effects on soil temperatures, tree canopies, and understory vegetation: a path analysis. Ecosphere 10, e02966. Available from: https://doi.org/10.1002/ecs2.2966.
Körner, C., 2012. Alpine Treelines: Functional Ecology of the Global High Elevation Tree Limits. Springer, Basel.
Kueppers, L.M., Conlisk, E., Castanha, C., Moyes, A.B., Germino, M.J., de Valpine, P., et al., 2017. Warming and provenance limit tree recruitment across and beyond the elevation range of subalpine forest. Glob. Change Biol. 23, 2383−2395.
Kurz, W.A., Shaw, C.H., Boisvenue, C., Stinson, G., Metsaranta, J., Leckie, D., et al., 2013. Carbon in Canada's boreal forest - a synthesis. Environ. Rev. 21, 260−292.
Kuuluvainen, T., 1994. Gap disturbance, ground microtopography, and the regeneration dynamics of boreal coniferous forests in Finland: a review. Ann. Zool. Fenn. 31, 35−51.
Laamrani, A., Valeria, O., Chehbouni, A., Bergeron, Y., 2020. Analysis of the effect of climate warming on paludification processes: will soil conditions limit the adaptation of northern boreal forests to climate change? A synthesis. Forests 11, 1176. Available from: https://doi.org/10.3390/f11111176.
Landhäusser, S.M., Wein, R.W., Lange, P., 1996. Gas exchange and growth of three arctic tree-line tree species under different soil temperature and drought preconditioning regimes. Can. J. Bot. 74, 686−693.
Lett, S., Nilsson, M.-C., Wardle, D.A., Dorrepaal, E., 2017. Bryophyte traits explain climate-warming effects on tree seedling establishment. J. Ecol. 105, 496−506.
Li, Y., Li, Z., Zhang, Z., Chen, L., Kurkute, S., Scaff, L., et al., 2019. High-resolution regional climate modeling and projection over western Canada using a weather research forecasting model with a pseudo-global warming approach. Hydrol. Earth Syst. Sci. 23, 4635−4659.
Liu, Q., Piao, S., Janssens, I.A., Fu, Y., Peng, S., Lian, X., et al., 2018. Extension of the growing season increases vegetation exposure to frost. Nat. Commun. 9, 426. Available from: https://doi.org/10.1038/s41467-017-02690-y.
Lorentzen, A.K., Austrheim, G., Solberg, E.J., De Vriendt, L., Speed, J.D.M., 2018. Pervasive moose browsing in boreal forests alters successional trajectories by severely suppressing keystone species. Ecosphere 9, e02458. Available from: https://doi.org/10.1002/ecs2.2458.
Måren, I.E., Vandvik, V., 2009. Fire and regeneration: the role of seed banks in the dynamics of northern heathlands. J. Veg. Sci. 20, 871−888.
Måren, I.E., Janovský, Z., Spindelböck, J.P., Daws, M.I., Kaland, P.E., Vandvik, V., 2010. Prescribed burning of northern heathlands: *Calluna vulgaris* germination cues and seed-bank dynamics. Plant Ecol. 207, 245−256.
McKee, J., Richards, A.J., 1998. The effect of temperature on reproduction in five *Primula* species. Ann. Bot. 82, 359−374.
Meyer, S.E., Monsen, S.B., 1991. Habitat-correlated variation in mountain big sagebrush (*Artemisia tridentata* ssp. *vaseyana*) seed germination patterns. Ecology 72, 739−742.
Meyer, S.E., McArthur, E.D., Jorgensen, G.L., 1989. Variation in germination response to temperature in rubber rabbitbrush (*Chrysothamnus nauseosus*: Asteraceae) and its ecological implications. Am. J. Bot. 76, 981−991.
Milbau, A., Graae, B.J., Shevtsova, A., Nijs, I., 2009. Effects of a warmer climate on seed germination in the subarctic. Ann. Bot. 104, 287−296.
Milbau, A., Shevtsova, A., Osler, N., Mooshammer, M., Graae, B.J., 2013. Plant community type and small-scale disturbances, but not altitude, influence the invasibility in subarctic ecosystems. New Phytol. 197, 1002−1011.
Molau, U., Larsson, E.-L., 2000. Seed rain and seed bank along an alpine altitudinal gradient in Swedish Lapland. Can. J. Bot. 78, 728−747.
Moles, A.T., Westoby, M., 2004a. Seedling survival and seed size: a synthesis of the literature. J. Ecol. 92, 372−383.
Moles, A.T., Westoby, M., 2004b. What do seedlings die from and what are the implications for evolution of seed size? Oikos 106, 193−199.
Mondoni, A., Probert, R., Rossi, G., Hay, F., Bonomi, C., 2008. Habitat-correlated seed germination behaviour in populations of wood anemone (*Anemone nemorosa* L.) from northern Italy. Seed Sci. Res. 18, 213−222.
Moreau, G., Chagnon, C., Auty, D., Caspersen, J., Achim, A., 2020. Impacts of climatic variation on the growth of black spruce across the forest-tundra ecotone: positive effects of warm growing seasons and heat waves are offset by late spring frosts. Front. For. Glob. Change 3, 145. Available from: https://doi.org/10.3389/ffgc.2020.613523.
Netherer, S., Schopf, A., 2010. Potential effects of climate change on insect herbivores in European forests—general aspects and the pine processionary moth as specific example. For. Ecol. Manage. 259, 831−838.
Nilsson, M.-C., Wardle, D.A., 2005. Understory vegetation as a forest ecosystem driver: evidence from the northern Swedish boreal forest. Front. Ecol. Environ. 3, 421−428.
Nystuen, K.O., Sundsdal, K., Opedal, Ø.H., Holien, H., Strimbeck, G.R., Graae, B.J., 2019. Lichens facilitate seedling recruitment in alpine heath. J. Veg. Sci. 30, 868−880.
Pardee, G.L., Inouye, D.W., Irwin, R.E., 2018. Direct and indirect effects of episodic frost on plant growth and reproduction in subalpine wildflowers. Glob. Change Biol. 24, 848−857.

Pardee, G.L., Jensen, I.O., Inouye, D.W., Irwin, R.E., 2019. The individual and combined effects of snowmelt timing and frost exposure on the reproductive success of montane forbs. J. Ecol. 107, 1970–1981.

Parisien, M.-A., Barber, Q.E., Hirsch, K.G., Stockdale, C.A., Erni, S., Wang, X., et al., 2020. Fire deficit increases wildfire risk for many communities in the Canadian boreal forest. Nat. Commun. 11, 2121. Available from: https://doi.org/10.1038/s41467-020-15961-y.

Pato, J., Obeso, J.R., 2012. Growth and reproductive performance in bilberry (*Vaccinium myrtillus*) along an elevation gradient. Ecoscience 19, 59–68.

Peng, Y.Y., Dang, Q.-L., 2003. Effects of soil temperature on biomass production and allocation in seedlings of four boreal tree species. For. Ecol. Manage. 180, 1–9.

Peterson, C.J., 2007. Consistent influence of tree diameter and species on damage in nine eastern North America tornado blowdowns. For. Ecol. Manage. 250, 96–108.

Pilsbacher, A.K., Lindgård, B., Reiersen, R., González, V.T., Bråthen, K.A., 2021. Interfering with neighbouring communities: allelopathy astray in the tundra delays seedling development. Funct. Ecol. 35, 266–276.

Price, D.T., Alfaro, R.I., Brown, K.J., Flannigan, M.D., Fleming, R.A., Hogg, E.H., et al., 2013. Anticipating the consequences of climate change for Canada's boreal forest ecosystems. Environ. Rev. 21, 322–365.

Prøsch-Danielsen, L., Simonsen, A., 2000. Palaeoecological investigations towards the reconstruction of the history of forest clearances and coastal heathlands in south-western Norway. Veg. Hist. Archaeobot. 9, 189–204.

Rayfield, B., Paul, V., Tremblay, F., Fortin, M.J., Hély, C., Bergeron, Y., 2021. Influence of habitat availability and fire disturbance on a northern range boundary. J. Biogeogr. 48, 394–404.

Rich, R.L., Frelich, L.E., Reich, P.B., 2007. Wind-throw mortality in the southern boreal forest: effects of species, diameter and stand age. J. Ecol. 95, 1261–1273.

Sayer, E.J., 2006. Using experimental manipulation to assess the roles of leaf litter in the functioning of forest ecosystems. Biol. Rev. 81, 1–31.

Schimmel, J., Granström, A., 1996. Fire severity and vegetation response in the boreal Swedish forest. Ecology 77, 1436–1450.

Shevtsova, A., Graae, B.J., Jochum, T., Milbau, A., Kockelbergh, F., Beyens, L., et al., 2009. Critical periods for impact of climate warming on early seedling establishment in subarctic tundra. Glob. Change Biol. 15, 2662–2680.

Simard, M.J., Bergeron, Y., Sirois, L., 1998. Conifer seedling recruitment in a southeastern Canadian boreal forest: the importance of substrate. J. Veg. Sci. 9, 575–582.

Sitters, J., Cherif, M., Egelkraut, D., Giesler, R., Olofsson, J., 2019. Long-term heavy reindeer grazing promotes plant phosphorus limitation in arctic tundra. Funct. Ecol. 33, 1233–1242.

Sommerville, K.D., Martyn, A.J., Offord, C.A., 2013. Can seed characteristics or species distribution be used to predict the stratification requirements of herbs in the Australian Alps? Bot. J. Linn. Soc. 172, 187–204.

Soudzilovskaia, N.A., Graae, B.J., Douma, J.C., Grau, O., Milbau, A., Shevtsova, A., et al., 2011. How do bryophytes govern generative recruitment of vascular plants? New Phytol. 190, 1019–1031.

Spindelböck, J.P., Cook, Z., Daws, M.I., Heegaard, E., Måren, I.E., Vandvik, V., 2013. Conditional cold avoidance drives between-population variation in germination behaviour in *Calluna vulgaris*. Ann. Bot. 112, 801–810.

Steijlen, I., Nilsson, M.-C., Zackrisson, O., 1995. Seed regeneration of Scots pine in boreal forest stands dominated by lichen and feather moss. Can. J. For. Res. 25, 713–723.

Stevens-Rumann, C.S., Morgan, P., 2019. Tree regeneration following wildfires in the western US: a review. Fire Ecol. 15, 1–17.

Stevens-Rumann, C.S., Kemp, K.B., Higuera, P.E., Harvey, B.J., Rother, M.T., Donato, D.C., et al., 2018. Evidence for declining forest resilience to wildfires under climate change. Ecol. Lett. 21, 243–252.

Steyaert, S., Frank, S., Puliti, S., Badia, R., Arnberg, M., Beardsley, J., et al., 2018. Special delivery: scavengers direct seed dispersal towards ungulate carcasses. Biol. Lett. 14, 20180388. Available from: https://doi.org/10.1098/rsbl.2018.0388.

Stinziano, J.R., Way, D.A., 2014. Combined effects of rising [CO_2] and temperature on boreal forests: growth, physiology and limitations. Botany 92, 425–436.

Stocks, B., Wotton, B., Flannigan, M., Fosberg, M., Cahoon, D., Goldammer, J., 2001. Boreal forest fire regimes and climate change. In: Beniston, M., Verstraete, M.M. (Eds.), Remote Sensing and Climate Modeling: Synergies and Limitations. Advances in Global Change Research. Springer, Dordrecht, pp. 233–246.

Stokes, A., 2002. Biomechanics of tree root anchorage. In: Waisel, Y., Eshel, A., Kafkafi, U. (Eds.), Plant Roots: The Hidden Half, Third ed. Marcel Dekker Inc, New York, pp. 269–286.

Ter Borg, S.J., 2005. Dormancy and germination of six Rhinanthus species in relation to climate. Folia Geobot. 40, 243–260.

Tingstad, L., Olsen, S.L., Klanderud, K., Vandvik, V., Ohlson, M., 2015. Temperature, precipitation and biotic interactions as determinants of tree seedling recruitment across the tree line ecotone. Oecologia 179, 599–608.

Usui, M., Kevan, P.G., Obbard, M., 2005. Pollination and breeding system of lowbush blueberries, *Vaccinium angustifolium* Ait. and *V. myrtilloides* Michx. (Ericaceae), in the boreal forest. Can. Field-Nat 119, 48–57.

Vandvik, V., 2004. Gap dynamics in perennial subalpine grasslands: trends and processes change during secondary succession. J. Ecol. 92, 86–96.

Vandvik, V., Vange, V., 2003. Germination ecology of the clonal herb *Knautia arvensis*: regeneration strategy and geographic variation. J. Veg. Sci. 14, 591–600.

Vandvik, V., Heegaard, E., Måren, I.E., Aarrestad, P.A., 2005. Managing heterogeneity: the importance of grazing and environmental variation on post-fire succession in heathlands. J. Appl. Ecol. 42, 139–149.

Vandvik, V., Töpper, J.P., Cook, Z., Daws, M.I., Heegaard, E., Måren, I.E., et al., 2014. Management-driven evolution in a domesticated ecosystem. Biol. Lett. 10, 20131082. Available from: https://doi.org/10.1098/rsbl.2013.1082.

Vandvik, V., Klanderud, K., Meineri, E., Måren, I.E., Töpper, J., 2016. Seed banks are biodiversity reservoirs: species—area relationships above versus below ground. Oikos 125, 218—228.

Vandvik, V., Elven, R., Töpper, J., 2017. Seedling recruitment in subalpine grassland forbs: predicting field regeneration behaviour from lab germination responses. Botany 95, 73—88.

van Zuijlen, K., Roos, R., Asplund, J., 2020. Mat-forming lichens affect microclimate and litter decomposition by different mechanisms. Fungal Ecol. 44, 100905. Available from: https://doi.org/10.1016/j.funeco.2019.100905.

Vikane, J.H., Vandvik, V., Vetaas, O.R., 2013. Invasion of *Calluna* heath by native and non-native conifers: the role of succession, disturbance and allelopathy. Plant Ecol. 214, 975—985.

Walck, J.L., Karlsson, L.M., Milberg, P., Hidayati, S.N., Kondo, T., 2012. Seed germination and seedling development ecology in world-wide populations of a circumboreal Tertiary relict. AoB Plants 2012, pls007. Available from: https://doi.org/10.1093/aobpla/pls007.

Wardle, D.A., Hörnberg, G., Zackrisson, O., Kalela-Brundin, M., Coomes, D.A., 2003. Long-term effects of wildfire on ecosystem properties across an island area gradient. Science 300, 972—975.

Weber, M.G., Flannigan, M.D., 1997. Canadian boreal forest ecosystem structure and function in a changing climate: impact on fire regimes. Environ. Rev. 5, 145—166.

Wookey, P.A., Parsons, A.N., Welker, J.M., Potter, J.A., Callaghan, T.V., Lee, J.A., et al., 1993. Comparative responses of phenology and reproductive development to simulated environmental change in sub-arctic and high arctic plants. Oikos 67, 490—502.

Zackrisson, O., Nilsson, M.-C., 1992. Allelopathic effects by *Empetrum hermaphroditum* on seed germination of two boreal tree species. Can. J. For. Res. 22, 1310—1319.

Zackrisson, O., Nilsson, M.-C., Steijlen, I., Hornberg, G., 1995. Regeneration pulses and climate-vegetation interactions in nonpyrogenic boreal Scots pine stands. J. Ecol. 469—483.

Zamorano, J.G., Hokkanen, T., Lehikoinen, A., 2018. Climate-driven synchrony in seed production of masting deciduous and conifer tree species. J. Plant Ecol. 11, 180—188.

Zasada, J., Viereck, L., 1975. The effect of temperature and stratification on germination in selected members of the *Salicaceae* in interior Alaska. Can. J. For. Res. 5, 333—337.

Zhang, T., 2005. Influence of the seasonal snow cover on the ground thermal regime: an overview. Rev. Geophys. 43, RG4002. Available from: https://doi.org/10.1029/2004RG000157.

Zhang, Y., Sherstiukov, A.B., Qian, B., Kokelj, S.V., Lantz, T.C., 2018. Impacts of snow on soil temperature observed across the circumpolar north. Environ. Res. Lett. 13, 044012. Available from: https://doi.org/10.1088/1748-9326/aab1e7.

Chapter 3

Effects of climate change on plant regeneration from seeds in the cold deserts of Central Asia

Juanjuan Lu[1], Dunyan Tan[1], Carol C. Baskin[1,2,3] and Jerry M. Baskin[1,2]

[1]College of Life Sciences, Xinjiang Agricultural University, Urümqi, Xinjiang, P.R. China, [2]Department of Biology, University of Kentucky, Lexington, KY, United States, [3]Department of Plant and Soil Sciences, University of Kentucky, Lexington, KY, United States

Introduction

Deserts cover $>19 \times 10^6$ km^2 or 15% of the global land surface (Ezcurra, 2006). Following the classification scheme of Laity (2008), about 23% of the global deserts are temperate (cold) deserts (Cowan, 2007; Zhang et al., 2016). Cold deserts are found at high latitudes in continental interiors, including the deserts of Central Asia, western USA, and Argentina (Laity, 2008; Mamtimin et al., 2011). They are characterized by low mean annual temperature, short frost-free season, frequent occurrence of frost at night in autumn and spring, very cold winters, and short, moderately warm summers (Mamtimin et al., 2011). Temperatures of cold deserts during winter can decrease to $-40°C$, but summers are hot. Precipitation in winter generally is in the form of snow (West, 1983a,b). Mean annual temperatures vary with location, and in Asia, for example, they range from 3.9°C at Dalanzadgad in the Gobi Desert (Mongolia) (Walter and Box, 1983a) to 16.7°C at Kerki in the Turanian Lowland (Turkmenistan) (Walter and Box, 1983b).

More than 80% of the world's cold deserts occur in Central Asia (CA, 34.3°−55.4°N, 46.5°−96.4°E) (Zhang et al., 2016; Bai et al., 2019). They cover a total area of 5.8×10^6 km^2 (Li et al., 2020) and include the arid lands of northwestern China (NW China), southwestern Mongolia Republic (SW Mongolia), and the five Central Asian states of the former Soviet Union (CAS5) (Kazakhstan, Kyrgyzstan, Tajikistan, Turkmenistan, and Uzbekistan) (Kottek et al., 2006; Cowan, 2007). The Tianshan Mountains divide CA into the western (the CAS5 countries) and the eastern (NW China and SW Mongolia) regions.

Due to its great distance from the ocean, CA has a typical continental arid and semiarid climate with low amount of precipitation and high evapotranspiration (Lioubimtseva and Cole, 2006). Annual precipitation ranges from 100 to 400 mm, and average annual precipitation (including rain and snow) is <150 mm, about one-fifth to one-third of which is snow that falls in winter and begins to melt in March or April. Rainfall is highly variable among seasons and years, but generally, it is higher in spring than in autumn. Furthermore, snowmelt in spring increases water availability for plants. Annual potential evaporation is >2000 mm (Zhang et al., 2016).

The dominant plants in cold deserts of CA are semishrubs and dwarf semishrubs, but many annuals and herbaceous perennials are present (Zhang et al., 2016; Li et al., 2020). The proportion of herbaceous annuals and perennials, including grasses, increases with increased precipitation (Zhang et al., 2016; Fang et al., 2019; Li et al., 2020). Additionally, the cold deserts of CA have salt lakes and salt pans with associated halophytic vegetation, as do cold deserts of other regions (Soriano, 1983; Walter and Box, 1983c; West, 1983c).

The herbaceous species of the cold deserts of CA mainly consist of perennial herbs such as many *Allium* and *Astragalus* species; various species of summer annual chenopods (Amaranthaceae), for example, *Salsola* and *Agriophyllum* species; winter annual/spring ephemeral (ephemeretum) species of Asteraceae, Boraginaceae, Brassicaceae, Fabaceae, Geraniaceae, Papaveraceae, and Poaceae; and perennial ephemerals (ephemeroids), for example, *Carex* spp. (Cyperaceae), *Eremurus* spp. (Xanthorrhoeaceae), and *Tragopogan* spp. (Asteraceae) (Wang et al., 2003; Zhang et al., 2016). Woody species (dwarf trees/shrubs/subshrubs) include *Artemisia* (Asteraceae),

Plant Regeneration from Seeds. DOI: https://doi.org/10.1016/B978-0-12-823731-1.00003-2
© 2022 Elsevier Inc. All rights reserved.

Calligonum (Polygonaceae), *Caragana* (Fabaceae), *Ephedra* (Ephedraceae) *Haloxylon* (Amaranthaceae), *Nitraria* (Nitrariaceae), *Reaumuria* (Tamaricaceae), and *Tamarix* (Tamaricaceae), among others (Wang et al., 2003; Zhang et al., 2016).

In this chapter, we review the research conducted on plant species in the cold deserts of CA, primarily NW China. Since much of the research on regeneration of plants from seeds has investigated the effects of changes in amount and timing of precipitation on annual species, this will be the primary focus of the chapter. Also, the limited information on the effect of climate change on shrubs is reviewed, especially with regard to changes in community composition.

Climate change in cold deserts of Central Asia

CA is a region that is being strongly affected by global warming (Solomon et al., 2007; Chen et al., 2010). During the last 100 years, the surface air temperature of CA has shown a significant warming trend of around $0.18°C$ $decade^{-1}$, which is higher than the global average and twice that of the Northern Hemisphere (Chen et al., 2009). This warming trend is predicted to continue throughout the 21st century, possibly with a rate well above the global mean (Parry et al., 2007; Huang et al., 2012; Hu et al., 2014). Furthermore, the substantial winter warming during the 20th century may change to a more prominent spring warming in the 21st century (He et al., 2014; Hu et al., 2014). Warming could enhance potential evapotranspiration, thus intensifying the aridity of the cold deserts of CA (Li et al., 2015). Also, increasing temperatures threaten the sustainability of glaciers in the Pamir and Tianshan mountains, which are the major summer water sources for many desert ecosystems in CA (Sorg et al., 2012; Zhang et al., 2016). In the cold deserts of CA, snowmelt increases water availability in spring, which is essential for the growth of wild as well as cultivated crop plants such as cotton and grapes (Fan et al., 2012; Bie et al., 2016).

Annual and seasonal precipitation increased from 1930 to 2009 (Chen et al., 2011). Annual precipitation increased at a rate of 0.13 mm $year^{-1}$. The increasing trend is as follows: winter (0.07 mm $year^{-1}$) > autumn (0.03 mm $year^{-1}$) > summer (0.02 mm $year^{-1}$) > spring (0.01 mm $year^{-1}$). This trend of increasing annual precipitation over most of CA is predicted to continue through the 21st century, with high rates of change in northern CA and the northeastern Tibetan Plateau (Lioubimtseva and Henebry, 2009; Chen et al., 2011).

Overall, the changes in both climatic mean and long-term trends of annual precipitation over CA are largely the result of changes in winter and spring. Additionally, atmospheric nitrogen (N) deposition is predicted to continue to increase in the cold desert of CA in the future (Liu et al., 2013; Xu et al., 2015). For example, N deposition has increased by 35.4 kg ha^{-1} $year^{-1}$ in the Gurbantunggut Desert (in Junggar Basin, NW China) during the past 30 years, and it is predicted to double by 2050 (Chang et al., 2012; Chen et al., 2019b).

Plant life history traits

Seed dormancy/germination

Many herbaceous species in the cold deserts of CA are facultative winter annuals/ephemerals, and physiological dormancy (PD) of their seeds is broken in summer during exposure to high temperatures and very dry conditions (Table 3.1) (Lu et al., 2010; Wang et al., 2010; Nur et al., 2014). Although PD is broken during summer, there are two possible germination seasons (Lu et al., 2010; Wang et al., 2010). (1) If the soil is moist in autumn, when temperatures are appropriate for germination, seeds dispersed/sown in the field germinate in autumn (Zeng et al., 2011). (2) If the soil is too dry during autumn for seeds to germinate, germination is delayed until snowmelt and rainfall in early spring (late March and early April) moisten the soil (Sun et al., 2016). Regardless of when germination occurs, the species usually complete their life cycle before early summer. Thus species whose seeds have PD have the potential to behave both as winter annuals and as spring ephemeral annuals, that is, facultative winter annuals. In fact, however, these species behave mostly as spring ephemeral annuals, because most nondormant seeds are prevented from germinating in autumn due to low soil moisture (Lu et al., 2010, 2017a,b; Wang et al., 2010; Nur et al., 2014; Zhou et al., 2015). Additionally, seeds of the perennial *Sterigmostemum fuhaiense* (Brassicaceae) require high temperatures to come out of PD, and cold stratification induced about half of the seeds into secondary dormancy (Lu et al., 2017a). Like seeds of facultative winter annuals, those of *S. fuhaiense* germinate in autumn and in spring (Lu et al., 2017a).

In the CA deserts, the germination behavior of seeds of these species allows them to germinate under natural precipitation in both seasons (i.e., autumn and spring). However, precipitation in late summer/autumn is only occasionally sufficient to stimulate seeds of the species, such as those of *Diptychocarpus strictus* (Brassicaceae) (Lu et al., 2010), *Echinops gmelinii* (Asteraceae) (Nur et al., 2014), and *Eremopyrum distans* (Poaceae) (Wang et al., 2010), to germinate

TABLE 3.1 Effects of watering treatment on seed dormancy and seedling growth of plant species in the cold deserts of Central Asia.

LF	Species	Family	PT	P PF	PA	ND/D	SB	GP	S SS	SeG	References
AH	*Chorispora sibirica*	Brassicaceae	SP/AU	–	↑	Nondeep PD	PSB	AG↑, SG↑	–	–	Lu et al. (2017a)
AH	*Diptychocarpus strictus*	Brassicaceae	AU	–	↑	Nondeep PD	–	AG↑	–	–	Lu et al. (2010)
AH	*D. strictus*	Brassicaceae	SP/AU	–	↑	Nondeep PD	–	AG↑, SG↑	SG↑[a]	–	Lu et al. (2014)
AH	*Echinops gmelinii*	Asteraceae	AU	–	↑	Type 6 nondeep PD	PSB	AG↑	–	–	Nur et al. (2014)
AH	*Epilasia acrolasia*	Asteraceae	AU	–	↑	Type 6 nondeep PD	TSB	AG↑	–	–	Nur et al. (2014)
AH	*Eremopyrum distans*	Gramineae	AU	–	↑	Nondeep PD	TSB	AG↑	↑	–	Wang et al. (2010)
AH	*Goldbachia laevigata*	Brassicaceae	SP/AU	–	↑	Nondeep PD	PSB	AG↑, SG↑	–	–	Lu et al. (2017a)
AH	*Isatis violascens*	Brassicaceae	AU	–	↑	Nondeep and intermediate PD	TSB	AG↑	–	–	Zhou et a.. (2015)
AH	*Koelpinia linearis*	Asteraceae	AU	–	↑	Type 6 nondeep PD	TSB	AG↑	–	–	Nur et al. (2014)
AH	*Leptaleum filifolium*	Brassicaceae	AU	–	↑	Nondeep PD	PSB	AG↑	–	–	Lu et al. (2017b)
AH	*Neotorularia korolkowii*	Brassicaceae	AU	–	↑	Nondeep PD	PSB	AG↑, SG↑	–	–	Lu et al. (2017b)
AH	*Spirorrhynchus sabulosus*	Brassicaceae	SP/AU	–	↑	Nondeep PD	TSB	AG↑, SG↑	–	–	Lu et al. (2017a)
AH	*Tauscheria lasiocarpa*	Brassicaceae	SP/AU	–	↑	Nondeep PD	TSB	AG↑, SG↑	–	–	Lu et al. (2017a)
AH	*Turgenia latifolia*	Apiaceae	AU	–	↑	Nondeep complex MPD	–	AG↑	–	–	Nurulla et al. (2014)
PH	*Astragalus lehmannianus*	Fabaceae	SU/AU	–	↑	PY	–	AG↑	↑	–	Han et al 2018
PH	*Sterigmostemum fuhaiense*	Brassicaceae	SP/AU	–	↑	Nondeep PD	PSB	AG↑, SG↑	–	–	Lu et al. (2017a)
S	*Eremosparton songoricum*	Fabaceae	–	–	–	PY	–	↑	–	↑	Liu et al. (2011)
SuS	*Reaumuria soongarica*	Tamaricaceae	SU/AU	–	↑	–	–	GP↑	–	–	Shan et al. (2018)
SuS	*R. soongarica*	Tamaricaceae	SU/AU	↓	–	–	–	GP↑	–	–	Shan et al. (2018)
SuS	*R. soongarica*	Tamaricaceae	SU	–	↑	–	–	GP↑	–	–	Xie et al. (2020)
SuS	*R. soongarica*	Tamaricaceae	SU	↓	–	–	–	GP↔	–	–	Xie et al. (2020)

Note: *D*, dormancy status of seeds: *ND/D*, nondormant or dormant; *MPD*, morphophysiological dormancy; *PD*, physiological dormancy; *PY*, physical dormancy; *GP*, germination percentage: *AG*, autumn germination; *SG*, spring germination; *LF*, life form: *H*, herb; *AH*, annual; *S*, shrub; *SuS*, subshrub; *P*, precipitation; *PA*, amount of precipitation; *PF*, frequency of precipitation; *PT*, timing of precipitation; *SP*, spring; *SU*, summer; *AU*, autumn; *WI*, winter; *S*, seedlings; *SeG*, seedling growth; *SS*, seedling survivorship; *SB*, soil seed bank; *PSB*, persistent soil seed bank; *TSB*, transient soil seed bank; ↑, increase; ↓, decrease; ↔, no change; –, no information.
[a] AG not observed.

(Table 3.1). For the annual grass *E. distans*, seed position in the spikelet [i.e., basal (group 1), middle (group 2), and distal (group 3)] had an effect on timing of germination (Wang et al., 2010). Under natural soil moisture conditions, none of the seeds germinated in autumn 2008; however, if the soil was hand-watered in autumn some seeds in all three groups germinated in late summer and autumn (August and September). Thus seeds of the winter annuals/ephemerals and perennials are more likely to germinate in spring than in late summer/autumn due to lack of soil moisture in autumn.

In some Brassicaceae species with indehiscent (Zhou et al., 2015; Lu et al., 2017a) or delayed-dehiscent (Lu et al., 2017b) fruits, seeds have PD (Table 3.1), which is enhanced by the (closed) pericarp. Some (or most) indehiscent fruits (Zhou et al., 2015; Lu et al., 2017a) and all delayed-dehiscent fruits open in moist soil in the field the following spring germination season (Lu et al., 2017b). However, some nondormant seeds remain nongerminated inside the open fruits, and this enhances retention of their viability. Thus the pericarp has a role in both dormancy/germination and retention of viability. As such, these species of Brassicaceae can form a transient (*Isatis violascens*, *Spirorrhynchus sabulosus*, and *Tauscheria lasiocarpa*) or persistent (*Chorispora sibirica*, *Goldbachia laevigata*, *Leptaleum filifolium*, *Neotorularia korolkowii*, and *Sterigmostemum fuhaiense*) seed bank. A decrease in soil moisture due to climate change is expected to delay pericarp opening, thereby prolonging the time seeds remain in the seed bank.

Seeds of the weedy annual species *Turgenia latifolia* (Apiaceae) have intermediate complex morphophysiological dormancy (MPD) (Table 3.1). Dormancy break and radicle emergence of this species occur if the soil is moist in autumn; however, the majority of seedlings do not emerge from the soil until spring (Nurulla et al., 2014). The probable reason for delay of emergence is that temperatures are too low for them to do so. Thus global warming may increase the seedling emergence of this species in autumn.

Physical dormancy [PY, water-impermeable (hard) seed coat] occurs in most species of Fabaceae in the cold deserts of CA, with many *Caragana* species being an exception (Abudureheman et al., 2014; Chen et al., 2019a). Seeds of the rare and endangered Central Asian endemic perennial sand dune species *Eremosparton songoricum* (Fabaceae) have PY (Table 3.1). Breaking PY in seeds of this species was more effective under constantly wet than under constantly dry conditions during a 1-year sequence of simulated natural temperature regimes, and germination percentage was significantly higher under dry-wet than under wet-dry conditions (Liu et al., 2017). Also, *E. songoricum* exhibits seasonal periodicity of seed germination in its cold desert sand dune habitat. PY break and germination are controlled by the sequence of winter-spring temperature and sand moisture conditions (Liu et al., 2017). That is, the seasonal cycle of natural temperature and moisture conditions in the habitat of *E. songoricum*, together with the seed dormancy-breaking requirements of this species, are such that germination occurs only in spring. Increased precipitation due to climate change might increase the effect of winter (via low temperatures) on dormancy break in seeds of *E. songoricum*.

In the sand dune perennial *Astragalus lehmannianus* (Fabaceae), PY of a portion of the seeds is broken by predispersal seed predation by *Etiella zinckenella* (Lepidoptera, Pyralididae) scarifying (via making exist holes in) the water-impermeable seed coat, in some cases without killing the embryo, thus allowing the seeds to imbibe water and germinate in autumn of the year of production if the soil/sand is moist (Han et al., 2018). In contrast, seeds that were not predated did not germinate in their first year following dispersal. If rainfall is sufficient in autumn, seedlings from predated seeds of *A. lehmannianus* that germinate in autumn can survive the winter as seedlings (Table 3.1). Thus increased precipitation due to climate change likely would increase seedling emergence and establishment of the seed predator-scarified seeds of this species in autumn.

Seedling survival and growth

An increase in precipitation can significantly increase (Zheng et al., 2005; Arfin-Khan et al., 2018), have no effect on (Weltzin and Mcpherson 2000) or decrease (Zhao et al., 2012) seedling emergence (germination). Increasing precipitation (hand-watering) in spring increased survival percentages of seedlings from autumn (AG) and spring (SG) germinating seeds of the winter annual/spring ephemeral *Erodium oxyrhinchum* (Geraniaceae), a species with PY seeds, although survival was lower for AG than SG (Table 3.1) (Chen et al., 2019c). Also, the different effects of increased precipitation on survival of SG and AG seedlings were due to the phenological stage and size of SG and AG when precipitation (by hand-watering) was increased in spring. Plants from SG seeds were still in the seedling stage with only a few leaves, while those from AG seeds were larger with more leaves. Thus plants from AG seeds were not as sensitive in their response to increased precipitation as those from SG seeds. Chen et al. (2019c) predicted that future increased precipitation in early spring may improve the survival of SG plants.

In arid and semiarid regions, water and N are the most important limiting factors for plant growth (Ehleringer et al., 1998; Allah et al., 2014). Compared with the natural condition (control), precipitation (hand-watering) plus N significantly improved survival of SG and AG plants of *Erodium oxyrhinchum* (Chen et al., 2019d). These results suggest that increasing precipitation and N due to climate change will increase seedling establishment of this species (Shan et al., 2018; Chen et al., 2019d).

No seedlings of *Diptychocarpus strictus* emerged in the not-watered treatment (natural precipitation) in autumn 2008 in an experimental garden (Lu et al., 2014). However, in the autumn-watered treatment (AW) seedlings from seeds that germinated in autumn 2008 survived the winter in a vegetative state under snow. Furthermore, there were more spring-germinating seedlings in the spring-watered (SW) than in the spring not-watered (SNW) treatment, but 98%–100% of SW and SNW survived and reproduced in late spring 2009 (Lu et al., 2014). That is to say, regardless of increasing precipitation (hand-watering), spring-germinating seedlings survived and completed their life cycle. Thus, in *D. strictus*, increases in precipitation due to climate change likely will increase the number of both autumn- and spring-germinating seedlings and their survival and reproduction.

Although a portion of intact, scarified seeds of *Astragalus lehmannianus* germinated (see above), no seedlings survived under natural precipitation or in the no-watering after germination treatment (Han et al., 2018). However, in the watered treatment (hand-watering) >50% of seedlings from scarified and low- and medium-insect predated seeds grew into juveniles and survived over winter. Survival of seedlings from high-predated seeds was significantly lower than that of seedlings from mechanically scarified and low-predated seeds but not from that of seedlings from medium-predated seeds. Clearly, increased precipitation in autumn would increase seedling establishment in this species.

Soil (sand) water content (SWC) for seedling survival of *Eremosparton songoricum* was ≥ 2.0%, and no seedlings survived at SWC <2.0% (Liu et al., 2011). Although seedling survival increased with an increase in SWC, a very low percentage (2.0) of seedlings survived in two dune sites. Also, growth rate of seedling roots was faster in the high (2.8 ± 0.6 cm d^{-1}) than in the low (1.6 ± 0.3 cm d^{-1}) moisture layer of the dunes. The inability of root extension to keep up with the retreating moisture-containing sand layer was the reason seedlings did not become established. Thus increased precipitation in spring/summer in the sand dune habitat of this species likely would increase seedling establishment.

Increasing the amount of precipitation (hand-watering) enhances seed germination (Zhou et al., 2015; Lu et al., 2017a) and seedling emergence (Wang et al., 2010; Lu et al., 2014) and growth (Liu et al., 2011) in the cold deserts of CA (Table 3.1). However, seedling emergence and growth are sensitive not only to precipitation amount but also to precipitation frequency and timing (Dalgleish et al., 2010; Zeppel et al., 2014; Zhu et al., 2014; Gao et al., 2015). In the salt-tolerant semishrub *Reaumuria soongarica* (Tamaricaceae), germination and seedling emergence and growth increased not only with increasing amount of precipitation (hand-watering) at the same precipitation frequency but also with a decrease in frequency of precipitation under the same precipitation amount (Shan et al., 2018). Moreover, germination and seedling emergence and growth were increased more by increased frequency of precipitation than by precipitation amount. Thus precipitation events characterized by low amounts and high frequency are required for seedling emergence of *Reaumuria soongarica* (Shan et al., 2018). However, Xie et al. (2020) reported that height and biomass of seedlings of this species were affected significantly by precipitation amount but nonsignificantly by precipitation frequency.

Plant growth, development, biomass accumulation/allocation, and seed production

The life cycle of winter annuals/spring ephemerals in the cold deserts of CA is flexible, which is an adaptation to the variation in timing and amount of precipitation in their cold desert habitats. Thus germination season [i.e., spring-germinating (SG) and autumn-germinating (AG)] and watering regime significantly affect the expression of most life history traits, including phenology (Lu et al., 2014, 2016; Chen et al., 2019b,c,d), plant size (Zhang et al., 2007; Wang et al., 2010; Lu et al., 2016; Chen et al., 2019b,c,d; Zang et al., 2020), fruit and seed production (Zhang et al., 2007; Wang et al., 2010; Lu et al., 2016; Chen et al., 2019b,c,d; Zang et al., 2020), accumulation and allocation of biomass to reproduction (Zhang et al., 2007; Lu et al., 2012, 2014, 2016; Chen et al., 2019b,c,d; Zang et al., 2020), and dormancy of seeds produced via maternal environmental effects (Wang et al., 2010; Chen et al., 2019b,c,d) (Table 3.2).

Phenology

In the cold deserts of NW China, increase in precipitation nonsignificantly delayed flowering, fruiting, and senescence of AG and SG plants of *Erodium oxyrhinchum* (Table 3.2) (Chen et al., 2019c). However, these phenological events were significantly delayed by an increase in precipitation (hand-watering) for AG and SG plants of *Eremopyrum distans*, *Nepeta micrantha* (Chen et al., 2019b), and *Diptychocarpus strictus* (Lu et al., 2014) and by an increase in precipitation plus N for AG and SG plants of *Erodium oxyrhinchum* (Chen et al., 2019d), AG plants of *Eremopyrum distans*, and SG plants of *Nepeta micrantha* (Chen et al., 2019b).

TABLE 3.2 Effects of watering treatment on life history traits of plant species in the cold deserts of Central Asia.

LF	Species	Family	P					Adults			OG	References
			PT	PA	PH	PB	BI	BA	DP			
AH	Atriplex aucheri	Amaranthaceae	W[a]	–	–	–	–	–	Seed yield↑ (each type of seed↑); seed size (A↔, B↔, C↓)	A↑, B↑, C↔	Wang et al. (2015)
AH	Ceratocarpus arenarius	Amaranthaceae	SP	↑	–	↑	RVA↔/↑, SVA↑, LVA↑, RA↑	RVA↔, SVA↑, LVA↔; RA↓	Number (DUa↔, DUc↔, DUf↑); Number percentage (DUa↔, DUc↓, DUf↑)	–	Gan et al. (2020)
AH	Diptychocarpus strictus	Brassicaceae	SP/AU	↑	FP↑, LS↑	↑	↑	RVA↑, SVA↑, LVA↓, RA↓ (upper siliques↑, lower siliques↓)	Total siliques↑, ratio of two siliques↑ (upper siliques↑, lower siliques↓)	–	Lu et al. (2014)
AH	D. strictus	Brassicaceae	SP	↑	–	↑	↑	RVA↑, SVA↑, LVA↔, RA (upper siliques↑, lower siliques↓)	Total siliques↑, ratio of two siliques↑ (upper siliques↑, lower siliques↓)	–	Lu et al. (2012)
AH	Eremopyrum distans	Gramineae	AU	↑	–	AU/SU↑	–	–	AU↑, SU↑	–	Wang et al. (2010)
AH	E. distans	Gramineae	W[a]	↑	ED↔, FD↑, LS↑	AG↑	RB↑	AG (RVA↑, SVA↓, LVA↓, RA↑)	↑	AG[b]↓	Chen et al. (2019b)
AH	Erodium oxyrhinchum	Geraniaceae	SP/SU	↑	FD↑, FD'↑, FP↑, LS↑	AG↑/SG↑, except for root length in SG	–	AG/SG [RVA↓, SVA↑, LVA↑, RA↓]	AG↑, SG↑	AG↓, SG↑	Chen et al. (2019c)
AH	E. oxyrhinchum	Geraniaceae	W[a]	↑ + N	ED↔, FD↑, FP↑, LS↑	AG↑/SG↑	–	AG/SG [RVA↓, SVA↑, LVA↔, RA↔]	AG↔, SG↑	AG↓, SG↑	Chen et al. (2019d)
AH	E. oxyrhinchum	Geraniaceae	SP	↓	–	–	SVB↓, LVB↓, RB↓	SVB↓, LVB↓, RB↑	–	–	Zang et al. (2020)
AH	Lachnoloma lehmannii	Brassicaceae	SP	↑	–	↑	↑	–	↑	–	Mamut et al. (2019)

AH	*Lappula duplicicarpa*	Boraginaceae	SP	↑	-	RVA↔, SVA↑, LVA↑, RA↔	RVA↓, SVA↑, LVA↔, RA↔	—	Lu et al. (2013)
AH	*Nepeta micrantha*	Labiatae	W[a]	↑	ED↔, FD↑, FD'↑, FP↑, LS↑	SG↓	RB↑	SG (RVA↑, SVA↓, LVA↓, RA↑)	Number proportion of nutlet combination (4L↑, 1S-3L↑, 2S-2L↓, 4S↓), total nutlets↑, ratio of two nutlets↑ (LN↑, SN↔)	SG[c]↑	Cher et al. (2019b)

Note: *BA*, biomass allocation; *AA*, biomass allocated to aboveground organs; *LVA*, biomass allocated to leaves; *RA*, biomass allocated to reproductive organs; *RVA*, biomass allocated to roots; *SVA*, biomass allocated to vegetative organs; *BI*, biomass; *PB*, plant biomass/size; *RB*, reproductive biomass; *VB*, vegetative biomass; *RVB*, biomass of roots; *SVB*, biomass of stems; *LVB*, biomass of leaves; *DP*, diaspore production; *DUa*, *Dub*, and *DUc* are diaspore units a, b, and c, respectively; *LF*, life form; *H*, herb *AH*, annual; *OG*, offspring seed germination; *A*, *B*, *C*, diaspore morphs of *Atriplex aucheri*; *P*, precipitation; *PA*, amount of precipitation; *PF*, frequency of precipitation; *PT*, timing of precipitation; *SP*, spring; *SU*, summer; *AU*, autumn; *WI*, winter; *PH*, phenology; *ED*, emergence date; *FD*, flowering date; *FD'*, fruiting date; *FP*, flowering period; *LS*, post-germination life span; ↑, increase; ↓, decrease; ↔, no change; -, no information.

[a]watered but season of watering not given.
[b]only AC plants were observed.
[c]only SG plants were observed.

Furthermore, in the watered treatment (hand-watering) the reproductive stage of SG plants of *Diptychocarpus strictus* (Lu et al., 2014) and *Erodium oxyrhinchum* (Chen et al., 2019c) was earlier than that of AG plants (Table 3.2). Additionally, time to reproduction of both species was accelerated under natural precipitation (no hand-watering) in spring. Consequently, the life cycle was shortened, which is adaptive in the cold desert, where precipitation is unpredictable in spring. Thus it is expected that increased temperatures in spring would shorten the life cycle of annual species but that increased precipitation would prolong it. In which case, these two factors would have contrasting effects on plant life history.

Plant morphological characters

With increasing precipitation (hand-watering), plant size (height) of AG and SG plants of *Diptychocarpus strictus* (Lu et al., 2014) and *Eremopyrum distans* (Wang et al., 2010) increased significantly (Table 3.2). When water was added in spring, plant height, leaf area, and number of leaves increased significantly in AG plants of *Eremopyrum distans* (Chen et al., 2019b) and in SG plants of *Lachnoloma lehmannii* (Mamut et al., 2019). With increasing precipitation, plant height, number and area of leaves, and number and length of branches, but not root length, increased in SG plants of *Erodium oxyrhinchum*, while in AG plants height increased significantly, leaf area increased nonsignificantly, and root length decreased significantly (Chen et al., 2019c). However, increased precipitation significantly decreased plant height, root length, and number of leaves, but not leaf area in SG plants of *Nepeta micrantha* (Chen et al., 2019b). With increasing precipitation plus N treatments, growth of SG (height, leaf area, and number of seeds) and AG (number and area of leaves) plants increased significantly in *Erodium oxyrhinchum* (Chen et al., 2019d). However, an increase in N (only) did not change or decreased the growth and seed production of SG and AG plants of *Erodium oxyrhinchum* (Chen et al., 2019d).

Chen et al. (2019b) reported that the root is probably the organ most sensitive to changes in precipitation and in precipitation plus N. *Eremopyrum distans* has a fibrous root system and traps sand around the roots, which is beneficial for holding moisture and for N uptake in plants (Verma et al., 2005). Thus an increase in precipitation, N, and precipitation plus N significantly increased plant height, leaf area, and leaf number in this species (Chen et al., 2019b). In contrast, *Nepeta micrantha* has a taproot system with a few lateral roots (Ding et al., 2016). Since a fibrous root system is more efficient than a taproot system in capturing mobile ions from spatially and temporally heterogeneous soils (Dunbabin et al., 2004), most morphological characters of *N. micrantha* showed no or a negative response to increased precipitation (Chen et al., 2019b). These results suggest that growth of annual plants in spring will increase, decrease, or not change with climate change in the CA, depending on the species.

Dry mass accumulation and allocation

With increased precipitation (hand-watering), dry mass accumulation increased in AG and SG plants of *Erodium oxyrhinchum* (Chen et al., 2019c), in AG (Lu et al., 2014) and SG (Lu et al., 2012, 2014) plants from dimorphic diaspores of *Diptychocarpus strictus* and *Lappula duplicicarpa* (Lu et al., 2013), and in SG plants from polymorphic diaspores of *Ceratocarpus arenarius* (Gan et al., 2020) (Table 3.2). Likewise, precipitation plus N treatments significantly increased biomass accumulation of SG and AG plants in *Erodium oxyrhinchum* (Chen et al., 2019d) and in AG plants of *Eremopyrum distans* (Chen et al., 2019b).

Furthermore, the increase in biomass of SG plants was significantly greater than that of AG plants in *Erodium oxyrhinchum* (Chen et al., 2019c) and *Diptychocarpus strictus* (Lu et al., 2014). This difference in biomass accumulation of SG and AG plants is also consistent with the response of height, number and area of leaves, and number and length of branches in SG and AG plants to increased precipitation. However, AG plants had a much longer period of time in which to grow than SG plants. With increased precipitation (hand-watering), the proportion of biomass allocated to stems and leaves increased, and the proportion allocated to reproduction and roots decreased in both SG and AG plants of *Erodium oxyrhinchum* (Chen et al., 2019c). In the watering treatment, the proportion of biomass allocated to roots and stems of *Diptychocarpus strictus* increased, but allocation to leaves and reproduction decreased in SG plants (Lu et al., 2012, 2014).

Reproductive allocation in hand-watered plants of *D. strictus* was lower in AG than in SG plants (Table 3.2) (Lu et al., 2014). The increase in proportion of biomass allocated to reproduction in a short growing season (AG watered vs SG watered) (Lu et al., 2014) and under limiting soil moisture (SG watered vs SG not watered) (Lu et al., 2012, 2014) can be interpreted as stress responses. Thus, in *D. strictus*, the increase in proportion of biomass allocated to reproduction in watered (vs not watered) SG plants was due to limiting soil moisture and in AG versus SG plants to a short growing season.

However, with increased precipitation (hand-watering) the proportion of biomass allocated to roots and reproduction increased in *Eremopyrum distans* and *Nepeta micrantha*, whereas the proportion of biomass allocated to stem and leaf decreased in AG plants of *E. distans* and in SG plants of *N. micrantha* (Table 3.2) (Chen et al., 2019b). Also, plants of *Erodium oxyrhinchum* under natural precipitation (vs hand-watering) allocated more aboveground biomass to reproductive organs, and the relative allocation to reproductive organs increased at the expense of allocation to stems and leaves (Zang et al., 2020). No significant effects on allocation to leaves or reproductive parts were observed in N or in precipitation plus N treatments for SG or AG plants of *Erodium oxyrhinchum* (Chen et al., 2019d) or for SG plants from the two nutlet morphs of *Lappula duplicicarpa* (Lu et al., 2013). It seems likely that increased precipitation will increase dry mass in AG and SG annual plants, whereas dry mass allocation to reproductive parts of AG and SG plants will increase, decrease, or not change with increasing precipitation in CA, depending on species.

Number of seeds produced

In the watering treatment (hand-watering), seed production significantly increased in AG and SG plants of *Diptychocarpus strictus* (Lu et al., 2014) and *Erodium oxyrhinchum* (Chen et al., 2019c) (Table 3.2). Furthermore, precipitation plus N treatments significantly increased the number of seeds produced by SG plants of *E. oxyrhinchum* but had no significant effect on number of seeds produced by AG plants (Chen et al., 2019d). These results suggest that the response of SG plants is more sensitive to increased precipitation than that of AG plants. Supplemental watering significantly increased seed production by SG plants of *Atriplex aucheri* (Wang et al., 2015), *Ceratocarpus arenarius* (Gan et al., 2020), *Diptychocarpus strictus* (Lu et al., 2012), *Lachnoloma lehmannii* (Mamut et al., 2019), *Lappula duplicicarpa* (Lu et al., 2013), and *Nepeta micrantha* (Chen et al., 2019b). Thus it is expected that seed production will be increased by increased precipitation.

Importance of heterodiaspory

Diaspore heteromorphism likely is a bet-hedging strategy in which an individual plant produces two or more distinct kinds of fruits and/or seeds (sometimes with accessory parts such as bracteoles, perianth, or phyllary) that differ in size/mass, shape, dispersal ability, and/or degree of dormancy. In addition, plants produced by the different morphs differ in many ways including survival, growth, competitive ability, life history, and demographic characteristics (reviewed in Baskin and Baskin, 2014). The cold deserts of CA are rich in the diaspore heteromorphic species due at least in part to the high number of annual heterodiasporous chenopods (Amaranthaceae) native to them. The seed dispersal/dormancy strategies of 20 annual diaspore heteromorphic species native to deserts of northern Xinjiang Province in northwest China were formulated by Baskin et al. (2014). These heteromorphic species are important components of plant communities in the cold deserts of CA.

Precipitation (hand-watering) significantly affected most life history traits of the heteromorphic species in the cold deserts of CA (Table 3.2) (Wang et al., 2015; Lu et al., 2012, 2013, 2014; Gan et al., 2020). The heterodiasporous Brassicaceae species *Diptychocarpus strictus* produces two dispersal unit morphs: (1) seeds from dehiscent upper siliques and (2) whole indehiscent lower siliques (Lu et al., 2010). Under natural rainfall conditions, germination of seeds from upper siliques was mostly delayed until the first spring following dispersal, primarily by drought in summer and autumn, and they did not form a long-lived persistent seed bank. In contrast, the thick, indehiscent pericarp of lower siliques prevented germination of the seeds inside them for >1 year, and they formed a seed bank that persisted for at least 7 years. This dormancy/germination/seed bank strategy in *D. strictus* likely enhances survival in its CA arid zone habitat with unpredictable rainfall (Lu et al., 2010, 2015; Lu, unpublished data). Seed production was higher for SG plants in the watered treatment than for those of SG plants in the nonwatered treatment for plants from the dimorphic seeds of *D. strictus* (Lu et al., 2012, 2014). Moreover, in the same watering treatment seed production was higher in AG than SG plants from the seeds of the two silique morphs (Lu et al., 2014).

Ecological differences between diaspores in heteromorphic species are usually explained as a bet-hedging strategy in highly variable (unpredictable) environments in time or space (Venable, 1985; Mandák, 1997). Thus the success of diaspore heteromorphic species in the cold desert of CA might be, at least partly, attributed to its heterocarpy and associated plastic responses to changing environmental conditions, which contribute to their ability to occupy variable environments. In the diaspore heteromorphic species *Atriplex aucheri* (Wang et al., 2015), *Ceratocarpus arenarius* (Quan et al., 2012; Gan et al., 2020), *Diptychocarpus strictus* (Lu et al., 2012, 2014), and *Lappula duplicicarpa* (Lu et al., 2013), the (high dispersal—low dormancy)/(low dispersal—high dormancy) [i.e., (HDi-LDo)/(LDi-HDo)] diaspore ratio can vary with growth conditions (site quality) of the mother plants.

In the high soil moisture condition, the (HDi-LDo)/(LDi-HDo) ratio for *Ceratocarpus arenarius* (Quan et al., 2012; Gan et al., 2020), *Diptychocarpus strictus* (Lu et al., 2012, 2014), and *Lappula duplicicarpa* (Lu et al., 2013) was relatively high, while in the low soil moisture condition it was relatively low. With increasing water stress, this ratio decreased due to a greater decrease in number of diaspores with high dispersal−low dormancy than in those with low dispersal−high dormancy. Furthermore, the relative allocation to HDi-LDo and LDi-HDo diaspores was significantly negatively correlated, suggesting that allocation to HDi-LDo diaspores in good growth conditions occurred at the expense of allocation to LDi-HDo diaspores (Lu et al., 2012, 2013, 2014; Gan et al., 2020). Thus the proportional increase in the LDi-HDo diaspores in low precipitation environments (or years) is a "low risk" strategy that increases the chance of retaining the mother site and of colonizing suitable sites nearby. On the other hand, a proportional increase in HDi-LDo diaspores in high precipitation environments (or years) is a "high risk" strategy that provides these desert winter annual/spring ephemerals with the chance of rapidly colonizing new sites and expanding the area occupied by the population (Sadeh et al., 2009). Thus, with climate change and especially increased precipitation during the growing season, we can expect a relatively high (HDi-LDo)/(LDi-HDo) ratio.

Vegetation and community dynamics

The effects of increased temperature and changes in precipitation on the dynamics of the natural vegetation of CA have been investigated using the normalized difference vegetation index (NDVI). Li et al. (2015) conducted a study of the natural vegetation dynamics of CA using NDVI data collected each year from 1982 to 2013. From 1982 to 1998, the NDVI for the natural vegetation in CA increased (0.004 decade^{-1}), but after 1998 it decreased (0.003 decade^{-1}). The trend for increased precipitation diminished in the early 21st century. However, intensified warming and slightly decreased precipitation caused increased negative values for the Palmer Drought Severity Index in surface moisture conditions during 1998−2013. Also, evapotranspiration has increased, thereby decreasing soil moisture, especially in the soil rooting zone of herbaceous species (Li et al., 2015). The annual NDVI was weakly negatively correlated with temperature and positively correlated with annual precipitation. Thus during the growing season the most important factor related to plant growth in most areas of CA is precipitation (Yin et al., 2016). Seasonally, the NDVI showed a significant increasing, nonsignificant decreasing, and significant decreasing trend in spring, autumn, and winter, respectively. At low elevations, a trend of increasing NDVI occurred in spring and summer, while in mountainous areas there was a significant increase in spring, summer, and autumn (Yin et al., 2016).

Compared to the 1980s and 1990s, the land area in CA occupied by shrubs increased from 2000 to 2013, and shrub encroachment into plant communities dominated by herbaceous species has occurred, resulting in a mosaic of herb- and shrub-dominated communities on the landscape. Shrub encroachment has been most significant in Uzbekistan, Turkmenistan, and the Tarim Basin in NW China (Li et al., 2015). One reason for shrub encroachment is that the effects of climate change are causing a decline in shallow groundwater levels and thus the shallow-rooted herbaceous plants die (Li et al., 2015).

With respect to the communities dominated by spring ephemerals, Jia et al. (2020) conducted a 5-year water-enhancement experiment in the (cold sandy) Gurbantunggut Desert of NW China. During the plant growing season (late March to late May) in each of the 5 years, they added 15 L of water to 1 m × 1 m plots in the desert four times. Each watering was equivalent to a 10 mm precipitation event. In one year (2006), they added 30 L of water four times, which is equivalent to 20 mm precipitation events. Between 2005 and 2009, precipitation during the growing season ranged from 29.9 to 88.2 mm. Compared to plots receiving only natural rainfall, watering significantly increased plant cover and density, but neither species richness nor diversity was affected. However, there was a higher relative increase in density of some rare than of common species. Annual primary production (ANPP) and seed production were increased by the addition of water. In 2006 the only year in which 80 mm of water was added to the plots, ANPP was 80 > 40 > 0 mm (control, natural precipitation) of water added, whereas seed production was (80 = 40) > 0. There was great variation in proportion of seeds produced by individual species in the control and in the water addition treatments between 2006 and 2007, the only years in which seed production was measured. Notably, in 2006 53%−72% of the seeds were produced by *Ceratocarpus arenarius* and 13%−21% by *Alyssum linifolium*, whereas in 2007 51%−53% of the seeds were produced by *A. linifolium* and 5%−6% by *C. arenarius*.

Zang et al. (2020) found that the cold desert annual plant community in the southern edge of the Gurbantunggut Desert did not change in species richness or composition after 2 years of extreme drought (imposed by rainout shelters). However, the extreme drought significantly decreased the biomass of the ephemeral plant community and increased biomass allocation to reproduction in the dominant species *Erodium oxyrhinchum*.

With an increase in amount of N applied to plots in the (cold) Tengger Desert of NW China, species richness of the herbaceous plants decreased in all four years (2007–10) of the study, during which time the annual precipitation ranged from 127 (2008) to 271.2 mm (2007) and significantly so in the two wettest years (2007 and 2009) (Su et al., 2013). Furthermore, species richness decreased with decrease in (natural) annual precipitation. Aboveground biomass increased with increased N addition in the wettest year (2007), decreased with increased N addition in 2008 and 2009, and did not change in 2010. Rainfall and N differentially affected production of aboveground biomass of the three plant functional groups: annual herbs (AS), perennial forbs (PF), and perennial grasses (PG). Notably, annuals produced a substantial amount of biomass in the wet year (2007) but little or no biomass in the 3 dry years. In general, higher N inputs inhibited biomass production of AS and PF and stimulated biomass production of PG. Overall, amount of annual rainfall and N application had an effect on community composition and biomass production.

Future research needs

Studies in arid areas (Reyer et al., 2013; Robinson et al., 2013) and on grassland and forest ecosystems (Dalgleish et al., 2010; Schneider et al., 2014; Didiano et al., 2016) outside CA have demonstrated that shifts in the precipitation pattern can have different and potentially greater ecological consequences on plants than changes in the amount of precipitation. Previous studies in the cold deserts of CA mainly have focused on effects of amount of precipitation on life history traits of annual plants (Nur et al., 2014; Lu et al., 2017a,b; Han et al., 2018; Chen et al., 2019b,c,d). An exception is a study on the shrub *Reaumuria soongarica* (Shan et al., 2018), which is discussed above. Thus more studies are needed to explore the effects of the amount, timing, and frequency of precipitation and their interactions on species and plant communities in the cold deserts of CA.

Previous research on diaspore heteromorphic species has focused mainly on comparisons of seed dormancy/germination (e.g., Lu et al., 2010, 2013, 2014) and biomass and reproduction allocation (e.g., Lu et al., 2012, 2014; Wang et al., 2015; Gan et al., 2020) in plants derived from the different diaspore morphs under different precipitation conditions (hand-watering) in the cold deserts of CA. However, an important aspect of the ecological functions of diaspore heteromorphism that has received little attention with regard to climate change is maintenance of a soil seed bank. Furthermore, most of the studies on diaspore heteromorphic species have focused on the effects of changes in precipitation. Thus additional studies need to test the effects of increased habitat temperatures (e.g., using open top chambers), N, and of N plus precipitation on the life history of the diaspore heteromorphic species, including morph ratio and generation-to-generation plasticity.

It generally is assumed that diaspore heteromorphism is a bet-hedging strategy. However, no studies have actually demonstrated that the production of two or more diaspore morphs is a bet-hedging strategy in any of the many heterodiasporous species in CA. To show that (diversifying) bet-hedging is an adaptive life history strategy in the temporally unpredictable rainfall environment of the cold deserts of CA, studies need to demonstrate a reduction in the arithmetic mean and variance in the number of offspring per year (or generation) and an increase in the geometric mean in the number of offspring across years. What might be the effect of climate change on the life history, ecology, and demography on diaspore heteromorphic species in CA, especially with regard to amount and frequency of precipitation?

Acknowledgments

This research was supported in part by the National Natural Science Foundation of China (U1803331, 32071668, 32160265), and Excellent Youth Foundation of Natural Science Foundation of Xinjiang Uygur Autonomous Region of China (2021D01E21).

References

Abudureheman, B., Liu, H., Zhang, D., Guan, K., 2014. Identification of physical dormancy and dormancy release patterns in several species (Fabaceae) of the cold desert, north-west China. Seed Sci. Res. 24, 133–145.

Allah, S., Khan, A.A., Fricke, T., Buerkert, A., Wachendorf, M., 2014. Fertilizer and irrigation effects on forage protein and energy production under semi-arid conditions of Pakistan. Field Crops Res. 159, 62–69.

Arfin-Khan, M.A.S., Vetter, V.M.S., Reshi, Z.A., Dar, P.A., Jentsch, A., 2018. Factors influencing seedling emergence of three global invaders in greenhouses representing major eco-regions of the world. Plant Biol. 20, 610–618.

Bai, J., Shi, H., Yu, Q., Xie, Z., Li, L., Luo, G., et al., 2019. Satellite-observed vegetation stability in response to changes in climate and total water storage in Central Asia. Sci. Total Environ. 659, 862–871.

Baskin, C.C., Baskin, J.M., 2014. Seed: Ecology, Biogeography, and Evolution of Dormancy and Germination, Second ed. Academic Press/Elsevier, San Diego.

Baskin, J.M., Lu, J.J., Baskin, C.C., Tan, D.Y., Wang, L., 2014. Diaspore dispersal ability and degree of dormancy in heteromorphic species of cold deserts of northwest China: a review. Persp. Plant Ecol. Evol. Syst. 16, 93–99.

Bie, B., Zhou, X., Zhang, Y., 2016. Effects of snow cover on seed germination of ten desert plant species. Chin. J. Ecol. 35, 2348–2354 (in Chinese with English abstract).

Chang, Y.H., Liu, X.J., Li, K.H., Lv, J.L., Song, W., 2012. Research progress in atmospheric nitrogen deposition. Arid Zone Res. 29, 972–979 (in Chinese with English abstract).

Chen, F.H., Chen, J.H., Holmes, J., Boomer, I., Austin, P., Gates, J.B., et al., 2010. Moisture changes over the last millennium in arid Central Asia: a review, synthesis and comparison with monsoon region. Quat. Sci. Rev. 29, 1055–1068.

Chen, F.H., Huang, W., Jin, L.Y., Chen, J.H., Wang, J.S., 2011. Spatiotemporal precipitation variations in the arid Central Asia in the context of global warming. Sci. China Earth Sci. 54, 1812–1821.

Chen, F., Wang, J., Jin, L., Zhang, Q., Li, J., Chen, J., 2009. Rapid warming in mid-latitude central Asia for the past 100 years. Front. Earth Sci. 3, 42–50.

Chen, D., Zhang, R., Baskin, C.C., Hu, X., 2019a. Water permeability/impermeability in seeds of 15 species of *Caragana* (Fabaceae). PeerJ 7, e6870. Available from: https://doi.org/10.7717/peerj.6870.

Chen, Y., Zhang, L., Shi, X., Liu, H., Zhang, D., 2019b. Life history responses of two ephemeral plant species to increased precipitation and nitrogen in the Gurbantunggut Desert. PeerJ 7, e6158. Available from: https://doi.org/10.7717/peerj.6158.

Chen, Y., Shi, X., Zhang, L., Baskin, J.M., Baskin, C.C., Liu, H., et al., 2019c. Effects of increased precipitation on the life history of spring- and autumn-germinated plants of the cold desert annual *Erodium oxyrhinchum* (Geraniaceae). AoB Plants 11, plz004. Available from: https://academic.oup.com/aobpla.

Chen, Y., Zhang, L., Shi, X., Ban, Y., Liu, H., Zhang, D., 2019d. Life history responses of spring-and autumn-germinated ephemeral plants to increased nitrogen and precipitation in the Gurbantunggut Desert. Sci. Total Environ. 659, 756–763.

Cowan, P.J., 2007. Geographic usage of the terms Middle Asia and Central Asia. J. Arid Environ. 69, 359–363.

Dalgleish, H.J., Koons, D.N., Adler, P.B., 2010. Can life-history traits predict the response of forb populations to changes in climate variability? J. Ecol. 98, 209–217.

Didiano, T.J., Johnson, M.T.J., Duval, T.P., 2016. Disentangling the effects of precipitation amount and frequency on the performance of 14 grassland species. PLoS One 11, e0162310. Available from: https://doi.org/10.1371/journal.pone.0162310.

Ding, J., Fan, L., Li, Y., Tang, L., 2016. Biomass allocation and allometric relationships of six desert herbaceous plants in the Gurbantunggut Desert. J. Desert Res. 36, 1323–1330 (in Chinese with English abstract).

Dunbabin, V., Rengel, Z., Diggle, A.J., 2004. Simulating form and function of root systems: efficiency of nitrate uptake is dependent on root system architecture and the spatial and temporal variability of nitrate supply. Funct. Ecol. 18, 204–211.

Ehleringer, J.R., Schwinning, S., Gebauer, R., 1998. Water use in arid land ecosystems. In: Scholes, J.D., Barker, M.G., et al., (Eds.) Plant Physiological Ecology. Blackwell Science, Oxford, pp. 347–365.

Ezcurra, E., 2006. Global Deserts Outlook. Renouf Publ. Co. Ltd., Ottawa.

Fang, X., Chen, Z., Guo, X., Zhu, S., Liu, T., Li, C., et al., 2019. Impacts and uncertainties of climate/CO_2 change on net primary productivity in Xinjiang, China (2000–2014): a modelling approach. Ecol. Model. 408, 108742. Available from: https://doi.org/10.1016/j.ecolmodel.2019.108742.

Fan, L.L., Ma, J., Wu, L.F., Xu, G.Q., Li, Y., Tang, L.S., 2012. Response of the herbaceous layer to snow variability at the south margin of the Gurbantonggut [Gurbantunggut] Desert of China. Chin. J. Plant Ecol. 36, 126–135 (in Chinese with English abstract).

Gan, L., Lu, J.J., Baskin, C.C., Baskin, J.M., Tan, D.Y., 2020. Phenotypic plasticity in diaspore production of a amphi-basicarpic cold desert annual that produces polymorphic diaspores. Sci. Rep. 10, 11142. Available from: https://doi.org/10.1038/s41598-020-67380-0.

Gao, R.R., Yang, X.J., Liu, G.F., Huang, Z.Y., Walck, J.L., 2015. Effects of rainfall pattern on the growth and fecundity of a dominant dune annual in a semi-arid ecosystem. Plant Soil 389, 335–347.

Han, Y.J., Baskin, J.M., Tan, D.Y., Baskin, C.C., Wu, M.Y., 2018. Effects of predispersal insect seed predation on the early life history stages of a rare cold sand-desert legume. Sci. Rep. 8, 3240. Available from: https://doi.org/10.1038/s41598-018-21487-7.

He, Y., Huang, J., Ji, M., 2014. Impact of land–sea thermal contrast on interdecadal variation in circulation and blocking. Clim. Dyn. 43, 3267–3279.

Huang, J., Guan, X., Ji, F., 2012. Enhanced cold-season warming in semi-arid regions. Atmos. Chem. Phys. 12, 5391–5398.

Hu, Z., Zhang, C., Hu, Q., Tian, H., 2014. Temperature changes in Central Asia from 1979 to 2011 based on multiple datasets. J. Clim. 27, 1143–1167.

Jia, Y., Sun, Y., Zhang, T., Zhang, T., Shi, Z., Maimaitiaili, B., et al., 2020. Elevated precipitation alters the community structure of spring ephemerals by changing dominant species density in Central Asia. Ecol. Evol. 10, 2196–2212.

Kottek, M., Grieser, J., Beck, C., Rudolf, B., Rubel, F., 2006. World map of the Köppen–Geiger climate classification updated. Meteorol. Zeits. 15, 259–263.

Laity, J., 2008. Deserts and Desert Environments. Wiley-Blackwell Publishes House, Oxford.

Lioubimtseva, E., Cole, R., 2006. Uncertainties of climate change in arid environments of Central Asia. Rev. Fish. Sci. 14, 29–49.

Lioubimtseva, E., Henebry, G., 2009. Climate and environmental change in arid Central Asia: impacts, vulnerability, adaptations. J. Arid. Environ. 73, 963–977.

Li, Z., Chen, Y., Li, W., Deng, H., Fang, G., 2015. Potential impacts of climate change on vegetation dynamics in Central Asia. J. Geophys. Res. Atmos. 120, 12345–12356.

Li, J., Chen, H., Zhang, C., 2020. Impacts of climate change on key soil ecosystem services and interactions in Central Asia. Ecol. Indic. 116, 106490. Available from: https://doi.org/10.1016/j.ecolind.2020.106490.

Liu, H., Abudureheman, B., Zhang, L., Baskin, J.M., Baskin, C.C., Zhang, D., 2017. Seed dormancy-breaking in a cold desert shrub in relation to sand temperature and moisture. AoB Plants 9, plx003. Available from: https://doi.org/10.1093/aobpla/plx003.

Liu, H.L., Shi, X., Wang, J.C., Yin, L.K., Huang, Z.Y., Zhang, D.Y., 2011. Effects of sand burial, soil water content and distribution pattern of seeds in sand on seed germination and seedling survival of *Eremosparton songoricum* (Fabaceae), a rare species inhabiting the moving sand dunes of the Gurbantunggut Desert of China. Plant Soil 345, 69–87.

Liu, X., Zhang, Y., Han, W., Tang, A., Shen, J., Cui, Z., et al., 2013. Enhanced nitrogen deposition over China. Nature 494, 459–462.

Lu, J.J., Ma, W.B., Tan, D.Y., Baskin, J.M., Baskin, C.C., 2013. Effects of environmental stress and nutlet morph on proportion and within-flower number-combination of morphs produced by the fruit-dimorphic species *Lappula duplicicarpa* (Boraginaceae). Plant Ecol. 214, 351–362.

Lu, J.J., Tan, D.Y., Baskin, C.C., Baskin, J.M., 2016. Effects of germination season on life history traits and on transgenerational plasticity in seed dormancy in a cold desert annual. Sci. Rep. 6, 25076. Available from: https://doi.org/10.1038/srep25076.

Lu, J.J., Tan, D.Y., Baskin, C.C., Baskin, J.M., 2017a. Role of indehiscent pericarp in formation of soil seed bank in five cold desert Brassicaceae species. Plant Ecol. 218, 1187–1200.

Lu, J.J., Tan, D.Y., Baskin, C.C., Baskin, J.M., 2017b. Delayed dehiscence of pericarp: role in germination and in retention of viability of seeds of two cold desert annual species of Brassicaceae. Plant Biol. 19, 14–22.

Lu, J.J., Tan, D.Y., Baskin, J.M., Baskin, C.C., 2010. Fruit and seed heteromorphism in the cold desert annual ephemeral *Diptychocarpus strictus* (Brassicaceae) and possible adaptive significance. Ann. Bot. 105, 999–1014.

Lu, J.J., Tan, D.Y., Baskin, J.M., Baskin, C.C., 2012. Phenotypic plasticity and bet-hedging in a heterocarpic winter annual/spring ephemeral cold desert species of Brassicaceae. Oikos 121, 357–366.

Lu, J.J., Tan, D.Y., Baskin, J.M., Baskin, C.C., 2014. Germination season and watering regime, but not seed morph, affect life history traits in a cold desert diaspore-heteromorphic annual. PLoS One 9, e102018. Available from: https://doi.org/10.1371/journal.pone.0102018.

Lu, J.J., Tan, D.Y., Baskin, J.M., Baskin, C.C., 2015. Post-maturity fates of seeds in dehiscent and indehiscent siliques of the diaspore heteromorphic species *Diptychocarpus strictus* (Brassicaceae). Persp. Plant Ecol. Evol. Syst. 17, 255–262.

Mamtimin, B., Et-Tantawi, A.M.M., Meixner, F.X., Domroes, M., 2011. Recent trends of temperature change under hot and cold desert climates: comparing the Sahara (Libya) and Central Asia (Xinjiang, China). J. Arid Environ. 75, 1105–1113.

Mamut, J., Tan, D.Y., Baskin, C.C., Baskin, J.M., 2019. Effects of water stress and NaCl stress on different life cycle stages of the cold desert annual *Lachnoloma lehmannii* in China. J. Arid Land 11, 774–784.

Mandák, B., 1997. Seed hetermorphism and the life cycle of plants: a literature review. Preslia 69, 129–159.

Nur, M., Baskin, C.C., Lu, J.J., Tan, D.Y., Baskin, J.M., 2014. A new type of non-deep physiological dormancy: evidence from three annual Asteraceae species in the cold deserts of Central Asia. Seed Sci. Res. 24, 301–314.

Nurulla, M., Baskin, C.C., Lu, J.J., Tan, D.Y., Baskin, J.M., 2014. Intermediate morphophysiological dormancy allows for life-cycle diversity in the annual weed, *Turgenia latifolia* (Apiaceae). Aust. J. Bot. 64, 930–937.

Parry, M.L., Canziani, O.F., Palutikof, J.P., van der Linden, P.J., Hanson, C.E., 2007. Climate Change 2007: Impacts, Adaptation and Vulnerability. Contribution of Working Group II to the Fourth Assessment Report of the Intergovernmental Panel on Climate Change. Cambridge University Press, Cambridge.

Quan, D.J., Wei, Y., Zhou, X.Q., Yan, C., 2012. Growth dynamics, biomass allocation and ecological adaptation in *Ceratocarpus arenarius* L. Acta Ecol. Sin. 32, 3352–3358 (in Chinese with English abstract).

Reyer, C.P.O., Leuzinger, S., Rammig, A., Wolf, A., Bartholomeus, R.P., Bonfante, A., 2013. A plant's perspective of extremes: terrestrial plant responses to changing climatic variability. Glob. Change Biol. 19, 75–89.

Robinson, T.M.P., La Pierre, K.J., Vadeboncoeur, M.A., Byrne, K.M., Thomey, M.L., Colby, S.E., 2013. Seasonal, not annual precipitation drives community productivity across ecosystems. Oikos 122, 727–738.

Sadeh, A., Guterman, H., Gersani, M., Ovadia, O., 2009. Plastic bet-hedging in an amphicarpic annual: an integrated strategy under variable conditions. Evol. Ecol. 23, 373–388.

Schneider, A.C., Lee, T.D., Kreiser, M.A., Nelson, G.T., 2014. Comparative and interactive effects of reduced precipitation frequency and volume on the growth and function of two perennial grassland species. Int. J. Plant. Sci. 175, 702–712.

Shan, L., Zhao, W., Li, Y., Zhang, Z., Xie, T., 2018. Precipitation amount and frequency affect seedling emergence and growth of *Reaumuria soongarica* in northwestern China. J. Arid Land 10, 574–587. Available from: https://doi.org/10.1007/s40333-018-0013-2.

Solomon, S., Qin, D., Manning, M., Chen, Z., Marquis, M., Averyt, K.B., et al., (Eds.), 2007. Climate Change 2007: The Physical Science Basis. Cambridge University Press, Cambridge.

Sorg, A., Bolch, T., Stoffel, M., Solomina, O., Beniston, M., 2012. Climate change impacts on glaciers and runoff in Tian Shan (Central Asia). Nat. Clim. Change 2, 725–731.

Soriano, A., 1983. Deserts and semi-deserts of Patagonia. In: West, N.E. (Ed.), Ecosystems of the World 5. Temperate Deserts and Semi Deserts. Elsevier, Amsterdam, pp. 423–460.

Sun, Y.-Y., Zhou, J., Liu, T., Liu, Z.-C., Hao, X.-R., Liu, H.-F., et al., 2016. Ecological implications and environmental dependence of the seed germination of common species in cold deserts. Pak. J. Bot. 48, 983–992.

Su, J., Li, X., Li, X., Feng, L., 2013. Effects of additional N on herbaceous species of desertified steppe in arid regions of China: a four-year field study. Ecol. Res. 28, 21–28.

Venable, D.L., 1985. The evolutionary ecology of seed heteromorphism. Am. Nat. 126, 577–595.

Verma, V., Worland, A.J., Sayers, E.J., Fish, L., Caligari, P.D.S., Snape, J.W., 2005. Identification and characterization of quantitative trait loci related to lodging resistance and associated traits in bread wheat. Plant Breed. 124, 234–241.
Walter, H., Box, E.O., 1983a. The deserts of Central Asia. In: West, N.E. (Ed.), Ecosystems of the World 5. Temperate Deserts and Semi-Deserts. Elsevier, Amsterdam, pp. 193–236.
Walter, H., Box, E.O., 1983b. Middle Asian deserts. In: West, N.E. (Ed.), Ecosystems of the World 5. Temperate Deserts and Semi-Deserts. Elsevier, Amsterdam, pp. 79–104.
Walter, H., Box, E.O., 1983c. Caspian lowland biome. In: West, N.E. (Ed.), Ecosystems of the World 5. Temperate Deserts and Semi-Deserts. Elsevier, Amsterdam, pp. 9–41.
Wang, X., Jiang, J., Lei, J., Zhang, W., Qian, Y., 2003. Distribution of ephemeral plants and their significance in dune stabilization in Gurbantunggut Desert. J. Geogr. Sci. 13, 323–330.
Wang, A.B., Tan, D.Y., Baskin, C.C., Baskin, J.M., 2010. Effect of seed position in spikelet on life history of *Eremopyrum distans* (Poaceae) from the cold desert of north-west China. Ann. Bot. 106, 95–105.
Wang, L., Wang, H.L., Zhang, K., Tian, C.Y., 2015. Effects of maternal nutrient and water availability on seed production, size and germination of heterocarpic *Atriplex aucheri*. Seed Sci. 43, 71–79.
Weltzin, J.F., Mcpherson, G.R., 2000. Implications of precipitation redistribution for shifts in temperate savanna ecotones. Ecology 81, 1902–1913.
West, N.E., 1983a. Approach. In: West, N.E. (Ed.), Ecosystems of the World 5. Temperate Deserts and Semi-Deserts. Elsevier, Amsterdam, pp. 1–2.
West, N.E., 1983b. Overview of North American temperate deserts and semi-deserts. In: West, N.E. (Ed.), Ecosystems of the World 5. Temperate Deserts and Semi-Deserts. Elsevier, Amsterdam, pp. 321–330.
West, N.E., 1983c. Comparisons and contrasts between the temperate deserts and semi-deserts of three continents. In: West, N.E. (Ed.), Ecosystems of the World 5. Temperate Deserts and Semi-Deserts. Elsevier, Amsterdam, pp. 461–472.
Xie, Y., Li, Y., Xie, T., Meng, R., Zhao, Z., 2020. Impact of artificially simulated precipitation patterns change on the growth and morphology of *Reaumuria soongarica* seedlings in Hexi Corridor of China. Sustainability 12, 2439. Available from: https://doi.org/10.3390/su12062439.
Xu, W., Luo, X.S., Pan, Y.P., Zhang, L., Tang, A.H., Shen, J.L., et al., 2015. Quantifying atmospheric nitrogen deposition through a nationwide monitoring network across China. Atmos. Chem. Phys. Disc 15, 18365–18405.
Yin, G., Hu, Z., Chen, X., Tiyip, T., 2016. Vegetation dynamics and its response to climate change in Central Asia. J. Arid Land 8, 375–388.
Zang, Y.X., Min, X.J., de Dios, V.R., Ma, J.Y., Sun, W., 2020. Extreme drought affects the productivity, but not the composition, of a desert plant community in Central Asia differentially across microtopographies. Sci. Total. Environ. 717, 137251. Available from: https://doi.org/10.1016/j.scitotenv.2020.137251.
Zeng, X., Tong, L., Shen, X., Niu, P., 2011. Environment-dependence of seed germination in autumn in Gurbantunggut Desert. Chin. J. Ecol. 30, 1604–1611 (in Chinese with English abstract).
Zeppel, M.J.B., Wilks, J.V., Lewis, J.D., 2014. Impacts of extreme precipitation and seasonal changes in precipitation on plants. Biogeosciences 11, 3083–3093.
Zhang, C., Lu, D., Chen, X., Zhang, Y., Maisupova, B., Tao, Y., 2016. The spatiotemporal patterns of vegetation coverage and biomass of the temperate deserts in Central Asia and their relationships with climate controls. Remote Sens. Environ. 175, 271–281.
Zhang, T., Sun, Y., Tian, C.Y., 2007. Ecological and biological differences between spring and autumn plants of two desert ephemerals. J. Plant Ecol. 31, 1174–1180.
Zhao, J., Song, Y., Sun, T., Mao, Z.J., Liu, C.Z., Liu, L., et al., 2012. Response of seed germination and seedling growth of *Pinus koraiensis* and *Quercus mongolica* to comprehensive action of warming and precipitation. Acta Ecol. Sin. 32, 7791–7800 (in Chinese with English abstract).
Zheng, Y.R., Xie, Z.X., Yu, Y., Jiang, L.H., Shimizu, H., Rimmington, G.M., 2005. Effects of burial in sand and water supply regime on seedling emergence of six species. Ann. Bot. 95, 1237–1245.
Zhou, Y.M., Lu, J.J., Tan, D.Y., Baskin, J.M., Baskin, C.C., 2015. Seed germination ecology of the cold desert annual *Isatis violascens* (Brassicaceae): two levels of physiological dormancy and role of the pericarp. PLoS One 10, e0140983. Available from: https://doi.org/10.1371/journal.pone.0140983.
Zhu, Y.J., Yang, X.J., Baskin, C.C., Baskin, J.M., Huang, Z.Y., 2014. Effects of amount and frequency of precipitation and sand burial on seed germination, seedling emergence and survival of the dune grass *Leymus secalinus* in semiarid China. Plant Soil 374, 399–409.

Chapter 4

Plant regeneration by seeds in hot deserts

Marina L. LaForgia[1], D. Lawrence Venable[2] and Jennifer R. Gremer[1]
[1]Department of Evolution and Ecology, University of California, Davis, CA, USA, [2]Department of Ecology and Evolutionary Biology, University of Arizona, Tucson, AZ, USA

Introduction

Drylands, including deserts, cover 41% of the Earth's terrestrial surface (Millennium Ecosystem Assessment, 2005; Reynolds et al., 2007) and are the largest single type of land by area (Laity et al., 2008). Although all deserts are characterized by high aridity, they vary in temperature, seasonality of precipitation, and vegetation. In general, deserts occur where evaporation exceeds precipitation; however, this can vary substantially depending on the geological, topographic, and floristic characteristics of the region and the interaction of these factors with precipitation and temperature. Current norms classify deserts as areas where the aridity index (ratio of mean annual precipitation to potential evapotranspiration) is <0.2 (UNEP, 1997).

Deserts typically are divided into hot, cold, and cold water (Meigs, 1953). In this chapter, we focus mainly on plant regeneration by seeds in hot deserts, but we also include the cold-water Atacama Desert in South America and the Namib Desert in southern Africa. Hot deserts are usually found near 30 degrees North and South, where there is dry subsiding air near the subtropical high-pressure belt (e.g., Sonoran and Chihuahuan deserts in North America, parts of Australia and Western Asia). They also occur in the interior of continents, where they are far from sources of moisture (e.g., interior of Australia) and in the rain shadow of mountains (e.g., Mojave Desert and parts of the Arabian Peninsula). Hot deserts have extreme diurnal changes in temperature with maximum temperatures commonly reaching $40°C-45°C$ during summer (Evenari, 1985). The lack of cloud cover leads to high levels of solar irradiance and limits the heat-holding capacity of deserts, allowing temperatures to decrease dramatically at night. They also have low total precipitation but with large variability in temporal and spatial distribution and intensity. Many hot deserts receive a substantial fraction of their total annual precipitation in the form of monsoonal rains (e.g., Australia, North America, Thar, and Southern Sahara).

Coastal cold-water deserts differ from those in the interior of continents owing to their proximity to the ocean. They typically occur on the western edges of continents, where cold upwelling suppresses convection. These areas have much higher humidity than hot interior deserts, lower maximum temperatures, and moisture inputs from fog. Rain shadows also may play a role in these deserts, for example, the Andes in South America prevent moisture-laden air from reaching the Atacama.

Deserts have an extraordinary diversity of flora (Fig. 4.1) and are sites of remarkable ecological and evolutionary convergence that address the challenges of life in the desert, including water limitation, extreme temperatures, strong intra- and interannual variability, and constrained biogeochemistry (Shreve and Wiggins, 1964; Orians and Solbrig, 1977; Knight and Ackerly, 2003). While convergence of form and function has resulted in resilience of desert ecosystems in the past (Axelrod, 1948,1979; Axelrod and Raven, 1978; Shmida and Whittaker, 1979; Evenari, 1985; Shmida, 1985), deserts are projected to experience significant effects due to climate change (Bates et al., 2008; Bayram and Öztürk, 2014; Hughes, 2011; Seager et al., 2007). Shifts in climate could overwhelm the ability of traits and strategies of desert biota to buffer individuals, populations, and communities from the negative effects of extreme and variable conditions. Thus understanding the effects of climate change on the regeneration of plants from seeds in these extreme environments is critical for predicting and mitigating the effects of climate change across the globe.

FIGURE 4.1 The flora in hot deserts around the world includes many annual herbs and long-lived perennials. (A) Flowers in Atacama Desert in western Chile, (B) shrubs in Simpson Desert in Australian interior, (C) flowers (note red tulips) in Negev Desert of Israel, (D) yuccas, shrubs, and grasses in Chihuahuan Desert in New Mexico, United States, (E) sand dunes and shrubs in Arabian Desert of the United Arab Emirates, (F) *Welwitschia mirabilis* in Namib Desert of coastal Namibia, (G) Joshua Trees (*Yucca brevifolia*) in Mojave Desert in California, USA, (H) iconic Saguaro cacti (*Carnegiea gigantea*) in Sonoran Desert in Arizona, United States. *(A) Photo by Javier Rubilar, CC-BY-SA 2.0. (B) Photo by Tandrew22, CC-BY-SA 4.0. (C) Photo by Gideon Pisanty, CC-BY 3.0. (D) Photo by Marina LaForgia. (E) Aidas U., CC-BY 3.0. (F) Photo by Derek Keats, CC-BY 2.0. (G) Photo by Marina LaForgia. (H) Photo by Joe Parks, CC-BY 2.0.*

Climate change in deserts

While specific deserts vary in the exact climatic changes they may experience, deserts are generally expected to increase in both temperature and rainfall variability, with an overall decrease of 5%−30% in mean annual rainfall (Bates et al., 2008; Bayram and Öztürk, 2014; Hughes, 2011; IPCC, 2013; Risbey, 2011; Seager et al., 2007). Increases in temperature will exacerbate drought regardless of changes in precipitation (Diffenbaugh et al., 2015; Pendergrass et al., 2017), although many desert regions also are expected to experience decreased precipitation (Thuiller et al., 2006; Risbey, 2011; Bayram and Öztürk, 2014). Finally, increases in precipitation variability, including changes in both intra- and inter-annual variability, as well as shifts in distribution and higher frequencies of extreme events, such as El Niño cycles, are all expected under climate change (Yoon et al., 2015; Pendergrass et al., 2017; Roque-Malo and Kumar, 2017).

There are relatively few studies assessing the effects of climate change in hot deserts compared to other biomes (Maestre et al., 2012; Tielbörger and Salguero-Gómez, 2014). Moreover, there is considerable geographic bias in the studies that do exist, with more in North America and Australia than elsewhere, likely due to proximity of universities and the geographic location of many hot deserts in remote and politically unstable regions (Maestre et al., 2012). Assessing the long-term impacts of climate change on these areas is further complicated by climate model disagreements on regional precipitation changes.

A large unknown for desert organisms is whether adaptation to the extreme climate will aid species in coping with future changes in climate (Tielbörger and Salguero-Gómez, 2014) or whether species are already living on the edge of their physiological limits and thus are highly vulnerable (Vale and Brito, 2015). Climate change will undoubtedly have effects across multiple plant life stages, affecting seed production, survival, dormancy, and germination, as well as seedling establishment and growth. In this chapter, we focus on the effects of increasing temperature and changing rainfall patterns in deserts on seed survival, dormancy, and germination.

Effects of increasing temperature

Air temperatures in deserts worldwide are expected to increase by at least 3°C by the end of the 21st century (IPCC, 2013). There already is evidence of increasing temperatures over the last 40 years in the Atacama Desert (Bennett et al., 2016) and in the Sonoran Desert (Kimball et al., 2010), where increasing temperatures have resulted in fewer days with temperatures below freezing (Weiss and Overpeck, 2005). For seeds of many species, the effects of increased air temperatures will predominantly be through increased soil temperatures. The degree to which soil temperatures change is determined by soil properties, but generally deserts are likely to experience more extreme soil temperature increases than mesic areas due to low vegetation cover (Harte et al., 1995). Furthermore, there is some evidence that increases in soil surface temperature will be nearly double that of air temperature (Ooi et al., 2009). Although temperatures deep in desert soils are relatively stable (Smith, 1986), the majority of seeds are often on or close to the surface (Gutterman, 1994; Pake and Venable, 1996), where maximum temperatures are the most extreme. For example, over the course of a typical day temperatures on the soil surface in the Sahara Desert can range from 20°C at sunrise to 70°C in the early evening (Larmuth, 1984). Warming will be greatest for seeds on the soil surface. Seeds in aerial seed banks will experience increases in air temperatures, while those buried in soil will be relatively buffered from temperature extremes. Regardless of where a seed is located, increased air and soil temperatures are likely to drive changes in seed survival, dormancy, and germination.

Increasing temperature effects on seed survival

Seed survival generally decreases with increasing temperatures (Ellis and Roberts, 1980), with high temperatures promoting seed mortality through rapid aging (Walters et al., 2010) and protein denaturation (Zhao et al., 2004). Seeds from desert-adapted species, however, may be resilient to future high-temperature exposure due to their adaptation to current temperature extremes. For example, a study on desert succulents in the Aizoaceae and Cactaceae found a significant positive correlation between the absolute maximum temperature at the seed collection site and the proportion of seed surviving exposure to 103°C for 17 h (Daws et al., 2007; Fig. 4.2). These results suggest that seeds of these species could withstand future temperature increases above what is typically experienced in their respective climates, at least for brief intervals. Conversely, species from relatively cool sites may lack sufficient adaptation to high-temperature extremes to survive. This may not be true for all species; however, as the same study found no correlation between seed survival and collection site temperatures for Crassulaceae.

In addition to experiencing short periods of extremely high temperatures, seeds also may experience long-term temperature increases. Few studies have looked at long-term exposure to increased temperatures. Ooi et al. (2009) assessed

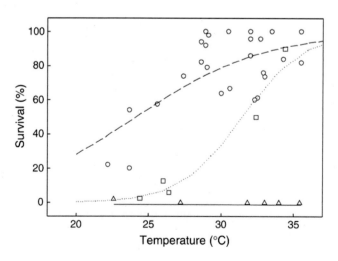

FIGURE 4.2 Relationship between seed survival after exposure to 103°C for 17 h and the maximum annual temperature at the collection location for 26 succulent species of Aizoaceae (*open circles*), Cactaceae (*open squares*), and Crassulaceae (*open triangles*). From Daws, M.I., Kabadajic, A., Manger, K., Kranner, I., 2007. Extreme thermo-tolerance in seeds of desert succulents is related to maximum annual temperature. S. Afr. J. Bot. 73, 262–265.

seed viability in eight short-lived species from an arid region of Australia after exposure to a daily temperature regime of 70°C/25°C for 70 days to mimic expected increases in summer temperatures for that region. They found decreased viability in only one of the eight species—the legume *Tephrosia sphaerospora*, again providing evidence that desert-adapted species may generally survive increases in temperature at the seed stage. It is important to note, however, that Australian deserts are relatively more fire prone than other hot deserts (Rice and Westoby, 1999), and many species have evolved to withstand high temperatures of fires. Thus information about fire tolerance may not generalize to desert habitats where fire is rare. Regardless, high heat tolerance in seeds of desert-adapted species may protect many species from heat-related seed death and instead may affect desert flora predominantly through changes in dormancy and germination.

Increasing temperature effects on dormancy and seed banks

Dormancy break is predominantly regulated by temperature (Mott, 1972; Bell, 1999; Baskin and Baskin, 2014). In hot deserts, physiological dormancy (PD) is the most prevalent (~60% of species), followed by physical dormancy (PY) (~25% of species), with only 10% of species having no dormancy and 5% morphophysiological dormancy (Baskin and Baskin, 2014). The majority of desert shrubs, herbs, and succulents (the most common types of desert plants) produce seeds with physiological dormancy (Baskin and Baskin, 2014). Species with PD that are adapted to germinating in the fall and winter often require a period of exposure to high (dry) temperatures to break dormancy (after-ripening). In Australian species, high temperatures reduced both PY and PD, thereby increasing germination (Mott, 1972; Ooi et al., 2009; Liyanage and Ooi, 2017).

Temperatures during seed development also can affect the depth of dormancy and thereby modulate final germination percentages (Ellis and Roberts, 1980; El-Keblawy and Al-Ansari, 2011; Donohue et al., 2012). In the Arabian desert annual grass *Brachypodium hybridum*, seeds produced under relatively low temperatures had higher final germination percentages than those produced under high temperatures (Elgabra et al., 2019). The authors posit that higher germination of seeds produced under low temperatures may be an adaptation to the hot climate, allowing seeds produced early in the season (under low temperatures) to germinate readily and produce more seeds before the end of the season. Those produced late in the season, under high temperatures, may have higher dormancy that would prevent germination when conditions were not favorable for establishment. More studies are needed to understand the general pattern of exposure to high temperatures during seed development and how this will affect germination under climate change.

Soil seed banks may contain both dormant and nondormant seeds that may be vulnerable to increased temperatures through seed mortality (see above) and in situ germination. While seed mortality lowers the buffering potential of a seed bank, germination only does so when it does not lead to seed replenishment through fecundity. Many desert species, especially annuals, rely on long-lived seed banks that spread the risk of germinating over multiple seasons (Bell, 1999; Ooi et al., 2009; Pake and Venable, 1996; Petrů and Tielbörger, 2008; Philippi, 1993; Smith et al., 2014; Sotomayor and Gutiérrez, 2015). Long-lived species on the other hand typically have relatively short seed longevity and lack persistent seed banks. Viability of *Yucca brevifolia* seeds in the soil declined by 50%−68% in 1 year (Reynolds et al., 2012). No seeds of *Stenocereus stellatus* (Cactaceae) survived more than 2 years (Álvarez Espino et al., 2014), and no seeds of the cacti *Carnegiea gigantea* or *Opuntia engelmannii* germinated after the first year (Bowers, 2005a). There is some evidence, however, of long-term seed survival in perennials, which often have episodic recruitment events. Bowers (2005a) found that seedlings of *Mammillaria grahamii* and *Ferocactus wislizeni* appeared 6 and 3 years postseeding, respectively, indicating the presence of a seed bank, and some cacti have long-lived aerial seed banks (Santini and Martorell, 2013).

Species with relatively long seed longevity and persistent seed banks may be at a disadvantage under future climate warming scenarios if increased temperatures break dormancy and promote germination when environmental factors are not favorable for seedling survival, thereby lowering the buffering capacity of the seed bank. Buried seeds in long-lived seed banks may be more buffered from temperature extremes than those on the soil surface, thus reducing heat-induced dormancy break. The majority of hot deserts, however, have a hard soil surface that precludes deep seed burial. Thus burial is likely to be beneficial only in sandy deserts. Similarly, seeds in aerial seed banks may be relatively more buffered from increases in temperature than those on the soil surface (El-Keblawy et al., 2018). Species with short seed longevity also may experience changes in dormancy under future warming; however, their ability to withstand warming may depend more on postgermination establishment and survival than on changes in dormancy.

Increasing temperature effects on germination

Although seeds of many desert species seem to be resistant to heat-related mortality, short-term exposure to high temperatures can affect germination percentages and rates. Seeds of *Yucca whipplei* had decreased germination after exposure to 110°C for 5 min (Keeley and Tufenkians, 1984), but the vast majority of desert species assessed showed either no change or an increase in germination in response to exposure to high temperatures (Table 4.1). For example, there was no change in germination response of *Mammillaria magnimamma* seeds after heating at 90°C for 12 h (Ruedas et al., 2000) or *Welwitschia mirabilis* seeds after exposure to 80°C for 48 h (Whitaker et al., 2004). Responses within groups of seemingly similar species vary widely. Pérez-Sánchez et al. (2011) investigated germination in multiple Chihuahuan desert perennial species (shrubs, succulents, and cacti) and found no consistent responses either within or across types of plants exposed to 70°C in 2-h windows for multiple days, with some species showing no response, others having increased germination, and still others having decreased germination.

Long-term temperature increases also may affect germination. After 70 days of warming at 70/25°C, Ooi et al. (2009) found no effect on the germination of four of eight species of short-lived Australian species, while three had increased germination. Similarly, three perennials in the Chihuahuan Desert either showed no change in germination (*Yucca filifera*) or increased germination (*Agave striata, Echinocactus platyacanthus*) after 1 year of warming at +1.7°C (Aragón-Gastélum et al., 2018). In a guild of eight desert annuals from the Mojave and Sonoran deserts, all species reached maximum germination following 5 weeks of exposure to 50°C (Capon and Van Asdall, 1967).

Many desert perennial shrubs and cacti germinate during summer monsoons (Steenbergh and Lowe, 1977; Mauchamp et al., 1993), when temperatures are relatively high. Species adapted to germinating at high temperatures such as *Ambrosia dumosa, Larrea tridentata*, and *Yucca brevifolia* may benefit from global warming (Ostler et al., 2002; Reynolds et al., 2012), although concurrent decreases in summer rainfall (Bachelet et al., 2016) may offset these beneficial effects. Furthermore, increases in winter temperature may limit dormancy-breaking cues in spring- and summer-germinating species. Thus even if nondormant seeds exhibit increased germination with warming, seed dormancy may limit their ability to benefit from increased temperatures. In slightly cooler coastal deserts like the Atacama, germination is more likely to be restricted at high temperatures. Germination of the perennial herb *Zephyra compacta* in the Atacama Desert was maximal at 20°C but decreased substantially at 25°C (Cuadra et al., 2017). Similarly, in six species of the Chilean endemic genus *Leucocoryne,* germination decreased when temperatures were above 20°C (Cuadra et al., 2016). Thus climate change could increase germination of species in hot deserts, but it may decrease germination of species in coastal cold-water deserts.

Generally, modest increases in temperature during germination events will improve germination success if temperatures are within the optimal temperature range for germination of the species. For two *Acacia* species in the Chihuahuan Desert, germination increased with temperature increases up to 7°C, above which germination decreased back to that of baseline temperatures (Cortés-Cabrera et al., 2018). In a study of 32 cactus species in the Americas, Seal et al. (2017) found that an increase in temperature of only 1°C would be above the optimal germination temperature for five species and +3.7°C would be above the optimal temperature for eight species, leading to decreased germination performance in these species under projected temperature increases (Fig. 4.3A). For the remaining 18 species, however, expected temperature increases were well below their respective optima. Additionally, at +1°C 24 species had faster germination rates, and at +3.7°C 27 species had faster germination rates (Fig. 4.3B).

Finally, desert soils are typically high in salts and related soil alkalinity due to high evaporation, and increased temperatures can exacerbate the effects of soil salinity (El-Keblawy and Al-Rawai, 2005; Bhatt et al., 2019). In the Arabian Desert, seeds of the invasive shrub *Prosopis juliflora* are tolerant of both high temperatures and high salinity when exposed to these factors independently, but seeds exhibited decreased germination when exposed to both high temperatures and high salinity levels at the same time (El-Keblawy and Al-Rawai, 2005). Increased temperatures under climate change may therefore amplify the negative effects of soil salinity.

Effects of decreasing precipitation

Predictions of how precipitation will change in deserts in response to climate change vary by region and climate change model, but the general consensus is that dry regions will become progressively drier (Bates et al., 2008; Risbey, 2011; Seager et al., 2007). Drying trends over the past century have already been observed in deserts across southwestern North America, western South America, Africa, and southwest Australia (Bates et al., 2008; Sarricolea et al., 2017). Largely, these decreases in precipitation primarily are expected to affect germination percentages and rates, with fewer studies and predictions for how decreased precipitation influences other components of regeneration from seeds, such

TABLE 4.1 Germination responses of various species following exposure of seeds to increased temperatures.

Species	Family	Life form	Region	Pregermination treatment	Effect	References
Acacia schaffneri	Fabaceae	Perennial shrub	Chihuahuan Desert	70°C 2 h (multiple days)	+	Pérez-Sánchez et al. (2011)
Agave lechuguilla	Agavaceae	Succulent	Chihuahuan Desert	70°C 2 h (multiple days)	−	Pérez-Sánchez et al. (2011)
Agave salmiana	Agavaceae	Succulent	Chihuahuan Desert	70°C 2 h (multiple days)	0	Pérez-Sánchez et al. (2011)
Agave striata	Agavaceae	Succulent	Chihuahuan Desert	+1.7°C (1 year)	+	Aragón-Gastélum et al. (2018)
Coreopsis bigelovii	Asteraceae	Annual forb	Mojave Desert	50°C (5 weeks)	+	Capon and Van Asdall (1967)
Echinocactus platyacanthus	Cactaceae	Succulent	Chihuahuan Desert	70°C 2 h (multiple days)	+	Pérez-Sánchez et al. (2011)
Echinocactus platyacanthus	Cactaceae	Succulent	Chihuahuan Desert	+1.7°C (1 year)	+	Aragón-Gastélum et al. (2018)
Enneapogon avenaceus	Poaceae	Perennial grass	Australian deserts	70°C/25°C 70 days	0	Ooi et al. (2009)
Enneapogon cylindricus	Poaceae	Perennial grass	Australian deserts	70°C/25°C 70 days	0	Ooi et al. (2009)
Eriophyllum wallacei	Asteraceae	Annual forb	Mojave Desert	50°C (5 weeks)	+	Capon and Van Asdall (1967)
Euphorbia polycarpa	Euphorbiaceae	Annual forb	Mojave Desert	50°C (5 weeks)	+	Capon and Van Asdall (1967)
Ferocactus histrix	Cactaceae	Succulent	Chihuahuan Desert	70°C 2 h (multiple days)	0	Pérez-Sánchez et al. (2011)
Geraea canescens	Asteraceae	Annual forb	Mojave Desert	50°C (5 weeks)	+	Capon and Van Asdall (1967)
Hibiscus krichauffianus	Malvaceae	Perennial shrub	Australian deserts	70°C/25°C 70 days	+	Ooi et al. (2009)
Isolatocereus dumortieri	Cactaceae	Succulent	Chihuahuan Desert	70°C 2 h (multiple days)	−	Pérez-Sánchez et al. (2011)
Larrea tridentata	Zygophyllaceae	Perennial shrub	North American deserts	71°C (7 days)	−	Barbour (1968)
Lepidium lasiocarpum	Brassicaceae	Annual forb	Sonoran Desert	50°C (5 weeks)	+	Capon and Van Asdall (1967)
Pachycereus pringlei	Cactaceae	Succulent	Sonoran Desert	70°C 3 weeks	−	Cancino et al. (1993)
Pachycereus pringlei	Cactaceae	Succulent	Sonoran Desert	40°C 1–3 weeks	+	Cancino et al. (1993)
Plantago insularis	Plantaginaceae	Annual forb	Sonoran Desert	50°C (5 weeks)	+	Capon and Van Asdall (1967)
Polycalymma stuartii	Asteraceae	Annual forb	Australian deserts	70°C/25°C 70 days	+	Ooi et al. (2009)
Portulaca oleracea	Portulacaceae	Annual forb	Australian deserts	70°C/25°C 70 days	+	Ooi et al. (2009)

(Continued)

TABLE 4.1 (Continued)

Species	Family	Life form	Region	Pregermination treatment	Effect	References
Prosopis laevigata	Fabaceae	Perennial shrub	Chihuahuan Desert	70°C 2 h (multiple days)	0	Pérez-Sánchez et al. (2011)
Salvia columbariae	Lamiaceae	Annual forb	Mojave Desert	50°C (5 weeks)	+	Capon and Van Asdall (1967)
Schismus arabicus	Poaceae	Annual grass	Negev Desert	40°C	+	Gutterman (1996)
Swainsona laxa	Fabaceae	Perennial forb	Australian deserts	70°C/25°C 70 days	0	Ooi et al. (2009)
Tephrosia sphaerospora	Fabaceae	Perennial shrub	Australian deserts	70°C/25°C 70 days	0	Ooi et al. (2009)
Welwitschia mirabilis	Welwitschiaceae	Succulent	Namib Desert	80°C 48 h	0	Whitaker et al. (2004)
Yucca decipiens	Agavaceae	Succulent	Chihuahuan Desert	70°C 2 h (multiple days)	−	Pérez-Sánchez et al. (2011)
Yucca filifera	Agavaceae	Succulent	Chihuahuan Desert	+1.7°C (1 year)	0	Aragón-Gastélum et al. (2018)
Yucca whipplei	Agavaceae	Succulent	Mojave Desert	110°C 5 min	−	Keeley and Tufenkians (1984)

Note: " − " indicates a significant negative effect on germination, "0" indicates no significant effect, and " + " indicates a significant positive effect.

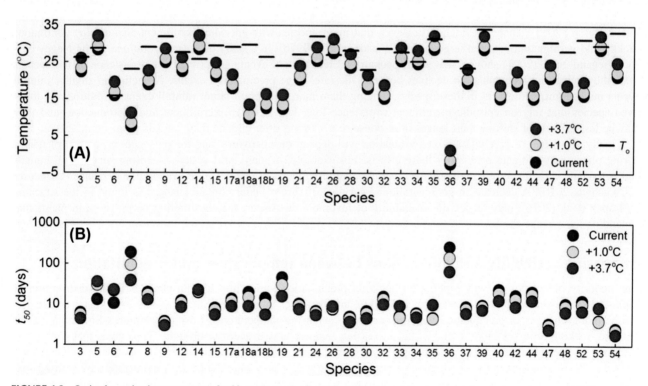

FIGURE 4.3 Optimal germination temperature for 32 cactus species from 33 seed lots in relation to expected temperature increases. (A) Even under temperature increases of 3.7°C (*red dots*), the majority of species were still below their optimal germination temperatures (T_o). (B) Germination rates increased in many species under temperature increases. *From Seal, C.E., Daws, M.I., Flores, J., Ortega-Baes, P., Galíndez, G., León-Lobos, P., et al., 2017. Thermal buffering capacity of the germination phenotype across the environmental envelope of the Cactaceae. Glob. Change Biol. 23, 5309–5317.*

as seed survival and dormancy. Here, we focus on germination, with a brief discussion of some potential responses of seed dormancy and seed banks.

Decreasing precipitation effects on dormancy and seed banks

Many desert annuals and short-lived perennials have seed banks that buffer against dry years, but how decreasing rainfall may affect dormancy has been little explored. There is some evidence that the maternal environment during seed development influences dormancy and subsequent germination. In a study of four annuals in the Negev Desert, Tielborger and Valleriani (2005) found that seeds produced under wet conditions had lower germination fractions the following growing season than those produced under dry conditions, suggesting that rainfall during seed production alters dormancy intensity. Thus, climate-driven decreases in precipitation could lead to the production of seeds with overall weaker dormancy.

Decreasing precipitation effects on germination

Decreasing rainfall likely will lead to decreased germination in a wide range of species, although responses will depend not only on the total amount of precipitation but also on how it is distributed in time. Rainfall events as small as 15 mm are enough to trigger germination in the Atacama (Vidiella et al., 1999), while 25 mm can trigger germination in North American deserts (Beatley, 1974). In some deserts, rainfall may occur only once each year, after which germination and seedling growth occur rapidly. For example, in 1989 roughly 85% of the total annual precipitation for the Carrizal Bajo region of the Atacama Desert fell in a single event of roughly 18.5 mm, triggering germination of annuals (Vidiella et al., 1999). Decreasing rainfall in these extremely arid systems may prohibit germination in a given year, increasing the importance of maintenance of seed longevity until a large rainfall event occurs.

Many desert species, especially perennials, display episodic recruitment (which often includes both germination and seedling survival) in which the majority of recruitment occurs during a series of above-average rainfall years (Ackerman, 1979; Watson et al., 1997a; Allen et al., 2008). For example, favorable recruitment events (i.e., multiple abnormally wet years) that occur once every century may be enough to sustain populations of the Saguaro cactus (Pierson et al., 2013). There is evidence, however, that many species with episodic recruitment also display continuous background germination between large events (Watson et al., 1997b). In a study on *Acacia raddiana* in the Negev desert, Wiegand et al. (2004) showed that small background recruitment events are critical for persistence of the species since they helped buffer population declines between the rare, large precipitation events. A decline in smaller rainfall events might limit recruitment in these species, making them more reliant on larger rainfall events. Additionally, long-lived species that rely on episodic recruitment experience long lags between recruitment and reproduction, and they may be less capable of shifting their ranges in response to a warmer, drier climate than annuals (Díaz et al., 2019).

Finally, the effect of precipitation on germination will depend on microsites, and the presence or absence of neighboring plants may mitigate or amplify these effects. For example, in many arid systems seedling survival of annuals and perennials is facilitated by the presence of shrubs that provide shade and decrease evaporative demands (Rolhauser and Pucheta, 2016). A decrease in germination and survival of shrubs under climate change, as found in the Atacama by López et al. (2016), may reduce the availability of favorable microsites, thereby limiting germination in plants that depend on these nurse shrubs.

Increasing variability and interactions between temperature and precipitation

The majority of hot deserts will experience increased precipitation variability through changes in the distribution of rainfall within a year, increased interannual variability, and increases in extreme climatic events over longer timescales. Furthermore, changes in precipitation will interact with increasing temperatures. Low temperatures limit evaporative demand, making droughts more tolerable, but increasing temperatures exacerbate drought. Thus areas that experience increases in precipitation may experience increased water stress as a result of increased temperatures (Gutzler and Robbins, 2011). Changes in the distribution of rainfall within a year also can interact with temperature by shifting seasonality of rainfall. Many deserts have both summer and winter rainfall. In southwestern North America, the proportion of rainfall that arrives in summer increases eastward, with the Mojave Desert receiving the least and the Chihuahuan receiving the most (Laity et al., 2008). Some models predict increases in winter precipitation in the Sonoran and Mojave deserts with less rainfall during spring and summer (Bachelet et al., 2016; Cui et al., 2020), while other models predict winter rainfall decreases across the southwestern USA deserts (Dominguez et al., 2012). Here, we focus on the effects of these factors on germination due to availability of current research, although the regeneration of plants from seeds also is likely to be affected via dormancy and seed bank dynamics.

Effects of increasing variability and temperature-precipitation interactions on germination

Germination responses to changing rainfall patterns and interactions with temperature under climate change are likely to be highly complex. In general, high temperatures become more stressful with decreasing rainfall, even among desert-adapted species. In a guild of eight perennial shrubs and cacti in the Chihuahuan Desert, Flores et al. (2017) found that many species tolerated low water potentials at current temperatures, but nearly all of them had decreased germination in response to decreased water potentials at high temperatures. However, seeds of *Prosopis laevigata* showed no change in germination in response to the interactive effects of decreased water potentials and increased temperatures. Thus species may tolerate increases in temperature and concurrent decreases in moisture inputs as long as these changes are within the normal range of variability they experience (e.g., Elgabra et al., 2019). More important than ability to germinate at low water potentials is how rapidly species can germinate given high water potential during a rainfall event and how long that high water potential remains afterward, both of which are influenced by temperature (Huxman et al., 2004). If conditions permit rapid germination, even a short window of high water potential may be enough for high germination, making postgermination seedling survival and establishment more critical than germination conditions. On the other hand, if germination is slow even under optimal conditions or if precipitation occurs in smaller amounts or outside optimal temperature windows due to climate change, germination fractions will decrease.

Temperatures during rainfall in hot deserts with bimodal precipitation drive two distinct plant communities: (1) summer annuals and many perennials germinate at relatively high temperatures during monsoonal summer rainfall, and (2) winter annuals germinate at low temperatures during winter rainfall (Went, 1948,1949; Bell, 1999). Reduced monsoonal rainfall combined with increased summer temperatures may decrease germination percentages in summer-germinating species, while increased mean temperatures could favor later germination of winter annuals. Shifting rainfall patterns within a season, however, may sometimes produce unexpected results. Although the Sonoran Desert has become both warmer and drier over a 25-year period (Weiss and Overpeck, 2005), the onset of germination-stimulating rains later in the winter favors cold-adapted species, which have higher germination fractions and germinate under cooler conditions than warm-adapted species (Kimball et al., 2010; Huang et al., 2016), driving an increase in abundance of these cold-adapted species. Thus the matching of timing of precipitation with appropriate temperatures will be most important in determining the germination response of desert species to climate change.

Finally, precipitation variability will increase under climate change driven by an increase in extreme events such as El Niño, which plays a large role in episodic recruitment of perennials and replenishment of annual seed banks in hot deserts (Gutiérrez and Meserve, 2003; Bowers, 2005b; León et al., 2011). Given an overall decrease in precipitation, extreme events like El Niño may become increasingly important for maintaining populations, possibly leading to large population increases during these events and large declines between events.

Other climatic changes

There are many other climatic changes not reviewed here that will affect regeneration of plants from seeds in hot deserts, including changes in fog in coastal deserts, changes in fire frequency, nitrogen deposition, and increased CO_2 concentration. Although difficult for current climate models to predict (Koračin, 2017), widespread decreases in fog over land have been recorded (Johnstone and Dawson, 2010; Van Schalkwyk, 2011; Klemm and Lin, 2016). Decreased fog may lead to negative effects on plant regeneration from seeds by limiting plant growth and biomass accumulation and therefore seed production (Seely and Henschel, 1998), but fog deposits may not supply enough water to induce germination (Southgate et al., 1996). Although fires in hot deserts are historically infrequent due to lack of continuous fuel on the soil surface (Humphrey, 1963), increasing human activity and species invasions have increased fuel loads, which subsequently have increased fire frequency in hot deserts, including in the Mojave (Bishop et al., 2020) and Sonoran (Schmid and Rogers, 1988) deserts. Nitrogen deposition also can drive changes in seed regeneration as nitrogen may break dormancy in desert annuals (El-Keblawy and Gairola, 2017). Nitrogen deposition also can promote the growth of invasive species and thus change the fire cycle. Finally, increased carbon dioxide concentrations have been shown to increase germination success in trees, herbs, and grasses but with different effect sizes between life forms (Marty and BassiriRad, 2014). In resource-limited deserts, positive effects of increased CO_2 on germination may only arise when other resource limitations are alleviated as was found in the Mojave Desert under increased precipitation (Smith et al., 2014). CO_2-driven germination may therefore be relatively less important than the effects of rainfall and temperature.

Summary and conclusions

Although desert climates are already among the most extreme and variable climates on the Earth, climate change is expected to amplify these conditions through increases in temperatures, changes in rainfall (including general declines, increases in variability, and shifts in distribution), and the interaction of temperature increases with rainfall changes. Seed regeneration of desert plants is likely to be affected through changes in seed survival, seed dormancy, seed bank dynamics, and seed germination.

Temperature increases may drive changes in seed regeneration across each of these stages, but current research suggests that seeds of species in hot deserts may be relatively insensitive to temperature increases in terms of seed mortality. Increased temperatures likely will have the greatest effect on seed dormancy and resultant germination. Increased temperatures, especially in regions where freezing temperatures currently limit plant establishment, as in Argentina (Mourelle and Ezcurra, 1996) and the Sonoran Desert (Steenbergh and Lowe, 1977), could lead to increased germination and overall plant productivity (Abella, 2020; McIntosh et al., 2020) by decreasing dormancy in species that require heat to break dormancy.

How shifting rainfall will affect seed regeneration is more complicated to dissect and has so far mainly been studied on germination dynamics and not necessarily on seed survival and dormancy. Furthermore, it is hard to separate the effects of increasing temperatures from the effects of changes in precipitation. If precipitation occurs in smaller amounts or outside optimal temperature windows, germination fractions are likely to decrease, as was found in warmer-germinating winter annuals in the Sonoran Desert (Kimball et al., 2010). Even small rainfall amounts during optimal temperature windows, however, may be sufficient to induce germination, provided evaporation rates are slower than germination rates. Given overall decreases in rainfall but increases in extreme rainfall events, population dynamics may exhibit more boom-and-bust cycles. How climate-driven changes in seed regeneration scale up to affect plant community composition will depend on postgermination factors. Although beyond the scope of this chapter, unpredictable postgermination precipitation combined with higher temperatures may result in lower seedling survival and establishment.

Future research

A more comprehensive understanding of the effects of climate change on regeneration of plants from seeds in hot deserts requires further study, including studies of how climate variability will influence seed dormancy and seed banking strategies, incorporating the role of altered biotic interactions under climate change on regeneration from seeds and a focus on lesser-studied desert ecosystems across the globe. While studies have focused on the effects of temperature or moisture, there is a need for more information on how shifting variability in these factors, including exposure to extreme temperatures as well as drought and floods, will interact to affect seed dormancy and seed survival. There is also a strong need to integrate across trophic levels to understand how changes in interacting organisms (i.e., critical dispersing animals and pollinators) will affect plant regeneration from seeds. Furthermore, understanding and predicting effects across systems hinges on our ability to sample ecosystems and life forms more broadly. Specifically, more research is needed in the Thar, Sahara, and Arabian deserts (Maestre et al., 2012). Finally, across deserts, we will benefit greatly from more research relating form to function and how functional differences influence plant regenerative processes from seeds to determine if groups of species sharing similar strategies have similar, predictable responses to climate change.

References

Abella, S.R., 2020. Cover–biomass relationships of an invasive annual grass, *Bromus rubens*, in the Mojave Desert. Inv. Plant Sci. Manage. 13, 288–292.

Ackerman, T.L., 1979. Germination and survival of perennial plant species in the Mojave Desert. Southw. Nat. 24, 399–408.

Allen, A.P., Pockman, W.T., Restrepo, C., Milne, B.T., 2008. Allometry, growth and population regulation of the desert shrub *Larrea tridentata*. Funct. Ecol. 22, 197–204.

Aragón-Gastélum, J.L., Flores, J., Jurado, E., Ramírez-Tobías, H.M., Robles-Díaz, E., Rodas-Ortiz, J.P., et al., 2018. Potential impact of global warming on seed bank, dormancy and germination of three succulent species from the Chihuahuan Desert. Seed Sci. Res. 28, 312–318.

Axelrod, D.I., 1948. Evolution of desert vegetation in western North America during Middle Pliocene time. Evolution 2, 127–144.

Axelrod, D.I., 1979. Age and origin of Sonoran desert vegetation. California Academy of Sciences Occasional Paper 132.

Axelrod, D.I., Raven, P.H., 1978. Late Cretaceous and Tertiary vegetation history of Africa. In: Werger, M.J.A. (Ed.), Biogeography and Ecology of Southern Africa, Monographiae Biologicae. Springer, Dordrecht, pp. 77–130.

Bachelet, D., Ferschweiler, K., Sheehan, T., Strittholt, J., 2016. Climate change effects on southern California deserts. J. Arid Environ. 127, 17–29.
Barbour, M.G., 1968. Germination requirements of the desert shrub *Larrea divaricata*. Ecology 49, 915–923.
Baskin, C.C., Baskin, J.M., 2014. Seeds: Ecology, Biogeography, and Evolution of Dormancy and Germination, Second Edition Academic Press/Elsevier, San Diego.
Bates, B., Kundzewicz, Z.W., Wu, S., Palutikof, J.P. (Eds.), 2008. Climate Change and Water: IPCC Technical Paper 6. IPCC.
Bayram, H., Öztürk, A.B., 2014. Global climate change, desertification, and its consequences in Turkey and the Middle East. In: Pinkerton, K.E., Rom, W.N. (Eds.), Global Climate Change and Public Health, Respiratory Medicine. Springer, New York, pp. 293–305.
Beatley, J.C., 1974. Phenological events and their environmental triggers in Mojave Desert ecosystems. Ecology 55, 856–863.
Bell, D.T., 1999. The process of germination in Australian species. Aust. J. Bot. 47, 475–517.
Bennett, M., New, M., Marino, J., Sillero-Zubiri, C., 2016. Climate complexity in the Central Andes: a study case on empirically-based local variations in the Dry Puna. J. Arid Environ. 128, 40–49.
Bhatt, A., Gairola, S., Carón, M.M., Santo, A., Murru, V., El-Keblawy, A., et al., 2019. Effect of light, temperature, salinity and maternal habitat on seed germination of *Aeluropus lagopoides* (Poaceae): an economically important halophyte of arid Arabian deserts. Botany 98, 117–125.
Bishop, T.B.B., Gill, R.A., McMillan, B.R., St. Clair, S.B., 2020. Fire, rodent herbivory, and plant competition: implications for invasion and altered fire regimes in the Mojave Desert. Oecologia 192, 155–167.
Bowers, J.E., 2005a. New evidence for persistent or transient seed banks in three Sonoran Desert cacti. Southw. Nat. 50, 482–487.
Bowers, J.E., 2005b. El Niño and displays of spring-flowering annuals in the Mojave and Sonoran deserts. J. Torrey Bot. Soc. 132, 38–49.
Cancino, J., de la Luz, J.L.L., Coria, R., Romero, H., 1993. Effect of heat treatment on germination of seeds of cardón [*Pachycereus pringlei* (S. Wats.) Britt. & Rose, Cactaceae]. J. Ariz.-Nev. Acad. Sci. 27, 49–54.
Capon, B., Van Asdall, W., 1967. Heat pre-treatment as a means of increasing germination of desert annual seeds. Ecology 48, 305–306.
Cortés-Cabrera, H.E., Pérez-Domínguez, R., Flores, J., González-Tagle, M., Cuéllar-Rodríguez, G., Jurado, E., 2018. Germination of two *Acacia* species at elevated temperatures simulating climate change. Rev. Mex. Cienc. For. 9, 304–322.
Cuadra, C.D.la, Vidal, A.K., Lefimil, S., Mansur, L., 2016. Temperature effect on seed germination in the genus *Leucocoryne* (Amaryllidaceae). HortScience 51, 412–415.
Cui, J., Piao, S., Huntingford, C., Wang, X., Lian, X., Chevuturi, A., et al., 2020. Vegetation forcing modulates global land monsoon and water resources in a CO_2-enriched climate. Nat. Commun. 11, 5184. Available from: https://doi.org/10.1038/s41467-020-18992-7.
Daws, M.I., Kabadajic, A., Manger, K., Kranner, I., 2007. Extreme thermo-tolerance in seeds of desert succulents is related to maximum annual temperature. S. Afr. J. Bot. 73, 262–265.
De la Cuadra, C., Vidal, A.K., Mansur, L.M., 2017. Optimal temperature for germination of *Zephyra compacta* (Tecophilaeaceae). HortScience 52, 432–435.
Diffenbaugh, N.S., Swain, D.L., Touma, D., 2015. Anthropogenic warming has increased drought risk in California. Proc. Natl. Acad. Sci. USA 112, 3931–3936.
Dominguez, F., Rivera, E., Lettenmaier, D.P., Castro, C.L., 2012. Changes in winter precipitation extremes for the western United States under a warmer climate as simulated by regional climate models. Geophys. Res. Lett. 39, L05803. Available from: https://doi.org/10.1029/2011GL050762.
Donohue, K., Barua, D., Butler, C., Tisdale, T.E., Chiang, G.C.K., Dittmar, E., et al., 2012. Maternal effects alter natural selection on phytochromes through seed germination. J. Ecol. 100, 750–757.
Díaz, F.P., Latorre, C., Carrasco-Puga, G., Wood, J.R., Wilmshurst, J.M., Soto, D.C., et al., 2019. Multiscale climate change impacts on plant diversity in the Atacama Desert. Glob. Change Biol. 25, 1733–1745.
El-Keblawy, A., Al-Ansari, F., 2011. Effects of site of origin, time of seed maturation, and seed age on germination behavior of *Portulaca oleracea* from the Old and New Worlds. Can. J. Bot. 78, 279–287.
El-Keblawy, A., Al-Rawai, A., 2005. Effects of salinity, temperature and light on germination of invasive *Prosopis juliflora* (Sw.) D.C. J. Arid Environ. 61, 555–565.
El-Keblawy, A., Gairola, S., 2017. Dormancy regulating chemicals alleviate innate seed dormancy and promote germination of desert annuals. J. Plant Growth Regul. 36, 300–311.
El-Keblawy, A., Shabana, H.A., Navarro, T., 2018. Seed mass and germination traits relationships among different plant growth forms with aerial seed bank in the sub-tropical arid Arabian deserts. Plant Ecol. Divers. 11, 393–404.
Elgabra, M., El-Keblawy, A., Mosa, K.A., Soliman, S., 2019. Factors controlling seed dormancy and germination response of *Brachypodium hybridum* growing in the hot arid mountains of the Arabian Desert. Botany 97, 371–379.
Ellis, R., Roberts, E., 1980. Influence of temperature and moisture on seed viability period in barley (*Hordeum distichum* L). Ann. Bot. 45, 31–37.
Evenari, M., 1985. The desert environment. In: Noy-Meir, I., Goodall, D.W., Evenari, M. (Eds.), Hot Deserts and Arid Shrublands, Ecosystems of the World. Vol. 12. Elsevier, Amsterdam, pp. 1–22.
Flores, J., Pérez-Sánchez, R.M., Jurado, E., 2017. The combined effect of water stress and temperature on seed germination of Chihuahuan Desert species. J. Arid. Environ. 146, 95–98.
Gutiérrez, J.R., Meserve, P.L., 2003. El Niño effects on soil seed bank dynamics in north-central Chile. Oecologia 134, 511–517.
Gutterman, Y., 1994. Strategies of seed dispersal and germination in plants inhabiting deserts. Bot. Rev. 60, 373–425.
Gutterman, Y., 1996. Temperatures during storage, light and wetting affecting caryopses germinability of *Schismus arabicus*, a common desert annual grass. J. Arid Environ. 33, 73–85.

Gutzler, D.S., Robbins, T.O., 2011. Climate variability and projected change in the western United States: regional downscaling and drought statistics. Clim. Dyn. 37, 835–849.

Harte, J., Torn, M.S., Chang, F.-R., Feifarek, B., Kinzig, A.P., Shaw, R., et al., 1995. Global warming and soil microclimate: results from a meadow-warming experiment. Ecol. Appl. 5, 132–150.

Huang, Z., Liu, S., Bradford, K.J., Huxman, T.E., Venable, D.L., 2016. The contribution of germination functional traits to population dynamics of a desert plant community. Ecology 97, 250–261.

Hughes, L., 2011. Climate change and Australia: key vulnerable regions. Reg. Environ. Change 11, 189–195.

Humphrey, R.R., 1963. The role of fire in the desert and desert grassland areas of Arizona. In: Proceedings of the Second Tall Timbers Fire Ecology Conference 17, 45–61.

Huxman, T.E., Snyder, K.A., Tissue, D., Leffler, A.J., Ogle, K., Pockman, W.T., et al., 2004. Precipitation pulses and carbon fluxes in semiarid and arid ecosystems. Oecologia 141, 254–268.

IPCC, 2013. Climate change 2013: the physical science basis. In: Stocker, T.F., Qin, D., Plattner, G.-K., Tignor, M., Allen, S.K., Boschung, J., et al., (Eds.). Contribution of Working Group I to the Fifth Assessment Report of the Intergovernmental Panel on Climate Change. Cambridge University Press, Cambridge.

Johnstone, J.A., Dawson, T.E., 2010. Climatic context and ecological implications of summer fog decline in the coast redwood region. Proc. Natl. Acad. Sci. USA 107, 4533–4538.

Keeley, J.E., Tufenkians, D.A., 1984. Garden comparison of germination and seedling growth of *Yucca whipplei* subspecies (Agavaceae). Madroño 31, 24–29.

Kimball, S., Angert, A.L., Huxman, T.E., Venable, D.L., 2010. Contemporary climate change in the Sonoran Desert favors cold-adapted species. Glob. Change Biol. 16, 1555–1565.

Klemm, O., Lin, N.-H., 2016. What causes observed fog trends: air quality or climate change? Aerosol Air Qual. Res. 16, 1131–1142.

Knight, C.A., Ackerly, D.D., 2003. Evolution and plasticity of photosynthetic thermal tolerance, specific leaf area and leaf size: congeneric species from desert and coastal environments. New Phytol. 160, 337–347.

Koračin, D., 2017. Modeling and forecasting marine fog. In: Koračin, D., Dorman, C.E. (Eds.), Marine Fog: Challenges and Advancements in Observations, Modeling, and Forecasting, Springer Atmospheric Sciences. Springer International Publishing, Cham, pp. 425–475.

Laity, J.J., Orme, A.J., Balachandran, B., 2008. Deserts and Desert Environments. John Wiley, Hoboken.

Larmuth, J., 1984. Microclimates. In: Cloudsley-Thompson, J.L. (Ed.), Key Environments: Sahara Desert. Pergamon Press, Oxford, pp. 57–66.

León, M.F., Squeo, F.A., Gutiérrez, J.R., Holmgren, M., 2011. Rapid root extension during water pulses enhances establishment of shrub seedlings in the Atacama Desert. J. Veg. Sci. 22, 120–129.

Liyanage, G.S., Ooi, M.K.J., 2017. Do dormancy-breaking temperature thresholds change as seeds age in the soil seed bank? Seed Sci. Res. 27, 1–11.

López, R.P., Squeo, F.A., Gutiérrez, J.R., 2016. Differential effect of shade, water and soil type on emergence and early survival of three dominant species of the Atacama Desert. Aust. Ecol. 41, 428–436.

Maestre, F.T., Salguero-Gómez, R., Quero, J.L., 2012. It is getting hotter in here: determining and projecting the impacts of global environmental change on drylands. Philos. Trans. R. Soc. B 367, 3062–3075.

Marty, C., BassiriRad, H., 2014. Seed germination and rising atmospheric CO_2 concentration: a *meta*-analysis of parental and direct effects. New Phytol. 202, 401–414.

Mauchamp, A., Montaña, C., Lepart, J., Rambal, S., 1993. Ecotone dependent recruitment of a desert shrub, *Flourensia cernua*, in vegetation stripes. Oikos 68, 107–116.

McIntosh, M.E., Boyd, A.E., Arnold, A.E., Steidl, R.J., McDade, L.A., 2020. Growth and demography of a declining, endangered cactus in the Sonoran Desert. Plant Species Biol. 35, 6–15.

Meigs, P., 1953. Arid and semiarid climate types of the world. Proceedings, International Geographical Union, 17th Congress, 8th General Assembly. Presented at the International Geographical Union. UNESCO, Washington, DC, pp. 135–138.

Millennium Ecosystem Assessment, 2005. Ecosystems and Human Well-being: Biodiversity Synthesis. Washington, DC.

Mott, J.J., 1972. Germination studies on some annual species from an arid region of Western Australia. J. Ecol. 60, 293–304.

Mourelle, C., Ezcurra, E., 1996. Species richness of Argentine cacti: a test of biogeographic hypotheses. J. Veg. Sci. 7, 667–680.

Ooi, M.K.J., Auld, T.D., Denham, A.J., 2009. Climate change and bet-hedging: interactions between increased soil temperatures and seed bank persistence. Glob. Change Biol. 15, 2375–2386.

Ostler, W K, Anderson, D C, Hansen, D J., 2002. Pre-treating seed enhance germination of desert shrubs. https://www.osti.gov/biblio/797467.

Orians, G.H., Solbrig, O.T., 1977. Convergent evolution in warm deserts. An Examination of Strategies and Patterns in Deserts of Argentina and the United States. Dowden, Hutchinson and Ross, Cambridge.

Pake, C.E., Venable, D.L., 1996. Seed banks in desert annuals: implications for persistence and coexistence in variable environments. Ecology 77, 1427–1435.

Pendergrass, A.G., Knutti, R., Lehner, F., Deser, C., Sanderson, B.M., 2017. Precipitation variability increases in a warmer climate. Sci. Rep. 7, 17966. Available from: https://doi.org/10.1038/s41598-017-17966-y.

Petrů, M., Tielbörger, K., 2008. Germination behaviour of annual plants under changing climatic conditions: separating local and regional environmental effects. Oecologia 155, 717–728.

Philippi, T., 1993. Bet-hedging germination of desert annuals: beyond the first year. Am. Nat. 142, 474–487.

Pierson, E.A., Turner, R.M., Betancourt, J.L., 2013. Regional demographic trends from long-term studies of saguaro (*Carnegiea gigantea*) across the northern Sonoran Desert. J. Arid Environ. 88, 57–69.

Pérez-Sánchez, R.M., Jurado, E., Chapa-Vargas, L., Flores, J., 2011. Seed germination of southern Chihuahuan Desert plants in response to elevated temperatures. J. Arid Environ. 75, 978–980.

Reynolds, M.B.J., DeFalco, L.A., Esque, T.C., 2012. Short seed longevity, variable germination conditions, and infrequent establishment events provide a narrow window for *Yucca brevifolia* (Agavaceae) recruitment. Am. J. Bot. 99, 1647–1654.

Reynolds, J.F., Smith, D.M.S., Lambin, E.F., Turner, B.L., Mortimore, M., Batterbury, S.P.J., et al., 2007. Global desertification: building a science for dryland development. Science 316, 847–851.

Rice, B., Westoby, M., 1999. Regeneration after fire in *Triodia* R. Br. Aust. J. Ecol. 24, 563–572.

Risbey, J.S., 2011. Dangerous climate change and water resources in Australia. Reg. Environ. Change 11, 197–203.

Rolhauser, A.G., Pucheta, E., 2016. Annual plant functional traits explain shrub facilitation in a desert community. J. Veg. Sci. 27, 60–68.

Roque-Malo, S., Kumar, P., 2017. Patterns of change in high frequency precipitation variability over North America. Sci. Rep. 7, 10853. Available from: https://doi.org/10.1038/s41598-017-10827-8.

Ruedas, M., Valverde, T., Castillo-Argüero, S., 2000. Respuesta germinativa y crecimiento de plántulas de *Mammillaria magnimamma* (Cactaceae) bajo diferentes condiciones ambientales. Bot. Sci. 25–35.

Santini, B.A., Martorell, C., 2013. Does retained-seed priming drive the evolution of serotiny in drylands? An assessment using the cactus *Mammillaria hernandezii*. Am. J. Bot. 100, 365–373.

Sarricolea, P., Meseguer Ruiz, O., Romero-Aravena, H., Sarricolea, P., Meseguer Ruiz, O., Romero-Aravena, H., 2017. Tendencias de la precipitación en el norte grande de Chile y su relación con las proyecciones de cambio climático. Diálogo Andino 54, 41–50.

Van Schalkwyk, L., 2011. Fog Forecasting at Cape Town International Airport: A Climatological Approach (Dissertation). University of Pretoria.

Schmid, M.K., Rogers, G.F., 1988. Trends in fire occurrence in the Arizona upland subdivision of the Sonoran Desert, 1955 to 1983. Southw. Nat. 33, 437–444.

Seager, R., Ting, M., Held, I., Kushnir, Y., Lu, J., Vecchi, G., et al., 2007. Model projections of an imminent transition to a more arid climate in southwestern North America. Science 316, 1181–1184.

Seal, C.E., Daws, M.I., Flores, J., Ortega-Baes, P., Galíndez, G., León-Lobos, P., et al., 2017. Thermal buffering capacity of the germination phenotype across the environmental envelope of the Cactaceae. Glob. Change Biol. 23, 5309–5317.

Seely, M., Henschel, J., 1998. The ecology of fog in Namib sand dunes. In: Schemenauer, R.S., Bridgman, H. (Eds.), Proceedings of the First International Conference on Fog and Fog Collection. International Development Research Centre, Vancouver, British Columbia, Canada, pp. 183–186.

Shmida, A., 1985. Biogeography of desert flora. In: Evenari, M., Noy-Meir, I., Goodall, D.W., Keast, A. (Eds.), Hot Deserts and Arid Shrublands, Ecosystems of the World. Vol. 12. Elsevier, Amsterdam, pp. 23–77.

Shmida, A., Whittaker, R.H., 1979. Convergent evolution of arid regions in the new and old worlds. In: Tuxen, R. (Ed.), Vegetation and History. Ver. Symp. Int. Ver. Vegetationskunde, Berlin, pp. 437–450.

Shreve, F., Wiggins, I.L., 1964. Vegetation and Flora of the Sonoran Desert. Stanford University Press, Stanford.

Smith, E.A., 1986. The structure of the Arabian heat low. Part I: surface energy budget. Mon. Weather. Rev. 114, 1067–1083.

Smith, S., Charlet, T.N., Titzer, S.F., Abella, S.R., Vanier, C.H., Huxman, T.E., 2014. Long-term response of a Mojave Desert winter annual plant community to a whole-ecosystem atmospheric CO_2 manipulation (FACE). Glob. Change Biol. 20, 879–892.

Sotomayor, D.A., Gutiérrez, J.R., 2015. Seed bank of desert annual plants along an aridity gradient in the southern Atacama coastal desert. J. Veg. Sci. 26, 1148–1158.

Southgate, R.I., Masters, P., Seely, M.K., 1996. Precipitation and biomass changes in the Namib Desert dune ecosystem. J. Arid. Environ. 33, 267–280.

Steenbergh, W.F., Lowe, C.H., 1977. Ecology of the Saguaro: II, Reproduction, Germination, Establishment, Growth, and Survival of the Young Plant, Monograph Series No. 8. National Park Service, Washington, DC.

Thuiller, W., Midgley, G.F., Hughes, G.O., Bomhard, B., Drew, G., Rutherford, M.C., et al., 2006. Endemic species and ecosystem sensitivity to climate change in Namibia. Glob. Change Biol. 12, 759–776.

Tielborger, K., Valleriani, A., 2005. Can seeds predict their future? Germination strategies of density-regulated desert annuals. Oikos 111, 235–244.

Tielbörger, K., Salguero-Gómez, R., 2014. Some like it hot: are desert plants indifferent to climate change? In: Lüttge, U., Beyschlag, W., Cushman, J. (Eds.), Progress in Botany. Vol. 75. Springer, Berlin, pp. 377–400.

UNEP, Middleton, N., Thomas, D. (Eds.), 1997. World Atlas of Desertification. United Nations, London.

Vale, C.G., Brito, J.C., 2015. Desert-adapted species are vulnerable to climate change: insights from the warmest region on Earth. Glob. Ecol. Conserv. 4, 369–379.

Vidiella, P.E., Armesto, J.J., Gutiérrez, J.R., 1999. Vegetation changes and sequential flowering after rain in the southern Atacama Desert. J. Arid Environ. 43, 449–458.

Walters, C., Ballesteros, D., Vertucci, V.A., 2010. Structural mechanics of seed deterioration: standing the test of time. Plant Sci. 179, 565–573.

Watson, I.W., Westoby, M., Holm, A.Mc.R., 1997a. Demography of two shrub species from an arid grazed ecosystem in Western Australia 1983-93. J. Ecol. 85, 815–832.

Watson, I.W., Westoby, M., Holm, A.Mc.R., 1997b. Continuous and episodic components of demographic change in arid zone shrubs: models of two *Eremophila* species from Western Australia compared with published data on other species. J. Ecol. 85, 833–846.

Weiss, J.L., Overpeck, J.T., 2005. Is the Sonoran Desert losing its cool? Glob. Change Biol. 11, 2065–2077.

Went, F.W., 1948. Ecology of desert plants. I. Observations on germination in the Joshua Tree National Monument, California. Ecology 29, 242–253.

Went, F.W., 1949. Ecology of desert plants. II. The effect of rain and temperature on germination and growth. Ecology 30, 1–13.

Whitaker, C., Berjak, P., Kolberg, H., Pammenter, N.W., Bornman, C.H., 2004. Responses to various manipulations, and storage potential, of seeds of the unique desert gymnosperm, *Welwitschia mirabilis* Hook. fil. S. Afr. J. Bot. 70, 622–630.

Wiegand, K., Jeltsch, F., Ward, D., 2004. Minimum recruitment frequency in plants with episodic recruitment. Oecologia 141, 363–372.

Yoon, J.H., Wang, S.Y., Gillies, R.R., Kravitz, B., Hipps, L., Rasch, P.J., 2015. Increasing water cycle extremes in California and in relation to ENSO cycle under global warming. Nat. Commun. 6, 8657. Available from: https://doi.org/10.1038/ncomms9657.

Zhao, Y., Mine, Y., Ma, C.-Y., 2004. Study of thermal aggregation of oat globulin by laser light scattering. J. Agric. Food Chem. 52, 3089–3096.

Álvarez Espino, R., Godínez-Álvarez, H., Almaráz, R., 2014. Seed banking in the columnar cactus *Stenocereus stellatus*: distribution, density and longevity of seeds. Seed Sci. Res. 24, 315–320.

Chapter 5

Effect of climate change on plant regeneration from seeds in steppes and semideserts of North America

Susan E. Meyer
US Forest Service Rocky Mountain Research Station Shrub Sciences Laboratory, Provo, UT, United States

Introduction

The steppes and semideserts of North America are located primarily in Intermountain Western United States, north of the warm deserts of the American Southwest, and extending into southern Canada on the north. The area is bounded on the west by the Sierra Nevada and Cascade Mountains and on the east by the Rocky Mountains. The region as defined here includes the Great Basin section of the Basin and Range physiographic province, the Colorado Plateau, and the Columbia Plateau (Fenneman, 1930). The Intermountain Western US is a region of immense topographic and climatic diversity. Most of the area currently has a temperate semiarid climate with predominantly winter precipitation, but the region is punctuated by mountain ranges with very strong climatic gradients over short distances.

In the Great Basin, in valley bottoms formerly occupied by huge Pleistocene lakes (e.g., Lake Bonneville), saline playas cover extensive areas and are fringed by halophytic vegetation. In fine-textured but less saline valley soils at a slightly higher elevation, semidesert communities dominated largely by chenopod shrubs (e.g., shadscale, *Atriplex confertifolia*) cover vast areas. These give way as elevation increases to steppe vegetation historically dominated by big sagebrush (*Artemisia tridentata*), with a high-diversity understory of bunchgrasses and perennial herbaceous species. Sagebrush steppe communities typically interdigitate with pinyon-juniper woodland or mountain brush communities as elevation increases further. On the Columbia Plateau of the Interior Northwest, sagebrush steppe historically transitioned to the cool season bunchgrass communities of the Palouse prairie. They may also transition directly to ponderosa pine (*Pinus ponderosa*) parkland. On the Colorado Plateau, the stair-step topography exposes diverse geological substrates that support a variety of semidesert communities, including extensive areas dominated by blackbrush (*Coleogyne ramosissima*). These give way to sagebrush steppe communities and to woodland and mountain brush communities at middle elevations. Forest and mountain meadow communities dominate the vegetation at higher elevations throughout the region.

Current and future climates

The semidesert and steppe biomes of the Intermountain West currently occur under a wide suite of climate scenarios (Table 5.1). One reason for this is the wide latitudinal range spanned, from approximately 35 degrees N at the edge of the Mogollon Rim, which marks the southern edge of the Colorado Plateau in Arizona, to approximately 50 degrees N in southern British Columbia, Canada. Another reason is the orographic effect of the many large mountain ranges that surround and lie within the region. Annual precipitation varies from <100 mm in the most xeric salt desert shrublands to >400 mm in the most mesic sagebrush steppe. Seasonal precipitation distribution also varies, with an increasing proportion of summer precipitation from northwest to southeast. Mean winter temperature varies as a function of both elevation and latitude, generally with lower values at higher elevations and further north, although there are exceptions, as in the Uinta Basin of eastern Utah.

Projected future climate scenarios under global warming also vary considerably across the region, particularly with regard to precipitation. At the global level, mean temperatures are expected to increase more in the north than in the south

Plant Regeneration from Seeds. DOI: https://doi.org/10.1016/B978-0-12-823731-1.00004-4
© 2022 Elsevier Inc. All rights reserved.

TABLE 5.1 Climate means (1981–2010) for 15 locations in semidesert and shrub-steppe communities of the US Intermountain West, ranked by total annual precipitation.

Location	Physiog. province[a]	Latitude longitude	Elev. (m)	January mean temp. (°C)	July mean temp. (°C)	Annual precip. (mm)	Precip. June–Sept. (%)
Malad ID	CmPl	42.18 −112.25	1381	−4.8	21.8	398	27
Milford UT	GB	38.39 −113.02	1531	−2.2	23.8	363	22
Pioche NV	GB	37.93 −114.43	1766	0.2	23.2	321	26
Reno NV	GB	39.41 −119.75	1388	1.1	22.5	290	11
Burns OR	CmPl	43.59 −119.00	1264	−3.0	19.2	287	17
Salina UT	CoPl	38.96 −111.87	1570	−2.8	22.8	269	31
Ely NV	GB	39.24 −114.87	1985	−3.3	19.9	262	30
Twin Falls ID	CmPl	42.58 −114.42	1127	−1.6	22.4	240	17
Moab UT	CoPl	38.57 −109.56	1219	−0.2	28.0	229	33
Vernal UT	CoPl	40.46 −109.54	1641	−6.6	22.4	225	35
Yakima WA	CmPl	46.62 −120.52	331	−0.2	22.0	223	18
Green River UT	CoPl	38.96 −110.21	1302	−2.3	26.4	178	37
Page AZ	CoPl	36.92 −111.46	1315	2.5	28.6	166	34
Lovelock NV	GB	40.17 −118.46	1211	−0.5	23.8	151	20

Note: Data accessed from the Prism Climate Group Data Explorer: https://prism.oregonstate.edu/explorer/.
[a]CmPl, *Columbia Plateau*; CoPl, *Colorado Plateau*; GB, *Great Basin*.

(Diffenbaugh and Field, 2013). In our region, mean temperature based on the "business as usual" emissions scenario during the period 2046–65 is expected to be 2°C–3°C higher than the 1986–2005 mean (Maloney et al., 2014). By the late 21st century (2081–2100), the mean temperature is expected to increase a total of 5°C–6°C. These changes will increase the frost-free period by approximately 40 days, with an earlier advent of spring and a later advent of winter.

Mean annual precipitation is expected to increase in the north, remain mostly the same in middle latitudes, and decrease in the south (Diffenbaugh and Field, 2013). In our region, the increase in total precipitation is expected to be about 10%–20% by the mid-21st century. By the end of the 21st century, the model projections of Diffenbaugh and Field (2013) predict no further increase in precipitation in the northern half of our region, but a decrease of about 10% in the more southern half. In the model projections of Maloney et al. (2014), a decrease in winter precipitation by the end of the century affects only the warm desert areas to the south, with little change in precipitation in the southern half of our region. Summer rainfall predictions are said to be less reliable because of complex convection dynamics, but in our region not much change is expected.

Overall variance around mean temperature and precipitation is also expected to increase as the climate warms, so that extreme conditions will occur more frequently (O'Gorman, 2015). These conditions may be of varying duration, either as events or as patterns that can persist through time. Thus, if the climate is generally warming there may still be periods of anomalously low temperature, but periods of anomalously high temperature are more likely. Similarly, even if the climate is becoming drier there will still be anomalously wet periods, although drought periods are more likely (Abatzoglou and Kolden, 2011; Donat et al., 2016). Precipitation events are predicted to become less frequent but to increase in intensity or duration, which could have positive or negative effects (Polley et al., 2013). One prediction is that even if the amount of precipitation remains constant there will be as much as a 20%–50% decrease in the proportion of winter precipitation that falls as snow (Abatzoglou and Kolden, 2011; Diffenbaugh and Field, 2013; Maloney et al., 2014).

Seed regeneration strategies and climate change

Most studies on effects of climate change on plant species and communities in our region have been based on examining current species distributions in relation to climate and how these might potentially shift geographically under global warming. They do not attempt to tease apart effects at different life stages and generally do not consider infraspecific genetic variation except in a very broad way (subspecies, regional ecotype). Bioclimatic envelope modeling (Araújo and Peterson, 2012) is the most widely used procedure. Effects of climate change on two important regional dominants, big sagebrush (*Artemisia tridentata*, Still and Richardson, 2015; Chaney et al., 2017; Richardson and Chaney, 2018) and blackbrush (*Coleogyne ramosissima*, Richardson et al., 2014) have been examined using these methods, and major extirpation along the trailing edges along with potential expansion to the north is predicted. Schlaepfer et al. (2015) created one of the few models for our region that attempts to model climate change effects on regeneration, in this case for big sagebrush. Their process-based model predicted decreased probability of successful establishment along the trailing southern edge of the distribution due to decreased spring soil moisture under climate change, although there was a predicted net increase in precipitation.

Most models that attempt to predict future plant species distributions as a function of climate change seem to be predicated on two simplifying assumptions. The first is that mean climatic conditions are the most important determinant of species distributions, while the second is that a species or subspecific taxon is a uniform entity whose populations and individuals will respond similarly to climate. These simplifying assumptions do not work very well when making predictions about how climate change will affect regeneration ecology, especially in semiarid and arid ecosystems. Many, if not most, years in such ecosystems are not favorable for seedling establishment even now, so that for a population to persist, seedling establishment must be integrated with other features of life history that permit persistence as seeds or plants during sometimes long intervals when establishment from seeds is not favored. Weather in an average year is much less important for seedling establishment in semiarid systems than the frequency of favorable above-average years. Population matrix models have shown that variation around the mean environmental condition can be necessary for persistence, especially for short-lived species, that is, if every year were average, the species could not survive (e.g., Meyer et al., 2006). As the changes in variance associated with shifts in mean temperature and especially precipitation during climate change are not well understood, it is not known how much this predicted increase in variability can compensate for a net negative shift in mean conditions at the seed zone level. The answer will likely be species-specific, and it will depend on other aspects of life history as well as on the existence of genetic variation in adaptive traits and/or adaptive phenotypic plasticity with regard to regeneration strategy.

The review of Walck et al. (2011) provides a broad framework for considering the interaction of regeneration niche with climate change. Very few of the studies considered in their review are on semidesert and steppe ecosystems, which are apparently understudied from this perspective. These authors draw a distinction between ecosystems where regeneration syndromes are controlled mainly by temperature and those that are controlled by moisture. However, in semidesert and steppe ecosystems these two sets of environmental controls are equally important. Winter cold and summer drought interact to make the window for establishment quite narrow in either fall or spring, and this interaction changes along elevational gradients, making selection for local adaptation to climate exceptionally strong.

Cochrane et al. (2015) discuss the role of among-population and within-population genetic variation in seed germination syndromes as well as the role of adaptive phenotypic plasticity in mediating species-level regeneration response to climate change. The authors point out the lack of population-level information for most species, the difficulty of making inferences from studies of a small number of populations, especially over a narrow ecological range, and the importance of separating genetic effects from phenotypic plasticity, including maternal effects related to seed maturation environment.

Studies with wild-collected seeds can be used to infer how a population might be able to adapt its regeneration strategy to climate change in situ through selection on standing within-population genetic variation in adaptively significant traits or through adaptive phenotypic plasticity. To distinguish between these two complementary modes of in situ adaptation, common environment studies that address differences, preferably at the individual plant level, are required. There have been numerous common garden studies demonstrating the existence of local adaptation for a wide range of species native to our region (Baughman et al., 2019). However, these studies almost never involve examination of variation in the germination syndrome to detect adaptively significant patterns or to determine whether these are genetically based.

In this review, I examine available evidence from a wide range of shrub-steppe and semidesert species for climate-correlated among-population genetic variation in regeneration syndrome and speculate on how this variation could be of potential adaptive significance under a changing climate. I also consider the role of within-population genetic variation and of potentially adaptive phenotypic plasticity.

The genus *Penstemon*

The genus *Penstemon* is now assigned to the family Plantaginaceae. It is comprised of approximately 250 species confined almost completely to temperate North America, with a primary center of diversity in the Intermountain West, which is home to over 80 species. The genus includes not only narrow endemics but also widely distributed species with broad ecological amplitude. It is best studied in terms of pollination syndromes and the evolution of floral morphology.

Germination syndromes of herbaceous perennial species of *Penstemon* have also been relatively well-studied and present many of the patterns observed in other species in our region (Meyer et al., 1995a,b; Meyer, 1992; Meyer and Kitchen, 1992, 1994a; Tilini et al., 2016). Penstemons set seed in spring or summer and are almost universally spring-germinating, necessitating predictive mechanisms for timing germination appropriately for spring emergence. Several species or groups of closely related species have wide ecological amplitude, setting the stage for climate-correlated variation.

Penstemon seeds are generally physiologically dormant at maturity and require moist chilling [cold (moist) stratification] to become germinable, and the length of the period of chilling required to break dormancy is closely related to the duration of winter conditions. Individuals and populations of many species are short-lived, making the ability to form a persistent seed bank another important feature of their regeneration strategy. They form a persistent seed bank by producing a variable fraction of seeds that are either nonresponsive to chilling regardless of its duration or that cannot be adequately chilled in the climate habitat they occupy. The chilling-responsive seeds of almost all species eventually germinate during prolonged chilling, but their germination under conditions simulating winter snowpack is tightly tied to winter severity, as shown for firecracker penstemon (*Penstemon eatonii*) in Fig. 5.1 (Meyer, 1992). This ensures that seeds will germinate just as snow cover melts in early summer at high elevation but allows for late winter emergence while moisture conditions are still favorable at warmer sites.

Variation in chilling requirement in penstemons has a strong genetic basis. For example, common garden-collected seeds of closely related species of the section *Glabri* from a range of habitats showed chilling responses that closely tracked the responses of their wild progenitors (Meyer and Kitchen, 1994a; Fig. 5.2). Variation among maternal plants in seed chilling response has been documented in firecracker penstemon, where among-plant differences within a wild population were nearly as large as among-population differences (Meyer, 1992). There were no differences between early and late seed collections from these plants.

Unlike most penstemons, Palmer penstemon (*P. palmeri*) mostly has nondormant seeds that are induced into dormancy by a short period of chilling (Meyer and Kitchen, 1992). In a study of response to a short period of chilling with common-garden seeds produced in 2 years, differences among individual maternal plants were large and correlated across years (Fig. 5.3). A major shift in overall response was observed the second year, with much lower postchilling dormancy. This demonstrates a large plastic response to seed maturation conditions.

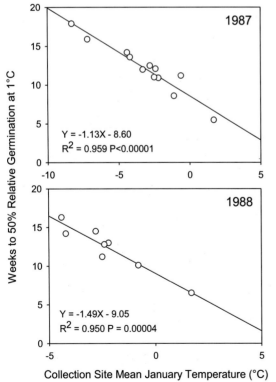

FIGURE 5.1 Weeks to 50% germination of the chilling-responsive fraction under simulated winter conditions (moist chilling at 1°C) for collections of firecracker penstemon (*Penstemon eatonii*) made in each of 2 years across a range of habitats that differed in winter severity as indicated by mean January temperature. Plotted regression equations are significant at the $P < .0001$ level. *From Meyer, S.E., 1992. Habitat correlated variation in firecracker penstemon (Penstemon eatonii Gray: Scrophulariaceae) seed germination response. Bull. Torrey Bot. Club 119, 268–279.*

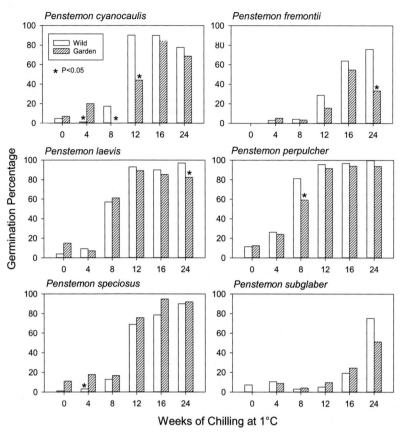

FIGURE 5.2 Germination response of wild-collected seeds of six species of *Penstemon* Section *Glabri* to chilling of 0–24 weeks compared with the germination response to chilling of their common garden-grown progeny. Pairs of bars headed with an asterisk are significantly different at $P < .05$ based on means separations from analysis of variance for each species. *From Meyer, S.E., Kitchen, S.G., 1994a. Habitat-correlated variation in seed germination response to chilling in Penstemon Section Glabri (Scrophulariaceae). Am. Midl. Nat. 32, 349–365.*

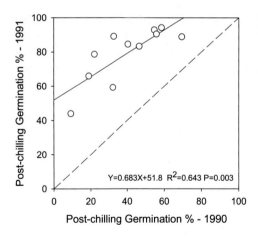

FIGURE 5.3 Germination percentage after a 4-week chilling period at 1°C followed by 4 weeks at 10/20°C for seeds of 11 individuals of *Penstemon palmeri* grown in a common garden and harvested in 2 consecutive years. The dotted line represents the regression line that would be obtained if seeds of each individual showed no change in response across years. *From Meyer, S.E., Kitchen, S.G., Carlson, S.L., 1995. Seed germination timing patterns in Intermountain Penstemon (Scrophulariaceae). Am. J. Bot. 82, 377–389.*

The combination of strong ecotypic variation within species, high within-population genetic variation, and potentially adaptive plastic responses to the maternal environment during seed development suggests that penstemons in our region are well-adapted to withstand climate change, at least in terms of germination and establishment.

Regionally important shrub species

Big sagebrush (Artemisia tridentata)

Big sagebrush is a very widely distributed autumn-flowering shrub that is a dominant species in steppe communities throughout the region. It is comprised of three principal subspecies, each of which occurs over a wide range of climates.

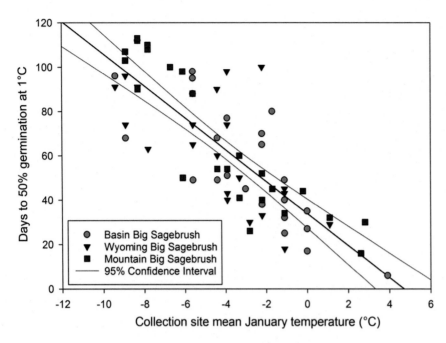

FIGURE 5.4 The relationship between days to 50% germination under conditions simulating winter snow cover for 69 seed collections of the three principal subspecies of big sagebrush (*Artemisia tridentata*) and mean January temperature, an index of winter severity at the collection site. Analysis of covariance indicated that the effect of mean January temperature was highly significant ($F_{1,63} = 124.8$, $P < .0001$; model $R^2 = 0.671$) but that there was no significant effect of subspecies and no significant interaction. *Adapted from Meyer, S. E., Monsen, S.B., 1992. Big sagebrush germination patterns: subspecies and population differences. J. Range Manage. 45, 87–93.*

In experiments with >20 populations from each subspecies, seeds of populations from colder montane habitats were more dormant at autumn temperature, more light-requiring, and up to 10 times slower to germinate at near-freezing temperatures than seeds of populations from warm desert fringe populations (Meyer et al., 1990; Meyer and Monsen, 1991, 1992). Germination syndrome variation in big sagebrush, for example, in germination speed at near-freezing temperature, is much more closely correlated with climate than with subspecific identity (Fig. 5.4). There were strong correlations among germination traits across subspecies and habitats, and all were significantly correlated with collection site mean January temperature, a surrogate variable for duration of winter snow cover (Meyer et al., 1990).

In a field seeding experiment with five populations from contrasting habitats at montane, mid-elevation, and valley planting sites, emergence and end-of-spring survival was highest for each population at the site that most closely matched its habitat of origin, demonstrating the ecological relevance of these differences in germination strategy (Meyer and Monsen, 1990). Germination regulation in this species functions predictively to time germination appropriately in the late winter or early spring. There is no strong mechanism for seed bank carryover. We have no information on the extent to which these germination strategies are genetically based, as all data were obtained from wild-collected seeds. The degree to which this species will be able to respond to selection on germination syndrome in situ is also not known. Big sagebrush seeds are dispersed only locally, which will likely limit its ability to move quickly into adjacent more upland habitats as climate warming progresses.

Rubber rabbitbrush (Ericameria nauseosa)

This autumn-flowering species is widely distributed and occurs over the full range of environments from warm desert fringes to montane meadows. It is comprised of numerous subspecies, some of which are early seral and widely distributed, while others are late-seral edaphic endemics of limited distribution. Mean germination time at near-freezing temperature (i.e., days to 50% germination) in this species is strongly negatively correlated with mean January temperature at the seed collection site, with warm desert fringe seed populations germinating as much as 10 times faster than montane seed populations (Meyer et al., 1989). As for big sagebrush, the germination syndrome in this species is much more strongly correlated with climate than with subspecific identity. Similar climate-related patterns have been observed in several subspecies of wide ecological amplitude (McArthur et al., 1987; Meyer and McArthur, 1987).

Among-population germination syndrome differences for rubber rabbitbrush have measurable ecological consequences. Seeds of three populations from contrasting environments showed germination time courses beneath winter snow cover at a mid-elevation site that closely tracked their germination time courses at 1°C in a laboratory experiment (Meyer, 1990). This species was included in the seeding study described earlier for big sagebrush (Meyer and Monsen, 1990). For five contrasting seed populations at lower, middle, and upper elevation planting sites, both maximum emergence and end-of-spring survival were highest for seed sources planted into habitats with climates that most closely matched climate at the site of seed origin. There is no evidence that this species forms persistent seed banks. The high dispersability of widely distributed, early seral subspecies may make it possible for them to migrate quickly to more climatically favorable habitats (Meyer, 1997, Meyer and Carlson, 2001). The distances that would need to be traversed to arrive at higher-elevation climate refugia in this region of high topographic relief would likely be relatively small.

*Blackbrush (*Coleogyne ramosissima*)*

This spring-flowering xerophytic shrub is the dominant species across a broad interface between our region and the warm deserts to the south. It is comprised of two climate ecotypes, a more cold-tolerant Colorado Plateau ecotype and a more drought-tolerant Mojave ecotype (Richardson and Meyer, 2012). These ecotypes are genetically differentiated and exhibit contrasting ecophysiological traits as well as differential survival in a common garden (Richardson et al., 2014). Experiments with seeds from populations from across the range showed that these ecotypes also exhibit contrasting germination strategies (Pendleton and Meyer, 2004). Seeds of all populations are generally dormant at dispersal and require moist chilling to break dormancy. Collections from the Colorado Plateau germinate more quickly under laboratory conditions simulating winter snow cover (1°C) than Mojave collections. In contrast, seeds of Mojave ecotype populations are less dormant at autumn/winter temperature (5/15°C) and germinate more quickly under conditions that simulate winter rainfall. Field retrieval experiments confirmed this result at two contrasting sites, and reciprocal seeding experiments at these two habitat sites demonstrated that the differences were ecologically meaningful. Each ecotype exhibited local advantage in establishment, with significantly higher emergence and end-of-spring survival at its site of origin than the nonlocal seed source (Meyer and Pendleton, 2005). The species is long-lived (Kitchen et al., 2015) and does not form persistent seed banks (Meyer and Pendleton, 2005). The seeds are dispersed by heteromyid rodents (Auger et al., 2016) and can be carried considerable distances in successive cache relocations. There is paleoecological evidence from packrat middens for local elevational migration of this species in response to short-term climate change during the Little Ice Age (Hunter and McAuliffe, 1994).

*Shadscale (*Atriplex confertifolia*)*

Seeds of this autumn-fruiting species often remain on the plant through the winter and spring. The seeds are enclosed in one-seeded utricles, which in turn are enveloped in indurated bracteoles. These diaspores are highly dormant and largely chilling-nonresponsive at maturity, often requiring an extended period of dry after-ripening to become chilling-responsive (Meyer et al., 1998). Dry after-ripening rate is an exponential function of temperature, thus some seeds become chilling-responsive relatively quickly during their first summer (Garvin and Meyer, 2003). Seeds from warm desert fringe populations were more chilling-responsive at maturity and after-ripened more rapidly under uniform conditions than those from populations at colder sites (Meyer et al., 1998; Garvin and Meyer, 2003). In a 5-year field retrieval experiment at a cold semidesert site, there was no germination the first spring, but warm desert fringe collections germinated to higher percentages than cold semidesert collections in the second spring. Cold semidesert collections also formed larger persistent seed banks than warm desert fringe collections under conditions at this retrieval site, with up to 50% of the seeds viable and nongerminated after 5 years. The ability to form persistent seed banks is a key life history feature for this relatively short-lived shrub.

*Antelope bitterbrush (*Purshia tridentata*)*

Antelope bitterbrush is an important and widely distributed species in our region. It has a relatively wide ecological amplitude, occurring from steppe communities up into montane communities, where it is an understory species in coniferous forests. In germination studies with collections from throughout its range, we failed to find any indication of habitat-correlated variation in germination syndrome for recently harvested seeds of this early summer-fruiting species (Meyer and Monsen, 1989). We examined whether there might be other germination timing mechanisms for ensuring a locally adaptive response to variation in duration of winter chilling across the range of this species. We found that warm dry conditions prior to chilling shortened the chilling requirement for germination, that is, dry after-ripening increases the response of seeds to chilling (Meyer, 1989). In contrast, under warm moist conditions prior to the initiation of winter chilling, the seeds became less responsive to subsequent chilling, so that they required a longer chilling period to become nondormant. The postulated effect of these two processes would be to make seeds from more xeric steppe sites, where summers are longer and drier and winters are shorter and less snowy, less dormant at the onset of winter. Seeds from more mesic sites, where summers are shorter, summer rainfall is more frequent, and winters are colder and snowier, would benefit from increased dormancy at the onset of winter. This hypothesis has yet to be tested in the field, but if confirmed it would suggest an environmentally regulated plastic dormancy loss response that could function in situ as an adaptation to a warming climate.

We also examined the genetic basis for seed dormancy in antelope bitterbrush through controlled crosses in a common garden (Meyer and Pendleton, 2000). Response to a suboptimal chilling period was mediated primarily through maternally derived tissues, whereas germination speed during prolonged chilling for both intact seeds and excised embryos was controlled at the embryo level. There were large genetically based differences among the nine individuals representing three populations in this study both as maternal and paternal parents, but these differences were not strongly correlated with habitat at the site of origin.

Other shrubs

Seed germination syndromes for several other shrub species in our region have been examined based on wild-collected seeds from a range of habitats. These include fourwing saltbush (*Atriplex canescens*, Meyer and Carlson, 2007), spiny hopsage (*Grayia spinosa*, Shaw and Haferkamp, 1994), spineless hopsage (*Zuckia brandegeei*, syn. *Grayia brandegeei*, Meyer and Pendleton, 1989), winterfat (*Krascheninnikovia lanata*, syn. *Eurotia lanata*, Moyer and Lang, 1976), and both curl-leaf and alder-leaf mountain mahogany (*Cercocarpus ledifolius* and *Cercocarpus montanus*, respectively, Kitchen and Meyer, 1991). In each case, apparently adaptive winter climate-correlated variation in seed germination syndrome was detected.

Native perennial grasses

Bottlebrush squirreltail (Elymus elymoides)

In contrast to native shrubs, most of which emerge and establish in spring, perennial grasses in our region are often more opportunistic in their germination phenology. Many spring and summer-flowering species are facultatively fall-emerging. These species may exhibit population differences, but they are usually not closely tied to variation in winter conditions. For example, bottlebrush squirreltail seeds demonstrated habitat-correlated differences in primary dormancy, which was higher in populations from lower elevation sites, where seeds ripen earlier in the summer and reliable autumn rains arrive later, than at higher elevation sites, where seeds ripen later in the summer (Beckstead et al., 1995). At the low elevation sites, a long after-ripening requirement likely functions to decrease the risk of premature germination in response to inadequate summer rainfall events. Local adaptation for regulation of germination phenology has also been reported for the related species big squirreltail (*Elymus multisetus*, Rowe and Leger, 2012).

Basin wildrye (Leymus cinereus)

This widely distributed bunchgrass exhibits a different pattern of apparent local adaptation. Although its seeds are relatively nondormant at dispersal, they germinate slowly and after-ripen very little under summer conditions (Meyer et al., 1995b). Field retrieval experiments demonstrated that slow germination makes them unlikely to emerge before winter in spite of lack of dormancy. Germination speed under winter conditions was strongly correlated with winter temperature and probable duration of winter snowpack. As for many of the shrubs discussed above, this is a mechanism for timing germination appropriately in early spring.

Bluebunch wheatgrass (Pseudoroegneria spicata)

Regulation of germination phenology was included by Kitchen and Monsen (1994) in a study of adaptive traits related to seedling vigor in a large germplasm collection of bluebunch wheatgrass. These authors did not formally examine the relationship between collection site climate and low-temperature germination speed, but they stated that the regional differences they observed in collections from across eight western states seemed to reflect climate patterns and that these differences might be related to the relative risk of winter frost and spring drought in different habitats.

Native herbaceous dicots

Lewis flax

One of the most widely distributed herbaceous perennial dicots in our region and throughout much of western North America is Lewis flax (*Linum lewisii*, syn. *Linum perenne* ssp. *lewisii*). A closely related species, *Linum perenne*, is native to northern Europe. We examined seed germination regulation for wild-produced and common-garden-produced seeds of 21 populations from habitats ranging from steppe and pinyon-juniper communities to montane meadows (Meyer and Kitchen, 1994b). We also included the cultivar "Appar," thought to be derived from native Lewis flax but later determined to be of European origin (Pendleton et al., 2008). The native populations had almost the full array of seed germination regulation mechanisms, including lack of dormancy, primary dormancy removed by after-ripening or by chilling, secondary dormancy induced by warm stratification or by chilling, and the presence of a chilling-nonresponsive fraction of variable size. The predicted net result of this complex set of mechanisms is appropriately timed spring germination coupled with a fraction that could carry over as a persistent seed bank in the habitat at the site of seed origin. The cultivar "Appar" did not have any of these regulatory mechanisms and germinated at the first opportunity in retrieval experiments at three contrasting sites, while seeds from mid- and high-elevation collections of native Lewis flax showed contrasting field germination patterns in different habitats predicted from laboratory experiments.

Common-garden progeny of 16 populations showed patterns of response similar to seeds of their parents but were generally less dormant, possibly because of inadvertent selection during propagation (Meyer and Kitchen, 1994b). There were major differences in germination response among the eight sibships representing each parental population, often larger than mean population differences, indicating a high level of within-population genetic variation for seed germination traits and consequently a high potential for selection, whether natural or artificial. The exception was "Appar," which showed no among-sibship variation, with either full germination or no germination in response to a given treatment. This likely represents the end-point of selection and explains why this cultivar, though easy to establish from seed, consistently fails to persist long-term in wildland seedings. The wealth of genetic variation for seed germination traits in Lewis flax may provide a buffer against extinction in the face of climate change.

Other herbaceous dicots

Several other genera of herbaceous dicots in our region demonstrate climate-correlated variation in the seed germination syndrome, including *Eriogonum* (Meyer and Paulsen, 2000), *Allium* (Phillips et al., 2010), *Erythronium* (Baskin et al., 1995), *Osmorhiza* (Baskin et al., 1995), and *Castilleja* (Meyer and Carlson, 2004). In contrast, Barga et al. (2017) failed to detect climate-correlated among-population variation in germination strategy in a study of 10 Great Basin herbaceous dicots, possibly because of the small number of populations used for each species (2–3) or the small range of environmental variation across collection sites. Seglias et al. (2018) used multivariate analysis of a complex experimental design and found a weak relationship between germination response in warm or cold stratification and precipitation seasonality for seven of eight herbaceous dicot species (2–4 populations per species) as well as weak relationships between temperature variables and overall bioclimatic variables with final germination across all treatments. These results are difficult to interpret but suggest that studies with more populations of these species might yield more definitive results.

An invasive annual grass

Cheatgrass or downy brome (*Bromus tectorum*) is the one invasive nonnative plant species in our region for which we have extensive information on genetic variation in the germination syndrome. This highly invasive facultative winter annual grass was introduced to our region from Eurasia before the turn of the 20th century (Mack, 1981) and has established postfire disclimax monocultures over large areas in our region (D'Antonio and Vitousek, 1992). This highly inbreeding species is comprised of numerous lineages that often possess contrasting germination syndromes that preadapt them to different climate environments in the invaded North American range (Meyer et al., 2013, 2016; Arnesen et al., 2017). Seeds are dispersed in early to mid-summer and after-ripen under dry conditions to become germinable in time for the first autumn or winter rains (Meyer et al., 1997). They exhibit conditional dormancy that protects them from summer germination, but after-ripening during summer renders them increasingly able to germinate at relatively warm autumn temperatures. Laboratory and field studies of warm desert, semidesert, and montane lineages showed that adaptively significant differences in initial dormancy and after-ripening rate were linked to climate differences at the site of seed origin. The late-ripening montane lineage showed low dormancy at dispersal, while the early-ripening warm desert lineage, which faces a prolonged hot summer with risk of monsoonal rain, was the most dormant at high temperature and the slowest to after-ripen (Bauer et al., 1998). These differences were later demonstrated to be genetically based in a common environment study (Meyer and Allen, 1999a). In addition, this latter study demonstrated that these lineages exhibited a plastic response to water stress during seed maturation, with increased water stress associated with more rapid maturation and decreased dormancy. The degree of phenotypic plasticity with regard to response to water stress during ripening was shown to be genetically based and lineage-dependent in these experiments (Meyer and Allen, 1999b).

The warm desert lineage used in these experiments was later demonstrated through population genetic studies to be confined to warm deserts in North America. It is the dominant lineage throughout the Mojave Desert and also occurs northward into the fringe area between the western Mojave and Great Basin deserts (Meyer et al., 2016). An establishment experiment at semidesert sites in western Nevada and central Utah showed that this lineage could not produce enough seeds to replace itself at the colder Utah site due to both poor emergence and survival and low seed production. It was surprisingly successful at the warmer western Nevada site. Another key difference between this warm desert lineage and common lineages of the Great Basin is that the warm desert lineage can flower without vernalization, which is clearly a preadaptation to warm desert that might place it at a phenological disadvantage at a colder-winter site (Meyer et al., 2004).

Potential changes in the geographic distribution of cheatgrass in response to global warming have been extensively examined using bioclimatic envelope modeling (Bradley 2009; Bradley et al., 2009; reviewed in Bradley et al., 2016). These models suggest that there will be loss of habitat for this species at the southern edge of its distribution in our region, which could create restoration opportunities in habitat that has become unsuitable. However, these models treat cheatgrass as a uniform entity and fail to take into account the strong ecotypic differentiation that we have observed in regulation of both flowering and seed dormancy loss. It seems more likely that warm desert-adapted ecotypes will be able to move northward as habitat is rendered unsuitable for Great Basin steppe-adapted ecotypes, particularly in the Lahontan Basin of western Nevada, where they already appear to be moving northward. Steppe ecotypes may in turn be able to move upward into areas now occupied by montane ecotypes. Cheatgrass is highly dispersable through the human agency, so it is plausible that these preadapted ecotypes will find their way to suitable habitat that they will quickly exploit. Although its population genetic structure is very different from that of most native species, cheatgrass possesses ample among-population and among-lineage genetic variation for seed dormancy and flowering regulation. This, coupled with its high dispersability, suggests that it will be able to cope with global warming both through in situ selection and through migration to adjacent habitats as they become suitable.

Conclusions and recommendations for future research

Many widely distributed species in our region have evolved contrasting regeneration strategies in response to diversifying selection across steep climatic gradients. Furthermore, many populations with adaptively advantageous mean germination responses to local climate can harbor considerable among-plant genetic diversity for relevant germination traits. This is likely because extreme and unpredictable interannual variation in weather patterns in semiarid environments leads to temporally varying selection on regeneration strategy, maintaining a range of genotypes that can cope with a range of contrasting conditions. Lastly, germination syndromes are not under rigid genetic control and can also respond to changes in local conditions through phenotypic plasticity. These features may preadapt species of our region to be better able to cope with climate change than species of mesic environments with more predictable weather patterns.

Future research should focus on the use of the many common gardens already in existence in our region for screening native plant populations for post-establishment traits. Laboratory germination and field retrieval experiments with seeds from multiple populations collected at gardens in contrasting environments could confirm and extend the findings outlined above. Establishment experiments in contrasting habitats with these garden-collected seeds could determine the potential ecological consequences of this variation in regeneration strategy in current and future climate environments.

References

Abatzoglou, J.T., Kolden, C.A., 2011. Climate change in western US deserts: potential for increased wildfire and invasive annual grasses. Rangel. Ecol. Manage 64, 471–478.

Araújo, M.B., Peterson, A.T., 2012. Uses and misuses of bioclimatic envelope modeling. Ecology 93, 1527–1539.

Arnesen, S., Coleman, C.E., Meyer, S.E., 2017. Population genetic structure of *Bromus tectorum* in the mountains of western North America. Am. J. Bot. 104, 879–890.

Auger, J., Meyer, S.E., Jenkins, S.H., 2016. A mast-seeding desert shrub regulates population dynamics and behavior of its heteromyid dispersers. Ecol. Evol. 6, 2275–2296.

Barga, S., Dilts, T.E., Leger, E.A., 2017. Climate variability affects the germination strategies exhibited by arid land plants. Oecologia 185, 437–452.

Baskin, C.C., Meyer, S.E., Baskin, J.M., 1995. Two types of morphophysiological dormancy in seeds of two genera (*Osmorhiza* and *Erythronium*) with an Arcto-Tertiary distribution pattern. Am. J. Bot. 82, 293–298.

Bauer, M.C., Meyer, S.E., Allen, P.S., 1998. A simulation model to predict seed dormancy loss in the field for *Bromus tectorum* L. J. Exp. Bot. 49, 1235–1244.

Baughman, O.W., Agneray, A.C., Forister, M.L., Kilkenny, F.F., Espeland, E.K., Fiegener, R., et al., 2019. Strong patterns of intraspecific variation and local adaptation in Great Basin plants revealed through a review of 75 years of experiments. Ecol. Evol. 9, 6259–6275.

Beckstead, J., Meyer, S.E., Allen, P.S., 1995. Effects of afterripening on cheatgrass (*Bromus tectorum*) and squirreltail (*Elymus elymoides*) germination. In: Roundy, B.A., McArthur, E.D., Haley, J.S., Manu, D.K. (comps.) (Eds.), Proceedings Wildland Shrub and Arid Land Restoration Symposium. General Technical Report INT-GTR-315, U.S. Department of Agriculture. Forest Service. Intermountain Research Station, Ogden, UT, pp. 165–172.

Bradley, B.A., 2009. Regional analysis of the impacts of climate change on cheatgrass invasion shows potential risk and opportunity. Glob. Change Biol. 15, 196–208.

Bradley, B.A., Curtis, C.A., Chambers, J.C., 2016. Bromus response to climate and projected changes with climate change. In: Germino, M.J., Chambers, J.C., Brown, C.S. (Eds.), Exotic Brome-Grasses in Arid and Semiarid Ecosystems of the Western US. Springer, Berlin, pp. 257–274.

Bradley, B.A., Oppenheimer, M., Wilcove, D.S., 2009. Climate change and plant invasions: restoration opportunities ahead? Glob. Change Biol. 15, 1511–1521.

Chaney, L., Richardson, B.A., Germino, M.J., 2017. Climate drives adaptive genetic responses associated with survival in big sagebrush (*Artemisia tridentata*). Evol. Appl. 10, 313–322.

Cochrane, A., Yates, C.J., Hoyle, G.L., Nicotra, A.B., 2015. Will among-population variation in seed traits improve the chance of species persistence under climate change? Glob. Ecol. Biogeogr. 24, 12–24.

Diffenbaugh, N.S., Field, C.B., 2013. Changes in ecologically critical terrestrial climate conditions. Science 341, 486–492.

Donat, M.G., Lowry, A.L., Alexander, L.V., O'Gorman, P.A., Maher, N., 2016. More extreme precipitation in the world's dry and wet regions. Nat. Clim. Change 6, 508–513.

D'Antonio, C.M., Vitousek, P.M., 1992. Biological invasions by exotic grasses, the grass/fire cycle, and global change. Annu. Rev. Ecol. Syst. 23, 63–87.

Fenneman, N.M., 1930. Physical divisions of the United States: U.S. Geological Survey map, scale 1:7,000,000, 1931, Physiography of Western United States. U.S. Geological Survey, Washington, D.C.

Garvin, S.C., Meyer, S.E., 2003. Multiple mechanisms for seed dormancy regulation in shadscale (*Atriplex confertifolia*: Chenopodiaceae). Can. J. Bot. 81, 601–610.

Hunter, K.L., McAuliffe, J.R., 1994. Elevational shifts of *Coleogyne ramossissima* in the Mojave Desert during the Little Ice Age. Quat. Res. 42, 216–221.

Kitchen, S.G., Meyer, S.E., Carlson, S.L., 2015. Mechanisms for maintenance of dominance in a nonclonal desert shrub. Ecosphere 6, 252. Available from: https://doi.org/10.1890/ES15-00083.1.

Kitchen, S.G., Meyer, S.E., 1991. Seed dormancy in two species of mountain mahogany (*Cercocarpus ledifolius* and *Cercocarpus montanus*). In: Johnson, K.L. (Ed.), Proceeding Fifth Utah Shrub Ecology Workshop. The genus Cercocarpus, Utah State University, Logan, pp. 27–42.

Kitchen, S.G., Monsen, S.B., 1994. Germination rate and emergence success in bluebunch wheatgrass. J. Range Manage. 47, 145–150.

Mack, R.N., 1981. Invasion of *Bromus tectorum* L. into western North America: an ecological chronicle. Agro-ecosystems 7, 145–165.

Maloney, E.D., Camargo, S.J., Chang, E., Colle, B., Fu, R., Geil, K.L., et al., 2014. North American climate in CMIP5 experiments: Part III: assessment of twenty-first-century projections. J. Clim. 27, 2230–2270.

McArthur, E.D., Meyer, S.E., Weber, D.J., 1987. Germination rate at low temperature: rubber rabbitbrush population differences. J. Range Manage. 40, 530–533.

Meyer, S.E., 1989. Warm pretreatment effects on antelope bitterbrush (*Purshia tridentata*) germination response to chilling. Northw. Sci. 63, 146–153.

Meyer, S.E., 1990. Seed source differences in germination under snowpack in northern Utah. In: Fifth Billings Symposium on Disturbed Land Rehabilitation. Volume I. Montana State University Reclamation Research Unit Publication No. 9003. Bozeman, MT, pp. 184–191.

Meyer, S.E., 1992. Habitat correlated variation in firecracker penstemon (*Penstemon eatonii* Gray: Scrophulariaceae) seed germination response. Bull. Torrey Bot. Club 119, 268–279.

Meyer, S.E., 1997. Ecological correlates of achene mass variation in *Chrysothamnus nauseosus* (Asteraceae). Am. J. Bot. 84, 471–477.

Meyer, S.E., Allen, P.S., 1999a. Ecological genetics of seed germination regulation in *Bromus tectorum* L. I. Phenotypic variance among and within populations. Oecologia 120, 27–34.

Meyer, S.E., Allen, P.S., 1999b. Ecological genetics of seed germination regulation in *Bromus tectorum* L. II. Reaction norms in response to a water stress gradient imposed during seed maturation. Oecologia 120, 35–43.

Meyer, S.E., Allen, P.S., Beckstead, J., 1997. Seed germination regulation in *Bromus tectorum* (Poaceae) and its ecological significance. Oikos 78, 475–485.

Meyer, S.E., Carlson, S.L., 2001. Achene mass variation in *Ericameria nauseosus* (Asteraceae) in relation to dispersal ability and seedling fitness. Funct. Ecol. 15, 274–281.

Meyer, S.E., Carlson, S.L., 2004. Comparative seed germination biology and seed propagation of eight intermountain species of Indian paintbrush. In: Hild, A.L., Shaw, N.L., Meyer, S.E., Booth, D.T., McArthur, E.D. (comps.), Seed and Soil Dynamics in Shrubland Ecosystems: Proceedings; 2002 August 12–16; Laramie, WY. Proceedings RMRS-P-31. U.S. Department of Agriculture, Forest Service, Rocky Mountain Research Station, Ogden UT, pp. 125–130.

Meyer, S.E., Carlson, S.L., 2007. Seed germination biology of Intermountain populations of fourwing saltbush (Atriplex canescens: Chenopodiaceae). In: McArthur, E.D., Kitchen, S.G. (comps.) Proceedings: Shrubland Dynamics—Fire and Water. TX: RMRS P-47. USDA Forest Service, Rocky Mountain Research Station, Fort Collins, CO, pp. 153–162.

Meyer, S.E., Carlson, S.L., Garvin, S.C., 1998. Seed germination regulation and field seed bank carryover in shadscale (*Atriplex confertifolia*: Chenopodiaceae). J. Arid. Environ. 38, 255–267.

Meyer, S.E., Ghimire, S., Decker, S., Merrill, K.R., Coleman, C.E., 2013. The ghost of outcrossing past in downy brome, an inbreeding annual grass. J. Hered. 104, 476–490.

Meyer, S.E., Kitchen, S.G., 1992. Cyclic seed dormancy in the short-lived perennial *Penstemon palmeri*. J. Ecol. 80, 115–122.

Meyer, S.E., Kitchen, S.G., 1994a. Habitat-correlated variation in seed germination response to chilling in *Penstemon* Section *Glabri* (Scrophulariaceae). Am. Midl. Nat. 32, 349–365.

Meyer, S.E., Kitchen, S.G., 1994b. Life history variation in blue flax (*Linum perenne*: Linaceae): seed germination phenology. Am. J. Bot. 81, 528–535.

Meyer, S.E., Kitchen, S.G., Carlson, S.L., 1995a. Seed germination timing patterns in Intermountain *Penstemon* (Scrophulariaceae). Am. J. Bot. 82, 377–389.

Meyer, S.E., Beckstead, J., Allen, P.S., Pullman, H., 1995b. Germination ecophysiology of *Leymus cinereus* (Poaceae). Int. J. Plant Sci. 156, 206–215.

Meyer, S.E., Leger, E.A., Eldon, D.R., Coleman, C.E., 2016. Strong genetic differentiation in the invasive annual grass *Bromus tectorum* across the Mojave–Great Basin ecological transition zone. Biol. Inv. 18, 1611–1628.

Meyer, S.E., McArthur, E.D., 1987. Studies on the seed germination biology of rubber rabbitbrush. In: Johnson, K.L. (Ed.), Proceedings of the Fourth Utah Shrub Ecology Workshop. Utah State University, Logan, pp. 19–25.

Meyer, S.E., McArthur, E.D., Jorgensen, G.L., 1989. Variation in germination response to temperature in rubber rabbitbrush (*Chrysothamnus nauseosus*: Asteraceae) and its ecological implications. Am. J. Bot. 76, 981–991.

Meyer, S.E., Monsen, S.B., 1989. Seed germination biology of antelope bitterbrush (*Purshia tridentata*). In: Wallace, A., McArthur, E.D., Haferkamp, M.R. (Eds.), Proceedings-Symposium on Shrub Ecophysiology and Biotechnology. USDA Forest Service. General Technical Report INT-256. Intermountain Research Station, Ogden UT, pp. 147–157.

Meyer, S.E., Monsen, S.B., 1990. Seed source differences in initial establishment for big sagebrush and rubber rabbitbrush. In: McArthur, E.D., Romney, E.M., Smith, S.D., Tueller, P.T. (comps.), Proceedings - Symposium on Cheatgrass Invasion, Shrub Die-Off, and Other Aspects of Shrub Biology and Management, Las Vegas, NV, April 5–7, 1989. USDA Forest Service. Intermountain Research Station General Technical Report INT-276., Ogden, UT, pp. 200–208.

Meyer, S.E., Monsen, S.B., McArthur, E.D., 1990. Germination response of *Artemisia tridentata* (Asteraceae) to light and chill: patterns of between-population variation. Bot. Gaz. 151, 176–183.

Meyer, S.E., Monsen, S.B., 1991. Habitat-correlated variation in mountain big sagebrush (*Artemisia tridentata* ssp. *vaseyana*) seed germination patterns. Ecology 72, 739–742.

Meyer, S.E., Monsen, S.B., 1992. Big sagebrush germination patterns: subspecies and population differences. J. Range Manage. 45, 87–93.

Meyer, S.E., Nelson, D.L., Carlson, S.L., 2004. Ecological genetics of vernalization response in *Bromus tectorum* L. (Poaceae). Ann. Bot. 93, 653–663.

Meyer, S.E., Paulsen, A., 2000. Chilling requirements for seed germination of 10 Utah species of perennial wild buckwheat (*Eriogonum* Michx. [Polygonaceae]). Native Plants J. 1, 18–24.

Meyer, S.E., Pendleton, B.K., 2005. Factors affecting seed germination and seedling establishment of a long-lived desert shrub (*Coleogyne ramosissima*: Rosaceae). Plant Ecol. 178, 171–187.

Meyer, S.E., Pendleton, R.L., 1989. Seed germination biology of spineless hopsage: between-population differences in dormancy and response to temperature. In: McArthur, E.D., Romney, E.M., Smith, S.D., Tueller, P.T. (comps.) Proceedings - Symposium on Cheatgrass Invasion, Shrub Die Off, and Other Aspects of Shrub Biology and Management, Las Vegas, NV, April 5–7, 1989. USDA Forest Service Intermountain Research Station General Technical Report INT-276, Ogden, UT, pp. 187–192.

Meyer, S.E., Pendleton, R.L., 2000. Genetic regulation of seed dormancy in *Purshia tridentata* (Rosaceae). Ann. Bot. 85, 521–529.

Meyer, S.E., Quinney, D., Weaver, J., 2006. A stochastic population model for *Lepidium papilliferum* (Brassicaceae), a rare desert ephemeral with a persistent seed bank. Am. J. Bot. 93, 891–902.

Moyer, J.L., Lang, R.L., 1976. Variable germination response to temperature for different sources of winterfat seed. J. Range Manage. 29, 320–321.

O'Gorman, P.A., 2015. Precipitation extremes under climate change. Curr. Clim. Change Rep. 1, 49–59.

Pendleton, R.L., Kitchen, S.G., Mudge, J., McArthur, E.D., 2008. Origin of the flax cultivar 'Appar' and its position within the *Linum perenne* complex. Int. J. Plant Sci. 169, 445–453.

Pendleton, B.K., Meyer, S.E., 2004. Habitat-correlated variation in blackbrush (*Coleogyne ramosissima*: Rosaceae) seed germination response. J. Arid. Environ. 59, 229–243.

Phillips, N.C., Drost, D.T., Varga, W.A., Shultz, L.M., Meyer, S.E., 2010. Germination characteristics along altitudinal gradients in three Intermountain *Allium* spp.(Amaryllidaceae). Seed Technol. 32, 15–25.

Polley, H.W., Briske, D.D., Morgan, J.A., Wolter, K., Bailey, D.W., Brown, J.R., 2013. Climate change and North American rangelands: trends, projections, and implications. Rangel. Ecol. Manage. 66, 493–511.

Richardson, B.A., Chaney, L., 2018. Climate-based seed transfer of a widespread shrub: population shifts, restoration strategies, and the trailing edge. Ecol. Appl 28, 2165–2174.

Richardson, B.A., Kitchen, S.G., Pendleton, R.L., Pendleton, B.K., Germino, M.J., Rehfeldt, G.E., et al., 2014. Adaptive responses reveal contemporary and future ecotypes in a desert shrub. Ecol. Appl. 24, 413–427.

Richardson, B.A., Meyer, S.E., 2012. Paleoclimate effects and geographic barriers shape regional population genetic structure of blackbrush (*Coleogyne ramosissima*: Rosaceae). Botany 90, 293–299.

Rowe, C.L., Leger, E.A., 2012. Seed source affects establishment of *Elymus multisetus* in postfire revegetation in the Great Basin. West. N. Am. Nat. 72, 543–553.

Schlaepfer, D.R., Taylor, K.A., Pennington, V.E., Nelson, K.N., Martyn, T.E., Rottler, C.M., et al., 2015. Simulated big sagebrush regeneration supports predicted changes at the trailing and leading edges of distribution shifts. Ecosphere 6, 3. Available from: https://doi.org/10.1890/ES14-00208.1.

Seglias, A.E., Williams, E., Bilge, A., Kramer, A.T., 2018. Phylogeny and source climate impact seed dormancy and germination of restoration-relevant forb species. PLoS One 13, e0191931. Available from: https://doi.org/10.1371/journal.pone.0191931.

Shaw, N.L., Haferkamp, M.R.Service, 1994. Spiny hopsage seed germination and seedling establishment. In: Kitchen, S.G., Monsen, S.B. (Eds.), Proceedings, Ecology and Management of Annual Rangelands. In: Intermountain Research Station Gen. Tech. Rep., INT-GTR-313. US Department of Agriculture, Forest, Ogden, UT, pp. 252–256.

Still, S.M., Richardson, B.A., 2015. Projections of contemporary and future climate niche for Wyoming big sagebrush (*Artemisia tridentata* subsp. *wyomingensis*): a guide for restoration. Nat. Areas J. 35, 30–43.

Tilini, K.L., Meyer, S.E., Allen, P.S., 2016. Breaking primary seed dormancy in Gibbens' beardtongue (*Penstemon gibbensii*) and blowout penstemon (*Penstemon haydenii*). Native Plants J. 17, 256–266.

Walck, J.L., Hidayati, S.N., Dixon, K.W., Thompson, K., Poschlod, P., 2011. Climate change and plant regeneration from seed. Glob. Change Biol. 17, 2145–2161.

Chapter 6

Effect of climate change on plant regeneration from seeds in steppes and semideserts of northern China

Xuejun Yang, Gaohua Fan, and Zhenying Huang
State Key Laboratory of Vegetation and Environmental Change, Institute of Botany, Chinese Academy of Sciences, Beijing, China

Introduction

The major biomes in northern China are temperate steppes (mainly in eastern and central Inner Mongolia), (semi-)deserts (from western Inner Mongolia to the west), and taiga (northern Heilongjiang Province). This area also has two large croplands, namely, the Songliao Plain and North China Plain. Steppes and semideserts, which are the focus of this chapter, are mainly located in Inner Mongolia (for a map of vegetation in China, see http://www.nsii.org.cn/mapvege). According to the Fifth Assessment Report of the Intergovernmental Panel on Climate Change (IPCC), this region has been experiencing a warming trend (including increase in number of warm days and decrease in number of cold days) and increasing temperature extremes over the past century. This warming trend is predicted to continue into the new millennium (Hijioka et al., 2014).

In this chapter, we review studies on the effect of climate change (warming, changes in precipitation patterns, nitrogen deposition, and salt stress) on plant regeneration from seeds in steppes and semideserts in northern China. Rising temperatures of the projected warming climate are likely to have different effects on population regeneration of different species by affecting seedling emergence in the local community. In addition, changes in soil moisture content resulting from climate warming and change in precipitation patterns will impact seed dormancy break and germination.

Projected climate changes and effects on natural ecosystems in northern China

Changes in mean annual temperature in northern China are predicted to exceed 2°C by the mid-21st century and 4°C − 6°C in the late-21st century above the late-20th-century baseline. Across northern China, precipitation is very likely to increase by the mid- and late-21st century, but this trend is characterized by strong variability (Hijioka et al., 2014).

The projected climate change during the 21st century is predicted to modify the distribution of vegetation across Asia (Tao and Zhang, 2010; Wang, 2014). However, limitations on seed dispersal, competition from established plants, rates of soil development, and habitat fragmentation will modify these responses (Corlett and Westcott, 2013). Responses of terrestrial systems to recent climate warming are reflected in shifts in phenology toward an earlier spring greening, a longer growing season, higher plant growth rates, and changes in the distribution of plant species, and the projected climate changes will increase these impacts (Hadano et al., 2013; Richardson et al., 2013; Ge et al., 2014). In semiarid and arid ecosystems, large changes also may occur due to changes in precipitation, but uncertainties in precipitation projections make these difficult to predict (Liancourt et al., 2012; Poulter et al., 2013).

In steppe vegetation, the future direction and rate of climate change are unclear because of the uncertainty in precipitation trends (Tchebakova et al., 2010; Hijioka et al., 2014). Steppe productivity in northern China already has been affected by climate change. For example, CENTURY ecosystem models show that typical and meadow steppes in Inner Mongolia are very sensitive to climate change, which will lead to the loss of soil organic C and a decrease of annual above-ground net primary production (ANPP) (Xiao et al., 1995). The dynamic land ecosystem model has shown that both total growing season precipitation and average temperature exert important controls on ANPP across grassland sites over a precipitation gradient in Mongolia (Dangal et al., 2016). Similarly, the annual mean NPP of the Hulunbuir Steppe in northern China is significantly positively correlated with the annual variation of precipitation (Zhang et al., 2019a,b). An analysis of the productivity of meadow and typical and desert steppes on the Inner Mongolian Plateau between 2011 and 2013 shows a significant positive correlation between precipitation and the ANPP of the three types of grasslands, and temperature has a significant impact on the ANPP of desert steppes (Su et al., 2020). In the Mu Us Desert of northern China, nitrogen (N) enrichment and precipitation addition significantly increased herbaceous productivity by improving water content and nutrient availability in the soil, but these effects were modified by the presence of shrubs (Bai et al., 2020). Climate change also affects functional traits of steppe species. As an example, functional traits (leaf area, specific leaf area, and root/shoot ratio) of four zonal *Stipa* species are sensitive to changing temperature and precipitation (Lv et al., 2016).

Climate change has led to steppe degradation in northern China. An in-depth empirical analysis of steppe degradation at the Xilinhot Plateau revealed that both natural and man-made causes had a significant influence on steppe degradation in this typical steppe region over the last 20 years. Measures taken by the government, such as fencing vulnerable areas, have played an important role in stopping this change (Li et al., 2012a). Steppe degradation also can have some feedback effects on regional climate change. The model-based analysis on grasslands of Mongolia shows that future grassland degradation will result in a significant change in regional climate. Thus, future grassland degradation could lead to an increasing trend in temperature in most areas and will cause a decreasing trend of precipitation (Zhang et al., 2013). Climate change also is changing the steppe community structure in northern China. For example, the increase in temperature has led to the invasion of shrub species into steppe grassland in arid and semiarid areas of the Mongolian Plateau, which may be intensified by future global warming (Zheng et al., 2019). Therefore, steppe degradation in this region caused by climate change is likely to be intensified in the future and will impair the functioning and services of the steppe ecosystems.

Effects of projected climate changes in northern China on regeneration of plants from seeds

Temperature and precipitation are critical for seed dormancy break and germination (radicle emergence) (Walck et al., 2011). Compared to adult stages, seed dormancy and germination are expected to be more sensitive to climate change and more critical for plant population recruitment (Lloret et al., 2004). Therefore, climate change (alterations in temperature and precipitation) will affect recruitment of plants and subsequent population dynamics in northern China.

Germination requirements vary greatly among species. For example, species with nondormant seeds can be germinated soon after dispersal, while species with dormant seeds delay germination to a period in which the probability for seedling survival and reproduction of adult plants is high (Baskin and Baskin, 2014). Therefore, responses of germination in a community to projected climate change depend on the species. Species-specific germination responses to environmental conditions are important in structuring the desert steppe community and have implications for predicting community structure under climate change in northern China. For example, although most species in the Siziwang Desert Steppe of Inner Mongolia are well-adapted to the desert by the ability to germinate in a habitat characterized by cold and dry conditions, the predicted warmer and dryer climate will favor germination of drought-tolerant species, such as *Artemisia scoparia*, *Bassia dasyphylla*, *Kochia prostrata*, and *Potentilla tanacetifolia*, thus altering the proportions of germinants of different species and subsequently changing community composition of the desert steppe (Yi et al., 2019).

As such, then, climate change could have differential effects on germination of species with various kinds and degrees of seed dormancy (Walck et al., 2011), which makes future dynamics of plant communities in this region under climate change difficult to predict. To understand such effects in northern China, some researchers have studied temperature and precipitation change on seed dormancy and germination.

Effect of climate warming

In northern China, various studies have examined the effect of increasing temperature on seed germination (Table 6.1). Seed germination of some species is high over a wide range of temperatures, thus a warming temperature is unlikely to significantly affect germination of these species. For example, an experiment in the Minqin Sandy Botanical Garden in Gansu Province found that the optimal temperature for germination of *Picea mongolica* seeds was 15°C–25°C and that the seeds did not germinate at temperature lower than 5°C (Zhu et al., 2011). Some species showed a decrease in germination percentage at low temperatures, and thus a warming climate may promote seed germination for these species.

TABLE 6.1 Studies showing effect of variable temperature regimes on seed germination of species in northern China.

Ecosystem/location	Species	Test temperatures	Response	References
Desert steppe				
Siziwang, IM	Amaranthus retroflexus	0/12°C, 7/19°C, 12/24°C, 15/27°C	↑	Yi et al. (2019)
Siziwang, IM	Allium polyrhizum	0/12°C, 7/19°C, 12/24°C, 15/27°C	↑	Yi et al. (2019)
Siziwang, IM	Allium ramosum	0/12°C, 7/19°C, 12/24°C, 15/27°C	↑	Yi et al. (2019)
Siziwang, IM	Allium tenuissimum	0/12°C, 7/19°C, 12/24°C, 15/27°C	↑	Yi et al. (2019)
Siziwang, IM	Artemisia annua	0/12°C, 7/19°C, 12/24°C, 15/27°C	↑	Yi et al. (2019)
Siziwang, IM	Artemisia frigida	0/12°C, 7/19°C, 12/24°C, 15/27°C	↑	Yi et al. (2019)
Siziwang, IM	Artemisia mongolica	0/12°C, 7/19°C, 12/24,°C 15/27°C	↑	Yi et al. (2019)
Siziwang, IM	Artemisia scoparia	0/12°C, 7/19°C, 12/24°C, 15/27°C	↑	Yi et al. (2019)
Siziwang, IM	Artemisia sieversiana	0/12°C, 7/19°C, 12/24°C, 15/27°C	↑	Yi et al. (2019)
Siziwang, IM	Bassia dasyphylla	0/12°C, 7/19°C, 12/24°C, 15/27°C	↑	Yi et al. (2019)
Siziwang, IM	Haplophyllum dauricum	0/12°C, 7/19°C, 12/24°C, 15/27°C	↑	Yi et al. (2019)
Siziwang, IM	Heteropappus altaicus	0/12°C, 7/19°C, 12/24°C, 15/27°C	↑	Yi et al. (2019)
Siziwang, IM	Kochia prostrata	0/12°C, 7/19°C, 12/24°C, 15/27°C	↑	Yi et al. (2019)
Siziwang, IM	Lagochilus ilicifolium	0/12°C, 7/19°C, 12/24°C, 15/27°C	↑	Yi et al. (2019)
Siziwang, IM	Linum stelleroides	0/12°C, 7/19°C, 12/24°C, 15/27°C	↑	Yi et al. (2019)
Siziwang, IM	Neopallasia pectinate	0/12°C, 7/19°C, 12/24°C, 15/27°C	↑	Yi et al. (2019)
Siziwang, IM	Plantago depressa	0/12°C, 7/19°C, 12/24°C, 15/27°C	↑	Yi et al. (2019)
Siziwang, IM	Potentilla multicaulis	0/12°C, 7/19°C, 12/24°C, 15/27°C	↑	Yi et al. (2019)
Siziwang, IM	Potentilla tanacetifolia	0/12°C, 7/19°C, 12/24°C, 15/27°C	↑	Yi et al. (2019)
Siziwang, IM	Stipa breviflora	0/12°C, 7/19°C, 12/24°C, 15/27°C	↑	Yi et al. (2019)
Semidesert				
Horqin Sandy Land, IM	Artemisia halodendron	10°C, 16°C, 22°C, 28°C, 34°C	↔	Li et al. (2012b)
Horqin Sandy Land, IM	Artemisia scoparia	10°C, 16°C, 22°C, 28°C, 34°C	↔	Li et al. (2012b)
Horqin Sandy Land, IM	Artemisia sieversiana	10°C, 16°C, 22°C, 28°C, 34°C	↔	Li et al. (2012b)
Horqin Sandy Land, IM	Artemisia wudanica	10°C, 16°C, 22°C, 28°C, 34°C	↔	Li et al. (2012b)
Hulun Buir and Bayin Oboo, IM	Picea mongolica	5°C, 10°C, 15°C, 20°C, 25°C, 30°C, 35°C, 40°C	↔	Li et al. (2011c)
Hulun Buir and Bayin Oboo, IM	Pinus sylvestris	5°C, 10°C, 15°C, 20°C, 25°C, 30°C, 35°C, 40°C	↔	Li et al. (2011c)
Minqin, Gansu Province	Ammopiptanthus mongolicus	10°C, 15°C, 20°C, 25°C, 30°C, 35°C, 40°C	↑	Li et al. (2011a)
Minqin, Gansu Province	Picea mongolica	10°C, 15°C, 20°C, 25°C, 30°C, 35°C, 40°C	↑	Zhu et al. (2011)
Mu Us Sandland, IM	Agriophyllum squarrosum	5/15°C, 10/20°C, 15/25°C, 20/30°C	↑	Gao et al. (2018)
Ordos Plateau, IM	Atriplex centralasiatica	5/15°C, 10/20°C, 15/25°C, 20/30°C	↔	Wang et al. (2020a,b)
Ordos Plateau, IM	Caragana korshinskii	5:15°C, 10:20°C, 15:25°C, 20:30°C, 25:35°C	↑	Lai et al. (2015)

(Continued)

TABLE 6.1 (Continued)

Ecosystem/location	Species	Test temperatures	Response	References
Qilian Mountain	*Amygdalus mongolica*	15/8°C, 20/10°C, 20/10°C, 25/15°C	↔	Wang et al. (2018a,b)
Tianzhu County, Qinghai	*Picea crassifolia*	5°C, 10°C, 15°C, 20°C, 25°C, 30°C	↑	Liu et al. (2019)
Tianzhu County, Qinghai	*Picea crassifolia*	25/15°C, 25/10°C, 30/10°C	↓	Liu et al. (2019)
Tianzhu County, Qinghai	*Picea mongolica*	5°C, 10°C, 15°C, 20°C, 25°C, 30°C	↑	Liu et al. (2019)
Tianzhu County, Qinghai	*Picea mongolica*	25/15°C, 25/10°C, 30/10°C	↓	Liu et al. (2019)
Steppe				
Baiyinxile steppe, IM	*Koeleria cristata*	25°C, 15°C	↓	Liu et al. (2013)
Daqingshan Mountain, IM	*Koeleria cristata*	25°C, 15°C	↑	Liu et al. (2013)
Shillamuren Steppe, IM	*Koeleria cristata*	25°C, 15°C	↑	Liu et al. (2013)
Xilingol, IM	*Koeleria cristata*	25°C, 15°C	↓	Liu et al. (2013)
Desert				
Dengkou, IM	*Ammopiptanthus mongolicus*	21.5°C, 40°C, 60°C, 80°C, 90°C	↑	Ding et al. (2006)
Tengger Desert, IM	*Cistanche deserticola*	15/5°C, 20/10°C, 25/15°C, 30/20°C	↔	Wang et al. (2017)

IM, Inner Mongolia.
Note: ↑, germination % increased with increasing temperature; ↓, germination % decreased with increasing temperature; ↔, germination % did not change significantly with temperature.

For example, *Koeleria cristata* seeds collected from four grasslands in Inner Mongolia germinated to the highest percentage at 25°C (88.2%), and germination was reduced significantly at low temperatures (Liu et al., 2013).

However, temperatures above the optimal for seed germination may cause adverse effects. For *Ammopiptanthus mongolicus* seeds collected from Dengkou, Inner Mongolia, the optimal temperature for initiation of seed germination was 30°C, but temperatures above 40°C had a detrimental effect on seedling growth, that is, hypocotyl and radical tissue damage and eventual decay of the seedlings (Ding et al., 2006). Seeds of *Pinus sylvestris* and *Picea mongolica* collected from Hulun Buir and Bayin Oboo in Inner Mongolia germinated to fairly high percentages at 10°C − 25°C; but some or all germinating seeds of both species rotted at temperatures above 30°C (Li et al., 2011b). Thus, species differ significantly in response to climate warming. Species-specific responses to changing temperature also have been demonstrated at the community level. In a study of 20 common species in a community of the Siziwang Desert Steppe, Inner Mongolia, seed germination percentage increased with increasing temperature regime for all species, but the magnitude of response was species-specific (Yi et al., 2019).

Studies have compared the responses of species of the same genus to different temperatures. Seeds of three congeneric shrub species, *Caragana korshinskii*, *C. intermedia*, and *C. microphylla* collected along the precipitation gradient across the Ordos Plateau in Inner Mongolia showed different germination responses to temperature. Seeds of *C. korshinskii* germinated to high percentages at 5/15°C − 25/35°C in both light and darkness, while those of *C. intermedia* and *C. microphylla* did so only at 15/25°C and 25/35°C, respectively (Lai et al., 2015). Temperatures for germination in *Picea crassifolia* and *P. mongolica* seeds collected from Chifeng, Inner Mongolia, were 15°C–30°C and 10°C–30°C, and the optimal temperatures were 25°C (72%) and 25/15°C (69%), respectively (Liu et al., 2019). Seeds of four *Artemisia* species were collected in the Horqin Sandy Land, Inner Mongolia. Seeds of *A. wudanica*, *A. halodendron*, and *A. sieversiana* were nondormant and germinated to a high percentage over a range of temperatures. However, seeds of *A. scoparia* had physiological dormancy, which was broken by cold stratification (Li et al., 2011a). Thus even congeneric species can have different germination responses to temperature, and a warming climate likely will have different effects on species in the same genus in the same habitat.

For species with physiologically dormant seeds that require low (moist) temperature during winter for dormancy break in the field, warming winter temperatures could have a significant effect on the ability of seeds to germinate in spring. For example, low temperatures (2°C − 5°C) improved (possibly by releasing physiological dormancy in seeds) germination of *Amygdalus mongolica* seeds collected from the Longchanghe Nature Protection Station of Domestic Nature Reserve of the Qilian Mountain in northern China (Wang et al., 2018a,b), suggesting an increased winter temperature could affect seed dormancy break in this species. In the Mu Us Sandland of Inner Mongolia, freshly harvested seeds of *Agriophyllum squarrosum* were in nondeep physiological (conditional) dormancy, which was alleviated by dry conditions at cold or warm temperatures but was re-induced by warm temperature (15°C/25°C) particularly at low soil moisture (Gao et al., 2018). For this annual species, warm winters could preclude the condition for seed dormancy release and thus cause seeds to be dormant in spring.

In sum, both seed dormancy and germination of species in northern China responded differently to changing temperatures. Thus, the projected warming climate is likely to have different effects on population regeneration of different species by affecting seedling emergence in the local community.

Effect of change in precipitation

Changes in rainfall pattern are another important aspect of climate change in northern China, which not only could impact survivorship, growth, and fecundity of plants but also the germination responses of offspring seeds. In this region, future changes in rainfall patterns—amount and frequency—may alter soil moisture and drought frequency, which in particular will impact seed germination of species in the steppe and semiarid ecosystems. Studies on the impact of variable precipitation patterns (or simulated treatments using PEG) on seed germination in northern China are shown in Table 6.2.

In general, seed germination of species in northern China is highly responsive to moisture and drought stress. An experiment on desert plants in this region showed that soil moisture had a significant effect on seedling emergence, with high moisture resulting in more seedlings than low moisture (Su et al., 2007). For seeds of *Pinus sylvestris* var. *mongolica* collected from the southern margin of Horqin Sandy Land, drought stress reduced germination percentage, delayed the beginning of germination, extended the duration of germination time, and slowed seed germination speed (Wang et al., 2020a,b). Under moderate drought stress (low concentrations of PEG-6000, 5%−10%), seeds of *Pinus densiflora*, *Pinus sylvestris* var. *mongolica*, and *Picea crassifolia* collected from Gonghe County in Qinghai Province showed an increase in germination percentage and radicle length, whereas high drought stress (concentrations of PEG-6000 > 10%) decreased germination (Heling et al., 2018). *Amygdalus mongolica* seeds collected from the Longchanghe Nature Protection Station of Domestic Nature Reserve in the Qilian Mountain showed a significant sequential decrease in germination percentage under increasing drought stress (Wang et al., 2018a,b). Similarly, with increasing drought stress (increase in PEG concentration, −0.177 to −1.25 MPa), germination percentage of *Ammopiptanthus mongolicus* seeds collected from Lanzhou in Gansu Province decreased (Duan et al., 2011).

The response of seed germination to rainfall can differ among species. For example, mild drought stress (low PEG concentration, −0.3 MPa) promoted germination of *Haloxylon ammodendron*, *Reaumuria soongarica*, *Ceratoides latens*, and *Suaeda glauca* seeds collected from the Alashan Desert in Inner Mongolia, and drought resistance differed among the four species (Yang et al., 2012). At the community level, seed germination of 20 species in the Siziwang Desert Steppe of Inner Mongolia decreased with increasing water stress, but the responses depended on species (Yi et al., 2019).

Difference in the response of seed germination also has been reported for congeneric species. Drought stress (high PEG-6000 concentrations, 4% − 24%) inhibited germination percentage, germination potential, and relative daily germination rate of seeds of *Ammopiptanthus mongolicus* and *A. nanus* collected from the Minqin Desert Botanical Garden in Gansu Province, but the degree of inhibition differed between the two species (Jiang et al., 2018). Similarly, the germination responses of seeds of *Caragana korshinskii*, *C. intermedia*, and *C. microphylla* from the Ordos Plateau in Inner Mongolia differed among the species (Lai et al., 2015). The germination response of seeds even has been reported to differ at the intraspecific level. Germination of *Agropyron mongolicum* seeds collected from seven populations in Yanchi of Ningxia Autonomous Region all decreased significantly with drought stress, but there were significant differences in the response to drought among the seven populations (Gao et al., 2015).

In addition, a change in precipitation pattern could affect germination behavior of the offspring seeds through the maternal environmental effect. Rainfall amount and frequency significantly affected all vegetative and reproductive traits of *Agriophyllum squarrosum*, a dominant annual in the Mu Us Sandland (Gao et al., 2015). This study suggested a maternal effect due to rainfall patterns, because germinability of offspring seeds tended to increase with increasing

TABLE 6.2 Studies showing effect of variable precipitation patterns (or simulated treatments) on seed germination of plants in northern China.

Ecosystem/location	Species	Precipitation (%) or (water potential, MPa)	Response	Reference
Semidesert				
Gonghe County, Qinghai	*Picea crassifolia*	5%, 10%, 15%, 20%, 25%, 30%	↔	Heling et al. (2018)
Gonghe County, Qinghai	*Pinus densiflora*	5%, 10%, 15%, 20%, 25%, 30%	↔	Heling et al. (2018)
Gonghe County, Qinghai	*Pinus sylvestris* var. *mongolica*	5%, 10%, 15%, 20%, 25%, 30%	↔	Heling et al. (2018)
Horqin Sandy Land, IM	*Pinus sylvestris* var. *mongolica*	0, −0.054, −0.177, −0.393, −0.735 MPa	↓	Wang et al. (2020a,b)
Ordos Plateau, IM	*Caragana korshinskii*	0, −0.2, −0.4, −0.6, −0.8, −1.0, −1.2, −1.4, −1.6, −1.8, −2.0 MPa	↓	Lai et al. (2015)
Ordos Plateau, IM	*C. intermedia*	0, −0.2, −0.4, −0.6, −0.8, −1.0, −1.2, −1.4, −1.6, −1.8, −2.0 MPa	↓	Lai et al. (2015)
Ordos Plateau, IM	*C. microphylla*	0, −0.2, −0.4, −0.6, −0.8, −1.0, −1.2, −1.4, −1.6, −1.8, −2.0 MPa	↓	Lai et al. (2015)
Ordos Sandland, IM	*Artemisia sphaerocephala*	0%, 2.5%, 5%, 10%, 20%	↓	Yang et al. (2010)
Ordos Sandland, IM	*Leymus secalinus*	1125, 2250, 4500, 6750, and 9000 mL	↔	Zhu et al. (2014)
Qilian Mountain	*Amygdalus mongolica*	5%, 10%, 15%, 20%, 25%, 30%	↓	Wang et al. (2018a,b)
Tianzhu County, Qinghai	*Picea crassifolia*	0, −0.3, −0.6, −0.9, −1.2, −1.5, −1.8, −2.1, −2.4, −2.7 MPa	↓	Liu et al. (2019)
Tianzhu County, Qinghai	*Picea mongolica*	0, −0.3, −0.6, −0.9, −1.2, −1.5, −1.8, −2.1, −2.4, −2.7 MPa	↓	Liu et al. (2019)
Desert				
Alxa Left Banner, IM	*Haloxylon ammodendron*	−0.3, −0.6, −0.9, −1.2, −1.5, −1.8, −2.1, −2.4, −2.7, −3.0 MPa	↓	Yang et al. (2012a)
Alxa Left Banner, IM	*Reaumuria soongarica*	−0.3, −0.6, −0.9, −1.2, −1.5, −1.8, −2.1, −2.4, −2.7, −3.0 MPa	↓	Yang et al. (2012a)
Alxa Left Banner, IM	*Ceratoides latens*	−0.3, −0.6, −0.9, −1.2, −1.5, −1.8, −2.1, −2.4, −2.7, −3.0 MPa	↓	Yang et al. (2012a)
Alxa Left Banner, IM	*Suaeda glauca*	−0.3, −0.6, −0.9, −1.2, −1.5, −1.8, −2.1, −2.4, −2.7, −3.0 MPa	↓	Yang et al. (2012a)
Dengkou, IM	*Nitraria tangutorum*	25%, 50%, 75%, 100%	↑	Zhang et al. (2019)
Jiuzhou, Gansu Province	*Ammopiptanthus mongolicus*	5%, 10%, 15%, 20%, 25%	↓	Duan et al. (2011)
Minqin, Gansu Province	*Ammopiptanthus mongolicus*	0%, 4%, 8%, 12%, 16%, 20%, 24%	↓	Jiang et al. (2018)
Minqin, Gansu Province	*Ammopiptanthus nanus*	0%, 4%, 8%, 12%, 16%, 20%, 24%	↓	Jiang et al. (2018)
Steppe				
Yanchi, Ningxia Province	*Agropyron mongolicum*	0, −0.3, −0.6, −0.9, −1.2 MPa	↓	Gao et al. (2015)

IM, Inner Mongolia.
Note: ↑, germination % increased with increasing precipitation (or simulated treatments); ↓, germination % decreased with increasing precipitation or increasing water stress; ↔, germination % did not change significantly with precipitation.

aridity to which the mother plant was exposed to during seed development. Similarly, an experiment on the effect of rainfall enhancement gradient on *Nitraria tangutorum* in Dengkou in Inner Mongolia reported that increased rainfall resulted in smaller seeds with an earlier time of germination (Zhang et al., 2019). However, increased precipitation on maternal *Potentilla tanacetifolia* plants in a temperate steppe in northern China stimulated seed production, but it had no effect on seed mass or germination percentage of offspring seeds (Li et al., 2011c). Therefore, a changing precipitation pattern is likely to affect plant regeneration from seeds in natural populations through maternal effects, but these effects can vary depending on the species.

Relatively few studies have examined the effect of the changing precipitation pattern on degree of seed dormancy. An experiment in the Mu Us Sandland in northern China showed that soil moisture, in addition to temperature, regulates dormancy cycling of the dominant pioneer species *Agriophyllum squarrosum* (seeds have physiological dormancy). Compared to high soil moisture, low soil moisture induced a deeper secondary dormancy in the dormancy cycle as evidenced by dormancy being overcome more easily in scarified seeds subjected to high moisture than in those subjected to low moisture (Gao et al., 2018). These results indicate that changes in soil moisture content resulting from climate warming and change in precipitation will impact the status of seed dormancy, at least for some species with physiological dormancy.

Effect of atmospheric nitrogen deposition

Atmospheric nitrogen (N) deposition is one of the important aspects of global change that also could play an important role in seed germination and seedling recruitment in natural populations. A pot incubation test in a semiarid grassland in the Mongolian Plateau of northern China revealed that N enrichment decreased seedling density and diversity of the 20 species tested and that responses to N addition differed among species, suggesting that seed germination is sensitive to atmospheric N deposition in the grassland ecosystems (Zhong et al., 2019). An experiment in a temperate steppe in northern China showed that N addition to maternal plants (F_o) stimulated seed production of *Potentilla tanacetifolia* but that it suppressed F_1 seed mass, germination percentage, and seedling biomass (Li et al., 2011d).

N deposition could also affect soil seed banks in plant communities, but we found only one such study that has been conducted in northern China. In a 6-year field experiment in a temperate steppe in Inner Mongolia, total soil seed density in the soil seed banks increased marginally, and total seed richness at a depth of 5–10 cm increased significantly following N addition (Miao et al., 2020).

Effect of salt stress

A consequence of the changes in rainfall pattern together with climate warming is change of soil salt concentration. The change of soil salinity is likely to have a large impact on plant growth and production, because salt is a major environmental stress that restricts the growth and yield of plants. Because seed germination and seedling growth are sensitive to salt stress, changing soil salt concentration would greatly impact seedling recruitment in northern China, especially in saline areas, which are common. Table 6.3 shows recent studies on the impact of salt stress on seed germination in northern China.

In general, species in northern China are negatively affected by salt stress. For example, germination percentage and germination index of *Morus mongolica* seeds collected from Inner Mongolia decreased significantly with salt stress (increase in NaCl concentration) (Yan et al., 2020). Seed germination of *Amygdalus mongolica* collected from Qilian Mountain in Gansu Province decreased with salt stress (0.2%–1.8% mixed salt) (Wang et al., 2018a,b). Salt stress decreased seed germination of *Picea mongolica* collected from Bai Yin Ao Bao of Inner Mongolia and of *Picea crassifolia* collected from Tianzhu County in the eastern Qilian Mountains (Wang et al., 2014). Therefore, future change in soil salt concentration is likely to have a significant effect on seed germination in northern China.

However, species can differ in their response to salt stress. For seeds of four species collected from Ningxia Province, low salt stress (50 and 100 mmol L^{-1} NaCl) had no significant effect on germination percentage, germination potential, germination index, or vigor index of *Melilotus suaveolens* or *Astragalus adsurgens* seeds but significantly inhibited germination of *Agriophyllum squarrosum* and *Agropyron mongolicum* seeds. Also, under high salt stress (150 and 300 mmol L^{-1} NaCl), seeds of *Agriophyllum squarrosum* and *Agropyron mongolicum* did not germinate (Huo et al., 2019). Germination of *Haloxylon ammodendron*, *Reaumuria soongarica*, *Ceratoides latens*, and *Suaeda glauca* seeds collected from the Alashan Desert in Inner Mongolia was not affected by low salt stress. However, high salt stress inhibited seed germination of all four species, with salt tolerance differing among them (Yang et al., 2012). For seeds of the six annual glycophytes *Artemisia sieversiana*, *A. scoparia*, *Chloris virgata*, *Eragrostis pilosa*, *Chenopodium acuminatum*, and *Chenopodium glaucum* collected in Horqin Sandy Land, Inner Mongolia, salt stress significantly reduced

TABLE 6.3 Studies showing effect of variable salt stress (NaCl) on seed germination of plants in northern China.

Ecosystem/location	Species	Salt stress	Response	References
Semidesert				
Bai Yin Ao Bao, IM	*Picea crassifolia*	0, 50, 100, 150, 200, 250, 300, 350, 400, 450 mmol L^{-1}	↓	Wang et al. (2014)
Bai Yin Ao Bao, IM	*Picea mongolica*	0, 50, 100, 150, 200, 250, 300, 350, 400, 450 mmol L^{-1}	↓	Wang et al. (2014)
Horqin Sandy Land, IM	*Artemisia sieversiana*	0, 50, 100, 200, 300 mmol L^{-1}	↓	Li et al. (2011a)
Horqin Sandy Land, IM	*A. scoparia*	0, 50, 100, 200, 300 mmol L^{-1}	↓	Li et al. (2011a)
Horqin Sandy Land, IM	*Chenopodium acuminatum*	0, 50, 100, 200, 300 mmol L^{-1}	↓	Li et al. (2011a)
Horqin Sandy Land, IM	*Chenopodium glaucum*	0, 50, 100, 200, 300 mmol L^{-1}	↔	Li et al. (2011a)
Horqin Sandy Land, IM	*Chloris virgata*	0, 50, 100, 200, 300 mmol L^{-1}	↓	Li et al. (2011a)
Horqin Sandy Land, IM	*Eragrostis pilosa*	0, 50, 100, 200, 300 mmol L^{-1}	↓	Li et al. (2011a)
IM	*Isatis indigotica*	0, 50, 100, 150 mmol L^{-1}	↓	Wang et al. (2018a,b)
IM	*Morus mongolica*	0, 20, 30, 50, 70, 100 mmol L^{-1}	↓	Yan et al. (2020)
Minqin, Gansu Province	*Ammopiptanthus mongolicus*	0%, 0.2%, 0.4%, 0.6%, 0.8%, 1.0%, 1.2%, 1.4%	↔	Li et al. (2011b)
Ordos Plateau, IM	*Atriplex centralasiatica*	0, 0.05, 0.1, 0.2, 0.4, 0.8, 1.5 mmol L^{-1}	↔	Wang et al. (2020a,b)
Ordos Plateau, IM	*Artemisia sphaerocephala*	5, 10, 50, 100, 200 mmol L^{-1}	↓	Yang et al. (2010)
Otog, IM	*Suaeda corniculata*	0.1, 0.2, 0.5, 0.8, 1.0 mol L^{-1}	↓	Yang et al. (2012b)
Qilian Mountain	*Amygdalus mongolica*	0.2%, 0.4%, 0.6%, 0.8%, 1.0%, 1.2%, 1.4%, 1.6%, 1.8%	↓	Wang et al. (2018a,b)
Yuansheng Lvyang, Ningxia	*Melilotus suaveolens*	50, 100, 150, 200, 300 mmol L^{-1}	↓	Huo et al. (2019)
Yuansheng Lvyang, Ningxia	*Astragalus adsurgens*	50, 100, 150, 200, 300 mmol L^{-1}	↓	Huo et al. (2019)
Yuansheng Lvyang, Ningxia	*Agriophyllum squarrosum*	50, 100, 150, 200, 300 mmol L^{-1}	↓	Huo et al. (2019)
Yuansheng Lvyang, Ningxia	*Agropyron mongolicum*	50, 100, 150, 200, 300 mmol L^{-1}	↓	Huo et al. (2019)
Desert				
Alxa Left Banner, IM	*Ceratoides latens*	0, 0.05, 0.1, 0.2, 0.3, 0.4, 0.5, 0.6, 0.7, 0.8, 0.9, 1.0, 1.2, 1.4, 1.6, 1.8 mol L^{-1}	↓	Yang et al. (2012a)
Alxa Left Banner, IM	*Haloxylon ammodendron*	0, 0.05, 0.1, 0.2, 0.3, 0.4, 0.5, 0.6, 0.7, 0.8, 0.9, 1.0, 1.2, 1.4, 1.6, 1.8 mol L^{-1}	↓	Yang et al. (2012a)
Alxa Left Banner, IM	*Reaumuria soongarica*	0, 0.05, 0.1, 0.2, 0.3, 0.4, 0.5, 0.6, 0.7, 0.8, 0.9, 1.0, 1.2, 1.4, 1.6, 1.8 mol L^{-1}	↓	Yang et al. (2012a)
Alxa Left Banner, IM	*Suaeda glauca*	0, 0.05, 0.1, 0.2, 0.3, 0.4, 0.5, 0.6, 0.7, 0.8, 0.9, 1.0, 1.2, 1.4, 1.6, 1.8 mol L^{-1}	↓	Yang et al. (2012a)
Minqin, Gansu Province	*Picea mongolica*	0%, 0.2%, 0.4%, 0.6%, 0.8%, 1.0%, 1.2%, 1.4%	↓	Zhu et al. (2011)

IM, Inner Mongolia.
Note: ↑, germination % increased with increasing salt stress; ↓, germination % decreased with increasing salt stress; ↔, germination % did not change significantly under salt stress.

germination percentages of *A. sieversiana*, *A. scoparia*, *C. virgata*, and *C. acuminatum*, but it had no effect on that of *E. pilosa*. Lower salinity (50 mM NaCl) significantly increased germination percentage of *C. glaucum*. In addition, >200 mM salinity led to some specific ion toxicity and reduced seed viability in *A. sieversiana*. For the other five species, an osmotic effect limited germination, but there was no specific ion toxicity effect (Li et al., 2011d).

Summing up, the impact of global climate change on seed germination and dormancy has received increasing attention in recent years. Studies reviewed here, although mostly conducted in the laboratory, suggest that seed germination and dormancy of species in northern China may be highly responsive to global change drivers, including global warming, changes in precipitation patterns, and increases in N deposition and soil salt concentration. Therefore, future climate change in northern China is likely to have a significant effect on regeneration of plants from seeds, through regulating seed dormancy break and germination, thus changing the timing of germination of species in this region. Such effects could result in changes in plant growth, biomass accumulation, seed production, and seed quality. In addition, these changes could result in modifications in plant community composition and relative abundance of various species in the natural habitats of northern China.

Future directions

Projected climate warming and changes in precipitation are likely to have a big influence on the semideserts and steppes in northern China. Such influence not only would lead to important changes in the structure of natural ecosystems but also be a major threat to human livelihood, because these natural systems support a rich flora and fauna that provide invaluable ecosystem services to human. Although the impacts of climate change on plant regeneration from seeds have received an increasing research attention in recent years, studies of observed climate changes and their impacts are still inadequate for many areas.

First, predicting future response of plant regeneration from seeds to climate change relies on an accurate projected change in temperature and precipitation. However, the projection of future changes in climate, especially in precipitation patterns, is highly uncertain in northern China. Therefore, improved projections for precipitation, and thus soil water supply, are most urgently needed for accurately predicting the effects of clime change on plant regeneration from seeds in this region.

Second, most studies (see above) used laboratory tests or greenhouse experiments to examine the effect of climate change on plant regeneration from seeds in northern China. Although these studies have provided some insight in understanding climate change impacts, results from them may be different from those of the real-world changes in natural ecosystems, because in nature climate change affects plant regeneration by interacting with a large number of biotic or abiotic environmental factors. Understanding how climate change impacts plant regeneration is currently limited by the lack of long-term studies in the natural ecosystems of northern China. Therefore studies on plant regeneration from seeds conducted in the natural ecosystems are urgently needed if we are to predict the future dynamics of plant communities more precisely in northern China.

Third, the effect of climate change on plant regeneration from seeds has mostly been examined for individual species, and the responses differed among them (see above). Therefore investigating how seed germination of the aggregate of multiple species in an ecosystem responds to environmental conditions is crucial for understanding the mechanisms for community structure and biodiversity maintenance. However, knowledge of seed germination response of species to environmental conditions is still scarce at the community level (Yi et al., 2019). Therefore field studies examining the responses of all species in a community are needed to determine the dynamics of the natural ecosystems in northern China under climate change.

References

Bai, Y., She, W., Zhang, Y., Qiao, Y., Fu, J., Qin, S., 2020. N enrichment, increased precipitation, and the effect of shrubs collectively shape the plant community in a desert ecosystem in northern China. Sci. Total Environ. 716, 135379. Available from: http://doi.org/10.1016/j.scitotenv.2019.135379.

Baskin, C.C., Baskin, J.M., 2014. Seeds: Ecology, Biogeography, and Evolution of Dormancy and Germination, second ed. Academic Press/Elsevier, San Diego, CA.

Corlett, R.T., Westcott, D.A., 2013. Will plant movements keep up with climate change? Trends Ecol. Evol. 28, 482–488.

Dangal, S.R.S., Tian, H., Lu, C., Pan, S., Pederson, N., Hessl, A., 2016. Synergistic effects of climate change and grazing on net primary production of Mongolian grasslands. Ecosphere 7, e01274. Available from: http://doi.org/10.1002/ecs2.1274.

Ding, Q., Wang, H., Jia, G., Hao, Y., 2006. Seed germination and seedling performance of *Ammopiptanthus mongolicus*. J. Plant Ecol. 30, 633–639 (in Chinese).

Duan, H.-R., Li, Y., Ma, Y.-J., 2011. Effects of PEG stress on seed germination of *Ammopiptanthus mongolicus*. Res. Soil Water Conserv. 18, 221–225 (in Chinese).

Gao, R., Yang, X., Liu, G., Huang, Z., Walck, J.L., 2015. Effects of rainfall pattern on the growth and fecundity of a dominant dune annual in a semi-arid ecosystem. Plant Soil 389, 335–347.

Gao, R., Zhao, R., Huang, Z., Yang, X., Wei, X., He, Z., et al., 2018. Soil temperature and moisture regulate seed dormancy cycling of a dune annual in a temperate desert. Environ. Exp. Bot. 155, 688–694.

Ge, Q., Wang, H., Dai, J., 2014. Simulating changes in the leaf unfolding time of 20 plant species in China over the twenty-first century. Int. J. Biometeorol. 58, 473–484.

Hadano, M., Nasahara, K.N., Motohka, T., Noda, H.M., Murakami, K., Hosaka, M., 2013. High-resolution prediction of leaf onset date in Japan in the 21st century under the IPCC A1B scenario. Ecol. Evol. 3, 1798–1807.

Heling, X.-Z., Jia, Z.-Q., Li, Q.-X., Zhang, Y.-Y., Feng, L.-L., Yang, K.-Y., et al., 2018. Seed germination upon drought stress of *Pinus densiflora* Sieb. et Zucc. and other two Pinaceae species. For. Res. 31, 173–179.

Hijioka, Y., Lin, E., Pereira, J.J., Corlett, R.T., Cui, X., Insarov, G.E., et al., 2014. Asia. In: Climate Change 2014: Impacts, adaptation, and vulnerability. Part B: Regional Aspects. In: Barros, V.R., Field, C.B., Dokken, D.J., Mastrandrea, M.D., Mach, K.J., Bilir, T.E., et al., Contribution of Working Group II to the Fifth Assessment Report of the Intergovernmental Panel on Climate Change. Cambridge University Press, Cambridge, pp. 1327–1370.

Huo, X.-T., Jia, H., Cao, B., 2019. Effects of NaCl treatment on seed germination of four herbage plants. Acta Agrestia Sin. 27, 1096–1101 (in Chinese).

Jiang, S.-X., Yan, Z.-Z., Wu, H., 2018. Effects of simulated droughts stress by PEG6000 on seed germination of the two *Ammopiptanthus* species. J. Northw. For. Univ. 33, 130–136 (in Chinese).

Lai, L.-M., Tian, Y., Wang, Y.-J., Zhao, X.-C., Jiang, L.-H., Baskin, J.M., et al., 2015. Distribution of three congeneric shrub species along an aridity gradient is related to seed germination and seedling emergence. AoB Plants 7, plv071. Available from: https://doi.org/10.1093/aobpla/plv071.

Liancourt, P., Spence, L.A., Boldgiv, B., Lkhagva, A., Helliker, B.R., Casper, B.B., et al., 2012. Vulnerability of the northern Mongolian steppe to climate change: insights from flower production and phenology. Ecology 93, 815–824.

Liu, Y.-J., Liu, S.-Z., Kang, C.-Z., Man, D.-Q., 2019. Comparative adaptation of seed germination and seedling growth to environmental factors in two *Picea* plant species. Acta Ecol. Sin. 39, 611–619 (in Chinese).

Liu, Z.-Y., Wang, Y.-R., Yang, J.-H., Hua, M., Liu, S.-S., 2013. Research of seed production capability and germination characteristics of *Koeleria cristata* (L.). Pers. Chin. J. Grassl. 35, 52–55 (in Chinese).

Li, S., Verburg, P.H., Lv, S., Wu, J., Li, X., 2012a. Spatial analysis of the driving factors of grassland degradation under conditions of climate change and intensive use in Inner Mongolia, China. Reg. Environ. Change 12, 461–474.

Li, X.-H., Jiang, D.-M., Alamusa, Zhou, Q.-L., Oshida, T., 2012b. Comparison of seed germination of four *Artemisia* species (Asteraceae) in northeastern Inner Mongolia. China J. Arid Land 4, 36–42.

Li, X.-H., Jiang, D.-M., Li, X.-L., Zhou, Q.-L., 2011a. Effects of salinity and desalination on seed germination of six annual weed species. J. For. Res. 22, 475–479.

Li, D.-L., Wei, Q.-S., Zhang, J.-H., He, F.-L., Yan, Z.-Z., 2011b. Seed germination and seedling culturing of *Ammopiptanthus mongolicus*. Chin. Agric. Sci. Bull. 27, 30–34 (in Chinese).

Li, Y., Liu, S., Kang, C., Man, D., Li, D., 2011c. Effects of temperature on germination characteristics of *Pinus sylvesiris* var. mongolica and *Picea mongolica* seed. Bull. Soil Water Conserv. 31, 73–77 (in Chinese).

Li, Y., Yang, H., Xia, J., Zhang, W., Wan, S., Li, L., 2011d. Effects of increased nitrogen deposition and precipitation on seed and seedling production of *Potentilla tanacetifolia* in a temperate steppe ecosystem. PLoS One 6, e28601. Available from: http://doi.org/10.1371/journal.pone.0028601.

Lloret, F., Peñuelas, J., Estiarte, M., 2004. Experimental evidence of reduced diversity of seedlings due to climate modification in a Mediterranean-type community. Glob. Change Biol. 10, 248–258.

Lv, X., Zhou, G., Wang, Y., Song, X., 2016. Sensitive indicators of zonal *Stipa* species to changing temperature and precipitation in Inner Mongolia grassland, China. Front. Plant. Sci. 7, 73. Available from: http://doi.org/10.3389/fpls.2016.00073.

Miao, R., Guo, M., Ma, J., Gao, B., Liu, Y., 2020. Shifts in soil seed bank and plant community under nitrogen addition and mowing in an Inner Mongolian steppe. Ecol. Eng. 153, 105900. Available from: http://doi.org/10.1016/j.ecoleng.2020.105900.

Poulter, B., Pederson, N., Liu, H., Zhu, Z., D'Arrigo, R., Ciais, P., et al., 2013. Recent trends in inner Asian forest dynamics to temperature and precipitation indicate high sensitivity to climate change. Agric. For. Meteorol. 178, 31–45.

Richardson, A.D., Keenan, T.F., Migliavacca, M., Ryu, Y., Sonnentag, O., Toomey, M., 2013. Climate change, phenology, and phenological control of vegetation feedbacks to the climate system. Agric. For. Meteorol. 169, 156–173.

Su, Y.-G., Li, X.-R., Cheng, Y.-W., Tan, H.-J., Jia, R.-L., 2007. Effects of biological soil crusts on emergence of desert vascular plants in North China. Plant Ecol. 191, 11–19.

Su, R., Yu, T., Dayananda, B., Bu, R., Su, J., Fan, Q., 2020. Impact of climate change on primary production of Inner Mongolian grasslands. Glob. Ecol. Conserv. 22, e00928. Available from: http://doi.org/10.1016/j.gecco.2020.e00928.

Tao, F., Zhang, Z., 2010. Adaptation of maize production to climate change in North China Plain: quantify the relative contributions of adaptation options. Eur. J. Agron. 33, 103–116.

Tchebakova, N.M., Rehfeldt, G.E., Parfenova, E.I., 2010. From vegetation zones to climatypes: effects of climate warming on Siberian ecosystems. In: Osawa, A., Zyryanova, O.A., Matsuura, Y., Kajimoto, T., Wein, R.W. (Eds.), Permafrost Ecosystems: Siberian Larch Forests. Springer, Berlin, pp. 427–446.

Walck, J.L., Hidayati, S.N., Dixon, K.W., Thompson, K., Poschlod, P., 2011. Climate change and plant regeneration from seed. Glob. Change Biol. 17, 2145–2161.

Wang, J., Baskin, J.M., Baskin, C.C., Liu, G., Yang, X., Huang, Z., 2017. Seed dormancy and germination of the medicinal holoparasitic plant *Cistanche deserticola* from the cold desert of northwest China. Plant Physiol. Biochem. 115, 279–285.

Wang, J., Yan, X., Li, J.-Y., Xie, Q.-G., Zhang, Y., Zhao, G., et al., 2018a. Response of the seed germination and seedling growth of *Amygdalus mongolica* to stress. J. Desert Res. 38, 140–148 (in Chinese).

Wang, Y., Zhou, R.-Y., Ma, L.-M., Bai, Y., Guan, J.-L., Tang, X.-Q., 2018b. Seed germination and seedling growth responses of *Isatis indigotica* in five populations from saline environments. Acta Pratacult. Sin. 27 (7), 145–154. in Chinese).

Wang, F., Liu, S.-Z., Liu, Y.-J., Li, D.-L., 2014. Response of *Picea mongolica* and *Picea crassifolia* seed germination and seedling growth to drought and salt stress. Acta Bot. Boreal.-Occid. Sin. 34, 2309–2316 (in Chinese).

Wang, D.-L., Zhang, R.-S., Fang, X., Wang, K., Wu, Y.-L., Qin, S.-Y., et al., 2020a. Seed germination and seedling growth response to drought stress and resistance evaluation for introduced *Pinus sylvestris var. mongolica* sandy-fixation plantation. J. Zhengjiang A&F Univ. 37, 60–68 (in Chinese).

Wang, Z., Baskin, J.M., Baskin, C.C., Yang, X., Liu, G., Huang, Z., 2020b. Dynamics of the diaspore and germination stages of the life history of an annual diaspore-trimorphic species in a temperate salt desert. Planta 251, 87. Available from: http://doi.org/10.1007/s00425-020-03380-8.

Wang, H., 2014. A multi-model assessment of climate change impacts on the distribution and productivity of ecosystems in China. Reg. Environ. Change 14, 133–144.

Xiao, X., Ojima, D.S., Parton, W.J., Chen, Z., Chen, D., 1995. Sensitivity of Inner Mongolia grasslands to climate change. J. Biogeogr. 22, 643–648.

Yan, J., Li, G.-T., Wang, Y.-L., Ma, Y.-X., Yang, Y., 2020. Effects of salt stress on seed germination and seedling physiological characteristics of *Morus mongolica*. J. Agric. Sci. Technol. 22, 28–37 (in Chinese).

Yang, F., Cao, D., Yang, X., Gao, R.-R., Huang, Z.-Y., 2012. Adaptive strategies of dimorphic seeds of the desert halophyte *Suaeda corniculata* in saline habitat. Chin. J. Plant Ecol. 36, 781–790 (in Chinese).

Yang, X., Dong, M., Huang, Z., 2010. Role of mucilage in the germination of *Artemisia sphaerocephala* (Asteraceae) achenes exposed to osmotic stress and salinity. Plant Physiol. Biochem. 48, 131–135.

Yi, F., Wang, Z., Baskin, C.C., Baskin, J.M., Ye, R., Sun, H., et al., 2019. Seed germination responses to seasonal temperature and drought stress are species-specific but not related to seed size in a desert steppe: implications for effect of climate change on community structure. Ecol. Evol. 9, 2149–2159.

Zhang, C., Zhang, Y., Li, J., 2019a. Grassland productivity response to climate change in the Hulunbuir steppes of China. Sustainability 11, 6760.

Zhang, J.-B., Dong, X., Xin, Z.-M., Liu, M.-H., Zhang, R.-H., Huang, Y.-R., et al., 2019b. Effects of artificial simulated precipitation on seed characters and germination of *Nitraria tangutorum*. Southw. China J. Agric. Sci. 32, 1181–1186 (in Chinese).

Zhang, F., Li, X., Wang, W., Ke, X., Shi, Q., 2013. Impacts of future grassland changes on surface climate in Mongolia. Adv. Meteorol. 2013, 263746. Available from: http://doi.org/10.1155/2013/263746.

Zheng, Y., Dong, L., Li, Z., Zhang, J., Li, Z., Miao, B., et al., 2019. Phylogenetic structure and formation mechanism of shrub communities in arid and semiarid areas of the Mongolian Plateau. Ecol. Evol. 9, 13320–13331.

Zhong, M., Miao, Y., Han, S., Wang, D., 2019. Nitrogen addition decreases seed germination in a temperate steppe. Ecol. Evol. 9, 8441–8449.

Zhu, G.-Q., Liu, S.Z., Li, D.-L., Kang, C.-Z., Yan, Z.-Z., Man, D.-Q., et al., 2011. Seed germination Seedl. Cult. Picea mongolica. Chin. Agric. Sci. Bull. 27 (16), 22–26 (in Chinese).

Zhu, Y., Yang, X., Baskin, C.C., Baskin, J.M., Dong, M., Huang, Z., 2014. Effects of amount and frequency of precipitation and sand burial on seed germination, seedling emergence and survival of the dune grass *Leymus secalinus* in semiarid China. Plant Soil 374, 399–409.

Chapter 7

Effect of climate change on regeneration of plants from seeds in grasslands

Eric Rae and Yuguang Bai
University of Saskatchewan, Saskatoon, SK, Canada

Introduction

Grasslands are natural or seminatural ecosystems with ≥ 10% cover of vascular plants usually dominated by C3 (cool season) and/or C4 (warm season) graminoids (grasses and grass-like herbs, e.g., sedges), and they cover about 40% of the earth's land surface, excluding Antarctica and Greenland (White et al., 2000; Dixon et al., 2014). They are diverse ecosystems that differ in composition along temperature, precipitation, and latitudinal and altitudinal gradients with 145 unique grassland ecoregions identified worldwide (Dixon et al., 2014). Grassland types of the world include savanna, shrubland, nonwoody grassland, and tundra (White et al., 2000). Thus, in addition to graminoids, grasslands can have broad-leaved herbaceous species (forbs) and subshrubs, shrubs, and trees. The grasslands in Asia, North America, and Oceania are mainly nonwoody (mostly <10% of the area covered by woody plants), while larger areas of those in Africa, South American, and Central America/Caribbean are covered by woody plants (White et al., 2000). In general, the relative biomass of C4 grasses decreases from tropics→temperate zone→arctic (Still et al., 2003; Winslow et al., 2003). Eurasian grasslands are mostly C3 species, African and Australian C4, and North American and South American both C3 and C4 (Collatz et al., 1998). C4 grasses are largely absent from Mediterranean regions and at high elevations and latitudes.

Much of the world's grasslands have been converted to cropland, especially in North America and South America (White et al., 2000). In North America, only about 20% of grasslands remain due to cultivation and other land use and development (Ceballos et al., 2010). Grasslands that have not been converted to cultivation are threatened by desertification, overgrazing, and climate change. Although grassland plants are adapted to disturbances such as grazing, drought, fire, and seasonal fluctuations in temperature and precipitation, plant species diversity, community structure, and ecosystem functions are being altered by climate change.

Experiments simulating climate change through CO_2 enrichment and warming were first conducted in growth chambers and greenhouses, although it was immediately acknowledged that generalizations from these controlled settings should be validated with field experiments (Drake et al., 1985). This prompted the use of Free Air CO_2 Enrichment (FACE) techniques that enabled field studies on seed germination, seedling recruitment, and population dynamics. Thus FACE experiments take into account the effects of interspecific interactions and factors such as irrigation, drought, nutrient addition, and soil microbial communities on plant responses in field settings.

While climate change effects on ecosystems are complex, there is a general understanding of the importance of differential responses among species leading to both altered community composition and function (Hovenden and Williams, 2010). However, to reach the goal of understanding altered community composition and function, we need to know how individual species will respond to climate change and in particular how regeneration of plants from seeds will be affected. In this chapter, we attempt to link results from controlled environments to field studies of plant regeneration to explore how climate change may affect sexual reproduction and plant population dynamics and community composition.

Climate change in grasslands

Changes in temperature and/or precipitation induced by climate change vary in magnitude or direction by region, limiting our ability to generalize responses of grasslands to global phenomena. Based on models for predicted global change

(Fischer et al., 2014), grasslands north of the Tropic of Cancer are expected to experience a greater degree of warming (1.2°C–2°C per 1°C of global warming) than those south of it (0.8°C–1.4°C per 1°C of global warming). Models by Fischer et al. (2014) also predict that annual precipitation in the prairies of North America, the Eurasia steppe, the grasslands of Pakistan and Western India, and the east African savannah will increase by 2%–14% per 1°C of global warming. Annual precipitation in the grasslands of Uruguay and northern Argentina is expected to increase by 2%–6% per 1°C degree of global warming. However, annual precipitation in all other major grassland regions (Australia, Central America, Colombia, Mediterranean, southern Africa, and Venezuela) is expected to decrease by 0%–8% per 1°C degree of global warming. We could not find similar models for CO_2. However, observations from NASA's Atmospheric Infrared Sounder system showed that CO_2 concentration varied by latitude, with northern latitudes and circumpolar regions experiencing higher CO_2 concentrations than those near the equator (Pagano et al., 2011). Such variation in grassland responses to climatic change should be considered when attempting to predict the future state of the grasslands.

The impact of future conditions on grasslands at a global scale is complicated not only by uncertainties in climate change projections but also by the diversity of species responses to climate and its interaction with other factors such as competition, species invasion, and biogeochemical processes (Jones, 2019). Bagne et al. (2012) summarized model predictions on the effects of climate change on North American grasslands and concluded that suitable conditions for grasslands are expected to increase. Climate conditions in the interior of North America (Great Plains) are expected to be favorable for northward expansion of grasslands in Canada but cause retraction of grasslands on their eastern and southern boundaries. Conditions in the semidesert grasslands are expected to be favorable for their expansion in southwestern USA but cause contraction in Mexico. Great Basin (USA) shrub-grasslands are expected to decline primarily along the eastern boundary, while California annual grasslands are likely to decline significantly. A poleward shift of forests in northern Eurasia and North America is expected according to a dynamic vegetation model (Yu et al., 2014), creating habitats suitable for grassland expansion along their southern boundaries. This same study also shows deciduous trees and shrubs replacing C3 grasses in Europe and broadleaf deciduous trees replacing C4 grasses in Central Africa and the Amazon. In general, climate change will facilitate expansion and/or invasion of naturalized and invasive species in grasslands around the world (Catford and Jones, 2019).

Grassland species composition and population dynamics

The effects of warming and CO_2 enrichment caused by anthropogenic activities on grassland species composition and the relative contribution of functional groups are mixed (Edwards et al., 2001a; Niklaus and Körner, 2004; Bloor et al., 2010). A *meta*-analysis by Ehrlén (2019) reviewed 13 long-term studies on the effects of climate change on demography and population dynamics of 24 grassland species and concluded that there were too few studies to reliably generalize among plant life histories, geographic regions, or climate factors. Compounding these concerns, studies by Andresen et al. (2018) and Reich et al. (2018) reported reversals of grassland functional group responses to CO_2 midway through long-term field experiments, challenging the utility of short-term studies as predictors of climate change effects. In Minnesota (USA), a 20-year FACE experiment found that CO_2 responses of C3 and C4 grasses reversed in 2008, a decade after the experiment had begun. Initially, C3 grasses responded more positively to CO_2 enrichment, but after the reversal C4 grasses responded more positively (Reich et al., 2018). In a 17-year FACE study on populations of C3 grasses, forbs, and legumes in Germany, forbs initially responded negatively to elevated CO_2. However, approximately a decade into the experiment, around 2007, forb relative abundance began to respond positively to CO_2 enrichment (Andresen et al., 2018).

These reversals in responses of C3 and C4 species may represent a response to changing global climate factors rather than an experimental effect. A 25-year study of the effects of shifting climate on an intermountain bunchgrass prairie in Montana (USA) revealed a shift in the dominant species from bluegrasses (primarily nonnative *Poa pratensis*), which began to decline around 2003, to wheatgrasses (primarily native *Pascopyrum smithii*), with wheatgrasses stabilizing as the new dominant species by 2010 (Belovsky and Slade, 2021). Despite the geographical distances between the sites in Germany, Minnesota, and Montana, there was a temporal correspondence in these reversals, suggesting that a common underlying environmental factor rather than experimental manipulation was responsible for the changes. Langley et al. (2018) analyzed data from 16 long-term climate change studies that included 791 plant species. For 57% of the species, magnitude of ambient change was greater than the treatment effects, which lends credence to the notion that ambient climatic factors are responsible for these trend reversals.

Experiments on C3 versus C4 species responses to climate change have yielded contradictory results. For instance, it is predicted that C4 plants will extend their range northward in a warming climate, yet their photosynthetic efficiency with increasing CO_2 levels is expected to restrict their range expansion (Epstein et al., 2002; Ainsworth and Long, 2005). Morgan et al. (2011) reported population growth of C4 grasses in response to experimental climate change manipulation (warming and increased CO_2). However, Zelikova et al. (2014) and Mueller et al. (2016) reported decreases in population density of C4 grasses in response to simulated climate change. Because many of these studies were relatively short-term and based on the observations of Reich et al. (2018) (see above), long-term trends may reveal different effects.

Despite the uncertainty raised by trend reversals and the high level of inconsistency in results, a few clear patterns have emerged from the decades of climate change experiments. Using data from 16 studies (Edwards et al., 2001a; Zavaleta et al., 2003; Marchi et al., 2004; Niklaus and Körner, 2004; Williams et al., 2007; Reich, 2009; Bloor et al., 2010; Dalgleish et al., 2011; Morgan et al., 2011; Adler et al., 2012; Compagnoni and Adler, 2014; Reich et al., 2014, 2018; Zelikova et al., 2014; Miranda-Apodaca et al., 2015; Mueller et al., 2016), we searched for reports of significant population-level changes in response to experimental climate manipulations.

Generally, species that reproduce both sexually and vegetatively responded more positively to climate change manipulations than those that reproduce predominately sexually or vegetatively. Four of the eight species (*Carex caryophyllea, C. flacca, Hypochaeris radicata,* and *Trifolium repens*) included in these studies that reproduced both by seeds and vegetatively had significant positive responses to climate change, while no species had significant negative responses. This contrasts with the 20 species that reproduce primarily by seeds, of which four had significant positive responses and four had significant negative responses. Of the six species that reproduce primarily vegetatively, only *Agrostis capillaris* responded significantly to climate change, that is, a negative response to drought/warming and a positive response to drought/warming/CO_2 enrichment. If we examine only the trends in graminoids rather than only the significant results, positive population trends in response to climate change were reported in 67% of the experiments for species that reproduce by seeds and vegetatively, 56% for species that reproduce by seeds, and 52% for species that reproduce vegetatively. However, 99% of the regeneration of grasses in the tallgrass prairies (North America) is via vegetative reproduction (Benson and Hartnett, 2006). These trends suggest that plant species with versatile reproductive strategies are well adapted to a changing climate.

Bunchgrasses are more likely to respond negatively to climate change manipulations than sod-forming grasses. Of the sod-forming graminoids among the 16 studies, *Agrostis capillaris, Carex caryophyllea,* and *C. flacca* had significant positive responses to climate change (Niklaus and Körner, 2004; Bloor et al., 2010). Although when CO_2 was removed as a factor in the experiment by Bloor et al. (2010) and only drought and warming were applied, the contribution of *A. capillaris* to community biomass decreased significantly. In contrast, among bunchgrasses only *Anthoxanthum odoratum* and *Lolium perenne* responded with significant enhancement in seed production, while population size and biomass were reduced significantly for *A. odoratum* (from a different study), *Dactylis glomerata,* and *Trisetum flavescens* under simulated climate change (Edwards et al., 2001a; Niklaus and Körner, 2004; Bloor et al., 2010). These results suggest that bunchgrass populations are more likely to decline than those of sod-forming grasses under future climate conditions, possibly due to a larger pool of reserves stored underground in rhizomes of the sod-formers that protect the plants against environmental stresses associated with warming (Derner and Briske, 2001). Intriguingly, subsequent studies by van Staalduinen and Anten (2005), Xu and Zhou (2011), and Zhang et al. (2018) found that caespitose grasses such as *Hesperostipa comata, H. curtiseta, Stipa grandis,* and *S. kryovii* are more resistant to drought than the rhizomatous grasses *Elymus lanceolatus, Leymus chinensis,* and *Pascopyrum smithii*. Coupled with the insights of Derner and Briske (2001), who noted that rhizomatous species in semiarid environments tended to thrive in microenvironments where water is relatively abundant, these studies suggest that the abundance of rhizomatous grasses increases in response to climate change. However, this effect may be reversed in regions that experience severe drought.

Biomass production and/or density of forbs exposed to artificial warming decreased (Zavaleta et al., 2003; Williams et al., 2007; Morgan et al., 2011; Buhrmann et al., 2016; Mueller et al., 2016). Belovsky and Slade (2021) cataloged the effects of (nonexperimental) climate change from 1975 to 2020 on intermountain grasslands of Montana (USA) and also found a reduction in the relative contribution of forbs to vegetation biomass as the climate became warmer and drier. These losses suggest that grassland communities with forbs and graminoids may shift more toward graminoid species under future climatic conditions. However, the contribution of seedling recruitment to these changes is complicated due to species-specific variation in regeneration strategies.

Reproductive phenology and seed production

It is generally understood that plant phenological stages are accelerated by warming (Parmesan and Yohe, 2003; Bertin, 2008; Wolkovich et al., 2012; Piao et al., 2019). In grasslands, this acceleration ranges from about 1 day (Parmesan,

2007; Piao et al., 2019) to more than 5 days (Ge et al., 2015) per decade, which suggests that location or grassland type may have a strong impact on shifts in phenology. Indeed, it has been suggested that artificial warming treatments underestimate the effect of warming on time of leafing-out and flowering compared to in situ changes (Wolkovich et al., 2012), emphasizing the need for field observations to confirm climate change impacts on phenology.

The effects of climate on grassland reproductive phenology are further complicated by interactions with species diversity. Wolf et al. (2017) demonstrated that flowering times in grasslands shifted to an earlier date as species diversity declined, meaning that experiments conducted on monocultures or planted polycultures are unlikely to accurately reflect the timing of phenological events in situ. However, the impact of this shift to an earlier flowering date in studies conducted on intact ecosystems may be minor since many studies have shown either no significant effect on species diversity or enhanced diversity in response to climate change in grasslands (Niklaus and Körner, 2004; Reich, 2009; Zelikova et al., 2015; Belovsky and Slade, 2021).

Warming in California grasslands resulted in earlier onset of flowering (2–5 days/1.5°C) for both forbs and C3 grasses (Cleland et al., 2006). In a Eurasian steppe field-heating experiment, faster flowering and fruiting were observed but mostly by less than 1 day among five forbs (*Allium bidentatum*, *Heteropappus altaicus*, *Potentilla acaulis*, *P. bifurca*, and *P. tanacetifolia*), two C3 grasses (*Agropyron cristatum* and *Stipa krylovii*), and one small shrub (*Artemisia frigida*) (Xia and Wan, 2013). An even greater shift in flowering phenology was reported in the tallgrass prairie (North America): up to 20 days with warming or warming plus a double amount of precipitation. However, doubling the amount of precipitation alone had no significant effect on flowering time (Sherry et al., 2007). The responses to warming reported by Sherry et al. (2007) varied by genus, with most forb genera and both C3 grass genera (*Bromus japonicus* and *Dichanthelium oligosanthes*) flowering earlier in response to artificial warming. Two C4 grasses (*Andropogon gerardii* and *Schizachyrium scoparium*) had delayed flowering, and one C4 grass (*Panicum virgatum*) had faster flowering. Under artificial warming, flowering times shifted to an earlier date for C3 grasses, the C4 grass *Themeda triandra*, and forbs, with an average shift of 11 days °C^{-1} (Hovenden et al., 2008b). Warming shifted flowering times to earlier for the C3 forb *Chenopodium album* and the C4 grass *Setaria viridis* (Lee, 2011).

A comprehensive observational study of the effect of temperature on timing of grass reproduction in the Western USA between 1895 and 2013 also found differential responses between C3 and C4 grasses (Munson and Long, 2017). C3 grasses reached the flowering stage faster with increase in mean annual temperature. Such effects were more pronounced for the three C3 annual species than for perennial species, while C4 grasses delayed flowering. Generally, precipitation had an inverse effect, with higher levels of mean annual precipitation associated with slower flowering times (+0.25 days cm^{-1}), particularly for C4 perennial grasses (Munson and Long, 2017). It was hypothesized that delay of flowering as mean annual precipitation increases was due to favorable conditions for vegetative growth. Location was a significant factor in the response. Many C3 annual and perennial grasses shifted from delayed flowering under warming conditions at more southerly latitudes to earlier flowering at more northerly latitudes (Munson and Long, 2017). The same study found no effect on C4 grasses.

A study on Tibetan Plateau grasslands from 1980 to 2014 found that warming was associated with an earlier and shorter growing phase, with both C3 grasses and forbs showing a significant shift (about 12 days °C^{-1}) but no effect on sedges (Wang et al., 2020b). Warming temperatures and decreasing December/January snowfall between 1995 and 2008 also were associated with earlier flowering (1.5 days °C^{-1}) of 31 grassland species in Montana (USA), with only *Hydrophyllum capitatum*, a perennial herb, exhibiting delayed flowering (Lesica and Kittelson, 2010). In general, observational studies of grasslands reported larger flowering time shifts than studies conducted using artificial climate change treatments, consistent with Wolkovich et al. (2012). This suggests that in grasslands, as in other ecoregions, experimental climate change treatments may underrepresent the predicted level of change. The magnitude of phenological shift appears to be affected by latitude and altitude, with more northerly and higher elevation grasslands experiencing larger shifts than those at more southerly latitudes or lower elevations (Sherry et al., 2007; Cleland et al., 2013; Munson and Long, 2017; Wang et al., 2020a).

Phenological responses of C3 grasses, C4 grasses, and forbs to climate change are shown by changes in the timing of flowering. Craine et al. (2012) demonstrated that C4 grasses in the North American tallgrass prairie flower significantly later than C3 grasses. Prior research had shown that the phenology of earlier flowering plants is more strongly affected by temperature change than that of late-flowering plants (Lesica and Kittelson, 2010; Cook et al., 2012; Wolkovich et al., 2012). Climate change has resulted in less mid-season flowering in montane meadows, with a shift to either earlier or later in the growing season (Aldridge et al., 2011). Similar results were reported by Sherry et al. (2007), who found that three late-flowering C4 genera delayed flowering in response to warming, while the relatively earlier flowering C4 *Panicum* grasses flowered even earlier in a warming climate. Dunnell and Travers (2011) compared flowering data from 1910 to 1961 with that from 2007 to 2010 in North Dakota (USA) and found that many spring-flowering species flowered earlier, while fall-flowering species flowered significantly later.

Research on the effects of CO_2 enrichment on flower initiation is sparse compared to that on warming effects. No significant effects of elevated CO_2 on the timing of flowering in grasslands were found in a mini-FACE experiment in France (Bloor et al., 2010), during a FACE experiment in an Australian grassland (Hovenden et al., 2008b), or for an invasive C3 forb or a C4 weedy grass grown in growth chambers (Lee, 2011). CO_2 enrichment delayed flowering in grasses and accelerated flowering in forbs. With very few studies reported, it is premature to generalize the effects of CO_2 enrichment on flowering phenology (Cleland et al., 2006).

Another factor complicating our understanding of climate change impacts on plant reproduction is the cumulative effect of temperature and precipitation for more than 1 year in perennial species (Bai and Romo, 1997; McKone et al., 1998; Dudney et al., 2017; Palit et al., 2017). *Festuca hallii* (C3 perennial grass) requires a specific combination of precipitation and temperature over more than 1 year to trigger flowering (Palit et al., 2017). For species with a mast seeding strategy that saturates predators, such as *Chionochloa* spp. (C3 perennial grasses) in New Zealand, previous January/February temperatures of >11°C trigger flowering the following year. This event occurred in about 20% of years between 1973 and 1995 but is expected to increase to 60% or 80% of years under 1°C or 2°C of warming (McKone et al., 1998). Such requirements tend to be species-specific and should be accounted for when attempting to predict species responses to climate change.

A *meta*-analysis on 79 species from published studies up to the year 2000 showed that in general, CO_2 enrichment resulted in more flowers (+19%), fruits (+18%), and seeds (+16%) and greater individual seed mass (+4%) and total seed mass (+25%), but that it was highly species-specific (Jablonski et al., 2002). A strong, positive correlation was found between germination percentage and seed mass with parental CO_2 enrichment in a *meta*-analysis (Marty and BassiriRad, 2014). Research on the effect of CO_2 enrichment on seed production showed similar effects in a *meta*-analysis by Jablonski et al. (2002), with enhancement in seed production and mass being common. For nine species in the North American mixed-grass prairie, parental CO_2 enrichment, heating, and irrigation significantly increased seed mass for *Centaurea diffusa*, *Grindelia squarrosa*, and *Hesperostipa comata*, while the other six species were not affected (Li, 2013, Li et al., 2018). Seed mass of the C4 grass *Andropogon gerardii* also was higher with parental CO_2 enrichment (Singh et al., 2019). After 5 years of CO_2 enrichment, seed production (+29%) and flowering (+24%) of graminoids and forbs increased in a calcareous grassland in Switzerland, with legumes unaffected (Thürig et al., 2003). Seeds from CO_2 enrichment-treated plants also tended to be heavier than those of controls. Elevated CO_2 significantly increased the number of seeds produced and dispersed (as measured by seed traps) for the majority of species in a New Zealand grassland (Edwards et al., 2001a).

However, other experiments have identified species for which CO_2 enrichment did not affect seed production. For example, a study on CO_2 effects on C3 grasses *Aegilops kotschyi*, *A. peregrine*, and *Hordeum spontaneum* found no significant effects on seed mass, and *Austrodanthonia caespitosa* (a perennial C3 grass) was similarly unaffected (Grünzweig and Körner, 2001; Hovenden et al., 2008a). Some species even responded negatively to elevated CO_2. Seed mass of *Poa annua* (C3 grass) was decreased significantly by parental exposure to elevated CO_2 (Bezemer and Jones, 2012).

While we found no equivalent *meta*-analysis on the effects of warming on seed production and mass, the results from individual experiments suggest that seed production in many species is affected by warming. Warming increased seed production per unit area of soil surface for the C3 annual grass *Bromus tectorum* and the C3 perennial grass *Leymus chinensis* (Gao et al., 2012; Blumenthal et al., 2016). For the C4 grass *Themeda triandra*, warming enhanced seed production in tetraploid plants, but the effect was reversed in diploid plants (Godfree et al., 2017). The annual weedy species, *Chenopodium album* and *Setaria viridis*, were not significantly affected by warming in a growth chamber experiment in terms of seed production and seed mass (Lee, 2011). Warming significantly reduced seed mass, seed production, flower number, and fruit number in nine species in alpine meadows on the Tibetan Plateau, including sedges, C3 grasses, and forbs (Liu et al., 2012). The negative response to climatic warming for species on the Tibetan Plateau may not be representative of nonmontane grasslands, as additional research in the region found that sensitivity to changes in temperature is strongly correlated with elevation (Li et al., 2019).

Generally, these results suggest that for many grassland species a changing global climate will result in advanced flowering and increased seed production, which should promote plant regeneration. However, many individual species may have atypical responses in terms of seed production that could result in shifts in ecosystem composition in favor of species that produce more seeds in response to climate change. For species that flower later in the season, especially C4 species, flowering appears to shift to even later, and while insufficient evidence exists this shift does not appear to be associated with reduced seed production. These concurrent earlier and later shifts in flowering may create a mid-season niche for invasive species that could alter grassland community composition. Additionally, a warming climate appears likely to have more pronounced effects on seed production in high elevation grasslands than in those at low elevations, which may make these ecosystems especially vulnerable to disturbance by climate change.

Seed germination

Interspecific variability in germination success can contribute to population dynamics and affect species diversity in plant communities (Harper, 1977; Bazzaz, 1998; Grime, 2006). Thus, climate change conditions as selection forces regulating seed germination can alter plant communities by differentially affecting the regeneration success of species. Climate change alters seed germination by both direct and parental environmental effects. Direct effects of climate change are primarily due to changes in environmental conditions during seed germination. Parental effects alter seed chemical composition, physiology, germination, and/or morphology as a result of environmental factors imposed on plants during seed production (Roach and Wulff, 1987; Fenner, 1991; Baskin and Baskin, 2014; Marty and BassiriRad, 2014). Studies on effects of parental climate change treatments on germination are limited by not being able to collect enough seeds from experimental plots, concern about removing seeds from plots, and year-to-year variability in seed production. However, field studies provide key insights on the contributions of parental environmental effects on germination success to ecosystem functions because they reflect the impact of complex ecosystem processes on plant regeneration success. Thus, it is crucial to compare the results of field experimentation with those conducted in growth chambers or greenhouses.

Marty and BassiriRad (2014) did a *meta*-analysis of germination data from 29 research papers up to the year 2012 on the effects of CO_2 enrichment on germination percentage and germination rate. They concluded that the responses to both direct and parental CO_2 enrichment are highly variable. In response to direct CO_2 enrichment, germination percentages for forb species including legumes increased significantly, while germination of grasses was not affected. In response to parental CO_2 enrichment, germination percentages increased significantly in six species and decreased significantly in nine species of wild grasses and forbs. They also found contrasting patterns between the effects of CO_2 enrichment on germination percentage and rate (speed). In the five papers that assessed the direct effects of CO_2 enrichment on germination rate, there was a significant enhancement for both grasses and forbs. Germination rate as a response to parental CO_2 enrichment was reported in 13 papers, and positive effects were found in two species and negative effects in five. However, forbs, C3 grasses, perennials, and wild (none-crop) species all showed significant increases in average germination rate, while significant increases in germination percentage were found for perennial and wild species across all species studied (Marty and BassiriRad, 2014). These authors suggested that elevated CO_2 levels may strongly interact with warming; however, their *meta*-analysis did not account for warming and included no C4 species.

Elevated CO_2 during germination significantly increased the germination percentage of three C4 grasses (Baruch and Jackson, 2005). However, the invasive species *Hyparrhenia rufa* and *Melinis minutiflora* exhibited a greater degree of enhancement than the native *Trachypogon plumosus*. de Faria et al. (2015) studied three invasive C4 grasses and found that elevated CO_2 or elevated CO_2 and temperature increased the germination capacity of *Urochloa brizantha* but had no effect on that of *Megathyrsus maximus* or *Urochloa decumbens*. Similar results were found for germination rate of *U. brizantha*; however, increasing temperatures increased germination rate of *U. decumbens* but had no effect on germination percentage. CO_2 enrichment, but not warming, significantly increased seedling emergence of three planted C4 grasses (Dodd et al., 2010). Seedlings of two weedy species *Chenopodium album* and *Setaria viridis* emerged faster with warming, and warming and CO_2 enrichment produced the same response as warming alone (Lee, 2011). Steinger et al. (2000) reported no significant effects of elevated CO_2 on germination percentage of *Bromus erectus*. In a study of C3 grasses and forbs, germination was significantly enhanced by elevated CO_2 only in the legume *Trifolium repens* (Edwards et al., 2001a). These results suggest that seed germination of C4 grasses is more responsive to the direct effects of CO_2 than that of C3 grasses.

Many studies have shown direct effects of warming on seed germination. Among species from grasslands in northern China, C4 species had higher optimal germination temperatures than C3 species (Zhang et al., 2015), and germination of C4 grasses was favored over that of C3 species under either symmetric (+5/+5 at day/night) or asymmetric (+3/+6 at day/night) warming (Zhang et al., 2014). For seven annual herbaceous species endemic to western Mediterranean grasslands, germination capacity for most species declined significantly as temperatures increased (Marques et al., 2007). In contrast, for 10 species of subalpine forbs germination of six species increased with increasing temperatures, seeds of three species did not germinate due to dormancy, and one species, *Viola biflora*, germinated to high percentages at all temperatures tested (Vandvik et al., 2017). A 3°C warming generally increased the average germination percentage of grasses, sedges, and forbs, but germination of legumes was not affected (Wang et al., 2020b). These results suggest that warming enhanced seed germination in most grassland species, particularly that of C4 species and species from cold environments, if moisture was not limiting.

Plants of *Poa annua* exposed to elevated parental CO_2 levels produced seeds with significantly lower germination (Bezemer and Jones, 2012). Germination of *Trifolium repens* seeds increased with parental CO_2 enrichment, and that of *Leontodon saxatilis* was reduced (Edwards et al., 2001b). However, when seed mass was treated as a covariate in the analysis the enhanced effect of parental CO_2 disappeared in *T. repens*, while decreased germination persisted in *L. saxatilis*. A study of Swiss grasslands also showed that while the fraction of germinating seeds did not differ under elevated CO_2, time to germination was shortened significantly for the forbs *Sanguisorba minor* and *Trifolium pratense* in elevated CO_2 plots but lengthened for the C3 perennial grass *Briza media* (Thürig et al., 2003). These contrasting effects of parental CO_2 enrichment on germination percentage and rate suggest that both should be considered when exploring subsequent seedling recruitment in field studies. The success of seedling recruitment from seeds was reduced by parental CO_2 enrichment in annual forbs in a Mediterranean grassland, while grasses and legumes were unaffected (Grünzweig and Dumbur, 2012). Overall, parental CO_2 enrichment tends to have a negative to neutral effect on germination percentage among C3 grasses, a negative impact on forbs, and a positive effect on legumes. However, there seems to be a tradeoff between germination percentage and germination rate in at least some species (Marty and BassiriRad, 2014).

The parental effect of warming has been demonstrated in a study by Zhang et al. (2020) in which species of *Stipa* (C3 grass) endemic to cool regions had higher optimal temperatures for germination than species from the warm regions. Germination percentage of *Leymus chinensis*, a C3 perennial grass in the meadow steppe in northeast China, was reduced after parental warming although seed mass increased (Gao et al., 2012). For the perennial forb *Solidago canadensis*, parental warming significantly increased germination percentage and rate in seeds collected from native populations in North America; however, the effect was reversed in seeds collected from invasive populations in China (Zhou and He, 2020). These authors suggested that the differences may be linked to climate of provenance, with seeds from cooler climates being more sensitive to warming than those from warmer regions. Therefore, local adaptations may complicate the generalizability of seed germination responses to climate change.

The combination of parental warming and CO_2 enrichment may affect seed germination differently than either factor alone, as demonstrated in *Bouteloua gracilis*, a C4 perennial grass for which germination was increased by the combination of warming and CO_2 enrichment but not by warming or CO_2 enrichment alone (Li et al., 2018). For *Andropogon gerardii*, the dominant C4 grass in the North American tallgrass prairie, there was a significant CO_2 × warming interaction on adaptability of seeds to temperature, despite there being a minimal effect of CO_2 enrichment (growth chamber) on germination and a significant effect of warming (Singh et al., 2019). These results challenge the generalization that CO_2 enrichment has minor effects on germination in C4 species. However, it seems that the effects of CO_2 enrichment on germination of some C4 species may depend on concurrent warming (Jaggard et al., 2010; Hampton et al., 2013).

Another example of the importance of accounting for interactions between elevated CO_2 and warming to accurately predict the effects of climate change is unpublished research by M. Cameron and Y. Bai (Table 7.1) on the germination of the small shrub *Artemisia frigida*. This species was unaffected by CO_2 enrichment alone, but warming alone increased both germinability (from 39% to 66%) and uniformity of germination (lower CVt). Germination uniformity disappeared when CO_2 and warming were applied concurrently. However, parental exposure to CO_2 and direct seed warming resulted in faster germination, that is, reduced time to 50% of maximum germination (T50m). Germination of *Austrodanthonia caespitosa* was reduced by elevated CO_2 or warming alone but not when applied in combination (Hovenden et al., 2008b). Collectively, these studies illustrate the importance of accounting for interactions among climatic factors.

Seed physiology

Parental exposure to climate change treatments induces physiological changes in seeds, which may explain the reported effects on germination. The most relevant and widely reported physiological parameter as a result of parental CO_2 enrichment is the carbon:nitrogen (C:N) ratio or N concentration of seeds. A *meta*-analysis showed an average of 14% decrease in seed N concentration among 79 crop and wild species subjected to parental CO_2 enrichment but only among nonlegumes (Jablonski et al., 2002). A higher C:N ratio may reduce protein content, thereby affecting seed functionality and chemical composition and lead to decreased seed viability (Andalo et al., 1996; Bai et al., 2003).

Several studies have explored the effects of parental CO_2 enrichment on grassland seeds. Seed N concentration decreased in eight of 13 species under parental CO_2 enrichment in a Swiss grassland, with all graminoids, most forbs,

TABLE 7.1 Effect of parental warming and CO_2 enrichment on (A) final germination percentage, (B) coefficient of variation of germination time (CVt), and (C) time (hours, h) to 50% maximum germination (T50m, h) of *Artemisia frigida* seeds incubated at 10°C, 15°C, and 20°C.

(A) Final germination (%)	Parental CO_2	Ambient CO_2	Enriched CO_2
Parental warming	No warming	39.20 ± 9.07 ab	42.25 ± 6.71 ab
	Warming	66.00 ± 6.67 a	33.56 ± 8.76 b
(B) CVt (%)	Parental warming	No warming	Warming
		34.49 ± 1.86 a	28.34 ± 2.13 b
(C) T50m (h)	Parental CO_2	Ambient CO_2	Enriched CO_2
Germination temperature	10°C	208.37 ± 16.55 a	236.92 ± 10.96 a
	15°C	170.57 ± 23.40 ab	186.92 ± 23.74 a
	20°C	216.00 ± 96.0 a	66.00 ± 12.49 b

Note: Data are means ± SE. Different letters across both rows and columns represent significant differences using Tukey's HSD test ($P \leq .05$). (M. Cameron and Y. Bai, unpublished data).

and the legume *Trifolium pratense* experiencing a significant decrease in seed N concentration (>C:N ratio) (Thürig et al., 2003). Huxman et al. (1998) also reported a higher C:N ratio for the C3 grass *Bromus rubens* seeds collected from plants grown under elevated CO_2, and Steinger et al. (2000) found a similar shift for *Bromus erectus*. However, when plants grown from seeds collected from wild populations of three Mediterranean C3 grasses (*Aegilops kotschyi*, *A. peregrine*, and *Hordeum spontaneum*) were exposed to elevated CO_2 seed N concentration decreased with increasing CO_2 only in *A. kotschyi* (Grünzweig and Körner, 2001). No significant effects of CO_2 enrichment were reported for seed nitrogen or carbon concentration for the C4 grass *Andropogon gerardii*; however, N concentration was significantly decreased by warming temperatures (Singh et al., 2019). In contrast, CO_2 enrichment, but not warming, decreased seed N concentration for the C3 perennial grass *Austrodanthonia caespitosa* (Hovenden et al., 2008b). This suggests that in terms of seed nitrogen content C4 species are less sensitive to CO_2 enrichment than C3 species (Wand et al., 1999; Barbehenn et al., 2004), while the reverse may be true for warming, although insufficient research has been conducted on C4 seed chemical composition in response to CO_2 enrichment or warming to draw definite conclusions.

In a comparison of effects of climate change on seed chemical composition in forbs and legumes, there were no changes in the C:N ratio for six New Zealand grassland species, which included legumes, C3 grasses, and forbs (Edwards et al., 2001b) or for *Poa annua* (Bezemer and Jones, 2012). However, Hikosaka et al. (2011) found that seed nitrogen content decreased much more in annual grasses than in annual forbs or legumes. A study of the C3 grasses *Lolium perenne* and *Poa pratensis* and the legumes *Medicago lupulina* and *Lotus corniculatus* contradicted reports of CO_2 enrichment primarily impacting nonlegumes (Jablonski et al., 2002; Hikosaka et al., 2011). These authors found that neither a drought/warming treatment nor a drought/warming/CO_2 treatment significantly affected the C:N ratio in plant tissue for either of the two grasses. However, for both N-fixing legumes the C:N ratio increased significantly in response to the drought/warming/CO_2 treatment but not the drought/warming treatment (AbdElgawad et al., 2014). These changes in the C:N ratio corresponded to a significant decrease in protein concentration in the leguminous species. Generally, the results of subsequent research have been consistent with the conclusions of Jablonski et al. (2002): elevated CO_2 appears to decrease seed N concentration. However, recent research challenges the conclusion that legumes are unaffected by elevated CO_2 (Thürig et al., 2003; AbdElgawad et al., 2014; Lamichaney et al., 2021).

Conclusions and future research

The species composition of grassland communities is shifting in response to climate change, and many of these changes are driven by species-specific responses. However, based on our review of the literature we expect an increase in graminoid dominance with climate warming, and where drought is not prominent we expect sod-forming species to dominate grasslands. Furthermore, we expect that as climate changes species with both sexual and vegetative reproduction will be dominant over those that reproduce only sexually or only vegetatively. Direction of change in community C3/C4 composition with climate change in grasslands with a mixture of cool and warm season grasses is not clear. However, for species that reproduces wholly or partially by seeds the evidence points toward C4 dominance. Compared to C3 species, germination of C4 species increases with elevated CO_2 and warming, likely because of differences in seed quality due to C4 seed nitrogen content being less affected by rising CO_2.

The changing climate may be creating new temporal niches, particularly in mid-growing season that could increase the success of invasive species able to exploit that gap (Sherry et al., 2007; Munson and Long, 2017). Nevertheless, despite shifts in species composition current evidence suggests that grassland community diversity is resilient to shifts in composition, with most studies finding no effect or increased diversity in response to climate change (Niklaus and Körner, 2004; Reich, 2009; Zelikova et al., 2015; Belovsky and Slade, 2021). Responses of seed production to climate change tend to be positive but highly species-specific. This may be one reason for variability in climate change response as interspecific differences in seed regeneration drive shifts in community composition.

Despite decades of research, the evidence is still insufficient to draw definitive conclusions about the effects of climate change on grassland community composition. More field-based research on the parental effects of climate change, particularly CO_2 enrichment and its interactions with regeneration traits of plants is needed. More *meta*-analyses on changes in regeneration capacity in response to annual temperature and precipitation variability at local, regional, and global scales would further our understanding of the contribution of seed regeneration to population dynamics and community composition in a changing climate. Discrepancies between field and laboratory results as well as between simulated and observed climate change require further study through comparative trials. Our understanding of climate change effects could be improved by spatial models that predict changing CO_2 concentration. Additional research focusing on the expected shifts highlighted in this chapter would increase our understanding of the effects of global climate change on grassland community composition, especially research to validate the impacts on the reproductive strategy of species to climate change.

References

AbdElgawad, H., Peshev, D., Zinta, G., Van Den Ende, W., Janssens, I.A., Asard, H., 2014. Climate extreme effects on the chemical composition of temperate grassland species under ambient and elevated CO_2: a comparison of fructan and non-fructan accumulators. PLoS One 9, e92044. Available from: https://doi.org/10.1371/journal.pone.0092044.

Adler, P.B., Dalgleish, H.J., Ellner, S.P., 2012. Forecasting plant community impacts of climate variability and change: when do competitive interactions matter? J. Ecol. 100, 478–487.

Ainsworth, E.A., Long, S.P., 2005. What have we learned from 15 years of free-air CO_2 enrichment (FACE)? A *meta*-analytic review of the responses of photosynthesis, canopy properties and plant production to rising CO2. New Phytol. 165, 351–372.

Aldridge, G., Inouye, D.W., Forrest, J.R.K., Barr, W.A., Miller-Rushing, A.J., 2011. Emergence of a mid-season period of low floral resources in a montane meadow ecosystem associated with climate change. J. Ecol. 99, 905–913.

Andalo, C., Godelle, B., Lefranc, M., Mousseau, M., Till-Bottraud, I., 1996. Elevated CO_2 decreases seed germination in *Arabidopsis thaliana*. Glob. Change Biol. 2, 129–135.

Andresen, L.C., Yuan, N., Seibert, R., Moser, G., Kammann, C.I., Luterbacher, J., et al., 2018. Biomass responses in a temperate European grassland through 17 years of elevated CO_2. Glob. Change Biol. 24, 3875–3885.

Bagne, K., Ford, P., Reeves, M., 2012. Grasslands. U.S. Department of Agriculture, Forest Service, Climate Change Resource Center. http://www.fs.usda.gov/ccrc/topics/grasslands/. (accessed 09.04.21). Available from: https://doi.org/10.1016/S0098-8472(03)00022-4.

Bai, Y., Romo, J.T., 1997. Seed production, seed rain, and the seedbank of fringed sagebrush. J. Range Manage 50, 151–155.

Bai, Y., Tischler, C.R., Booth, D.T., Taylor Jr., E.M., 2003. Variations in germination and grain quality within a rust-resistant common wheat germplasm as affected by parental CO_2 conditions. Environ. Exp. Bot. 50, 159–168.

Barbehenn, R.V., Chen, Z., Karowe, D.N., Spickard, A., 2004. C3 grasses have higher nutritional quality than C4 grasses under ambient and elevated atmospheric CO2. Glob. Change Biol. 10, 1565–1575.

Baruch, Z., Jackson, R.B., 2005. Responses of tropical native and invader C4 grasses to water stress, clipping and increased atmospheric CO_2 concentration. Oecologia 145, 522–532.

Baskin, C.C., Baskin, J.M., 2014. Seeds: Ecology, Biogeography, and Evolution of Dormancy and Germination, second ed. Academic Press/Elsevier, San Diego.

Bazzaz, F.A., 1998. Tropical forests in a future climate: changes in biological diversity and impact on the global carbon cycle. Clim. Change 39, 317–336.

Belovsky, G.E., Slade, J.B., 2021. Climate change and primary production: forty years in a bunchgrass prairie. PLoS One 15, e0243496. Available from: https://doi.org/10.1371/journal.pone.0243496.

Benson, E.J., Hartnett, D.C., 2006. The role of seed and vegetative reproduction in plant recruitment and demography in tallgrass prairie. Plant Ecol. 187, 163–178.

Bertin, R.I., 2008. Plant phenology and distribution in relation to recent climate change. J. Torrey Bot. Soc. 135, 126–146.
Bezemer, T.M., Jones, T.H., 2012. The effects of CO_2 and nutrient enrichment on photosynthesis and growth of *Poa annua* in two consecutive generations. Ecol. Res. 27, 873–882.
Bloor, J.M.G., Pichon, P., Falcimagne, R., Leadley, P., Soussana, J.F., 2010. Effects of warming, summer drought, and CO_2 enrichment on aboveground biomass production, flowering phenology, and community structure in an upland grassland ecosystem. Ecosystems 13, 888–900.
Blumenthal, D.M., Kray, J.A., Ortmans, W., Ziska, L.H., Pendall, E., 2016. Cheatgrass is favored by warming but not CO_2 enrichment in a semi-arid grassland. Glob. Change Biol. 22, 3026–3038.
Buhrmann, R.D., Ramdhani, S., Pammenter, N.W., Naidoo, S., 2016. Grasslands feeling the heat: the effects of elevated temperatures on a subtropical grassland. Bothalia 46, a2122. Available from: https://doi.org/10.4102/abc.v46i2.2122.
Catford, J.A., Jones, L.P., 2019. Grassland invasion in a changing climate. In: Gibson, D.J., Newman, J.A. (Eds.), Grassland and Climate Change. Cambridge University Press, Cambridge, pp. 149–171.
Ceballos, G., Davidson, A., List, R., Pacheco, J., Manzano-Fischer, P., Santos-Barrera, G., et al., 2010. Rapid decline of a grassland system and its ecological and conservation implications. PLoS One 5, e8562. Available from: https://doi.org/10.1371/journal.pone.0008562.
Cleland, E.E., Chiariello, N.R., Loarie, S.R., Mooney, H.A., Field, C.B., 2006. Diverse responses of phenology to global changes in a grassland ecosystem. Proc. Natl. Acad. Sci. USA 103, 13740–13744.
Cleland, E.E., Collins, S.L., Dickson, T.L., Farrer, E.C., Gross, K.L., Gherardi, L.A., et al., 2013. Sensitivity of grassland plant community composition to spatial vs. temporal variation in precipitation. Ecology 94, 1687–1696.
Collatz, G.J., Berry, J.A., Clark, J.S., 1998. Effects of climate and atmospheric CO_2 partial pressure on the global distribution of C_4 grasses: present, past, and future. Oecologia 114, 441–454.
Compagnoni, A., Adler, P.B., 2014. Warming, competition, and *Bromus tectorum* population growth across an elevation gradient. Ecosphere 5, 121. Available from: https://doi.org/10.1890/ES14-00047.1.
Cook, B.I., Wolkovich, E.M., Davies, T.J., Ault, T.R., Betancourt, J.L., Allen, J.M., et al., 2012. Sensitivity of spring phenology to warming across temporal and spatial climate gradients in two independent databases. Ecosystems 15, 1283–1294.
Craine, J.M., Wolkovich, E.M., Gene Towne, E., Kembel, S.W., 2012. Flowering phenology as a functional trait in a tallgrass prairie. New Phytol. 193, 673–682.
Dalgleish, H.J., Koons, D.N., Hooten, M.B., Moffet, C.A., Adler, P.B., 2011. Climate influences the demography of three dominant sagebrush steppe plants. Ecology 92, 75–85.
de Faria, A.P., Fernandes, G.W., França, M.G.C., 2015. Predicting the impact of increasing carbon dioxide concentration and temperature on seed germination and seedling establishment of African grasses in Brazilian Cerrado. Austral Ecol. 40, 962–973.
Derner, J.D., Briske, D.D., 2001. Below-ground carbon and nitrogen accumulation in perennial grasses: a comparison of caespitose and rhizomatous growth forms. Plant Soil 237, 117–127.
Dixon, A.P., Faber-Langendoen, D., Josse, C., Morrison, J., Loucks, C.J., 2014. Distribution mapping of world grassland types. J. Biogeogr. 41, 2003–2019.
Dodd, M.B., Newton, P.C.D., Lieffering, M., Luo, D., 2010. The responses of three C4 grasses to elevated temperature and CO_2 in the field. Proc. New Zeal. Grassl. Assoc. 72, 61–66.
Drake, B.G., Rogers, H.H., Allen, L.H., 1985. Methods of exposing plants to elevated carbon dioxide. In: Strain, B., Cure, J. (Eds.), Direct Effects of Increasing Carbon Dioxide on Vegetation. U.S. Department of Energy, Washington, DC, pp. 11–32.
Dudney, J., Hallett, L.M., Larios, L., Farrer, E.C., Spotswood, E.N., Stein, C., et al., 2017. Lagging behind: have we overlooked previous-year rainfall effects in annual grasslands? J. Ecol. 105, 484–495.
Dunnell, K.L., Travers, S.E., 2011. Shifts in the flowering phenology of the northern Great Plains: patterns over 100 years. Am. J. Bot. 98, 935–945.
Edwards, G.R., Clark, H., Newton, P.C.D., 2001a. The effects of elevated CO_2 on seed production and seedling recruitment in a sheep-grazed pasture. Oecologia 127, 383–394.
Edwards, G.R., Newton, P.C.D., Tilbrook, J.C., Clark, H., 2001b. Seedling performance of pasture species under elevated CO_2. New Phytol. 150, 359–369.
Ehrlén, J., 2019. Climate change in grasslands—demography and population dynamics. In: Gibson, D.J., Newman, J.A. (Eds.), Grassland and Climate Change. Cambridge University Press, Cambridge, pp. 172–187.
Epstein, H.E., Gill, R.A., Paruelo, J.M., Lauenroth, W.K., Jia, G.J., Burke, I.C., 2002. The relative abundance of three plant functional types in temperate grasslands and shrublands of North and South America: effects of projected climate change. J. Biogeogr. 29, 875–888.
Fenner, M., 1991. The effects of the parent environment on seed germinability. Seed Sci. Res. 1, 75–84.
Fischer, E.M., Sedláček, J., Hawkins, E., Knutti, R., 2014. Models agree on forced response pattern of precipitation and temperature extremes. Geophys. Res. Lett. 41, 8554–8562.
Gao, S., Wang, J., Zhang, Z., Dong, G., Guo, J., 2012. Seed production, mass, germinability, and subsequent seedling growth responses to parental warming environment in *Leymus chinensis*. Crop Pasture Sci. 63, 87–94.
Ge, Q., Wang, H., Rutishauser, T., Dai, J., 2015. Phenological response to climate change in China: a *meta*-analysis. Glob. Change Biol. 21, 265–274.
Godfree, R.C., Marshall, D.J., Young, A.G., Miller, C.H., Mathews, S., 2017. Empirical evidence of fixed and homeostatic patterns of polyploid advantage in a keystone grass exposed to drought and heat stress. R. Soc. Open Sci. 4, 1270934. Available from: https://doi.org/10.1098/rsos.170934.
Grime, J.P., 2006. Plant Strategies, Vegetation Processes, and Ecosystem Properties, second ed. John Wiley, Chichester.

Grünzweig, J.M., Dumbur, R., 2012. Seed traits, seed-reserve utilization and offspring performance across pre-industrial to future CO_2 concentrations in a Mediterranean community. Oikos 121, 579–588.

Grünzweig, J.M., Körner, C., 2001. Growth, water and nitrogen relations in grassland model ecosystems of the semi-arid Negev of Israel exposed to elevated CO_2. Oecologia 128, 251–262.

Hampton, J.G., Boelt, B., Rolston, M.P., Chastain, T.G., 2013. Effects of elevated CO_2 and temperature on seed quality. J. Agric. Sci. 151, 154–162.

Harper, J.L., 1977. Population Biology of Plants. Academic Press, London.

Hikosaka, K., Kinugasa, T., Oikawa, S., Onoda, Y., Hirose, T., 2011. Effects of elevated CO_2 concentration on seed production in C3 annual plants. J. Exp. Bot. 62, 1523–1530.

Hovenden, M.J., Williams, A.L., 2010. The impacts of rising CO_2 concentrations on Australian terrestrial species and ecosystems. Austral Ecol. 35, 665–684.

Hovenden, M.J., Williams, A.L., Pedersen, J.K., Vander Schoor, J.K., Wills, K.E., 2008a. Elevated CO_2 and warming impacts on flowering phenology in a southern Australian grassland are related to flowering time but not growth form, origin or longevity. Aust. J. Bot. 56, 630–643.

Hovenden, M.J., Wills, K.E., Chaplin, R.E., Vander Schoor, J.K., Williams, A.L., Osanai, Y., et al., 2008b. Warming and elevated CO_2 affect the relationship between seed mass, germinability and seedling growth in *Austrodanthonia caespitosa*, a dominant Australian grass. Glob. Change Biol. 14, 1633–1641.

Huxman, T.E., Hamerlynck, E.P., Jordan, D.N., Salsman, K.J., Smith, S.D., 1998. The effects of parental CO_2 environment on seed quality and subsequent seedling performance in *Bromus rubens*. Oecologia 114, 202–208.

Jablonski, L.M., Wang, X., Curtis, P.S., 2002. Plant reproduction under elevated CO_2 conditions: a *meta*-analysis of reports on 79 crop and wild species. New Phytol. 156, 9–26.

Jaggard, K.W., Qi, A., Ober, S., 2010. Possible changes to arable crop yields by 2050. Philos. Trans. R. Soc. B 365, 2835–2851.

Jones, M.B., 2019. Projected climate change and the global distribution of grasslands. In: Gibson, D.J., Newman, J.A. (Eds.), Grassland and Climate Change. Cambridge University Press, Cambridge, pp. 67–81.

Lamichaney, A., Tewari, K., Basu, P.S., Katiyar, P.K., Singh, N.P., 2021. Effect of elevated carbon-dioxide on plant growth, physiology, yield and seed quality of chickpea (*Cicer arietinum* L.) in Indo-Gangetic plains. Physiol. Mol. Biol. Plants 27, 251–263.

Langley, J.A., Chapman, S.K., La Pierre, K.J., Avolio, M., Bowman, W.D., Johnson, D.S., et al., 2018. Ambient changes exceed treatment effects on plant species abundance in global change experiments. Glob. Change Biol. 24, 5668–5679.

Lee, J.S., 2011. Combined effect of elevated CO_2 and temperature on the growth and phenology of two annual C3 and C4 weedy species. Agric. Ecosyst. Environ. 140, 484–491.

Lesica, P., Kittelson, P.M., 2010. Precipitation and temperature are associated with advanced flowering phenology in a semi-arid grassland. J. Arid Environ. 74, 1013–1017.

Li, J., 2013. Germination Thresholds of the Mixed-grass Prairie Species as Affected by Global Climate Change: A FACE Study (M.Sc. thesis). University of Saskatchewan, Saskatoon, Canada.

Liu, Y., Mu, J., Niklas, K.J., Li, G., Sun, S., 2012. Global warming reduces plant reproductive output for temperate multi-inflorescence species on the Tibetan plateau. New Phytol. 195, 427–436.

Li, J., Ren, L., Bai, Y., Lecain, D., Blumenthal, D., Morgan, J., 2018. Seed traits and germination of native grasses and invasive forbs are largely insensitive to parental temperature and CO_2 concentration. Seed Sci. Res. 28, 303–311.

Li, L., Zhang, Y., Wu, J., Li, S., Zhang, B., Zu, J., et al., 2019. Increasing sensitivity of alpine grasslands to climate variability along an elevational gradient on the Qinghai-Tibet Plateau. Sci. Total Environ. 678, 21–29.

Marchi, S., Tognetti, R., Vaccari, F.P., Lanini, M., Kaligarič, M., Miglietta, F., et al., 2004. Physiological and morphological responses of grassland species to elevated atmospheric CO_2 concentrations in FACE-systems and natural CO_2 springs. Funct. Plant Biol. 31, 181–194.

Marques, I., Draper, D., Martins-Loução, M.A., 2007. Influence of temperature on seed germination in seven Mediterranean grassland species from SE Portugal. Bocconea 21, 367–372.

Marty, C., BassiriRad, H., 2014. Seed germination and rising atmospheric CO_2 concentration: a *meta*-analysis of parental and direct effects. New Phytol. 202, 401–414.

McKone, M.J., Kelly, D., Lee, W.G., 1998. Effect of climate change on mast-seeding species: frequency of mass flowering and escape from specialist insect seed predators. Glob. Change Biol. 4, 591–596.

Miranda-Apodaca, J., Pérez-López, U., Lacuesta, M., Mena-Petite, A., Muñoz-Rueda, A., 2015. The type of competition modulates the ecophysiological response of grassland species to elevated CO_2 and drought. Plant Biol. 17, 298–310.

Morgan, J.A., Lecain, D.R., Pendall, E., Blumenthal, D.M., Kimball, B.A., Carrillo, Y., et al., 2011. C4 grasses prosper as carbon dioxide eliminates desiccation in warmed semi-arid grassland. Nature 476, 202–206.

Mueller, K.E., Blumenthal, D.M., Pendall, E., Carrillo, Y., Dijkstra, F.A., Williams, D.G., et al., 2016. Impacts of warming and elevated CO_2 on a semi-arid grassland are non-additive, shift with precipitation, and reverse over time. Ecol. Lett. 19, 956–966.

Munson, S.M., Long, A.L., 2017. Climate drives shifts in grass reproductive phenology across the western USA. New Phytol. 213, 1945–1953.

Niklaus, P.A., Körner, C., 2004. Synthesis of a six-year study of calcareous grassland responses to in situ CO_2 enrichment. Ecol. Monogr. 74, 491–511.

Pagano, T.S., Olsen, E.T., Chahine, M.T., Ruzmaikin, A., Nguyen, H., Jiang, X., 2011. Monthly representations of mid-tropospheric carbon dioxide from the atmospheric infrared sounder. Proc. SPIE 8158, 81580C–1. Available from: https://doi.org/10.1117/12.894960.

Palit, R., Bai, Y., Romo, J., Coulman, B., Warren, R., 2017. Seed production in *Festuca hallii* is regulated by adaptation to long-term temperature and precipitation patterns. Rangel. Ecol. Manage. 70, 238–243.

Parmesan, C., 2007. Influences of species, latitudes and methodologies on estimates of phenological response to global warming. Glob. Change Biol. 13, 1860–1872.

Parmesan, C., Yohe, G., 2003. A globally coherent fingerprint of climate change impacts across natural systems. Nature 421, 37–42.

Piao, S., Liu, Q., Chen, A., Janssens, I.A., Fu, Y., Dai, J., et al., 2019. Plant phenology and global climate change: current progresses and challenges. Glob. Change Biol. 25, 1922–1940.

Reich, P.B., 2009. Elevated CO_2 reduces losses of plant diversity caused by nitrogen deposition. Science 326, 1399–1402.

Reich, P.B., Hobbie, S.E., Lee, T.D., 2014. Plant growth enhancement by elevated CO_2 eliminated by joint water and nitrogen limitation. Nat. Geosci. 7, 920–924.

Reich, P.B., Hobbie, S.E., Lee, T.D., Pastore, M.A., 2018. Unexpected reversal of C3 vs C4 grass response to elevated CO_2 during a 20-year field experiment. Science 360, 317–320.

Roach, D.A., Wulff, R.D., 1987. Maternal effects in plants. Annu. Rev. Ecol. Syst. 18, 209–235.

Sherry, R.A., Zhou, X., Gu, S., Arnone, J.A., Schimel, D.S., Verburg, P.S., et al., 2007. Divergence of reproductive phenology under climate warming. Proc. Natl. Acad. Sci. USA 104, 198–202.

Singh, B., Singh, S.K., Matcha, S.K., Kakani, V.G., Wijewardana, C., Chastain, D., et al., 2019. Parental environmental effects on seed quality and germination response to temperature of *Andropogon gerardii*. Agronomy 9, 304. Available from: https://doi.org/10.3390/agronomy9060304.

Steinger, T., Gall, R., Schmid, B., 2000. Maternal and direct effects of elevated CO_2 on seed provisioning, germination and seedling growth in *Bromus erectus*. Oecologia 123, 475–480.

Still, C.J., Berry, J.A., Collatz, G.J., DeFries, R.S., 2003. Global distribution of C3 and C4 vegetation: carbon cycle implications. Glob. Biogeochem. Cycles 17, 6–14.

Thürig, B., Körner, C., Stöcklin, J., 2003. Seed production and seed quality in a calcareous grassland in elevated CO_2. Glob. Change Biol. 9, 873–884.

Vandvik, V., Elven, R., Töpper, J., 2017. Seedling recruitment in subalpine grassland forbs: predicting field regeneration behaviour from lab germination responses. Botany 95, 73–88.

van Staalduinen, M.A., Anten, N.P.R., 2005. Differences in the compensatory growth of two co-occurring grass species in relation to water availability. Oecologia 146, 190–199.

Wand, S.J.E., Midgley, G.F., Jones, M.H., Curtis, P.S., 1999. Responses of wild C4 and C3 grass (Poaceae) species to elevated atmospheric CO_2 concentration: a *meta*-analytic test of current theories and perceptions. Glob. Change Biol. 5, 723–741.

Wang, H., Liu, H., Cao, G., Ma, Z., Li, Y., Zhang, F., et al., 2020a. Alpine grassland plants grow earlier and faster but biomass remains unchanged over 35 years of climate change. Ecol. Lett. 23, 701–710.

Wang, X., Niu, B., Zhang, X., Yongtao He, Y., Shi, P., Miao, Y., et al., 2020b. Seed germination in alpine meadow steppe plants from Central Tibet in response to experimental warming. Sustainability 12, 1884. Available from: https://doi.org/10.3390/su12051884.

White, R., Murray, S., Mark, R., 2000. Pilot Analysis of Global Ecosystems: Grassland Ecosystems. World Resources Institute, Washington, DC.

Williams, A.L., Wills, K.E., Janes, J.K., Vander Schoor, J.K., Newton, P.C.D., Hovenden, M.J., 2007. Warming and free-air CO_2 enrichment alter demographics in four co-occurring grassland species. New Phytol. 176, 365–374.

Winslow, J.C., Hunt Jr., E.R., Piper, S.C., 2003. The influence of seasonal water availability on global C3 vs C4 grassland biomass and its implications for climate change research. Ecol. Model. 163, 153–173.

Wolf, A.A., Zavaleta, E.S., Selmants, P.C., 2017. Flowering phenology shifts in response to biodiversity loss. Proc. Natl. Acad. Sci. USA 114, 3463–3468.

Wolkovich, E.M., Cook, B.I., Allen, J.M., Crimmins, T.M., Betancourt, J.L., Travers, S.E., et al., 2012. Warming experiments underpredict plant phenological responses to climate change. Nature 485, 494–497.

Xia, J., Wan, S., 2013. Independent effects of warming and nitrogen addition on plant phenology in the Inner Mongolian steppe. Ann. Bot. 111, 1207–1217.

Xu, Z., Zhou, G., 2011. Responses of photosynthetic capacity to soil moisture gradient in perennial rhizome grass and perennial bunchgrass. BMC Plant Biol. 11, 21. Available from: https://doi.org/10.1186/1471-2229-11-21.

Yu, M., Wang, G., Parr, D., 2014. Future changes of the terrestrial ecosystem based on a dynamic vegetation model driven with RCP8.5 climate projections from 19 GCMs. Clim. Change 127, 257–271.

Zavaleta, E.S., Shaw, M.R., Chiariello, N.R., Mooney, H.A., Field, C.B., 2003. Additive effects of simulated climate changes, elevated CO_2, and nitrogen deposition on grassland diversity. Proc. Natl. Acad. Sci. USA 100, 7650–7654.

Zelikova, T.J., Blumenthal, D.M., Williams, D.G., Souza, L., LeCain, D.R., Morgan, J., et al., 2014. Long-term exposure to elevated CO_2 enhances plant community stability by suppressing dominant plant species in a mixed-grass prairie. Proc. Natl. Acad. Sci. USA 111, 15456–15461.

Zelikova, T.J., Williams, D.G., Hoenigman, R., Blumenthal, D.M., Morgan, J.A., Pendall, E., 2015. Seasonality of soil moisture mediates responses of ecosystem phenology to elevated CO_2 and warming in a semi-arid grassland. J. Ecol. 103, 1119–1130.

Zhang, R., Luo, K., Chen, D., Baskin, J., Baskin, C., Wang, Y., et al., 2020. Comparison of thermal and hydrotime requirements for seed germination of seven *Stipa* species from cool and warm habitats. Front. Plant Sci. 11, 560714. Available from: https://doi.org/10.3389/fpls.2020.560714.

Zhang, R., Schellenberg, M.P., Han, G., Wang, H., Li, J., 2018. Drought weakens the positive effects of defoliation on native rhizomatous grasses but enhances the drought-tolerance traits of native caespitose grasses. Ecol. Evol. 8, 12126–12139.

Zhang, H., Tian, Y., Zhou, D., 2015. A modified thermal time model quantifying germination response to temperature for C3 and C4 species in temperate grassland. Agriculture 5, 412–426.

Zhang, H., Yu, Q., Huang, Y., Zheng, W., Tian, Y., Song, Y., et al., 2014. Germination shifts of C3 and C4 species under simulated global warming scenario. PLoS One 9, e105139. Available from: https://doi.org/10.1371/journal.pone.0105139.

Zhou, X.H., He, W.M., 2020. Climate warming facilitates seed germination in native but not invasive *Solidago canadensis* populations. Front. Ecol. Evol. 8, 595214. Available from: https://doi.org/10.3389/fevo.2020.595214.

Chapter 8

Climate change and plant regeneration from seeds in Mediterranean regions of the Northern Hemisphere

Efisio Mattana[1], Angelino Carta[2,3], Eduardo Fernández-Pascual[4], Jon E. Keeley[5,6] and Hugh W. Pritchard[1]

[1]*Royal Botanic Gardens, Kew, Wakehurst, West Sussex, United Kingdom,* [2]*Department of Biology, Botany Unit, University of Pisa, Pisa, Italy,* [3]*CIRSEC - Centre for Climate Change Impact, University of Pisa, Pisa, Italy,* [4]*IMIB – Biodiversity Research Institute, University of Oviedo, Mieres, Spain,* [5]*US Geological Survey, Western Ecological Research Center, Sequoia-Kings Canyon Field Station, Three Rivers, CA, United States,* [6]*Department of Ecology and Evolutionary Biology, University of California, Los Angeles, CA, United States*

Introduction

Mediterranean regions host approximately one-sixth of the global flora (Cowling et al., 1996; Rundel et al., 2018) but cover only around 2% of the land area. For this reason, they are considered to be global biodiversity hotspots (Myers et al., 2000). These regions are characterized by mild-wet winters and hot-dry summers. This type of climate occurs on the west side of continents between approximately 30 degrees and 40 degrees of latitude (Lionello et al., 2006), notably in the Mediterranean Basin, California, central Chile, the Cape Region of South Africa, and southern Australia. The most distinctive feature of this kind of climate is the concentration of rainfall in the winter half of the year (i.e., November–April in the Northern Hemisphere and May–October in the Southern Hemisphere). Although winter frost may occur, especially in continental areas and at high elevations, its duration usually is short (Aschmann, 1973). In this context, landscapes are dominated by evergreen sclerophyllous-leaved shrubs but also have a wide diversity of herbaceous plant species (Mooney and Dunn, 1970).

The Mediterranean Basin includes more than one-half of the total area of the Mediterranean climate regions (Aschmann, 1973). The region is both a refuge and an area of active plant speciation (Quézel, 1978, 1985; Thompson, 2005) and an area of floristic exchange between the Holarctic and Holotropical kingdoms (Morrone, 2015; Carta et al., 2022). Since the Neolithic, the spatial configuration and size of natural habitats in this region have been driven by intense human impact on the landscape, which has created new opportunities for colonization of some plant species and caused others to retract into isolated patches (Thompson, 2005).

In North America, the Mediterranean climate region is centered in California (United States), which has the most extreme summer drought among the Mediterranean regions (Aschmann, 1973) and receives most of its precipitation later in the year than the Mediterranean Basin, typically peaking in January (Iacobellis et al., 2016). Major gradients in topography, climate, and edaphic conditions, as well as geographical isolation, have contributed to high plant diversification in California (Rundel et al., 2018). The Mediterranean region of California has a diverse endemic flora associated with serpentine soils (Harrison et al., 2006). As with the Mediterranean Basin, California has a long history of human perturbations that have greatly altered much of the native vegetation (Syphard et al., 2018).

Projections of how climate change will affect Mediterranean climate regions differ according to the model chosen and timeframe considered. However, it is possible to summarize the most likely scenarios as: (1) a decrease in amount of annual precipitation and changes in its timing and frequency, (2) a pronounced warming, and (3) an increase in frequency of extreme temperature events (Giorgi and Lionello, 2008; Berg and Hall, 2015; Polade et al., 2017) (Fig. 8.1). The two Mediterranean climate regions of the Northern Hemisphere are located in areas where projected warming is strongest (Polade et al., 2017) and where frequencies of extreme precipitation events are projected to increase, especially in California (Polade et al., 2017). However, the overall amount of annual precipitation is projected to decrease

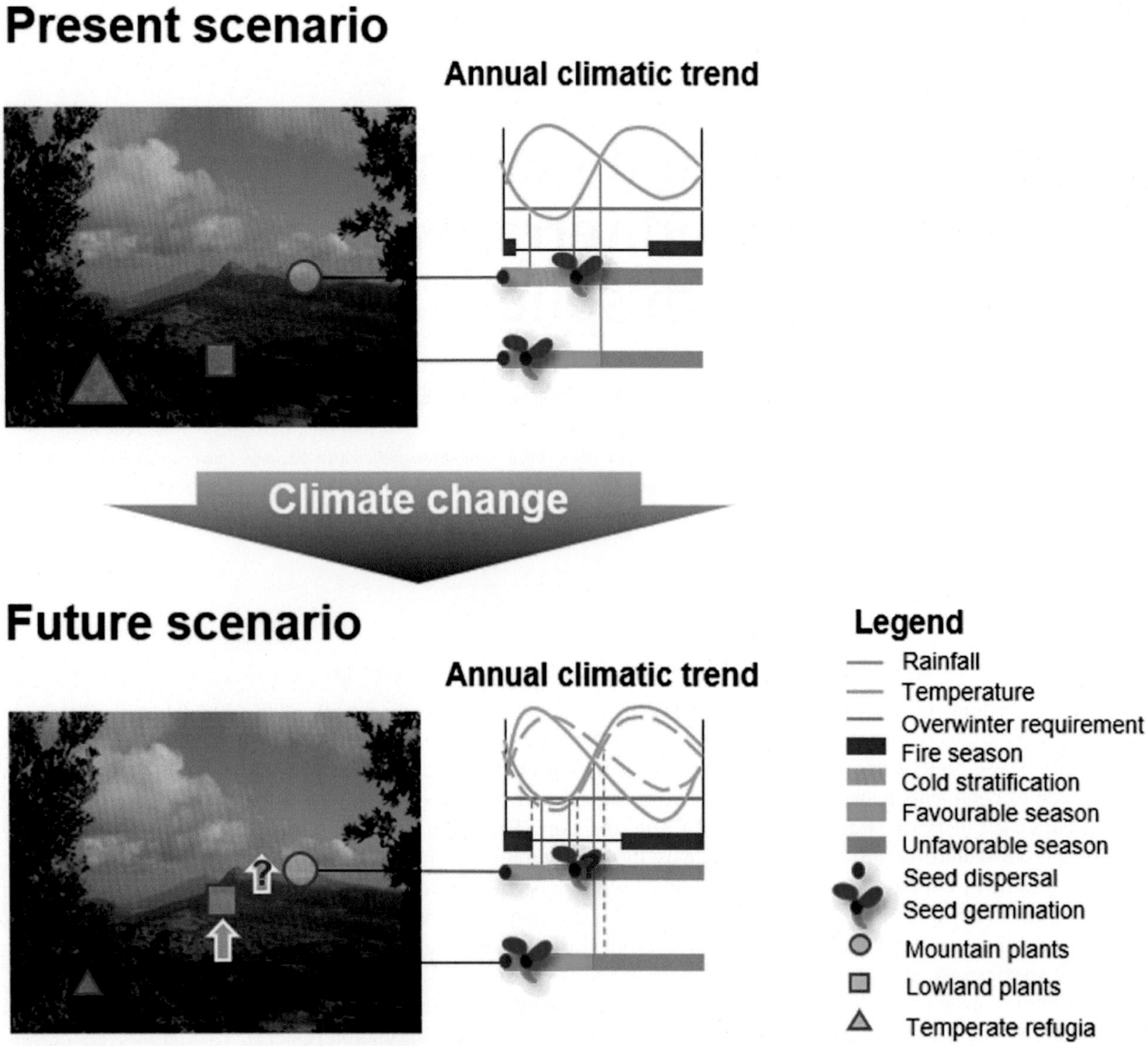

FIGURE 8.1 Temporal and spatial effects of climate change on reproduction by seeds of lowland and mountain plants under a Mediterranean climate. Mediterranean lowland plants exhibit the typical "Mediterranean germination syndrome" with seed germination occurring mainly in the autumn following dispersal (Box 8.1). As a result of climate change, the favorable season for germination will decrease, while length of the summer drought (adverse season) will increase, narrowing the favorable window for seedling establishment. However, these lowland plants may migrate to higher elevations or latitudes. Mediterranean mountain plants usually have seeds that need overwintering to break dormancy and germinate the first spring following seed dispersal. In addition to the decrease in the favorable germination season, the overwintering period may become too short and mild to break dormancy, negatively affecting (or even inhibiting) germination and seedling establishment. Further migration to higher elevations may not be possible for mountain plants. Temperate refugia, such as riparian woodlands or patches of mountain vegetation in an evergreen sclerophyllous scrubland matrix, may be an alternative escape route, but these refugia likely will be reduced in size and number, thus increasing interspecific competition. Annual climate plots start from the end of the seed dispersal season (late summer). Vertical lines link the changes of seasons with the annual trends of rainfall and temperature. In the "Future scenario," dashed lines correspond to former conditions under the "Present scenario."

in the Mediterranean Basin (Polade et al., 2017). Climate warming will be coupled with a pronounced rise in sea level, which in coastal habitats will result in greater seawater incursion into estuaries than at present (Callaway et al., 2007). The subsequent increase in salinity could have devastating effects on water resources, terrestrial and marine ecosystems, and human wellbeing (Giorgi and Lionello, 2008).

The seasonal alternation of mild-wet winters and hot-dry summers typical of the Mediterranean climate is conducive to wildfires each summer (Fig. 8.1). The mild-wet winter-spring season leads to moderate plant productivity that generates broad landscapes of contiguous fuels, and during the annual summer drought this biomass becomes highly flammable fuel (Keeley, 2012). Wildfire intensity and frequency have been hypothesized to increase in Mediterranean climates as a result of increasing aridity and increasing anthropic activities (Piñol et al., 1998). Increased fire intensity and frequency do not allow ecosystems to fully recover between wildfire events, greatly altering the biota (Keeley et al., 2009). In the Mediterranean Basin, climate change is predicted to decrease the time interval between successive fires, leading to shrub-dominated landscapes (Mouillot et al., 2002). In California, there has been a twofold increase in size of area burned since 2000 compared to that burned in the previous two decades. Climatologists contend that this change in fire regime correlates with an increase in temperature (Williams et al., 2019), but anthropic activities are involved (Keeley and Syphard, 2019). Projecting future fire regimes requires consideration of climate, human-caused ignitions, and land management decisions. For example, in the montane forested landscape of California fires are strongly climate-limited, whereas at lower elevations they are ignition-limited. Therefore, in some areas increased human population growth ultimately may be more important in driving future fire regimes than climate (Keeley and Syphard, 2016).

In this chapter, we review studies dealing with effects of climate change on the different steps of plant regeneration from seeds (seed production to seedling survival) in the Mediterranean Basin and California. We provide a synthesis of the complex interactions between plant reproduction and climate change and identify future research directions for assessing the importance of early plant life-cycle stages in the Northern Hemisphere Mediterranean landscape under a changing climate.

Plant regeneration from seeds under a changing climate

Seed dormancy-breaking and germination requirements ensure that the timing of reproduction from seeds is synchronized with the most favorable season for seedling establishment (Baskin and Baskin, 2014). This is of pivotal importance under the highly seasonal Mediterranean climate (Box 8.1). Here, the temporal window for regeneration is restricted by drought in summer and by low temperature and frost (with increasing elevation) in winter (McClain, 2016; Picciau et al., 2019; Gremer et al., 2020) (Fig. 8.1). At the same time, seedlings are the most vulnerable stage of the plant life cycle, being subjected to abiotic and biotic constraints that affect their emergence, survival, and establishment (Harper, 1977; Moles and Westoby, 2004; Leck et al., 2008; Cogoni et al., 2012). Climatic changes predicted for Mediterranean regions in the Northern Hemisphere (Giorgi and Lionello, 2008; Polade et al., 2017) will affect seed dormancy, germination, and seedling establishment.

Seed production

Drought and warming treatments simulating climate change have reduced flower production and fruit and seed set for the Mediterranean species, *Dorycnium pentaphyllum*, *Erica multiflora*, and *Helianthemum syriacum* (Del Cacho et al., 2013b). Under future warming, it is predicted that *Acer platanoides* will produce smaller seeds with lower nutrient concentration and reduced viability. However, drier conditions in southern Europe may actually increase the seed quality of *A. platanoides* and thereby partly compensate for the negative effect of warmer conditions on seed production (Carón et al., 2014). In contrast, seed nutrient concentration, mass, and size and early seedling establishment are not expected to change for *A. pseudoplatanus* (Carón et al., 2014). However, Daws et al., (2006) found that increased temperature likely had a positive effect on seed quality traits of this species during development across Europe (from Italy to Scotland).

Mild winters, decreased cold stratification, and seed dormancy break

Mild winters, with a decrease in duration of the cold stratification period, can negatively affect species that require winter cold to break physiological dormancy of the seeds (Walck et al., 2011) (Fig. 8.1). Under projected climate change predictions, there is a risk for reduction in dormancy-break in seeds of *Vitis vinifera* subsp. *sylvestris* in Sardinia (Italy), with lowland populations more threatened than high elevation populations due to winter temperatures in the lowlands being too high to break seed dormancy (Orrù et al., 2012). Similarly, according to climate change predictions the number of days with temperatures low enough for seed dormancy release in *Gentiana lutea* subsp. *lutea* in the Sardinian mountains will be less than the 30-day threshold required, thereby preventing high percentages of germination (Cuena-Lombraña et al., 2020).

> **BOX 8.1 Seed germination ecology of Mediterranean plants from the Northern Hemisphere.**
>
> The "Mediterranean germination syndrome" is characterized by a cold-cued and slow germination (Thanos et al., 1989, 1991, 1995; Doussi and Thanos, 2002). This delay mechanism is an ecological adaptation to the unpredictable rainfall pattern of the Mediterranean climate (Doussi and Thanos, 2002). In practice, it limits germination to autumn/winter, maximizing the length of the growing season before the onset of summer drought (Thanos et al., 1995). In addition, Mediterranean plants often rely on a depth-sensing mechanism for seedling establishment, via photoinhibition of seed germination, which prevents germination on and close to the surface (Thanos et al., 1989, 1991; Carta et al., 2017), often balanced by a diurnally fluctuating temperature detection mechanism that prevents seeds from germinating too deep in the soil (Saatkamp et al., 2011).
>
> During the Mediterranean summer, seeds experience dry after-ripening, that is, warm-dry conditions after dispersal (Baskin and Baskin, 2014), which can release dormancy in seeds of coastal and lowland species, thereby helping to synchronize seed germination with the onset of autumn precipitation (Picciau et al., 2019).
>
> Warm-wet stratification is needed for embryo growth in species that disperse seeds before embryo development is complete, that is, species with morphological or morphophysiological seed dormancy (Baskin and Baskin, 2014). In the Mediterranean climate, where the warm season is also the dry season, warm-wet stratification is associated with riparian woods and patches of mesophilous mountain vegetation (Mattana et al., 2012; Carta et al., 2014; Porceddu et al., 2017, 2020), which act as "temperate refugia" within an evergreen and sclerophyllous landscape (Sanz et al., 2011).
>
> Cold-wet stratification has been reported to have no effect or to be detrimental for seed germination of typical lowland Mediterranean species (Skordilis and Thanos, 1995; Luna et al., 2008). However, some California chaparral species require cold-wet stratification for germination, but the duration may be relatively brief (Keeley, 1987), suggesting this may be a critical cue to indicate winter conditions, the time of most seedling establishment (Keeley, 1987).
>
> Mediterranean mountain plants do not seem to differ very much in their germination characteristics from cold-adapted temperate plants, suggesting that elevation extremes outweigh geographical patterns. Similar to temperate alpine plants (Fernández-Pascual et al., 2021), Mediterranean mountain species have higher germination temperatures and germinate faster than Mediterranean lowland species (Giménez-Benavides et al., 2005; Picciau et al., 2019). Nevertheless, seeds of a high number of Mediterranean mountain species are nondormant at dispersal and can germinate without any pretreatment (Giménez-Benavides et al., 2005; Lorite et al., 2007; Picciau et al., 2019).
>
> The ability of seeds to survive desiccation plays an important role in the regeneration ecology of Mediterranean species (Tweddle et al., 2003). Although species with desiccation-sensitive seeds are most common in tropical forest systems, seed desiccation sensitivity may also occur in trees from temperate regions (Tweddle et al., 2003). Indeed, the most widespread broadleaf forests in the Mediterranean Basin are dominated by *Quercus* species, which produce desiccation-sensitive seeds despite the drought that characterizes the Mediterranean climate (Joët et al., 2013), as is also the case with a number of California sclerophyllous species (Keeley, 1997).
>
> As a response to fire, some plant species, that is, obligate seeders (versus obligate resprouters) (Keeley, 1987), have evolved delayed regeneration, whereby, despite continuous seed production, seedling recruitment occurs mainly in one pulse after a fire (Keeley, 2012). Regeneration after fire requires response to physical (heat) and physiological (smoke) fire cues. Physically dormant seeds, such as those in the Cistaceae (Thanos et al., 1992), have water-impermeable seed or fruit coats that may be rendered permeable by heat. Seeds with a physiological inhibiting mechanism in the embryo (Keeley and Fotheringham, 1997, 1998) require a fire-induced cue, that is, charcoal, and nitrogenous compounds collectively termed "smoke" (Keeley and Pausas, 2018), to break physiological seed dormancy and trigger germination. Also, some species have seeds with little or no dormancy that are produced in serotinous cones or fruits that open only after fire, for example, some Pinaceae species (Moya et al., 2013).
>
> The role of fire in the selection of physical dormancy in Mediterranean Basin ecosystems recently has been questioned (Santana et al., 2020). In the northern Mediterranean Basin, postfire seeders include species of gymnosperms (Pinaceae) and species in various angiosperm families, especially those belonging to the rosid clade (*sensu* APG-IV, Chase et al., 2016), such as Cistaceae, Fabaceae, Rosaceae, and those in the asterid clade, including Ericaceae and Lamiaceae (Paula et al., 2009). In California, postfire seeders include species in the families mentioned above plus *Ceanothus* spp. in the Rhamnaceae (Keeley, 2012). However, fire-independent recruitment is widespread in floras of these two Mediterranean regions, and postfire seeding tends to dominate at the arid end of the gradient (Keeley, 2012).

Increased germination temperatures

After dormancy is broken, seeds of many species have specific temperature requirements for germination, and for Mediterranean species the optimum temperatures are relatively low, for example, 20°C, 17.7°C, and 21.4°C for trees, shrubs, and herbs, respectively (Baskin and Baskin, 2014). Due to global warming, it is expected that temperatures will become high enough during the rainy season to impact cold-cued germination of Mediterranean species (Box 8.1; Fig. 8.1). Temperatures above 20°C reduced germination percentage in eight perennial species of *Euphorbia* from Sicily (Cristaudo et al., 2019) and of *Stipa tenacissima* from North Africa (Krichen et al., 2014, 2017). In the northeastern Mediterranean region (Italy), rising temperatures are predicted to alter the structure of foredune plant communities by favoring regeneration from seeds of exotic and semifixed dune species to the detriment of important foredune species (Del Vecchio et al., 2021).

On the other hand, a milder winter also could mean earlier emergence for spring germinating species. A reduced-snowpack experiment in the mountains of Eldorado National Forest in California, following a stand-replacing fire, resulted in earlier emergence of conifer seedlings (Werner et al., 2019). However, identifying a single pattern of seed germination responses to warming is not possible due to inter- and intraspecific variability in dormancy-breaking and germination requirements. For example, germination of four Cistaceae species with physical dormancy (water-impermeable seed coat) from the Iberian Peninsula responded differently to increasing temperatures: some were affected negatively, some benefitted, and some were not affected (Chamorro et al., 2017).

If species with small geographic ranges also have a narrower regeneration niche breadth than widespread species (Brown, 1984; Thompson, 2005), endemic species should be more threatened by climate change than common ones (Walck et al., 2011). Seed germination of Iberian endemics is more sensitive to high incubation temperatures than widespread Mediterranean species, making them more vulnerable to increases in temperature (Luna et al., 2012). The same pattern was detected in four *Limonium* species from the east coast of Spain, with the endemic species *L. girardianum* and *L. santapolense* being more sensitive to increasing temperatures than the widespread Mediterranean species *L. virgatum* and *L. narbonense* (Monllor et al., 2018). However, more studies are needed to confirm the link between species range size and seed germination niche breadth.

Altered precipitation regimes and germination

Seed germination phenology in drought-adapted Mediterranean annual grasses is driven mainly by rainfall events (Jiménez-Alfaro et al., 2018), which could be altered by the projected changes in the seasonality of precipitation. Germination and normal seedling development of the acorn-recalcitrant species *Quercus ilex* are tightly linked to the water content of the acorn after winter, suggesting that seed desiccation sensitivity also may influence recruitment of this species in Mediterranean forests (Joët et al., 2013). However, in a field experiment that simulated projected decreases in precipitation in southern France the proportion of *Q. ilex* acorns that germinated was not affected significantly (García De Jalón et al., 2020). On the other hand, an increase in frequency and severity of flash floods is predicted for the Mediterranean Basin (Kyselý et al., 2012), and an increase in soil water levels during winter reduced germination and emergence of *Q. canariensis*, *Q. pyrenaica*, and *Q. suber* (Urbieta et al., 2008).

Sea level rise and salinity stress

Many studies have evaluated the effects of salinity on seed germination of Mediterranean plants from coastal and arid environments (e.g., Weber and D'Antonio, 1999; Gorai and Neffati, 2007; Santo et al., 2014; Estrelles et al., 2015; Al Hassan et al., 2017; Raddi et al., 2019). However, very few studies have explicitly identified seed germination responses to salt stress in a changing climate. Increasing temperature, drought stress (PEG-6000), and salinity (NaCl) inhibited seed germination of *Stipa tenacissima* in North Africa. According to climate predictions, the window of time during the year for recruitment of new individuals of this species will be narrower than that in the present climate, due to a shorter period of suitable water availability and temperature (Krichen et al., 2014, 2017). Sea level rise will directly affect the structure of coastal vegetation. In the San Francisco Bay area, sea level is expected to rise by 1.4 m during the next 90 years, and inundation will affect the population structure of the dominant halophyte *Sarcocornia pacifica*. Seedling survival of this species is estimated to decrease by about one-half in response to increased salinity (Woo and Takekawa, 2012).

Altered fire regimes

Fire is a natural ecosystem factor in all Mediterranean regions except Chile (Keeley, 2012). However, this natural disturbance is increasingly being modified by human intervention that alters different fire regime parameters, depending on the ecosystem (Keeley and Pausas, 2019). In some cases, perturbations have resulted in increased fire frequency and in others decreased fire frequency and subsequently increased fuel accumulation that leads to higher intensity fires.

The predicted increase in fire frequency could inhibit natural regeneration of species not adapted to fire. This may be the case for *Abies cephalonica* in the Greek mountains, where no seedling recruitment was observed three years after the fire (Ganatsas et al., 2012). Increasing fire frequency already has been implicated in the loss of many postfire-seeding shrubs in California (Syphard et al., 2018). Increase in wildfire frequency also will accelerate extinction of rare relict chaparral populations, such as those of the endemic shrub *Adenostoma sparsifolium* in California, whose relatively few viable seeds do not survive wildfires (Wiens et al., 2012). Some species also may benefit from increasing fire frequency, such as *Quercus trojana* in Italy, whose seedlings are more tolerant to the combined effect of drought and fire than those of the co-occurring species *Q. ilex* and *Q. virgiliana* (Chiatante et al., 2015).

The high yearly variability in amount of rainfall in these Mediterranean ecosystems (Moreno et al., 2011) means that interactions of fire with other climatic features in Mediterranean shrublands will be important. This is particularly the case for seeder species (see Box 8.1) since they regenerate only from seeds after fire. Changes in seedling emergence and recruitment in the seeder species *Cistus ladanifer*, *Erica umbellata*, and *Salvia rosmarinus* have been evaluated after experimental burns conducted both early and late during the fire season of three consecutive years. While fire season generally was not an important factor in controlling seedling emergence and recruitment, seedling emergence of *E. umbellata* and *S. rosmarinus*, but not of *C. ladanifer*, was positively correlated with precipitation in the autumn and winter immediately after fire. *E. umbellata* is the species most sensitive to reduced rainfall and thus will be the most threatened of the three species under projected climate change scenarios (Moreno et al., 2011).

Seedling survival

Overall, hotter and drier summers will decrease seedling survival (Werner et al., 2019). In the mountains of the Mediterranean region of California, higher summer temperatures and temperature anomalies are associated with lower survival and growth for most seedlings of forest trees, while rainfall and snow have weaker and more variable effects (Moran et al., 2019). Under projected future climate scenarios, survival and growth of most forest trees are predicted to decrease at their present sites, leading to a more open forest structure—or potentially to a transition to nonforest vegetation types (Moran et al., 2019). Similarly, seedling establishment windows for *Quercus douglasii* and *Q. kelloggii* in California are projected to decrease by 50%−95%, with populations becoming smaller and restricted to higher elevations and north-facing slopes by the end of the 21st century, depending on species and climate scenario (Davis et al., 2016). Sporadic rainfall events may be critical for woody species recruitment in Mediterranean climate communities (Matías et al., 2012). Overall, shrub species have higher seedling survival and growth and are less affected by severe drought than trees, whereas some tree species (mainly pines) are extremely dependent on wet summer conditions. Therefore the predicted reduction in the frequency of wet summers likely will alter species composition and dominance of future plant communities (Matías et al., 2012).

Local edaphic conditions may interact with changes in the climate to affect seedling survival, thus limiting species to specific soils, for example, sandy versus clayish soils for the cork oak (*Quercus suber*) (Ibáñez et al., 2014). Apart from drought, seedling survival of four Spanish oaks (*Q. canariensis*, *Q. ilex*, *Q. suber*, and *Q. pyrenaica*) also decreased with risk of waterlogging in wet sites (Gómez-Aparicio et al., 2008). Interactions of climatic factors and presence or absence of ectomycorrhizal−fungal colonization on seedling survival also should be considered. Indeed, drier conditions in spring could increase seedling survival of *Q. ilex* in summer via a synergistic effect of drought-adapted ectomycorrhizal−fungal colonization and less favorable conditions for root pathogens (García De Jalón et al., 2020).

Seedling establishment is controlled by climate, especially at the trailing edge of the distribution of a species, that is, at lower latitudes and elevations in the case of Mediterranean species. Mediterranean populations of species with a Eurasian distribution such as yew (*Taxus baccata*), which are at the southern limit of the species' distribution, strongly depend on water availability for recruitment (García et al., 2000). Seedling emergence and recruitment of yew are consistently enhanced in localities with high soil moisture (Sanz et al., 2009). These authors observed a trend of decreasing yew regeneration in the Mediterranean Basin, and they hypothesized that dry conditions due to climate change will further decrease the southern range of this species (Sanz et al., 2009). The same trend also was identified for *Betula alba*, for which the close dependence on moist microsites for germination could make low germination percentages an important bottleneck in its regeneration in the Mediterranean mountains (Sanz et al., 2011). Increasing temperatures also reduce seedling survival of Scots pine (*Pinus sylvestris*) at the southern edge of its distribution (Spain) (Matías and Jump, 2014). At the southern margin of the distribution of beech (*Fagus sylvatica*) in the Mediterranean region of France, high soil moisture, precipitation, and temperature during the growing season increased seedling density, while late spring and early autumn frosts decreased seedling density (Silva et al., 2012). Thus, projected increases in temperatures and drought severity will jeopardize future regeneration of the species at its trailing edge of distribution

(Silva et al., 2012). Intensification of summer drought also will hinder establishment of beech seedlings on the Iberian Peninsula (Robson et al., 2009). However, at the southern edge of distribution of beech in north-eastern Greece seedlings of this species exhibit differences among provenances, including drought tolerance and duration of growing period (timing of bud burst), that may enable them to survive under climate change (Varsamis et al., 2019).

What will happen in the new sites to which species might migrate, either to higher latitudes or elevations, that is, at the leading edge of the species distribution? Lenoir et al. (2009) provided evidence on how global warming already is having an impact on the altitudinal distribution of forest tree species in mountains in the Mediterranean region. In eight of 10 tree species, they found seedlings growing at a higher elevation than adults (mean difference in elevation 69 m) in the mountains of France. In an experiment involving three common gardens, one each at oceanic, continental, and Mediterranean sites in France, Mediterranean and southern European tree species (*Castanea sativa, Pinus halepensis,* and *Quercus ilex*) had high seed germination percentages and good seedling performance when placed in the oceanic and continental sites (Merlin et al., 2018). Since these three species are expected to migrate northward under climate warming, these results help confirm the projection that these species will increase their distribution range by shifting northward. On the other hand, relatively low seedling survival of the cold-adapted tree species *Pinus sylvestris, Picea abies, Abies alba*, and *Larix decidua* in the southern Mediterranean site suggested that these four species may retreat northward from southern France and to higher elevations (Merlin et al., 2018). Similarly, under warmer and drier conditions than at present the lowland Mediterranean pines *Pinus halepensis* and *P. pinaster* had higher survival and growth than the mountain species *P. nigra, P. sylvestris*, and *P. uncinata* (Matías et al., 2017).

Facilitation and drought stress

Presence of a vegetation canopy can facilitate seedling establishment in Mediterranean-type ecosystems limited by water availability (Mendoza et al., 2009). At the southern margin of its distribution in central and southern Spain, seedling emergence of holly (*Ilex aquifolium*) is strongly canopy-dependent, and seedlings reach their highest density in closed holly woodland but were absent in grasslands (Arrieta and Suárez, 2006). However, in a drought year no holly seedlings survived the first year even in holly woodlands, highlighting a bottleneck in population recruitment due to summer drought that is predicted to become more frequent in many climate change scenarios (Arrieta and Suárez, 2006). Similarly, under increasing temperatures it is predicted that seedling recruitment in Mediterranean populations of *Helleborus foetidus* likely will be restricted to areas with woody plant cover and fail to persist in open areas (Ramírez et al., 2006). Seedling recruitment in Mediterranean populations of *Quercus ilex* (Pérez-Ramos et al., 2013) and *Sarcopoterium spinosum* (Rysavy et al., 2016) also was facilitated by canopy shade of shrubs under dry conditions simulating climate change.

Undoubtedly, canopy effect and climate interactions are species-specific and vary significantly between years. Under experimental warming and drying in Mediterranean shrublands, the presence of a canopy increased survival of the dominant species *Erica multiflora* in stands that were warmed, but a canopy decreased survival of *Globularia alypum* in stands subjected to drought stress (Lloret et al., 2005). That is, under drought conditions canopy-forming species may compete with seedlings for water to the detriment of the seedlings. For example, survival of *E. multiflora* and *G. alypum* seedlings decreased when they were growing close to competing neighbors, which would have increased drought stress. Furthermore, seedlings of *G. alypum* established in long-term (9 years) drought plots lived for a longer period of time than those in warmed plots, probably because the drought plots had reduced canopy cover (Del Cacho et al., 2013a). The relative influence of drought, herbivory, and canopy facilitation on regeneration of *Pinus jeffreyi* varies by year, highlighting the importance of climatic and microhabitat factors, as well as that of species interactions, in the context of ecosystem responses to climate change (Alpert and Loik, 2013).

Canopy dieback caused by severe drought in Mediterranean shrublands in southern Spain was followed by germination and establishment of the woody species *Halimium halimifolium, Salvia rosmarinus, Lavandula stoechas*, and *Thymus mastichina* in the year following the drought (Del Cacho and Lloret, 2012). In the subsequent year, emergence of seedlings of these species decreased, apparently due to depletion of the soil seed bank and reduced seed rain; many seed-producing plants were killed by the drought. Seed germination of *Cistus libanotis* was delayed after drought, while seedling emergence of *Juniperus phoenicea* and *Pinus pinea* decreased significantly. Thus, increased frequency of recurrent severe droughts, which likely will increase under future climatic conditions, has the potential to induce permanent changes in composition of shrubland communities that need more time to replenish their seed banks between plant-killing drought events (Del Cacho and Lloret, 2012).

Moderate shading reduced the growth of *P. sylvestris*, but not of *P. nigra*, thereby increasing the vulnerability of *P. sylvestris* seedlings to drought (Bachofen et al., 2019). At its southernmost distribution in southern Spain, the shade of the shrub canopy facilitated both seedling emergence and survival of *P. sylvestris*. Thus, the ecological factors controlling seedling establishment in the southern *P. sylvestris* forests differ greatly from those operating in its main distribution area (Castro et al., 2004), where shading reduces seedling growth and increases vulnerability to drought (Bachofen et al., 2019).

Local adaptation and phenotypic plasticity in the face of climate change

The California chaparral species *Ceanothus cuneatus* and *Arctostaphylos pungens* also occur in the Sierra Madre of Mexico under a summer rain climate. These two species germinate under winter temperatures in California and under summer temperatures in Mexico (Keeley et al., 2012), highlighting how some Mediterranean species seem to be capable of adapting to changes in temperature during the germination period and thus may not be threatened by global warming. Climate-linked genetic variation in natural populations indicates that an evolutionary response is possible, as detected by seedling establishment of the Mediterranean shrub *Fumana thymifolia* in response to climate manipulations. In this species, single-locus genetic divergence occurred in drought and warming treatments, suggesting that rapid evolution in response to climate change may be widespread in natural populations, based on genetic variation already present within the population (Jump et al., 2008).

Furthermore, species may be able to track climate changes through phenotypic plasticity (Carta et al., 2016b). Plants from seeds of *Streptanthus tortuosus* in California that germinate in response to rain in early autumn were likely to flower the following spring and die, while those from seeds that germinated late in autumn did not flower in the first year and mostly behaved as biennials; a few were iteroparous perennials (Gremer et al., 2020). Thus, timing of germination-triggering rainfall events strongly affect plant life history. Richter et al. (2012) compared seedling growth performance of *Pinus sylvestris* seedlings from the Alps with that of *P. sylvestris* and *P. nigra* from a Mediterranean seed source and found that Mediterranean seedlings had lower phenotypic plasticity than alpine seedlings under high precipitation. Their results suggest that local provenances have the potential to cope with changes in climatic conditions as a function of both phenotypic plasticity and genotypic variation (Richter et al., 2012). Higher germination percentage of typical Mediterranean shrubland species, such as *Phillyrea angustifolia*, is associated with populations growing in habitats with more severe summers, suggesting high plasticity in the species, which would allow it to adapt rapidly to the future climate (Mira et al., 2017). Seedlings of *Quercus ilex* tolerate multiple stresses and physiologically acclimate to heat waves and cold snaps, suggesting that this species could cope with the increasingly stressful conditions imposed by climate change (Gimeno et al., 2009), although desiccation sensitivity of the acorn could still constrain natural regeneration.

Future directions

Further studies are needed to better understand the interactions of rapid climate change and plant reproduction from seeds in the Mediterranean Basin and California. Comparative studies among the world's Mediterranean climate regions could help us identify common responses to climate change and highlight specific threats to the plant communities of each region. Although some specific habitats such as the Mediterranean temporary ponds are still under-explored (see Box 8.2), there is a vast literature from the Mediterranean Basin and California that analyzes seed ecology in relation to one or more factors that could be related to climate change, such as increasing temperatures, salinization, drought stress, and fire. While only a few of these studies were explicitly designed to replicate current and projected climate scenarios, the wealth of information that has been accumulated in the past decades presents the opportunity to produce a global synthesis of the climatic drivers of plant regeneration by seeds in Mediterranean climates [but see recommendations in Körner and Hiltbrunner (2018)].

As highlighted by Del Cacho et al. (2013b), experimental work is needed to determine the effects of climate change on all stages in the reproductive life cycle of plants and the relationship between the stages. Furthermore, how climate change will affect plant production of specialized metabolites (Hashoum et al., 2020) and plant responses to pathogens (Homet et al., 2019) need to be investigated. There is limited knowledge about the ability of Mediterranean species to cope with climate change through phenotypic plasticity (Richter et al., 2012) and local adaptation (Petrů and Tielbörger, 2008). This knowledge could be gained by performing common garden and reciprocal transplant experiments with species differing in life-history traits (Giménez-Benavides et al., 2018). In addition, further studies on climate-linked genetic variation and seed germination (De Vitis et al., 2014, 2018) should be carried out to determine if rapid evolution of plant reproduction from seeds in response to climate change is common (Jump et al., 2008).

BOX 8.2 Seed ecology of plant species of Mediterranean temporary ponds under a changing climate.

Mediterranean temporary ponds (MTPs) are very specialized and threatened habitats (Keeley and Zedler, 1998; Deil, 2005). MTPs have received considerable attention over the last two decades, because of the multiple ecological roles they play in the landscape (Casanova and Brock, 2000; Deil, 2005), especially in the Mediterranean Basin, where MTPs are associated with traditional rural activities (Rhazi et al., 2012). In these habitats, the regeneration niche of seeds is very dynamic and diverse (Bliss and Zedler, 1997; Carta, 2016), with some species exhibiting specialized life-history strategies (Tuckett et al., 2010c; Cross et al., 2015), including germination requirements related to fungi and anaerobic conditions (Keeley, 1988), and others showing a wider and opportunistic germination niche breadth (Tuckett et al., 2010b; Carta et al., 2013). Nevertheless, species are usually finely tuned to the soil moisture gradients (Casanova and Brock, 2000), even within the same genus (Carta et al., 2016a).

In the Northern Hemisphere and especially in the Mediterranean Basin, MTPs are already threatened by pollution, abandonment of traditional land use (Calhoun et al., 2017), and climate change via a decrease in annual precipitation (Lefebvre et al., 2019). Although water deficits are expected to impact all Mediterranean regions, the countries at highest risk of wetland degradation are those in Southern Europe and Northern Africa (Lefebvre et al., 2019). Nevertheless, we also may expect an increase in MTPs via habitat shifting from permanent water bodies.

Plants living in MTPs are adapted to the large intra- and interannual variability in water availability (Deil, 2005). However, harsher spring and summer drought conditions due to climate change can adversely affect seed production and increase competition with other species living in drier habitats, resulting in an overall reduction of the specialized pond communities (Deil, 2005). The effects of climate change on seed desiccation tolerance and longevity of taxa living in MTPs are poorly investigated (Tuckett et al., 2010a; Carta et al., 2018), particularly in the Northern Hemisphere, and such studies should be enhanced not only for their *ex situ* conservation potential but also to predict seed persistence in the soil (Faist et al., 2013; Metzner et al., 2017; Gioria et al., 2020).

The regenerative phase of many rare and endemic plant species living in these unique habitats has not been well studied. Thus, there is a risk that some species will disappear before we understand their specialized reproductive biology.

Concluding remarks

Plant reproduction from seeds is a complex process that integrates past and present climatic inputs to achieve successful establishment outputs (Fernández-Pascual et al., 2019). Climate change resulting from anthropogenic processes also is a complex phenomenon, the scale of which now justifies the identification of a new epoch in the planet's history, the "Anthropocene." In this epoch, the human imprint rivals the impact of natural forces on the functioning of the Earth's systems (Steffen et al., 2011). How these two complex phenomena interact needs to be understood in terms of changing rates and thresholds for the physiological processes that underlie reproduction of plants by seeds (Fernández-Pascual et al., 2019). Notwithstanding all of these considerations, it is possible to identify some main climate-change driven effects on plant reproduction from seeds in the Northern Hemisphere Mediterranean regions, as detailed below and schematized in Fig. 8.1.

Global warming clashes with the cold-cued germination of the Mediterranean germination syndrome (Thanos et al., 1989, 1991, 1995; Doussi and Thanos, 2002) (Box 8.1). Species with a narrow thermal niche for seed germination will be more at risk from climate change than those with a broad thermal niche (Luna et al., 2012, Monllor et al., 2018). Typical Mediterranean species from coastal and lowland environments are already adapted to warm and arid conditions (Keeley et al., 2012; Matías et al., 2017). On the other hand, peripheral populations of species with a more temperate distribution, which usually occupy wetter micro-sites (Carta et al., 2016b; Orrù et al., 2012; Porceddu et al., 2020) or higher elevations (Cuena-Lombraña et al., 2020; McClain, 2016), could be more at risk due to a compromised seed dormancy release as a result of milder winters.

Independent of the effects on seed dormancy breaking and germination processes, the real bottleneck for plant reproduction by seeds under a warmer and drier Mediterranean climate will be seedling survival. Harsh summer drought already represents the main cause of seedling mortality in these regions under the current climate (Arrieta and Suárez, 2006). This is particularly true in Mediterranean high-mountains, where summer drought and extreme heatwaves are the most limiting factors for the plant regeneration process (Giménez-Benavides et al., 2018). Altered and intensified fire regimes add another level of threat (Keeley et al., 2009; Keeley, 2012).

In conclusion, warmer winters primarily will affect the seed germination phase of the plant life cycle, suppressing seed germination for "typical" Mediterranean species that require low temperatures to germinate in autumn/winter and inhibiting seed dormancy release for species that need overwintering to germinate in spring. Harsher summers will be more detrimental for the establishment phase of the plant life cycle, compromising seedling survival and plant regeneration. Thus, the regeneration of Mediterranean plants from seeds faces a double challenge under a continuously changing climate.

Acknowledgments

Efisio Mattana is supported by the Kew Future Leaders Fellowship, from the Royal Botanic Gardens, Kew. Eduardo Fernández-Pascual was supported by the Jardín Botánico Atlántico [SV-20-GIJON-JBA]. The authors thank V. Thomas Parker (San Francisco State University) and Lesley DeFalco (US Geological Survey) for reviewing the manuscript.

References

Al Hassan, M., Estrelles, E., Soriano, P., López-Gresa, M.P., Bellés, J.M., Boscaiu, M., et al., 2017. Unraveling salt tolerance mechanisms in halophytes: a comparative study on four Mediterranean *Limonium* species with different geographic distribution patterns. Front. Plant Sci. 8, 1438. Available from: https://doi.org/10.3389/fpls.2017.01438.

Alpert, H., Loik, M.E., 2013. *Pinus jeffreyi* establishment along a forest-shrub ecotone in eastern California, USA. J. Arid Environ. 90, 12−21.

Arrieta, S., Suárez, F., 2006. Marginal holly (*Ilex aquifolium* L.) populations in Mediterranean central Spain are constrained by a low-seedling recruitment. Flora 201, 152−160.

Aschmann, H., 1973. Distribution and peculiarity of Mediterranean ecosystems. In: di Castri, F., Mooney, H.A. (Eds.), Mediterranean Type Ecosystems: Origin and Structure. Springer-Verlag, Berlin, pp. 11−19.

Bachofen, C., Wohlgemuth, T., Moser, B., 2019. Biomass partitioning in a future dry and CO_2 enriched climate: shading aggravates drought effects in Scots pine but not European black pine seedlings. J. Appl. Ecol. 56, 866−879.

Baskin, C.C., Baskin, J.M., 2014. Seeds: Ecology, Biogeography, and Evolution of Dormancy and Germination, Second ed. Academic Press/Elsevier, San Diego.

Berg, N., Hall, A., 2015. Increased interannual precipitation extremes over California under climate change. J. Clim. 28, 6324−6334.

Bliss, S.A., Zedler, P.H., 1997. The germination process in vernal pools: sensitivity to environmental conditions and effects on community structure. Oecologia 113, 67−73.

Brown, J.H., 1984. On the relationship between abundance and distribution of species. Am. Nat. 124, 255−279.

Calhoun, A.J.K., Mushet, D.M., Bell, K.P., Boix, D., Fitzsimons, J.A., Isselin-Nondedeu, F., 2017. Temporary wetlands: challenges and solutions to conserving a 'disappearing' ecosystem. Biol. Conserv. 211, 3−11.

Callaway, J.C., Thomas Parker, V., Vasey, M.C., Schile, L.M., 2007. Emerging issues for the restoration of tidal marsh ecosystems in the context of predicted climate change. Madroño 54, 234−248.

Carta, A., 2016. Seed regeneration in Mediterranean temporary ponds: germination ecophysiology and vegetation processes. Hydrobiologia 782, 23−35.

Carta, A., Bedini, G., Müller, J.V., Probert, R.J., 2013. Comparative seed dormancy and germination of eight annual species of ephemeral wetland vegetation in a Mediterranean climate. Plant Ecol. 214, 339−349.

Carta, A., Bottega, S., Spanò, C., 2018. Aerobic environment ensures viability and anti-oxidant capacity when seeds are wet with negative effect when moist: implications for persistence in the soil. Seed Sci. Res. 28, 16−23.

Carta, A., Hanson, S., Müller, J.V., 2016a. Plant regeneration from seeds responds to phylogenetic relatedness and local adaptation in Mediterranean *Romulea* (Iridaceae) species. Ecol. Evol. 6, 4166−4178.

Carta, A., Peruzzi, L., Ramírez-Barahona, S., 2022. A global phylogenetic regionalization of vascular plants reveals a deep split between Gondwanan and Laurasian biotas. New Phytol. 233, 1494−1504.

Carta, A., Probert, R., Moretti, M., Peruzzi, L., Bedini, G., 2014. Seed dormancy and germination in three *Crocus* ser. *Verni* species (Iridaceae): implications for evolution of dormancy within the genus. Plant Biol. 16, 1065−1074.

Carta, A., Probert, R., Puglia, G., Peruzzi, L., Bedini, G., 2016b. Local climate explains degree of seed dormancy in *Hypericum elodes* L. (Hypericaceae). Plant Biol. 18, 76−82.

Carta, A., Skourti, E., Mattana, E., Vandelook, F., Thanos, C.A., 2017. Photoinhibition of seed germination: occurrence, ecology and phylogeny. Seed Sci. Res. 27, 131−153.

Carón, M.M., De Frenne, P., Brunet, J., Chabrerie, O., Cousins, S.A.O., De Backer, L., et al., 2014. Latitudinal variation in seeds characteristics of *Acer platanoides* and *A. pseudoplatanus*. Plant Ecol. 215, 911−925.

Casanova, M.T., Brock, M.A., 2000. How do depth, duration and frequency of flooding influence the establishment of wetland plant communities? Plant Ecol. 147, 237−250.

Castro, J., Zamora, R., Hódar, J.A., Gómez, J.M., 2004. Seedling establishment of a boreal tree species (*Pinus sylvestris*) at its southernmost distribution limit: consequences of being in a marginal Mediterranean habitat. J. Ecol. 92, 266−277.

Chamorro, D., Luna, B., Moreno, J.M., 2017. Germination responses to current and future temperatures of four seeder shrubs across a latitudinal gradient in western Iberia. Am. J. Bot. 104, 83−91.

Chase, M.W., Christenhusz, M.J.M., Fay, M.F., Byng, J.W., Judd, W.S., Soltis, D.E., et al., 2016. An update of the Angiosperm Phylogeny Group classification for the orders and families of flowering plants: APG IV. Bot. J. Linn. Soc. 181, 1–20.

Chiatante, D., Tognetti, R., Scippa, G.S., Congiu, T., Baesso, B., Terzaghi, M., et al., 2015. Interspecific variation in functional traits of oak seedlings (*Quercus ilex, Quercus trojana, Quercus virgiliana*) grown under artificial drought and fire conditions. J. Plant Res. 128, 595–611.

Cogoni, D., Mattana, E., Fenu, G., Bacchetta, G., 2012. From seed to seedling: a critical transitional stage for the Mediterranean psammophilous species *Dianthus* morisianus (Caryophyllaceae). Plant Biosyst. 146, 910–917.

Cowling, R.M., Rundel, P.W., Lamont, B.B., Arroyo, M.K., Arianoutsou, M., 1996. Plant diversity in Mediterranean-climate regions. Trends Ecol. Evol. 11, 362–366.

Cristaudo, A., Catara, S., Mingo, A., Restuccia, A., Onofri, A., 2019. Temperature and storage time strongly affect the germination success of perennial *Euphorbia* species in Mediterranean regions. Ecol. Evol. 9, 10984–10999.

Cross, A.T., Turner, S.R., Renton, M., Baskin, J.M., Dixon, K.W., Merritt, D.J., 2015. Seed dormancy and persistent sediment seed banks of ephemeral freshwater rock pools in the Australian monsoon tropics. Ann. Bot. 115, 847–859.

Cuena-Lombraña, A., Porceddu, M., Dettori, C.A., Bacchetta, G., 2020. Predicting the consequences of global warming on *Gentiana lutea* germination at the edge of its distributional and ecological range. PeerJ 8, e8894. Available from: https://doi.org/10.7717/peerj.8894.

Davis, F.W., Sweet, L.C., Serra-Diaz, J.M., Franklin, J., Mccullough, I., Flint, A., et al., 2016. Shrinking windows of opportunity for oak seedling establishment in southern California mountains. Ecosphere 7, e01573. Available from: https://doi.org/10.1002/ecs2.1573.

Daws, M.I., Cleland, H., Chmielarz, P., Gorian, F., Leprince, O., Mullins, C.E., et al., 2006. Variable desiccation tolerance in *Acer pseudoplatanus* seeds in relation to developmental conditions: a case of phenotypic recalcitrance? Funct. Plant Biol. 33, 59–66. Available from: https://doi.10.1071/FP04206.

Deil, U., 2005. A review on habitats, plant traits and vegetation of ephemeral wetlands—a global perspective. Phytocoenologia 35, 533–706.

Del Cacho, M., Estiarte, M., Peñuelas, J., Lloret, F., 2013a. Inter-annual variability of seed rain and seedling establishment of two woody Mediterranean species under field-induced drought and warming. Popul. Ecol. 55, 277–289.

Del Cacho, M., Lloret, F., 2012. Resilience of Mediterranean shrubland to a severe drought episode: the role of seed bank and seedling emergence. Plant Biol. 14, 458–466.

Del Cacho, M., Peñuelas, J., Lloret, F., 2013b. Reproductive output in Mediterranean shrubs under climate change experimentally induced by drought and warming. Perspect. Plant Ecol. Evol. Syst. 15, 319–327.

Del Vecchio, S., Mattana, E., Ulian, T., Buffa, G., 2021. Functional seed traits and germination patterns predict species coexistence in NE Mediterranean foredune communities. Ann. Bot. 127, 361–370.

De Vitis, M., Mattioni, C., Mattana, E., Pritchard, H.W., Seal, C.E., Ulian, T., et al., 2018. Integration of genetic and seed fitness data to the conservation of isolated subpopulations of the Mediterranean plant *Malcolmia littorea*. Plant Biol. 20, 203–213.

De Vitis, M., Seal, C.E., Ulian, T., Pritchard, H.W., Magrini, S., Fabrini, G., et al., 2014. Rapid adaptation of seed germination requirements of the threatened Mediterranean species *Malcolmia littorea* (Brassicaceae) and implications for its reintroduction. S. Afr. J. Bot. 94, 46–50.

Doussi, M.A., Thanos, C.A., 2002. Ecophysiology of seed germination in Mediterranean geophytes. 1. *Muscari* spp. Seed Sci. Res. 12, 193–201.

Estrelles, E., Biondi, E., Galiè, M., Mainardi, F., Hurtado, A., Soriano, P., 2015. Aridity level, rainfall pattern and soil features as key factors in germination strategies in salt-affected plant communities. J. Arid Environ. 117, 1–9.

Faist, A.M., Ferrenberg, S., Collinge, S.K., 2013. Banking on the past: seed banks as a reservoir for rare and native species in restored vernal pools. AoB Plants 5, plt043. Available from: https://doi.org/10.1093/aobpla/plt043.

Fernández-Pascual, E., Carta, A., Mondoni, A., Cavieres, L., Rosbakh, S., Venn, S., et al., 2021. The seed germination spectrum of alpine plants: a global meta-analysis. New Phytol. 229, 3573–3586.

Fernández-Pascual, E., Mattana, E., Pritchard, H.W., 2019. Seeds of future past: climate change and the thermal memory of plant reproductive traits. Biol. Rev. 94, 439–456.

Ganatsas, P., Daskalakou, E., Paitaridou, D., 2012. First results on early post-fire succession in an *Abies cephalonica* forest (Parnitha National Park, Greece). IForest 5, 6–12.

García De Jalón, L., Limousin, J.M., Richard, F., Gessler, A., Peter, M., Hättenschwiler, S., et al., 2020. Microhabitat and ectomycorrhizal effects on the establishment, growth and survival of *Quercus ilex* L. seedlings under drought. PLoS One 15, e0229807. Available from: https://doi.org/10.1371/journal.pone.0229807.

García, D., Zamora, R., Hódar, J.A., Gómez, J.M., Castro, J., 2000. Yew (*Taxus baccata* L.) regeneration is facilitated by fleshy-fruited shrubs in Mediterranean environments. Biol. Conserv. 95, 31–38.

Gimeno, T.E., Pas, B., Lemos-Filho, J.P., Valladares, F., 2009. Plasticity and stress tolerance override local adaptation in the responses of Mediterranean holm oak seedlings to drought and cold. Tree Physiol. 29, 87–98.

Giménez-Benavides, L., Escudero, A., García-Camacho, R., García-Fernández, A., Iriondo, J.M., Lara-Romero, C., et al., 2018. How does climate change affect regeneration of Mediterranean high-mountain plants? An integration and synthesis of current knowledge. Plant Biol. 20, 50–62.

Giménez-Benavides, L., Escudero, A., Pérez-García, F., 2005. Seed germination of high mountain Mediterranean species: altitudinal, interpopulation and interannual variability. Ecol. Res. 20, 433–444.

Giorgi, F., Lionello, P., 2008. Climate change projections for the Mediterranean region. Glob. Planet Change 63, 90–104.

Gioria, M., Pyšek, P., Baskin, C.C., Carta, A., 2020. Phylogenetic relatedness mediates persistence and density of soil seed banks. J. Ecol. 108, 2121–2131.

Gorai, M., Neffati, M., 2007. Germination responses of *Reaumuria vermiculata* to salinity and temperature. Ann. Appl. Biol. 151, 53–59.

Gremer, J.R., Chiono, A., Suglia, E., Bontrager, M., Okafor, L., Schmitt, J., 2020. Variation in the seasonal germination niche across an elevational gradient: the role of germination cueing in current and future climates. Am. J. Bot. 107, 350–363.

Gómez-Aparicio, L., Pérez-Ramos, I.M., Mendoza, I., Matías, L., Quero, J.L., Castro, J., et al., 2008. Oak seedling survival and growth along resource gradients in Mediterranean forests: implications for regeneration in current and future environmental scenarios. Oikos 117, 1683–1699.

Harper, J.L., 1977. Population Biology of Plants. Academic Press, London.

Harrison, S., Safford, H.D., Grace, J.B., Viers, J.H., Davies, K.F., 2006. Regional and local species richness in an insular environment: serpentine plants in California. Ecol. Monogr. 76, 41–56.

Hashoum, H., Saatkamp, A., Gauquelin, T., Ruffault, J., Fernandez, C., Bousquet-Mélou, A., 2020. Mediterranean woody plant specialized metabolites affect germination of *Linum perenne* at its dry and upper thermal limits. Plant Soil 446, 291–305.

Homet, P., González, M., Matías, L., Godoy, O., Pérez-Ramos, I.M., García, L.V., et al., 2019. Exploring interactive effects of climate change and exotic pathogens on *Quercus suber* performance: damage caused by *Phytophthora cinnamomi* varies across contrasting scenarios of soil moisture. Agric. For. Meteror. 276–277, 107605. Available from: https://doi.org/10.1016/j.agrformet.2019.060.004.

Iacobellis, S.F., Cayan, D.R., Abatzoglou, J.T., Mooney, H.A., 2016. Climate. In: Mooney, H.A., Zavaleta, E.S. (Eds.), Ecosystems of California. University of California Press, Oakland, pp. 9–26.

Ibáñez, B., Ibáñez, I., Gómez-Aparicio, L., Ruiz-Benito, P., García, L.V., Marañón, T., 2014. Contrasting effects of climate change along life stages of a dominant tree species: the importance of soil-climate interactions. Divers. Distrib. 20, 872–883.

Jiménez-Alfaro, B., Hernández-González, M., Fernández-Pascual, E., Toorop, P., Frischie, S., Gálvez-Ramírez, C., 2018. Germination ecology of winter annual grasses in Mediterranean climates: applications for soil cover in olive groves. Agric. Ecosyst. Environ. 262, 29–35.

Joët, T., Ourcival, J.M., Dussert, S., 2013. Ecological significance of seed desiccation sensitivity in *Quercus ilex*. Ann. Bot. 111, 693–701.

Jump, A.S., Peñuelas, J., Rico, L., Ramallo, E., Estiarte, M., Martínez-Izquierdo, J.A., et al., 2008. Simulated climate change provokes rapid genetic change in the Mediterranean shrub *Fumana thymifolia*. Glob. Change Biol. 14, 637–643.

Keeley, J.E., 1987. Role of fire in seed germination of woody taxa in California chaparral. Ecology 68, 434–443.

Keeley, J.E., 1988. Anaerobiosis as a stimulus to germination in two vernal pool grasses. Am. J. Bot. 75, 1086–1089.

Keeley, J.E., 1997. Seed longevity of non-fire recruiting chaparral shrubs. Four Seasons 10 (3), 36–42.

Keeley, J.E., 2012. Fire in mediterranean climate ecosystems—a comparative overview. Isr. J. Ecol. Evol. 58, 123–135.

Keeley, J.E., Bond, W.J., Bradstock, R.A., Pausas, J.G., Rundel, P.W., 2012. Fire in Mediterranean Ecosystems: Ecology, Evolution and Management. Cambridge University Press, Cambridge.

Keeley, J.E., Fotheringham, C.J., 1997. Trace gas emissions and smoke-induced seed germination. Science 276, 1248–1250.

Keeley, J.E., Fotheringham, C.J., 1998. Mechanism of smoke-induced seed germination in a post-fire chaparral annual. J. Ecol. 86, 27–36.

Keeley, J.E., Pausas, J.G., 2018. Evolution of 'smoke' induced seed germination in pyroendemic plants. S. Afr. J. Bot. 115, 251–255.

Keeley, J.E., Pausas, J.G., 2019. Distinguishing disturbance from perturbations in fire-prone ecosystems. Int. J. Wildl. Fire 28, 282–287.

Keeley, J.E., Safford, H., Fotheringham, C.J., Franklin, J., Moritz, M., 2009. The 2007 Southern California wildfires: lessons in complexity. J. For. 107, 287–296.

Keeley, J.E., Syphard, A., 2016. Climate change and future fire regimes: examples from California. Geosciences 6, 37. Available from: https://doi.org/10.3390/geosciences6030037.

Keeley, J.E., Syphard, A.D., 2019. Twenty-first century California, USA, wildfires: fuel-dominated vs. wind-dominated fires. Fire Ecol. 15, 24. Available from: https://doi.org/10.1186/s42408-019-0041-0.

Keeley, J.E., Zedler, P.H., 1998. Characterization and global distribution of vernal pools. In: Witham, C.W., Bauder, E.T., Belk, D., Ferren Jr., W.R., Ornduff, R. (Eds.), Ecology, Conservation, and Management of Vernal Pool Ecosystems—Proceedings from a 1996 Conference. California Native Plant Society, Sacramento, CA, pp. 1–14.

Krichen, K., Ben Mariem, H., Chaieb, M., 2014. Ecophysiological requirements on seed germination of a Mediterranean perennial grass (*Stipa tenacissima* L.) under controlled temperatures and water stress. S. Afr. J. Bot. 94, 210–217.

Krichen, K., Vilagrosa, A., Chaieb, M., 2017. Environmental factors that limit *Stipa tenacissima* L. germination and establishment in Mediterranean arid ecosystems in a climate variability context. Acta Physiol. Plant 39, 175. Available from: https://doi.org/10.1007/s11738-017-2475-9.

Kyselý, J., Beguería, S., Beranová, R., Gaál, L., López-Moreno, J.I., 2012. Different patterns of climate change scenarios for short-term and multi-day precipitation extremes in the Mediterranean. Glob. Planet. Change 98–99, 63–72.

Körner, C., Hiltbrunner, E., 2018. The 90 ways to describe plant temperature. Perspect. Plant Ecol. Evol. Syst. 30, 16–21.

Leck, M.A., Parker, V.T., Simpson, R.L., Simpson, R.S., 2008. Seedling Ecology and Evolution. Cambridge University Press, Cambridge.

Lefebvre, G., Redmond, L., Germain, C., Palazzi, E., Terzago, S., Willm, L., et al., 2019. Predicting the vulnerability of seasonally-flooded wetlands to climate change across the Mediterranean Basin. Sci. Total Environ. 692, 546–555.

Lenoir, J., Gégout, J.C., Pierrat, J.C., Bontemps, J.D., Dhôte, J.F., 2009. Differences between tree species seedling and adult altitudinal distribution in mountain forests during the recent warm period (1986–2006). Ecography 32, 765–777.

Lionello, P., Malanotte-Rizzoli, P., Boscolo, R., Alpert, P., Artale, V., Li, L., et al., 2006. The Mediterranean climate: an overview of the main characteristics and issues. In: Lionello, P., Malanotte-Rizzoli, P., Boscolo, R. (Eds.), Developments in Earth and Environmental Sciences, vol 4. Elsevier, pp. 1–26.

Lloret, F., Peñuelas, J., Estiarte, M., 2005. Effects of vegetation canopy and climate on seedling establishment in Mediterranean shrubland. J. Veg. Sci. 16, 67–76.

Lorite, J., Ruiz-Girela, M., Castro, J., 2007. Patterns of seed germination in Mediterranean mountains: study on 37 endemic or rare species from Sierra Nevada, SE Spain. Candollea 62, 5–16.

Luna, B., Pérez, B., Céspedes, B., Moreno, J.M., 2008. Effect of cold exposure on seed germination of 58 plant species comprising several functional groups from a mid-mountain Mediterranean area. Ecoscience 15, 478–484.

Luna, B., Pérez, B., Torres, I., Moreno, J.M., 2012. Effects of incubation temperature on seed germination of Mediterranean plants with different geographical distribution ranges. Folia Geobot. 47, 17–27.

Mattana, E., Pritchard, H.W., Porceddu, M., Stuppy, W.H., Bacchetta, G., 2012. Interchangeable effects of gibberellic acid and temperature on embryo growth, seed germination and epicotyl emergence in *Ribes multiflorum* ssp. *sandalioticum* (Grossulariaceae). Plant Biol. 14, 77–87.

Matías, L., Castro, J., Villar-Salvador, P., Quero, J.L., Jump, A.S., 2017. Differential impact of hotter drought on seedling performance of five ecologically distinct pine species. Plant Ecol. 218, 201–212.

Matías, L., Jump, A.S., 2014. Impacts of predicted climate change on recruitment at the geographical limits of Scots pine. J. Exp. Bot. 65, 299–310.

Matías, L., Zamora, R., Castro, J., 2012. Sporadic rainy events are more critical than increasing of drought intensity for woody species recruitment in a Mediterranean community. Oecologia 169, 833–844.

McClain, K., 2016. Seed germination requirements for four fire-recruiter chaparral shrubs. Senior Honors Projects, Paper 91. John Carroll University, Ohio. <http://collected.jcu.edu/honorspapers/91>.

Mendoza, I., Zamora, R., Castro, J., 2009. A seeding experiment for testing tree-community recruitment under variable environments: implications for forest regeneration and conservation in Mediterranean habitats. Biol. Conserv. 142, 1491–1499.

Merlin, M., Duputié, A., Chuine, I., 2018. Limited validation of forecasted northward range shift in ten European tree species from a common garden experiment. For. Ecol. Manage. 410, 144–156.

Metzner, K., Gachet, S., Rocarpin, P., Saatkamp, A., 2017. Seed bank, seed size and dispersal in moisture gradients of temporary pools in Southern France. Basic Appl. Ecol. 21, 13–22.

Mira, S., Arnal, A., Pérez-Garciá, F., 2017. Habitat-correlated seed germination and morphology in populations of *Phillyrea angustifolia* L. (Oleaceae). Seed Sci. Res. 27, 50–60.

Moles, A.T., Westoby, M., 2004. What do seedlings die from and what are the implications for evolution of seed size? Oikos 106, 193–199.

Monllor, M., Soriano, P., Llinares, J.V., Boscaiu, M., Estrelles, E., 2018. Assessing effects of temperature change on four *Limonium* species from threatened Mediterranean salt-affected habitats. Not. Bot. Horti Agrobot. Cluj-Napoca 46, 286–291.

Mooney, H.A., Dunn, E.L., 1970. Convergent evolution of Mediterranean-climate evergreen sclerophyll shrubs. Evolution 24, 292–303.

Moran, E.V., Das, A.J., Keeley, J.E., Stephenson, N.L., 2019. Negative impacts of summer heat on Sierra Nevada tree seedlings. Ecosphere 10, e02776. Available from: https://doi.org/10.1002/ecs2.2776.

Moreno, J.M., Zuazua, E., Pérez, B., Luna, B., Velasco, A., Resco De Dios, V., 2011. Rainfall patterns after fire differentially affect the recruitment of three Mediterranean shrubs. Biogeosciences 8, 3721–3732.

Morrone, J.J., 2015. Biogeographic regionalisation of the world: a reappraisal. Aust. Syst. Bot. 28, 81–90.

Mouillot, F., Rambal, S., Joffre, R., 2002. Simulating climate change impacts on fire frequency and vegetation dynamics in a Mediterranean-type ecosystem. Glob. Change Biol. 8, 423–437.

Moya, D., De Las Heras, J., Salvatore, R., Valero, E., Leone, V., 2013. Fire intensity and serotiny: response of germination and enzymatic activity in seeds of *Pinus halepensis* Mill. from southern Italy. Ann. For. Sci. 70, 49–59.

Myers, N., Mittermeier, R.A., Mittermeier, C.G., Da Fonseca, G.A., Kent, J., 2000. Biodiversity hotspots for conservation priorities. Nature 403, 853–858.

Orrù, M., Mattana, E., Pritchard, H.W., Bacchetta, G., 2012. Thermal thresholds as predictors of seed dormancy release and germination timing: altitude-related risks from climate warming for the wild grapevine *Vitis vinifera* subsp. *sylvestris*. Ann. Bot. 110, 1651–1660.

Paula, S., Arianoutsou, M., Kazanis, D., Tavsanoglu, Ç., Lloret, F., Buhk, C., et al., 2009. Fire-related traits for plant species of the Mediterranean Basin. Ecology 90, 1420. Available from: https://doi.org/10.1890/08-1309.1.

Petrů, M., Tielbörger, K., 2008. Germination behaviour of annual plants under changing climatic conditions: separating local and regional environmental effects. Oecologia 155, 717–728.

Picciau, R., Pritchard, H.W., Mattana, E., Bacchetta, G., 2019. Thermal thresholds for seed germination in Mediterranean species are higher in mountain compared with lowland areas. Seed Sci. Res. 29, 44–54.

Piñol, J., Terradas, J., Lloret, F., 1998. Climate warming, wildfire hazard, and wildfire occurrence in coastal eastern Spain. Clim. Change 38, 345–357.

Polade, S.D., Gershunov, A., Cayan, D.R., Dettinger, M.D., Pierce, D.W., 2017. Precipitation in a warming world: assessing projected hydro-climate changes in California and other Mediterranean climate regions. Sci. Rep. 7, 10783. Available from: https://doi.org/10.1038/s41598-017-11285-y.

Porceddu, M., Mattana, E., Pritchard, H.W., Bacchetta, G., 2017. Dissecting seed dormancy and germination in *Aquilegia barbaricina*, through thermal kinetics of embryo growth. Plant Biol. 19, 983–993.

Porceddu, M., Pritchard, H.W., Mattana, E., Bacchetta, G., 2020. Differential interpretation of mountain temperatures by endospermic seeds of three endemic species impacts the timing of in situ germination. Plants 9, 1382. Available from: https://doi.org/10.3390/plants9101382.

Pérez-Ramos, I.M., Rodríguez-Calcerrada, J., Ourcival, J.M., Rambal, S., 2013. *Quercus ilex* recruitment in a drier world: a multi-stage demographic approach. Perspect. Plant Ecol. Evol. Syst. 15, 106–117.

Quézel, P., 1978. Analysis of the flora of Mediterranean and Saharan Africa. Ann. Mo. Bot. Gard. 65, 479–534.

Quézel, P., 1985. Definition of the Mediterranean region and the origin of its flora. In: Gomez-Campo, C. (Ed.), Plant Conservation in the Mediterranean Area. Junk, Dordrecht, pp. 9–24.

Raddi, S., Mariotti, B., Martini, S., Pierguidi, A., 2019. Salinity tolerance in *Fraxinus angustifolia* Vahl.: seed emergence in field and germination trials. Forests 10, 940. Available from: https://doi.org/10.3390/f10110940.

Ramírez, J.M., Rey, P.J., Alcántara, J.M., Sánchez-Lafuente, A.M., 2006. Altitude and woody cover control recruitment of *Helleborus foetidus* in a Mediterranean mountain area. Ecography 29, 375–384.

Rhazi, L., Grillas, P., Saber, E.-R., Rhazi, M., Brendonck, L., Waterkeyn, A., 2012. Vegetation of Mediterranean temporary pools: a fading jewel? Hydrobiologia 689, 23–36.

Richter, S., Kipfer, T., Wohlgemuth, T., Guerrero, C.C., Ghazoul, J., Moser, B., 2012. Phenotypic plasticity facilitates resistance to climate change in a highly variable environment. Oecologia 169, 269–279.

Robson, T.M., Rodríguez-Calcerrada, J., Sánchez-Gómez, D., Aranda, I., 2009. Summer drought impedes beech seedling performance more in a sub-Mediterranean forest understory than in small gaps. Tree Physiol. 29, 249–259.

Rundel, P.W., Arroyo, M.T.K., Cowling, R.M., Keeley, J.E., Lamont, B.B., Pausas, J.G., et al., 2018. Fire and plant diversification in Mediterranean-climate regions. Front. Plant Sci. 9, 851. Available from: https://doi.org/10.3389/fpls.2018.00851.

Rysavy, A., Seifan, M., Sternberg, M., Tielbörger, K., 2016. Neighbour effects on shrub seedling establishment override climate change impacts in a Mediterranean community. J. Veg. Sci. 27, 227–237.

Saatkamp, A., Affre, L., Dutoit, T., Poschlod, P., 2011. Germination traits explain soil seed persistence across species: the case of Mediterranean annual plants in cereal fields. Ann. Bot. 107, 415–426.

Santana, V.M., Alday, J.G., Adamo, I., Alloza, J.A., Baeza, M.J., 2020. Climate, and not fire, drives the phylogenetic clustering of species with hard-coated seeds in Mediterranean Basin communities. Perspect. Plant Ecol. Evol. Syst. 45, 125545. Available from: https://doi.org/10.1016/j.ppees.2020.125545.

Santo, A., Mattana, E., Frigau, L., Bacchetta, G., 2014. Light, temperature, dry after-ripening and salt stress effects on seed germination of *Phleum sardoum* (Hackel) Hackel. Plant Species Biol. 29, 300–305.

Sanz, R., Pulido, F., Julio Camarero, J., 2011. Boreal trees in the Mediterranean: recruitment of downy birch (*Betula alba*) at its southern range limit. Ann. For. Sci. 68, 793–802.

Sanz, R., Pulido, F., Nogués-Bravo, D., 2009. Predicting mechanisms across scales: amplified effects of abiotic constraints on the recruitment of yew *Taxus baccata*. Ecography 32, 993–1000.

Silva, D.E., Rezende Mazzella, P., Legay, M., Corcket, E., Dupouey, J.L., 2012. Does natural regeneration determine the limit of European beech distribution under climatic stress? For. Ecol. Manage. 266, 263–272.

Skordilis, A., Thanos, C.A., 1995. Seed stratification and germination strategy in the Mediterranean pines *Pinus brutia* and *P. halepensis*. Seed Sci. Res. 5, 151–160.

Steffen, W., Grinevald, J., Crutzen, P., Mcneill, J., 2011. The Anthropocene: conceptual and historical perspectives. Philos. Trans. R. Soc. A 369, 842–867.

Syphard, A.D., Brennan, T.J., Keeley, J.E., 2018. Chaparral landscape conversion in southern California. In: Underwood, E., Safford, H., Molinari, N., Keeley, J.E. (Eds.), Springer Series on Environmental Management. Springer, Cham, pp. 323–346.

Thanos, C.A., Georghiou, K., Douma, D.J., Marangaki, C.J., 1991. Photoinhibition of seed germination in Mediterranean maritime plants. Ann. Bot. 68, 469–475.

Thanos, C.A., Georghiou, K., Kadis, C., Pantazi, C., 1992. Cistaceae: a plant family with hard seeds. Isr. J. Plant Sci. 41, 251–263.

Thanos, C.A., Georghiou, K., Skarou, F., 1989. *Glaucium flavum* seed germination—an ecophysiological approach. Ann. Bot. 63, 121–130.

Thanos, C.A., Kadis, C.C., Skarou, F., 1995. Ecophysiology of germination in the aromatic plants thyme, savory and oregano (Labiatae). Seed Sci. Res. 5, 161–170.

Thompson, J.D., 2005. Plant Evolution in the Mediterranean. Oxford University Press, Oxford.

Tuckett, R.E., Merritt, D.J., Hay, F.R., Hopper, S.D., Dixon, K.W., 2010a. Comparative longevity and low-temperature storage of seeds of Hydatellaceae and temporary pool species of south-west Australia. Aust. J. Bot. 58, 327–334.

Tuckett, R.E., Merritt, D.J., Hay, F.R., Hopper, S.D., Dixon, K.W., 2010b. Dormancy, germination and seed bank storage: a study in support of ex situ conservation of macrophytes of southwest Australian temporary pools. Freshw. Biol. 55, 1118–1129.

Tuckett, R.E., Merritt, D.J., Rudall, P.J., Hay, F., Hopper, S.D., Baskin, C.C., et al., 2010c. A new type of specialized morphophysiological dormancy and seed storage behaviour in Hydatellaceae, an early-divergent angiosperm family. Ann. Bot. 105, 1053–1061.

Tweddle, J.C., Dickie, J.B., Baskin, C.C., Baskin, J.M., 2003. Ecological aspects of seed desiccation sensitivity. J. Ecol. 91, 294–304.

Urbieta, I.R., Pérez-Ramos, I.M., Zavala, M.A., Marañón, T., Kobe, R.K., 2008. Soil water content and emergence time control seedling establishment in three co-occurring Mediterranean oak species. Can. J. For. Res. 38, 2382–2393.

Varsamis, G., Papageorgiou, A.C., Merou, T., Takos, I., Malesios, C., Manolis, A., et al., 2019. Adaptive diversity of beech seedlings under climate change scenarios. Front. Plant Sci. 9, 1918. Available from: https://doi.org/10.3389/fpls.2018.01918.

Walck, J.L., Hidayati, S.N., Dixon, K.W., Thompson, K., Poschlod, P., 2011. Climate change and plant regeneration from seed. Glob. Change Biol. 17, 2145–2161.

Weber, E., D'Antonio, C.M., 1999. Germination and growth responses of hybridizing *Carpobrotus* species (Aizoaceae) from coastal California to soil salinity. Am. J. Bot. 86, 1257–1263.

Werner, C.M., Young, D.J.N., Safford, H.D., Young, T.P., 2019. Decreased snowpack and warmer temperatures reduce the negative effects of interspecific competitors on regenerating conifers. Oecologia 191, 731–743.

Wiens, D., Allphin, L., Wall, M., Slaton, M.R., Davis, S.D., 2012. Population decline in *Adenostoma sparsifolium* (Rosaceae): an ecogenetic hypothesis for background extinction. Biol. J. Linn. Soc. 105, 269–292.

Williams, A.P., Abatzoglou, J.T., Gershunov, A., Guzman-Morales, J., Bishop, D.A., Balch, J.K., et al., 2019. Observed impacts of anthropogenic climate change on wildfire in California. Earth's Future 7, 892–910.

Woo, I., Takekawa, J.Y., 2012. Will inundation and salinity levels associated with projected sea level rise reduce the survival, growth, and reproductive capacity of *Sarcocornia pacifica* (pickleweed)? Aquat. Bot. 102, 8–14.

Chapter 9

Plant regeneration from seeds in the southern Mediterranean regions under a changing climate

Jennifer A. Cochrane[1,2] and Sarah Barrett[3]

[1]*Department of Biodiversity, Conservation and Attractions, Kensington, WA, Australia,* [2]*The Australian National University, Canberra, Australia,*
[3]*Department of Biodiversity, Conservation and Attractions, Centennial Park, WA, Australia*

Southern Hemisphere Mediterranean-type ecosystems

The Mediterranean-type ecosystems (MTEs) of the Southern Hemisphere encompass the southern parts of Australia, Central Chile, and the Cape Region of South Africa (Fig. 9.1) and contain some 20,000 native species (Keeley et al., 2012; Rundel et al., 2018). These regions are located along the western and southern sides of the continents between 30 and 40 degrees south of the equator (Köppen, 1936). Ancient landscapes with infertile soils characterize the regions in South Africa and Australia, while the landscape of Central Chile is younger, more fertile, and topographically complex (Hopper, 2009). All three regions support global biodiversity hotspots with high levels of endemism (Klausmeyer and Shaw, 2009), but disproportionate conversion to agriculture, development, and other human uses makes them some of the most vulnerable ecosystems to climate change in the world (Underwood et al., 2009). It is predicted that these regions will be hardest hit by climate change (Lavorel et al., 1998).

The native vegetation of the three regions has been characterized as "sclerophyllous woodlands with winter rain" (Walter, 1985), but the considerable natural variability in climate and rainfall reliability contributes to a sizeable variation in vegetation and ecosystem processes within and between the regions (Cowling et al., 1996). Broadleaf evergreen sclerophyll woodlands are abundant in Australia, of lesser importance in Central Chile, and rather depauperate in the Cape Region (Keeley et al., 2012). Also present in the Central Chile region are relict rainforest, succulent, montane, and alpine floras (Mooney et al., 2001). The Cape Region has a predominance of broad-leaved shrubs, particularly Proteaceae (Allsopp et al., 2014), as well as restiod rushes and ericoid shrubs (Mooney et al., 2001). At the drier end of this region (<250 mm of annual precipitation) are nonflammable succulent shrublands. Native annuals are not abundant in these southern MTEs (Boucher and Moll, 1981; Rundel, 1981; Pignatti et al., 2002). Shrublands and heathlands are a very important component of the southern MTEs, and these are typically evergreen with broad or small, stiff, sclerophyllous leaves. These shrublands harbor the greatest diversity of plant species and are called *kwongan* in Australia, *fynbos* in South Africa, and *matorral* in Chile (Rundel et al., 2016). These vegetation types contain species with a wide range of adaptations that allow them to cope with selection forces such as summer drought, low soil fertility, and periodic fires (Klausmeyer and Shaw, 2009; Keeley et al., 2012).

Seed types, seed storage syndromes, and dispersal mechanisms differ amongst the regions. Fleshy fruits are more common in Chile than in Australia, and these are frequently bird-dispersed. More small dry propagules are produced in Australia than in Chile, and large propagules are mostly absent in Australia but present in Chile. Wind dispersed propagules are more evident in Chile than in Australia, and small arillate seeds are frequent in Australia and rare in Chile (Hoffmann and Armesto, 1995). Kwongan and fynbos have a high degree of convergence, with the greatest global occurrence of obligate postfire seeders, canopy seed storage (serotiny), and ant seed dispersal (myrmecochory) (Keeley, 1992; Le Maitre and Midgley, 1992; Cowling et al., 1994). In South Africa and Australia, plant migration is limited by relatively short dispersal distances and lack of colonization ability (Hammill et al., 1998; Fitzpatrick et al., 2008; Schurr et al., 2007).

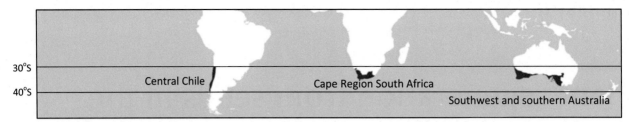

FIGURE 9.1 Mediterranean climate regions of the Southern Hemisphere. *Adapted from Watson, A., Judd, S., Watson, J., Lam, A., Mackenzie, D., 2008. The Extraordinary Nature of the Great Western Woodlands. The Wilderness Society of WA Inc., Perth with permission.*

Seed germination in southern MTEs is usually restricted to autumn and spring, when soils are moist and temperatures are mild. For most life forms, mean optimal temperature for germination is usually between 15°C and 25°C with varying light or dark requirements (Baskin and Baskin, 2014). Over half of the trees in these MTEs are considered to have nondormant seeds, whilst 10% of shrubs are nondormant at dispersal (Baskin and Baskin, 2014), germinating if soil moisture is nonlimiting and temperatures are favorable. A full review of dormancy in life forms in the winter rainfall temperate zones can be found in Baskin and Baskin (2014). Fire plays a significant role in recruitment from seeds for many taxa in the Australian and South African MTEs (Keeley et al., 2012), in particular the many species of Proteaceae and Myrtaceae that retain seeds in woody serotinous fruits that open after fire and release their nondormant seeds. Smoke-stimulated seed germination has been reported for many species in the South African fynbos (De Lange et al., 2018) and in the Australian kwongan (Dixon et al., 1995; Roche et al., 1997; Tieu et al., 2001).

Current and predicted environmental changes due to global warming

MTEs are considered transitional climate regions (Lavorel et al., 1998), and the El Niño Southern Oscillation (ENSO) is the main interannual variable climate phenomenon that impacts the climate in Southern Hemisphere Mediterranean ecosystems. MTEs have marked climate seasonality defined by mild wet winters with low solar irradiance and warm, dry summers (Dallman, 1998). The coastal influence moderates winter temperatures so that rains coincide with suitable growing temperatures. Average monthly summer temperatures (October to March) can exceed 22°C, depending on region, latitude, and altitude. Annual rainfall ranges from 100 to 2000 mm, with at least 65% of it falling in the winter, whilst mean temperature for 1 month during winter is below 15°C (Aschmann, 1973). Fog is a distinctive feature of the coastal areas of Central Chile, a narrow strip of land bounded on the west by the Pacific Ocean and on the east by the Andean Cordillera.

It is generally accepted that these regions will experience a hotter, drier climate with severe consequences for biodiversity (Malcolm et al., 2006). Relatively minor decreases in rainfall and increases in temperature may lead to the expansion of adjacent semiarid and arid ecosystems at the expense of Mediterranean ecosystems (Fischlin et al., 2007; IPCC, 2014). Winter growing conditions followed by summer drought contribute to making MTEs some of the most fire-prone in the world (Keeley et al., 2012). Alterations to the fire regimes of the Cape Region and southern Australia are likely to occur (Klausmeyer and Shaw, 2009). As fire regimes shift due to the warming and drying trend, timing and severity of fire will result in a shorter period during which conditions will be suitable for postfire seed germination [termed "interval squeeze" (Enright et al., 2015)] with long-term consequences for seedling density and community composition.

Southwestern and southern Australia

Until recently, the climate of southern Australia was considered typically seasonal and predictable (Cowling et al., 2004). Annual rainfall varied from 1500 mm in the southwest to around 250 mm in the east. However, since 1950 rainfall in south-west Western Australia has decreased significantly, with a marked decline in autumn and winter (April to October). The drying in recent decades across southern Australia is the most sustained large-scale change in rainfall since national records began in 1900. The upsurge in heatwaves and summer storms, plus a decline in rainfall and increasing fire frequency and severity are predicted to become the new norm (Steffen et al., 2014). Overall, the region is predicted to have higher temperatures, hotter and more frequent hot days, less frost, and less rainfall in winter and spring under various climate change scenarios. Intensity of heavy rainfall events will increase, but the duration of drought periods also will increase. There will be a decrease in mean winter wind speed and increased solar irradiance coupled with reduced relative humidity in winter and spring. Evapotranspiration rates will increase, and soil moisture and runoff will decline. Declining rainfall and extended drought in southern Australia may cause the loss of some temporary wetlands (Nielsen and Brock, 2009). Fire regimes will change due to changes in temperature, rainfall and humidity, and atmospheric CO_2 will increase, resulting in increased risk of severe fire weather (Williams et al., 2009). The size of this Mediterranean biome is expected to contract to 77%−49% of its current size by the end of the 21st century (Klausmeyer and Shaw, 2009).

Central Chile

Climate-model simulations for the 21st-century project a reduction in precipitation of up to 50% relative to the present-day conditions, mainly in Central Chile (Neukom and Luterbacher, 2011; Stocker et al., 2013). The region has already faced a dramatic rainfall decline and persistent drought (2010−15), resulting in a precipitation deficit of approximately 30% (Centre for Climate and Resilience Research, 2015). These changes are of the same magnitude as those projected for Mediterranean areas worldwide (IPCC, 2014). Temperatures will increase by 2°C−4°C by the late 21st century, reaching nearly 5°C during summer in some Andean regions. This trend is consistent with historically observed temperature increases in this part of the country (Falvey and Garreaud, 2009). Temperature-related effects are expected to change snow cover and seasonality of runoff from snowmelt-dominated basins. Climate change will result in a greater difference in temperature between land and sea and is likely to increase the incidence of fog (Mooney et al., 2001). These predicted changes could have important impacts on species distribution, but predictions vary greatly. Klausmeyer and Shaw (2009) predicted that the Mediterranean biome of Central Chile would expand by 129%−153%, but more recent modeling of species distributions by Bambach et al. (2013) concluded that under climate change scenarios, assuming nonmigration, distribution area for all the species they investigated would decrease. The area with the highest reduction in suitable environment was along the coastline, where high temperature increases have been projected. These more recent models indicate that woodlands and shrublands are the plant communities most likely to change in both structure and composition under environmental change (Bambach et al., 2013). However, invasive species will also become more common and widespread if rainfall seasonality shifts.

Cape Region of South Africa

The climate in the Cape Region has warmed over the last century, but rainfall trends are less clear across the Cape (Altwegg et al., 2014). An analysis of rainfall in Cape Town from 1841 to 2016 shows a long-term decrease from 1900 to 2016 (Ndebele et al., 2020), and interannual variability in rainfall has increased since the late 1960s, with droughts becoming more widespread and intense (Fauchereau et al., 2003). Maximum temperatures have increased significantly throughout South Africa, with increases in minimum temperature for most of the country from 1960 to 2010 (Mackellar et al., 2014). Extreme temperature indices for 1962−2009 show strong increases in warm but decreases in cold extremes (Kruger and Sekele, 2013). Mean annual temperature increases of 1.3°C−4.5°C have been predicted by the year 2080 in the south-western Cape (Hulme et al., 2001). More recent modeling predicts temperature increases of 0.5°C−1.5°C to 1.5°C−2.5°C (Maúre et al., 2018). Decreases in precipitation will be accompanied by increases in number of consecutive dry (rainless) days and decreases in consecutive wet days. An anticipated increase in aridity of the Cape Region is associated with declines in the reliability and frequency of small rainfall events during the winter growing season (Midgley et al., 2003; Cowling et al., 2004). Droughts are likely to become more intense and widespread, and extreme rainfall events are likely to increase (Wilson et al., 2020). High impact climate events such as heatwaves and high-fire danger days also are consistently projected to increase in frequency (Engelbrecht et al., 2015). On average, fire frequency has already increased by about 4 years over the past three decades across the region (Wilson et al., 2010). A 51%−65% reduction in the Mediterranean biome in South Africa is projected by 2050 (Midgley et al., 2002) under warming of 1.8°C, with the likely extinction of 23% of the Cape fynbos flora (Thomas et al., 2004).

Current knowledge on plant regeneration from seeds

Southwestern and southern Australia

Cochrane (2020a) synthesized information on the potential impact of global warming on seed germination of 102 native species in south-west Western Australia (SW MTE) using a temperature gradient plate. Seeds were subjected to sub- and supraoptimal temperatures, and their germination responses were evaluated in the light of global warming scenarios. Twenty-six common, threatened, and geographically restricted *Eucalyptus* species (Myrtaceae) demonstrated high ability to germinate outside current and predicted future autumn—winter wet season temperatures, suggesting that climatic distribution is a poor proxy for thermal tolerance in *Eucalyptus* seeds (Cochrane, 2017b). While germination for six species was predicted to decline under future climate conditions, germination of the majority would be maintained or improved, particularly during cooler winter months. There was no evidence of local adaptation to thermal conditions, and rare species were not less tolerant of increased temperatures than common species. Temperature sensitivity of seeds of 38 endemic obligate seeder species of *Banksia* (Proteaceae) demonstrated that many have wide physiological tolerance for high germination temperatures, although a number of geographically restricted species such as *B. praemorsa*, *B. oreophila*, and *B. quercifolia* have narrower temperature windows for germination (Cochrane et al., 2014a). Only germination of *B. dryandroides* is expected to decline in the future. However, optimal germination timing for many species is predicted to shift under climate warming. In conjunction with declining rainfall, this delay in germination will decrease the time for seedling establishment and growth prior to the onset of summer drought.

Eight herbaceous species of Haemodoraceae, Asteraceae, Boryaceae, Cyperaceae, and Xyridaceae in the SW MTE exhibited a range of germination strategies (Cochrane, 2019a). Two germinated at temperatures exceeding those in their natural habitats, suggesting that the environmental space currently occupied underestimates their physiological tolerances. *Xerochrysum macranthum* (Asteraceae) strongly preferred alternating temperatures, and five species demonstrated at least some requirement for fluctuating temperatures for optimal germination. Little is known about germination timing in the field for these species, but the ability to germinate at low alternating temperatures may indicate a propensity to recruit when seasonal temperatures are likely to be changing most or when temperatures are low and soil is moist. Like the woody species assessed, none of these herbaceous species are expected to undergo major collapses due to changes in temperature; however, shifts in timing of optimal germination events are likely. Modeling of empirical results revealed that at least for some species the opportunity for germination may decline due to increasing temperatures. Seeds of 20 small-range endemic woody perennial species of Casuarinaceae, Fabaceae, Myrtaceae, and Proteaceae were tested for germination and the data subsequently modeled against current and predicted (2070) mean monthly minimum and maximum temperatures (Cochrane, 2020b). The data suggested only minimal changes in germination percentages, despite a predicted increase in diurnal temperatures over the next 50 years. Germination of nine species was predicted to decline <1% to 7% and that of 11 species to increase <1% to 3%. Overall, speed of germination was predicted to increase, but timing of germination for most species shifted seasonally (both advances and delays) due to changing diurnal temperatures.

High temperatures were not limiting for germination of several species restricted to the low altitude mountains of south-west Western Australia (Cochrane et al., 2011), with exception of the obligate seeding species *Sphenotoma drummondii* (Ericaceae), which only germinated over a narrow range of cool temperatures. Increasing temperature above the optimum constant temperature had a pronounced negative effect on germination, making this rare species highly vulnerable to climate warming. Seeds of a lowland population of *Andersonia echinocephala* (Ericaceae) were more dormant and germinated at higher temperatures than those from a montane population (Cochrane et al., 2011).

Germination speed and percentage of multiple populations of *Neurachne alopecuroidea* (Poaceae) from Western Australia declined with decreasing moisture availability, with greater impact at higher temperatures. Tolerance to temperature and moisture availability was higher than expected, suggesting local adaptation across its Western Australian distribution (Gray et al., 2019). Seed age and air temperature (in particular the diurnal range) in combination with rainfall, number of rainy days, and rain intensity influenced germination in the rare *Verticordia staminosa* (Myrtaceae) (Simpson, 2011). Guerin et al. (2010) investigated the impact of climate change on seed germination of 13 South Australian threatened species over a range of constant and alternating temperatures. Four species of Liliaceae, Scrophulariaceae, and Poaceae had low temperature thresholds and thus are vulnerable to increasing temperatures. For three species of Apiaceae, the negative effects of increased temperatures were alleviated by a diurnal regime. Seeds of herbs, forbs, and grasses appeared to have lower thermal thresholds than shrubs and trees. In contrast, Dwyer and Erickson (2016) investigated germination of six winter annuals of Goodeniaceae, Poaceae, Asteraceae, and Araliaceae from small, isolated habitat remnants and found that seeds of most species would be able to germinate to high percentages under future climate conditions due to the cumulative effects of after-ripening and warmer maternal and germination environments. In a study of South Australian *Brachyscome* species (Asteraceae), populations from warmer climates germinated under very high temperatures whilst those from cooler climates germinated under lower temperatures. However, for some both temperature extremes had negative effects on seedling development (Aleman, 2014). Guerin et al. (2008/2009) screened seven species of Asteraceae, Fabaceae, Myrtaceae, and Scrophulariaceae from South Australia for germination over a range of temperatures and moisture stress regimes and reported that highest germination percentages occurred at 10°C and 15°C and was completely inhibited at a water stress of −1.0 MPa. Germination of the relictual *Cephalotus follicularis* (Cephalotaceae) on water agar occurred after 8−16 weeks of embryo development within a thermal range of 15°C−20°C (Just et al., 2019). Seeds appeared to have limited capacity for persistence through warm, dry seasonal periods or following dispersal into less mesic habitats. Alteration to natural hydrological regimes has been implicated in the decline of populations of this species. Under predicted future climate and land-use scenarios for the SW MTE, recruitment of *C. follicularis* from seeds may become increasingly episodic and unpredictable.

Results from studies using open-top chambers and rainout shelters showed that seed germination, seedling survival and growth, and adult survival, flowering, and fruiting in kwongan shrubland in Western Australia were reduced by drought and warming (Williams, 2014). Cochrane et al. (2015) reported that soil warming of less than 2°C led to delays and declines in germination of four obligate seeder *Banksia* species in a common garden experiment. Furthermore, moisture stress during germination of these species indicated that the threshold water potential for a significant decline and delay in germination was high (−0.25 MPa). In the absence of moisture stress, germination was uniformly high, but increasing drought stress led to reduced and delayed germination in all four species, with *Banksia coccinea* the most vulnerable to germination failure under predicted changes in rainfall patterns (Cochrane et al., 2014b).

Serotinous species are good climate change indicators. For example, rainfall reductions and temperature increases between 1987 and 2012 resulted in a >50% reduction in total seed production in *Banksia hookeriana* in Western Australia (Enright et al., 2015). Dry years also lowered postfire recruitment in this species (Enright and Lamont, 1992) and increased the fire interval required for stand self-replacement (Enright et al., 1996, 2014). Seed production in 54% of 11 serotinous species of Proteaceae and Myrtaceae was reduced by 4%−76% in 2018 compared with 1983−2004 (Cowan, 2018). Cone opening was more rapid at mesic sites than at xeric sites. Although seed viability remained unchanged, climate change potentially reduced the overall size of canopy seed store, with decreases in fruit production and more rapid cone opening being most significant. Other studies also reported changes in serotiny across a climatic gradient, with degrees of serotiny lower at more mesic sites, where fire is less frequent (Cowling and Lamont, 1985). Premature opening of serotinous fruits of *Banksia* and *Hakea* (Proteaceae) may occur due to exposure to the hot dry conditions of extreme summer drought (Lamont et al., 1991; Causley et al., 2016). In this case, seeds may be released into an interfire environment with unfavorable conditions for germination (Lamont and Enright, 2000). Species of *Hakea* have shown almost complete fruit opening from drought stress, increasing the risk of recruitment failure due to climate change.

Seeds of the rare *Acacia awestoniana* (Fabaceae), collected in different years at the same site, differed in thermal resilience to heat shock treatments (Cochrane, 2019b). Strong interannual responses were attributed to the maternal environment during seed development, as well as to storage conditions. High rainfall and cool conditions during seed maturation are related to production of larger, less thermally tolerant seeds. The extreme tolerance of seeds of *A. awestoniana* exposed to extended heat pulse treatment may represent adaptation to the longer, hotter conditions predicted for the region. Two range-restricted *Eucalyptus* taxa exhibited narrower thermal tolerance ranges for germination than their widespread congenerics, but they exhibited higher physiological plasticity for thermal and drought stress tolerance (Rajapakshe et al., 2020).

In a study of thermal thresholds for physical dormancy break in *Acacia*, Cochrane (2017a) found that species whose seeds had a low dormancy-break threshold responded to sudden heatwave conditions or the sustained dry heat of summer. More than 40% of *A. nigricans* seeds lost dormancy during a simulated long hot summer period and germinated in conditions unfavorable for seedling establishment. In an assessment of after-ripening in seeds of *Acanthocarpus preissii* (Asparagaceae), warm dry conditions promoted alleviation of physiological dormancy (Turner et al., 2006). Baker et al. (2005) proposed that seasonal dormancy cycling may be disrupted by changing temperature and moisture conditions during seed burial in the soil, with high temperatures and wet−dry cycles increasing dormancy release in *Actinotus leucocephalus* (Apiaceae).

In an investigation of the response of native and nonnative woodland species to burial depth and fire regime, seeds of serotinous species were among the most tolerant to elevated soil temperatures, surviving temperatures under all fuel treatments and emerging from 2 to 7 cm soil depth (Tangney et al., 2020). However, plant communities in seasonally dry Mediterranean climates may be highly sensitive to fire seasonality given the close relationship between plant phenology and environmental conditions (He et al., 2016; Miller et al., 2019). Although some species may survive future changes in temperature and rainfall, their persistence may be compromised by changed fire regimes (He et al., 2016). For example, smoke applied to nonburnt sites in autumn promoted a significantly greater germination response for the majority of 37 species investigated than treatment in winter or spring (Roche et al., 1998).

Establishment of seedlings from seeds planted into degraded *Banksia* woodlands was not affected by a 50% experimental rainfall reduction, but water addition significantly improved seedling survival and growth (Standish et al., 2012).

Holloway-Phillips et al. (2015) demonstrated that species differences in water-use strategies of two co-occurring species of *Banksia* may start to shift community dynamics as the climate changes.

Water-stressed native species are susceptible to infection by stem canker fungi (Shearer and Crane, 2014), with branch decline and death reducing reproductive potential. Newly emerging diseases such as the rust pathogen *Puccinia psidii* can impact reproduction (Glen et al., 2007), and myrtaceous species from southern MTEs will be vulnerable to this disease (Morin et al., 2012). Seeds of invasive plant species such as *Malva parvifolia* (Malvaceae) that have optimal germination temperatures associated with winter rainfall may be less dormant than native species under lower rainfall conditions (Michael et al., 2006). Little is known about the life cycle and resilience of insect pollinators to drought and fire, but pollination studies suggest that honeyeater bird communities in southern Australia will exhibit some resilience to environmental change, although with some local changes to fruit set and pollen movement (Phillips et al., 2010).

A mechanistic model developed from empirical data evaluated germination phenology of three co-occurring *Eucalyptus* species from the MTE of Victoria under a range of climate change scenarios (Rawal et al., 2015). Germination of all species declined when soil moisture potential was below field capacity, and all exhibited shifts in germination timing. Only *E. microcarpa* had high germination and establishment percentages under the projected climate change conditions. Changes to future community composition may occur in this woodland as shifts in timing and success of recruitment occur. Fitzpatrick et al. (2008) projected that 5%–25% of *Banksia* species would result in range losses of 100% by 2080, depending on climate scenarios. Hughes et al. (1996) predicted that 53% of *Eucalyptus* species had annual temperature envelopes of only $3°C$ and projected that climate warming would completely displace these species.

Central Chile

There are only a few seed germination studies in the Chilean matorral in the context of global warming (but see Muñoz and Fuentes, 1989; Gómez-González et al., 2008, 2011b). However, the implications of climate change can be construed from many published studies, particularly those on fire ecology and invasive species. Lightning-ignited (natural) fires are rare under current environmental conditions (Mooney, 1977; Fuentes et al., 1994), although anthropogenic fires are more frequent (Gómez-González et al., 2008). Unlike the MTEs of South Africa and Australia, the woody flora of the matorral consists predominantly of resprouters (Montenegro et al., 2004), possibly due to environmental stress (cold or drought) (Mooney, 1977). Nevertheless, the vegetation of the Chilean matorral exhibits fire-adapted regeneration strategies (Montenegro et al., 1983). It has been surmised that abrupt changes in historical fire regime, high levels of resprouter species, and low recruitment of seeder species may have resulted in a loss of seeder species in the Chilean flora (Keeley et al., 2012). Thus, there is little similarity between the soil seed bank and the aboveground vegetation in the Chilean matorral, and very few viable seeds of native woody species are present in the seed bank (Jiménez and Armesto, 1992; Figueroa et al., 2004). Consequently, postfire germination and seedling recruitment are low (Muñoz and Fuentes, 1989; Segura et al., 1998). Seeds of Chilean shrubs cannot survive high-intensity fires, and viable seeds and naturally established seedlings occur only in patches burned at low intensity (Segura et al., 1998). Gómez-González et al. (2008) provide what appears to be the first result on smoke-related germination in the region. Smoke significantly stimulated germination in three species and decreased it in eight. The three smoke-stimulated species are colonizers, whilst smoke-inhibited species tended to be major dominants in the matorral vegetation. These results suggest that current and escalating human-caused fires could drastically change the structure of the matorral in the future (Gómez-González et al., 2008). Anthropogenic fires have also been implicated in selection for seed traits such as shape, pubescence, and pericarp thickness (Gómez-González et al., 2011a), which can have far-reaching consequences for plant regeneration in the future. Although several studies that have investigated the impact of fire-associated triggers (heat, smoke, and their combination) on germination of native and exotic seeds from the Chilean matorral (e.g., Figueroa and Cavieres, 2012; Figueroa et al., 2009), apparently none have considered the impact global warming might have on changes to community composition as a result of changes to fire frequency and severity.

Placea (Amaryllidaceae), endemic to the Andes of Central Chile, has a narrow range of low temperatures for germination (Guerrero et al., 2007), suggesting that increasing temperatures may result in germination failure. The effect of heat shock on seed survival and germination in 21 native woody species of the Chilean matorral was evaluated to investigate the importance of fire-adaptive responses in the context of other MTEs (Gómez-González et al., 2017). Ten species were adversely affected by fire cues, seven stimulated by high-temperature heat shock, and four unaffected. All species stimulated to germinate by heat shock were resprouters in the hard-seeded families Fabaceae, Anacardiaceae, and Rhamnaceae, and thus they may have had physical dormancy.

A study of thermal buffering capacity of 55 species of Cactaceae, including four native to the Mediterranean climate region of Central Chile (*Maihueniopsis glomerata* and three subspecies of *Echinopsis chiloensis*) predicted changes in germination under global warming (Seal et al., 2017). For *M. glomerata*, time to 50% germination was expected to be more rapid under a +3.7°C climate change scenario by the end of the 21st century and around 10 days faster for the subspecies of *Echinopsis chiloensis*. *Maihueniopsis glomerata* had a 61°C temperature range for commencement of germination, leading to the suggestion that the seasonality of germination could change significantly for this species. In contrast, the three *Echinopsis* taxa were only able to germinate over an ca. 30°C range.

Similar to other MTEs, germination can be inhibited during rainless winters followed by hot summers, and this could be a determining factor for regeneration of the native *Leucocoryne dimorphopetala* (Amaryllidaceae) (De La Cuadra et al., 2016). Retention of leaf litter promoted earlier and higher germination percentages of the endangered tree, *Beilschmiedia miersii* (Lauraceae), particularly during dry years (Becerra et al., 2004). Extreme drought events enhanced the susceptibility of a semiarid plant community in north-Central Chile to invasion by exotic annuals, but moderately high precipitation events changed species composition by favoring regeneration of native annuals (Jiménez et al., 2011).

Seed availability rather than seedling emergence may be the greatest limiting factor for initiating succession in disturbed patches of Chilean matorral (Holmgren et al., 2000). Seed densities were much higher during ENSO years than non-ENSO years, showing the importance of this phenomenon for seed bank replenishment in the arid region of Chile (Gutiérrez and Meserve, 2003). In some cases, nurse plants create milder microclimatic conditions for seed regeneration and seedling establishment, but in other cases invasive species occupy these microsites (Lenz and Facelli, 2003), resulting in the potential for changes to community structure and composition. The facilitative effects of the cushion plant *Azorella madreporica* (Apiaceae) on seedling establishment of *Hordeum comosum* did not appear to decrease under experimentally applied warmer conditions (Cavieres and Sierra-Almeida, 2012). The facilitation effect that nurse shrubs have on seedling establishment is particularly crucial in drier areas of the matorral, with recolonization of open places in wetter sites inprobable without the protection of a nurse shrub (Holmgren et al., 2000).

The lower elevations of the Central Chilean Andes experience the long summer drought that characterizes MTE, and drought is an important factor in seedling mortality. With duration of snow cover decreasing due to climate change (Burger et al., 2018), which impacts hydrology, evapotranspiration, and snowmelt, the presence of refugia may be reduced.

Climate change not only causes direct effects on plant regeneration but also increases the potential for a multitude of indirect impacts such as the disruption of important mutualisms. The region contains a high number of plants with fleshy fruits dispersed by birds, and avian-gut passage increases woodland seed regeneration (Reid and Armesto, 2010). The onset of precipitation mediates plant–avian dispersal interactions such that regeneration of *Cryptocarya alba* (Lauraceae) was maximized by dispersal only at the beginning of the wet season (Bustamante et al., 2012).

South Africa Cape Region

In one of a few studies simulating climate warming, seed germination of 11 Proteaceae species was reduced in 91% of the species with soil temperatures increases of 3.5°C above ambient and in 64% of the species by 1.4°C above ambient. However, some seeds germinated at these temperatures, and a gradual temperature increase may allow some species to have sufficient variability to rapidly adapt to projected temperature increases (Arnolds et al., 2015). Germination speed also declined, making these species susceptible to competition from faster regenerating fynbos species. Reductions in seedling recruitment due to seed quiescence were more closely related to diurnal soil temperature minima than to the diurnal maxima. Seed dormancy enforced by elevated night temperatures was considered to be the postfire recruitment stage most sensitive to global warming in the South African MTE (Bond and Van Wilgen, 1996). Soil-stored seeds of *Leucospermum* (Proteaceae) are thermo-inhibited at even moderately high temperatures and have highly specific alternating temperature requirements (Brits et al., 2014).

Several experimental studies have assessed the impact of drought on seedling emergence and survival in the Cape Region. In a laboratory experiment, soil desiccation prior to radicle emergence of 23 Proteaceae species from a wide variety of fynbos habitats had no impact on germination percentages (Mustart et al., 2012). However, desiccation imposed after radicle emergence significantly reduced seedling emergence after subsequent rewetting. The longer the desiccation period, the greater the impact, and as few as 6 days of drying resulted in major declines for some species. Species had highly individualistic responses to desiccation and variability in responses that extended across species occurring in both the western winter rainfall and eastern bimodal rainfall parts of the Cape. Low tolerance to desiccation may have evolved in response to reliably moist habitats during the winter rainfall germination period. Proteoid species in high rainfall environments, reliably wet seepages, or marshes have low desiccation tolerance after radicle emergence and are at high risk of local extinction with aridification of their habitats. In another simulated drought experiment, projected drought periods leading to the complete cessation of transpiration and death in 1-year-old seedlings of 16 Proteaceae species greatly exceeded the number of days without rain per month during summer in the current distribution ranges of those species (Arnolds et al., 2015).

In the Cape Region, *Erica* is characterized by high diversity and endemism and a high proportion (90%) of obligate seeders. Within this genus, summer drought strongly influences effectiveness of postfire regeneration and growth and is a major selective force in the distribution of seeders in this region (Ojeda, 1998). West et al. (2012) found considerable variability in drought response both between and within growth forms. Shallow-rooted anisohydric (continue to transpire when soil moisture is low) ericoid shrubs showed reduced growth and flowering and increased mortality in contrast to shallow-rooted isohydric (stop transpiring when soil moisture is low) restiods and deep-rooted isohydric proteoid shrubs that were more resilient to drought. Even nonlethal drought may be of great importance for seeder species due to diminished reproductive output. Data from permanent plots in the fynbos monitored for 44 years showed a significant decline in species diversity, which was driven by increasingly severe postfire summer weather events and the legacy effects of woody alien plant densities (Slingsby et al., 2017). However, microsite characteristics may influence the impact of drought, and shallow water tables of high-altitude sites with summer clouds may buffer some elements of this flora from increased summer drought. Lotter et al. (2014) investigated the physiological response of the economically significant *Aspalathus linearis* (Fabaceae) to drought stress. This species partially offsets the negative effects of water deficit stress by increasing water use efficiency and sclerophylly, which may enable seedlings to survive and persist, albeit with reduced biomass.

The Proteaceae are an example of an iconic biodiverse suite of species projected to be vulnerable to climate change (Midgley et al., 2002, 2003; Schurr et al., 2012; Cabral et al., 2013). Increases in the threat status of Proteaceae have been predicted as well as species extinctions due to future change in land use and climate change, with climate change having the most severe impact (Bomhard et al., 2005). Only 5% of the 330 endemic Proteaceae species modeled are predicted to retain more than two-thirds of their range (Midgley et al., 2002), and about 10% of endemic Proteaceae have ranges restricted to the area of this biome at greatest risk. One-third of the species could undergo complete range dislocation by 2050. It is likely that climate change will exceed the potential of Proteaceae populations to track climate change by migrating because of seed dispersal limitations and the limitation of regeneration opportunities immediately postfire (Midgley et al., 2002). Species-level modeling for 28 Proteaceae species from areas of high risk of biome loss predicted range contractions in 17, with five showing range elimination, and 11 potential range expansions (Midgley et al., 2003). For those species predicted to have no future range, microrefuges may still mitigate this impact. Current geographical distributions of species do not always match their ecological niches, and species with poor dispersal may be absent from suitable sites. On the other hand, species with high persistence ability may be present in sites currently unsuitable for them (Pagel et al., 2020). Demographic distribution models for *Protea repens* predict higher population growth rates in the core of its range under projected climates for 2050, with declines along arid range margins due to the interaction of more frequent fire and drying climate (Merow et al., 2014). Modeling the distribution of the fynbos legume *Aspalathus linearis* under the A2 greenhouse gas emissions scenario suggests that this species may lose up to 57% of its climatically suitable range, especially in lowland areas (Nakicenovic et al., 2000).

For several functional groups characteristic of the fynbos at their climatic limits (e.g., members of Proteaceae and Restionaceae), global warming or rainfall reduction has been predicted to lead to a retreat of fynbos elements to higher, cooler elevations on mesic southerly slope aspects (Mooney et al., 2001). Although much of the protected area in the Cape Region is mountainous, the area available for colonization by species decreases with elevation as mountain peaks taper (Agenbag et al., 2008). Low-lying areas often are fragmented, which limits opportunities for range expansion. Thus, high-altitude specialists have nowhere to go, and south coastal species have no potential for expansion poleward. Furthermore, low-lying areas north of Cape Town are not protected from human activities and cannot support species in the future (Agenbag et al., 2008; Van Wilgen et al., 2016). In addition, the intrinsic adaptation potential of many endemic species is limited by short seed dispersal distances and lack of colonization ability (Klausmeyer and Shaw, 2009).

The succulent karoo community is also sensitive to change in rainfall seasonality. Since this community occupies a flat landscape with no topographic refugia, it is threatened with extinction (Mooney et al., 2001). The response of succulents from the region to experimental warming demonstrated that current thermal regimes are likely to closely approximate tolerable extremes for many endemic succulents in the region and that warming could significantly exceed their thermal thresholds. Under experimental warming, plant and canopy mortalities of these specialized-dwarf and shrubby succulents increased by 2.1- to 4.9-fold (Musil et al., 2005). Young et al. (2016) identified the risks of climate change to dwarf succulents in the region, although this life form is among the most resilient plant groups in terms of drought tolerance. The distribution of *Conophytum* (Aizoaceae) in southern Africa is controlled primarily by amount of winter and summer rainfall and geology. The predicted effect of climate change scenarios for this genus is a severe contraction in the bioclimatic envelope and significant range dislocation. On the other hand, modeling has predicted that climate change may lead to invasion of the fynbos by succulent species from the karoo (Agenbag, 2006). Fire is likely to facilitate invasion of marginal habitats by succulents because of the greater sensitivity of fynbos regeneration stages to high temperatures and drought. Phenological monitoring of fynbos species across the fynbos-succulent karoo boundary showed that growth is negatively affected by high temperatures and that low but regular rainfall is required to sustain growth during the dry summer. Faster growing species were more sensitive to interannual climate variation than slower growing species with more conservative water use strategies (Agenbag, 2006).

Seeds of *Oxalis* from the Cape flora are classified as either orthodox, recalcitrant and intermediate, the latter two strategies confined to the winter rainfall region. Their occupation of specialized niches suggests that they may face adverse impacts under predicted climate change (hotter and drier winters) (Watson et al., 2008). Phenological studies of *Oxalis* by Dreyer et al. (2006) showed that warmer and drier years compress the flowering period, with the start of flowering dependent on temperature decline and onset of winter rains. This sensitivity suggests that the seeds of species that are nondormant and must germinate directly after dispersal are potentially vulnerable to climate change. A shortened flowering period negatively affects the potential for sexual reproduction, which is limiting in the long term.

In this MTE, it is predicted that invasive species will reduce the opportunities for native species to shift their range (Wilson et al., 2020). Alien species, especially those that tap into groundwater, may have significant hydrological impacts, worsening the effect of droughts and affecting river flow. Extreme weather events such as floods create disturbance, including the release of nutrients into the system and opening the habitat for invasion. Seed regeneration of native riparian species is not disturbance triggered, in contrast to invasive species that benefit from disturbance. Fire in the fynbos favors exotic woody plant invasion (Kruger and Bigalke, 1984), and seed size is one of the factors influencing naturalization and invasion success of woody plants (Moodley et al., 2013). Invasion success of naturalized species of Proteaceae was associated with small seed size and serotiny, although large mammal-dispersed seeds were associated with greater naturalization success.

Heat and drought impacts associated with climate change may induce high levels of stress on fynbos species by reducing resistance to insect pests and introduced and native pathogens. Some of the more important fungal pathogens associated with the Proteaceae in South Africa are species of Botryosphaeriaceae that cause stem canker, dieback, or leaf blight when plants are stressed (Marincowitz et al., 2008). Although numerous canker pathogens are capable of infecting vigorous trees, canker-causing fungi are more likely to reach epidemic levels and cause substantial damage to trees that are weakened by heat and drought stress thereby creating invasion windows (Schoeneweiss, 1975; Sturrock et al., 2011).

Future research needs

In light of the extreme conditions of rising temperatures and increasing moisture deficiencies that have been predicted for the coming decades, additional empirical research is required to better understand how such climatic variables impact flowering, pollination, seed production, seed bank dynamics, seed viability, dormancy, germination, seedling establishment, and survival of plant groups in Southern Hemisphere MTEs. Understanding and quantifying demographic processes are key to more accurate modeling of population range shift and persistence under environmental change (Malcolm et al., 2006; Yates et al., 2010). More specifically, data are required on climate change impacts on MTEs:

- relationship between time to reproductive age and changing climates and fire regimes
- potential mismatches in climate tolerances between fundamental and realized niches
- plant pollinator interactions and potential phenological mismatches
- comparisons of past and present seed production
- seed longevity and seed bank persistence under altered temperature and moisture regimes
- restoration seeding in the context of a warming, drying climate
- realistic data on seed dispersal capability
- identification of refugial areas and their role in regeneration from seeds
- within-species variation in seed traits driving regeneration
- tolerance of seeds and seedlings to temperature and moisture stress

- impact of germination cues (e.g., smoke, heat, and volatile inhibitors) and their interaction with fire, water, and radiation regimes
- seed biology and climatic tolerances of invasive species and the role of fire in exotic plant invasions
- prevalence and spread of introduced and native plant pathogens in relation to global warming

There is clear evidence of concern about global warming, changing fire regimes, and biological invasions in these Southern Hemisphere MTEs. However, future conservation of these highly significant biodiversity hotspots is challenged by our limited understanding of seed regeneration ecology in relation to disturbance regimes and their interaction with a changing climate.

References

Agenbag, L., 2006. A Study on an Altitudinal Gradient Investigating the Potential Effects of Climate Change on Fynbos and the Fynbos-Succulent Karoo Boundary. Master of Science, Stellenbosch University, Stellenbosch, South Africa.

Agenbag, L., Elser, K.J., Midgley, G.F., Boucher, C., 2008. Diversity and species turnover on an altitudinal gradient in Western Cape, South Africa: baseline data for monitoring range shifts in response to climate change. Bothalia 38, 161–191.

Aleman, R., 2014. Seed Biology and Germination Requirements of *Brachyscome* Species in South Australia. Ph.D. dissertation. University of South Australia, Adelaide.

Allsopp, N., Colville, J.F., Verboom, G.A. (Eds.), 2014. Fynbos: Ecology, Evolution and Conservation of a Megadiverse Region. Oxford University Press, Oxford.

Altwegg, R., West, A., Gillson, L., Midgley, G.F., 2014. Impacts of climate change in the Greater Cape Floristic Region. In: Allsopp, N., Colville, J.F., Verboom, G.A. (Eds.), Fynbos: Ecology, Evolution, and Conservation of a Megadiverse Region. Oxford University Press, Oxford, pp. 299–320.

Arnolds, L., Musil, C.F., Rebelo, A.G., Kruger, G.H.J., 2015. Experimental climate warming enforces seed dormancy in South African Proteaceae but seedling drought resilience exceeds summer drought periods. Oecologia 177, 1103–1116.

Aschmann, H., 1973. Distribution and peculiarity of Mediterranean ecosystems. In: Di Castri, F., Mooney, H.A. (Eds.), Mediterranean Type Ecosystems: Origin and Structure. Springer, Berlin, pp. 11–19.

Baker, K.S., Steadman, K.J., Plummer, J.A., Merritt, D.J., Dixon, K.W., 2005. The changing window of conditions that promote germination of two fire ephemerals, *Actinotus leucocephalus* (Apiaceae) and *Tersonia cyathiflora* (Gyrostemonaceae). Ann. Bot. 96, 1225–1236.

Bambach, N., Meza, F.J., Gilabert, H., Miranda, M., 2013. Impacts of climate change on the distribution of species and communities in the Chilean Mediterranean ecosystem. Reg. Environ. Change 13, 1245–1257.

Baskin, C.C., Baskin, J.M., 2014. Seeds: Ecology, Biogeography, and Evolution of Dormancy and Germination, Second ed. Academic Press/Elsevier, San Diego.

Becerra, P.I., Celis-Diez, J.L., Bustamante, R.O., 2004. Effects of leaf litter and precipitation on germination and seedling survival of the endangered tree *Beilschmiedia miersii*. Appl. Veg. Sci. 7, 253–257.

Bomhard, B., Richardson, D.M., Donaldson, J.S., Hughes, G.O., Midgley, G.F., Raimondo, D.C., et al., 2005. Potential impacts of future land use and climate change on the Red List status of the Proteaceae in the Cape Floristic Region, South Africa. Glob. Change Biol. 11, 1452–1468.

Bond, W.J., Van Wilgen, B.W., 1996. Surviving fires—vegetative and reproductive responses. In: Bond, W.J., Van Wilgen, B.W. (Eds.), Fire and Plants. Chapman and Hall, London, pp. 34–51.

Boucher, C., Moll, D.J., 1981. South African Mediterranean shrublands. In: Allsopp, N., Colville, J.F., Verboom, G.A. (Eds.), Fynbos: Ecology, Evolution, and Conservation of a Megadiverse Region. Oxford University Press, Oxford, pp. 543–571.

Brits, G.J., Brown, N.A.C., Calitz, F.J., 2014. Alternating temperature requirements in *Leucospermum* R.Br. seed germination and ecological correlates in fynbos. S. Afr. J. Bot. 92, 112–119.

Burger, F., Brock, B., Montecinos, A., 2018. Seasonal and elevational contrasts in temperature trends in Central Chile between 1979 and 2015. Glob. Planet. Change 162, 136–147.

Bustamante, R.O., Vásquez, R.A., Grez, A.A., Moreira, D., 2012. The onset of precipitation mediates plant–avian disperser interaction in recalcitrant seeds: the case of *Cryptocarya alba* (MOL) Looser, in Mediterranean ecosystems. Cent. Chile. Plant. Ecol. Divers. 5, 75–79.

Cabral, J.S., Jeltsch, F., Thuiller, W., Higgins, S., Midgley, G.F., Rebelo, A.G., et al., 2013. Impacts of past habitat loss and future climate change on the range dynamics of South African Proteaceae. Divers. Distrib. 19, 363–376.

Causley, C.L., Fowler, W.M., Lamont, B.B., He, T., 2016. Fitness benefits of serotiny in fire- and drought-prone environments. Plant Ecol. 217, 773–779.

Cavieres, L.A., Sierra-Almeida, A., 2012. Facilitative interactions do not wane with warming at high elevations in the Andes. Oecologia 170, 575–584.

Centre for Climate and Resilience Research, Chile, 2015. Report to the Nation. The 2010–2015 mega-drought: a lesson for the future. Centre for Climate and Resilience Research (CR) 2, Chile.

Cochrane, A., 2017a. Are we underestimating the impact of rising summer temperatures on dormancy loss in hard-seeded species? Aust. J. Bot. 65, 248–256.

Cochrane, A., 2017b. Modelling seed germination response to temperature in *Eucalyptus* L'Her. (Myrtaceae) species in the context of global warming. Seed Sci. Res. 27, 99–109.

Cochrane, A., 2019a. Effects of temperature on germination in eight Western Australian herbaceous species. Folia Geobot. 54, 29–42.

Cochrane, A., 2019b. Multi-year sampling provides insight into the bet-hedging capacity of the soil-stored seed reserve of a threatened *Acacia* species from Western Australia. Plant Ecol. 220, 241–253.

Cochrane, A., 2020a. Thermal requirements underpinning germination allude to risk of species decline from climate warming. Plants 9, 796. Available from: https://doi.org/10.3390/plants9060796.

Cochrane, A., 2020b. Temperature thresholds for germination in 20 short-range endemic plant species from a Greenstone Belt in southern Western Australia. Plant Biol. 22, 103–112.

Cochrane, A., Daws, M.I., Hay, F.R., 2011. Seed-based approach for identifying flora at risk from climate warming. Austral Ecol. 36, 923–935.

Cochrane, A., Hoyle, G.L., Yates, C.J., Wood, J., Nicotra, A.B., 2014a. Predicting the impact of increasing temperatures on seed germination among populations of Western Australian *Banksia* (Proteaceae). Seed Sci. Res. 24, 195–205.

Cochrane, J.A., Hoyle, G.L., Yates, C.J., Wood, J., Nicotra, A.B., 2014b. Evidence of population variation in drought tolerance during seed germination in four *Banksia* (Proteaceae) species from Western Australia. Aust. J. Bot. 62, 481–489.

Cochrane, A., Hoyle, G.L., Yates, C.J., Wood, J., Nicotra, A.B., 2015. Climate warming delays and decreases seedling emergence in a Mediterranean ecosystem. Oikos 124, 150–160.

Cowan, E., 2018. Evidence for Demographic Shift in Woody Plant Species under a Changing Climate in Southwest Australian Mediterranean-type Ecosystems. Honours thesis. Murdoch University, Perth.

Cowling, R.M., Lamont, B.B., 1985. Variation in serotiny of three Banksia species along a climatic gradient. Aust. J. Ecol. 10, 345–350.

Cowling, R.M., Ojeda, F., Lamont, B.B., Rundel, P.W., 2004. Climate stability in Mediterranean-type ecosystems: implications for the evolution and conservation of biodiversity. In: Arianoutsou, M., (Ed.), MEDECOS 10th Conference, 2004. Rhodes, Greece. Mill Press, Rotterdam, pp. 1–11.

Cowling, R.M., Pierce, S.M., Stock, W.D., Cocks, M., 1994. Why are there so many myrmecochorous species in the Cape fynbos? In: Arianoutsou, M., Groves, R.H. (Eds.), Plant-Animal Interactions in Mediterranean-Type Ecosystems. Kluwer Academic Publishers, Amsterdam, pp. 159–168.

Cowling, R.M., Rundel, P.W., Lamont, B.B., Arroyo, M.K., Arianoutsou, M., 1996. Plant diversity in Mediterranean-climate regions. Trends Ecol. Evol. 11, 362–366.

Dallman, P.R., 1998. Plant Life in the World's Mediterranean Climates. University of California Press, Berkeley.

De La Cuadra, C., Vidal, A.K., Lefimil, S., Mansur, L., 2016. Temperature effect on seed germination in the genus *Leucocoryne* (Amaryllidaceae). Hortscience 51, 412–415.

De Lange, J.H., Brown, N.A.C., Van Staden, J., 2018. Perspectives on the contributions by South African researchers in igniting global research on smoke-stimulated seed germination. S. Afr. J. Bot. 115, 219–222.

Dixon, K.W., Roche, S., Pate, J.S., 1995. The promotive effect of smoke derived from burnt native vegetation on seed germination of Western Australian plants. Oecologia 101, 185–192.

Dreyer, L.L., Esler, K.J., Zietsman, J., 2006. Flowering phenology of South African *Oxalis*—possible indicator of climate change? S. Afr. J. Bot. 72, 150–156.

Dwyer, J.M., Erickson, T.E., 2016. Warmer seed environments increase germination fractions in Australian winter annual plant species. Ecosphere 7, e01497. Available from: https://doi.org/10.1002/ecs2.1497.

Engelbrecht, C.J., Landman, W.A., Engelbrecht, F.A., Malherbe, J., 2015. A synoptic decomposition of rainfall over the Cape south coast of South Africa. Clim. Dyn. 44, 2589–2607.

Enright, N.J., Fontaine, J.B., Bowman, D.M.J.S., Bradstock, R.A., Williams, R.J., 2015. Interval squeeze: altered fire regimes and demographic responses interact to threaten woody species persistence as climate changes. Front. Ecol. Environ. 13, 265–272.

Enright, N.J., Fontaine, J.B., Lamont, B.B., Miller, B.P., Westcott, V.C., 2014. Resistance and resilience to changing climate and fire regime depend on plant functional traits. J. Ecol. 102, 1572–1581.

Enright, N.J., Lamont, B.B., 1992. Survival, growth and water relations of *Banksia* seedlings on a sand mine rehabilitation site and adjacent scrub-heath sites. J. Appl. Ecol. 29, 663–671.

Enright, N.J., Lamont, B.B., Marsula, R., 1996. Canopy seed bank dynamics and optimum fire regime for the highly serotinous shrub, *Banksia hookeriana*. J. Ecol. 84, 9–17.

Falvey, M., Garreaud, R.D., 2009. Regional cooling in a warming world: recent temperature trends in the southeast Pacific and along the west coast of subtropical South America (1979–2006). J. Geophy. Res. Atmos. 114. Available from: https://doi.org/10.1029/2008JD010519.

Fauchereau, N., Trzaska, S., Rouault, M., Richard, Y., 2003. Rainfall variability and changes in Southern Africa during the 20th Century in the global warming context. Nat. Haz 29, 139–154.

Figueroa, J.A., Cavieres, L.A., 2012. The effect of heat and smoke on the emergence of exotic and native seedlings in a Mediterranean fire-free matorral of central Chile. Rev. Chil. Hist. Nat. 85, 101–111.

Figueroa, J.A., Cavieres, L.A., Gómez-González, S., Montenegro, M.M., Jaksic, F.M., 2009. Do heat and smoke increase emergence of exotic and native plants in the matorral of central Chile? Acta Oecol. 35, 335–340.

Figueroa, J.A., Teillier, S., Jaksic, F.M., 2004. Composition, size and dynamics of the seed bank in a Mediterranean shrubland of Chile. Austral Ecol. 29, 574–584.

Fischlin, A., Midgley, G.F., Price, J.T., Leemans, R., Gopal, B., Turley, C., et al., 2007. Ecosystems, their properties, goods and services. In: Parry, M.L., Canziani, O.F., Palutikof, J.P., Van Der Linden, P.J., Hanson, C.E. (Eds.), Climate Change 2007: Impacts, Adaptation and Vulnerability. Contribution of Working Group II to the Fourth Assessment Report of the Intergovernmental Panel of Climate Change (IPCC). Cambridge University Press, Cambridge, pp. 211–272.

Fitzpatrick, M.C., Gove, A.D., Sanders, N.J., Dunn, R.R., 2008. Climate change, plant migration, and range collapse in a global biodiversity hotspot: the *Banksia* (Proteaceae) of Western Australia. Glob. Change Biol. 14, 1337–1352.

Fuentes, E.R., Segura, A.M., Holmgren, M., 1994. Are the responses of matorral shrubs different from those in an ecosystem with a reputed fire history? In: Moreno, J.M., Oechal, W.C. (Eds.), The Role of Fire in Mediterranean-Type Ecosystems. Ecological Studies (Analysis and Synthesis). Springer, New York, pp. 16–25.

Glen, M., Alfenas, A.C., Zauza, E.A.V., Wingfield, M.J., Mohammed, C., 2007. *Puccinia psidii*: a threat to the Australian environment and economy—a review. Australas. Plant Pathol. 36, 1–16.

Gray, F., Cochrane, A., Poot, P., 2019. Provenance modulates sensitivity of stored seeds of the Western Australian native grass *Neurachne alopecuroidea* to temperature and moisture availability. Aust. J. Bot. 67, 106–115.

Guerin, J., Te, T., Thorpe, M., Duval, D., Ainsley, P., 2008/2009. Developing a Tool to Identify Plant Species at Risk of Climate Change. A report to the Wildlife Fund Grants Program 2008/2009. South Australian Seed Conservation Centre, Botanic Gardens Directorate, Department of Environment and Natural Resources, Adelaide.

Guerin, J., Te, T., Thorpe, M., Duval, D., Ainsley, P., 2010. Developing a Screening Procedure to Determine the Impact of Climate Change on Seed Germination in Threatened Native Plant Species. Final Report to the Australian Flora Foundation. South Australian Seed Conservation Centre, Botanic Gardens Directorate, Department of Environment and Natural Resources, Adelaide.

Guerrero, P.C., Sandoval, A.C., León-Lobos, P., 2007. The effect of chilling on seed germination of *Placea* species (Asparagales: Amaryllidaceae), an endemic genus to central Chile. Gayana Bot. 64, 40–45.

Gutiérrez, J.R., Meserve, P.L., 2003. El Niño effects on soil seed bank dynamics in north-central Chile. Oecologia 134, 511–517.

Gómez-González, S., Paula, S., Cavieres, L.A., Pausas, J.G., 2017. Postfire responses of the woody flora of Central Chile: insights from a germination experiment. PLoS One 12, e0180661. Available from: https://dx.doi.org/10.1371/journal.pone.0180661.

Gómez-González, S., Sierra-Almeida, A., Cavieres, L.A., 2008. Does plant-derived smoke affect seed germination in dominant woody species of the Mediterranean matorral of central Chile? For. Ecol. Manage. 255, 1510–1515.

Gómez-González, S., Torres-Díaz, C., Bustos-Schindler, C., Gianoli, E., 2011a. Anthropogenic fire drives the evolution of seed traits. Proc. Natl. Acad. Sci. U.S.A. 108, 18743–18747.

Gómez-González, S., Torres-Díaz, C., Gianoli, E., 2011b. The effects of fire signals on germination and seed viability of *Helenium aromaticum* (Hook.) LH Bailey (Asteraceae). Gayana Bot. 68, 86–88.

Hammill, K.A., Bradstock, R.A., Allaway, W.G., 1998. Post-fire seed dispersal and species re-establishment in proteaceous heath. Aust. J. Bot. 46, 407–419.

He, T., D'agui, H., Lim, S.L., Enright, N.J., Luo, Y., 2016. Evolutionary potential and adaptation of *Banksia attenuata* (Proteaceae) to climate and fire regime in southwestern Australia, a global biodiversity hotspot. Sci. Rep. 6, 26315. Available from: https://doi.org/10.1038/srep26315.

Hoffmann, A.J., Armesto, J.J., 1995. Modes of seed dispersal in the Mediterranean regions in Chile, California, and Australia. In: Arroyo, M.T.K., Zedler, P.H., Fox, M.D. (Eds.), Ecology and Biogeography of Mediterranean Ecosystems in Chile, California, and Australia. Springer-Verlag, New York, pp. 289–310.

Holloway-Phillips, M.M., Huai, H., Cochrane, A., Nicotra, A.B., 2015. Differences in seedling water-stress response of two co-occurring *Banksia* species. Aust. J. Bot. 63, 647–656.

Holmgren, M., Segura, A.M., Fuentes, E.R., 2000. Limiting mechanisms in the regeneration of the Chilean matorral — experiments on seedling establishment in burned and cleared mesic sites. Plant Ecol. 147, 49–57.

Hopper, S.D., 2009. OCBIL theory: towards an integrated understanding of the evolution, ecology and conservation of biodiversity on old, climatically buffered, infertile landscapes. Plant Soil. 322, 49–86.

Hughes, L., Cawsey, E.M., Westoby, M., 1996. Climatic range sizes of *Eucalyptus* species in relation to future climate change. Glob. Ecol. Biogeogr. Lett. 5, 23–29.

Hulme, M., Doherty, R., Ngara, T., New, M., Lister, D., 2001. African climate change: 1900–2100. Clim. Res. 17, 145–168.

IPCC, 2014. Summary for policymakers. In: Field, C.B., Barros, V.R., Dokken, D.J., Mach, K.J., Mastrandrea, M.D., Bilir, T.E. et al. (Eds.), Climate Change 2014: Impacts, Adaptation, and Vulnerability.Part A: Global and Sectoral Aspects. Contribution of Working Group II to the Fifth Assessment Report of the Intergovernmental Panel on Climate Change. Cambridge University Press, Cambridge, UK and New York, NY.

Jiménez, H.E., Armesto, J.J., 1992. Importance of the soil seed bank of disturbed sites in Chilean matorral in early secondary succession. J. Veg. Sci. 3, 579–586.

Jiménez, M.A., Jaksic, F.M., Armesto, J.J., Gaxiola, A., Meserve, P.L., Kelt, D.A., et al., 2011. Extreme climatic events change the dynamics and invasibility of semi-arid annual plant communities. Ecol. Lett. 14, 1227–1235.

Just, M.P., Merritt, D.J., Turner, S.R., Conran, J.G., Cross, A.T., 2019. Seed germination biology of the Albany pitcher plant, *Cephalotus follicularis*. Aust. J. Bot. 67, 480–489.

Keeley, J.E., Bond, W.J., Bradstock, R.A., Pausas, J.G., Rundell, P.W., 2012. Fire in Mediterranean Ecosystems: Ecology, Evolution and Management. Cambridge University Press, Cambridge.

Keeley, J.E., 1992. A Californian's view of fynbos. In: Cowling, R.M. (Ed.), The Ecology of Fynbos: Nutrients, Fire and Diversity. Oxford University Press, Cape Town, pp. 372–388.

Klausmeyer, K.R., Shaw, M.R., 2009. Climate change, habitat loss, protected areas and the climate adaptation potential of species in Mediterranean ecosystems worldwide. PLoS One 4, e6392. Available from: https://doi.org/10.1371/journal.pone.0006392.

Kruger, F.J., Bigalke, R.C., 1984. Fire and fynbos. In: Booysen, P.D.V., Tainton, N.M. (Eds.), Ecological Effects of Fire in South Africa Ecosystem. Springer-Verlag, Berlin, pp. 220–240.

Kruger, A.C., Sekele, S.S., 2013. Trends in extreme temperature indices in South Africa: 1962–2009. Int. J. Climatol. 33, 661–676.

Köppen, W., 1936. Das Geographische System der Klimate. Gebrüder Borntraeger, Berlin.

Lamont, B.B., Enright, N.J., 2000. Adaptive advantages of aerial seed banks. Plant Species Biol. 15, 157–166

Lamont, B.B., Lemaitre, D.C., Cowling, R.M., Enright, N.J., 1991. Canopy seed storage in woody plants. Bot. Rev. 57, 277–317.

Lavorel, S., Canadell, J., Rambal, S., Terradas, J., 1998. Mediterranean terrestrial ecosystems: research priorities on global change effects. Glob. Ecol. Biogeogr. Lett. 7, 157–166.

Lenz, T.I., Facelli, J.M., 2003. Shade facilitates an invasive stem succulent in a chenopod shrubland in South Australia. Austral Ecol. 28, 480–490.

Lotter, D., Valentine, A.J., Archer Van Garderen, E., Tadross, M., 2014. Physiological responses of a fynbos legume, *Aspalathus linearis* to drought stress. S. Afr. J. Bot. 98, 214–223.

Mackellar, N., New, M., Jack, C., 2014. Observed and modelled trends in rainfall and temperature for South Africa: 1960–2010. S. Afr. J. Sci. 110, 13. Available from: https://doi.org/10.1590/sajs.2014/20130353.

Le Maitre, D., Midgley, J., 1992. Plant reproductive ecology. In: Cowling, R.M. (Ed.), The Ecology of Fynbos: Nutrients, Fire and Diversity. Oxford University Press, Cape Town, pp. 135–174.

Malcolm, J., Liu, C., Neilson, R., Hansen, L., Hannah, L., 2006. Global warming and extinctions of endemic species from biodiversity hotspots. Conserv. Biol. 20, 538–548.

Marincowitz, S., Groenewald, J.Z., Wingfield, M.J., Crous, P.W., 2008. Species of Botryosphaeriaceae occurring on Proteaceae. Persoonia 21, 111–118.

Maúre, G., Pinto, I., Ndebele-Murisa, M., Muthige, M., Lennard, C., Nikulin, G., et al., 2018. The southern African climate under 1.5°C and 2°C of global warming as simulated by CORDEX regional climate models. Environ. Res. Lett. 13, 065002. Available from: https://doi.org/10.1088/1748-9326/aab190.

Merow, C., Latimer, A.M., Wilson, A.M., Mcmahon, S.M., Rebelo, A.G., Silander Jr, J.A., 2014. On using integral projection models to generate demographically driven predictions of species' distributions: development and validation using sparse data. Ecography 37, 1167–1183.

Michael, P.J., Steadman, K.J., Plummer, J.A., 2006. Climatic regulation of seed dormancy and emergence of diverse *Malva parviflora* populations from a Mediterranean-type environment. Seed Sci. Res. 16, 273–281.

Midgley, G.F., Hannah, L., Millara, D., Thuiller, W., Booth, A., 2003. Developing regional and species-level assessments of climate change impacts on biodiversity in the Cape Floristic Region. Biol. Conserv. 112, 87–97.

Midgley, G.F., Hannah, L., Millar, D., Rutherford, M.C., Powrie, L.W., 2002. Assessing the vulnerability of species richness to anthropogenic climate change in a biodiversity hotspot. Glob. Ecol. Biogeogr. Lett. 11, 445–451.

Miller, R.G., Tangney, R., Enright, N.J., Fontaine, J.B., Merritt, D.J., Ooi, M.K.J., et al., 2019. Mechanisms of fire seasonality effects on plant populations. Trends Ecol. Evol. 34, 1104–1117.

Montenegro, G., Avila, G., Schatte, P., 1983. Presence and development of lignotubers in shrubs of the Chilean matorral. Can. J. Bot. 61, 1804–1808.

Montenegro, G., Ginocchio, R., Segura, A., Keely, J.E., Gomez, M., 2004. Fire regimes and vegetation responses in two Mediterranean-climate regions. Rev. Chil. Hist. Nat. 77, 455–464.

Moodley, D., Geerts, S., Richardson, D.M., Wilson, J.R.U., 2013. Different traits determine introduction, naturalization and invasion success in woody plants: Proteaceae as a test case. PLoS One 8, e75078. Available from: https://doi.org/10.1371/journal.pone.0075078.

Mooney, H.A., 1977. Convergent Evolution in Chile and California: Mediterranean Climate Ecosystems. Dowden, Hutchinson and Ross, Inc, Stroudsburg, Pennsylvania.

Mooney, H.A., Arroyo, M.T.K., Bond, W.J., Canadell, J., Hobbs, R.J., Lavorel, S., et al., 2001. Mediterranean-climate ecosystems. In: Chapin, F.S., Sala, O.E., Huber-Sannwald, E. (Eds.), Global Biodiversity in a Changing Environment. Springer, New York, pp. 157–199.

Morin, L., Aveyard, R., Lidbetter, J.R., Wilson, P.G., 2012. Investigating the host-range of the rust fungus *Puccinia psidii* sensu lato across tribes of the family Myrtaceae present in Australia. PLoS One 7, e35434. Available from: https://doi.org/10.1371/journal.pone.0035434.

Musil, C.F., Schmiedel, U., Midgley, G.F., 2005. Lethal effects of experimental warming approximating a future climate scenario on southern African quartz-field succulents: a pilot study. New Phytol. 165, 539–547.

Mustart, P.J., Rebelo, A.G., Juritz, J., Cowling, R.M., 2012. Wide variation in post-emergence desiccation tolerance of seedlings of fynbos proteoid shrubs. S. Afr. J. Bot. 80, 110–117.

Muñoz, M.R., Fuentes, E.R., 1989. Does fire induce shrub germination in the Chilean matorral? Oikos 56, 177–181.

Nakicenovic, N., Alcamo, J., Grubler, A., Riahi, K., Roehrl, R.A., Rogner, H.-H., et al., 2000. Special Report on Emissions Scenarios (SRES). A Special Report of Working Group III of the Intergovernmental Panel on Climate Change. Cambridge University Press, Cambridge.

Ndebele, N.E., Grab, S., Turasie, A., 2020. Characterizing rainfall in the south-western Cape, South Africa: 1841–2016. Int. J. Climatol. 40, 1992–2014.

Neukom, R., Luterbacher, J., 2011. Climate variability in the Southern Hemisphere. Glob. Change Mag. 76, 26–29.

Nielsen, D.L., Brock, M.A., 2009. Modified water regime and salinity as a consequence of climate change: prospects for wetlands of Southern Australia. Clim. Change 95, 523–533.

Ojeda, F., 1998. Biogeography of seeder and resprouter *Erica* species in the Cape Floristic Region - where are the resprouters? Biol. J. Linn. Soc. 63, 331–347.

Pagel, J., Treurnicht, M., Bond, W.J., Kraaij, T., Nottebrock, H., Schutte-Vlok, A., et al., 2020. Mismatches between demographic niches and geographic distributions are strongest in poorly dispersed and highly persistent plant species. Proc. Natl. Acad. Sci. USA 117, 3663–3669.

Phillips, R.D., Hopper, S.D., Dixon, K.W., 2010. Pollination ecology and the possible impacts of environmental change in the Southwest Australian Biodiversity Hotspot. Philos. Trans. R. Soc. B: Biol. Sci. 365, 517–528.

Pignatti, E., Pignatti, S., Ladd, P.G., 2002. Comparison of ecosystems in the Mediterranean Basin and Western Australia. Plant Ecol. 163, 177–186.

Rajapakshe, R.P.V.G.S.W., Turner, S.R., Cross, A.T., Tomlinson, S., 2020. Hydrological and thermal responses of seeds from four co-occurring tree species from southwest Western Australia. Conserv. Physiol. 8, coaa021. Available from: https://doi.org/10.1093/conphys/coaa021.

Rawal, D.S., Kasel, S., Keatley, M.R., Nitschke, C.R., 2015. Environmental effects on germination phenology of co-occurring eucalypts: implications for regeneration under climate change. Int. J. Biometeorol. 59, 1237–1252.

Reid, S., Armesto, J., 2010. Avian gut-passage effects on seed germination of shrubland species in Mediterranean central Chile. Plant Ecol. 212, 1–10.

Roche, S., Dixon, K.W., Pate, J.S., 1997. Seed ageing and smoke: partner cues in the amelioration of seed dormancy in selected Australian native species. Aust. J. Bot. 45, 783–815.

Roche, S., Dixon, K.W., Pate, J.S., 1998. For everything a season: smoke-induced seed germination and seedling recruitment in a Western Australian *Banksia* woodland. Aust. J. Ecol. 23, 111–120.

Rundel, P.W., Arroyo, M.T.K., Cowling, R.M., Keeley, J.E., Lamont, B.B., Pausas, J.G., et al., 2018. Fire and plant diversification in Mediterranean-climate regions. Front. Plant Sci. 9, 851. Available from: https://doi.org/10.3389/fpls.2018.00851.

Rundel, P.W., Arroyo, M.T.K., Cowling, R.M., Keeley, J.E., Lamont, B.B., Vargas, P., 2016. Mediterranean biomes: evolution of their vegetation, floras, and climate. Annu. Rev. Ecol. Evol. Syst. 47, 383–407.

Rundel, P.W., 1981. The matorral zone of Central Chile. In: Di Castri, F., Goodall, D., Specht, R.L. (Eds.), Mediterranean-Type Shrublands. Vol. 11. Elsevier, The Hague, pp. 175–201.

Schoeneweiss, D.F., 1975. Predisposition, stress, and plant disease. Annu. Rev. Phytopathol. 13, 193–211.

Schurr, F.M., Esler, K.J., Slingsby, J.A., Allsopp, N., 2012. Fynbos Proteaceae as model organisms for biodiversity research and conservation. S. Afr. J. Sci. 108, 12–16.

Schurr, F.M., Midgley, G.F., Rebelo, A.G., Reeves, G., Poschlod, P., Higgins, S.I., 2007. Colonization and persistence ability explain the extent to which plant species fill their potential range. Glob. Ecol. Biogeogr. 16, 449–459.

Seal, C.E., Daws, M.I., Flores, J., Ortega-Baes, P., Galíndez, G., León-Lobos, P., et al., 2017. Thermal buffering capacity of the germination phenotype across the environmental envelope of the Cactaceae. Glob. Change Biol. 23, 5309–5317.

Segura, A.M., Holmgren, M., Anabalón, J.J., Fuentes, E.R., 1998. The significance of fire intensity in creating local patchiness in the Chilean matorral. Plant Ecol. 139, 259–264.

Shearer, B.L., Crane, C.E., 2014. Host range of the stem canker pathogen *Luteocirrhus shearii* mainly limited to *Banksia*. Australas. Plant Pathol. 43, 257–266.

Simpson, G., 2011. Cracking the Niche: An Investigation Into the Impact of Climatic Variables on Germination of the Rare Shrub *Verticordia staminosa* subspecies *staminosa* (Myrtaceae). Bachelor of Science In: Murdoch University, Perth.

Slingsby, J.A., Merow, C., Aiello-Lammens, M., Allsopp, N., Hall, S., Kilroy Mollmann, H., et al., 2017. Intensifying postfire weather and biological invasion drive species loss in a Mediterranean-type biodiversity hotspot. Proc. Natl. Acad. Sci. U.S.A. 114, 4697. Available from: https://doi.org/10.1073/pnas.1619014114.

Standish, R.J., Fontaine, J.B., Harris, R.J., Stock, W.D., Hobbs, R.J., 2012. Interactive effects of altered rainfall and simulated nitrogen deposition on seedling establishment in a global biodiversity hotspot. Oikos 121, 2014–2025.

Steffen, W., Hughes, L., Perkins, L., 2014. Heatwaves: Hotter, Longer, More Often. Climate Council of Australia Limited.

Stocker, T.F., Qin, D., Plattner, G.-K., Alexander, L.V., Allen, S.K., Bindoff, N.L., et al., 2013. 2013: technical summary. In: Stocker, T.F., Qin, D., Plattner, G.-K., Tignor, M., Allen, S.K., Boschung, J. et al. (Eds.), Climate Change 2013: The Physical Science Basis. Contribution of Working Group I to the Fifth Assessment Report of the Intergovernmental Panel on Climate Change. Cambridge University Press, Cambridge.

Sturrock, R.N., Frankel, S.J., Brown, A.V., Hennon, P.E., Kliejunas, J.T., Lewis, K.J., et al., 2011. Climate change and forest diseases. Plant Pathol. 60, 133–149.

Tangney, R., Merritt, D.J., Callow, J.N., Fontaine, J.B., Miller, B.P., 2020. Seed traits determine species responses to fire under varying soil heating scenarios. Funct. Ecol. 34, 1967–1978.

Thomas, C.D., Cameron, A., Green, R.E., Bakkenes, M., Beaumont, L.J., Collingham, Y.C., et al., 2004. Extinction risk from climate change. Nature 427, 145–148.

Tieu, A., Dixon, K.W., Meney, K.A., Sivasithamparam, K., 2001. The interaction of heat and smoke in the release of seed dormancy in seven species from southwestern Western Australia. Ann. Bot. 88, 259–265.

Turner, S.R., Merritt, D.J., Ridley, E.C., Commander, L.E., Baskin, J.M., Baskin, C.C., et al., 2006. Ecophysiology of seed dormancy in the Australian endemic species *Acanthocarpus preissii* (Dasypogonaceae). Ann. Bot. 98, 1137–1144.

Underwood, E.C., Viers, J.H., Klausmeyer, K.R., Cox, R.L., Shaw, M.R., 2009. Threats and biodiversity in the Mediterranean biome. Divers. Distrib. 15, 188–197.

Van Wilgen, B.W., Carruthers, J., Cowling, R.M., Esler, K.J., Forsyth, A.T., Gaertner, M., et al., 2016. Ecological research and conservation management in the Cape Floristic Region between 1945 and 2015: history, current understanding and future challenges. Trans. R. Soc. S. Afr. 71, 207–303.

Walter, H., 1985. Vegetation of the Earth and Ecological Systems of the Geo-Biosphere, third ed. Springer-Verlag, Berlin.

Watson, A., Judd, S., Watson, J., Lam, A., Mackenzie, D., 2008. The Extraordinary Nature of the Great Western Woodlands. The Wilderness Society of WA Inc, Perth.

West, A.G., Dawson, T.E., February, E.C., Midgley, G.F., Bond, W.J., Aston, T.L., 2012. Diverse functional responses to drought in a Mediterranean-type shrubland in South Africa. New Phytol. 195, 396–407.

Williams, A., 2014. Climate Change in Southwest Australian Shrublands: Response to Altered Rainfall and Temperature (Ph.D. thesis). Murdoch University, Perth.

Williams, R.J., Bradstock, R.A., Cary, G.J., Enright, N.J., Gill, A.M., Leidloff, A.C., et al., 2009. Interactions Between Climate Change, Fire Regimes and Biodiversity in Australia - A Preliminary Assessment. Department of Climate Change and Department of the Environment, Heritage and Arts, Canberra.

Wilson, J.R., Foxcroft, L.C., Geerts, S., Hoffman, M.T., Macfadyen, S., Measey, J., et al., 2020. The role of environmental factors in promoting and limiting biological invasions in South Africa. In: Van Wilgen, B.W., Measey, J., Richardson, D.M., Wilson, J.R., Zengeya, T.A. (Eds.), Biological Invasions in South Africa. Springer International Publishing, Cham, Switzerland, pp. 355–385.

Wilson, A.M., Latimer, A.M., Silander, J.A., Gelfand, A.E., De Klerk, H., 2010. A hierarchical Bayesian model of wildfire in a Mediterranean biodiversity hotspot: implications of weather variability and global circulation. Ecol. Model. 221, 106–112.

Yates, C.J., Elith, J., Latimer, A.M., Maitre, D.L., Midgley, G.F., Schurr, F.M., et al., 2010. Projecting climate change impacts on species distributions in megadiverse South African Cape and Southwest Australian Floristic Regions: opportunities and challenges. Austral Ecol. 35, 374–391.

Young, A.J., Guo, D., Desmet, P.G., Midgley, G.F., 2016. Biodiversity and climate change: risks to dwarf succulents in Southern Africa. J. Arid Environ. 129, 16–24.

Chapter 10

Plant regeneration from seeds in the temperate deciduous forest zone under a changing climate

Jeffrey L. Walck and Siti N. Hidayati
Department of Biology, Middle Tennessee State University, Murfreesboro, TN, United States

Introduction

Twenty-five percent of the land area on earth that supports (or potentially can support) forests occurs within the temperate region of the Northern Hemisphere, and this includes a wide range of forest types (Tyrrell et al., 2012). The exact boundaries of temperate forests are not always clear-cut, but generally boreal forests (taiga) are found to the north, semitropical/tropical forest, or sclerophyllous woodlands to the south and temperate grasslands (steppes) to the east or west. Of the forest types in the temperate region, the deciduous (nemoral) forest (Walter, 1985) has the greatest extent, occurring primarily in eastern North America, western and central Europe, and eastern Asia (see Fig. 1.1). Winter deciduous trees are the dominant feature of this forest, and forest structure consists of three strata: tree (canopy), shrub and small tree (subcanopy), and herbaceous ground cover. Originally the deciduous forest with interspersed grasslands, savannas, and rock outcrop communities dominated the landscape in the three deciduous forest regions (Anderson et al., 1999; Vasseur, 2012; Noss, 2013). Today, forested regions are highly fragmented due to agriculture and urbanization, and areas abandoned from agriculture are in various stages of succession. Thus, we will consider plant species growing in the region of the temperate deciduous forest and not just in the forest per se.

Seeds of plants that grow in the temperate deciduous forest zone are dispersed primarily from late summer through autumn (mostly) into early winter and from late spring to early summer (Willson and Traveset, 2000). Most of these seeds are dormant or conditionally dormant when dispersed, with physiological dormancy being the prevalent kind (class) of dormancy found in them (Baskin and Baskin, 2014). Physical dormancy is the second most important class of dormancy followed by morphophysiological, combinational (physiological + physical), and morphological dormancy. Some species produce nondormant seeds, which is about equally important as physical dormancy. To overcome dormancy with a physiological component, warm (summer, moist) and/or cold (winter, moist) stratification is/are required for temperate species (Walck et al., 2005; Baskin and Baskin, 2014). Also, seeds of some species undergo after-ripening during the warm dry conditions of summer and are nondormant by autumn. As such, species mostly can be divided into "spring germinators," seeds dispersed in autumn and require cold stratification for dormancy break, and "autumn germinators," seeds dispersed in late spring/early summer and require warm stratification (or after-ripening) for dormancy break.

Temperature and precipitation will continue to change with global warming due to increased atmospheric carbon dioxide (CO_2) concentration, which is the major global greenhouse gas. The concentration of CO_2 in the atmosphere has increased steadily since the 1960s and is predicted to continue to do so (Dunn et al., 2020). Temperatures are expected to increase by up to 5°C by 2100 (under RCP8.5) in eastern North America, western and central Europe, and eastern Asia (IPCC, 2013). Between October and March, the amount of precipitation falling in these regions is expected to increase by 10%–20% by 2100, but between April and September it is predicted to remain about the same in eastern North America, decrease by 10% in central Europe and increase by 10% in eastern Asia. With warming, any predicted increase in precipitation may be offset by increased evaporation, leading to decreases in soil moisture and increases in droughts (Costanza et al., 2016). With increased winter temperatures since the 1970s, the amount of precipitation falling as snow and snow-coverage area has decreased and snowmelt increased; these trends are expected to continue (McCabe and Wolock, 2010). Without a snow cover, the duration of frozen soil will increase, causing winter cooling for subnivium-dependent organisms (Zhu et al., 2019).

Plant regeneration from seeds results from a complexity of interacting traits starting with the maternal environment (Huang et al., 2018). In addition to changes in the climate, several other anthropogenic changes such as atmospheric nitrogen deposition and land-use disturbances are simultaneously occurring with global warming (Danneyrolles et al., 2019). The main purpose of this chapter is to summarize results from research on the effects of climate change and its interaction with other anthropogenic changes on plant regeneration from seeds in the temperate deciduous forest zone. First, we review studies on warming and changes in precipitation (moisture) and their interactions with biotic and non-climatic abiotic factors on dormancy break and germination. Second, we examine germination responses to snow reduction and elevated CO_2. Third, effects of climate change on seed production, soil seed banks, and geographical range shifts of species are considered. Lastly, we suggest aspects of climate change that need to be studied in the future to increase our understanding of plant regeneration from seeds.

Changes in temperature and precipitation

Temperatures in the temperate deciduous forest are distinctly seasonal: a warm (summer) season and a cold to mild (winter) season, with seasons (spring and autumn) transitioning between summer and winter (Walter, 1985). On average, a distinct dry season does not occur. Thus, adequate rainfall occurs every month, and it supports plant growth during the warm season. As such, plant regeneration from seeds is predominantly driven by temperature and secondarily by soil moisture in the temperate deciduous forest (Ibáñez et al., 2007; De Frenne et al., 2009; Staehlin and Fant, 2015; Canham and Murphy, 2016; Gao et al., 2017). The effects of temperature change on seed dormancy and germination have been addressed in several studies—more so than the effects of precipitation changes.

Changes in temperature

The effects of warming during the dormancy break and germination seasons have been examined for seeds from a broad array of species. Artificial warming in the field and laboratory have led to decreases (Solarik et al., 2016; Footitt et al., 2018) or increases (Thompson and Naeem, 1996; Park et al., 2019) in germination and seedling emergence. However, warming also will affect the timing of germination. To illustrate the complexity of seedling emergence with climate change among ecologically similar forest geophytes, warmer (+5°C) autumns and winters delayed time of germination of *Narcissus pseudonarcissus* seeds during the first autumn following dispersal but advanced germination time of *Galanthus nivalis* seeds (Newton et al., 2020). Germination phenology due to increased temperatures between the normal (present) dispersal and germination periods may shift from spring predominantly to autumn (mostly), with some still occurring in spring (Gosling et al., 2009; Bandara et al., 2019; Finch et al., 2019; Kondo et al., 2019; Flanigan et al., 2020) or earlier in spring (De Frenne et al., 2011; Footitt et al., 2018; Bandara et al., 2019; Albrecht et al., 2020). On the other hand, not all species will experience a change in their germination timing with warming. For *Quercus robur*, which has epicotyl physiological dormancy, emergence time of roots and shoots is predicted to remain about the same in a future climate as they are at present (McCartan et al., 2015). While onset of germination did not change with soil warming for seeds of *Acer rubrum*, seedling growth was accelerated (Wheeler et al., 2017).

Shifts in germination phenology from spring to autumn may occur for seeds with conditional (relative) dormancy that are dispersed in autumn and if temperatures increase enough in autumn so that the germination requirements overlap with the warmed temperatures. This type of response will occur primarily in seeds of species that show a Type 2 pattern of changes to temperature requirements for germination as seeds come out of nondeep physiological dormancy (e.g., Walck et al., 1997). That is, in the early stages of dormancy break seeds with Type 2 nondeep physiological dormancy germinate to moderate to high percentages at moderate to high temperatures, and as dormancy break proceeds they gain the ability to also germinate at low temperatures. As such, seeds dispersed in autumn do not germinate under present climate conditions since temperatures in the habitat are too low for germination (Fig. 10.1A). Since the seeds gain the ability to germinate to high percentages at low temperatures with loss of dormancy during winter, they can germinate at low temperatures in spring (Fig. 10.1B).

The shift in germination phenology to autumn (mostly) was shown in laboratory experiments on seeds of herbaceous perennial (Bandara et al., 2019; Finch et al., 2019; Kondo et al., 2019) and woody (Flanigan et al., 2020) species. Although shifts in flowering and seed dispersal phenologies could offset autumn germination if dispersal is delayed until later in autumn (Baskin and Baskin, 1984; Galloway and Burgess, 2009), the germination requirements may not

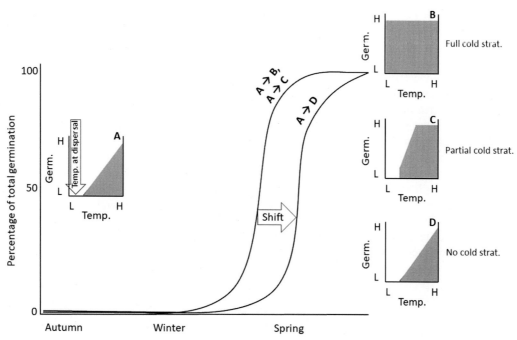

FIGURE 10.1 Effect of a shortened winter on germination of conditionally dormancy seeds with Type 2 nondeep physiological dormancy (see Soltani et al., 2017). At time of dispersal in autumn, seeds can germinate at high (but not at low) temperatures (inset A, shaded triangle); thus, germination in autumn is prevented by low temperatures of the environment. Following full cold stratification (present climate conditions), a high percentage of germination occurs across the temperature range possible for the species (inset B, shaded square); thus, seeds germinate in spring at temperatures that were too low for germination in autumn. If the length of the cold stratification period in winter is not long enough for full dormancy break to occur (including no lowering of the minimum temperature for germination), germination will still occur at the normal time in spring due to climate warming and early arrival of spring (inset C, shaded area). Thus, the predicted climate warming will have no effect on timing of germination. In the extreme case that little or no cold stratification occurs during winter, there would be no dormancy break; thus, timing of germination would be delayed until temperatures increase (inset D). *Germ*, germination; *L*, low temperature; *H*, high temperature.

overlap with the warmed temperatures. Such changes in germination phenology may lead to large repercussions in population dynamics if seedlings die overwinter, thereby depleting soil seed banks, or if seedlings survive overwinter giving them increased preemptive competitive advantage (Bandara et al., 2019; Finch et al., 2019; Newton et al., 2020; Flanigan et al., 2020).

Depending on the amount and timing of warming during the year and temperature requirements for dormancy break and germination, the time of seedling emergence in spring may shift to an earlier time than at present, remain the same, or be delayed. Using seeds with Type 2 nondeep physiological dormancy as a model (Fig. 10.1), we can examine the effects of changes in length of the cold stratification period in winter on time of germination. Under present climate conditions, seeds receive enough cold stratification to become fully nondormant, and thus in spring they can germinate to high percentages over the widest range of temperatures possible for the species, including those in early spring (Fig. 10.1B).

Shifts to earlier spring germination have been shown for the herbs *Alliaria petiolata* (Footitt et al., 2018) and *Penstemon digitalis* (Bandara et al., 2019). Although a cold stratification requirement may not be fully satisfied with shortening of winters and winter warming, emergence will occur from seeds in the population that have a low degree of dormancy. Consequently, there may be selection for seeds with low dormancy and reduction in the amount of cold stratification required; the nongerminated seeds remain in the soil seed bank and may have increased mortality with soil warming (Footitt et al., 2018). With advanced (earlier) emergence in spring, susceptibility to late frosts may occur (Souther and McGraw, 2011). While seedling emergence varied with warming among six species of broad-leaf trees, the more rapid growth of the seedlings was associated with increased mortality from frost (Fisichelli et al., 2014a). On the other hand, tree seedlings that emerged early were exposed to late frosts, but they benefitted from a longer growing period that resulted in increased overall survival (Bianchi et al., 2019).

However, it is likely that the increased temperatures in spring due to warming may compensate for the lack of full dormancy break in winter, resulting in little, or no, change in the timing of germination in spring (Fig. 10.1C). That is, with the onset of warmer temperatures earlier in spring seeds that are not fully nondormant may germinate at the same time as usual, that is, no change in germination phenology would occur.

Timing of germination also could be delayed in spring. If the cold stratification period in winter is not long enough to reduce the minimum temperature for germination, seeds would germinate later in spring when temperatures increased (Fig. 10.1D). We do not know of any germination studies that have directly addressed the effects of decreased amount of cold stratification on delayed germination timing. The delay in simulated spring germination in a laboratory warming experiment on the woody vine *Parthenocissus quinquefolia* in Flanigan et al.'s (2020) study might be an example. A similar phenomenon has been documented for buds, since they require chilling to release winter dormancy and a critical thermal time (growing degree days) in spring for budburst (Murray et al., 1989; Körner and Basler, 2010; Nanninga et al., 2017). With a decrease in number of chilling days and insufficient fulfillment of the chilling requirement, a delayed green-up has been reported with earlier springs south of 40°N in eastern North America (Zhang et al., 2007). In contrast, north of 40°N buds receive sufficient chilling, and earlier springs result in advanced green-up.

It is important to determine niche breadths (or tolerance ranges) for dormancy loss and germination since they determine, in part, how species will respond to changes in climate (Walck et al., 2011). Populations of a species may persist at a particular location if the niche breadth is as wide as a future change in environmental conditions, whereas species must adapt, migrate, or risk extinction if the niche breadth does not overlap with the changed conditions. Niche breadths, determined from temperature responses and calculated as Levins' niche breadth index (B_n), varied among perennial herbaceous species related to their ecological characteristics. *Asclepias incarnata* and *A. verticillata*, with limited dispersal and relatively high habitat specificity had narrower niche breadths than *A. syriaca*, with broad dispersal and relatively low habitat specificity (Finch et al., 2019). *Penstemon digitalis* with limited genetic neighborhood size and gene dispersal range had a narrower niche breadth than *Asclepias syriaca* and *Physalis longifolia* with broad genetic neighborhood sizes and gene dispersal ranges (Bandara et al., 2019). Species with narrow niche breadths are predicted to be more sensitive to climate change than those with wide breadths. However, Bandara et al. (2019) found that the germination phenology of *P. digitalis* with a narrow niche breadth was mostly unaffected by warming but that of *A. syriaca* and *P. longifolia* with wide breadths shifted dramatically, which possibly is maladaptive.

Changes in temperature during the dormancy breaking period—for example, warm periods during cold—rarely have been studied for deciduous forest species. Flanigan et al. (2020) examined the effects of normal and extreme warm periods during cold stratification on seeds of woody plants in southeastern United States, an area that typically experiences warm periods during winter. Although interrupting cold stratification with warm periods shifted germination to earlier in spring for some species, the effects of warm periods on germination responses were not consistent across nine native and nonnative woody plants, perhaps indicating that selection had occurred that prevented erratic responses to such warm interruptions (Flanigan et al., 2020). On the other hand, seeds of the exotic *Lonicera maackii* germinated in the field during a warm period in winter. Although the seedlings died following a late-winter freeze, warming winters might enable them to survive, giving the species a preemptive competitive advantage.

Interaction between warming and changes in precipitation

Changes in precipitation along with warming undoubtedly will influence moisture content of the soil and seeds. In addition, moisture interacts with temperature to overcome dormancy. A moist environment is required for effective dormancy break at cold temperatures (cold stratification) but either a moist (warm stratification) or dry (after-ripening) environment is required for dormancy break at warm temperatures (Baskin and Baskin, 2014). Most studies examining the effects of warming and soil moisture on regeneration of plants from seeds have focused on germination instead of the dormancy break.

A few studies on trees have tested the effects of both temperature and moisture on recruitment via seeds. Among several tree species in eastern USA, warming advanced germination by 2 weeks and seedling leafing-out by 10 days during a 3-year field experiment (Kaye and Wagner, 2014). However, a 20% increase in precipitation had no effect on germination or leafing-out phenology. The authors concluded that southern species that might be expected to move northward did not exhibit responses that would place them at a competitive advantage over current resident (northern) species. For two secondary forest trees in Europe, *Acer pseudoplatanus* and *A. platanoides*, germination and survival decreased with warming and/or soil moisture reduction, while early growth of seedlings decreased with soil moisture reduction (Carón et al., 2015a). Increases in precipitation frequency stimulated germination of both species and warming reduced seedling survival and growth (Carón et al., 2015b).

Studies on herbaceous, mostly annual, species had mixed and surprising results for effects of temperature and moisture changes on their recruitment. Distinct differences occurred among arable (annual and perennial) weeds in Europe that made germination of rare weeds more sensitive to changes in climate than common weeds due to their low-temperature optima, narrow moisture requirements, and stronger declines with decreased water availability (Rühl et al., 2015, 2016).

Among North American winter annuals, a warm period during after-ripening across a relative humidity gradient had no effect on germination, whereas a decrease in the diurnal temperature range had mixed results, the most extreme being no germination in the nonnative *Ranunculus parviflorus* (Hicks et al., 2019).

At the community level, only one study has investigated the effects of warming and changes in moisture on recruitment of species within the region of the temperate deciduous forest. An English calcareous grassland was subjected to simulated warmer winters and increased summer rainfall and to warmer winters and summer drought (Sternberg et al., 1999). With increased summer rainfall, seedling establishment was limited by microsite availability due to prolonged life span of perennial grasses. With summer drought, establishment was reduced due to presence of dense leaf litter and reduced seed production of annual species. The authors suggested that succession in grasslands may be influenced under both climate scenarios due primarily to microsite availability.

Interactions of warming and/or changes in moisture with biotic and nonclimatic abiotic factors

Although climate has a strong influence on plant regeneration, other factors interact with climate change and affect recruitment. The interaction of warming and changes in moisture with biotic factors such as competition and herbivory and abiotic factors such as leaf litter and nitrogen deposition have been investigated in a few studies. In Europe, two secondary forest trees showed widely divergent responses to warming and competition with natural understory vegetation but not to reduced precipitation (Carón et al., 2015c). Competition reduced germination and seedling survival of *A. platanoides*, whereas warming enhanced germination and seedling survival of *A. pseudoplatanus*. Reducing competition strongly increased growth of *A. platanoides* seedlings. The authors concluded that with increased competition projected under future climates *A. platanoides* will be more strongly impacted than *A. pseudoplatanus*. With regards to leaf litter, warming mostly increased seedling emergence for several temperate trees regardless of presence of litter, but emergence was much less with litter (Fisichelli et al., 2014b). The effects of leaf litter on emergence across a precipitation gradient varied greatly among the same species. Seedling growth was increased by warming, but conditions too dry or too wet limited the positive response to warming. Survivorship of *Quercus alba* seedlings was independent of warming in a field experiment using open-top chambers (OTCs) and had a complex relationship to herbivory (Burt et al., 2014). As temperatures warmed, levels of insect herbivory on seedlings decreased, perhaps due to decreased herbivore abundance. However, seedlings with higher levels of herbivory had higher survivorship than those without herbivory regardless of warming treatments, which might have been related to a positive relationship observed between soil moisture and herbivory.

Since nitrogen deposition has increased over time (Ackerman et al., 2019), it makes sense to test the effects of nitrogen addition concomitantly with warming. In a temperate old field, moisture availability played a role in driving warming responses by reducing seedling survival in some successional tree species, whereas nitrogen addition increased survivorship for the nonnative *Elaeagnus umbellata* (McWhirter and Henry, 2015). However, these responses of tree seedlings were influenced by grasses: increased grass cover facilitated seedling establishment in a dry year, probably by providing shade and retaining soil moisture. In another study on nitrogen addition combined with warming, seedling emergence time of the forest understory grass *Milium effusum* increased with warming, but nitrogen addition had no effect (Maes et al., 2014).

Snow cover reduction

Snow provides an effective blanket that protects seeds and seedlings from freezing temperatures, since it keeps soil at temperatures slightly above freezing and reduces soil temperature fluctuations (Decker et al., 2003; Simons et al., 2010). Snow cover is especially important in northern temperate deciduous forests, where snow remains on the ground for extended periods, unlike in southern forests (Hughes et al., 1996). As such, germination of some species occurs beneath snow cover and that of other species coincides with snowmelt providing ample soil moisture for seedling establishment (Phartyal et al., 2009; Kondo et al., 2011). The importance of snow cover on seed and seedling biology is shown by climate change studies that manipulate snow depth via removing or adding snow.

Snow conditions have been shown to affect seed germination and seedling survivorship, mostly of trees in a few climate-change studies in temperate deciduous forests. Postwinter germination varied among eastern North American trees differing in dispersal phenology in response to decreased snow cover. Germination was decreased for five tree species that mostly disperse seeds from summer to autumn but increased for three other species that disperse seeds mostly in winter (Drescher and Thomas, 2013). For all eight species, sapling survival decreased with decreasing snow cover. With the addition of passive warming (using OTCs), germination for three of these species did not vary across snow cover manipulations; the other five species were not tested (Drescher, 2014). On the other hand, survival may depend on the interaction of snow with other factors. Survival of *Acer saccharum* seedlings in northern Minnesota declined with snow removal only in forests lacking conifers, and leafing-out was slowed under a shallow snow cover due to small-mammal herbivory (Guiden et al., 2019). Studies investigating snow cover on herbaceous plants rarely have been done. In the monocarpic perennial *Lobelia inflata*, reducing snow cover decreased survivorship of overwintering rosettes (Simons et al., 2010).

Elevated CO_2

In addition to global warming, increased CO_2 concentration in the atmosphere may have a direct effect on regeneration of plants from seeds. In a *meta*-analysis by Marty and BassiriRad (2014), elevated CO_2 had varied parental and direct effects on percentage and rate (speed) of seed germination among several weedy herb and tree species common in temperate biomes. Often, studies on the effects of elevated CO_2 on germination are combined with other climate variables. Four studies illustrate the complexity of germination results with elevated CO_2, and they usually show that CO_2 is not as important as other climate factors influencing germination and emergence. Elevated temperature affected germination of 17 temperate Korean trees more so than CO_2, and most of the species were unaffected by either of these factors (Kim and Han, 2018). While elevated temperature decreased germination percentage and rate in the common temperate weed *Silene noctiflora*, elevated CO_2 decreased percentage but increased rate (Qaderi and Reid, 2008). When elevated CO_2 was tested along with elevated temperature and moisture stress, temperature interacting with moisture stress had the largest effects and was unrelated to CO_2 for germination of the wild-type *Arabidopsis thaliana* (Abo Gamar and Qaderi, 2019). In an old-field ecosystem, air temperature and soil moisture explained most of the variation in seedling emergence and establishment of the early successional trees *Acer saccharinum*, *Pinus taeda*, and *Liquidambar styraciflua*, while CO_2 explained the least amount (Classen et al., 2010).

Seed production

Changes in seed production due to global warming will influence the amount of seed input into the soil and the number of potential seedlings established. Climate change studies have shown that seed production monitored over many years is related to warming. In France, *Quercus petraea* and *Q. robur* have increased seed production over the last decade with increasing spring temperature; no relationship was found with annual or seasonal precipitations (Caignard et al., 2017). Moreover, acorn mass for *Q. petraea* increased over the same period, which may enhance seedling establishment. Masting is a common feature of temperate trees, especially oaks. As such, not only has seed production increased, but masting interval has decreased with increased temperature over the past 38 years for *Quercus crispula* in Japan (Shibata et al., 2020). However, we must be cautious about conclusions drawn from observational studies on masting behavior over time since demographic effects such as aging may be more important than climatic variables (Pesendorfer et al., 2020).

In at least a few studies, experimental manipulation of climate variables did not have a direct influence on seed set at the species or community level. In the spring ephemeral *Gagea lutea*, experimental warming decreased seed set, probably through reduced pollen germination (Sunmonu and Kudo, 2015). In the English heathlands, species composition had a larger influence on seed production of target species than precipitation manipulations of drought and heavy rainfall events (Gellesch et al., 2015). However, offspring from mother plants grown under these extreme manipulations had advanced germination (Walter et al., 2016).

Seed predation may offset benefits to increased production caused by warming. For example, while seed production increased and masting variability decreased over 39 years for *Fagus sylvatica* in England, less effective pollination and greater seed predation reduced the benefits of increased seed set (Bogdziewicz et al., 2020a,b). Another example of how increased seed set may not be positively related to seedling establishment is seen in the study by Touzot et al. (2020). With modeling, they found that populations of wild boar (*Sus scrofa*), a widespread and abundant acorn-consumer species, increased with an increase in mast seeding of *Quercus petraea* related to spring warming.

Predators can act in subtle ways with climate change to influence seed set. In the biennial herb *Oenothera biennis*, total lifetime seed production was highest for plants with Japanese beetle (*Popillia japonica*) herbivory in warmed conditions (Lemoine et al., 2017). Insect pollinators—both the number of species and their foraging activities—responding to changes in climate may impact seed production. An extremely warm spring in Japan decreased seed set in bee-pollinated spring ephemerals but did not affect seed set in fly-pollinated species (Kudo et al., 2004). Phenological mismatches also may occur. In the spring ephemeral *Corydalis ambigua*, flowering tended to occur before pollinator (bumble bee) activity when spring came early during 10–14 years of study, resulting in reduced seed production (Kudo and Ida, 2013). Early snowmelt and slow soil warming increased the risk of this mismatch (Kudo and Cooper, 2019). For the herbaceous perennials, *Asclepias incarnata* and *Tradescantia ohiensis*, manipulating flowering onset in greenhouses to cause earlier flowering decreased seed set compared to that in historical times (Rafferty and Ives, 2012). The cause for the decreased seed set was changes in the composition of the pollinator assemblage as well as effectiveness of some pollinators.

Soil seed banks

Depending on the species, any changes in seed input (production), seed persistence, or seed output (seedling emergence or seed death) either directly or indirectly in response to climate change may influence soil seed bank dynamics (Plue et al., 2013). Although soil seed banks have been suggested to serve as important sources to maintain species diversity in communities altered by global warming, their capacity to buffer change will be limited and mostly temporary (Plue et al., 2020). Changes in the soil environment caused by climate change that might influence seed survivorship and longevity in the soil—like deleterious fungi—have been investigated. Manipulating temperature, moisture, and CO_2 yielded no consistent response, but studies have been limited to grasslands in the temperate deciduous forest zone. Winter warming with supplemented summer rainfall in a calcareous grassland at Oxfordshire, UK, influenced neither fungal pathogens nor seed persistence in the soil over 1–2 years (Leishman et al., 2000). Only two of 69 species showed a response to climate manipulations after 6–7 years in a calcareous grassland in Derbyshire, UK. Number of seeds increased with either summer drought or winter heating, but this response was unrelated to their persistence (Akinola et al., 1998a,b). Seventeen years of drought in the Derbyshire grassland reduced the size and species richness of the soil seed bank, and the effects were greater for the seed bank than the standing vegetation (Basto et al., 2018). In a long-term (18 years) manipulative study in a grassland in Germany, elevated CO_2 concentration caused seed density of long-term persistent species to increase in the soil seed bank but that of short-term persistent species to decrease (Seibert et al., 2019).

Changes in environmental conditions also may affect seedling emergence from soil seed banks. For example, longer intervals (12 vs 8 or 4 days) between watering events associated with a low-water treatment increased emergence of the old-field weed *Setaria faberi*, but watering interval length had no effect on emergence in the average-water treatment (Robinson and Gross, 2010). In contrast, emergence of the weed *Chenopodium album* in the same study increased with long watering intervals as well as with high amounts of water. For the common forest understory grass *Elymus canadensis*, soil freeze–thaw cycles affected seed persistence by limiting seedling emergence, delaying emergence, and providing opportunities for fungal pathogens (Connolly and Orrock, 2015).

Geographical range shifts

Although species specific, current evidence shows that a northward shift in temperate tree species is occurring but lagging the rate of warming (Evans and Brown, 2017; Sittaro et al., 2017). However, this shift is not consistent across the temperate deciduous forest zonobiome (Merlin et al., 2018). Recruitment in response to warming among several trees species was (and will continue to be) greater at northern than at southern locations (Boisvert-Marsh et al., 2019; Gao et al., 2017; Matías and Jump, 2014), with treefall gaps being particularly important for establishment of northward migrating trees (Leithead et al., 2010). In contrast, Fei et al. (2017) reported that eastern USA tree species have experienced a stronger westward than poleward shift, which is related to changes in moisture availability.

Tracking climate change across a highly fragmented landscape will be especially problematic for species with limited dispersal capacity and a long generation time (Wang et al., 2018). The discrepancy between modeled future shift rates (24–200 m year^{-1}), which is comparable to pollen-reconstructed rates, is far less than the velocity of climate change (>1000 m year^{-1}, Loarie et al., 2009). In fact, seed dispersal for some species may not occur in the direction predicted by warming. Instead of migrating upslope on mountains to escape warming, seeds of some animal-dispersed autumn-fruiting plants are moved toward the foot of mountains by Asian black bears (*Ursus thibetanus*), Japanese martens (*Martes melampus*), and Japanese macaques (*Macaca fuscata*) (Naoe et al., 2016, 2019). These mammals descend mountains to track the phenology of their food plants and to visit lowlands to consume agricultural crops. In contrast, animal-dispersed spring- and summer-fruiting plants can escape warming since their seeds are dispersed upslope by bears and martens. Although models predict that species with poor dispersal ability will have difficulty keeping pace with warming, historical evidence has shown that these species exhibited fast migration in the past, for example, during the early Holocene in Europe. This discrepancy between historical and future migration rates may be partly explained by germination responses. Among several forest herbs, Graae et al. (2009) found that germination of poor colonizers was positively related to temperatures favoring increased recruitment ability with warming. In contrast, germination of fast-colonizing species was independent of temperature.

For some plant species, dispersal limitation may prevent them from migrating fast enough to keep pace with climate change (Bellemare and Moeller, 2014). Such might be the case for seeds of forest herbs that are primarily ant dispersed. However, Vellend et al. (2003) documented long-distance dispersal of *Trillium grandiflorum* by white-tailed deer (*Odocoileus virginianus*). Their deer-generated seed shadow model predicted that seeds of this species occasionally could be dispersed >3 km from parent plants compared to up to 70 m via ants. Suitable habitat may not limit seedling establishment if seeds disperse long distances. For another ant-dispersed species, *Jeffersonia diphylla*, seeds germinated and seedlings established in forested habitats 200 km beyond its natural edge of the range in the northeastern USA (Bellemare and Moeller, 2014).

Acer saccharum has been the focus of several studies to understand how range shifts of temperate trees occur into boreal forests. In Canada, seedling survival of *A. saccharum* was higher in adjacent boreal forest than that in temperate forest, probably due to better illumination caused by canopy-opening disturbances in the boreal forest (Kellman, 2004). While establishment of *A. saccharum* was highest within the species' range, survival was comparable to that at the range margin and beyond in Canada (Solarik et al., 2018). Thus, there should be no impediment for seedling establishment and survival in the boreal forest. However, establishment and survivorship may be influenced by changes in the climate. In the mixed deciduous–boreal forest in central Minnesota (USA), emergence and survival among several temperate tree species (including *A. saccharum*) were unaffected by warming and simulated drought conditions, but seedling growth increased with warming when combined with drought (Wright et al., 2018).

Several studies have shown that lack of suitable microsites may constrain recruitment during migration of temperate species. For *Acer saccharum*, soil chemistry, particularly low base cation concentrations, was a major limitation for seedling establishment in boreal forest soils, and an insufficient arbuscular mycorrhizal fungal inoculum strongly reduced seedling performance in these soils (Carteron et al., 2020). Soil fungal pathogens and seed predation by small mammals were suggested to prevent establishment of *A. saccharum* on boreal forest soil beyond the range of the species in Quebec (Canada) (Brown and Vellend, 2014). The substrate underneath a boreal forest canopy is an unfavorable environment for seedling establishment of several temperate trees (including *A. saccharum*) due to needle cover and decayed wood (Solarik et al., 2019). On the other hand, species could experience increased favorable conditions during migration. For the grass *Milium effusum*, seedling emergence and growth were accelerated more on soils from colder sites than on soil from the site where seeds were collected (De Frenne et al., 2014).

Seed predation also may act alone or in conjunction with microsites to limit the range expansion of a species. For *Acer saccharum*, seed predation reduced seedling establishment beyond the current edge of its range (Brown and Vellend, 2014). Over time, however, seedling density for *A. saccharum* was highest at and beyond the edge compared to that in the core of the range, which is consistent with the hypothesis of release from natural enemies (Urli et al., 2016). Predation and substrates are important factors limiting seedling emergence not only for *A. saccharum* but also for a few other temperate species in the boreal forest. The large seeds produced by temperate species such as *A. saccharum* and *Fraxinus nigra* were vulnerable to granivorous vertebrates within the boreal forest, whereas the small seeds of *Betula alleghaniensis* and *Thuja occidentalis* were prevented from germinating by leaf litter (Evans et al., 2020).

Herbaceous plants have been less studied than trees in terms of mechanisms underlying range shifts. For the European forest herbs *Milium effusum* and *Stachys sylvatica*, soil biota emerged as an important driver in seedling emergence (Ma et al., 2019). In nonsterilized soil, seedling emergence increased when seeds of *M. effusum* were sown in northern soils, while that of *S. sylvatica* showed no response. When foreign soil microorganisms were inoculated into soils, emergence of *S. sylvatica*, but not *M. effusum*, increased in southern soils that simulated future warming. In the United Kingdom, sowing seeds of southern species into a northern experimental plot subjected to artificial winter warming and summer drought and following seedling establishment over 6 years showed that most of the species had poor establishment (Buckland et al., 2001). However, the timing of seedling emergence was important. With warming alone, earlier emergence of seeds from the southern invaders *Brachypodium pinnatum* and *Bromus erectus* at the northern site resulted in increased biomass by the end of the growing season (Moser et al., 2011). Summer drought did not affect *B. erectus* but negated the benefits of warming for *B. pinnatum*. Thus, northward migration of these two species is not limited by substrate conditions but apparently by dispersal limitation due to low seed production (Moser and Thompson, 2014). Since both species are self-incompatible, founder populations in isolated grasslands may be unable to produce seeds, thus slowing northward migration.

Future considerations

Plant regeneration results from a complexity of interacting traits. For example, in the endangered *Cirsium pitcheri* temperature (and not precipitation) was the most critical factor impacting seedling growth under future climate conditions.

However, seed mass, a trait influenced by maternal environmental effects, was related to seedling emergence and survival (Staehlin and Fant, 2015). For many traits and their interactions, we lack an understanding of how they will function and interact within a climate change framework (Saatkamp et al., 2019).

Partitioning the effects of climate-change variables into seed development, seed dispersal, seed dormancy loss, seed germination, and seedling emergence/establishment life-cycle stages would allow a better understanding of the underlying mechanisms that control responses. For example, although studies have shown that a shift in timing of germination from spring to autumn may occur, a delay in dispersal may prevent the germination requirements from overlapping with habitat temperatures in autumn and maintain the predominantly spring emergence. One group of species in need of study is herbaceous species occurring in mesic deciduous forests (Bellemare and Moeller, 2014). Many of these species have seeds with morphophysiological dormancy in which the breaking of the morphological (embryo growth) and physiological components of dormancy require specific temperatures (Baskin and Baskin, 1988, 2014). As such, investigations into how mismatched temperatures between the timing of embryo growth and physiological dormancy break are needed to understand how seedling emergence and establishment may be affected by climate change.

For many aspects of climate change, we do not know how they will affect plant regeneration, especially extreme events. Key examples include increasing precipitation downpour events (Kendon et al., 2014) that could wash seeds out of the habitat or bury them deeper in the soil than normal thus affecting seedling emergence; increased frequency and/or intensity of winter warm spells and freeze/thaw cycles (Tomczyk et al., 2019) interrupting cold stratification and/or increasing seed mortality, respectively; increased wildfires, especially in the Coastal Plain of the southeastern USA (Fill et al., 2019), influencing seed mortality and/or seed germinability in the soil seed bank; and interactions between heat waves and precipitation (De Boeck et al., 2010) affecting warm stratification/after-ripening and mortality of recalcitrant seeds. Changes in large-scale climatic events like the frequency and intensity of hurricanes, which are major disturbances in temperate deciduous forests of the eastern USA that influence seedling establishment (Carlton and Bazzaz, 1998), have not been studied as well.

Variation among populations has been documented for various seed traits from small to large spatial scales (Lemke et al., 2015; Bandara et al., 2019). However, understanding the range of variation in extreme climate events (e.g., heat waves, droughts) that match predicted changes in climatic variables would greatly assist in predicting the response of plant population's climate change. We also are far from understanding the roles that local adaptation and phenotypic plasticity will play on plant regeneration from seeds of diverse species over large scales, especially at range edges or in a diversity of environments (De Frenne et al., 2012; Varsamis et al., 2019).

References

Abo Gamar, M.I., Qaderi, M.M., 2019. Interactive effects of temperature, carbon dioxide and watering regime on seed germinability of two genotypes of *Arabidopsis thaliana*. Seed Sci. Res. 29, 12–20.

Ackerman, D., Millet, D.B., Chen, X., 2019. Global estimates of inorganic nitrogen deposition across four decades. Glob. Biogeochem. Cycl. 33, 100–107.

Akinola, M.O., Thompson, K., Buckland, S.M., 1998a. Soil seed bank of an upland calcareous grassland after 6 years of climate and management manipulations. J. Appl. Ecol. 35, 544–552.

Akinola, M.O., Thompson, K., Hillier, S.H., 1998b. Development of soil seed banks beneath synthesized meadow communities after seven years of climate manipulations. Seed Sci, Res. 8, 493–500.

Albrecht, M.A., Dell, N.D., Long, Q.G., 2020. Seed germination traits in the rare sandstone rockhouse endemic *Solidago albopilosa* (Asteraceae). J. Torrey Bot. Soc. 147, 172–184.

Anderson, R.C., Fralish, J.S., Baskin, J.M. (Eds.), 1999. Savannas, Barrens, and Rock Outcrop Plant Communities of North America. Cambridge University Press, Cambridge.

Bandara, R.G., Finch, J., Walck, J.L., Hidayati, S.N., Havens, K., 2019. Germination niche breadth and potential response to climate change differ among three North American perennials. Folia Geobot. 54, 5–17.

Baskin, J.M., Baskin, C.C., 1984. The ecological life cycle of *Campanula americana* in northcentral Kentucky. Bull. Torrey Bot. Club 111, 329–337.

Baskin, C.C., Baskin, J.M., 1988. Germination ecophysiology of herbaceous plant species in a temperate region. Am. J. Bot. 75, 286–305.

Baskin, C.C., Baskin, J.M., 2014. Seeds: Ecology, Biogeography, and Evolution of Dormancy and Germination, second ed. Academic Press/Elsevier, San Diego.

Basto, S., Thompson, K., Grime, J.P., Jason, D., 2018. Severe effects of long-term drought on calcareous grassland seed banks. Npj Clim. Atmos. Sci. 1, 1. Available from: https://doi.org/10.1038/s41612-017-0007-3.

Bellemare, J., Moeller, D.A., 2014. Climate change and the herbaceous layer of temperate deciduous forests. In: Gilliam, F.S. (Ed.), The Herbaceous Layer in Forests of Eastern North America, second ed. Oxford University Press, Oxford, pp. 460–480.

Bianchi, E., Bugmann, H., Bigler, C., 2019. Early emergence increases survival of tree seedlings in Central European temperate forests despite severe late frost. Ecol. Evol. 9, 8238–8252.

Bogdziewicz, M., Kelly, D., Thomas, P.A., Lageard, J.G.A., Hacket-Pain, A., 2020a. Climate warming disrupts mast seeding and its fitness benefits in European beech. Nat. Plants 6, 88–94.

Bogdziewicz, M., Kelly, D., Tanentzap, A.J., Thomas, P.A., Lageard, J.G.A., Hacket-Pain, A., 2020b. Climate change strengthens selection for mast seeding in European beech. Curr. Biol. 30, 3477–3483.

Boisvert-Marsh, L., Périé, C., De Blois, S., 2019. Divergent responses to climate change and disturbance drive recruitment patterns underlying latitudinal shifts of tree species. J. Ecol. 107, 1956–1969.

Brown, C.D., Vellend, M., 2014. Non-climatic constraints on upper elevational plant range expansion under climate change. Proc. R. Soc. B 281, 20141779. Available from: https://doi.org/10.1098/rspb.2014.1779.

Buckland, S.M., Thompson, K., Hodgson, J.G., Grime, J.P., 2001. Grassland invasions: effects of manipulations of climate and management. J. Appl. Ecol. 38, 301–309.

Burt, M.A., Dunn, R.R., Nichols, L.M., Sanders, N.J., 2014. Interactions in a warmer world: effects of experimental warming, conspecific density, and herbivory on seedling dynamics. Ecosphere 5, 9. Available from: https://doi.org/10.1890/ES13-00198.1.

Caignard, T., Kremer, A., Firmat, C., Nicolas, M., Venner, S., Delzon, S., 2017. Increasing spring temperatures favor oak seed production in temperate areas. Sci. Rep. 7, 8555. Available from: https://doi.org/10.1038/s41598-017-09172-7.

Carteron, A., Parasquive, V., Blanchard, F., Guilbeault-Mayers, X., Turner, B.L., Vellend, M., et al., 2020. Soil abiotic and biotic properties constrain the establishment of a dominant temperate tree into boreal forests. J. Ecol. 108, 931–944.

Canham, C.D., Murphy, L., 2016. The demography of tree species response to climate: seedling recruitment and survival. Ecosphere 7 (8), e01424. Available from: https://doi.org/10.1002/ecs2.1424.

Classen, A.T., Norby, R.J., Company, C.E., Sides, K.E., Weltzin, J.F., 2010. Climate change alters seedling emergence and establishment in an old-field ecosystem. PLoS One 5, e13476. Available from: https://doi.org/10.1371/journal.pone.0013476.

Carlton, G.C., Bazzaz, F.A., 1998. Resource congruence and forest regeneration following an experimental hurricane blowdown. Ecology 79, 1305–1319.

Carón, M.M., De Frenne, P., Brunet, J., Chabrerie, O., Cousins, S.A.O., De Backer, L., et al., 2015a. Interacting effects of warming and drought on regeneration and early growth of *Acer pseudoplatanus* and *A. platanoides*. Plant Biol. 17, 52–62.

Carón, M.M., De Frenne, P., Chabrerie, O., Cousins, S.A.O., De Backer, L., Decocq, G., et al., 2015b. Impacts of warming and changes in precipitation frequency on the regeneration of two *Acer* species. Flora 214, 24–33.

Carón, M.M., De Frenne, P., Brunet, J., Chabrerie, O., Cousins, S.A.O., Decocq, G., et al., 2015c. Divergent regeneration responses of two closely related tree species to direct abiotic and indirect biotic effects of climate change. For. Ecol. Manage 342, 21–29.

Connolly, B.M., Orrock, J.L., 2015. Climatic variation and seed persistence: freeze–thaw cycles lower survival via the joint action of abiotic stress and fungal pathogens. Oecologia 179, 609–616.

Costanza, J., Beck, S., Pyne, M., Terando, A., Rubino, M., White, R., et al., 2016. Assessing climate-sensitive ecosystems in the Southeastern United States. U.S. Geological Survey Open-File Report 2016-1073. Available from: https://doi.org/10.3133/ofr20161073.

Danneyrolles, V., Dupuis, S., Fortin, G., Leroyer, M., De Römer, A., Terrail, R., et al., 2019. Stronger influence of anthropogenic disturbance than climate change on century-scale compositional changes in northern forests. Nat. Commun. 10, 1265. Available from: https://doi.org/10.1038/s41467-019-09265-z.

De Boeck, H.J., Dreesen, F.E., Janssens, I.A., Nijs, I., 2010. Climatic characteristics of heat waves and their simulation in plant experiments. Glob. Change Biol. 16, 1992–2000.

Decker, K.L.M., Wang, D., Waite, C., Scherbatskoy, T., 2003. Snow removal and ambient air temperature effects on forest soil temperatures in northern Vermont. Soil Sci. Soc. Am. J. 67, 1234–1242.

De Frenne, P., Kolb, A., Verheyen, K., Brunet, J., Chabrerie, O., Decocq, G., et al., 2009. Unravelling the effects of temperature, latitude and local environment on the reproduction of forest herbs. Glob. Ecol. Biogeogr. 18, 641–651.

De Frenne, P., Brunet, J., Shevtsova, A., Kolb, A., Graae, B.J., Chabrerie, O., et al., 2011. Temperature effects on forest herbs assessed by warming and transplant experiments along a latitudinal gradient. Glob. Change Biol. 17, 3240–3253.

De Frenne, P., Graae, B.J., Brunet, J., Shevtsova, A., De Schrijver, A., Chabrerie, O., et al., 2012. The response of forest plant regeneration to temperature variation along a latitudinal gradient. Ann. Bot. 109, 1037–1046.

De Frenne, P., Coomes, D.A., De Schrijver, A., Staelens, J., Alexander, J.M., Bernhardt-Römermann, M., et al., 2014. Plant movements and climate warming: intraspecific variation in growth responses to nonlocal soils. New Phytol. 202, 431–441.

Drescher, M., 2014. Snow cover manipulations and passive warming affect post-winter seed germination: a case study of three cold-temperate tree species. Clim. Res. 60, 175–186.

Drescher, M., Thomas, S.C., 2013. Snow cover manipulations alter survival of early life stages of cold-temperate tree species. Oikos 122, 541–554.

Dunn, R.J.H., Stanitski, D.M., Gobron, N., Willett, K.M., 2020. Global climate. In: Blunden, J., Arndt, D.S. (Eds.), State of the Climate in 2019. Bulletin of the American Meteorological Society, vol. 101, pp. S9–S127.

Evans, P., Brown, C.D., 2017. The boreal-temperate forest ecotone response to climate change. Environ. Rev. 25, 423–431.

Evans, P., Crofts, A.L., Brown, C.D., 2020. Biotic filtering of northern temperate tree seedling emergence in beyond-range field experiments. Ecosphere 11, e03108. Available from: https://doi.org/10.1002/ecs2.3108.

Fei, S., Desprez, J.M., Potter, K.M., Jo, I., Knott, J.A., Oswalt, C.M., 2017. Divergence of species responses to climate change. Sci. Adv. 3, e1603055. Available from: https://doi.org/10.1126/sciadv.1603055.

Fill, J.M., Davis, C.N., Crandall, R.M., 2019. Climate change lengthens southeastern USA lightning-ignited fire seasons. Glob. Change Biol. 25, 3562–3569.

Finch, J., Walck, J.L., Hidayati, S.N., Kramer, A.T., Lason, V., Havens, K., 2019. Germination niche breadth varies inconsistently among three *Asclepias* congeners along a latitudinal gradient. Plant Biol. 21, 425–438.

Fisichelli, N., Vor, T., Ammer, C., 2014a. Broadleaf seedling responses to warmer temperatures "chilled" by late frost that favors conifers. Eur. J. For. Res. 133, 587–596.

Fisichelli, N., Wright, A., Rice, K., Mau, A., Buschena, C., Reich, P.B., 2014b. First-year seedlings and climate change: species-specific responses of 15 North American tree species. Oikos 123, 1331–1340.

Flanigan, N., Bandara, R., Wang, F., Jastrzębowski, S., Hidayati, S.N., Walck, J.L., 2020. Germination responses to winter warm spells and warming vary widely among woody plants in a temperate forest. Plant Biol. 22, 1052–1061.

Footitt, S., Huang, Z., Ölcer-Footitt, H., Clay, H., Finch-Savage, W.E., 2018. The impact of global warming on germination and seedling emergence in *Alliaria petiolata*, a woodland species with dormancy loss dependent on low temperature. Plant Biol. 20, 682–690.

Galloway, L.F., Burgess, K.S., 2009. Manipulation of flowering time: phenological integration and maternal effects. Ecology 90, 2139–2148.

Gao, W.-Q., Ni, Y.-Y., Xue, Z.-M., Wang, X.-F., Kang, F.-F., Hu, J., et al., 2017. Population structure and regeneration dynamics of *Quercus variabilis* along latitudinal and longitudinal gradients. Ecosphere 8, e01737. Available from: https://doi.org/10.1002/ecs2.1737.

Gellesch, E., Wellstein, C., Beierkuhnlein, C., Kreyling, J., Walter, J., Jentsch, A., 2015. Plant community composition is a crucial factor for heath performance under precipitation extremes. J. Veg. Sci. 26, 975–984.

Gosling, P.G., McCartan, S.A., Peace, A.J., 2009. Seed dormancy and germination characteristics of common alder (*Alnus glutinosa* L.) indicate some potential to adapt to climate change in Britain. Forestry 82, 573–582.

Graae, B.J., Verheyen, K., Kolb, A., Van Der Veken, S., Heinken, T., Chabrerie, O., et al., 2009. Germination requirements and seed mass of slow- and fast- colonizing temperate forest herbs along a latitudinal gradient. Écoscience 16, 248–257.

Guiden, P.W., Connolly, B.M., Orrock, J.L., 2019. Seedling responses to decreased snow depend on canopy composition and small-mammal herbivore presence. Ecography 42, 780–790.

Hicks, R.N., Wang, F., Hidayati, S.N., Walck, J.L., 2019. Seed germination of exotic and native winter annuals differentially responds to temperature and moisture, especially with climate change scenarios. Plant Species Biol. 34, 174–183.

Huang, Z., Footitt, S., Tang, A., Finch-Savage, W.E., 2018. Predicted global warming scenarios impact on the mother plant to alter seed dormancy and germination behaviour in *Arabidopsis*. Plant Cell Environ. 41, 187–197.

Hughes, M.G., Frei, A., Robinson, D.A., 1996. Historical analysis of North American snow cover extent: merging satellite and station derived snow cover observations. In: 53rd Eastern Snow Conference, Williamsburg, Virginia, USA, pp. 21–31.

Ibáñez, I., Clark, J.S., LaDeau, S., Lambers, J.H.R., 2007. Exploiting temporal variability to understand tree recruitment response to climate change. Ecol. Monogr. 77, 163–177.

IPCC, 2013. Annex I: atlas of global and regional climate projections. In: van Oldenborgh, G.J., Collins, M., Arblaster, J., Christensen, J.H., Marotzke, J., Power, S.B., et al.,Climate Change 2013: The Physical Science Basis. Contribution of Working Group I to the Fifth Assessment Report of the Intergovernmental Panel on Climate Change. Cambridge University Press, Cambridge [Stocker, T.F., Qin, D., Plattner, G.-K., Tignor, M., Allen, S.K., Boschung, J., et al. (Eds.)].

Kaye, M.W., Wagner, R.J., 2014. Eastern deciduous tree seedlings advance spring phenology in response to experimental warming, but not wetting, treatments. Plant Ecol. 215, 543–554.

Kellman, M., 2004. Sugar maple (*Acer saccharum* Marsh.) establishment in boreal forest: results of a transplantation experiment. J. Biogeogr. 31, 1515–1522.

Kendon, E.J., Roberts, N.M., Fowler, H.J., Roberts, M.J., Chan, S.C., Senior, C.A., 2014. Heavier summer downpours with climate change revealed by weather forecast resolution model. Nat. Clim. Change 4, 570–576.

Kim, D.H., Han, S.H., 2018. Direct effects on seed germination of 17 tree species under elevated temperature and CO_2 conditions. Open Life Sci. 13, 137–148.

Kondo, T., Mikubo, M., Yamada, K., Walck, J.L., Hidayati, S.N., 2011. Seed dormancy in *Trillium camschatcense* (Melanthiaceae) and the possible roles of light and temperature requirements for seed germination in forests. Am. J. Bot. 98, 215–226.

Kondo, T., Walck, J.L., Hidayati, S.N., 2019. Radicle emergence with increased temperatures following summer dispersal in *Trillium camschatcense*: a species with deep simple double morphophysiological dormancy in seeds. Plant Species Biol. 34, 45–52.

Körner, C., Basler, D., 2010. Phenology under global warming. Science 327, 1461–1462.

Kudo, G., Cooper, E.J., 2019. When spring ephemerals fail to meet pollinators: mechanism of phenological mismatch and its impact on plant reproduction. Proc. R. Soc. B 286, 20190573. Available from: https://doi.org/10.1098/rspb.2019.0573.

Kudo, G., Ida, T.Y., 2013. Early onset of spring increases the phenological mismatch between plants and pollinators. Ecology 94, 2311–2320.

Kudo, G., Nishikawa, Y., Kasagi, T., Kosuge, S., 2004. Does seed production of spring ephemerals decrease when spring comes early? Ecol. Res. 19, 255–259.

Leishman, M.R., Masters, G.J., Clarke, I.P., Brown, V.K., 2000. Seed bank dynamics: the role of fungal pathogens and climate change. Funct. Ecol. 14, 293–299.

Leithead, M.D., Anand, M., Silva, L.C.R., 2010. Northward migrating trees establish in treefall gaps at the northern limit of the temperate–boreal ecotone, Ontario, Canada. Oecologia 164, 1095–1106.

Lemke, I.H., Kolb, A., Graae, B.J., De Frenne, P., Acharya, K.P., Blandino, C., et al., 2015. Patterns of phenotypic trait variation in two temperate forest herbs along a broad climatic gradient. Plant Ecol. 216, 1523–1536.

Lemoine, N.P., Doublet, D., Salminen, J.-P., Burkepile, D.E., Parker, J.D., 2017. Responses of plant phenology, growth, defense, and reproduction to interactive effects of warming and insect herbivory. Ecology 98, 1817–1828.

Loarie, S.R., Duffy, P.B., Hamilton, H., Asner, G.P., Field, C.B., Ackerly, D.D., 2009. The velocity of climate change. Nature 462, 1052−1057.

Ma, S., De Frenne, P., Wasof, S., Brunet, J., Cousins, S.A.O., Decocq, G., et al., 2019. Plant−soil feedbacks of forest understorey plants transplanted in nonlocal soils along a latitudinal gradient. Plant Biol. 21, 677−687.

Maes, S.L., De Frenne, P., Brunet, J., De La Peña, E., Chabrerie, O., Cousins, S.A.O., et al., 2014. Effects of enhanced nitrogen inputs and climate warming on a forest understorey plant assessed by transplant experiments along a latitudinal gradient. Plant Ecol. 215, 899−910.

Marty, C., BassiriRad, H., 2014. Seed germination and rising atmospheric CO_2 concentration: a *meta*-analysis of parental and direct effects. New Phytol. 202, 401−414.

Matías, L., Jump, A.S., 2014. Impacts of predicted climate change on recruitment at the geographical limits of Scots pine. J. Exp. Bot. 65, 299−310.

McCabe, G.J., Wolock, D.M., 2010. Long-term variability in Northern Hemisphere snow cover and associations with warmer winters. Clim. Change 99, 141−153.

McCartan, S.A., Jinks, R.L., Barsoum, N., 2015. Using thermal time models to predict the impact of assisted migration on the synchronization of germination and shoot emergence of oak (*Quercus robur* L.). Ann. For. Sci. 72, 479−487.

McWhirter, B.D., Henry, H.A.L., 2015. Successional processes and global change: tree seedling establishment in response to warming and N addition in a temperate old field. Plant Ecol. 216, 17−26.

Merlin, M., Duputié, A., Chuine, I., 2018. Limited validation of forecasted northward range shift in ten European tree species from a common garden experiment. For. Ecol. Manage. 410, 144−156.

Moser, B., Thompson, K., 2014. Self-incompatibility will slow climate driven northward shift of two dominants of calcareous grasslands. Biol. Conserv. 169, 297−302.

Moser, B., Fridley, J.D., Askew, A.P., Grime, J.P., 2011. Simulated migration in a long-term climate change experiment: invasions impeded by dispersal limitation, not biotic resistance. J. Ecol. 99, 1229−1236.

Murray, M.B., Cannell, M.G.R., Smith, R.I., 1989. Date of budburst of fifteen tree species in Britain following climatic warming. J. Appl. Ecol. 26, 693−700.

Nanninga, C., Buyarski, C.R., Pretorius, A.M., Montgomery, R.A., 2017. Increased exposure to chilling advances the time to budburst in North American tree species. Tree Physiol. 37, 1727−1738.

Naoe, S., Tayasu, I., Sakai, Y., Masaki, T., Kobayashi, K., Nakajima, A., et al., 2016. Mountain-climbing bears protect cherry species from global warming through vertical seed dispersal. Curr. Biol. 26, R315−R316.

Naoe, S., Tayasu, I., Sakai, Y., Masaki, T., Kobayashi, K., Nakajima, A., et al., 2019. Downhill seed dispersal by temperate mammals: a potential threat to plant escape from global warming. Sci. Rep. 9, 14932. Available from: https://doi.org/10.1038/s41598-019-51376-6.

Newton, R.J., Hay, F.R., Ellis, R.H., 2020. Temporal patterns of seed germination in early spring-flowering temperate woodland geophytes are modified by warming. Ann. Bot. 125, 1013−1023.

Noss, R.F., 2013. Forgotten Grasslands of the South: Natural History and Conservation. Island Press, Washington, DC.

Park, S., Kim, T., Shim, K., Kong, H.-Y., Yang, B.-G., Suh, S., et al., 2019. The effects of experimental warming on seed germination and growth of two oak species (*Quercus mongolica* and *Q. serrata*). Korean J. Ecol. Environ. 52, 210−220.

Pesendorfer, M.B., Bogdziewicz, M., Szymkowiak, J., Borowski, Z., Kantorowicz, W., Espelta, J.M., et al., 2020. Investigating the relationship between climate, stand age, and temporal trends in masting behavior of European forest trees. Glob. Change Biol. 26, 1654−1667.

Phartyal, S.S., Kondo, T., Baskin, J.M., Baskin, C.C., 2009. Temperature requirements differ for the two stages of seed dormancy break in *Aegopodium podagraria* (Apiaceae), a species with deep complex morphophysiological dormancy. Am. J. Bot. 96, 1086−1095.

Plue, J., De Frenne, P., Acharya, K., Brunet, J., Chabrerie, O., Decocq, G., et al., 2013. Climate-controlled seed bank patterns. Glob. Ecol. Biogeogr. 22, 1106−1117.

Plue, J., Van Calster, H., Auestad, I., Basto, S., Bekker, R.M., Bruun, H.H., et al., 2020. Buffering effects of soil seed banks on plant community composition in response to land use and climate. Glob. Ecol. Biogeogr. 30, 128−139.

Qaderi, M.M., Reid, D.M., 2008. Combined effects of temperature and carbon dioxide on plant growth and subsequent seed germinability of *Silene noctiflora*. Int. J. Plant Sci. 169, 1200−1209.

Rafferty, N.E., Ives, A.R., 2012. Pollinator effectiveness varies with experimental shifts in flowering time. Ecology 93, 803−814.

Robinson, T.M.P., Gross, K.L., 2010. The impact of altered precipitation variability on annual weed species. Am. J. Bot. 97, 1625−1629.

Rühl, A.T., Eckstein, R.L., Otte, A., Donath, T.W., 2015. Future challenge for endangered arable weed species facing global warming: low temperature optima and narrow moisture requirements. Biol. Conserv. 182, 262−269.

Rühl, A.T., Eckstein, R.L., Otte, A., Donath, T.W., 2016. Distinct germination response of endangered and common arable weeds to reduced water potential. Plant Biol. 18, 83−90.

Saatkamp, A., Cochrane, A., Commander, L., Guja, L., Jimenez-Alfaro, B., Larson, J., et al., 2019. A research agenda for seed-trait functional ecology. New Phytol. 221, 1764−1775.

Seibert, R., Grünhage, L., Müller, C., Otte, A., Donath, T.W., Zobel, M., 2019. Raised atmospheric CO_2 levels affect soil seed bank composition of temperate grasslands. J. Veg. Sci. 30, 86−97.

Shibata, M., Masaki, T., Yagihashi, T., Shimada, T., Saitoh, T., 2020. Decadal changes in masting behaviour of oak trees with rising temperature. J. Ecol. 108, 1088−1100.

Simons, A.M., Goulet, J.M., Bellehumeur, K.F., 2010. The effect of snow depth on overwinter survival in *Lobelia inflata*. Oikos 119, 1685−1689.

Sittaro, F., Paquette, A., Messier, C., Nock, C.A., 2017. Tree range expansion in eastern North America fails to keep pace with climate warming at northern range limits. Glob. Change Biol. 23, 3292−3301.

Solarik, K., Gravel, D., Ameztegui, A., Bergeron, Y., Messier, C., 2016. Assessing tree germination resilience to global warming: a manipulative experiment using sugar maple (*Acer saccharum*). Seed Sci. Res. 26, 153–164.

Solarik, K.A., Messier, C., Ouimet, R., Bergeron, Y., Gravel, D., 2018. Local adaptation of trees at the range margins impacts range shifts in the face of climate change. Glob. Ecol. Biogeogr. 27, 1507–1519.

Solarik, K.A., Cazelles, K., Messier, C., Bergeron, Y., Gravel, D., 2019. Priority effects will impede range shifts of temperate tree species into the boreal forest. J. Ecol. 108, 1155–1173.

Soltani, E., Baskin, C.C., Baskin, J.M., 2017. A graphical method for identifying the six types of nondeep physiological dormancy in seeds. Plant Biol. 19, 673–682.

Souther, S., McGraw, J.B., 2011. Vulnerability of wild American ginseng to an extreme early spring temperature fluctuation. Popul. Ecol. 53, 119–129.

Staehlin, B.M., Fant, J.B., 2015. Climate change impacts on seedling establishment for a threatened endemic thistle, *Cirsium pitcheri*. Am. Midl. Nat. 173, 47–60.

Sternberg, M., Brown, V.K., Masters, G.J., Clarke, I.P., 1999. Plant community dynamics in a calcareous grassland under climate change manipulations. Plant Ecol. 143, 29–37.

Sunmonu, N., Kudo, G., 2015. Warm temperature conditions restrict the sexual reproduction and vegetative growth of the spring ephemeral *Gagea lutea* (Liliaceae). Plant Ecol. 216, 1419–1431.

Thompson, L.J., Naeem, S., 1996. The effects of soil warming on plant recruitment. Plant Soil 182, 339–343.

Tomczyk, A.M., Sulikowska, A., Bednorz, E., Półrolniczak, M., 2019. Atmospheric circulation conditions during winter warm spells in Central Europe. Nat. Hazards 96, 1413–1428.

Touzot, L., Schermer, É., Venner, S., Delzon, S., Rousset, C., Baubet, É., et al., 2020. How does increasing mast seeding frequency affect population dynamics of seed consumers? Wild boar as a case study. Ecol. Appl. 30, e02134. Available from: https://doi.org/10.1002/eap.2134.

Tyrrell, M.L., Ross, J., Kelty, M., 2012. Carbon dynamics in the temperate forest. In: Ashton, M.S., Tyrrell, M.L., Spalding, D., Gentry, B. (Eds.), Managing Forest Carbon in a Changing Climate. Springer, New York, pp. 77–107.

Urli, M., Brown, C.D., Narváez Perez, R., Chagnon, P.-L., Vellend, M., 2016. Increased seedling establishment via enemy release at the upper elevational range limit of sugar maple. Ecology 97, 3058–3069.

Varsamis, G., Papageorgiou, A.C., Merou, T., Takos, I., Malesios, C., Manolis, A., et al., 2019. Adaptive diversity of beech seedlings under climate change scenarios. Front. Plant Sci. 9, 1918. Available from: https://doi.org/10.3389/fpls.2018.01918.

Vasseur, L., 2012. Restoration of deciduous forests. Nature Edu. Knowl. 3 (12), 1.

Vellend, M., Myers, J.A., Gardescu, S., Marks, P.L., 2003. Dispersal of Trillium seeds by deer: implications for long-distance migration of forest herbs. Ecology 84, 1067–1072.

Walck, J.L., Baskin, J.M., Baskin, C.C., 1997. A comparative study of the seed germination biology of a narrow endemic and two geographically-widespread species of *Solidago* (Asteraceae). 1. Germination phenology and effect of cold stratification on germination. Seed Sci. Res. 7, 47–58.

Walck, J.L., Baskin, J.M., Baskin, C.C., Hidayati, S.N., 2005. Defining transient and persistent seed banks in species with pronounced seasonal dormancy and germination patterns. Seed Sci. Res. 15, 189–196.

Walck, J.L., Hidayati, S.N., Dixon, K.W., Thompson, K., Poschlod, P., 2011. Climate change and plant regeneration from seed. Glob. Change Biol. 17, 2145–2161.

Walter, H., 1985. Vegetation of the Earth and Ecological Systems of the Geo-biosphere. Springer-Verlag, Berlin, third revised and enlarged edition. Translated from the fifth, revised German edition by O. Muise.

Walter, J., Harter, D.E., Beierkuhnlein, C., Jentsch, A., 2016. Transgenerational effects of extreme weather: perennial plant offspring show modified germination, growth and stoichiometry. J. Ecol. 104, 1032–1040.

Wang, W.J., He, H.S., Thompson, F.R., Spetich, M.A., Fraser, J.S., 2018. Effects of species biological traits and environmental heterogeneity on simulated tree species distribution shifts under climate change. Sci. Total. Environ. 634, 1214–1221.

Wheeler, J.A., Gonzalez, N.M., Stinson, K.A., 2017. Red hot maples: *Acer rubrum* first-year phenology and growth responses to soil warming. Can. J. For. Res. 47, 159–165.

Willson, M.F., Traveset, A., 2000. The ecology of seed dispersal. In: Fenner, M. (Ed.), Seeds: The Ecology of Regeneration in Plant Communities, second ed. CABI Publishing, New York, pp. 85–110.

Wright, A.J., Fisichelli, N.A., Buschena, C., Rice, K., Rich, R., Stefanski, A., et al., 2018. Biodiversity bottleneck: seedling establishment under changing climatic conditions at the boreal–temperate ecotone. Plant Ecol. 219, 691–704.

Zhang, X., Tarpley, D., Sullivan, J.T., 2007. Diverse responses of vegetation phenology to a warming climate. Geophys. Res. Lett. 34, L19405. Available from: https://doi.org/10.1029/2007GL031447.

Zhu, L., Ives, A.R., Zhang, C., Guo, Y., Radeloff, V.C., 2019. Climate change causes functionally colder winters for snow cover-dependent organisms. Nat. Clim. Change 9, 886–893.

Chapter 11

Plant regeneration from seeds: Tibet Plateau in China

Kun Liu[1], Miaojun Ma[1], Carol C. Baskin[2,3] and Jerry M. Baskin[2]
[1]State Key Laboratory of Grassland and Agro-ecosystems, School of Life Sciences, Lanzhou University, Lanzhou, Gansu, P.R. China, [2]Department of Biology, University of Kentucky, Lexington, KY, United States, [3]Department of Plant and Soil Sciences, University of Kentucky, Lexington, KY, United States

Introduction

The Tibetan Plateau (hereafter Plateau) is a unique landform with an average elevation of >4000 m a.s.l.; it is the "roof of the world" (Latif et al., 2019). Many Asian rivers such as Yangtze, Yellow, Brahmaputra, Lantsang (known as the Mekong River downstream), Ganges, and Indus originate on the Plateau, which is known as the "Asian Water Tower" (Xu et al., 2008). Parts of the Plateau occur in Bhutan, China, India, Kyrgyzstan, Nepal, Pakistan, and Tajikistan, but the majority of it is in China. We will cover only the part of the Plateau in China in this chapter.

Vegetation types on the Plateau are very diverse due to the influence of topography, climate, and human activities (Zhao et al., 2011). In China, approximately, 8.6% of the Plateau is covered by forest: broad-leaved forest on the southern edge and coniferous forest, mixed broad-leaved and coniferous forest and shrub on the eastern edge. Seventy percent of the Plateau in China is covered by vegetation dominated by herbaceous plants; about 3% by lakes, wetlands, and farmlands; and about 18%–19% by sparse alpine vegetation and desert (Fig. 11.1). Mean monthly annual temperature ranges from −15°C in the northwest to 10°C in the southeast, and mean monthly annual precipitation ranges from 50 to 150 mm in the northwest to 300–450 mm in the southeast (You et al., 2013; Li et al., 2020). The spatial heterogeneity of precipitation on the Plateau results in a diversity of grassland types. From southeast to northwest, the type of grassland changes from humid alpine meadow to semiarid alpine steppe to arid alpine desert steppe (Li et al., 2020).

The wide range in elevation, temperature and precipitation, diversity of vegetation, and topographic complexity contribute to the Plateau being a plant species biodiversity hotspot (Myers et al., 2000; Huang et al., 2016). About 3764 of the 12,058 species of seed plants (mostly angiosperms) on the Plateau are endemic, and they belong to 519 genera and 113 families (Zhang et al., 2016a,b; Yu et al., 2018). Most (76.3%) of the endemics are herbaceous, and 67.5% are temperate species, with most of them occurring at intermediate elevations (2800–4300 m) (Yu et al., 2018).

Low temperatures, a long frost period, and a short growing season are the main climate features on the Plateau. They result in individual plants and plant communities on the Plateau being vulnerable to global climate change (Klein et al., 2004; Elmendorf et al., 2012). In such a climate, an accurate determination of the suitable time for dormancy breaking and seedling establishment and growth is critical for plant survival. Thus the temperature requirements for seed production, seed dormancy breaking, and seed germination are well adapted to current temperature regimes, but they may be less well adapted to the warming conditions due to future climate change. As such, plant regeneration from seeds is likely to be significantly affected by global warming, which may cause some changes in distribution of species on the Plateau and ultimately affect community composition, vegetation distribution, and ecosystem function. Therefore it is necessary to evaluate the effect of climate change on each aspect of plant regeneration from seeds, that is, seed production, soil seed bank, seed dormancy, seed germination, and seedling establishment, and try to predict future changes in distribution of species and community composition.

In this chapter, we address the following questions about the effects of climatic change on the Plateau. (1) What is known about climate warming? (2) What are the effects of increased temperatures on seed production, dormancy, and germination? (3) What is known about the effect of changes in precipitation on regeneration of species from seeds? (4) What are the effects of climate change on community structure and composition?

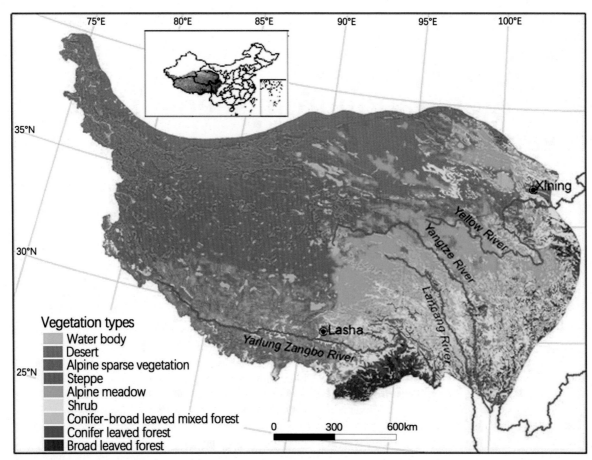

FIGURE 11.1 Vegetation types on the Tibetan Plateau in China. *From Zhao, D., Wu, S., Yin., Y., Yin, Z., 2011. Vegetation distribution on Tibetan Plateau under climate change scenario. Regul. Environ. Change 11, 905–915, with permission.*

Climate change

Due to the tendency for climate warming to increase with elevation (Giorgi et al., 1997; Kuang and Jiao, 2016), temperature increase on the Plateau is two times as high as that of the average global temperature (IPCC, 2007). From 1955 to 1996, increases in mean annual and mean winter temperatures on the Plateau were about 0.16°C decade^{-1} and 0.32°C decade^{-1}, respectively (Liu and Chen, 2000). Both of these increases in temperature exceed those for the Northern Hemisphere as a whole and for the same latitudinal zone in the same period (Beniston et al., 1996; Beniston, 2003; Cui and Hans, 2009). Climate warming on the Plateau is caused mainly by an increase in the minimum temperature (Fig. 11.2) and especially by the strong night and winter warming (Harvey, 1995; IPCC, 2007; You et al., 2016). Thus increase in temperature is higher in winter than in summer and higher at night than during the day (Liu and Chen, 2000; Duan et al., 2006; You et al., 2016). Data from 1961 to 2005 show that the diurnal temperature range exhibits a statistically decreasing trend at a rate of 0.20°C decade^{-1} (You et al., 2008).

Climate warming can cause significant changes in other aspects of the environment on the Plateau. For example, global warming decreased the number of soil freeze–thaw events and increased the number of growing-degree days, frost-free days, and amount of time leaf surface temperatures were in the optimal range for photosynthesis (Pu and Xu, 2009; Chen et al., 2011; Latif et al., 2019). In the context of global warming, decreases in depth and duration of snow cover on the Plateau have occurred over the last few decades (Qin et al., 2006; Xu, et al., 2017).

Warming is one of several factors that can cause considerable change in precipitation on the Plateau (IPCC, 2007). Unlike temperature, there is no uniform pattern of change in precipitation on the Plateau (Kuang and Jiao, 2016), with an increase in the amount in some places and a decrease in others. Also, it is predicted that the frequency of heavy precipitation and drought events will increase on the Plateau (IPCC, 2007). Zhao et al. (2015) found that precipitation is increasing in most regions of the Plateau, except in the southern part, where it is decreasing. However, although there has been a slight increase in precipitation in some parts of the Plateau there may be a decrease of soil moisture

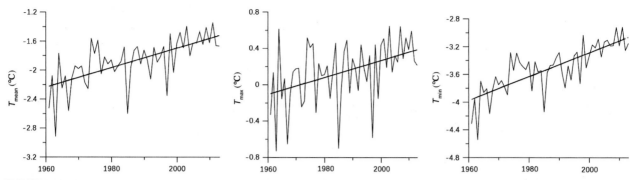

FIGURE 11.2 Changes in annual mean (T_{mean}), maximum (T_{max}), and minimum (T_{min}) temperatures above 4500 m a.s.l. on the Tibet Plateau during 1961–2012. *From You, Q.L., Min, J.Z., Kang, S.C., 2016. Rapid warming in the Tibetan Plateau from observations and CMIP5 models in recent decades. Int. J. Climatol. 36, 2660–2670, with permission.*

especially during winter because of warming. That is, warming increases evapotranspiration, causing more water vapor to escape into the atmosphere, thus counteracting and even exceeding the increased precipitation (Xie et al., 2010).

Due to human activities, the concentration of atmospheric nitrogen (N) and N deposition is increasing (Gruber and Galloway, 2008; Ackerman et al., 2019). Since many ecosystems, such as those in the alpine zone, are N-limited (Kou et al., 2020), N deposition has the potential to impact plant growth (Zong et al., 2016). However, global warming can cause N loss from ecosystems, which may or may not be balanced by N deposition (Bai et al., 2013; Zhang et al., 2015; Lin et al., 2016) In addition, Lin et al. (2016) found that increased precipitation may lead to loss of dissolved N pools in the soil on the Plateau.

Seed production

The effects of warming on seed production can be difficult to distinguish from those of elevated CO_2 because both CO_2 and temperature are critical determinants of photosynthetic rates (Sage and Kubien, 2007). According to the IPCC (2013), atmospheric CO_2 concentration increased by about 40% between 1750 and 2011. A simulation study on the Plateau showed that elevated CO_2 increased the annual mean gross primary productivity by 23 g C m^{-2} year^{-1} during 1981–2010 (Luo et al., 2020). Furthermore, elevated CO_2 can enhance plant water use efficiency, which also can result in increased plant growth and seed production (Jackson et al., 1994; Morgan et al., 2004).

For 12 herbaceous species from an alpine meadow on the eastern Plateau, warming significantly reduced the number of flowers and total number of seeds per plant for nine multiinflorescence species but not for three single-inflorescence species (Liu et al., 2012). Furthermore, reduction in seed production was largely attributable to decline in number of flowers per plant (Liu et al., 2012). However, number of seeds produced by the annual plant *Koenigia islandica* from a desertified alpine meadow on the central Plateau increased significantly with warming (Cui et al., 2017). Thus the effect of warming on seed production may differ from region to region, and more cross-region studies are needed to test it.

Warming can significantly affect seed mass, and this effect is species-specific. For example, Zhang et al. (2019a,b) tested the effect of long-term warming on seed mass of species from a lightly grazed and a heavily grazed alpine meadow on the eastern Plateau. In the lightly grazed meadow, warming significantly increased seed mass of four of 19 species and significantly decreased seed mass of six species; seed mass of the remaining nine species was not significantly affected. In the heavily grazed meadow, warming significantly increased seed mass of six of 20 species and significantly decreased seed mass of six species; seed mass of the remaining eight species was not significantly affected.

An increase in precipitation promotes plant growth and seed production, while a decrease in precipitation decreases seed production. An increase in precipitation promoted an increase in number of seeds and seed mass of *Kobresia humilis*, which is the dominant plant in alpine meadows on the eastern Plateau, but both number of seeds and seed mass were decreased by a decrease in rainfall (Peng et al., 2020b). In the dry northwestern part of the Plateau, seed production may be more sensitive to changes in precipitation than in the wet southeastern part. However, the influence of changes in precipitation on seed production of plants on the northwestern part of the Plateau has not been investigated.

Although theoretically increased precipitation on the Plateau is beneficial to plant growth, increased evaporation of soil moisture caused by increased temperatures and increased plant water demand could decrease or even offset the favorable effect of increased precipitation on plants (Shen et al., 2011; Ganjurjav et al., 2020). Thus there may be an interaction effect of increased temperature and precipitation on seed production, but this has not been well studied.

Additionally, although there are a few studies demonstrating that increased precipitation promotes seed production of plants on the eastern part of the Plateau (Peng et al., 2020b), we have not found any such studies on seed production by plants from the northwestern part of the Plateau.

As the climate on the Plateau warms, the frequency of warm events and wet events increases (Liu et al., 2006; Qin et al., 2011). When climatic extremes occur, variation in environmental factors such as temperature and soil moisture can exceed the tolerance range of plants for seed production (Chiluwal et al., 2019; Wang et al., 2020b). Therefore climate extremes can negatively affect plants at specific developmental stages, such as temperature thresholds during flowering. Furthermore, climatic extremes may increase plant disease and promote pest outbreaks (Alig et al., 2002; Gan, 2004). As a result, the effect of extreme climatic events on seed production is greater than that of the increase in mean temperature (Antle et al., 2004; Porter and Semenov, 2005; Tubiello et al., 2007). Thus the increase in frequency of climatic extremes on the Plateau might result in a decrease in seed production. However, there are very few studies on the effect of climate extremes on plant sexual reproduction on the Plateau (Hou et al., 2018).

Soil seed bank

The general categories of alpine grasslands on the Plateau are alpine steppes, marshy meadows, saline meadows, and alpine meadows (Feng and Squires, 2020). Of these categories, soil seed banks of alpine meadows (Ma et al., 2010, 2018, 2019, 2020; An et al., 2020) and alpine wetlands/marshes (marshy meadows) (Ma et al., 2011, 2014, 2017, 2018; He et al., 2021) have been studied most extensively. Also, seed banks of alpine steppe, alpine desert (Zhao et al., unpublished data), and subalpine meadows (Ma et al., 2009, 2013, 2019) have been studied. The alpine grasslands investigated have high species richness and seed density (seeds m^{-2} of ground surface). Furthermore, as expected, alpine meadows have higher species richness and seed density than alpine deserts: species richness 18.1 ± 1.1 and 7.0 ± 0.6, respectively, and seed density 3459.1 ± 486.1 and 647.6 ± 138.2 m^{-2}, respectively (Zhao et al., unpublished data).

An et al. (2020) used an elevational gradient (3158–4002 m a.s.l.) on the Plateau as a surrogate for climate change (space-for-time, sensu Pickett, 1989) to explore the effect of climate change on species composition and seed density of seed banks and standing vegetation in alpine meadows. With increased elevation, temperature decreased and precipitation increased, with little effect on aboveground species richness and abundance. With increased elevation, however, species richness and seed density in the soil seed bank, as well as similarity between species in the aboveground vegetation and seed bank, decreased. The decrease in similarity between species in aboveground vegetation and the seed bank was correlated with increase in mean annual precipitation but not with decrease in mean annual temperature.

On the northeastern part of the Plateau, Ma et al. (2020) also used an elevational gradient (i.e., space-for-time) to investigate the effects of climate change on seed banks on a regional scale. Across an elevation gradient of 2039–4002 m a.s.l., the authors compared seed banks and standing vegetation in six types of meadows: alpine, swamp, alpine scrub, subalpine, forest edge, and upland. With increase in mean annual temperature and mean annual precipitation, species richness, but not seed density, of the seed bank decreased. Abundance of aboveground vegetation increased with precipitation. However, species richness of the aboveground vegetation decreased with increased temperature, resulting in decreased species richness and seed density of the seed bank.

It is likely that increased N deposition would affect plant growth of some species more than that of others, thereby having an effect of the composition and density of the soil seed bank. In a 9-year fertilization experiment in an alpine meadow on the Plateau, addition of $(NH4)_2HPO_4$ decreased species richness of both the soil seed bank and aboveground vegetation. However, the effect of fertilization was greater on the aboveground vegetation than on the seed bank (Ma et al., 2014; Zhang et al., 2019a,b). The effects on the seed bank and aboveground vegetation of adding only N to alpine meadows on the Plateau need to be determined.

Seed dormancy and germination

Seed dormancy that is broken by cold-wet conditions of winter prevents germination of many species on the Plateau in autumn after seed dispersal, and thus seedling death due to exposure to low temperature in winter is avoided (Liu et al., 2011, 2018a). On the other hand, if cold-wet conditions break seed dormancy germination will occur in spring. For example, the nondeep physiological dormancy in seeds of *Primula alpicola* and *Pedicularis fletcheri* is broken by cold-wet conditions during winter, and seeds germinate soon after snowmelt in spring (Wang et al., 2017). However, if seeds of these species fail to germinate in spring they reenter dormancy, come out of dormancy in winter and have the potential to germinate the following spring, that is, they undergo dormancy cycling.

Seeds of most of the 489 species from an alpine meadow on the eastern Plateau studied by Liu et al. (2011) required the wet-cold conditions in winter to germinate. Compared with seeds stored at dry-warm conditions, mean community germination of seeds stored in wet-cold conditions increased by 17.93%. Storing seeds in dry-warm versus dry-cold conditions decreased mean community germination by 4.61%. Thus wet conditions in the winter are more favorable for dormancy break than dry conditions. The increased frequency of extreme weather events due to global warming will increase the probability of a warmer and/or drier winter, which might inhibit complete breaking of seed dormancy of some species.

The effect of wet-cold conditions on dormancy break of seeds from the eastern Plateau differs among species. Seeds of about 35% of the species from alpine meadows on the eastern Plateau require a wet winter to break dormancy and thus to germinate the following spring. However, wet-cold conditions do not promote dormancy break of other species. Germination of seeds in which dormancy is broken only by wet-cold conditions may be negatively affected by dry-warm conditions in winter. Thus the dry-warm trends in some regions of the Plateau could have an indirect effect on the community structure due to lack of seed germination of some species and promotion of germination of other species. However, this has not been tested in the field.

A simulation study indicated that during 1981–2010 there was a linear delay of 4.0 days decade^{-1} of the date when the ground on the Plateau was first frozen in winter and that the date of thawing was advanced linearly by 4.6 days decade^{-1}. As a result, the freeze duration was shortened by 8.6 days decade^{-1} for seasonally frozen ground (Guo and Wang, 2013). Hence, insufficient cold stratification due to shortening of winter potentially could result in a delay or failure of dormancy break.

Promotion of seed germination by fluctuating temperatures is a mechanism by which seeds detect vegetation gaps and depth of burial in soil (Fenner and Thompson, 2005). This mechanism ensures that germination occurs only in gaps and at/near the soil surface (Liu et al., 2013, 2020). Seed germination of about 36% of 445 species from the eastern Plateau was increased by temperature fluctuation (Liu et al., 2013). Thus a lowering of the magnitude of diurnal temperature fluctuations on the Plateau via climate change might suppress seed germination of some species (You et al., 2008), which could have important consequences on plant community composition.

Data from 2004–13 indicate that precipitation in the northwest and central parts of the Plateau is increasing with global warming (Zhao et al., 2015), which likely promotes seed germination by increasing soil moisture. Increased precipitation could be important in decreasing salt stress in areas with high soil salinity such as in the desert steppes on the northwestern part of the Plateau. It is well documented that germination in saline habitats is correlated with precipitation that decreases soil salinity (Baskin and Baskin, 2014). However, the effects of leaching salts by rainwater on germination of species growing in saline soil on the Plateau have not been investigated. In the southern part of the Plateau, precipitation is predicted to decrease with global warming (Zhao et al., 2015), and this along with increased evaporation of water from the soil caused by increased temperature may inhibit seed germination in various habitats.

Seedling emergence and establishment

The main seed germination season on the Plateau differs among plant species. For 144 species from the eastern Plateau, the proportion of species with seedlings emerging in autumn (33%), spring (44%), and summer (23%) differed significantly (Cao et al., 2018). Due to global warming, the beginning of the growing season on the Plateau has advanced and the end delayed (Dong et al., 2010; Wang et al., 2020a; Sun et al., 2020). However, fruiting time of plants on the Plateau is not changing as much as time of other phenological events (Jiang et al., 2016; Hu et al., 2020). Changes in length of the growing season may increase the proportion of species whose seeds germinate in autumn and the proportion with bi-seasonal (autumn and spring) seedling emergence. However, this needs to be documented. For species that normally germinate in spring, a shift in emergence to autumn may have a significant effect on seedling establishment, but further research is needed to determine what the effect is, if any.

Results of observations and of modeling indicate that during 1961–2005 the rate of increasing night temperature was two times that of the day temperature (Duan et al., 2006; You et al., 2016). It seems likely that the warming of nights may decrease seedling mortality due to freezing. Moreover, during 1961–2005, the number of extreme cold days and nights decreased by 0.85 and 2.38 days decade^{-1}, respectively, and the number of frost days has decreased by 4.32 days decade^{-1} (You et al., 2008). These decreases in number of extreme cold days, extreme cold nights, and frost days may decrease seedling death due to freezing, but this needs to be investigated.

On the southern part of the Plateau, precipitation is decreasing with global warming (Zhao et al., 2015), suggesting that seedling mortality due to desiccation could become an increasingly important factor in plant regeneration from seeds. However, increasing precipitation in other parts of the Plateau potentially will promote seedling establishment.

In addition, an increase in extreme weather events is one of the consequences of climate warming (IPCC, 2007; Jiang et al., 2012). During the last 50 years, the frequency of wet events increased and that of drought events decreased noticeably (Qin et al., 2011), implying that the mortality of seedlings caused by drought events will decrease in the future if regional warming continues.

An increase in snowfall is a characteristic of a warming climate in cold regions (Karl et al., 1993; Leathers et al., 1993; Houghton et al., 1996). However, an increase in snowfall did not result in an increase in snow cover on the Plateau. Annual cumulative daily snow depth on the Plateau increased from the 1960s to 1980s (Qin et al., 2006; Xu et al., 2017), but it has decreased since the 1980s (Xu et al., 2017). The duration of snow cover exhibited a significant decreasing trend (3.5 days decade^{-1}) due to a delay in the date when the first snow falls and an earlier date when the snow melted, which is consistent with a response to climate change (Xu et al., 2017). Another change in snow cover on the Plateau is that its interannual variation in depth is increasing due to increased interannual variation of snowfall and temperature (Qin et al., 2006).

The insulating effect of snow cover on the ground surface is of great benefit to seedling survival during winter (Walck et al., 2005). However, the decreasing trends in snow depth and duration of snow cover on the Plateau since the 1980s have reduced the insulating effect of snow cover. A decrease in insulation by snow may result in the soil being colder during winter. Seeds of species on the Plateau that are not dormant at maturity usually germinate immediately after dispersal in autumn, and seedling survival in winter depends on snow cover (Wang et al., 2018; Guiden et al., 2019). Thus decreasing snow depth in winter is expected to adversely affect the seedling survival of species whose seeds germinate in autumn. Furthermore, shortening the duration of snow cover and its increasing interannual variation in depth likely will increase the risk of damage by spring frost, which also is expected to have adverse effects on seedling survival (Liu et al., 2018b; Richardson et al., 2018).

Data from satellite and from field observations indicate a significant advancement in alpine spring phenology over decades of climate warming on the Plateau (Bibi et al., 2018; Piao et al., 2019; Wang et al., 2020a). A 35-year (1980–2014) observation period on an alpine grassland on the Plateau indicates that the start of the fast-growing phase of plants in spring has advanced 5 days decade^{-1} (Wang et al., 2020a). This advancement of the growing season of plants likely will increase competition, which might result in increased seedling mortality.

Change in community structure and composition

In a warming experiment conducted in an alpine meadow on the eastern Plateau using open-top chambers (OTCs), total species richness had decreased by 16%–30% after 2 years of warming (Klein et al., 2004) and by 26%–40.2% after 4 years of warming (Klein et al., 2004; Zhang et al., 2017). In a 5-year warming study (started in spring of 2006) using OTCs in an alpine fen on the Plateau, species richness and diversity of the community, especially graminoids, had not changed in 2006 but had decreased by 2010 (Yang et al., 2015). In a 4-year warming study using OTCs in an alpine meadow and alpine swamp (marshy meadow) on the Plateau, warming increased vegetation height in both communities. Warming marginally increased cover of legumes in the meadow but not in the swamp (Peng et al., 2020a). The authors attributed the increase in legumes in the meadow to decreased soil moisture associated with increased soil temperatures.

In a long-term warming experiment in an alpine meadow on the eastern Plateau, plant species diversity recovered following an initial decrease (Zhang et al., 2017). However, species composition of the experimentally warmed communities was significantly altered. That is, some species that disappeared in the early part of the experiment did not reappear and were replaced by species that were not present in the original community. These results suggest that some plant species are not adapted to cope with rapid climate warming and that immigration from elsewhere is important for reassembly of plant communities.

The importance of migration for species survival on the Plateau under rapid climate change was shown by modeling research. Yan and Tang (2019) tested the effect of climate change on range shifts of 993 endemic species on the Plateau for 2050–70, using an ensemble modeling method for changes in species distribution. Under a full-dispersal scenario, the distribution of most species would shift to the west, 72%–81% would expand their distribution, 6%–20% would experience >30% range loss, and net species richness would increase across the region. However, under a no-dispersal scenario 15%–59% of the species would lose >30% of their current habitat by 2070, and severe species loss may occur in the southeastern and the eastern peripheral parts of the Plateau. Thus under the rapid climate change scenario migration is very important for maintenance of biodiversity on the Plateau.

Changes in plant community composition and structure can occur in response to deposition of atmospheric N (Bowman et al., 2006; Zong et al., 2016). Since alpine ecosystems such as those on the Plateau are N-limited, it is predicted that increased N deposition will have an effect on them (Zong et al., 2016). In an alpine meadow on the Plateau dominated by the sedge *Kobresia pygmaea*, Zong et al. (2016) found that the level of experimental N addition at which the vegetation begins to change in species composition is 8.8–12.7 kg ha^{-1} year^{-1}. Minimum, maximum, and mean total levels of N deposition in Qinghai (on the Plateau) for 1990 to 2003 were 1.08, 17.81, and 7.55 kg ha^{-1} year^{-1}, respectively (Lü and Tian, 2007). Thus there is reason to be concerned about increase in N deposition and community stability.

In a 4-year experiment in an alpine meadow on the mid-south part of the Plateau, 0, 10, 20, 40, and 80 kg N ha^{-1} year^{-1} (i.e., 0, 1.5, 3, 6, and 12 times more N than ambient atmospheric deposition, respectively) were added during the growing season (Zong et al., 2016). By the fourth year of the experiment, percentage of plant cover had increased significantly for all levels of N addition, with percent coverage of the grasses increasing significantly but not that of sedges or forbs. In terms of biomass production in response to N addition, grasses were also the most sensitive. During the 4-year study, species richness and diversity index did not change, but due to differences in species cover and biomass production the authors predicted that there would be long-term changes in community composition.

In a 4-year experiment that combined warming and N addition in an alpine meadow on the Plateau, Zong et al. (2018) monitored N in the soil and evaluated the N resorption efficiency (NRE) in leaves of the dominant plant species in the community. Warming increased the release of N from the soil and increased community NRE. Furthermore, NRE increased with moderate but not high levels of N addition. Although N addition did not change the overall NRE of the community, it promoted phosphorus uptake by the grass *Stipa capillacea*, suggesting that the relative importance of this species in the community would increase.

Changes in community structure and composition due to climate change may be related to the effects of climate change on plant regeneration from seeds. First, the effect of warming on seed production is species-specific (Liu et al., 2012). Thus some species will increase in the community due to enhanced seed production and seedling establishment, and others may decrease due to suppression of seed production by warming, resulting in their regeneration in the community being seed limited. Second, the effect of warming on dormancy break also differs among species (Liu et al., 2011). Dormancy breaking of seeds of some species in winter may be inhibited by increased temperatures and drying. In which case, seeds of some species cannot germinate in spring, and thus the species ultimately will disappear from the community. Third, the dependence of population regeneration on seeds differs among species. The abundance of species that rely almost exclusively on clonal/vegetative reproduction, such as sedges and grasses (Ma et al., 2013, 2017), is unlikely to be affected by the positive or negative influence of climate on plant regeneration from seeds (An et al., 2020; Ma et al., 2020). However, for species whose regeneration depends mainly on seeds, such as annuals, their abundance in the community is sensitive to the influence of climate on regeneration from seeds.

Research needs

The rapidly changing climate on the Tibet Plateau in China likely will have a significant impact on all aspects of plant regeneration from seeds. As we have indicated throughout the chapter, there are many knowledge gaps about how climate change on the Plateau will affect plant regeneration from seeds. Thus in many cases we can only speculate about the specifics of the effects of climate change on plant species and communities. Clearly, much research remains to be done on the effect of changes in temperature, precipitation, and snow cover, as well as the effect of climatic extremes and of atmospheric N deposition on seed production, dormancy-break, germination, seedling emergence, and establishment. The results of such research will help us predict the fate of individual species, changes in community composition and structure, and species migration.

References

Ackerman, D., Millet, D.B., Chen, X., 2019. Global estimates of inorganic nitrogen deposition across four decades. Glob. Biogeochem. Cycles 33, 100–107.

Alig, R.J., Adams, D.M., McCarl, B.A., 2002. Projecting impacts of global climate change on the US forest and agriculture sectors and carbon budgets. For. Ecol. Manage. 169, 3–14.

An, H., Zhao, Y., Ma, M., 2020. Precipitation controls seed bank size and its role in alpine meadow community regeneration with increasing altitude. Glob. Change Biol. 26, 5767–5777.

Antle, J.M., Capalbo, S.M., Elliott, E.T., Paustian, K.H., 2004. Adaptation, spatial heterogeneity, and the vulnerability of agricultural systems to climate change and CO_2 fertilization: an integrated assessment approach. Clim. Change 64, 289–315.

Bai, E., Li, S., Xu, W., Li, W., Dai, W., Jiang, P., 2013. A *meta*-analysis of experimental warming effects on terrestrial nitrogen pools and dynamics. New Phytol. 119, 441–451.

Baskin, C.C., Baskin, J.M., 2014. Seeds: Ecology, Biogeography, and Evolution of Dormancy and Germination, 2nd ed. Academic Press/Elsevier, San Diego.

Beniston, M., 2003. Climatic change in mountain regions: a review of possible impacts. Clim. Change 59, 5–31.

Beniston, M., Fox, D.G., Adhikary, S., Andressen, R., Guisan, A., Holten, J., et al., 1996. The impacts of climate change on mountain regions. In: Watson, R.T., Zinyowera, M.C., Moss, R.H. (Eds.), Second Assessment Report of Intergovernmental Panel on Climate Change (IPCC). Cambridge University Press, Cambridge, pp. 191–213.

Bibi, S., Wang, L., Li, X., Zhou, J., Chen, D., Yao, T., 2018. Climatic and associated cryospheric, biospheric, and hydrological changes on the Tibetan Plateau: a review. Int. J. Climatol. 38, e1–e17.

Bowman, W.D., Gartner, J.R., Holland, K., Widermann, M., 2006. Nitrogen critical loads for alpine vegetation and terrestrial ecosystem responses: are we there yet? Ecol. Appl. 16, 1183–1193.

Cao, S., Liu, K., Du, G., Baskin, J.M., Baskin, C.C., Bu, H., et al., 2018. Seedling emergence of 144 subalpine meadow plants: effects of phylogeny, life cycle type and seed mass. Seed Sci. Res. 28, 93–99.

Chen, H., Zhu, Q., Wu, N., Wang, Y., Peng, C.H., 2011. Delayed spring phenology on the Tibetan Plateau may also be attributable to other factors than winter and spring warming. Proc. Nat. Acad. Sci. USA 108, E93. Available from: https://doi.org/10.1073/pnas.1100091108.

Chiluwal, A., Bheemanahalli, R., Kanaganahalli, V., Boyle, D., Perumal, R., Pokharel, M., et al., 2019. Deterioration of ovary plays a key role in heat stress-induced spikelet sterility in sorghum. Plant Cell Environ. 43, 448–462.

Cui, X., Hans, F., 2009. Recent land cover changes on the Tibetan Plateau: a review. Clim. Change 94, 47–61.

Cui, S., Meng, F., Suonan, J., Wang, Q., Li, B., Liu, P., et al., 2017. Responses of phenology and seed production of annual *Koenigia islandica* to warming in a desertified alpine meadow. Agric. For. Meteorol. 247, 376–384.

Dong, W., Jiang, Y., Yang, S., 2010. Response of the starting dates and the lengths of seasons in Mainland China to global warming. Clim. Change 99, 81–91.

Duan, A.M., Wu, G.X., Zhang, Q., Liu, Y.M., 2006. New proofs of the recent climate warming over the Tibetan Plateau as a result of the increasing greenhouse gases emissions. Chin. Sci. Bull. 51, 1396–1400.

Elmendorf, S.C., Henry, G.H.R., Hollister, R.D., Björk, R.G., Bjorkman, A.D., Callaghan, T.V., et al., 2012. Global assessment of experimental climate warming on tundra vegetation: heterogeneity over space and time. Ecol. Lett. 15, 164–175.

Feng, H., Squires, V.R., 2020. Socio-environmental dynamics of alpine grasslands, steppes and meadows of the Qinghai-Tibetan Plateau, China: a commentary. Appl. Sci. 10, 6488. Available from: https://doi.org/10.3390/app10186488.

Fenner, M., Thompson, K., 2005. The Ecology of Seeds. Cambridge University Press, Cambridge.

Gan, J., 2004. Risk and damage of southern pine beetle outbreaks under global climate change. For. Ecol. Manage. 191, 61–71.

Ganjurjav, H., Gornish, E., Hu, G., Schwartz, M., Wan, Y., Li, Y., et al., 2020. Warming and precipitation addition interact to affect plant spring phenology in alpine meadows on the central Qinghai-Tibetan Plateau. Agric. For. Meteorol. 287, 107943. Available from: https://doi.org/10.1016/j.agrformet.2020.107943.

Giorgi, F., Hurrell, J., Marinucci, M.R., Beniston, M., 1997. Elevation dependency of the surface climate change signal: a model study. J. Clim. 10, 288–296.

Gruber, N., Galloway, J.N., 2008. An earth-system perspective of the global nitrogen cycle. Nature 451, 293–296.

Guiden, P.W., Connolly, B.M., Orrock, J.L., 2019. Seedling responses to decreased snow depend on canopy composition and small-mammal herbivore presence. Ecography 42, 780–790.

Guo, D., Wang, H., 2013. Simulation of permafrost and seasonally frozen ground conditions on the Tibetan Plateau, 1981–2010. J. Geophys. Res. Atmos. 118, 5216–5230.

Harvey, D.L., 1995. Warm days, hot nights. Nature 377, 15–16.

He, M., Xin, C., Baskin, C.C., Li, J., Zhao, Y.P., An, H., et al., 2021. Different response of transient and persistent seed bank of alpine wetland to grazing disturbance on the Tibetan Plateau. Plant Soil 459, 93–107.

Hou, W., Wang, J., Hu, D., Feng, X., 2018. Effects of drought in post-flowering on leaf water potential, photosynthetic physiology, seed phenotype and yield of hulless barley in Tibet Plateau. Sci. Agric. Sin. 51, 2675–2688.

Houghton, J.T., Meira Filho, L.G., Callender, B.A., Harris, N., Kattenberg, A., Maskell, K. (Eds.), 1996. Climate Change 1995: The Science of Climate Change. Cambridge University Press, Cambridge.

Hu, X., Zhou, W., Sun, S., 2020. Responses of plant reproductive phenology to winter-biased warming in an alpine meadow. Front. Plant Sci. 11, 534703. Available from: https://doi.org/10.3389/fpls.2020.534703.

Huang, J.H., Huang, J.H., Liu, C.R., Zhang, J.L., Lu, X.H., Ma, K.P., 2016. Diversity hotspots and conservation gaps for the Chinese endemic seed flora. Biol. Conserv. 198, 104–112.

IPCC, 2007. Core Writing Team (Intergovernmental Panel on Climate Change) In: Pachauri, R.K., Reisinger, A. (Eds.), Climate Change 2007: Synthesis report. Contribution of Working Groups I, II and III to the Fourth Assessment Report of the Intergovernmental Panel on Climate Change. Cambridge University Press, Cambridge.

IPCC, 2013. (Intergovernmental Panel on Climate Change) Climate Change 2013. The Physical Science Basis. Working Group I. Contribution to the Fifth Assessment Report of the Intergovernmental Panel on Climate Change. Cambridge University Press, Cambridge.

Jackson, R.B., Sala, O.E., Field, C.B., Mooney, H.A., 1994. CO_2 alters water use, carbon gain, and yield for the dominant species in a natural grassland. Oecologia 98, 257–262.

Jiang, Z., Song, J., Li, L., Chen, W., Wang, Z., 2012. Extreme climate events in China: IPCC-AR4 model evaluation and projection. Clim. Change 110, 385–401.

Jiang, L., Wang, S., Meng, F., Duan, T., Niu, H., Xu, G., et al., 2016. Relatively stable response of fruiting stage to warming and cooling relative to other phenological events. Ecology 97, 1961–1969.

Karl, T.R., Groisman, P.Y., Knight, R.W., Heim, R.R., 1993. Recent variations of snow cover and snowfall in North America and their relation to precipitation and temperature variations. J. Clim. 6, 1327–1344.

Klein, J.A., Harte, J., Zhao, X., 2004. Experimental warming causes large and rapid species loss, dampened by simulated grazing, on the Tibetan Plateau. Ecol. Lett. 7, 1170–1179.

Kou, D., Yang, G., Li, F., Feng, X., Zhang, D., Mao, C., et al., 2020. Progressive nitrogen limitation across the Tibetan alpine permafrost region. Nat. Commun. 11, 3331. Available from: https://doi.org/10.1038/s41467-020-17169-6.

Kuang, X., Jiao, J.J., 2016. Review of climate change on the Tibetan Plateau during the last half century. J. Geophys. Res. Atmos. 121, 3979–4007.

Latif, A., Ilyas, S., Zhang, Y., Xin, Y., Zhou, L., Zhou, Q., 2019. Review on global change status and its impacts on the Tibetan Plateau environment. J. Plant Ecol. 6, 917–930.

Leathers, D.J., Mote, T.L., Kaivinen, K.C., McFeeters, S., 1993. Temporal characteristics of USA snowfall 1945/1946 through to 1984/1985. Int. J. Climatol. 13, 65–76.

Li, M., Zhang, X., He, Y., Niu, B., Wu, J., 2020. Assessment of the vulnerability of alpine grasslands on the Qinghai Tibetan Plateau. PeerJ 8, e8513. Available from: https://doi.org/10.7717/peerj.8513.

Lin, L., Zhu, B., Chen, C., Zhang, Z., Wang, Q.-B., He, J.-S., 2016. Precipitation overrides warming in mediating soil nitrogen pools in an alpine grassland ecosystem on the Tibetan Plateau. Sci. Rep. 6, 31438. Available from: https://doi.org/10.1038/srep31438.

Liu, X., Chen, B., 2000. Climatic warming in the Tibetan Plateau during recent decades. Int. J. Climatol. 20, 1729–1742.

Liu, X., Yin, Z., Shao, X., Qin, N., 2006. Temporal trends and variability of daily maximum and minimum, extreme temperature events, and growing season length over the eastern and central Tibetan Plateau during 1961–2003. J. Geophys. Res. Atmos. 111, D19109. Available from: https://doi.org/10.1029/2005JD006915.

Liu, Y., Mu, J., Niklas, K.J., Li, G., Sun, S., 2012. Global warming reduces plant reproductive output for temperate multi-inflorescence species on the Tibetan Plateau. New Phytol. 195, 427–436.

Liu, K., Baskin, J.M., Baskin, C.C., Bu, H., Liu, M., Liu, W., et al., 2011. Effect of storage conditions on germination of seeds of 489 species from high elevation grasslands of the eastern Tibet Plateau and some implications for climate change. Am. J. Bot. 98, 12–19.

Liu, K., Baskin, J.M., Baskin, C.C., Bu, H., Du, G., Ma, M., 2013. Effect of diurnal fluctuating vs constant temperatures on germination of 445 species from the eastern Tibet Plateau. PLoS One 8, e69364. Available from: https://doi.org/10.1371/journal.pone.0069364.

Liu, K., Cao, S., Du, G., Baskin, J.M., Baskin, C.C., Bu, H., et al., 2018a. Linking seed germination and plant height: a case study of a wetland community on the eastern Tibet Plateau. Plant Biol. 30, 886–893.

Liu, Q., Piao, S., Janssens, I.A., Fu, Y., Peng, S., Lian, X., et al., 2018b. Extension of the growing season increases vegetation exposure to frost. Nat. Commun. 9, 426. Available from: https://doi.org/10.1038/s41467-017-02690-y.

Liu, K., Liang, T., Qiang, W., Du, G., Baskin, J.M., Baskin, C.C., et al., 2020. Changes in seed germination strategy along the successional gradient from abandoned cropland to climax grassland in a subalpine meadow and some implications for rangeland restoration. Agric. Ecosyst. Environ. 289, 106746. Available from: https://doi.org/10.1016/j.agee.2019.106746.

Lü, C., Tian, H., 2007. Spatial and temporal patterns of nitrogen deposition in China: synthesis of observational data. J. Geophys. Res. Atmos. 112, D22S05. Available from: https://doi.org/10.1029/2006JD007990.

Luo, X., Jia, B., Lai, X., 2020. Contributions of climate change, land use change and CO_2 to changes in the gross primary productivity of the Tibetan Plateau. Atmos. Ocean Sci. Lett. 13, 8–15.

Ma, M., Dalling, J.W., Ma, Z., Zhou, X., 2017. Soil environmental factors drive seed density across vegetation types on the Tibetan Plateau. Plant Soil 419, 349–361.

Ma, M., Baskin, C.C., Li, W., Zhao, Y.P., Zhao, Y., Zhao, L., et al., 2019. Seed banks trigger ecological resilience in subalpine meadows abandoned after arable farming on the Tibetan Plateau. Ecol. Appl. 29, e01959. Available from: https://doi.org/10.1002/eap.1959.

Ma, M., Collins, S.L., Du, G., 2020. Direct and indirect effects of temperature and precipitation on alpine seed banks in the Tibetan Plateau. Ecol. Appl. 30, e02096. Available from: https://doi.org/10.1002/eap.2096.

Ma, M., Du, G., Zhou, X., 2009. Role of the soil seed bank during succession in a subalpine meadow on the Tibetan plateau. Arct. Antarct. Alp. Res. 41, 469–477.

Ma, M., Zhou, X., Wang, G., Ma, Z., Du, G., 2010. Seasonal dynamics in alpine meadow seed banks along an altitudinal gradient on the Tibetan Plateau. Plant Soil 336, 291–302.

Ma, M., Zhou, X., Du, G., 2011. Soil seed bank dynamics in alpine wetland succession on the Tibetan Plateau. Plant Soil 346, 19–28.

Ma, M., Zhou, X., Du, G., 2013. Effects of disturbance intensity on seasonal dynamics of alpine meadow soil seed banks on the Tibetan Plateau. Plant Soil 369, 283–295.

Ma, Z., Ma, M., Baskin, J.M., Baskin, C.C., Li, J., Du, G., 2014. Responses of alpine meadow seed bank and vegetation to nine consecutive years of soil fertilization. Ecol. Eng. 70, 92–101.

Ma, M., Walck, J.L., Ma, Z., Wang, L., Du, G., 2018. Grazing disturbance increases transient but decreases persistent soil seed bank. Ecol. Appl. 28, 1020–1031.

Morgan, J.A., Mosier, A.R., Milchunas, D.G., LeCain, D.R., Nelson, J.A., Parton, W.J., 2004. CO_2 enhances productivity, alters species composition, and reduces digestibility of shortgrass steppe vegetation. Ecol. Appl. 14, 208–219.

Myers, N., Mittermeier, R.A., Mittermeier, C.G., da Fonseca, G.A.B., Kent, J., 2000. Biodiversity hotspots for conservation priorities. Nature 403, 853–858.

Peng, A., Klanderud, K., Wang, G., Zhang, L., Xiao, Y., Yang, Y., 2020a. Plant community responses to warming modified by soil moisture in the Tibetan Plateau. Arct. Antarct. Alp. Res. 52, 60–69.

Peng, Z., Xiao, H., He, X., Xu, C., Pan, T., Yu, X., 2020b. Different levels of rainfall and trampling change the reproductive strategy of *Kobresia humilis* in the Qinghai-Tibet Plateau. Rangel. J. 42, 143–152.

Piao, S., Liu, Q., Chen, A., Janssens, I.A., Fu, Y., Dai, J., et al., 2019. Plant phenology and global climate change: current progresses and challenges. Glob. Change Biol. 25, 1922–1940.

Pickett, S.T.A., 1989. Space-for-time substituion as an alternative to long-term studies. In: Likens, G.E. (Ed.), Long-term Studies in Ecology. Approaches and Alternatives. Springer-Verlag, New York, pp. 110–135.

Porter, J.R., Semenov, M.A., 2005. Crop responses to climatic variation. Philos. Trans. R. Soc. B 360, 2021–2035.

Pu, Z., Xu, L., 2009. MODIS/Terra observed snow cover over the Tibet Plateau: distribution, variation and possible connection with the East Asian Summer Monsoon (EASM). Theor. Appl. Climatol. 97, 265–278.

Qin, D., Liu, S., Li, P., 2006. Snow cover distribution, variability, and response to climate change in western China. J. Clim. 19, 1820–1833.

Qin, C., Yang, B., Brauning, A., Sonechkin, D.M., Huang, K., 2011. Regional extreme climate events on the northeastern Tibetan Plateau since AD 1450 inferred from tree rings. Glob. Planet Change 75, 143–154.

Richardson, A.D., Hufkens, K., Milliman, T., Aubrecht, D.M., Furze, M.E., Seyednasrollah, B., et al., 2018. Ecosystem warming extends vegetation activity but heightens vulnerability to cold temperatures. Nature 560, 368–371.

Sage, R.F., Kubien, D.S., 2007. The temperature response of C_3 and C_4 photosynthesis. Plant Cell Environ. 30, 1086–1106.

Shen, M., Tang, Y., Chen, J., Zhu, X., Zheng, Y., 2011. Influences of temperature and precipitation before the growing season on spring phenology in grasslands of the central and eastern Qinghai-Tibetan Plateau. Agric. For. Meteorol. 151, 1711–1722.

Sun, Q., Li, B., Zhou, G., Jiang, Y., Yuan, Y., 2020. Delayed autumn leaf senescence date prolongs the growing season length of herbaceous plants on the Qinghai–Tibetan Plateau. Agric. For. Meteorol. 284, 107896. Available from: https://doi.org/10.1016/j.agrformet.2019.107896.

Tubiello, F.N., Soussana, J.F., Howden, S.M., 2007. Crop and pasture response to climate change. Proc. Nat. Acad. Sci. USA 104, 19686–19690.

Walck, J.L., Baskin, J.M., Baskin, C.C., Hidayati, S.N., 2005. Defining transient and persistent seed banks in species with pronounced seasonal dormancy and germination patterns. Seed Sci. Res. 15, 189–196.

Wang, G., Baskin, C.C., Baskin, J.M., Yang, X., Liu, G., Zhang, X., et al., 2017. Timing of seed germination in two alpine herbs on the southeastern Tibetan Plateau: the role of seed dormancy and annual dormancy cycling in soil. Plant Soil 421, 465–476.

Wang, G., Baskin, C.C., Baskin, J.M., Yang, X., Liu, G., Zhang, X., et al., 2018. Effects of climate warming and prolonged snow cover on phenology of the early life history stages of four alpine herbs on the southeastern Tibetan Plateau. Am. J. Bot. 105, 967–976.

Wang, H., Liu, H., Cao, G., Ma, Z., Li, Y., Zhang, F., et al., 2020a. Alpine grassland plants grow earlier and faster but biomass remains unchanged over 35 years of climate change. Ecol. Lett. 23, 701–710.

Wang, Y., Tao, H., Zhang, P., Hou, X., Sheng, D., Tian, B., et al., 2020b. Reduction in seed set upon exposure to high night temperature during flowering in maize. Physiol. Plant 169, 73–82.

Xie, H., Ye, J., Liu, X., E, C., 2010. Warming and drying trends on the Tibetan Plateau (1971–2005). Theor. Appl. Climatol. 101, 241–253.

Xu, X., Lu, C., Shi, X., Gao, S., 2008. World water tower: an atmospheric perspective. Geophys. Res. Lett. 35, L20815. Available from: https://doi.org/10.1029/2008GL035867.

Xu, W., Ma, L., Ma, M., Zhang, H., Yuan, W., 2017. Spatial-temporal variability of snow cover and depth in the Qinghai–Tibetan Plateau. J. Clim. 30, 1521–1533.

Yan, Y., Tang, Z., 2019. Protecting endemic seed plants on the Tibetan Plateau under future climate change: migration matters. J. Plant Ecol. 12, 962–971.

Yang, Y., Wang, G., Klanderud, K., Wang, J., Liu, G., 2015. Plant community responses to five years of simulated climate warming in an alpine fen of the Qinghai-Tibetan Plateau. Plant Ecol. Divers. 82, 211–218.

You, Q.L., Kang, S.C., Aguilar, E., Yan, Y.P., 2008. Changes in daily climate extremes in the eastern and central Tibetan Plateau during 1961–2005. J. Geophys. Res. Atmos. 113, D07101. Available from: https://doi.org/10.1029/2007JD009389.

You, Q., Fraedrich, K., Ren, G., Pepin, N., Kang, S., 2013. Variability of temperature in the Tibetan Plateau based on homogenized surface stations and reanalysis data. Int. J. Climatol. 33, 1337–1347.

You, Q.L., Min, J.Z., Kang, S.C., 2016. Rapid warming in the Tibetan Plateau from observations and CMIP5 models in recent decades. Int. J. Climatol. 36, 2660–2670.

Yu, H.B., Zhang, Y.L., Liu, L.S., Chen, Z., Qi, W., 2018. Floristic characteristics and diversity patterns of seed plants endemic to the Tibetan Plateau. Biodivs. Sci. 26, 130–137 (in Chinese).

Zhang, X.Z., Shen, Z.X., Fu, G., 2015. A *meta*-analysis of the effects of experimental warming on soil carbon and nitrogen dynamics on the Tibetan Plateau. Appl. Soil Ecol. 87, 32–38.

Zhang, D., Ye, J., Sun, H., 2016a. Quantitative approaches to identify floristic units and centres of species endemism in the Qinghai-Tibetan Plateau, south-western China. J. Biogeogr. 43, 2465–2476.

Zhang, K., Shi, Y., Jing, X., Jing, S., He, J.-S., Sun, R., et al., 2016b. Effects of short-term warming and altered precipitation on soil microbial communities in alpine grassland of the Tibetan Plateau. Front. Microbiol. 7, 1032. Available from: https://doi.org/10.3389/fmicb.2016.01032.

Zhang, C., Willis, C.G., Klein, J.K., Ma, Z., Li, J.Y., Zhou, H., et al., 2017. Recovery of plant species diversity during long-term experimental warming of a species-rich alpine meadow community on the Qinghai-Tibet plateau. Biol. Conserv. 213, 218–224.

Zhang, C., Ma, Z., Zhou, H., Zhao, X., 2019a. Long-term warming results in species-specific shifts in seed mass in alpine communities. PeerJ 7, e7416. Available from: https://doi.org/10.7717/peerj.7416.

Zhang, C., Willis, C.G., Klein, J.K., Ma, Z., Ma, M., Csontos, P., et al., 2019b. Direct and indirect effects of long-term fertilization on the stability of the persistent seed bank. Plant Soil 438, 239–250.

Zhao, D., Wu, S., Yin, Y., Yin, Z., 2011. Vegetation distribution on Tibetan Plateau under climate change scenario. Regul. Environ. Change 11, 905–915.

Zhao, X., Wang, W., Wan, W., Li, H., 2015. Influence of climate change on potential productivity of naked barley in the Tibet Plateau in the past 50 years. Chin. J. Eco-Agric 23, 1329–1338.

Zong, N., Shi, P., Song, M., Zhang, X., Jiang, J., Chai, X., 2016. Nitrogen critical loads for an alpine meadow ecosystem on the Tibetan Plateau. Environ. Manage. 57, 531–542.

Zong, N., Shi, P., Chai, X., 2018. Effects of warming and nitrogen addition on nutrient resorption efficiency in an alpine meadow on the northern Tibetan Plateau. Soil Sci. Plant Nutr 64, 482–490.

Chapter 12

Effect of climate change on regeneration of plants from seeds in tropical wet forests

James Dalling[1,2], Lucas A. Cernusak[3], Yu-Yun Chen[4,5], Martijn Slot[2], Carolina Sarmiento[2,6] and Paul-Camilo Zalamea[2,6]

[1]*Department of Plant Biology, University of Illinois at Urbana-Champaign, Champaign, IL, United States,* [2]*Smithsonian Tropical Research Institute, Panama City, Panama, Republic of Panama,* [3]*College of Science and Engineering, James Cook University, Cairns, QLD, Australia,* [4]*Department of Natural Resources and Environmental Studies, National Dong Hwa University, Hualien, Taiwan, Republic of China,* [5]*Center for Interdisciplinary Research on Ecology and Sustainability, National Dong Hwa University, Hualien, Taiwan, Republic of China,* [6]*Department of Integrative Biology, University of South Florida, Tampa, FL, United States*

Introduction

Tropical forests play an essential role in the global carbon cycle. On the one hand, they are an important component of the global carbon sink that is estimated to have absorbed 55% of anthropogenic carbon emissions over the last 60 years. On the other hand, degradation and deforestation in the tropics account for a significant part of the 20% of anthropogenic carbon emissions attributable to land-use change (Mitchard, 2018). While estimating the contributions of tropical forests to the global carbon budget has been exceptionally challenging, the emerging consensus is that tropical forests are transitioning from being a sink or net-neutral carbon pool to becoming a net carbon source, as forest cover is degraded and drought and rising temperatures reduce the ability of trees to absorb more CO_2 via increased growth rates (Brienen et al., 2015; Liu et al., 2017; Qie et al., 2017).

Continued climate change and deforestation in the tropics have the potential to radically alter the course of warming on planet earth. Tropical forests are estimated to account for 55% of total above-ground carbon and 30% of global soil carbon stocks (reviewed in Cusack et al., 2016). Nonetheless, our understanding of the sensitivity of these pools to climate change remains very limited. In part, this reflects a historical focus on montane, alpine, and high latitude ecosystems, where increases in temperature are predicted to be greatest. However, the impacts of global warming are a function not only of the magnitude, rate, and variance in changes in temperature and precipitation patterns but also of the sensitivity of ecosystems to these changes (Sheldon, 2019). Tropical forest species may be the most sensitive to temperature changes because unlike boreal and temperate species their constituent species experience relatively little seasonal temperature variation and therefore may have adapted to a narrow temperature range (Janzen, 1967; Ghalambor et al., 2006; Feeley and Silman, 2010). On the other hand, tropical forests may be resilient. The fossil record shows that diverse neotropical forests existed during the Paleocene Eocene Thermal Maximum c. 55 million years ago, when mean annual temperatures were 4°C–7°C higher than current (Jaramillo et al., 2010). Furthermore, the high local diversity of tropical tree communities may allow for the maintenance of some ecosystem functions through compositional change.

Widespread increases in tree mortality following drought and extreme temperature events highlight the importance of changes in community composition in determining how climate change will affect tropical forest diversity and carbon storage. Species composition in tropical forests is highly variable at small spatial scales due to successional stages following treefall gap formation, dispersal limitation, and environmental filtering associated with variation in soil moisture and nutrient availability. Thus, mixing of plant functional types can result in large variation in standing biomass within forests and large projected changes in future carbon storage depending on the favored species groups. For example, Bunker et al. (2005) showed that extinction scenarios influencing species with different functional traits could reduce carbon storage on Barro Colorado Island (BCI), Panama by up to 70%. Clearly, climate change effects on species regeneration processes play a critical role in determining future carbon cycling and the fate of the biome.

The critical regeneration stages of seeds and seedlings largely determine the future composition of tropical forest canopies. The distribution of maternal seed sources and the dispersal of their seeds in part shape the habitat associations of tropical trees (Hubbell et al., 1999), which are further influenced by differential mortality rates that occur primarily from the seedling emergence to the small sapling stage (Baldeck et al., 2013). For most tropical trees, high mortality rates at these juvenile stages in turn reflect the actions of herbivores and pathogens and availability of abiotic resources (light, moisture, and nutrient availability, Paine et al., 2008). In this chapter, we highlight how global change can alter recruitment success of tropical tree species, starting with effects on reproductive phenology and then impacting seed traits and seedling establishment via effects of elevated CO_2 and temperature on germination, growth and nutrient stoichiometry and of elevated temperature and altered rainfall regimes on seedling drought and shade tolerance (Fig. 12.1).

Observed and predicted climate change in tropical forests

Average temperatures in tropical rain forests worldwide increased 0.26°C per decade from the 1970s to 1990s (Malhi and Wright, 2004) and by 0.7°C from 1980 to 2014 in the Amazon Basin (Gloor et al., 2015). At the same time, daily temperature fluctuations also increased by 0.29°C (Wang and Dillon, 2014). Predicted future temperature increases in tropical forests are further linked to an increased frequency of extreme temperature events (>2 standard deviations of mean, Beaumont et al., 2011) and with increased intensity and duration of heatwaves (Giorgi et al., 2014) that can cause heat-induced tree mortality (Cusack et al., 2016). Warming trends are further exacerbated by El Niño Southern Oscillation (ENSO) events (Rifai et al., 2019; Wigneron et al., 2020). Warming in the lowlands also impacts tropical montane forests. Increased sea surface temperatures are accompanied by an increase in the elevation at which freezing occurs and with the retreat of glaciers impacting hydrological cycles in the Andes and elsewhere (Diaz and Graham, 1996; Bradley et al., 2009). Furthermore, the composition and physiognomy of montane forests are strongly tied to fog water inputs and the frequency of cloud cover. Both paleo and contemporary data provide evidence for rising cloud-base heights associated with warming (Pounds et al., 1999; Los et al., 2019). Under climate change, cloud-base heights are predicted to increase through a combination of global warming effects on sea surface temperatures and regional warming effects resulting from the conversion of lowland forest to other land uses (Still et al., 1999; Foster, 2001; Nair et al., 2010; van der Molen et al, 2010).

Changes in precipitation have been more variable than temperature across tropical forest regions, with an average of 1% ($\pm 0.8\%$; mean $\pm$ 2 SE) decrease from 1960 to 1998 (Malhi and Wright, 2004). Although precipitation trends are

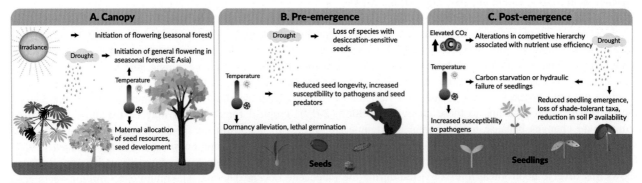

FIGURE 12.1 Critical climate change effects on tropical forest regeneration from flowering to seedling establishment. In the canopy (A), changes in solar irradiance patterns through the year are important flowering cues for many tree species in seasonal climates. In aseasonal forests in SE Asia, a combination of drought and low temperatures cue community-wide general flowering (GF) events. For all species, alterations in temperature and moisture availability can impact seed quality via effects on maternal seed provisioning, thereby potentially affecting seed longevity and degree of dormancy. After dispersal and before seedling emergence (B), increasing temperatures have direct effects on seed persistence via loss of dormancy and lethal germination. Increased productivity under a warmer climate also may increase herbivore populations and thus losses to seed predators and seed/seedling pathogens. Even relatively short dry spells immediately following seed dispersal may result in loss of species with desiccation-sensitive seeds. After seedling emergence (C), community composition may be affected strongly by species differences in tolerance to extreme temperatures and to altered competitive hierarchies resulting from different levels of nutrient limitation under elevated CO_2. Increased drought may contribute to loss of strongly shade-tolerant species due to trade-offs between drought and shade tolerance, while increasing temperature and decreasing moisture availability may result in carbon starvation or hydraulic failure under drought conditions and contribute to nutrient limitation.

region-specific, it is likely that all tropical forests will experience an increase in seasonal and interannual variation in rainfall, resulting in an increase in frequency of extreme droughts and extreme rainfall events. For example, in the Amazon there has been a greater intensification of the hydrological cycle, with an overall increase in wet season precipitation, a slight decrease in dry season precipitation (Gloor et al., 2015), and a 25% reduction in rainfall from 2000 to 2012 in the SE Amazon (Hilker et al., 2014). While links between the frequency of ENSO events and anthropogenic climate change currently are uncertain, extreme drought and warming periods associated with ENSO provide insights into tropical forest responses to predicted future climate conditions. A series of recent ENSO events have created intense drought conditions in eastern Amazonia (Jiménez-Muñoz et al., 2016) that are associated with increased rates of tree mortality (Brienen et al., 2015; Feldpausch et al., 2016; Zuleta et al., 2017), a shift in species composition, increased mortality of wet-affiliated taxa, and increased abundance of dry-affiliated taxa (Esquivel-Muelbert et al., 2019). Similar effects have been reported for earlier ENSO events in Panama (Condit et al., 1996).

Effects of climate change on reproductive phenology

Seed production is affected by factors influencing flowering and fruit development. Temporal fluctuation of environmental factors is the primary determinant for flowering (Yasuda et al., 1999; Wich and van Schaik, 2000). In addition, irradiance (van Schaik et al., 1993) and resource availability such as photosynthetic assimilates and mineral nutrients also may constrain flowering (Ichie and Nakagawa, 2013). On the other hand, fruiting is affected by environmental conditions during fruit development and by biotic interactions such as seed predation. Since the future climate in tropical wet forests may become warmer with stronger seasonality in rainfall, identifying factors that determine flowering and fruiting is important to understanding seeding and therefore the onset of forest regeneration.

In the tropics, temporal patterns of ambient temperature and rainfall in wet forests exhibit little seasonality compared to seasonally dry forests. Nonetheless, flowering and fruiting in most tree species of wet tropical forests exhibit seasonality. In La Selva, Costa Rica, most species flower subannually or annually (Newstrom et al., 1994). Similarly, species in El Verde, Puerto Rico, exhibit annual patterns with mean flowering dates scattered throughout the year (Zimmerman et al., 2007). Although flowers are produced regularly, fruiting tends to be more intermittent than in Costa Rica due to failure of pollination and fruit development and to seed predation. In contrast to the importance of water availability for flowering in seasonal forests (Wright et al., 2019), flowering of tree species in the wet forest of El Verde is highly correlated with seasonal variation in irradiance (Zimmerman et al., 2007), which explains the annual flowering pattern in most species in the absence of strong temperature and rainfall seasonality.

Taking the seemingly unmatched phenomena of climate and flowering schedules to an extreme, hundreds of forest species in Southeast Asian wet forests produce flowers in a supraannual fashion at varying intervals of 2−10 years (Sakai, 2002). Such supraannual flowering events involve species in dozens of families and occur in the same years; thus, they are referred to as general flowering (GF) events (Appanah, 1985; Ashton et al., 1988). Changes in zenithal sun angles do not explain supraannual GF in the wet forests of SE Asia. Instead, recent studies provide evidence that the synergistic effect of cool temperature and drought trigger synchronized blooming in GF events at the species (Yeoh et al., 2017; Chen et al., 2018) and community (Azmy et al., 2016; Ushio et al., 2019) levels. The occurrence of strong cooling and severe drought is rare and unpredictable in SE Asian wet forests. GF species may have high survival of the seed and seedling stages because herbivore populations decrease to a low level between GF events.

A detailed analysis of five *Shorea* (Dipterocarpaceae) species in Malaysia indicates that temperature and rainfall thresholds that best simulate flowering patterns differ among species (Chen et al., 2018). For example, *S. macroptera* has the lowest temperature requirement for flowering among the five species but the highest drought requirement for flowering. On the other hand, *S. leprosula* requires very cool temperatures but a mild drought to trigger flowering (Chen et al., 2018). Therefore, the warmer climates predicted for SE Asia may create a stronger barrier for flowering in *S. leprosula*, while *S. macroptera* may receive flowering cues more often if rainfall patterns become more seasonal in wet forests. Furthermore, if climate patterns are altered significantly from those at present we may expect decoupling of synchronized flowering events and a reduced magnitude of reproductive output during GF events. For example, species such as *S. macroptera* may produce flowers and fruits more frequently and other species less frequently than they do at present. The ecological consequences of decoupling synchronized flowering events may be dramatic. According to the predator satiation hypothesis (Janzen, 1971), seed survivorship in masting species is increased when they produce a surplus of fruits, thereby allowing a higher proportion of seeds to escape seed predation (Sun et al., 2007; Hosaka et al., 2017). Loss of synchronized flowering and fruiting in SE Asian forests may change forest composition by reducing survivorship of GF seeds and seedlings. Furthermore, if some species reproduce more frequently the continued provisioning of fruit may maintain higher densities of natural enemies, leading to lower seed survivorship regardless of when they are produced.

Climate change also may create a barrier to seedling establishment of GF species. The environmental prediction hypothesis suggests that the time when conditions favoring seedling establishment occur is an important driver of flowering events (Williamson and Ickes, 2002). A recent analysis revealed that environmental triggers for GF events at the Pasoh Forest Reserve (West Malaysia) serve as a good predictor of wet conditions on the forest floor (Satake et al., 2021). Warming in SE Asia may lead to more or to less frequent flowering or even to a shift of flowering time, depending on the species. The present link between environmental conditions that trigger flowering and seeding time may or may not remain the same with climate change as it is at present. In the latter case, seeds of some species may be ready for dispersal before the forest floor becomes wet.

Seed responses to elevated temperature and drought

By 2100, mean surface temperatures are predicted to have increased by 1.4°C and 4.8°C according to the IPCC's RCP4.5 and RCP8.5 scenarios, respectively Stocker et al. 2013. In tropical rain forests, extreme climatic events such as droughts, storms, and heatwaves are expected to become longer, more intense, and more frequent. The decadal-to-century-long lifespan of seeds (Dalling and Brown, 2009) and seedlings (Green and Harms, 2018; Chang-Yang et al., 2021) in tropical forests means that the conditions under which seeds developed may not coincide with those that seedlings currently experience. However, strong connections between climate and seed development, persistence, and germination have long been accepted, and they have been correlated to changes in seed and seedling morphology (Everingham et al., 2021). Thus, it is expected that future climatic conditions will significantly affect plant species distributions and population dynamics (Ooi, 2012).

The environmental conditions of the maternal plant can strongly influence seed development. For example, higher temperatures experienced by the parent plant lead to a decrease in dormancy and shorter seed longevity (Long et al., 2015). Stress caused by elevated temperatures before seeds reach maturity can decrease germination and subsequent seedling establishment by reducing the maternal allocation of seed reserves (Kitajima and Fenner, 2000). For example, soybean plants exposed to a 5°C increase in temperature during the seed filling period had an increased proportion of abnormal seeds and reduced seed germination relative to control plants (Spears et al., 1997). However, surprisingly little is known about how elevated temperatures affect the accumulation of seed reserves of noncrop species under natural conditions. Everingham et al. (2021) used modern and historic seed collections of 43 native Australian plant species (ranging from herbs to trees) that had experienced different levels of climatic change and found that two-thirds of them showed substantial differences in seed size between modern and historical collections. Nonetheless, there was little consistency in the direction of these changes, with seed size increasing in modern seeds (subject to warmer conditions) for 32% of the species and decreasing in 28%. This apparent inconsistency in seed size responses may have important implications for future ecosystem dynamics and forest composition. Furthermore, most species included in the Everingham et al. (2021) study have subtropical and temperate distributions, leaving a substantial knowledge gap about how traits such as seed size, viability, and germination respond to climate change in natural conditions for purely tropical species.

The time that elapses between dispersal and germination is referred to as seed persistence (Long et al., 2015). Seeds can spend this time in the soil seed bank either in a dormant or nondormant state. Many, but not all, nondormant seeds germinate rapidly and generally do not persist for long periods of time in the soil, while dormant seeds can persist in moist surface layers of tropical soils for decades (Dalling and Brown, 2009). By improving the chances of germinating when environmental conditions are favorable for seedling recruitment, growth, and survival, dormancy in seeds serves as a mechanism to reduce the risk of reproductive failure (Long et al., 2015). Traits such as germination, persistence, dormancy, and aging are strongly influenced by the temperature and moisture conditions seeds experience in the soil. In laboratory settings, elevated temperatures can alleviate physical (Baskin, 2003) and physiological (Iglesias-Fernández et al., 2011) dormancy and accelerate seed aging (Ellis and Roberts, 1981). Therefore, the warmer and drier conditions predicted in future climate scenarios may influence seed dormancy alleviation and germination, thereby contributing to net seed losses from the seed bank for some but not all species (Ooi, 2012). Although warmer conditions can reduce seed viability in the seed bank, it is unclear how close tropical seeds are to lethal temperature thresholds. Seed dormancy occurs in 51% and 57% of species in continuously moist tropical and seasonal tropical forests, respectively (Baskin and Baskin, 2014). On BCI, Panama, more than half of 157 species studied had delayed germination (>4 weeks from dispersal to emergence, Garwood, 1983). In a similar study in the Panama Canal Watershed forests, 45 of 94 tree species exhibited seed dormancy, with most of the 45 species having either physiological (23 species) or physical (13 species) dormancy. Physical dormancy was more common among species in the strongly seasonal forests on the Pacific side of the canal watershed (Sautu et al., 2007). If climate change results in greater seasonal soil moisture deficit, then species with physical rather than physiological dormancy may have an advantage in persisting through the dry season. However, Ooi et al. (2009) have shown that sustained higher temperatures can increase the mortality of at least some species with physical dormancy in arid ecosystems in Australia. After exposure to predicted increases in soil temperature, seed viability decreased by more than 40% in seeds of one of the five physically dormant species studied.

With climate change, short droughts are likely to increase in frequency as variance in precipitation regimes increases. If short dry spells of a few weeks coincide with the timing of seedling emergence, they may have strong effects on aseasonal forests. Seedlings of light-demanding species, particularly those from small seeds in surface soil layers in exposed treefall gaps, may be disproportionately negatively affected by drought at the emergence stage. For example, after only 4 days seedling survival for six pioneer species was significantly lower in rain-excluded than in irrigated gaps. All plots were protected from rain, and the irrigated plots received 18 mm of rain-collected water each day (540 mm per month). In contrast, seeds that germinate in the understory may be relatively well-buffered from short-term droughts by canopy shade that prevents large declines in soil water potential that would affect seedling emergence (Engelbrecht et al., 2006). However, longer term droughts may strongly affect species with desiccation-sensitive (recalcitrant) seeds. In the seasonally moist forest on BCI, 16% of the species have desiccation-sensitive seeds (Daws et al., 2005). This trait appears to reflect a cost associated with low investment in physical defense and rapid germination—traits that maximize investment in seed resource provisioning and reduce opportunities for seed losses to predators. Desiccation-sensitive seeds are common in aseasonal tropical forests and include ecologically and economically important species in families such as the Dipterocarpaceae and Sapotaceae (Tweddle et al., 2003). Under more intense and longer dry periods due to climate change, these taxa may be the first ones lost from wet tropical forests. Although most attention has been given to woody plants, epiphytes contribute to the physiognomy of aseasonal tropical forests and may be particularly vulnerable to climate change because most of their life cycle occurs at the interface of vegetation and atmosphere, where future changes are predicted to be strong (Zotz and Bader, 2009).

In addition to direct effects on seeds and seedlings, climate change also may impact interactions with microbes. Recent work has highlighted the importance of fungal infections in determining both seed and seedling survival in tropical forests (Mangan et al., 2010; Sarmiento et al., 2017; Hazelwood et al., 2021; Zalamea et al., 2021). Shade-tolerant species with short seed persistence times are most susceptible to fungal infection and damping-off disease from oomycetes (Augspurger, 1984; Augspurger and Wilkinson, 2007). Seeds also are susceptible to a broad range of ascomycete fungal infections in the soil (Sarmiento et al., 2017). The prevalence and virulence of these infections also may increase under warmer and wetter conditions (Spear et al., 2015). Rapid seedling emergence, which could be accelerated under warmer conditions, would reduce opportunities to invade poorly defended tissues (Dalling et al., 2020). Seeds of the annual grass *Bromus tectorum* that germinated fast often escaped fungal pathogen (*Pyrenophora semeniperda*) attacks compared to seeds of the same annual grass that germinated more slowly (Beckstead et al., 2007). However, in cool montane forests seedlings also may escape infection if colder temperatures favor seedling growth versus fungal or bacterial growth (Leach, 1947), in which case warmer conditions may result in greater losses to pathogens.

Seedling responses to increased temperatures

A key component of the response of plants to an increase in temperature is the parallel increase in vapor pressure deficit (VPD), which is perceived by plants as atmospheric drought. Leaf stomata close when VPD is high, thereby greatly decreasing water loss from the plant but also reducing the potential for carbon fixation by photosynthesis (Slot and Winter, 2017; Smith et al., 2020). Increased VPD associated with elevated temperature can significantly decrease survival of temperate tree species during drought (Adams et al., 2009, 2017; Will et al., 2013), either because reduced photosynthesis causes carbon starvation or because increased transpiration demands cause hydraulic failure. Similar experiments have not been conducted with tropical species.

Whether increased temperature and VPD in response to global warming will have negative effects on plant regeneration from seeds depends on the plasticity of tropical seedlings to adjust to the changed environment. A 3°C daytime warming did not negatively impact seedling growth in mixed tropical species mesocosms, apparently due to thermal acclimation of photosynthetic parameters (Slot and Winter, 2018). However, it appears that late-successional species performed more poorly under warming conditions than early-successional species. Likewise, in a combined 5°C warming and elevated CO_2 experiment late-successional species exposed to full sunlight had high mortality. Survival was improved when shade netting was used to reduce light to levels that better reflected the regeneration niche of late succession species; however, seedlings of these species did not perform as well as those of early-successional species (K. Winter, M. Slot, unpublished results).

Carbon loss via mitochondrial respiration strongly affects the whole-plant light compensation point of seedlings (Baltzer and Thomas, 2007). Thus understory seedlings are particularly vulnerable to global warming, since respiration is highly temperature sensitive. Further, the amount of carbon lost by respiration at elevated temperatures is mitigated by thermal acclimation of leaves of tropical tree seedlings (e.g., Cheesman and Winter, 2013; Zhu et al., 2021), saplings (Slot and Winter, 2018; Mujawamariya et al., 2021), and canopy trees (Slot et al., 2014). However, acclimation does not always result in perfect homeostasis. That is, even when plants acclimate to higher temperatures, they often respire more carbon than they would at cooler temperatures (but see Mujawamariya et al., 2021). Furthermore, acclimation varies among species (Slot and Kitajima, 2015). Thus, with small increases in temperature in the forest understory seedlings of some species may require increased light levels to survive.

A closed canopy has a strong buffering effect on maximum temperature and maximum VPD in tropical forests (Jucker et al., 2018). Thus, even with moderate warming understory seedlings will be exposed to less atmospheric drought stress than seedlings growing in forest gaps. Differences in the severity of extreme conditions for carbon fixation across habitats and regeneration niches potentially create conditions for nonrandom impacts of climate change on tropical forest seedling communities. Interestingly, however, a recent study that analyzed tree rings from four temperate species across forest strata found that high temperatures had a significantly greater negative effect on the growth of understory plants than of canopy trees of the same species (Rollinson et al., 2021). While these results do not refute the hypothesis that understory-regenerating species may be affected differentially by warming compared to species that regenerate in open sites, they do suggest that the buffering effect of canopies does not protect understory plants from the negative effects of warming. Closed canopies have a particularly strong buffering effect on maximum temperatures and VPD in the understory and a more moderate effect on mean temperature and VPD (Jucker et al., 2018). Extremes can have disproportionately large effects on plant performance if they lead to irreversible damage, but mean conditions are more likely to affect long-term growth patterns. Ultimately, the effects of global climate change on tropical seedlings will depend on the microclimate seedlings experience and on the capacity to acclimate to higher temperatures and episodic drought within the constraints of their habitat. The need for simultaneous adjustments of seedlings to multiple potential stressors (temperature, drought, shade, and altered species interactions) poses unique challenges to plants because of trade-offs in optimal allocation of resources that maximize growth and survival. Furthermore, lack of information on these simultaneous adjustments to environmental factors limits our ability to predict plant responses to climate change.

Drought tolerance and shade tolerance trade-offs for seedlings

Trade-offs in strategies for growth and survival occur in seedlings of wet tropical forest species (Huston, 1994). Adaptations that confer a survival advantage for seedlings in one environment may be disadvantageous in another environment, which is especially true along environmental gradients. One hypothesized trade-off is between adaptations for surviving dry spells and those for surviving and growing in the low light conditions that dominate wet forests in the absence of treefall gaps. That is, there is a trade-off between drought tolerance and shade tolerance (Smith and Huston, 1989; Brenes-Arguedas et al., 2013). Although drought typically is not associated with wet tropical forests, rainless periods that can significantly affect seedling survival are not uncommon (Engelbrecht et al., 2005; Engelbrecht et al., 2006; Poorter and Markesteijn, 2008). Dry spells of several days can strongly reduce water potentials in the top 20 cm of soil, potentially resulting in the death of seedlings (Engelbrecht and Kursar, 2003). Drought tolerance of seedlings is significantly associated with distribution patterns of adult tropical forest trees at both local and regional scales across the Isthmus of Panama (Engelbrecht et al., 2007).

Attempts to test for a trade-off between drought and shade tolerance among tropical rainforest tree species have yielded mixed results. In part, this likely is due to the challenge of designing experiments that can evaluate the drought and shade tolerances of a range of species. Evidence for trade-off was found among 24 species distributed across the Isthmus of Panama (Brenes-Arguedas et al., 2013). Here, the trade-off appeared to be due mostly to variation in photosynthetic capacity (higher in more drought-tolerant species). In contrast, observations on seedlings of 62 Bolivian tropical tree species indicated no direct trade-off between traits associated with drought and shade tolerance (Markesteijn and Poorter, 2009). However, photosynthetic capacity was not assessed in the Bolivian study. Global climate change clearly has the potential to shift the distribution and intensity of drought and shade within wet tropical forests by altering rainfall patterns and disturbance regimes. Understanding and predicting species responses to such shifts as a result of seedling survival and growth is an important undertaking for conservation planning and forest management.

Effects of climate change on nutrient availability

In addition to light and soil moisture, productivity and distribution of tropical trees are affected by soil nutrients (John et al., 2007; Muller-Landau et al., 2021). Global climate change phenomena can directly and indirectly affect plant-nutrient relations that impact seedling growth and survival. Anthropogenic nitrogen (N) deposition is predicted to increase in most tropical forests, and long-term increases in leaf and wood $\delta^{15}N$ have been documented (Hietz et al., 2011). The greatest impact of N deposition is expected in N-limited montane forests (Tanner et al., 1998) and not in phosphorus (P)-limited lowland forests (Vitousek, 1984). Soil warming in tropical forests also is likely to increase N availability by stimulating microbial-decomposer activity, as has been observed in temperate and high latitude ecosystems (Cusack et al., 2016; Wood et al., 2019). Similar effects might be expected for P, although experimental warming indicates that P availability may be reduced if the production of microbial enzymes that release P is diminished (Cusack et al., 2010). However, increased temperature also is associated with soil drying. While little effect of drying has been detected on soil N availability, experimental drying treatments reduced extractable soil P (Wood and Silver, 2012), despite greater leaching losses of P and base cations in wetter soils (Posada and Schuur, 2011, but see Turner and Engelbrecht, 2011). Thus, much still needs to be learned about the effects of global climate change on soil nutrient availability and plant regeneration from seeds.

Interactions between elevated CO_2 and nutrient limitation

To date, the relatively small number of CO_2 manipulations conducted in wet tropical forests has been done on seedling and sapling responses in closed or open-top chambers, rather than in large free-air enrichment experiments. These studies show that productivity of tropical trees in general responds positively to elevated CO_2. However, growth chamber experiments found high variation in the magnitude of growth enhancement among species, with stronger growth response for pioneer than for shade-tolerant species (Cusack et al., 2016), and for gymnosperms than for angiosperms (Dalling et al., 2016). These responses may be further constrained by nutrient limitation, in particular the supply rate of P, which is thought to limit productivity in highly weathered soils across much of the lowland tropics (Vitousek, 1984; Paoli and Curran, 2007; Cleveland et al., 2011). Currently, predictions on how P availability will impact projected future biomass growth in tropical forests are not well developed, but models that incorporate P limitation suggest that ecosystem resilience to climate change may be much lower than previously assumed (Fleischer et al., 2019).

Tropical forests across the Panama Canal watershed show large differences in species composition associated with gradients of dry season intensity and soil P availability (Condit et al., 2013). Species sorting across the soil P gradient reflects adaptations to maximize seedling growth under different levels of nutrient availability. In a pot experiment, species from high P sites grew fastest in high P soils, while those from low P sites grew fastest in low P soils—a clear performance trade-off (Zalamea et al., 2016). In a follow-up experiment, this species sorting was disrupted under elevated CO_2 (800 ppmv) such that relative growth rates were no longer related to species habitat associations (Thompson et al., 2019). Therefore, species-specific responses to elevated CO_2 pose challenges for predicting community-wide responses to changes in temperature and precipitation related to global climate change.

Nitrogen-fixing legume trees are common in tropical forests, and they are predicted to have a strong positive response to elevated CO_2. P acquisition is via fine root growth and phosphatase production, and both require large amounts of N. Thus N-fixing trees could gain a competitive advantage if additional photosynthate under elevated CO_2 increases N_2 fixation (Houlton et al., 2008). Alternatively, N_2 fixation, ecosystem N availability, and continued carbon fixation by tropical forests may be increasingly constrained by low P availability. In a pot experiment, growth of seedlings of the N-fixing tree legume *Ormosia macrocalyx* was increasingly limited by P [and molybdenum (Mo)—a cofactor of the nitrogenase enzyme] applied at preindustrial, ambient, and double-ambient concentrations of CO_2 (Trierweiler et al., 2018). Thus, although the N-fixer in this study could greatly increase N_2 fixation when P and Mo were available, N-fixation rates may not be able to keep up with demand as CO_2 rises.

Concluding remarks and future research challenges

A consistent theme throughout this chapter has been the low confidence in predicted responses of seeds and seedlings to climate change in wet tropical forests. Understanding how climate change will impact regeneration is one of the most challenging, yet significant, problems involved in modeling the carbon cycle. In other biomes, inferences from observations and experiments carried out in one ecosystem can reasonably be translated to others. Tropical forests, however, have several unique characteristics that demand much more attention from in situ studies. This chapter has highlighted the presence of key features of wet tropical forests that lack parallels elsewhere: synchronized mass flowering events triggered by climatic cues; presence of desiccation-sensitive seeds susceptible to increased frequency of short-term droughts; adaptations of seedlings to survive in deep shade that potentially constrains their ability to tolerate moisture deficits and elevated temperatures; and nutrient limitation imposed by P availability. Climate alterations impacting each of these features have the potential to strongly impact forest species and trait composition. Overlaying the direct climate-change effects are alterations to complex species webs involving seed predators, herbivores, pathogens, and mycorrhizae. Due to logistical and cost constraints, most climate change studies in tropical forests have been conducted in closed greenhouse or small open-top chambers (Cernusak et al., 2013). While these approaches are well-suited to experimental work on seed and seedling physiology, they largely exclude biotic interactions.

More information also is needed about the seed and seedling stages of the life cycle of wet tropical forest species. For example, will particular seed dormancy and persistence traits buffer more or buffer less increased temperature and moisture fluctuations in the soil, and will these traits provide protection against lethal germination (i.e., germination during conditions unfavorable for subsequent survival) as the climate changes? Laboratory studies have been successful in providing insight into sensitivities of physiological processes to climate change. However, predictions of how projected future climates, including changes in temperature and precipitation regimes, will alter recruitment patterns of tropical plants require field experiments in which seeds and seedlings are exposed to the full range of natural hazards that shape survivorship curves. Such studies need to include possible climate-related changes in rates of seed predation. For example, using a seed translocation experiment along an elevation gradient in the Peruvian Andes, Hillyer and Silman (2010) showed that seed survival rates generally increased with elevation from 1500 to 3000 m a.s.l. and suggested that warming may affect productivity, thereby increasing seed predator abundance and reducing recruitment.

A greater array of well-replicated long-term translocation experiments and seed survival analyses such as those carried out in Queensland, Australia (Green et al., 2014) are needed to predict not only direct physiological effects of climate change on seeds and seedlings but also broader impacts on the species interactions that affect tropical forest regeneration. Furthermore, a trade-off between seedling drought and shade tolerance is a long-standing hypothesis grounded in consideration of physiological constraints. Additional testing of this hypothesis would help better understand the extent to which a trade-off between drought and shade tolerance is manifested in wet tropical forest communities and whether it forms a cornerstone of forest demography models as has been suggested (Huston, 1994).

We eagerly await the integration of seed and seedlings stages into ongoing and proposed field manipulations of climate change drivers, including soil warming (Carter et al., 2020; Nottingham et al., 2020) and long-anticipated CO_2 enrichment studies (Tollefson, 2013).

References

Adams, H.D., Guardiola-Claramonte, M., Barron-Gafford, G.A., Villegas, J.C., Breshears, D.D., Zou, C.B., et al., 2009. Temperature sensitivity of drought-induced tree mortality portends increased regional die-off under global-change-type drought. Proc. Natl. Acad. Sci. USA 106, 7063–7066.

Adams, H.D., Barron-Gafford, G.A., Minor, R.L., Gardea, A.A., Bentley, L.P., Law, D.J., et al., 2017. Temperature response surfaces for mortality risk of tree species with future drought. Environ. Res. Lett. 12, 115014. Available from: https://doi.org/10.1088/1748-9326/aa93be.

Appanah, S., 1985. General flowering in the climax rain forests of Southeast Asia. J. Trop. Ecol. 1, 225–240.

Ashton, P.S., Givnish, T.J., Appanah, S., 1988. Staggered flowering in the Dipterocarpaceae: new insights into floral induction and the evolution of mast fruiting in the seasonal tropics. Am. Nat. 132, 44–66.

Augspurger, C.K., 1984. Seedling survival of tropical tree species: interactions of dispersal distance, light-gaps, and pathogens. Ecology 65, 1705–1712.

Augspurger, C.K., Wilkinson, H.T., 2007. Host specificity of pathogenic *Pythium* species: implications for tree species diversity. Biotropica 39, 702–708.

Azmy, M.M., Hashim, M., Numata, S., Hosaka, T., Noor, S.M.N., Fletcher, C., 2016. Satellite-based characterization of climatic conditions before large-scale general flowering events in Peninsular Malaysia. Sci. Rep. 6, 32329. Available from: https://doi.org/10.1038/srep32329.

Baldeck, C.A., Harms, K.E., Yavitt, J.B., John, R., Turner, B.L., Valencia, R., et al., 2013. Habitat filtering across tree life stages in tropical forest communities. Proc. R. Soc. B 280, 20130548. Available from: https://doi.org/10.1098/rspb.2013.0548.

Baltzer, J.L., Thomas, S.C., 2007. Determinants of whole-plant light requirements in Bornean rain forest tree saplings. J. Ecol. 95, 1208–1221.

Baskin, C.C., 2003. Breaking physical dormancy in seeds–focussing on the lens. New Phytol. 158, 229–232.

Baskin, C.C., Baskin, J.M., 2014. Seeds: Ecology, Biogeography and Evolution of Dormancy and Germination, Second ed. Academic Press/Elsevier, San Diego.

Beaumont, L.J., Pitman, A., Perkins, S., Zimmermann, N.E., Yoccoz, N.G., Thuiller, W., 2011. Impacts of climate change on the world's most exceptional ecoregions. Proc. Natl. Acad. Sci. USA 108, 2306–2311.

Beckstead, J., Meyer, S.E., Molder, C.J., Smith, C., 2007. A race for survival: can *Bromus tectorum* seeds escape *Pyrenophora semeniperda*-caused mortality by germinating quickly? Ann. Bot. 99, 907–914.

Brenes-Arguedas, T., Roddy, A.B., Kursar, T.A., 2013. Plant traits in relation to the performance and distribution of woody species in wet and dry tropical forest types in Panama. Funct. Ecol. 27, 392–402.

Brienen, R.J., Phillips, O.L., Feldpausch, T.R., Gloor, E., Baker, T.R., Lloyd, J., et al., 2015. Long-term decline of the Amazon carbon sink. Nature 519, 344–348.

Bunker, D.E., DeClerck, F., Bradford, J.C., Colwell, R.K., Perfecto, I., Phillips, O.L., et al., 2005. Species loss and aboveground carbon storage in a tropical forest. Science 310, 1029–1031.

Bradley, R.S., Keimig, F.T., Diaz, H.F., Hardy, D.R., 2009. Recent changes in freezing level heights in the tropics with implications for the deglacierization of high mountain regions. Geophys. Res. Lett. 36, L17701. Available from: https://doi.org/10.1029/2009GL037712.

Carter, K.R., Wood, T.E., Reed, S.C., Schwartz, E.C., Reinsel, M.B., Yang, X., et al., 2020. Photosynthetic and respiratory acclimation of understory shrubs in response to in situ experimental warming of a wet tropical forest. Front. For. Glob. Change 3, 765–785.

Cernusak, L.A., Winter, K., Dalling, J.W., Holtum, J.A., Jaramillo, C., Körner, C., et al., 2013. Tropical forest responses to increasing atmospheric CO_2: current knowledge and opportunities for future research. Funct. Plant Biol. 40, 531–551.

Chang-Yang, C.-H., Needham, J., Lu, C.-L., Hsieh, C.-F., Sun, I.-F., McMahon, S.M., 2021. Closing the life cycle of forest trees: the difficult dynamics of seedling to sapling transitions in a subtropical rainforest. J. Ecol. 109, 2705–2716. Available from: https://doi.org/10.1111/1365-2745.13677.

Cheesman, A.W., Winter, K., 2013. Growth response and acclimation of CO_2 exchange characteristics to elevated temperatures in tropical tree seedlings. J. Exp. Bot. 64, 3817–3828.

Chen, Y.-Y., Satake, A., Sun, I.F., Kosugi, Y., Tani, M., Numata, S., et al., 2018. Species specific flowering cues among general flowering *Shorea* species at the Pasoh Research Forest, Malaysia. J. Ecol. 106, 586–598.

Cleveland, C.C., Townsend, A.R., Taylor, P., Alvarez-Clare, S., Bustamante, M.M.C., Chuyong, G., et al., 2011. Relationships among net primary productivity, nutrients and climate in tropical rain forest: a pan-tropical analysis. Ecol. Lett. 14, 939–947.

Condit, R., Hubbell, S.P., Foster, R.B., 1996. Changes in tree species abundance in a neotropical forest: impact of climate change. J. Trop. Ecol. 12, 231–256.

Condit, R., Engelbrecht, B.M., Pino, D., Perez, R., Turner, B.L., 2013. Species distributions in response to individual soil nutrients and seasonal drought across a community of tropical trees. Proc. Natl. Acad. Sci. USA 110, 5064–5068.

Cusack, D.F., Torn, M.S., McDowell, W.H., Silver, W.L., 2010. The response of heterotrophic activity and carbon cycling to nitrogen additions and warming in two tropical soils. Glob. Change Biol. 16, 2555–2572.

Cusack, D.F., Karpman, J., Ashdown, D., Cao, Q., Ciochina, M., Halterman, S., et al., 2016. Global change effects on humid tropical forests: evidence for biogeochemical and biodiversity shifts at an ecosystem scale. Rev. Geophys. 54, 523–610.

Dalling, J.W., Brown, T.A., 2009. Long-term persistence of pioneer species in tropical rain forest soil seed banks. Am. Nat. 173, 531–535.

Dalling, J.W., Cernusak, L.A., Winter, K., Aranda, J., Garcia, M., Virgo, A., et al., 2016. Two tropical conifers show strong growth and water-use efficiency responses to altered CO_2 concentration. Ann. Bot. 118, 1113–1125.

Dalling, J.W., Davis, A.S., Arnold, A.E., Sarmiento, C., Zalamea, P.C., 2020. Extending plant defense theory to seeds. Annu. Rev. Ecol. Evol. Syst. 51, 123–141.

Daws, M.I., Garwood, N.C., Pritchard, H.W., 2005. Traits of recalcitrant seeds in a semi-deciduous tropical forest in Panama: some ecological implications. Funct. Ecol. 19, 874–885.

Diaz, H.F., Graham, N.E., 1996. Recent changes in tropical freezing heights and the role of sea surface temperature. Nature 383, 152–155.

Ellis, R.H., Roberts, E.H., 1981. Improved equations for the prediction of seed longevity. Ann. Bot. 45, 13–30.

Engelbrecht, B.M.J., Kursar, T.A., 2003. Comparative drought-resistance of seedlings of 28 species of co-occurring tropical woody plants. Oecologia 136, 383–393.

Engelbrecht, B.M.J., Kursar, T.A., Tyree, M.T., 2005. Drought effects on seedling survival in a tropical moist forest. Trees 19, 312–321.

Engelbrecht, B.M., Dalling, J.W., Pearson, T.R., Wolf, R.L., Galvez, D.A., Koehler, T., et al., 2006. Short dry spells in the wet season increase mortality of tropical pioneer seedlings. Oecologia 148, 258–269.

Engelbrecht, B.M.J., Comita, L.S., Condit, R., Kursar, T.A., Tyree, M.T., Turner, B.L., et al., 2007. Drought sensitivity shapes species distribution patterns in tropical forests. Nature 447, 80–82.

Esquivel-Muelbert, A., Baker, T.R., Dexter, K.G., Lewis, S.L., Brienen, R.J., Feldpausch, T.R., et al., 2019. Compositional response of Amazon forests to climate change. Glob. Change Biol. 25, 39–56.

Everingham, S.E., Offord, C.A., Sabot, M.E., Moles, A.T., 2021. Time travelling seeds reveal that plant regeneration and growth traits are responding to climate change. Ecology 102, e03272. Available from: https://doi.org/10.1002/ecy.3272.

Feeley, K.J., Silman, M.R., 2010. Biotic attrition from tropical forests correcting for truncated temperature niches. Glob. Change Biol. 16, 1830–1836.

Feldpausch, T.R., Phillips, O.L., Brienen, R.J.W., Gloor, E., Lloyd, J., Lopez-Gonzalez, G., et al., 2016. Amazon forest response to repeated droughts. Glob. Biogeochem. Cycles 30, 964–982.

Fleischer, K., Rammig, A., De Kauwe, M.G., Walker, A.P., Domingues, T.F., Fuchslueger, L., et al., 2019. Amazon forest response to CO_2 fertilization dependent on plant phosphorus acquisition. Nat. Geosci. 12, 736–741.

Foster, P., 2001. The potential negative impacts of global climate change on tropical montane cloud forests. Earth-Sci. Rev. 55, 73–106.

Garwood, N.C., 1983. Seed germination in a seasonal tropical forest in Panama: a community study. Ecol. Monogr. 53, 159–181.

Ghalambor, C.K., Huey, R.B., Martin, P.R., Tewksbury, J.J., Wang, G., 2006. Are mountain passes higher in the tropics? Janzen's hypothesis revisited. Integr. Comp. Biol. 46, 5–17.

Giorgi, F., Coppola, E., Raffaele, F., Diro, G.T., Fuentes-Franco, R., Giuliani, G., et al., 2014. Changes in extremes and hydroclimatic regimes in the CREMA ensemble projections. Clim. Change 125, 39–51.

Gloor, M., Barichivich, J., Ziv, G., Brienen, R., Schöngart, J., Peylin, P., et al., 2015. Recent Amazon climate as background for possible ongoing and future changes of Amazon humid forests. Glob. Biogeochem. Cycles 29, 1384–1399.

Green, P.T., Harms, K., 2018. How old is an understory seedling? Long-term records reveal the great age of suppressed, juvenile rainforest trees in north Queensland, Australia. In: Proceedings of the 2018 ESA Annual Meeting, New Orleans.

Green, P.T., Harms, K.E., Connell, J.H., 2014. Nonrandom, diversifying processes are disproportionately strong in the smallest size classes of a tropical forest. Proc. Natl. Acad. Sci. USA 111, 18649–18654.

Hazelwood, K., Beck, H., Timothy Paine, C.E., 2021. Negative density dependence in the mortality and growth of tropical tree seedlings is strong, and primarily caused by fungal pathogens. J. Ecol. 109, 1909–1918.

Hietz, P., Turner, B.L., Wanek, W., Richter, A., Nock, C.A., Wright, S.J., 2011. Long-term change in the nitrogen cycle of tropical forests. Science 334, 664–666.

Hilker, T., Lyapustin, A.I., Tucker, C.J., Hall, F.G., Myneni, R.B., Wang, Y., 2014. Vegetation dynamics and rainfall sensitivity of the Amazon. Proc. Natl. Acad. Sci. USA 111, 16041–16046.

Hillyer, R., Silman, M.R., 2010. Changes in species interactions across a 2.5 km elevation gradient: effects on plant migration in response to climate change. Glob. Change Biol. 16, 3205–3214.

Hosaka, T., Yumoto, T., Chen, Y.Y., Sun, I.F., Wright, S.J., Numata, S., 2017. Responses of pre-dispersal seed predators to sequential flowering of Dipterocarps in Malaysia. Biotropica 49, 177–185.

Houlton, B.Z., Wang, Y.P., Vitousek, P.M., Field, C.B., 2008. A unifying framework for dinitrogen fixation in the terrestrial biosphere. Nature 454, 327–330.

Hubbell, S.P., Foster, R.B., O'Brien, S.T., Harms, K.E., Condit, R., Wechsler, B., et al., 1999. Light-gap disturbances, recruitment limitation, and tree diversity in a neotropical forest. Science 283, 554–557.

Huston, M.A., 1994. Biological Diversity: The Coexistence of Species on Changing Landscapes. Cambridge University Press, Cambridge.

Ichie, T., Nakagawa, M., 2013. Dynamics of mineral nutrient storage for mast reproduction in the tropical emergent tree *Dryobalanops aromatica*. Ecol. Res. 28, 151–158.

Iglesias-Fernández, R., del Carmen Rodriguez-Gacio, M., Matilla, A.J., 2011. Progress in research on dry after ripening. Seed Sci. Res. 21, 69–80.

Janzen, D.H., 1967. Why mountain passes are higher in the tropics. Am. Nat. 101, 233–249.

Janzen, D.H., 1971. Seed predation by animals. Annu. Rev. Ecol. Syst. 2, 465–492.

Jaramillo, C., Ochoa, D., Contreras, L., Pagani, M., Carvajal-Ortiz, H., Pratt, L.M., et al., 2010. Effects of rapid global warming at the Paleocene-Eocene boundary on neotropical vegetation. Science 330, 957–961.

Jiménez-Muñoz, J.C., Mattar, C., Barichivich, J., Santamaría-Artigas, A., Takahashi, K., Malhi, Y., et al., 2016. Record-breaking warming and extreme drought in the Amazon rainforest during the course of El Niño 2015–2016. Sci. Rep. 6, 33130. Available from: https://doi.org/10.1038/srep33130.

John, R., Dalling, J.W., Harms, K.E., Yavitt, J.B., Stallard, R.F., Mirabello, M., et al., 2007. Soil nutrients influence spatial distributions of tropical tree species. Proc. Natl. Acad. Sci. USA 104, 864–869.

Jucker, T., Hardwick, S.R., Both, S., Elias, D.M., Ewers, R.M., Milodowski, D.T., 2018. Canopy structure and topography jointly constrain the microclimate of human-modified tropical landscapes. Glob. Change Biol. 24, 5243–5258.

Kitajima, K., Fenner, M., 2000. Ecology of seedling regeneration. In: Fenner, M. (Ed.), Seeds: The Ecology of Regeneration in Plant Communities. CABI Publishing, Wallingford, pp. 331–359.

Leach, L.D., 1947. Growth rates of host and pathogen as factors determining the severity of pre-emergence damping off. J. Agric. Res. 75, 161–179.

Liu, J., Bowman, K.W., Schimel, D.S., Parazoo, N.C., Jiang, Z., Lee, M., et al., 2017. Contrasting carbon cycle responses of the tropical continents to the 2015–2016 El Niño. Science 358. Available from: https://doi.org/10.1126/science.aam5690.

Long, R.L., Gorecki, M.J., Renton, M., Scott, J.K., Colville, L., Goggin, D.E., 2015. The ecophysiology of seed persistence: a mechanistic view of the journey to germination or demise. Biol. Rev. 90, 31–59.

Los, S.O., Street-Perrott, F.A., Loader, N.J., Froyd, C.A., Cuní-Sanchez, A., Marchant, R.A., 2019. Sensitivity of a tropical montane cloud forest to climate change, present, past and future: Mt. Marsabit, N. Kenya. Quat. Sci. Rev. 218, 34–48.

Malhi, Y., Wright, J., 2004. Spatial patterns and recent trends in the climate of tropical rainforest regions. Philos. Trans. R. Soc. B 359, 311–329.

Mangan, S.A., Schnitzer, S.A., Herre, E.A., Mack, K.M.L., Valencia, M.C., Sanchez, E.I., et al., 2010. Negative plant-soil feedback predicts tree-species relative abundance in a tropical forest. Nature 466, 752–755.

Markesteijn, L., Poorter, L., 2009. Seedling root morphology and biomass allocation of 62 tropical tree species in relation to drought- and shade-tolerance. J. Ecol. 97, 311–325.

Mitchard, E.T., 2018. The tropical forest carbon cycle and climate change. Nature 559, 527–534.

Mujawamariya, M., Wittemann, M., Manishimwe, A., Ntirugulirwa, B., Zibera, E., Nsabimana, D., et al., 2021. Complete or overcompensatory thermal acclimation of leaf dark respiration in African tropical trees. New Phytol. 229, 2548–2561.

Muller-Landau, H.C., Cushman, K.C., Arroyo, E.E., Martinez Cano, I., Anderson-Teixeira, K.J., Backiel, B., 2021. Patterns and mechanisms of spatial variation in tropical forest productivity, woody residence time, and biomass. New Phytol. 229, 3065–3087.

Nair, U.S., Ray, D.K., Lawton, R.O., Welch, R.M., Pielke, R.A., Calvo-Alvarado, J., 2010. The impact of deforestation on orographic cloud formation in a complex tropical environment. I. In: Bruijnzeel, L.A., Scatena, F.N., Hamilton, L.S. (Eds.), Tropical Montane Cloud Forests. Cambridge University Press, Cambridge, pp. 538–548.

Newstrom, L., Frankie, G.W., Baker, H.G., 1994. A new classification for plant phenology based on flowering patterns in lowland tropical rain forest trees at La Selva, Costa Rica. Biotropica 26, 141–159.

Nottingham, A.T., Meir, P., Velasquez, E., Turner, B.L., 2020. Soil carbon loss by experimental warming in a tropical forest. Nature 584, 234–237.

Ooi, M.K.J., 2012. Seed bank persistence and climate change. Seed Sci. Res. 22, S53–S60.

Ooi, M.K.J., Auld, T.D., Denham, A.J., 2009. Climate change and bet-hedging: interactions between increased soil temperatures and seed bank persistence. Glob. Change Biol. 15, 2375–2386.

Paine, C.E., Harms, K.E., Schnitzer, S.A., Carson, W.P., 2008. Weak competition among tropical tree seedlings: implications for species coexistence. Biotropica 40, 432–440.

Paoli, G.D., Curran, L.M., 2007. Soil nutrients limit fine litter production and tree growth in mature lowland forest of southwestern Borneo. Ecosystems 10, 503–518.

Poorter, L., Markesteijn, L., 2008. Seedling traits determine drought tolerance of tropical tree species. Biotropica 40, 321–331.

Posada, J.M., Schuur, E.A., 2011. Relationships among precipitation regime, nutrient availability, and carbon turnover in tropical rain forests. Oecologia 165, 783–795.

Pounds, J.A., Fogden, M.P., Campbell, J.H., 1999. Biological response to climate change on a tropical mountain. Nature 398, 611–615.

Qie, L., Lewis, S.L., Sullivan, M.J., Lopez-Gonzalez, G., Pickavance, G.C., Sunderland, T., et al., 2017. Long-term carbon sink in Borneo's forests halted by drought and vulnerable to edge effects. Nat. Commun. 8, 1966. Available from: https://doi.org/10.1038/s41467-017-01997-0.

Rifai, S.W., Li, S., Malhi, Y., 2019. Coupling of El Niño events and long-term warming leads to pervasive climate extremes in the terrestrial tropics. Environ. Res. Lett. 14, 105002. Available from: https://doi.org/10.1088/1748-9326/ab402f.

Rollinson, C.R., Alexander, M.R., Dye, A.W., Moore, D.J., Pederson, N., Trouet, V., 2021. Climate sensitivity of understory trees differs from overstory trees in temperate mesic forests. Ecology 102, e03264. Available from: https://doi.org/10.1002/ecy.3264.

Sakai, S., 2002. General flowering in lowland mixed dipterocarp forests of South-east Asia. Biol. J. Linn. Soc. 75, 233–247.

Sarmiento, C., Zalamea, P.C., Dalling, J.W., Davis, A.S., Stump, S.M., U'Ren, J.M., et al., 2017. Soilborne fungi have host affinity and host-specific effects on seed germination and survival in a lowland tropical forest. Proc. Natl. Acad. Sci. USA 114, 11458–11463.

Satake, A., Yao, T.L., Kosugi, Y., Chen, Y.-Y., 2021. Testing the environmental prediction hypothesis for community-wide mass flowering in South-East Asia. Biotropica 53, 608–618.

Sautu, A., Baskin, J.M., Baskin, C.C., Deago, J., Condit, R., 2007. Classification and ecological relationships of seed dormancy in a seasonal moist tropical forest, Panama, Central America. Seed Sci. Res. 17, 127–140.

Sheldon, K.S., 2019. Climate change in the tropics: ecological and evolutionary responses at low latitudes. Annu. Rev. Ecol. Evol. Syst. 50, 303–333.

Slot, M., Kitajima, K., 2015. General patterns of acclimation of leaf respiration to elevated temperatures across biomes and plant types. Oecologia 177, 885–900.

Slot, M., Winter, K., 2017. In situ temperature relationships of biochemical and stomatal controls of photosynthesis in four lowland tropical tree species. Plant Cell Environ. 40, 3055–3068.

Slot, M., Winter, K., 2018. High tolerance of tropical sapling growth and gas exchange to moderate warming. Funct. Ecol. 32, 599–611.

Slot, M., Rey-Sánchez, C., Gerber, S., Lichstein, J.W., Winter, K., Kitajima, K., 2014. Thermal acclimation of leaf respiration of tropical trees and lianas: response to experimental canopy warming, and consequences for tropical forest carbon balance. Glob. Change Biol. 20, 2915–2926.

Smith, T., Huston, M., 1989. A theory of the spatial and temporal dynamics of plant communities. Vegetatio 83, 49–69.

Smith, M.N., Taylor, T.C., van Haren, J., Rosolem, R., Restrepo-Coupe, N., Adams, J., et al., 2020. Empirical evidence for resilience of tropical forest photosynthesis in a warmer world. Nat. Plants 6, 1225–1230.

Spear, E.R., Coley, P.D., Kursar, T.A., 2015. Do pathogens limit the distributions of tropical trees across a rainfall gradient? J. Ecol. 103, 165–174.

Spears, J.F., TeKrony, D.M., Egli, D.B., 1997. Temperature during seed filling and soybean seed germination and vigour. Seed Sci. Technol. 25, 233–244.

Still, C.J., Foster, P.N., Schneider, S.H., 1999. Simulating the effects of climate change on tropical montane cloud forests. Nature 398, 608–610.

Stocker, T.F., Qin, D., Plattner, G.K., Tignor, M.M.B., Allen, S.K., Boschung, J., et al., 2013. IPCC 2013: Climate change 2013: The physical science basis Working Group I Contribution to the Fifth Assessment Report of the Intergovernmental Panel on Climate Change. Cambridge University Press, Cambridge.

Sun, I.F., Chen, Y.Y., Hubbell, S.P., Wright, S.J., Noor, S.M.N., 2007. Seed predation during general flowering events of varying magnitude in a Malaysian rain forest. J. Ecol. 95, 818–827.

Tanner, E.V.J., Vitousek, P.A., Cuevas, E., 1998. Experimental investigation of nutrient limitation of forest growth on wet tropical mountains. Ecology 79, 10–22.

Thompson, J.B., Slot, M., Dalling, J.W., Winter, K., Turner, B.L., Zalamea, P.C., 2019. Species-specific effects of phosphorus addition on tropical tree seedling response to elevated CO_2. Funct. Ecol. 33, 1871–1881.

Tollefson, J., 2013. Experiment aims to steep rainforest in carbon dioxide. Nature 496, 405–406.

Trierweiler, A.M., Winter, K., Hedin, L.O., 2018. Rising CO_2 accelerates phosphorus and molybdenum limitation of N_2-fixation in young tropical trees. Plant Soil 429, 363–373.

Tweddle, J.C., Dickie, J.B., Baskin, C.C., Baskin, J.M., 2003. Ecological aspects of seed desiccation sensitivity. J. Ecol. 91, 294–304.

Turner, B.L., Engelbrecht, B.M., 2011. Soil organic phosphorus in lowland tropical rain forests. Biogeochemistry 103, 297–315.

Ushio, M., Osada, Y., Kumagai, T., Kume, T., Pungga, R.S., Nakashizuka, T., et al., 2019. Dynamic and synergistic influences of air temperature and rainfall on general flowering in a Bornean lowland tropical forest. Ecol. Res. 35, 17–29.

van der Molen, M.K., Vugts, H.F., Bruijnzeel, L.A., Scatena, F.N., Pielke, R.A., Kroon, L.J.M., 2010. Meso-scale climate change due to lowland deforestation in the maritime tropics. In: Bruijnzeel, L.A., Scatena, F.N., Hamilton, L.S. (Eds.), Tropical Montane Cloud Forests. Cambridge University Press, Cambridge, pp. 527–537.

van Schaik, C.P., Terborgh, J.W., Wright, S.J., 1993. The phenology of tropical forests: adaptive significance and consequence for primary consumers. Annu. Rev. Ecol. Syst. 24, 353–377.

Vitousek, P.M., 1984. Litterfall, nutrient cycling, and nutrient limitation in tropical forests. Ecology 65, 285–298.

Wang, G., Dillon, M.E., 2014. Recent geographic convergence in diurnal and annual temperature cycling flattens global thermal profiles. Nat. Clim. Change 4, 988–992.

Wich, S., van Schaik, C.P., 2000. The impact of El Niño on mast fruiting in Sumatra and elsewhere in Malesia. J. Trop. Ecol. 16, 563–577.

Wigneron, J.P., Fan, L., Ciais, P., Bastos, A., Brandt, M., Chave, J., 2020. Tropical forests did not recover from the strong 2015–2016 El Niño event. Sci. Adv. 6, eaay4603. Available from: https://doi.org/10.1126/sciadv.aay4603.

Will, R.E., Wilson, S.M., Zou, C.B., Hennessey, T.C., 2013. Increased vapor pressure deficit due to higher temperature leads to greater transpiration and faster mortality during drought for tree seedlings common to the forest–grassland ecotone. New Phytol. 200, 366–374.

Williamson, G.B., Ickes, K., 2002. Mast fruiting and ENSO cycles – does the cue betray a cause? Oikos 97, 459–461.

Wood, T.E., Cavaleri, M.A., Giardina, C.P., Khan, S., Mohan, J.E., Nottingham, A.T., et al., 2019. Soil warming effects on tropical forests with highly weathered soils. In: Mohan, J.E. (Ed.), Ecosystem Consequences of Soil Warming: Microbes, Vegetation, Fauna and Soil Biogeochemistry. Elsevier Science & Technology, Amsterdam, pp. 385–439.

Wood, T.E., Silver, W.L., 2012. Strong spatial variability in trace gas dynamics following experimental drought in a humid tropical forest. Glob. Biogeochem. Cycles 26, GB3005. Available from: https://doi.org/10.1029/2010GB004014.

Wright, S.J., Calderon, O., Muller-Landau, H.C., 2019. A phenology model for tropical species that flower multiple times each year. Ecol. Res. 34, 20–29.

Yasuda, M., Matsumoto, J., Osada, N., Ichikawa, S., Kachi, N., Tani, M., et al., 1999. The mechanism of general flowering in Dipterocarpaceae in the Malay Peninsula. J. Trop. Ecol. 15, 437–449.

Yeoh, S.H., Satake, A., Numata, S., Ichie, T., Lee, S.L., Basherudin, N., et al., 2017. Unraveling proximate cues of mass flowering in the tropical forests of Southeast Asia from gene expression analyses. Mol. Ecol. 26, 5074–5085.

Zalamea, P.C., Turner, B.L., Winter, K., Jones, F.A., Sarmiento, C., Dalling, J.W., 2016. Seedling growth responses to phosphorus reflect adult distribution patterns of tropical trees. New Phytol. 212 (2), 400–408.

Zalamea, P.C., Sarmiento, C., Arnold, A.E., Davis, A.S., Ferrer, A., Dalling, J.W., 2021. Closely related tree species support distinct communities of seed-associated fungi in a lowland tropical forest. J. Ecol. 109, 1858–1872.

Zhu, L., Bloomfield, K.J., Asao, S., Tjoelker, M.G., Egerton, J.J., Hayes, L., 2021. Acclimation of leaf respiration temperature responses across thermally contrasting biomes. New Phytol. 229, 1312–1325.

Zimmerman, J.K., Wright, S.J., Calderon, O., Pagan, M.A., Paton, S., 2007. Flowering and fruiting phenologies of seasonal and aseasonal neotropical forests: the role of annual changes in irradiance. J. Trop. Ecol. 23, 231–251.

Zotz, G., Bader, M.Y., 2009. Epiphytic plants in a changing world-global: change effects on vascular and non-vascular epiphytes. In: Lüttge, U., Beyschlag, W., Büdel, B., Francis, D. (Eds.), Progress in Botany, vol. 70. Springer, Berlin, pp. 147–170.

Zuleta, D., Duque, A., Cardenas, D., Muller-Landau, H.C., Davies, S.J., 2017. Drought-induced mortality patterns and rapid biomass recovery in a terra firme forest in the Colombian Amazon. Ecology 98, 2538–2546.

Chapter 13

Climate change and plant regeneration from seeds in tropical dry forests

Guillermo Ibarra-Manríquez[1], Jorge Cortés-Flores[2], María Esther Sánchez-Coronado[3], Diana Soriano[4], Ivonne Reyes-Ortega[4], Alma Orozco-Segovia[3], Carol C. Baskin[5,6] and Jerry M. Baskin[5]

[1]*Instituto de Investigaciones en Ecosistemas y Sustentabilidad. Universidad Nacional Autónoma de México. Antigua carretera a Pátzcuaro No. 8701. Col. San José de la Huerta. C. P. Morelia, Michoacán, México,* [2]*Jardín Botánico, Instituto de Biología, Sede Tlaxcala, Universidad Nacional Autónoma de México. Ex Fábrica San Manuel S/N. Colonia San Manuel. C. P. Santa Cruz Tlaxcala, Tlaxcala, México,* [3]*Instituto de Ecología, Universidad Nacional Autónoma de México, Av. Universidad 3000, C. P. Ciudad de México, México,* [4]*Departamento de Ecología y Recursos Naturales, Facultad de Ciencias, Universidad Nacional Autónoma de México, Av. Universidad 3000, C. P. Ciudad de México, México,* [5]*Department of Biology, University of Kentucky, Lexington, KY, United States,* [6]*Department of Plant and Soil Sciences, University of Kentucky, Lexington, KY, United States*

Introduction

Tropical dry forests (TDF) occur from Mexico to Argentina and Brazil; on the Caribbean Islands; in Africa, Madagascar, India, Sri Lanka, central Indochina, and Australia; and on the island chain east of Java, Indonesia (Miles et al., 2006; Portillo-Quintero and Sánchez-Azofeifa, 2010). Distribution maps of TDF had been published previously (Miles et al., 2006; Pennington et al., 2006; Portillo-Quintero and Sánchez-Azofeifa, 2010). Two-thirds of the TDF occur in the Americas (Miles et al., 2006). These forests are delineated globally based on a combination of several climatic variables (Murphy and Lugo, 1986; Meir and Pennington, 2011; FAO (Food and Agriculture Organization of the United Nations), 2012, 2020; Siyum, 2020): (1) mean annual temperature $>17°C$, without frost; (2) mean annual precipitation $600-2000$ mm; (3) an annual potential evapotranspiration:precipitation ratio >1; and (4) a 3- to 8-month dry season. Variation in the annual amount of precipitation and duration of the wet and dry periods in TDF determine the patterns of seed germination, seedling establishment, and plant growth and reproduction (Murphy and Lugo, 1986; Singh and Kushwaha, 2005; Vieira and Scariot, 2006; Shumba et al., 2010; Allen et al., 2017).

TDF vary in species composition, and structural and functional characteristics (Rzedowski, 1978; Gentry, 1995; Lott and Atkinson, 2006; Lebrija-Trejos et al., 2010; FAO, 2012, 2020; Hulshof et al., 2013; Banda-R et al., 2016). Unlike savannas with one tree stratum, an open canopy, and a grassy understory, TDF are dominated by trees that grow in two to three strata with a closed (or nearly closed) canopy. Most trees are deciduous during the dry season and produce new leaves during the wet season (Rzedowski, 1978; Murphy and Lugo, 1986; FAO, 2012, 2020; Ibarra-Manríquez et al., 2021). However, an exception to this phenology occurs in Australia, where evergreen eucalyptus trees maintain their leaves during the dry season (Murphy and Lugo, 1986; Bowman and Prior, 2005). The proportion of deciduous species in the landscape is influenced by environmental and biotic factors such as length of the dry season, geological substrate, topography, forest structure, and land-use history (Rzedowski, 1978; Murphy and Lugo, 1986; Gentry, 1995; Sánchez-Azofeifa et al., 2005; FAO, 2012, 2020; Hulshof et al., 2013). TDF are rich in tree, shrub, herb, and climber species, but epiphytes usually are not abundant (Gentry, 1995; Lott and Atkinson, 2006; Ibarra-Manríquez et al., 2021). Cacti occur in TDF on dry sites in the New World (Godinez-Alvarez and Valiente-Banuet, 1998; Ibarra-Manríquez et al., 2021), and succulent species of Euphorbiaceae in those in Africa (Pennington et al., 2009, 2018). TDF have high β diversity values due to differences among sites or regions in factors such as elevation, slope, insolation, potential of evapotranspiration, and soil characteristics (Balvanera et al., 2002; Trejo and Dirzo, 2002; Gallardo-Cruz et al., 2009; Apgaua et al., 2014).

TDF cover about 40% of the land area occupied by subtropical and tropical forests (Murphy and Lugo, 1986), and millions of people (an estimated 100 million in Africa) live in or near them (Djoudi et al., 2015). Thus, TDF are important for the livelihood of a significant proportion of the human population (Murphy and Lugo, 1986; Trejo and Dirzo, 2000; Miles et al., 2006; Agarwala et al., 2016; Siyum, 2020). TDF have relatively high soil fertility (Pennington et al., 2006). Consequently, much deforestation and habitat fragmentation of this forest type have occurred via use of the land for agriculture (crops and pasture) and other purposes (Murphy and Lugo, 1986; Miles et al., 2006; Klemens et al., 2011; Meir and Pennington, 2011; Armenteras and Rodríguez-Eraso, 2014; Portillo-Quintero and Sánchez-Azofeifa, 2010; Portillo-Quintero et al., 2015). TDF are severely threatened in all the regions in which they occur (Janzen, 1988; Miles et al., 2006). In addition to threats due to direct human activities, including increased fire frequency, TDF are predicted to be seriously impacted by climate change in the future (Hasnat, 2020; IPBES (Intergovernmental Science-Policy Platform on Biodiversity and Ecosystem Services), 2019; Meir and Pennington, 2011; Siyum, 2020; IPCC, 2014), and more in the Americas than in other regions (Miles et al., 2006).

This chapter reviews the effects of climate change on regeneration of plants from seeds in the TDF. In particular, key events in the life cycle of plants, including flowering/pollination, seed production, seed dispersal, seed dormancy-break and germination, seed banks, and establishment of seedlings, are evaluated in the context of predicted climatic changes in TDF (Allen et al., 2017; IPCC, 2014). We also briefly discuss possible changes in community composition and shifts in distribution patterns of TDF species in response to climate change and future research needs.

Predicted climate changes

Climate change is caused by emission of greenhouse gases (primarily CO_2) as a result of economic activities and growth of the human population (Allen et al., 2017; Sentinella et al., 2020; Siyum, 2020). The IPCC, 2014 estimates that by the end of the 21st-century global temperature will increase by $0.3°C-1.7°C$ under RCP2.6 (with stringent greenhouse gas mitigation) and by $2.6°C-4.8°C$ under RCP8.5 (very high greenhouse gas emissions). Under the RCP8.5 scenario, tropical and subtropical regions generally are predicted to become warmer and drier than at present. Although the amount of precipitation is predicted to decrease, the occurrence of extreme precipitation events probably will be higher (IPCC, 2014). Wright (2005) predicted that the duration of the dry season in TDF will increase.

Several environmental factors in TDF display an important variation, including latitude, aspect, topographic position, slope, microtopography, plant cover, organic matter in and on the soil, percentage of rock exposure, soil moisture, and soil temperature, that could affect the regeneration of plants from seeds (Gallardo-Cruz et al., 2009; Lebrija-Trejos et al., 2010; Cao and Sanchez-Azofeifa, 2017; Chaturvedi and Raghubanshi, 2018; Méndez-Toribio et al., 2020; Bradford et al., 2020). Although habitat variation is not a part of climate change models, it needs to be considered in evaluating the ability of species to persist in TDF. Furthermore, high-accuracy climate change models for TDF are not available, and one reason for this is that the climate of TDF varies between regions of a continent and between continents (Albuquerque et al., 2012; Stan and Sánchez-Azofeifa, 2019; WMO (World Meteorological Organization), 2019).

BOX 13.1 Literature survey.

We surveyed papers in the Web of Science, Schoolar Google, and Scopus databases searching for the following combination of terms in the title, abstract or keywords of papers: (Tropical dry forest* OR seasonal tropical dry forest* or tropical deciduous forest) AND (seed* OR germination OR dormancy OR seed storage behavior or seedling* OR regeneration) AND (drought OR climate change OR temperature OR global warming). After removing duplicates, 561 references were screened. To be included in our review, papers must have addressed the effect of at least one environmental driver related to climate change (temperature or drought) in a species from the tropical dry forest. We also recorded studies that analyzed and predicted changes in abiotic factors associated with climate change in the TDF. In addition, we included articles in which phenological changes in flowering and their consequences on seed production were analyzed. Review papers were used to screen for further references that we could have missed. A very important reference for the determination seed dormancy was the book by Baskin and Baskin (2014), and for the determination of seed storage behavior we mainly consulted the database of the Royal Botanic Gardens, Kew (RBGK (Royal Botanic Gardens, Kew), 2021). Each study was classified into four regeneration stages: seed germination, seed dormancy, lag time, soil seed banks, and seedlings. We also recorded whether the study assessed changes in seed viability after exposure to treatments related to the environmental drivers. Regarding climate change-related variables, we classified the studies into two categories: temperature and drought. Under "temperature," we included studies that used experimental temperatures or those measured in natural habitats. Under "drought," we included studies that addressed the effect of desiccation (reduction in seed water content) and environmental water deficit (soil water potentials) on seeds.

Another reason is due to the lack of regional climate data with which to make models (Boko et al., 2007). Nonetheless, the prediction from general models for dry regions in Africa (Boko et al., 2007), Latin America (Margin et al., 2007), and Australia (Hennessy et al., 2007) is that the temperature will increase, and precipitation will decrease. It also is predicted that TDF could become savannas (Hasnat, 2020; IPCC, 2007; Siyum, 2020) and that in some regions in Brazil tropical rainforests could become TDF or savannas (Rodrigues et al., 2015). Dexter et al. (2018) concluded that soil moisture would be the most important factor in tropical biome transitions.

Flowering, seed production, and dispersal

During periods of drought, an increase in temperature in combination with changes in the rainfall patterns can affect the optimal range of conditions for photosynthesis. If photosynthesis is reduced, plants will have decreased energy reserves, which may negatively affect the number of flowers, fruits, and seeds produced (Becknell et al., 2012; Maza-Villalobos et al., 2013; Slot and Winter, 2017). It is expected that flowering and seed production will be negatively affected by climate change. During the dry season, high temperatures can diminish nectar production (Mu et al., 2015) and pollen viability (Ejsmond et al., 2015), negatively impacting attraction for pollinators. In wind-pollinated plants, atypical precipitation events during the dry season related to climate change would negatively affect pollen viability (Cobert and Plumridge, 1985). For tree species of TDF, a decrease in amount of precipitation and/or duration of the wet season can modify onset, synchrony, and duration of flowering (Shrestha et al., 2018; Stan and Sánchez-Azofeifa, 2019). In pollinator-dependent plants, variations in flowering phenology can have a negative impact on pollination vectors (Quesada et al., 2009; Watanabe, 2014).

A mismatch between timing of flowering and pollinator activity could lead to a decrease in plant reproductive success (i.e., fruit and seed set), since more than 90% of species depend on biotic vectors for their pollination (Bawa, 1990; Quesada et al., 2011; Parmesan and Hanley, 2015). Disruptions of these plant–animal interactions because of climate change could have negative effects on regeneration of plants from seeds. For other growth forms of TDF such as herbs, the life cycle is closely related to the availability of water, and flowering occurs during the wet season. However, if precipitation decreases after plants are established it is expected that flowering will either not occur or the flowering period will be very short, both of which will decrease seed production.

If viable seeds are produced, dispersal per se will not be critical since it is mainly via anemochory, autochory, or epizoochory (Griz and Machado, 2001; Cortés-Flores et al., 2017). Seeds dispersed at the beginning of the rainy season are likely to germinate during this period. However, it is predicted that rains will be more erratic and that there will be a decrease in the amount of precipitation and in duration of the wet season, potentially resulting in changes in germination phenology that will negatively impact success of species establishment (Marod et al., 2002; Donohue et al., 2010; Walck et al., 2022, Ashton 2016; Siyum, 2020).

The timing of seed dispersal varies with continent, class of dormancy, and season (Fig. 13.1). Relatively more species in Africa and America have seeds with physical (PY) than physiological (PD), or nondormancy (ND), while relatively more species in Asia have PD than PY or ND. In Africa, dispersal of seeds with PY is strongly associated with the dry season, but some seeds with PY are dispersed in the wet season (Fig. 13.1). In America and Asia, seeds with PY also were dispersed in both the wet and dry seasons, with slightly fewer species dispersing seeds in the wet than in the dry season. Seeds with PD and those that were ND are dispersed in the wet and dry seasons in Africa, America, and Asia. It seems likely that changes in precipitation regime could affect not only seed production but also timing of dispersal, and indirectly have a major impact on germination phenology and successful seedling establishment.

Seed dormancy and germination

Seed germination and the establishment of seedlings are critical events in the life cycle of plants (Donohue et al., 2010), especially in forests characterized by a marked seasonality in water availability such as TDF (Bhadouria et al., 2016). On a global scale, seed traits, such as size, moisture content, storage behavior, and germination responses, of TDF species have been evaluated mainly for trees. Furthermore, the amount/kind of available information on seed traits varies with the species. Seed attributes for other growth forms, such as herbs and lianas, are scarcely known despite their importance in the structure, dynamic, and diversity of TDF (Gentry, 1995; Lott and Atkinson, 2006; Ibarra-Manríquez et al., 2015; Ibarra-Manríquez et al., 2021). The main classes of seed dormancy in TDF trees are PY and PD. However, seeds of a few species have morphophysiological dormancy and some ND seeds. Additionally, some seed lots of a species may consist of seeds with PY and ND or of PD and ND (Baskin and Baskin, 2014).

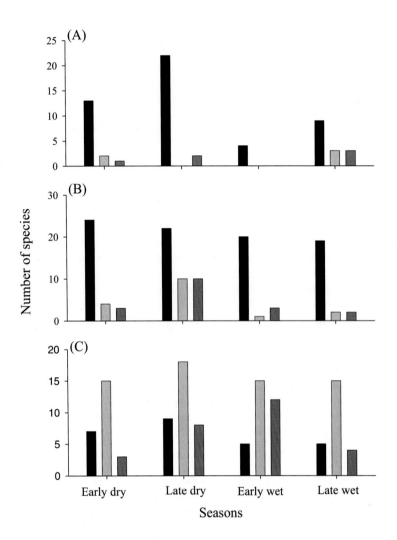

FIGURE 13.1 Number of species with physically dormant (*black bars*), physiologically dormant (*light gray bars*), and nondormant seeds (*dark gray bars*) at time of seed dispersion in tropical dry forests in Africa (A), America (B), and Asia (C). (See Box 13.1).

Although most TDF trees have desiccation-tolerant (orthodox) seeds, a few species have desiccation-sensitive (recalcitrant) seeds (Khurana and Singh, 2001; Tweddle et al., 2003; Pritchard et al., 2004). Of the 290 TDF tree species in our review of seed storage behavior (see Box 13.1), 3%, 1%, and 7% of the species that have recalcitrant seeds occur in Africa, America, and Asia, respectively. These species may be at risk if climate change results in a delay of arrival and/or duration of the wet season. Orthodox seeds are not only desiccation-tolerant, but they may have the protection of a thick water-impermeable seed coat (e.g., Fabaceae), an indehiscent fruit or stony endocarp (e.g., Anacardiaceae and Myrtaceae). Thus, orthodox seeds produced during the dry season remain viable until the onset of the wet season in the next year (Khurana and Singh, 2001).

Physical and physiological dormancy

Our review of seed dormancy in 302 TDF tree species (see Box 13.1) revealed that 58.3% have PY. Seeds with PY dispersed in the dry season or late in the wet season would remain in the soil and not germinate until the wet season of the following year (Singh and Singh, 1992; Khurana and Singh, 2001). In Asia, 11 of 34 species were reported to have both PY and ND. This could explain an increase in number of seedlings coming from seeds considered to have PY. That is, ND seeds would germinate immediately in the wet season, while the seeds with PY require dormancy-breaking cues, which may delay germination until the subsequent wet season. In this sense, Gutterman (2000) and Hudson et al. (2015) report contrasting examples of the effect of rainfall and temperature on the acquisition of PY during seed maturation. In species whose seeds have the potential to develop PY, seeds first gain physiological maturity, after which seed drying results in development of seed coat impermeability to water (Baskin and Baskin, 2014). Thus, changes in timing of precipitation events could have an impact on the proportion of seeds in a given seed cohort with PY versus ND.

Our review of seed dormancy in 302 TDF tree seeds (see Box 13.1) revealed that 26.7% had PD. Thus, after dispersal seeds cannot germinate until after they have had a period during which dormancy is broken. Under nursery or field conditions in Thailand, seeds of *Glochidion kerrii* and *Heynea trijuga* began to germinate 39 and 83 days, respectively, after sowing, and the last ones to germinate did so 134 and 203 days, respectively, after sowing. Seed dispersal of both species occurs near the end of the wet season or the beginning of the dry season (Blakesley et al., 2000, 2002; Elliott et al., 2002); thus, seeds remained nongerminated for 6 months or longer in/on the soil before the onset of the wet season. Seeds of *Diospyros melanoxylon* in India mature during the dry season and germinate in the following wet season (Ghosh et al., 1976). On the other hand, seeds of *Mesua ferrea* in India are dispersed during the dry season, but they do not germinate until the beginning of the next dry season (Richards, 1957). Thus seeds of some trees in TDF can persist in the soil during the dry season and germinate rather quickly after the wet season begins. The effect of a delay in onset of the wet season predicted by climate change models on seed viability and germination needs to be investigated. Furthermore, seed storage behavior of *H. trijuga* indicates that the seeds probably are desiccation sensitive (Agustin et al., 2018), which suggests that decreased precipitation could have a negative impact on plant regeneration from seeds of some species.

Effect of temperature on seed germination

From our review of the temperatures at which germination tests were conducted in 169 TDF species (see Box 13.1), 20°C, 25°C, and 30°C were the most frequently used, and they are optimal for the germination of many tropical species (Baskin and Baskin, 2014). The most frequently used alternating temperature regime in germination studies is 20/30°C.

Daily accumulated temperatures (measured at 10-min intervals) in growth chambers at 30°C and 20/30°C (12/12 h, day/night) were 4320°C and 3600°C, respectively, which can be higher or lower than accumulated field temperatures. For example, accumulated temperatures in open sites at a soil depth of 2–3 cm in Veracruz (México) and in Ciudad de México (México) were 4777.2°C and 3894.7°C, respectively (Peraza-Villarreal, Sánchez-Coronado, and Orozco-Segovia, unpublished data). Nevertheless, seeds of a few TDF tree species have been germinated at controlled temperatures higher than 30°C. Among them *Dalbergia retusa* germinated to 60% and 15% at 40°C and 45°C, respectively (García and Di Stéfano, 2000); *Pterocarpus macrocarpus* to about 40% at 45/40°C (Liengsiri and Hellum, 1988); and *Stylosanthes viscosa* to 58% and 31% at 38°C and 46°C, respectively (Gomes and Kretschmer, 1978). Seeds of *S. viscosa* also germinated to 69% at 46°C after receiving five wet–dry cycles. Thus, for these and probably many more TDF species the predicted increase in temperature due to climate change is not expected to decrease germination unless water is limiting. In fact, increased temperatures due to climate change may promote germination of some species. However, although seeds of TDF species may be tolerant of relatively high field temperatures germination at high temperatures does not guarantee seedling survival (Orozco-Segovia and Sánchez-Coronado, 2009).

Decreased and erratic precipitations and shortening of the wet season may negatively affect germination and seedling establishment. Seeds of *Myracrodruon urundeuva* from TDF (Caatinga) in northeastern Brazil can germinate to low percentages at 40°C and −0.7 MPa in the laboratory. However, climate change models (RCP8.5) for northeastern Brazil predict a negative effect of decreased precipitation on germination and seedling recruitment of this species in the field (Oliveira et al., 2019). Thermal and hydrothermal time models for germination of *Cenostigma microphyllum* in northeastern Brazil predict that decreases in length of the wet season would significantly decrease germination and seedling establishment (Gomes et al., 2019). In the wet season of the Chamela TDF (Mexico), Soriano et al. (2014) found that soil moisture at a soil depth of 5 cm in an open site varied between 1.3% and 2.9% (close to −40 MPa), while at 15 cm it reached 5.36%–6.6% (−0.035 MPa, Gouge et al., 2000; soil field capacity −0.03 MPa, Taiz and Zeiger, 1998).

Seeds of *Terminalia chebula* from Asia germinated to 20.8% at −0.5 MPa (Khurana and Singh, 2004). However, seedling roots need to grow and reach water quickly if seedlings are going to survive. In closed sites of the Chamela TDF, field capacity of the soil can be reached at a depth of 5 cm, probably due to the protection from evaporation by the canopy and litter layer. In this Mexican TDF, the highest accumulation of litter occurs in May–July (9–10.4 mg ha^{-1} yr^{-1}), and it decreases 54%–67% by the end of the wet season due to decomposition (Martínez-Yrízar and Sarukhán, 1993). However, litter decomposition depends on frequency of precipitation, and events ≥ 10 mm are positively correlated with decomposition rate (Anaya et al., 2012). Thus, decreases in precipitation due to climate change might promote litter accumulation, which not only protects the soil from temperature changes and creates safe sites for germination and establishment but also promotes occurrence of fire, which is increasing in TDF, thereby affecting forest regeneration (Stan and Sánchez-Azofeifa, 2019).

Seedling growth and survival

Seedlings from seeds that germinate at the beginning of the wet season in tropical forests have a higher probability of surviving than those from seeds that germinate in the middle or at the end of this season (Garwood, 1983; Marod et al., 2002). However, establishment success also depends on seedling growth rate. Species in the Chamela TDF with a high relative growth rate (RGR), for example, *Apoplanesia paniculata*, *Cordia alliodora*, and *Cochlospermum vitifolium*, grew rapidly during a short period of drought in the wet season, when water and solar irradiance were high (Huante and Rincón, 1997). Seedlings of species with intermediate RGR (e.g., *Caesalpinia eriostachys*) grew well at the beginning of the wet season and in variable levels of solar irradiance and lower water availability. On the other hand, seedlings of the shade-tolerant *Celaenodendron mexicanum* had a low RGR. In view of likely decreases in precipitation due to climate change, research is needed to determine the relationship between RGR and the percentage of seedlings that become established.

In addition to timing of germination in relation to the wet season, morphological and functional traits of seedlings also are important for survival during drought (Kitajima, 1994; Marod et al., 2002; Poorter and Markesteijn, 2008). For example, seedlings of more than half of the tree species in a Mexican TDF have reserve-storing cotyledons (Cortés-Flores et al., 2020), which can increase the probability of seedling survival since the reserves could help compensate for predation or slow rates of carbon fixation (photosynthesis) due to unpredictable droughts (Baraloto and Forget, 2007). It also is suggested that the loss of leaves and the presence of taproots, as in *Amburana cearensis* and *Spondias mombin* (Poorter and Markesteijn, 2008), or roots with modifications to store water (e.g., Burseraceae) may be important for the seedling establishment and tolerance of drought (Cortés-Flores et al., 2020). Thus, we might expect these drought-adapted species to become more abundant if the amount of precipitation decreases in TDF due to climate change. Another consideration is that precipitation events outside the wet season can promote germination, but seedlings die due to drought stress. A contrasting scenario that may result from climate change is an increase in amount of precipitation that can increase growth of fungi that kill seedlings (Marod et al., 2002).

At the beginning of the wet season, seeds of herbaceous species in TDF of Mexico can germinate in 6.2 ± 4.4 days, and germination of most seedlings is epigeal, with photosynthetic cotyledons that allow them to grow rapidly (Soriano et al., 2011; Cortés-Flores et al., 2020). Hence, predicted changes in the timing and amount of precipitation are expected to reduce relative recruitment success of seedlings (Ray and Brown, 1994; Morris et al., 2008; Ooi, 2012; Hudson et al., 2015) since they usually do not tolerate desiccation. In this sense, the time within the wet season when seeds germinate not only determines seedling survival, but it also can affect how well they grow. That is, early germination allows herbs to reach a larger size before reproduction and/or to lengthen their reproductive period. Thus, from a climate change perspective a decrease in duration of the wet season would negatively affect seedling survival and decrease seed production.

Soil seed bank

Reserves of viable, nongerminated seeds in the soil can play an important role in stabilizing species population dynamics in plant communities (del Cacho et al., 2012; Basto et al., 2018; Plue et al., 2021). Thus, the role of soil seed banks in maintaining plant communities affected by climate change is of considerable interest to plant ecologists (Ooi, 2012; Basto et al., 2018; Plue et al., 2021). Seed bank studies of TDF have been conducted in many countries, including Argentina (Lipoma et al., 2020), Brazil (Costa and Araújo, 2003; Mendes et al., 2015; Menezes et al., 2019; Souza et al., 2020), Costa Rica (Wijdeven and Kuzee, 2000), Ecuador (Jara-Guerrero et al., 2020), Ethiopia (Teketay and Granström, 1995; Reubens et al., 2007), Mexico (Rico-Gray and García-Franco, 1992; Meave et al., 2012; Álvarez-Aquino et al., 2014), Nicaragua (Uasuf et al., 2009), Sri Lanka (Madawala, et al., 2016), and Thailand (Marod et al., 2002; Chalermsri et al., 2020).

The soil seed bank of TDF usually does not contain seeds of very many tree species (Teketay and Granström, 1995; Uasuf et al., 2009; Álvarez-Aquino et al., 2014; Chalermsri et al., 2020), and the tree species found in seed banks are those that occur in the early stages of succession (Khurana and Singh, 2001; Ceccon et al., 2006). However, most seeds found in seed banks of TDF are those of the numerous herbaceous species, one of the most important growth forms in terms of composition and structure (Lott and Atkinson, 2006; Ibarra-Manríquez et al., 2021). In southern Mexico, the seed bank in TDF is mostly (80%) composed of herbaceous species of Asteraceae, Acanthaceae, Euphorbiaceae, and Poaceae (Álvarez-Aquino et al., 2014; Meave et al., 2012). The low density of seeds of tree species in the seed bank indicates that restoration of TDF cannot be supported only by the soil seed bank (Lemenih and Teketay, 2006; Uasuf et al., 2009; Álvarez-Aquino et al., 2014; Chalermsri et al., 2020). However, the seed bank is a source of propagules for regeneration of early successional tree species and herbaceous species of TDF (Vieira and Scariot, 2006; Madawala et al., 2016; Menezes et al., 2019).

It is likely that climate change will impact the soil seed banks of TDF in several ways. First, increased air temperature is correlated with increased soil temperature, which is predicted to decrease seed viability (Ooi et al., 2009, 2012) and/or promote germination (Ooi, 2012). In fire-prone woodlands in New South Wales Australia, Ooi et al. (2012) found that for each 1°C increase in air temperature soil temperature increased 1.5°C. These temperature increases significantly increased PY-break of seeds of Fabaceae. In the rocky Reserva Ecológica del Pedregal de San Ángel, Ciudad de México (México), soil temperature increased 0.76°C to 1.38°C with each 1°C increase in air temperature, depending on microsite (Peraza-Villarreal, Sánchez-Coronado, and Orozco-Segovia, unpublished data). In contrast, soil temperatures registered beneath litter in open (nonshaded) and understory sites the TDF had no relationship with mean air temperature (Soriano et al., 2014).

Second, changes in the pattern of precipitation in TDF could have an impact on seedling establishment and on survival of plants to maturity. If precipitation events and amounts are not favorable for replacement of plants of a particular species that have died in the TDF, there will be an overall decrease in the number of individuals that produce seeds. Thus, the number of seeds added to the seed bank each year will decrease. Without replenishment, the seed bank of a particular species could be depleted after several years of failed seedling establishment (Teketay and Granström, 1995). Lack of replenishment leads to low similarity between the seed bank and aboveground vegetation, which does occur in TDF (Uasuf et al., 2009; Chalermsri et al., 2020).

Third, changes in temperature and/or amount and timing of precipitation due to climate change could result in variation in seed traits of a species when they enter the seed bank. For example, seeds of 12 of 19 tree species in TDF in Mexico produced in a dry year (384 mm of rain during the wet season) had a higher concentration of nitrogen than those produced in a wet year (652 mm of rainfall during the wet season). Furthermore, dry mass of the seed coat (and presumably its thickness) was higher in seeds with high nitrogen than in those with low nitrogen (Soriano et al., 2011).

Fourth, germination in open- versus closed-canopy sites in TDF may provide some insight into germination responses of seeds in the soil seed bank to increased temperature and decreased soil moisture that may result from climate change. That is, the open-canopy sites in some respects would simulate climate change. For example, seeds of 18 TDF tree species buried 3 cm deep in open and closed sites in a TDF in Mexico for about 3 years had varying responses, depending on species and site of burial (Soriano et al., 2014). Fresh seeds of the 18 species were buried in mesh bags at a depth of 3 cm in soil in open and closed sites in the field for about 1 and 2 years. Prior to the first and second wet seasons, some seeds were exhumed and buried again at 3 cm in boxes of native soil in their respective open or closed sites. Seeds of seven of the 18 species had water-impermeable seed coats (PY), and some seeds of four of these seven species germinated in the field in the wet season of the year in which they were dispersed. Seeds of two of the four species germinated only in the open, and those of two species germinated equally well in the open and closed sites. In the second wet season following seed maturation, seeds of four of the seven species with PY germinated: one only in the open site, two equally well in open and closed sites, and one better in closed than in open sites (Soriano et al., 2014). These results suggest that increased temperature and decreased soil moisture due to climate change are not likely to eliminate most species with PY from the short-term persistent soil seed bank.

Regarding the 11 species with permeable seeds in the study by Soriano et al. (2014), seeds of 10 species germinated during the wet season of the year in which they were dispersed: five equally well in open and closed sites, three better in open, one better in closed, one only in open and one only in closed. In the wet season of the following year, seeds of only four species germinated: one equally well in open and closed sites, two only in open, and one better in open than closed. These results suggest that permeable seeds do not persist in the TDF seed bank as well as those with PY and that increased temperatures associated with open sites (and climate change) may promote germination and thus depletion of the seed bank. However, permeable and impermeable seeds of most of the species buried continuously did not germinate in the soil but remained viable, suggesting that in some microsites undisturbed soil would maintain a seed bank for more than 1 year. Soil movement by animals and runoff from rainfall (erosion) could expose seeds to conditions that promote germination.

A possible explanation for the relatively low number of seeds in the TDF soil seed bank is seed predation by insects during the dry season. Briones-Salas et al. (2006) carried out an animal-exclusion experiment on 10 species and showed that postdispersal removal (and presumably predation) by arthropods was higher for *Crescentia alata* (53.3%) than for the other species (0%–30.0%). In the nonexclusion treatments, seed removal was 75%–90% for all species except *Albizia occidentalis* (46.4%), *Amphipterygium adstringens* (12.2%), and *Guazuma ulmifolia* (38.0%). Barnes (2001) reported that in *Acacia erioloba* larvae of bruchid beetles damaged the embryo in a high proportion of the seeds. However, in seeds with nondamaged embryos, the exit holes made by the beetles allowed the water-impermeable seeds to imbibe and germinate. In any case, the seeds were not incorporated into the soil seed bank, suggesting that effects of climate change on seed predation could be an important factor in plant regeneration from seeds.

Community composition and shifts in species distribution

Changes in temperature and precipitation predicted by climate change will have different impacts on regeneration from seeds, depending on the species (Allen et al., 2017; Dantas et al., 2020). Variation in water availability has been considered the most important abiotic factor influencing the growth, reproduction, and evolution of TDF species (Murphy and Lugo, 1986; Becerra, 2005; De-Nova et al., 2012; Chaturvedi and Raghubanshi, 2018). However, both temperature and water availability could act as environmental filters for seed dormancy, germination, seedling recruitment and survival and growth to reproductive maturity (Allen et al., 2010, 2017; Walck et al., 2011; Cochrane, 2017; Venier et al., 2017), resulting in changes in population dynamics. Thus, some species will remain in a particular area, while others migrate or disappear. Consequently, major changes are expected in the composition, structure, and diversity of the TDF (Walck et al., 2011; Feeley et al., 2012; Ooi, 2012). Furthermore, it is predicted that climate change will cause a decrease in growth of some tree species of TDF (Feng et al., 2018), which could affect plant–plant interactions and lead to changes in community composition.

One of the results from predictive models is that the land area suitable for TDF will change due to climate change. For example, models predict that the future distribution of four of five species evaluated in TDF in southern Ecuador will decrease due to narrowing of their fundamental niche even under the optimistic RCP2.6 climate change scenario (Aguirre et al., 2017). Modeling for northeastern Brazil showed that habitat for endemic species of TDF would be reduced by 10% for trees and 13% for nonarboreal species (Silva et al., 2019).

Using ecological niche modeling for 16 tree species in the TDF of Brazil, Collevatti et al. (2013) obtained a potential distribution of each species at the end of the present century. Models predict that climatically suitable areas for all 16 species will expand southeast, negatively affecting their occurrence in protected areas of TDF. Prieto-Torres et al. (2016) modeled the ecological niche for 15 species of plants in Mexican TDF and concluded that the area in western Mexico favorable for TDF would decrease. However, for eastern Mexico they predicted that the area favorable for TDF would increase due to relocation to new areas at higher elevations, such as in the states of Tamaulipas and Veracruz. The models indicated that due to climate change TDF would shift upward about 200 m in elevation. Currently, TDF occur below 1500 m elevation and rarely at 1900 m (Rzedowski, 1978) and 2300 m (Zacarias-Eslava et al., 2011; Cornejo-Tenorio et al., 2013). Various studies support the possibility of migration of TDF to higher elevations. For example, this forest can form ecotones with temperate communities dominated by oak (*Quercus*) species (Zacarias-Eslava et al., 2011; Cornejo-Tenorio et al., 2013; Prieto-Torres and Rojas-Soto, 2016). Considering the predictions of climate change, seasonal humid forests would be the sites with climatic conditions similar to those in current TDF. Therefore TDF species that currently coexist in ecotones with other more humid seasonal forests (e.g., seasonal tropical moist forest) could migrate to the adjacent more humid forests.

Future research needs

More research is needed to increase our understanding of how climate change will impact regeneration of TDF species from seeds. Research on seed dormancy and germination and seedling establishment needs to be carried out under field conditions in each country where this forest occurs. These investigations must encompass the diversity of the growth forms found in these forests and not be restricted to the arboreal component. The resilience of TDF to climatic change may reside in seed and seedling functional diversity. Since predicted changes in temperature (increase) and water availability (decrease) will occur simultaneously, it is important to design experiments in which the response of species to these conditions can be studied separately and in combination. However, due to the high diversity of species and germination responses and the imminent climate change, it is advisable to evaluate germination in groups of species that share several traits related to seed germination and the establishment and survival of seedlings. To achieve this objective, available information on germination needs to be integrated with that for other regeneration traits (e.g., seed dispersal, seed size, seedling types) to further define functional groups. With this information and controlling for phylogenetic relations among taxa, it will be easier to define the species groups most vulnerable or resilient to climate change. Also, it is necessary to identify the germination variables that need to be evaluated in order to develop meaningful germination experiments that will facilitate the synthesis of information and the search for ecological patterns. Furthermore, a framework of information needs to be created that identifies species vulnerability to the main impacts of climate change, including water availability and temperature in a changing and heterogeneous environment. Such information will be helpful in developing sound management strategies and in promoting sustainable management, conservation, and restoration of TDF in the context of climate change.

Acknowledgments

We thank to Ma. Guadalupe Cornejo-Tenorio, Irma Acosta Calixto, Alejandro González Ponce de León, José Miguel Baltazar, Jimena Rey Loaiza, María Graciela García-Guzmán, Humberto Peraza-Villarreal, and Pedro Eloy Mendoza Hernández for technical support. Software support was provided by Laboratorio Nacional de Análisis y Síntesis Ecológica, Escuela Nacional de Estudios Superiores, Unidad Morelia, Universidad Nacional Autónoma de México.

References

Agarwala, M., DeFries, R.S., Qureshi, Q., Jhala, Y.V., 2016. Factors associated with long-term species composition in dry tropical forests of central India. Environ. Res. Lett. 11, 105008. Available from: https://doi.org/10.1088/1748-9326/11/10/105008.

Aguirre, N., Eguiguren, P., Maita, J., Ojeda, T., Sanamiego, N., Furniss, et al., 2017. Potential impacts to dry forest species distribution under two climate change scenarios in southern Ecuador. Neotrop. Biodivers. 3, 18−29.

Agustin, E.K., Wawangningrum, H., Wanda, I.F., 2018. Seed storage behavior and morphological characterization fruit and seed of pasat fruits (*Heynea trijuga*). Biodivers. Indonesia 4, 83−86.

Albuquerque, U.V., Araújo, E.L., El-Deir, A.C.A., de Lima, A.L.A., Souto, A., Bezerra, B.M., et al., 2012. Caatinga revisited: ecology and conservation of an important seasonal dry forest. Sci. World J. 2012, 205182.

Allen, C.D., Macalady, A.K., Chenchouni, H., Bachelet, D., McDowell, N., Vennetier, M., et al., 2010. A global overview of drought and heat-induced tree mortality reveals emerging climate change risks for forests. For. Ecol. Manage. 259, 660−684.

Allen, K., Dupuy, J.M., Gei, M.G., Hulshof, C., Medvigy, D., Pizano, C., et al., 2017. Will seasonally dry tropical forests be sensitive or resistant to future changes in rainfall regimes? Environ. Res. Lett. 12, 023001. Available from: https://doi.org/10.1088/1748-9326/aa5968.

Álvarez-Aquino, C., Barradas-Sánchez, L., Ponce-González, O., Williams-Linera, G., 2014. Soil seed bank, seed removal, and germination in a seasonally dry tropical forest in a seasonally dry tropical forest in Veracruz, Mexico. Bot. Sci. 92, 111−121.

Anaya, C.A., Jaramillo, V.J., Martínez-Yrízar, A., García-Oliva, F., 2012. Large rainfall pulses control litter decomposition in a tropical dry forest: evidence from an 8-year study. Ecosystems 15, 652−663.

Armenteras, D., Rodríguez-Eraso, N., 2014. Forest deforestation dynamics and drivers in Latin America: a review since 1990. Colombia For. 17, 233−246.

Apgaua, D.M.G., dos Santos, R.M., Pereira, D.G.S., de Oliveira Menino, G.C., Pires, G.G., Fontes, M.A.L., et al., 2014. Beta-diversity in seasonally dry tropical forests (SDTF) in the Caatinga Biogeographic Domain, Brazil, and its implications for conservation. Biodivers. Conserv. 23, 217−232.

Ashton, P., 2016. Towards a common understanding of the lowland deciduous forests of tropical Asia: south Asia and Indo-Burma compared. Nat. Hist. Bull. Siam. Soc. 61, 59−70.

Balvanera, P., Lott, E., Segura, G., Siebe, C., Islas, A., 2002. Patterns of β-diversity in a Mexican tropical dry forest. J. Veg. Sci. 13, 145−158.

Banda-R, K., Delgado-Salinas, A., Dexter, K.G., Linares-Palomino, R., Oliveira-Filho, A., Prado, D., et al., 2016. Plant diversity patterns in neotropical dry forests and their conservation implications. Science 353, 1383−1387.

Baraloto, C., Forget, P.M., 2007. Seed size, seedling morphology, and response to deep shade and damage in neotropical rain forest trees. Am. J. Bot. 94, 901−911.

Barnes, M.E., 2001. Seed predation, germination and seedling establishment of *Acacia erioloba* in northern Botswana. J. Arid Environ. 49, 541−554.

Baskin, C.C., Baskin, J.M., 2014. Seeds: Ecology, Biogeography, and Evolution of Dormancy and Germination, Second ed. Academic Press/Elsevier, San Diego.

Basto, S., Thompson, K., Grime, J.P., Fridley, J.D., Calhim, S., Askew, A.P., et al., 2018. Severe effects of long-term drought on calcareous grassland seed banks. Clim. Atmos. Sci. 1, 1. Available from: https://doi.org/10.1038/s41612-017-0007-3.

Bawa, K.S., 1990. Plant-pollinator interactions in tropical rain forests. Annu. Rev. Ecol. Syst. 21, 399−422.

Becerra, J.X., 2005. Timing the origin and expansion of the Mexican tropical dry forest. Proc. Natl. Acad. Sci. USA 102, 10919−10923.

Becknell, J.M., Kissing, L.K., Powers, J.S., 2012. Aboveground biomass in mature and secondary seasonally dry tropical forests: a literature review and global synthesis. For. Ecol. Manage. 276, 88−95.

Bhadouria, R., Sing, R., Srivastava, P., Raghubanshi, A.S., 2016. Understanding the ecology of tree-seedling growth in dry tropical environment: a management perspective. Energ. Ecol. Environ. 1, 296−309.

Blakesley, D., Anusarnsunthorn, V., Kerby, J., Navakitbumrung, P., Kuarak, C., Zangkum, S., et al., 2000. Nursery technology and tree species selection for restoring forest biodiversity in northern Thailand. In: Elliott, S., Kerby, J., Blakesley, D., Hardwick, K., Woods, K., Anusarnsunthorn, V. (Eds.), Forest Restoration for Wildlife Conservation. International Tropical Timber Organization and The Forest Restoration Research Unit. Chiang Mai University, Thailand, pp. 207−220.

Blakesley, D., Elliott, S., Kuarak, C., Navakitbumrung, P., Zangkum, S., Anusarnsunthorn, V., 2002. Propagating framework tree species to restore seasonally dry tropical forest: implications of seasonal seed dispersal and dormancy. For. Ecol. Manage. 164, 31−38.

Boko, M., Niang, I., Nyong, A., Vogel, C., Githeko, A., Medany, M., et al., 2007. Africa. Climate change 2007: adaptation and vulnerability. In: Parry, M.L., Canziani, O.F., Palutikof, J.P., van der Linden, P.J., Hanson, C.E. (Eds.), Working Group II to the Fourth Assessment Report of the Intergrovernmental Panel on Climate Change. Cambridge University Press, Cambridge, pp. 433−467.

Bowman, D.M.J.S., Prior, L.D., 2005. Why do evergreen trees dominate the Australian seasonal tropics? Aust. J. Bot. 53, 379−399.

Bradford, J.B., Schlaepfer, D.R., Lauenroth, W.K., Palmquist, K.A., 2020. Robust ecological drought projections for drylands in the 21st century. Glob. Change Biol. 26, 3906–3919.

Briones-Salas, M., Sánchez-Cordero, V., Sánchez-Rojas, G., 2006. Multi-species fruit and seed removal in a tropical deciduous forest in Mexico. Botany 84, 433–442.

Cao, S., Sanchez-Azofeifa, A., 2017. Modeling seasonal surface temperature variations in secondary tropical dry forests. Int. J. Appl. Earth Observ. Geoinform. 62, 122–134.

Ceccon, E., Huante, P., Rincón, E., 2006. Abiotic factors influencing tropical dry forests regeneration. Brazil. Arch. Biol. Technol. 49, 305–312.

Chalermsri, A., Ampornpan, L.-A., Purahong, W., 2020. Seed rain, soil seed bank, and seedling emergence indicate limited potential for self-recovery in a highly disturbed tropical, mixed deciduous forest. Plants 9, 1391. Available from: https://doi.org/10.3390/plants9101391.

Chaturvedi, R.K., Raghubanshi, A.S., 2018. Soil water availability influences major ecosystem processes in tropical dry forest. Int. J. Hydrol. 2, 14–15.

Cobert, S.A., Plumridge, J.R., 1985. Hydrodynamics and the germination of oil-seed rape pollen. J. Agric. Sci. 104, 445–451.

Cochrane, A., 2017. Modelling seed germination response to temperature in *Eucalyptus* L'Her. (Myrtaceae) species in the context of global warming. Seed Sci. Res. 27, 99–109.

Collevatti, R., Lima-Ribeiro, M., Diniz-Filho, J., Oliveira, G., Dobrovolski, R., Terribile, L., 2013. Stability of Brazilian seasonally dry forests under climate change: inferences for long-term conservation. Am. J. Plant Sci. 4, 792–805.

Cornejo-Tenorio, G., Sánchez-García, E., Flores-Tolentino, M., Santana-Michel, F.J., Ibarra-Manríquez, G., 2013. Estudio florístico del cerro El Águila, Michoacán, México. Bot. Sci. 91, 155–180. Available from: https://doi.org/10.17129/botsci.411.

Cortés-Flores, J., Cornejo-Tenorio, G., Sánchez-Coronado, M.E., Orozco-Segovia, A., Ibarra-Manríquez, G., 2020. Disentangling the influence of ecological and historical factors on seed germination and seedling types in a Neotropical dry forest. PLoS One 15, e0231526. Available from: https://doi.org/10.1371/journal.pone.0231526.

Cortés-Flores, J., Hernández-Esquivel, K.B., González-Rodríguez, A., Ibarra-Manríquez, G., 2017. Flowering phenology, growth forms and pollination syndromes in a tropical dry forest species: influence of phylogeny and abiotic factors. Am. J. Bot. 104, 39–49.

Costa, R.C., Araújo, F.S., 2003. Densidade, germinação e flora do banco de sementes no solo, no final da estação seca, em uma área de caatinga, Quixadá, CE. Acta Bot. Bras. 17, 259–264.

Dantas, B.F., Moura, M.S.B., Pelacani, C.R., Angelotti, F., Taura, T.A., Oliveira, G.M., et al., 2020. Rainfall, not soil temperature, will limit the seed germination of dry forest species with climate change. Oecologia 192, 529–541.

del Cacho, M., Saura-Mas, S., Estiarte, M., Peñuelas, Lloret, F., 2012. Effect of experimentally induced climate change on the seed bank of a Mediterranean shrubland. J. Veg. Sci. 23, 280–291.

De-Nova, J.A., Medina, R., Montero, J.C., Weeks, A., Rosell, J.A., Olson, M.E., et al., 2012. Insights into the historical construction of species-rich Mesoamerican seasonally dry tropical forests: the diversification of *Bursera* (Burseraceae, Sapindales). New Phytol. 193, 276–287.

Dexter, K.G., Pennington, R.T., Oliveira-Filho, A.T., Bueno, M.L., Miranda, P.L.S., Bueno, M.L., 2018. Inserting tropical dry forests into the discussion on biome transitions in the tropics. Front. Ecol. Evol. 6, 104. Available from: https://doi.org/10.3389/fevo.2018.00104.

Djoudi, H., Vergles, E., Blackie, R.R., Koame, C.K., Gauthier, D., 2015. Dry forests, livelihoods and poverty alleviation: understanding currents trends. Int. For. Rev. 17, 54–69.

Donohue, R., Rubio de Cases, R., Burghardt, L., Kovach, K., Willis, C.G., 2010. Germination, postgermination adaptation, and species ecological ranges. Annu. Rev. Ecol. Evol. Syst. 41, 293–319.

Ejsmond, M.J., Ejsmond, A., Banasiak, Ł., Karpińska-Kołaczek, M., Kozłowski, J., Kołaczek, P., 2015. Large pollen at high temperature: an adaptation to increased competition on the stigma? Plant Ecol. 216, 1407–1417.

Elliott, S., Kuarak, C., Navakitbumrung, P., Zangkum, S., Anusarnsunthorn, V., Blakesley, D., 2002. Propagating framework trees to restore seasonally dry tropical forest in northern Thailand. New For. 23, 63–70.

FAO (Food and Agriculture Organization of the United Nations), 2012. Global ecological zones for FAO forest reporting: 2010. Forest Resources Assessment Working Paper 179. Update. Rome. <http://www.fao.org/3/ap861e/ap861e00.pdf>.

FAO (Food and Agriculture Organization of the United Nations), 2020. Forest resources assessment programme. <http://www.fao.org/3/ad652e/ad652e05.htm#TopOfPage>.

Feeley, K.J., Rehm, E.M., Machovina, B., 2012. Perspective: the responses of tropical forest species to global climate change: acclimate, adapt, migrate, or go extinct? Front. Biogeogr. 4, 69–84.

Feng, X., Uriarte, M., González, G., Reed, S., Thompson, J., Zimmerman, J.K., et al., 2018. Improving predictions of tropical forest response to climate change through integration of field studies and ecosystem modeling. Glob. Change Biol. 24, e213–e232.

Gallardo-Cruz, J.A., Pérez-García, E.A., Meave, J.A., 2009. β-Diversity and vegetation structure as influenced by slope aspect and altitude in a seasonally dry tropical landscape. Landsc. Ecol. 24, 473–482.

García, E.G., Di Stéfano, J.F., 2000. Temperatura y germinación de las semillas de *Dalbergia retusa* (Papilionaceae), árbol en peligro de extinción. Rev. Biol. Trop. 48, 43–45.

Garwood, N.C., 1983. Seed germination in a seasonal tropical forest in Panama: a community study. Ecol. Monogr. 53, 159–181.

Gentry, A.H., 1995. Diversity and floristic composition of neotropical dry forests. In: Bullock, S.H., Mooney, H.A., Medina, E. (Eds.), Seasonally Dry Tropical Forests. Cambridge University Press, Cambridge, pp. 146–194.

Ghosh, R.C., Mathur, N.K., Singh, R.P., 1976. *Diospyros melanoxylon* - its problems and cultivation. Indian. For. 102, 326–336.

Godinez-Alvarez, A., Valiente-Banuet, A., 1998. Germination and early seedling growth of Tehuacan Valley cacti species: the role of soils and seed ingestion by dispersers on seedling growth. J. Arid Environ. 38, 21–31.

Gomes, D.T., Kretschmer Jr., A.E., 1978. Effect of three temperature regimes on tropical legume seed germination. Soil Crop. Sci. Soc. Fla. Proc. 37, 61–63.

Gomes, S.E.V., Oliveira, G.M., Araujo, M.N., Seal, C.E., Dantas, B.F., 2019. Influence of current and future climate on the seed germination of *Cenostigma microphyllum* (Mart. ex G. Don) E. Gagnon & G.P. Lewis. Folia Geobot. 54, 19–28.

Gouge, D.H., Smith, K.A., Lee, L.L., Henneberry, T.J., 2000. Effect of soil depth and moisture on the vertical distribution of *Steinernema riobrave* (Nematoda: Steinernematidae). J. Nematol. 32, 223–228.

Griz, L.M.S., Machado, I.C.S., 2001. Fruiting phenology and seed dispersal syndromes in Caatinga, a tropical dry forest in the northeast of Brazil. J. Trop. Ecol. 17, 303–321.

Gutterman, Y., 2000. Maternal effects on seeds during development. In: Fenner, M. (Ed.), Seeds: The ecology of Regeneration in Plant Communities. CABI Publishing, Wallingford, pp. 59–84.

Hasnat, G.N.T., 2020. Climate change effects, adaptation, and mitigation techniques in tropical dry forests. In: Bhadouria, R., Tripathi, S., Srivastava, P., Singh, P. (Eds.), Handbook of Research on the Conservation and Restoration of Tropical Dry Forests. IGI Global, Hershey, PA, pp. 42–64.

Hennessy, K., Fitzharris, B., Bates, B.C.,, Harvey, N., Howden, S.M., Hughes, L., et al., 2007. Australia and New Zealand. climate change 2007: adaptation and vulnerability. In: Parry, M.L., Canziani, O.F., Palutikof, J.P., van der Linden, P.J., Hanson, C.E. (Eds.), Working Group II to the Fourth Assessment Report of the Intergrovermental Panel on Climate Change. Cambridge University Press, Cambridge, pp. 507–540.

Huante, P., Rincón, E., 1997. Responses to light changes in tropical deciduous woody seedlings with contrasting growth rates. Oecologia 113, 53–66.

Hudson, A.R., Ayre, D.J., Ooi, M.K., 2015. Physical dormancy in a changing climate. Seed Sci. Res. 25, 66–81.

Hulshof, C.M., Martínez-Yrízar, A., Burquez, A., Boyle, B., Enquist, B.J., 2013. Plant functional trait variation in tropical dry forests: a review and synthesis. In: Sánchez-Azofeifa, A., Powers, J.S., Fernandes, G.W., Quesada, M. (Eds.), Tropical Dry Forests in the Americas: Ecology, Conservation, and Management. CRC Press, Boca Raton, pp. 129–140.

Ibarra-Manríquez, G., Carrillo-Reyes, P., Rendón-Sandoval, F.J., Cornejo-Tenorio, G., 2015. Diversity and distribution of lianas in Mexico. In: Schnitzer, S.A., Bongers, F., Putz, F., Burnham, R. (Eds.), Ecology of Lianas. Wiley-Blackwell, Oxford, pp. 91–103.

Ibarra-Manríquez, G., Cornejo-Tenorio, G., Hernández-Esquivel, K.B., Rojas-López, M., Sánchez-Sánchez, L., 2021. Vegetación y flora vascular del ejido Llano de Ojo de Agua, Depresión del Balsas, municipio de Churumuco, Michoacán, México. Rev. Mex. Biodivers. 92, e923482. Available from: https://doi.org/10.22201/ib.20078706e.2021.92.3482.

IPBES (Intergovernmental Science-Policy Platform on Biodiversity and Ecosystem Services), 2019. In: Díaz, S.J., Settele, E.S., Brondízio, E.S., Ngo, H.T., Guèze, M., Agard, J. (Eds.), Summary for Policymakers of the Global Assessment Report on Biodiversity and Ecosystem Services of the Intergovernmental Science-Policy Platform on Biodiversity and Ecosystem Services. IPBES Secretariat, Bonn, p. 56.

IPCC (Intergovernmental Panel on Climate Change), 2007. Climate Change 2007: Impacts, Adaptation and Vulnerability. Working Group II Contribution to the Intergovernmental Panel on Climate Change Fourth Assessment Report Summary for Policymakers. IPCC, Brussels.

IPCC (Intergovernmental Panel on Climate Change)., 2014. Climate Change 2014: Synthesis Report. Contribution of Working Groups I, II and III to the Fifth Assessment Report of the Intergovernmental Panel on Climate Change [Core Writing Team, R.K. Pachauri and L.A. Meyer (Eds.)], IPCC, Geneva.

Janzen, D.H., 1988. Tropical dry forests. The most endangered major forest ecosystem. In: Wilson, E.O. (Ed.), Biodiversity. National Academy Press, Washington, DC, pp. 130–137.

Jara-Guerrero, A., Espinosa, C.I., Méndez, M., De la Cruz, M., Escudero, A., 2020. Dispersal syndrome influences the match between seed rain and soil seed bank of woody species in a Neotropical dry forest. J. Veg. Sci. 31, 995–1005.

Khurana, E., Singh, J.S., 2001. Ecology of seed and seedling growth for conservation and restoration of tropical dry forest: a review. Environ. Conserv. 28, 39–52.

Khurana, E., Singh, J.S., 2004. Germination and seedling growth of five tree species from tropical dry forest in relation to water stress: impact of seed size. J. Trop. Ecol. 20, 385–396.

Kitajima, K., 1994. Relative importance of photosynthetic traits and allocation patterns as correlates of seedling shade tolerance of 13 tropical trees. Oecologia 98, 419–428.

Klemens, J.A., Deacon, N.J., Cavender-Bares, J., 2011. Pasture recolonization by a tropical oak and the regeneration ecology of seasonally dry tropical forests. In: Dirzo, R., Young, H.S., Mooney, H.A., Ceballos, G. (Eds.), Seasonally Dry Tropical Forests. Island Press, Washington, DC, pp. 221–237.

Lebrija-Trejos, E., Pérez-García, E.A., Meave, J.A., Bongers, F., Poorter, L., 2010. Functional traits and environmental filtering drive community assembly in a species-rich tropical system. Ecology 91, 386–398.

Lemenih, M., Teketay, D., 2006. Changes in soil seed bank composition and density following deforestation and subsequent cultivation of a tropical dry Afromontane forest in Ethiopia. Trop. Ecol. 47, 1–12.

Liengsiri, C., Hellum, A.K., 1988. Effects of temperature on seed germination in *Pterocarpus macrocarpus*. J. Seed Technol. 12, 66–75.

Lipoma, M.L., Fortunato, V., Enrico, L., Díaz, S., 2020. Where does the forest come back from? Soil and litter seed banks and the juvenile bank as sources of vegetation resilience in a semiarid Neotropical forest. J. Veg. Sci. 31, 1017–1027.

Lott, E.J., Atkinson, T.H., 2006. Mexican and Central American seasonally dry tropical forests: Chamela-Cuixmala, Jalisco, as a focal point for comparison. In: Pennington, T., Lewis, G.P., Ratter, J.A. (Eds.), Neotropical Savannas and Seasonally Dry Forests. Plant Diversity, Biogeography, and Conservation. CRC Press, Boca Raton, pp. 315–342.

Madawala, H.M.S.P., Ekanayake, S.K., Perera, G.A.D.,, 2016. Diversity, composition and richness of soil seed banks in different forest communities at Dotalugala Man and Biosphere Reserve, Sri Lanka. Ceylon J. Sci. 45, 43–55.

Margin, G., Gary Garcia, C., Cruz Choque, D., Giménez, J.C., Moreno, A.R., Nagy, G.J., et al., 2007. Latin America. Climate Change 2007: adaptation and vulnerability. In: Parry, M.L., Canziani, O.F., Palutikof, J.P., van der Linden, P.J., Hanson, C.E. (Eds.), Working Group II to the Fourth Assessment Report of the Intergovernmental Panel on Climate Change. Cambridge University Press, Cambridge, pp. 581–615.

Marod, D., Kutintara, U., Tanaka, H., Nakashizuka, T., 2002. The effects of drought and fire on seed and seedling dynamics in a tropical seasonal forest in Thailand. Plant Ecol. 161, 41–57.

Martínez-Yrízar, A., Sarukhán, J., 1993. Cambios estacionales del mantillo en el suelo de un bosque tropical caducifolio y uno subcaducifolio en Chamela, Jalisco, México. Acta Bot. Mex. 21, 1–6.

Maza-Villalobos, S., Poorter, L., Martínez-Ramos, M., 2013. Effects of ENSO and temporal rainfall variation on the dynamics of successional communities in old-field succession of a tropical dry forest. PLoS One 8, 008204. Available from: https://doi.org/10.1371/journal.pone.0082040.

Meave, J.A., Flores-Rodríguez, C., Pérez-García, E.A., Romero-Romero, M.A., 2012. Edaphic and seasonal heterogeneity of seed banks in agricultural fields of a tropical dry forest region in southern Mexico. Bot. Sci. 90, 313–329.

Meir, P., Pennington, R.T., 2011. Climate change and seasonally dry tropical forests. In: Dirzo, R., Young, H.S., Mooney, H.A., Ceballos, G. (Eds.), Seasonally Dry Tropical Forests: Ecology and Conservation. Island Press, Washington, DC, pp. 279–299.

Mendes, L.B., da Silva, K.A., dos Santos, D.M., dos Santos, J.M.F.F., Albuquerque, U.P., Araújo, E.L., 2015. What happens to the soil seed bank 17 years after clear cutting of vegetations? Rev. Biol. Trop. 63, 321–332.

Méndez-Toribio, M., Ibarra-Manríquez, G., Paz, H., Lebrija-Trejos, E., 2020. Atmospheric and soil drought risks combined shape community assembly of trees in a tropical dry forest. J. Ecol. 108, 1347–1357.

Menezes, J.C., Neto, O.C.C., Azevedo, I.F.P., Machado, A.O., Nunes, Y.R.F., 2019. Soil seed bank at different depths and light conditions in a dry forest in northern Minas Gerais. Flor. Amb. 26, e20170314. Available from: https://doi.org/10.1590/2179-8087.031417.

Miles, L., Newton, A.C., DeFries, R.S., Ravilious, C., May, I., Blyth, S., et al., 2006. A global overview of the conservation status of tropical dry forests. J. Biogeogr. 33, 491–505.

Morris, W.F., Pfister, C.A., Tuljapurkar, S., Haridas, C.V., Boggs, C.L., Boyce, M.S., et al., 2008. Longevity can buffer plant and animal populations against changing climatic variability. Ecology 89, 19–25.

Mu, J., Peng, Y., Xi, X., Wu, X., Li, G., Niklas, K.J., et al., 2015. Artificial asymmetric warming reduces nectar yield in a Tibetan alpine species of Asteraceae. Ann. Bot. 116, 899–906.

Murphy, P.G., Lugo, A.E., 1986. Ecology of tropical dry forest. Annu. Rev. Ecol. Syst. 17, 67–88.

Oliveira, G.M., Silva, F.F.S., Araujo, M.N., Costa, D.C.C., Gomes, S.E.V., Matias, J.R., et al., 2019. Environmental stress, future climate, and germination of *Myracrodruon urundeuva* seeds. J. Seed Sci. 41, 32–43.

Ooi, M., 2012. Seed bank persistence and climate change. Seed Sci. Res. 22, S53–S60.

Ooi, M.K., Auld, T.D., Denham, A.J., 2009. Climate change and bet-hedging: interactions between increased soil temperatures and seed bank persistence. Glob. Change Biol. 15, 2375–2386.

Ooi, M.K., Auld, T.D., Denham, A.J., 2012. Projected soil temperature increase and seed dormancy response along an altitudinal gradient: implications for seed bank persistence under climate change. Plant Soil 353, 289–303.

Orozco-Segovia, A., Sánchez-Coronado, M.E., 2009. Functional diversity in seeds and its implications for ecosystem functionality and restoration ecology. In: Gamboa de Buen, A., Orozco-Segovia, A., Cruz-García, F. (Eds.), Functional Diversity of Plant Reproduction. Research Signpost, Kerala, pp. 175–216.

Parmesan, C., Hanley, M.E., 2015. Plants and climate change: complexities and surprises. Ann. Bot. 116, 849–864.

Pennington, R.T., Lewis, G.P., Ratter, J.A., 2006. An overview of the plant diversity, biogeography and conservation of neotropical savannas and seasonally dry forests. In: Pennington, T., Lewis, G.P., Ratter, J.A. (Eds.), Neotropical Savannas and Seasonally Dry Forests. Plant Diversity, Biogeography, and Conservation. CRC Press, Boca Raton, pp. 1–29.

Pennington, R.T., Lavin, M., Oliveira-Filho, A., 2009. Woody plant diversity, evolution, and ecology in the tropics: perspectives from seasonally dry tropical forests. Annu. Rev. Ecol. Evol. Syst. 40, 437–457.

Pennington, R.T., Lehmann, C.E.R., Rowland, L.M., 2018. Tropical savannas and dry forests. Curr. Biol. 38, R527–R548.

Plue, J., Van Calster, H., Auestand, I., Basto, S., Bekker, R.M., Bruun, H.H., et al., 2021. Buffering effects of soil seed banks on plant community composition in response to land use and climate. Glob. Ecol. Biogeogr. 30, 128–139.

Poorter, L., Markesteijn, L., 2008. Seedling traits determine drought tolerance of tropical tree species. Biotropica 40, 321–331.

Portillo-Quintero, C.A., Sánchez-Azofeifa, G.A., 2010. Extent and conservation of tropical dry forests in the Americas. Biol. Conserv. 143, 144–155.

Portillo-Quintero, C., Sánchez-Azofeifa, G.A., Calvo-Alvarado, J., Quesada, M., Santo, M.M.E., 2015. The role of tropical dry forests for biodiversity, carbon and water conservation in the neotropics: lessons learned and opportunities for its sustainable management. Reg. Environ. Change 15, 1039–1049.

Prieto-Torres, D.A., Rojas-Soto, O.R., 2016. Reconstructing the Mexican tropical dry forests via an autoecological niche approach: reconsidering the ecosystem boundaries. PLoS One 11, e0150932. Available from: https://doi.org/10.1371/journal.pone.0150932.

Prieto-Torres, D.A., Navarro-Sigüenza, A.G., Santiago-Alarcon, D., Rojas-Soto, O.R., 2016. Response of the endangered tropical dry forest to climate change and the role of Mexican protected areas for their conservation. Glob. Change Biol. 22, 364–379.

Pritchard, H.W., Daws, M.I., Fletcher, B.J., Gaméné, C.S., Msanga, H.P., Omondi, W., 2004. Ecological correlates of seed desiccation tolerance in tropical African dryland trees. Am. J. Bot. 91, 863–870.

Quesada, M., Rosas, F., Aguilar, R., Ashworth, L., Rosas-Guerrero, V.M., Sayago, R., et al., 2011. Human impacts on pollination, reproduction, and breeding systems in tropical forest plants. In: Dirzo, R., Young, H.S., Mooney, H.A., Ceballos, G. (Eds.), Seasonally Dry Tropical Forests: Ecology and Conservation. Island Press, Washington, DC, pp. 173–194.

Quesada, M., Sánchez-Azofeifa, G.A., Alvarez-Añorve, M., Stoner, K.E., Avila-Cabadilla, L., Calvo-Alvarado, J., et al., 2009. Succession and management of tropical dry forests in the Americas: review and new perspectives. For. Ecol. Manage. 258, 1014–1024.

Ray, G.J., Brown, B.J., 1994. Seed ecology of woody species in a Caribbean dry forest. Restor. Ecol. 2, 156–163.

RBGK (Royal Botanic Gardens, Kew), 2021. <https://data.kew.org/sid>. January 13, 2021.

Reubens, B., Gebrehiwot, K., Hermy, M., Muys, B., 2007. Persistent soil seed banks for natural rehabilitation of dry tropical forests in northern Ethiopia. Tropicultura 25, 204–214.

Richards, P.W., 1957. The Tropical Rainforest: An Ecological Study. The University Press, Cambridge.

Rico-Gray, V., García-Franco, J.G., 1992. Vegetation and soil seed bank of successional stages in tropical lowland deciduous forest. J. Veg. Sci. 3, 617–624.

Rodrigues, P.M.S., Silva, J.O., Eisenlohr, P.V., Schaefer, C.E.G.R., 2015. Climate change effects on the geographic distribution of specialist tree species of the Brazilian tropical dry forests. Braz. J. Biol. 75, 679–684.

Rzedowski, J., 1978. Vegetación de México. Limusa, México.

Sánchez-Azofeifa, G.A., Quesada, M., Rodríguez, J.P., Nassar, J.M., Stoner, K.E., Castillo, A., et al., 2005. Research priorities for neotropical dry forests. Biotropica 37, 477–485.

Sentinella, A.T., Warton, D.I., Sherwin, W.B., Offord, C.A., Moles, A.T., 2020. Tropical plants do not have narrower temperature tolerances, but are more at risk from warming because they are close to their upper thermal limits. Glob. Ecol. Biogeogr. 29, 1387–1398.

Shrestha, M., Garcia, J.E., Bukovac, Z., Dorin, A., Dyer, A.G., 2018. Pollination in a new climate: assessing the potential influence of flower temperature variation on insect pollinator behaviour. PLoS One 13, e0200549. Available from: https://doi.org/10.1371/journal.pone.0200549.

Shumba, E., Chidumayo, E., Gumbo, D., Kambole, C., Chishaleshale, M., 2010. Biodiversity of plants. In: Chidumayo, E.N., Gumbo, D.J. (Eds.), The Dry Forests and Woodlands of Africa. Managing for Products and Services. Earthscan, Washington, DC, pp. 43–61.

Silva, J.L.S., Cruz-Neto, O., Peres, C.A., Tabarelli, M., Lopes, A.V., 2019. Climate change will reduce suitable Caatinga dry forest habitat for endemic plants with disproportionate impacts on specialized reproductive strategies. PLoS One 14, e0217028. Available from: https://doi.org/10.1371/journal.pone.0217028.

Singh, K.P., Kushwaha, C.P., 2005. Emerging paradigms of tree phenology in dry tropics. Curr. Sci. 89, 964–975.

Singh, J.S., Singh, V.K., 1992. Phenology of seasonally dry tropical forest. Curr. Sci. 63, 684–689.

Siyum, Z.G., 2020. Tropical dry forest dynamics in the context of climate change: syntheses of drivers, gaps, and management perspectives. Ecol. Process. 9, 25. Available from: https://doi.org/10.1186/s13717-020-00229-6.

Slot, M., Winter, K., 2017. Photosynthetic acclimation to warming in tropical forest tree seedlings. J. Exp. Bot. 68, 2275–2284.

Soriano, D., Orozco-Segovia, A., Márquez-Guzmán, J., Kitajima, K., Gamboa-de Buen, A., Huante, P., 2011. Seed reserve composition in 19 tree species of a tropical deciduous forest in Mexico and its relationship to seed germination and seedling growth. Ann. Bot. 107, 939–951.

Soriano, D., Huante, P., Gamboa-deBuen, A., Orozco-Segovia, A., 2014. Effects of burial and storage on germination and seed reserves of 18 tree species in a tropical deciduous forest in Mexico. Oecologia 174, 33–44.

Souza, J.D., Aguiar, B.A.S., Santos, D.M., Araujo, V.K.R., Simões, J.A., Andrade, J.R., et al., 2020. Dynamics in the emergence of dormant and nondormant herbaceous species from the soil seed bank from a Brazilian dry forest. J. Plant Ecol. 13, 256–265.

Stan, K., Sánchez-Azofeifa, A., 2019. Tropical dry forest diversity, climatic response, and resilience in a changing climate. Forests 10, 443. Available from: https://doi.org/10.3390/f10050443.

Taiz, L., Zeiger, E., 1998. Plant Physiology, Second ed. Sinauer Associates, Sunderland, MA.

Teketay, D., Granström, A., 1995. Soil seed banks in dry Afromontane forests of Ethiopia. J. Veg. Sci. 6, 777–786.

Trejo, I., Dirzo, R., 2000. Deforestation of seasonally dry tropical forest: a national and local analysis in Mexico. Biol. Conserv. 94, 133–142.

Trejo, I., Dirzo, R., 2002. Floristic diverity of Mexican seasonally dry tropical forests. Biodivers. Conserv. 11, 2063–2084.

Tweddle, J.C., Dickie, J.B., Baskin, C.C., Baskin, J.M., 2003. Ecological aspects of seed desiccation sensitivity. J. Ecol. 91, 294–304.

Uasuf, A., Tigabu, M., Odén, P.C., 2009. Soil seed banks and regeneration of neotropical dry deciduous and gallery forests in Nicaragua. Bois For. Trop. 299, 49–62.

Venier, P., Cabido, M., Funes, G., 2017. Germination characteristics of five coexisting neotropical species of *Acacia* in seasonally dry Chaco forests in Argentina. Plant Species Biol. 32, 134–146.

Vieira, D.L.M., Scariot, A., 2006. Principles of natural regeneration of tropical dry forests for restoration. Restor. Ecol. 14, 11–20.

Walck, J.L., Hidayati, S.N., Dixon, K.W., Thompson, K., Poschlod, P., 2011. Climate change and plant regeneration from seed. Glob. Change Biol. 17, 2145–2161.

Watanabe, M.E., 2014. Pollinators at risk: human activities threaten key species. BioScience 64, 5–10.

Wijdeven, S.M.J., Kuzee, M.E., 2000. Seed availability as a limiting factor in forest recovery processes in Costa Rica. Restor. Ecol. 8, 414–424.

WMO (World Meteorological Organization), 2019. WMO provisional statement on the state of the global climate in 2019. <https://public.wmo.int/en/resources/library/wmo-provisional-statement-state-of-global-climate-2019>.

Wright, S.J., 2005. Tropical forests in a changing environment. Trends Ecol. Evol. 20, 553–560.

Zacarias-Eslava, L.E., Cornejo-Tenorio, G., Cortés-Flores, J., González-Castañeda, N., Ibarra-Manríquez, G., 2011. Composición, estructura y diversidad del cerro El Águila, Michoacán, México. Rev. Mex. Biodivers. 82, 854–869.

Chapter 14

Regeneration from seeds in South American savannas, in particular the Brazilian Cerrado

L. Felipe Daibes[1], Carlos A. Ordóñez-Parra[2], Roberta L.C. Dayrell[3], and Fernando A.O. Silveira[2]

[1]*Departamento de Biodiversidade, Instituto de Biociências, Universidade Estadual Paulista (Unesp), Rio Claro, Brazil,* [2]*Center for Ecological Synthesis and Conservation, Department of Genetics, Ecology and Evolution, Universidade Federal de Minas Gerais, Belo Horizonte, Brazil,* [3]*Faculty of Biology and Preclinical Medicine, University of Regensburg, Universitätsstraße 31, Regensburg, Germany*

Introduction

South American savannas are fire-prone ecosystems dominated by a continuous herbaceous layer and seasonal climates (Bond, 2019). They contain a large fraction of the rich biodiversity of Latin America and are threatened by human activities, including extensive agriculture, livestock production, and industry development (Salazar et al., 2015; Velazco et al., 2019). Savannas cover large parts of the Neotropics (Borghetti et al., 2020) and include the Cerrado in central Brazil, Bolivia, and Paraguay (Strassburg et al., 2017), the Beni Savanna in northeastern Bolivia (Larrea-Alcázar et al., 2011), the Llanos in eastern Colombia and Venezuela (Huber et al., 2006), and scattered patches of Amazonian savannas (de Carvalho and Mustin, 2017; Devecchi et al., 2020) (Fig. 14.1). We focus on the effects of climate change on the Brazilian Cerrado, for which the most data on regeneration from seeds are available.

Our goal is to synthesize the available evidence on how climate change will affect regeneration from seeds in the Cerrado. For this purpose, we focus on the effect of three environmental cues known to affect seed germination and dormancy in this ecosystem: temperature, fire, and water availability. First, we describe the Cerrado vegetation and discuss how climate change is expected to alter these key environmental cues. Then, we provide an overview of the current knowledge on the germination ecology of Cerrado species in response to the three environmental drivers and discuss how predicted changes might cascade into shifts in species distribution, population dynamics, and community composition. We conclude by providing a list of key research areas that need to be addressed to enhance our predictive ability on the effects of climate change on the regeneration of plants from seeds in the Cerrado.

The Cerrado vegetation

The Cerrado is the largest Neotropical savanna, covering more than 2 million km^2, and most of it occurs in central Brazil (Oliveira-Filho and Ratter, 2002). It extends over 20 degrees of latitude and at elevations ranging from 100 to almost 2000 m a.s.l. (Ribeiro and Walter, 2008; Borghetti et al., 2020). Cerrado savannas have assembled in the last 6–8 million years, coinciding with the expansion of C4 grasslands and reoccurrence of frequent fires (Beerling and Osborne, 2006; Simon et al., 2009), but significant expansions and retractions occurred during the Quaternary (Bueno et al., 2017). Such ancient savanna formations are classified as fire-prone "old-growth" grassy ecosystems (Veldman et al., 2015). Despite its simple characterization as a savanna, the Cerrado contains considerable vegetation heterogeneity driven by differences in fire regime, water availability, and soil fertility (Bueno et al., 2018). "Open formations" have a continuous herbaceous layer, such as wet grasslands (locally known as *campo úmido*), grassy savannas (*campo sujo*), *campo rupestre*, and *cerrado sensu stricto*. "Closed formations" in the Cerrado, include woodlands (*cerradão*), riparian forests, and seasonally dry forests.

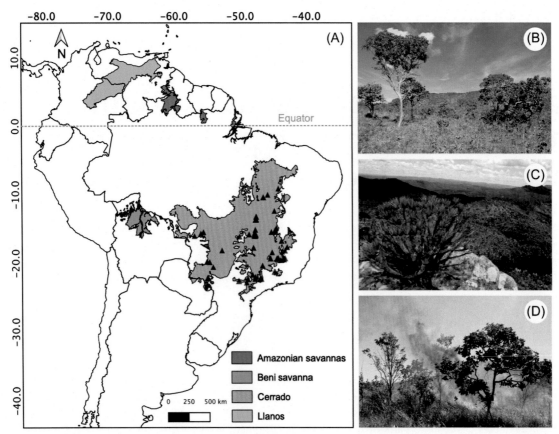

FIGURE 14.1 Distribution of South American savannas from Olson et al. (2001) and Cerrado landscape. (A) Geographic location of studies (▲) addressing the effect of climate change-related environmental cues on germination of Cerrado species. (B) Typical landscape of a Cerrado open savanna in Central Brazil. (C) Montane rocky outcrop vegetation (*campo rupestre*) in southeastern Brazil. (D) Experimental fire in invaded Cerrado (Gorgone-Barbosa et al., 2015). *Photos by L. Felipe Daibes (B and D) and Rafael Oliveira (C).*

The Cerrado climate is mostly seasonal, with rainy summers and dry winters. Climate is variable across the region. Mean monthly annual temperatures range from 18°C to 28°C, and rainfall ranges from 800 to 2000 mm yr^{-1}, with a long dry season during the austral winter (Bueno et al., 2018). Such climatic variability defines distinct biogeographic units with unique species composition (Françoso et al., 2020), resulting in the world's richest savannas, with at least 13,000 plant species, nearly half of them endemics (BFG, 2018). The typical Cerrado vegetation grows on acidic, dystrophic, and clayish soils, and the region provides key ecosystem services such as belowground carbon storage and freshwater reserves (Strassburg et al., 2017). Threats to Cerrado biodiversity and ecosystem services include changes in land use for agriculture, cattle ranching, and mining, leading to decreased soil moisture, increased albedo, and fragmentation of vegetation (Carvalho et al., 2009; Salazar et al., 2015; Lambers et al., 2020). The current network of protected areas is insufficient to halt biodiversity loss in South American savannas (Velazco et al., 2019). Due to its high species richness, a high proportion of endemic species, and current conservation status, the Cerrado is considered a global biodiversity hotspot (Myers et al., 2000). Fire is a natural component of the Cerrado vegetation dynamics, including relatively small and frequent burns, ignited both by lightning in the transition between dry and wet seasons and anthropic fires annually set by local farmers (Miranda et al., 2009; Schmidt and Eloy, 2020).

Climate change effects in the Cerrado

The increasing input of CO_2 into the atmosphere is predicted to affect community responses, nutrient availability, fire regimes, and ecosystem functioning in the Cerrado (Bustamante et al., 2012; Franco et al., 2014; Feeley et al., 2020). Despite the United Nations' intention to keep the global temperature increase to <2°C, predictions are that average temperatures could rise by as much as 4°C in Brazil (Marengo et al., 2009). In addition to the increase in temperature, a 20%—40% decrease in precipitation is also expected in the next decades (Vera et al., 2006; Marengo et al., 2009, 2010).

Recently, Hofmann et al. (2021) have shown consistent increases in air temperature and reductions in relative humidity in the Cerrado, suggesting high impacts on biodiversity and vegetation.

Future fire-related trends are hard to predict due to spatial variability and complex interactions between fire regime, climate, vegetation, and anthropic factors (Flannigan et al., 2009), but they are likely to affect seed germination cues. The global climate affects fire-vegetation relationships, determining the occurrence of wildfires (Jolly et al., 2015). Cerrado areas naturally have a high fire frequency, mainly in the open vegetation, with natural fire return intervals between 3 and 4 years (Coutinho, 1990; Pereira Júnior et al., 2014). CO_2 emissions have been suggested to increase resprouting and growth, both for native species and invasive grasses (Hoffmann et al., 2000; Baruch and Jackson, 2005), thus increasing flammable biomass that may lead to increased fire occurrence in the following decades (Hoffmann et al., 2002). Such an increase in fire frequency is directly related to greenhouse gas emissions, forming a positive feedback loop that will result in larger areas burned and increased fire severity (Silva et al., 2019). Moreover, eastern Brazil, which is predicted to experience the driest climate in the Cerrado region, coincides with areas of increased fire occurrence, although other areas of the Cerrado might experience a fire retreat (Krawchuk et al., 2009; Feeley et al., 2020).

Fire regimes in the Cerrado are altered by anthropic activities, such as deforestation and fire suppression (Schmidt and Eloy, 2020). For instance, fire suppression leads to an expansion of woody plant encroachment into savannas, resulting in significant decreases in the diversity of herbaceous species (Abreu et al., 2017). Such activities cause harmful side effects, given the accumulation of flammable biomass through time, thus causing unwanted burns of higher proportions than natural or managed fires (Durigan and Ratter, 2016). Recently, an increasing incidence of megafires ($>50,000$ km^2 burned) has been reported not only in the fire-prone savannas but also in fire-sensitive vegetation such as the Amazon rainforest (Fidelis et al., 2018). Moreover, human activity favors the occurrence of fires during the mid-dry months (June to August) (Ramos-Neto and Pivello, 2000), while protected areas tend to burn later in the rainy period (October to November) (Alvarado et al., 2018).

Germination ecology in the Cerrado: how much do we know?

Plant regeneration traits in the Cerrado have been shaped by multiple environmental drivers in the last several million years, including frequent fires, high temperatures, and seasonal droughts. According to our literature survey (see Box 14.1), there has been an increase in the number of published papers that assessed the effect of fire, temperature, or drought on Cerrado seeds, with 37% of the studies published between 2015 and 2020 (Fig. 14.2A). Most studies addressed the effect of fire-related cues and temperature, especially the impact on seed germination and viability (Fig. 14.2B). On the other hand, studies on drought and soil seed banks are scarce, with only 15 and five studies, respectively. Our survey recovered studies dealing with 264 species belonging to 45 families, but Fabaceae (20%) and Poaceae (19%) contain nearly 40% of the studied species (Fig. 14.3). Additionally, we found that the effect of only one environmental driver has been assessed for most species, and in only 23% of the species have studies addressed two or three drivers (Fig. 14.2C).

BOX 14.1 Literature survey.

On October 31, 2020, we surveyed papers in the Web of Science and Scopus databases searching for the following combination of terms in the title, abstract, or keywords of papers: (Cerrado OR Neotropical savanna*) AND (seed* OR germination OR dormancy OR regeneration) AND (fire OR drought OR climate change OR temperature OR global warming). After removing duplicates, 665 references were screened. To be included in our review, papers must have addressed the effect of at least one environmental driver related to climate change (temperature, fire, or drought) on a species from the Cerrado. Review papers were used to screen for further references that we could have missed. In the end, we were left with 112 studies.

Each study was classified into three regeneration stages: seed germination, seed dormancy, and soil seed banks. We also recorded whether the study assessed changes in seed viability after exposure to treatments related to the environmental drivers. Regarding climate change-related variables, we classified the studies into three categories: temperature, fire, and drought. Under "Fire," we included studies that addressed the direct (i.e., using natural or experimental fires or exposing seeds to heat shock or smoke treatments) or indirect effects of fire (i.e., daily alternating temperatures). Under "Drought," we included studies that addressed the effect of desiccation (i.e., reduction in seed water content) and environmental water deficit (i.e., soil water potentials) on seeds. We also recorded whether the study was conducted in the field or laboratory. For each study, species identification was checked against The Plant List (http://www.theplantlist.org/, last access on November 2, 2020). Finally, we classified the type of vegetation into "open savannas" and "closed formations."

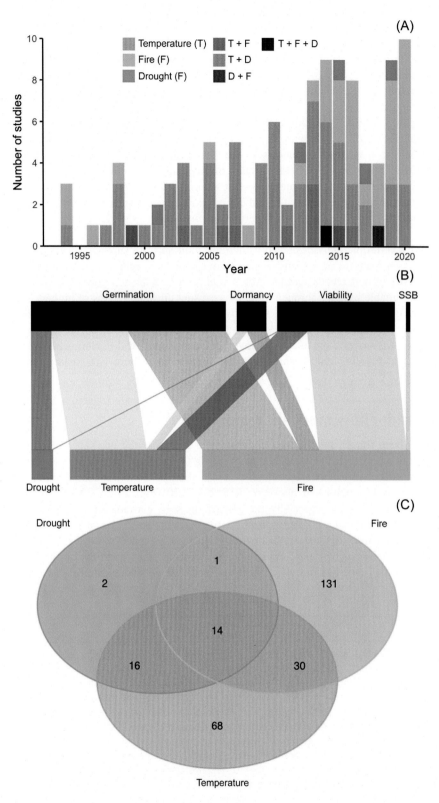

FIGURE 14.2 Overview of our knowledge of climate change-related environmental cues on regeneration of plants from seeds in the Cerrado. (A) Temporal variation in the number of published papers regarding environmental cues. (B) Bipartite network of studies illustrating the number of species assessed in each combination of a given driver and regeneration stage. Bar width is proportional to the number of species. *SSB*, Soil Seed Bank. (C) Venn diagram showing number of species studied for multiple environmental drivers.

Regeneration traits in the Cerrado also are linked with habitat heterogeneity and seasonality. Fruit types, seed dispersal strategies, and growth forms are intermingled, with the frequency of different dispersal modes, seed dormancy, seed sizes, and heat tolerance, depending on vegetation structure and dispersal season (Kuhlmann and Ribeiro, 2016;

FIGURE 14.3 Families with the highest number of species reported in our literature survey. A. *Senna* (Fabaceae), B. *Echinolaena* (Poaceae), C. *Mikania* (Asteraceae), D. *Lavoisiera* (Melastomataceae), E. *Vellozia* (Velloziaceae), and F. *Vochysia* (Vochysiaceae). *Photos by L. Felipe Daibes (A), Maria G.G. Camargo (B and E), Roberta L.C. Dayrell (C), Fernando A.O. Silveira (D), and Roberta Grillo (F).*

Daibes et al., 2019; Escobar et al., 2020). In general, Cerrado seeds are desiccation-tolerant (Ribeiro and Borghetti, 2014; Escobar et al., 2018), and germination is not affected by fire (Daibes et al., 2019). Most species from open Cerrado communities produce small seeds that are self- or wind-dispersed (Escobar et al., 2020; Garcia et al., 2020). In contrast, species from closed communities produce larger, vertebrate-dispersed seeds with higher heat tolerance and faster germination (Escobar et al., 2018, 2020; Daibes et al., 2019). The reason for higher heat tolerance in plants from closed vegetation is that they tend to have large seeds that provide insulation of the embryo from the heat of fires (Ribeiro et al., 2015; Daibes et al., 2019). The most common dormancy classes are physiological dormancy and physical dormancy, but contrary to global vegetation patterns (Willis et al., 2014), a considerable number of Cerrado species produce non-dormant seeds (Dayrell et al., 2017; Escobar et al., 2018). Overall, species in many families produce high percentages of embryoless seeds (Dayrell et al., 2017; Dairel and Fidelis, 2020), relying on long-term survival of adult plants, resprouting, or vegetative propagation for persistence (de Moraes et al., 2016; Pausas et al., 2018; Pilon et al., 2021; Zupo et al., 2021).

Environmental drivers of Cerrado regeneration from seeds under a changing climate

Increased fire frequency

Precise future trends in fire regimes are hard to predict (Flannigan et al., 2009), but it is generally expected that climate change will lead to increases in fire frequency, size and number of areas burned, and fire severity in the Cerrado (Silva et al., 2019). Cerrado plants are adapted to disturbance and exhibit different post-fire regeneration strategies, including resprouting and producing heat-tolerant propagules (Pilon et al., 2021; Zupo et al., 2021).

Many species tolerate fire passage (Daibes et al., 2019); hence, no major changes in regeneration from seeds are expected under changing fire regimes. Therefore, heat tolerance, the maintenance of seed viability during exposure to high temperatures, is a key trait that acts as a proxy for seed survival to fire-related heat shocks. In the Cerrado, seed survival is positively associated with seed mass and dormancy (Ribeiro et al., 2015; Ramos et al., 2016; Daibes et al., 2019). Nevertheless, available data are taxonomically biased, with species in Fabaceae and Poaceae comprising nearly half of all studies on seeds of Cerrado species (Fig. 14.3). Recently, Gawryszewski et al. (2020) showed that an increase in fire frequency decreased seed production, but not germination, of *Vochysia alata*. Although rare, field experimental burns provide reliable and realistic data on the role of fire on regeneration traits, showing that most seeds directly exposed on the soil surface are killed during Cerrado fires (Daibes et al., 2017, 2018). On the other hand, seeds incorporated into soil seed banks are insulated from high temperatures and usually survive fire events. Thus, seasonal burns do not seem to affect seedling emergence in the field (de Andrade and Miranda, 2014; Fontenele et al., 2020b).

Irrespective of fires, many Cerrado species seem to synchronize dispersal and seedling emergence patterns with the onset of the rainy season (Salazar et al., 2011; Ramos et al., 2017; Escobar et al., 2018). However, once established, most Cerrado seedlings survive fires, having the capacity to resprout at a young age (<1 year), a trait lacking in most forest species (Hoffmann, 2000). Thus, enhanced fire frequency is a strong environmental filter, selecting species whose seedlings survive via belowground-protected buds even if their aerial parts are removed (fire topkill). Wildfires might also affect the diversity of invertebrate dispersal agents (ants, for instance), mainly in closed formations (Vasconcelos et al., 2017), hindering regeneration from seeds in fire-sensitive vegetation.

In contrast to Mediterranean-type ecosystems, fire has little or no effect on seed germination in the Cerrado. This conclusion is based on studies that assessed seed responses to fire-related cues by performing laboratory heat shock (Ribeiro et al., 2013; Ribeiro and Borghetti, 2014; Daibes et al., 2019) and smoke (Le Stradic et al., 2015; Fichino et al., 2016; Ramos et al., 2019; Zirondi et al., 2019; Fernandes et al., 2021) treatments. For heat shock treatments, exposure times of up to 5 min have been applied, considering that Cerrado savannas experience fast-moving surface fires (Miranda et al., 1993). Fire has little effect on germination via stimulation by smoke (Le Stradic et al., 2015; Zirondi et al., 2019) or via alleviation of physical dormancy (e.g., heat treatments broke physical dormancy of seeds of only six of 46 legumes; Daibes et al., 2019). However, smoke-stimulated effects can increase the rate (speed) of germination (Fernandes et al., 2021) and enhance the root growth of grass seedlings (Ramos et al., 2019). Cerrado fires stimulate flowering in some species (Le Stradic et al., 2015; Pilon et al., 2018; Fidelis et al., 2019), although post-fire seed production can be relatively poor with regard to seed quality and germination in typically resprouting savanna grasses and sedges (Le Stradic et al., 2015; Fontenele et al., 2020a).

Increase in mean temperature

Many tropical plants are predicted to experience temperatures exceeding their optimum or even maximum germination temperature with climate warming, which will put them at great risk (Sentinella et al., 2020). Regeneration from seeds in many Cerrado species is related to seed shed in the post-fire environment; thus, seeds often are exposed to increased daily temperature fluctuations and dry soils (Daibes et al., 2017). In studies addressing temperature-dependent germination, seeds usually are incubated over a range of constant temperatures in the laboratory to define the breadth of the thermal niche of germination (Ranieri et al., 2012; Marques et al., 2014; Correa et al., 2021; Oliveira et al., 2021). Germination of small-seeded species, such as those of *Vellozia* and *Xyris*, commonly is light-dependent, but higher environmental temperatures (35°C–40°C) promote germination in the dark in *Vellozia* (Soares da Mota and Garcia, 2013). Seeds tolerant of high incubation temperatures, such as those of Velloziaceae, may retain high germination percentages with climate change in contrast to other endemic plant groups, such as Xyridaceae (Giorni et al., 2018) and some Bromeliaceae (Marques et al., 2014), whose seeds mostly are sensitive to high temperatures.

In *Mimosa*, temperatures around 35°C are high enough to alleviate physical dormancy without the need for heat shock (Silveira and Fernandes, 2006; Dayrell et al., 2015). Under field conditions, increased temperature fluctuations partially alleviated physical dormancy in *Mimosa leiocephala* but showed little to no effect on seeds of other Cerrado legume shrubs (Daibes et al., 2017). Other studies have shown that high temperatures mostly alleviate physical dormancy under wet rather than dry conditions in legumes from fire-prone ecosystems (Van Klinken et al., 2006; Wiggers et al., 2017). In Brazil, dormancy break in one forest legume *Senna multijuga* was mediated by high incubation temperatures, becoming more "sensitive" during soil storage (Rodrigues-Junior et al., 2018). A similar pattern has been found in three of four Australian legumes (Liyanage and Ooi, 2017), and it has been recently shown that high temperatures are predicted to exceed germination thresholds for many Cerrado trees (Correa et al., 2021). Legume recruitment would be enhanced by increased environmental temperatures if combined with water availability during the wet season; in a dry environment, most seeds would remain nongerminated (but viable) until a suitable rainfall event.

Decrease in mean annual rainfall

While seasonality of precipitation traditionally has been recognized as a critical factor in the natural regeneration from seeds in the Cerrado (Escobar et al., 2018), our literature review shows that drought is the least studied environmental driver in this ecosystem. Predicted increases in the length of the dry season (Feeley et al., 2020) suggest significant decreases in the length of the establishment season in the Neotropics. Surprisingly, 73% of the studies for this environmental driver addressed seed responses to desiccation (i.e., a reduction in seed water content) rather than the effect of soil moisture on germination. Therefore, it is unclear how an increase in length and severity of the dry seasons could affect regeneration from seeds in the Cerrado.

Current evidence suggests that species from open savannas produce seeds that tolerate desiccation better than those of forest species (Ribeiro and Borghetti, 2014). For instance, Oliveira et al. (2018) showed that the ability of seeds of two *Xyris* species from wet savannas to germinate was not affected by wetting and drying, suggesting they can survive isolated rains in the dry season or dry spells in the rainy season. As a result, species from open savannas are potentially more resilient to decreased rainfall than those from closed savannas.

In the Cerrado, more species produce non-dormant than dormant seeds (Dayrell et al., 2017); thus, non-dormant seeds could germinate following a short rainfall episode, but seedlings would die due to the extended length of the dry season. For instance, seedlings from species dispersing seeds during the mid-rainy season would have much less time to attain a sufficient size to withstand the dry season than those from seeds dispersed at the beginning of the rainy season. Soil seed bank assessments reveal that Cerrado species have a seasonal recruitment pattern, independent of fire, and short-lived seed banks (Medina and Fernandes, 2007; Salazar et al., 2011; de Andrade and Miranda, 2014; Garcia et al., 2020). Dormancy cycling has been documented in only a few endemic species of Eriocaulaceae and Xyridaceae. The tiny non-dormant seeds of species in these two families are dispersed during the winter dry season, and if they are buried and fail to germinate in the spring wet season, they are induced into dormancy. Then, they come out of dormancy during the subsequent dry season and dormancy cycling continues (Garcia et al., 2014, 2020; Oliveira et al., 2017). Daily temperature fluctuation may partially alleviate the physiological dormancy of grass species under dry conditions (Musso et al., 2015; Dairel and Fidelis, 2020). Species producing desiccation-tolerant, dormant seeds may be favored by decreasing rainfall, possibly forming longer-lived seed banks due to climate change. However, this hypothesis remains to be formally tested.

Predicted consequences of climate change on plant populations and communities

Changes in species distribution

Shifts in environmental cues induced by climate change are expected to alter the distribution of Cerrado species (Siqueira and Peterson, 2003; Ribeiro et al., 2019), including losses of up to 78% of the original distribution area of economic and culturally relevant species (Simon et al., 2013). Reductions in the geographic range of species can be mediated by multiple drivers, and the consequence of such shifts likely results in changes in community structure and composition (Correa et al., 2021). First, the lack of specialized means of seed dispersal in many Cerrado endemics suggests limited opportunities for migration to track suitable future climatic conditions, especially for edaphic specialists (Corlett and Tomlinson, 2020). Landscape fragmentation makes migration even more challenging for dispersal-limited species (Carvalho et al., 2009). Thus, expected shifts in species distribution mediated by seed dispersal should result in filtering autochoric species out of future communities. On the other hand, zoochoric, fleshy fruited species, such as those of *Miconia* (Melastomataceae), could increase in dominance in communities, given their role as potential keystone taxa (Kuhlmann and Ribeiro, 2016; Messeder et al., 2021).

Second, increasing temperatures and decreasing soil water potential will decrease tolerance thresholds below the point where germination is possible (Correa et al., 2021). In eurythermic species, for example, Velloziaceae with a wide germination niche breadth (Soares da Mota and Garcia, 2013), suboptimum germination is expected, whereas in stenothermic species, for example, some bromeliads with a narrow germination niche breadth, germination may be greatly reduced by high temperatures (Duarte et al., 2018). For instance, in the bromeliad *Racinaea aerisincola* germination percentage decreased by one-third at temperatures >20°C (Marques et al., 2014). Interspecific variation in thermal niche breadth explains differences in species responses to projected temperature increases (Oliveira et al., 2021).

Third, desiccation sensitivity is rare in the Cerrado, but the seeds of some savanna trees are recalcitrant (Mayrinck et al., 2019). Thus, an increase in dry season length might limit recruitment to moist sites, such as gallery forests. Nevertheless, a few species with recalcitrant seeds have features such as a persistent pericarp around the seeds of *Swartzia langsdorffii* that maintain a moist microenvironment and survival until rainfall occurs (Vaz et al., 2016).

Fourth, fire-sensitive species from open savannas likely will be negatively affected by increasing fire frequency. Such species may shift their habitats to closed savannas, where fire frequency is lower. However, such a shift in habitat is unlikely owing to low shade tolerance of herbaceous species in the Cerrado, thereby reducing species abundance and geographic distribution of fire-sensitive species from open savannas.

Future decreases in the geographic range of Cerrado species are supported by both thermal-time models (Correa et al., 2021) and ecological niche modeling studies. In *Eugenia* (Myrtaceae) species from the Cerrado, which typically produce recalcitrant seeds (Delgado and Barbedo, 2007), geographic range and species richness are predicted to be limited to moist vegetation formations in southeastern Brazil near the Atlantic forest (Oliveira et al., 2019). To date, few studies have combined information on seed responses to environmental drivers and modeling the ecological niche of a Cerrado species in future climatic scenarios. High temperatures are detrimental to seedlings of *Dipteryx alata* (Fabaceae), and the geographic distribution of this widespread species is expected to decrease in the next 50 years (Ribeiro et al., 2019). A compilation of studies on seed germination showed that higher temperatures and fire frequency generally will reduce the germinability of savanna species (Borghetti et al., 2021). However, high fire tolerance typical of many Cerrado species coupled with recruitment models suggest a potential for expansion of Cerrado into eastern Amazonia over the next several decades (Borghetti et al., 2021).

Changes in species abundance and community composition

Species abundance and community composition of native savanna communities potentially will be strongly affected by climate warming (Feeley et al., 2020). Populations of stenothermic, recalcitrant, and fire-sensitive species are predicted to decline. This does not mean that eurythermic, desiccation-tolerant, and fire-tolerant species will increase in abundance since they are expected to experience moderate levels of negative effects from the shortened establishment seasons. "Risk-taking" and "risk-avoidance" strategies (*sensu* Duncan et al., 2019) likely will cause changes in community structure and composition under scenarios of unpredictable and shortened establishment seasons. If mid-dry season fires, such as those ignited by humans (Schmidt and Eloy, 2020), trigger germination in species with physical dormancy, a mismatch between germination and optimum conditions for seedling establishment is expected.

Significant changes in the structure and composition of Cerrado communities are expected, further decreasing the resilience of these communities to man-made impacts. Moreover, the impacts of humans on plant communities may differ between open and closed vegetation formations in the Cerrado. Open savannas would be exposed to increased fire frequency, favoring regeneration via vegetative resprouting (Bond and Midgley, 2001; Zupo et al., 2021). Moist patches in closed vegetation types would protect seeds from fire and provide sufficient moisture for germination (Fig. 14.4). Heat-tolerant species such as those of Fabaceae and Velloziaceae would be favored, whereas heat-sensitive species of Bromeliaceae and Xyridaceae would not be favored (Fig. 14.5).

Climate change is predicted to increase biological invasions, further aggravating negative impacts on natural communities. The distribution of invasive fire-adapted grasses (e.g., *Melinis minutiflora* and *Urochloa brizantha*) are expected to increase with the changing climate, impacting the structure and composition of Cerrado communities (Gorgone-Barbosa et al., 2015; Damasceno et al., 2018). Although fire-related heat shocks may negatively affect their germination (Paredes et al., 2018; Gorgone-Barbosa et al., 2020), these invasive African grasses exhibit fast biomass recovery, outcompeting native vegetation in the post-fire environment (Damasceno et al., 2018). The abundance of invasive grasses is determined by both fire regime and seasonal climatic factors, thereby decreasing the cover of native species (Damasceno and Fidelis, 2020). The presence of invasive grasses enhances fire intensity and flame height, and fire improves the habitat for fast-recovering grasses (Gorgone-Barbosa et al., 2015). Thus, adequate ecological policies should be applied to conserve Cerrado species and ecosystem services (Strassburg et al., 2017; Velazco et al., 2019). Among such policies, integrated fire management is essential to control fuel load accumulation and thus prevent the risk of catastrophic wildfires (Durigan and Ratter, 2016; Schmidt and Eloy, 2020).

Knowledge gaps and research needs

There is a huge knowledge gap in studies addressing climate change effects on the regeneration of plants from seeds in tropical savannas (Walck et al., 2011). Despite recent efforts, our review shows that this knowledge gap still persists for Neotropical savannas. Thus, we have little capacity to predict how changes in climate and fire regimes will affect the regeneration of plants from seeds and subsequently changes in species range and composition of ecological communities. Notably, our current knowledge on the impacts of climate change on regeneration from seeds in the Cerrado is shaped by pervasive taxonomic and geographic biases. Below, we propose some research avenues that can enhance our ability to predict the effect of climate change on regeneration from seeds in the Cerrado and improve the accuracy of data for informed decision-making.

1. More data are needed on how seeds of Cerrado species respond to increasing temperatures, decreasing substrate water potentials, and changes in fire-related cues. Hydrothermal time models, which can be used to predict germination thresholds, are available for plants in arid zones (Duncan et al., 2019; Rajapakshe et al., 2020) and tropical dry

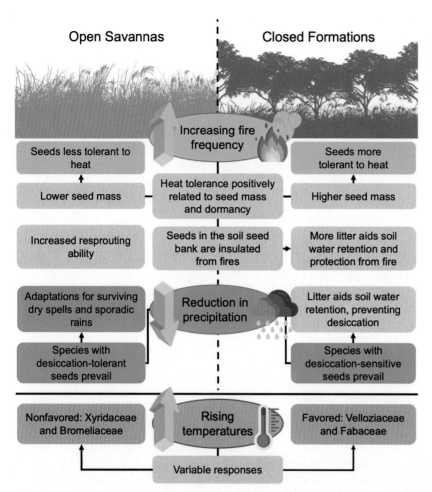

FIGURE 14.4 Overview of the predicted changes in regeneration from seeds in the Cerrado for each of the environmental drivers assessed, contrasting expected responses to increased fires and reduced precipitation in open and closed plant formations. *This figure was designed with resources from Flaticon.com.*

forests (Dantas et al., 2020). Still, little effort has been made for Cerrado species (but see Duarte et al., 2018; Correa et al., 2021; Oliveira et al., 2021). Future studies should focus on understudied clades and regions to reduce taxonomic and geographical biases (Ribeiro et al., 2016). Phylogenetic relationships are a weak predictor of heat tolerance in adult plants (Perez and Feeley, 2021), but many seed traits have been shown to be phylogenetically conserved, including response to temperatures (Arène et al., 2017), desiccation tolerance (Wyse and Dickie, 2017), and persistence in soil (Gioria et al., 2020). Therefore, modeling seed responses to environmental cues for taxa across phylogenetic trees may help improve our ability to understand how different clades will be affected by global warming.

2. We need experiments across a wide phylogenetic diversity of plants from different ecosystems to identify easy-to-measure traits correlated with seed responses to germination cues. By doing so, it will be possible to determine whether we can use easy-to-measure or readily available traits in trait databases (e.g., TRY; Kattge et al., 2020) as surrogates for heat tolerance, drought resistance, and dormancy-breaking cues. For example, the positive correlation between heat tolerance and seed mass suggests that we can use seed mass to predict heat tolerance (Daibes et al., 2019). However, the study by Daibes et al. was restricted to Fabaceae, and we need data that are distributed across the phylogenetic tree of the Cerrado.
3. Temperature, precipitation, and elevation gradients have been widely used to understand community assembly patterns and recruitment from seeds (Ooi et al., 2012; Rosbakh and Poschlod, 2015; Wang et al., 2021). Mountains in the Cerrado have large climatic variation over short distances and are widely distributed across the landscape (Vidal Jr. et al., 2019). Thus, they can serve as natural laboratories to increase our understanding of how changing abiotic conditions affect regeneration from seeds.
4. Intraspecific variation is important in community assembly (Violle et al., 2012). Several species occur over large areas of the Cerrado, but few of them have been assessed for genetic diversity or intraspecific variation in seed traits. The few studies available on intraspecific variation show that seed dormancy differs among populations

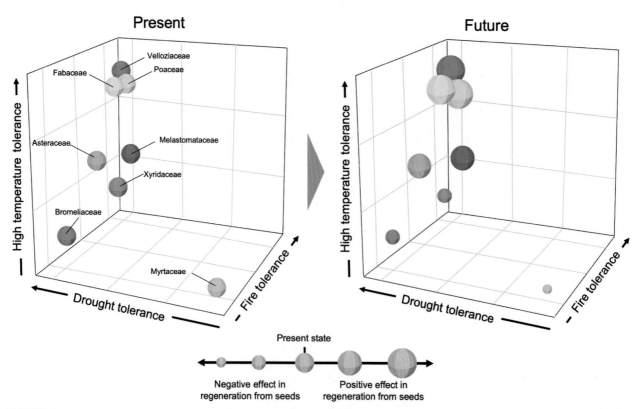

FIGURE 14.5 Projected effects of seed regeneration in selected plant families based on their overall responses to high temperatures, drought, and fire.

(Lacerda et al., 2004) and that germination traits respond to environmental changes (Sales et al., 2013). Moreover, seed desiccation tolerance varies between individuals of some species from contrasting climates and/or vegetation physiognomies (Pereira et al., 2012, 2017). Advancing our understanding of intraspecific variation in seed traits is important to understanding shifts in geographic distribution of species and how community composition might change in the future (Valladares et al., 2014).
5. Finally, we need a broader understanding of the seed dispersal of Cerrado species. Seed germination ecology frequently is linked to dispersal phenology, fruit traits, and species growth forms (Salazar et al., 2011, 2012; Escobar et al., 2018, 2020). Understanding how habitat loss and fragmentation affect seed dispersal ecology and fruit–frugivore interactions is important in predicting dispersal, seed limitation, and regeneration patterns.

Conclusions

A combination of increased fire frequency, increased temperatures, decreased air humidity and an extended dry season (Hofmann et al., 2021) due to climate change likely will affect the regeneration of plants from seeds in the Cerrado. Because germination of Cerrado seeds is highly influenced by seasonality (Salazar et al., 2011; Escobar et al., 2018), changes in annual precipitation could lead to major changes in emergence from soil seed banks, especially in open savannas. Our study identified drought tolerance and emergence from soil seed banks as the least studied environmental drivers under a changing climate. In contrast to other fire-prone vegetation types, regeneration from seeds in the Cerrado shows little to no effect of fire-related cues (Ribeiro et al., 2013; Fichino et al., 2016; Daibes et al., 2019). However, differences in thermal niche breadth (Soares da Mota and Garcia, 2013; Marques et al., 2014; Dayrell et al., 2015; Oliveira et al., 2021) might filter out some species from future plant communities in the Cerrado. Thus, changes in species distribution and the structure and composition of communities are expected.

Locally, anthropic factors related to land use contribute to Cerrado fragmentation and loss of ecosystem services. Neotropical savannas have been studied less than fire-prone temperate ecosystems, and a better picture of how a changing climate will affect regeneration from seeds will emerge if persistent phylogenetic and geographical knowledge gaps are filled. Such knowledge will better inform the development of management and conservation policies such as fire management, translocation of species, and exotic species control.

Acknowledgments

FAOS acknowledges support from CNPq and CAO-P a scholarship from CAPES.

References

Abreu, R.C.R., Hoffmann, W.A., Vasconcelos, H.L., Pilon, N.A., Rossatto, D.R., Durigan, G., 2017. The biodiversity cost of carbon sequestration in tropical savanna. Sci. Adv. 3, e1701284. Available from: https://doi.org/10.1126/sciadv.1701284.

Alvarado, S.T., Silva, T.S.F., Archibald, S., 2018. Management impacts on fire occurrence: a comparison of fire regimes of African and South American tropical savannas in different protected areas. J. Environ. Manage. 218, 79–87.

Arène, F., Affre, L., Doxa, A., Saatkamp, A., 2017. Temperature but not moisture response of germination shows phylogenetic constraints while both interact with seed mass and lifespan. Seed Sci. Res. 27, 110–120.

Baruch, Z., Jackson, R.B., 2005. Responses of tropical native and invader C4 grasses to water stress, clipping and increased atmospheric CO_2 concentration. Oecologia 145, 522–532.

Beerling, D.J., Osborne, C.P., 2006. The origin of the savanna biome. Glob. Change Biol. 12, 2023–2031.

BFG (Brazilian Flora Group), 2018. Brazilian flora 2020: Innovation and collaboration to meet target 1 of the global strategy for plant conservation (GSPC). Rodriguesia 69, 1513–1527.

Bond, W.J., 2019. Open Ecosystems: Ecology and Evolution Beyond the Forest Edge. Oxford University Press, Oxford.

Bond, W.J., Midgley, J.J., 2001. Ecology of sprouting in woody plants: the persistence niche. Trends Ecol. Evol. 16, 45–51.

Borghetti, F., Barbosa, E.R.M., Ribeiro, L.C., Ribeiro, J.F., Walter, B.M.T., 2020. South American savannas. In: Scogings, P.F., Sankaran, M. (Eds.), Savanna Woody Plants and Large Herbivores. Wiley, Hoboken, pp. 77–122.

Borghetti, F., de Oliveira Caetano, G.H., Colli, G.R., Françoso, R., Sinervo, B.R., 2021. The firewall between Cerrado and Amazonia: interaction of temperature and fire govern seed recruitment in a neotropical savanna. J. Veg. Sci. 32, e12988. Available from: https://doi.org/10.1111/jvs.12988.

Bueno, M.L., Pennington, R.T., Dexter, K.G., Kamino, L.H.Y., Pontara, V., Neves, D.M., et al., 2017. Effects of Quaternary climatic fluctuations on the distribution of Neotropical savanna tree species. Ecography 40, 403–414.

Bueno, M.L., Dexter, K.G., Pennington, R.T., Pontara, V., Neves, D.M., Ratter, J.A., et al., 2018. The environmental triangle of the Cerrado Domain: ecological factors driving shifts in tree species composition between forests and savannas. J. Ecol. 106, 2109–2120.

Bustamante, M.M.C., Nardoto, G.B., Pinto, A.S., Resende, J.C.F., Takahashi, F.S.C., Vieira, L.C.G., 2012. Potential impacts of climate change on biogeochemical functioning of Cerrado ecosystems. Brazil. J. Biol. 72, 655–671.

Carvalho, F.M.V., De Marco, P., Ferreira, L.G., 2009. The Cerrado into-pieces: habitat fragmentation as a function of landscape use in the savannas of central Brazil. Biol. Conserv. 142, 1392–1403.

Corlett, R.T., Tomlinson, K.W., 2020. Climate change and edaphic specialists: irresistible force meets immovable object? Trends Ecol. Evol. 35, 367–376.

Correa, A.R., Silva, A.M.P., Arantes, C.R.A., Guimarães, S.C., Camili, E.C., Coelho, M.F.B., 2021. Quantifying seed germination based on thermal models to predict global climate change impacts on Cerrado species. Seed Sci. Res. 31, 126–135.

Coutinho, L.M., 1990. Fire in the ecology of the Brazilian Cerrado. In: Goldamer, J.G. (Ed.), Fire in the Tropical Biota. Springer-Verlag, Berlin, pp. 82–105.

Daibes, L.F., Zupo, T., Silveira, F.A.O., Fidelis, A., 2017. A field perspective on effects of fire and temperature fluctuation on Cerrado legume seeds. Seed Sci. Res. 27, 74–83.

Daibes, L.F., Gorgone-Barbosa, E., Silveira, F.A.O., Fidelis, A., 2018. Gaps critical for the survival of exposed seeds during Cerrado fires. Aust. J. Bot. 66, 116–123.

Daibes, L.F., Pausas, J.G., Bonani, N., Nunes, J., Silveira, F.A.O., Fidelis, A., 2019. Fire and legume germination in a tropical savanna: ecological and historical factors. Ann. Bot. 123, 1219–1229.

Dairel, M., Fidelis, A., 2020. How does fire affect germination of grasses in the Cerrado? Seed Sci. Res. 30, 275–283.

Damasceno, G., Fidelis, A., 2020. Abundance of invasive grasses is dependent on fire regime and climatic conditions in tropical savannas. J. Environ. Manage. 271, 111016. Available from: https://doi.org/10.1016/j.jenvman.2020.111016.

Damasceno, G., Souza, L., Pivello, V.R., Gorgone-Barbosa, E., Giroldo, P.Z., Fidelis, A., 2018. Impact of invasive grasses on Cerrado under natural regeneration. Biol. Inv. 20, 3621–3629.

Dantas, B.F., Moura, M.S.B., Pelacani, C.R., Angelotti, F., Taura, T.A., Oliveira, G.M., et al., 2020. Rainfall, not soil temperature, will limit the seed germination of dry forest species with climate change. Oecologia 192, 529–541.

Dayrell, R.L.C., Gonçalves-Alvim, S.J., Negreiros, D., Fernandes, G.W., Silveira, F.A.O., 2015. Environmental control of seed dormancy and germination of *Mimosa calodendron* (Fabaceae): implications for ecological restoration of a highly threatened environment. Brazil. J. Bot. 38, 395–399.

Dayrell, R.L.C., Garcia, Q.S., Negreiros, D., Baskin, C.C., Baskin, J.M., Silveira, F.A.O., 2017. Phylogeny strongly drives seed dormancy and quality in a climatically buffered hotspot for plant endemism. Ann. Bot. 119, 267–277.

de Andrade, L.A.Z., Miranda, H.S., 2014. The dynamics of the soil seed bank after a fire event in a woody savanna in central Brazil. Plant Ecol. 215, 1199–1209.

de Carvalho, W.D., Mustin, K., 2017. The highly threatened and little known Amazonian savannahs. Nat. Ecol. Evol. 1, 0100. Available from: https://doi.org/10.1038/s41559-017-0100.

de Moraes, M.G., de Carvalho, M.A.M., Franco, A.C., Pollock, C.J., Figueiredo-Ribeiro, R., de, C.L., 2016. Fire and drought: soluble carbohydrate storage and survival mechanisms in herbaceous plants from the Cerrado. Bioscience 66, 107–117.

Delgado, L.F., Barbedo, C.J., 2007. Tolerância à dessecação de sementes de espécies de *Eugenia*. Pesq. Agropec. Bras. 42, 265–272.

Devecchi, M.F., Lovo, J., Moro, M.F., Andrino, C.O., Barbosa-Silva, R.G., Viana, P.L., et al., 2020. Beyond forests in the Amazon: biogeography and floristic relationships of the Amazonian savannas. Bot. J. Linn. Soc. 193, 478–503.

Duarte, A.A., Lemos Filho, J.P., Marques, A.R., 2018. Seed germination of bromeliad species from the campo rupestre: thermal time requirements and response under predicted climate-change scenarios. Flora 238, 119–128.

Duncan, C., Schultz, N.L., Good, M.K., Lewandrowski, W., Cook, S., 2019. The risk-takers and -avoiders: germination sensitivity to water stress in an arid zone with unpredictable rainfall. AoB Plants 11, plz066. Available from: https://doi.org/10.1093/aobpla/plz066.

Durigan, G., Ratter, J.A., 2016. The need for a consistent fire policy for Cerrado conservation. J. Appl. Ecol. 53, 11–15.

Escobar, D.F.E., Silveira, F.A.O., Morellato, L.P.C., 2018. Timing of seed dispersal and seed dormancy in Brazilian savanna: two solutions to face seasonality. Ann. Bot. 121, 1197–1209.

Escobar, D.F.E., Silveira, F.A.O., Morellato, L.P.C., 2020. Do regeneration traits vary according to vegetation structure? A case study for savannas. J. Veg. Sci. 32, e12940. Available from: https://doi.org/10.1111/jvs.12940.

Feeley, K.J., Bravo-Avila, C., Fadrique, B., Perez, T.M., Zuleta, D., 2020. Climate-driven changes in the composition of New World plant communities. Nat. Clim. Change 10, 965–970.

Fernandes, A.F., Oki, Y., Fernandes, G.W., Moreira, B., 2021. The effect of fire on seed germination of campo rupestre species in the South American Cerrado. Plant Ecol. 222, 45–55.

Fichino, B.S., Dombroski, J.R.G., Pivello, V.R., Fidelis, A., 2016. Does fire trigger seed germination in the neotropical savannas? Experimental tests with six Cerrado species. Biotropica 48, 181–187.

Fidelis, A., Alvarado, S.T., Barradas, A., Pivello, V., 2018. The year 2017: megafires and management in the Cerrado. Fire 1, 49. Available from: https://doi.org/10.3390/fire1030049.

Fidelis, A., Rosalem, P., Zanzarini, V., Camargos, L.S., Martins, A.R., 2019. From ashes to flowers: a savanna sedge initiates flowers 24 h after fire. Ecology 100, e02648. Available from: https://doi.org/10.1002/ecy.2648.

Flannigan, M.D., Krawchuk, M.A., de Groot, W.J., Wotton, B.M., Gowman, L.M., 2009. Implications of changing climate for global wildland fire. Int. J. Wildl. Fire 18, 483–507.

Fontenele, H.G.V., Cruz-Lima, L.F.S., Pacheco-Filho, J.L., Miranda, H.S., 2020a. Burning grasses, poor seeds: post-fire reproduction of early-flowering Neotropical savanna grasses produces low-quality seeds. Plant Ecol. 221, 1265–1274.

Fontenele, H.G.V., Figueirôa, R.N.A., Pereira, C.M., Nascimento, V.T., Musso, C., Miranda, H.S., 2020b. Protected from fire, but not from harm: seedling emergence of savanna grasses is constrained by burial depth. Plant Ecol. Divers. 13, 189–198.

Franco, A.C., Rossatto, D.R., Ramos Silva, L.de C., da Silva Ferreira, C., 2014. Cerrado vegetation and global change: the role of functional types, resource availability and disturbance in regulating plant community responses to rising CO_2 levels and climate warming. Theor. Exp. Plant Physiol. 26, 19–38.

Françoso, R.D., Dexter, K.G., Machado, R.B., Pennington, R.T., Pinto, J.R.R., Brandão, R.A., et al., 2020. Delimiting floristic biogeographic districts in the Cerrado and assessing their conservation status. Biodivers. Conserv. 29, 1477–1500.

Garcia, Q.S., Oliveira, P.G., Duarte, D.M., 2014. Seasonal changes in germination and dormancy of buried seeds of endemic Brazilian Eriocaulaceae. Seed Sci. Res. 24, 113–117.

Garcia, Q.S., Barreto, L.C., Bicalho, E.M., 2020. Environmental factors driving seed dormancy and germination in tropical ecosystems: a perspective from *campo rupestre* species. Environ. Exp. Bot. 178, 104164. Available from: https://doi.org/10.1016/j.envexpbot.2020.104164.

Gawryszewski, F.M., Sato, M.N., Miranda, H.S., 2020. Frequent fires alter tree architecture and impair reproduction of a common fire-tolerant savanna tree. Plant Biol. 22, 106–112.

Gioria, M., Pyšek, P., Baskin, C.C., Carta, A., 2020. Phylogenetic relatedness mediates persistence and density of soil seed banks. J. Ecol. 108, 2121–2131.

Giorni, V.T., Bicalho, E.M., Garcia, Q.S., 2018. Seed germination of *Xyris* spp. from Brazilian campo rupestre is not associated to geographic distribution and microhabitat. Flora 238, 102–109.

Gorgone-Barbosa, E., Pivello, V.R., Bautista, S., Zupo, T., Rissi, M.N., Fidelis, A., 2015. How can an invasive grass affect fire behavior in a tropical savanna? A community and individual plant level approach. Biol. Inv. 17, 423–431.

Gorgone-Barbosa, E., Daibes, L.F., Novaes, R.B., Pivello, V.R., Fidelis, A., 2020. Fire cues and germination of invasive and native grasses in the Cerrado. Acta Bot. Brasil. 34, 185–191.

Hoffmann, W.A., 2000. Post-establishment seedling success in the Brazilian Cerrado: a comparison of savanna and forest species. Biotropica 32, 62–69.

Hoffmann, W.A., Bazzaz, F.A., Chatterton, N.J., Harrison, P.A., Jackson, R.B., 2000. Elevated CO_2 enhances resprouting of a tropical savanna tree. Oecologia 123, 312–317.

Hoffmann, W.A., Schroeder, W., Jackson, R.B., 2002. Positive feedbacks of fire, climate, and vegetation and the conversion of tropical savanna. Geophys. Res. Lett. 29, 2052. Available from: https://doi.org/10.1029/2002GL015424.

Hofmann, G.S., Cardoso, M.F., Alves, R.J.V., Weber, E.J., Barbosa, A.A., Toledo, P., et al., 2021. The Brazilian Cerrado is getting hotter and drier. Global Change Biol. 27, 4060–4073.

Huber, O., Stefano, R.D., Aymard, G., Riina, R., 2006. Flora and vegetation of the Venezuelan Llanos: a review. In: Pennington, R.T., Lewis, G.P., Ratter, J.A. (Eds.), Neotropical Savannas and Seasonally Dry Forests: Plant Diversity, Biogeography, and Conservation. CRC Press, Boca Raton, pp. 95–120.

Jolly, W.M., Cochrane, M.A., Freeborn, P.H., Holden, Z.A., Brown, T.J., Williamson, G.J., et al., 2015. Climate-induced variations in global wildfire danger from 1979 to 2013. Nat. Commun. 6, 7537. Available from: https://doi.org/10.1038/ncomms8537.

Kattge, J., Bönisch, G., Díaz, S., Lavorel, S., Prentice, I.C., Leadley, P., et al., 2020. TRY plant trait database – enhanced coverage and open access. Glob. Change Biol. 26, 119–188.

Krawchuk, M.A., Moritz, M.A., Parisien, M.-A., Van Dorn, J., Hayhoe, K., 2009. Global pyrogeography: the current and future distribution of wildfire. PLoS One 4, e5102. Available from: https://doi.org/10.1371/journal.pone.0005102.

Kuhlmann, M., Ribeiro, J.F., 2016. Evolution of seed dispersal in the Cerrado biome: ecological and phylogenetic considerations. Acta Bot. Brasil. 30, 271–282.

Lacerda, D.R., Lemos Filho, J.P., Goulart, M.F., Ribeiro, R.A., Lovato, M.B., 2004. Seed-dormancy variation in natural populations of two tropical leguminous tree species: *Senna multijuga* (Caesalpinoideae) and *Plathymenia reticulata* (Mimosoideae). Seed Sci. Res. 14, 127–135.

Lambers, H., de Britto Costa, P., Oliveira, R.S., Silveira, F.A.O., 2020. Towards more sustainable cropping systems: lessons from native Cerrado species. Theor. Exp. Plant Physiol. 32, 175–194.

Larrea-Alcázar, D.M., Embert, D., Aguirre, L.F., Ríos-Uzeda, B., Quintanilla, M., Vargas, A., 2011. Spatial patterns of biological diversity in a neotropical lowland savanna of northeastern Bolivia. Biodivers. Conserv. 20, 1167–1182.

Le Stradic, S., Silveira, F.A.O., Buisson, E., Cazelles, K., Carvalho, V., Fernandes, G.W., 2015. Diversity of germination strategies and seed dormancy in herbaceous species of *campo rupestre* grasslands. Austral Ecol. 40, 537–546.

Liyanage, G.S., Ooi, M.K.J., 2017. Do dormancy-breaking temperature thresholds change as seeds age in the soil seed bank? Seed Sci. Res. 27, 1–11.

Marengo, J.A., Jones, R., Alves, L.M., Valverde, M.C., 2009. Future change of temperature and precipitation extremes in South America as derived from the PRECIS regional climate modeling system. Int. J. Climatol. 29, 2241–2255.

Marengo, J.A., Ambrizzi, T., da Rocha, R.P., Alves, L.M., Cuadra, S.V., Valverde, M.C., et al., 2010. Future change of climate in South America in the late twenty-first century: intercomparison of scenarios from three regional climate models. Clim. Dyn. 35, 1089–1113.

Marques, A.R., Atman, A.P.F., Silveira, F.A.O., Lemos-Filho, J.P., 2014. Are seed germination and ecological breadth associated? Testing the regeneration niche hypothesis with bromeliads in a heterogeneous neotropical montane vegetation. Plant Ecol. 215, 517–529.

Mayrinck, R.C., Vilela, L.C., Pereira, T.M., Rodrigues-Junior, A.G., Davide, A.C., Vaz, T.A.A., 2019. Seed desiccation tolerance/sensitivity of tree species from Brazilian biodiversity hotspots: considerations for conservation. Trees 33, 777–785.

Medina, B.M.O., Fernandes, G.W., 2007. The potential of natural regeneration of rocky outcrop vegetation on rupestrian field soils in "Serra do Cipó,". Brazil. Rev. Brasil. Bot. 30, 665–678.

Messeder, J.V.S., Silveira, F.A.O., Cornelissen, T.G., Fuzessy, L.F., Guerra, T.J., 2021. Frugivory and seed dispersal in a hyperdiverse plant clade and its role as a keystone resource for the Neotropical fauna. Ann. Bot. 127, 577–595.

Miranda, A.C., Miranda, H.S., de Fátima Oliveira Dias, I., de Souza Dias, B.F., 1993. Soil and air temperatures during prescribed cerrado fires in Central Brazil. J. Trop. Ecol. 9, 313–320.

Miranda, H.S., Sato, M.N., Neto, W.N., Aires, F.S., 2009. Fires in the cerrado, the Brazilian savanna. In: Cochrane, M.A. (Ed.), Tropical Fire Ecology. Springer, Berlin, pp. 427–450.

Musso, C., Miranda, H.S., Aires, S.S., Bastos, A.C., Soares, A.M.V.M., Loureiro, S., 2015. Simulated post-fire temperature affects germination of native and invasive grasses in cerrado (Brazilian savanna). Plant Ecol. Divers. 8, 219–227.

Myers, N., Mittermeier, R.A., Mittermeier, C.G., da Fonseca, G.A.B., Kent, J., 2000. Biodiversity hotspots for conservation priorities. Nature 403, 853–858.

Oliveira-Filho, A.T., Ratter, J.A., 2002. Vegetation physiognomies and woody flora of the Cerrado biome. In: Oliveira, P., Marquis, R. (Eds.), The Cerrados of Brazil. Columbia University Press, New York, pp. 91–120.

Oliveira, T.G.S., Diamantino, I.P., Garcia, Q.S., 2017. Dormancy cycles in buried seeds of three perennial *Xyris* (Xyridaceae) species from the Brazilian *campo rupestre*. Plant Biol. 19, 818–823.

Oliveira, T.G.S., Souza, M.G.M., Garcia, Q.S., 2018. Seed tolerance to environmental stressors in two species of *Xyris* from Brazilian campo rupestre: effects of heat shock and desiccation. Flora 238, 210–215.

Oliveira, H.R., Staggemeier, V.G., Quintino Faria, J.E., de Oliveira, G., Diniz-Filho, J.A.F., 2019. Geographical ecology and conservation of *Eugenia* L. (Myrtaceae) in the Brazilian Cerrado: past, present and future. Austral Ecol. 44, 95–104.

Oliveira, T.G.S., Duarte, A.A., Diamantino, I.P., Garcia, Q.S., 2021. Thermal niche for seed germination of *Xyris* species from Brazilian montane vegetation: implications for climate change. Plant Species Biol. 36, 284–294.

Olson, D.M., Dinerstein, E., Wikramanayake, E.D., Burgess, N.D., Powell, G.V.N., Underwood, E.C., et al., 2001. Terrestrial ecoregions of the world: a new map of life on Earth. Bioscience 51, 933–938.

Ooi, M.K.J., Auld, T.D., Denham, A.J., 2012. Projected soil temperature increase and seed dormancy response along an altitudinal gradient: implications for seed bank persistence under climate change. Plant Soil 353, 289–303.

Paredes, M.V.F., Cunha, A.L.N., Musso, C., Aires, S.S., Sato, M.N., Miranda, H.S., 2018. Germination responses of native and invasive Cerrado grasses to simulated fire temperatures. Plant Ecol. Divers. 11, 193–203.

Pausas, J.G., Lamont, B.B., Paula, S., Appezzato-da-Glória, B., Fidelis, A., 2018. Unearthing belowground bud banks in fire-prone ecosystems. New Phytol. 217, 1435–1448.

Pereira Júnior, A.C., Oliveira, S.L.J., Pereira, J.M.C., Turkman, M.A.A., 2014. Modelling fire frequency in a Cerrado savanna protected area. PLoS One 9, e102380. Available from: https://doi.org/10.1371/journal.pone.0102380.

Pereira, W.V.S., Faria, J.M.R., Tonetti, O.A.O., Silva, E.A.A.da, 2012. Desiccation tolerance of *Tapirira obtusa* seeds collected from different environments. Rev. Brasil. Sem. 34, 388–396.

Pereira, W.V.S., Faria, J.M.R., José, A.C., Tonetti, O.A.O., Ligterink, W., Hilhorst, H.W.M., 2017. Is the loss of desiccation tolerance in orthodox seeds affected by provenance? South African J. Bot. 112, 296–302.

Perez, T.M., Feeley, K.J., 2021. Weak phylogenetic and climatic signals in plant heat tolerance. J. Biogeogr. 48, 91–100.

Pilon, N.A.L., Hoffmann, W.A., Abreu, R.C.R., Durigan, G., 2018. Quantifying the short-term flowering after fire in some plant communities of a cerrado grassland. Plant Ecol. Divers. 11, 259–266.

Pilon, N.A.L., Cava, M.G.B., Hoffmann, W.A., Abreu, R.C.R., Fidelis, A., Durigan, G., 2021. The diversity of post-fire regeneration strategies in the cerrado ground layer. J. Ecol. 109, 154–166.

Rajapakshe, R.P.V.G.S.W., Turner, S.R., Cross, A.T., Tomlinson, S., 2020. Hydrological and thermal responses of seeds from four co-occurring tree species from southwest Western Australia. Conserv. Physiol. 8, coaa021. Available from: https://doi.org/10.1093/conphys/coaa021.

Ramos-Neto, M.B., Pivello, V.R., 2000. Lightning fires in a Brazilian savanna national park: rethinking management strategies. Environ. Manage. 26, 675–684.

Ramos, D.M., Liaffa, A.B.S., Diniz, P., Munhoz, C.B.R., Ooi, M.K.J., Borghetti, F., et al., 2016. Seed tolerance to heating is better predicted by seed dormancy than by habitat type in Neotropical savanna grasses. Int. J. Wildl. Fire 25, 1273. Available from: https://doi.org/10.1071/WF16085.

Ramos, D.M., Diniz, P., Ooi, M.K.J., Borghetti, F., Valls, J.F.M., 2017. Avoiding the dry season: dispersal time and syndrome mediate seed dormancy in grasses in Neotropical savanna and wet grasslands. J. Veg. Sci. 28, 798–807.

Ramos, D.M., Valls, J.F.M., Borghetti, F., Ooi, M.K.J., 2019. Fire cues trigger germination and stimulate seedling growth of grass species from Brazilian savannas. Am. J. Bot. 106, 1190–1201.

Ranieri, B.D., Pezzini, F.F., Garcia, Q.S., Chautems, A., França, M.G.C., 2012. Testing the regeneration niche hypothesis with Gesneriaceae (tribe Sinningiae) in Brazil: implications for the conservation of rare species. Austral Ecol. 37, 125–133.

Ribeiro, J.F., Walter, B.M.T., 2008. As principais fitofisionomias do bioma Cerrado. In: Sano, S.M., Almeida, S.P., Ribeiro, J.F. (Eds.), Cerrado: Ecologia e Flora. Embrapa Cerrados. Embrapa Inovação Tecnológica, Brasília, pp. 151–212.

Ribeiro, L.C., Pedrosa, M., Borghetti, F., 2013. Heat shock effects on seed germination of five Brazilian savanna species. Plant Biol. 15, 152–157.

Ribeiro, L.C., Borghetti, F., 2014. Comparative effects of desiccation, heat shock and high temperatures on seed germination of savanna and forest tree species. Austral Ecol. 39, 267–278.

Ribeiro, L.C., Barbosa, E.R.M., van Langevelde, F., Borghetti, F., 2015. The importance of seed mass for the tolerance to heat shocks of savanna and forest tree species. J. Veg. Sci. 26, 1102–1111.

Ribeiro, G.V.T., Teixido, A.L., Barbosa, N.P.U., Silveira, F.A.O., 2016. Assessing bias and knowledge gaps on seed ecology research: implications for conservation agenda and policy. Ecol. Appl. 26, 2033–2043.

Ribeiro, R.M., Tessarolo, G., Soares, T.N., Teixeira, I.R., Nabout, J.C., 2019. Global warming decreases the morphological traits of germination and environmental suitability of *Dipteryx alata* (Fabaceae) in Brazilian Cerrado. Acta Bot. Brasil. 33, 446–453.

Rodrigues-Junior, A.G., Baskin, C.C., Baskin, J.M., Garcia, Q.S., 2018. Sensitivity cycling in physically dormant seeds of the Neotropical tree *Senna multijuga* (Fabaceae). Plant Biol. 20, 698–706.

Rosbakh, S., Poschlod, P., 2015. Initial temperature of seed germination as related to species occurrence along a temperature gradient. Funct. Ecol. 29, 5–14.

Salazar, A., Goldstein, G., Franco, A.C., Miralles-Wilhelm, F., 2011. Timing of seed dispersal and dormancy, rather than persistent soil seed-banks, control seedling recruitment of woody plants in Neotropical savannas. Seed Sci. Res. 21, 103–116.

Salazar, A., Goldstein, G., Franco, A.C., Miralles-Wilhelm, F., 2012. Differential seedling establishment of woody plants along a tree density gradient in Neotropical savannas. J. Ecol. 100, 1411–1421.

Salazar, A., Baldi, G., Hirota, M., Syktus, J., McAlpine, C., 2015. Land use and land cover change impacts on the regional climate of non-Amazonian South America: a review. Glob. Planet. Change 128, 103–119.

Sales, N.M., Pérez-García, F., Silveira, F.A.O., 2013. Consistent variation in seed germination across an environmental gradient in a Neotropical savanna. S. African J. Bot. 87, 129–133.

Schmidt, I.B., Eloy, L., 2020. Fire regime in the Brazilian savanna: recent changes, policy and management. Flora 268, 151613. Available from: https://doi.org/10.1016/j.flora.2020.151613.

Sentinella, A.T., Warton, D.I., Sherwin, W.B., Offord, C.A., Moles, A.T., 2020. Tropical plants do not have narrower temperature tolerances, but are more at risk from warming because they are close to their upper thermal limits. Glob. Ecol. Biogeogr. 29, 1387–1398.

Silva, P.S., Bastos, A., Libonati, R., Rodrigues, J.A., DaCamara, C.C., 2019. Impacts of the 1.5°C global warming target on future burned area in the Brazilian Cerrado. For. Ecol. Manage. 446, 193–203.

Silveira, F.A.O., Fernandes, G.W., 2006. Effect of light, temperature and scarification on the germination of *Mimosa foliolosa* (Leguminosae) seeds. Seed Sci. Technol. 34, 585–592.

Simon, M.F., Grether, R., de Queiroz, L.P., Skema, C., Pennington, R.T., Hughes, C.E., 2009. Recent assembly of the Cerrado, a neotropical plant diversity hotspot, by in situ evolution of adaptations to fire. Proc. Natl. Acad. Sci. USA 106, 20359–20364.

Simon, L.M., Oliveira, G., Barreto, B.de S., Nabout, J.C., Rangel, T.F.L.V.B., Diniz-Filho, J.A.F., 2013. Effects of global climate changes on geographical distribution patterns of economically important plant species in Cerrado. Rev. Árv. 37, 267–274.

Siqueira, M.F., Peterson, A.T., 2003. Consequences of global climate change for geographic distributions of Cerrado tree species. Biota Neotrop. 3, 1–14.

Soares da Mota, L.A., Garcia, Q.S., 2013. Germination patterns and ecological characteristics of *Vellozia* seeds from high-altitude sites in southeastern Brazil. Seed Sci. Res. 23, 67–74.

Strassburg, B.B.N., Brooks, T., Feltran-Barbieri, R., Iribarrem, A., Crouzeilles, R., Loyola, R., et al., 2017. Moment of truth for the Cerrado hotspot. Nat. Ecol. Evol. 1, 0099. Available from: https://doi.org/10.1038/s41559-017-0099.

Valladares, F., Matesanz, S., Guilhaumon, F., Araújo, M.B., Balaguer, L., Benito-Garzón, M., et al., 2014. The effects of phenotypic plasticity and local adaptation on forecast of species range shifts under climate change. Ecol. Lett. 17, 1351–1364.

Van Klinken, R.D., Flack, L.K., Pettit, W., 2006. Wet-season dormancy release in seed banks of a tropical leguminous shrub is determined by wet heat. Ann. Bot. 98, 875–883.

Vasconcelos, H.L., Maravalhas, J.B., Cornelissen, T., 2017. Effects of fire disturbance on ant abundance and diversity: a global *meta*-analysis. Biodivers. Conserv. 26, 177–188.

Vaz, T.A.A., Davide, A.C., Rodrigues-Junior, A.G., Nakamura, A.T., Tonetti, O.A.O., da Silva, E.A.A., 2016. *Swartzia langsdorffii* Raddi: morphophysiological traits of a recalcitrant seed dispersed during the dry season. Seed Sci. Res. 26, 47–56.

Velazco, S.J.E., Villalobos, F., Galvão, F., De Marco Júnior, P., 2019. A dark scenario for Cerrado plant species: effects of future climate, land use and protected areas ineffectiveness. Divers. Distrib. 25, 660–673.

Veldman, J.W., Buisson, E., Durigan, G., Fernandes, G.W., Le Stradic, S., Mahy, G., et al., 2015. Toward an old-growth concept for grasslands, savannas, and woodlands. Front. Ecol. Environ. 13, 154–162.

Vera, C., Silvestri, G., Liebmann, B., González, P., 2006. Climate change scenarios for seasonal precipitation in South America from IPCC-AR4 models. Geophys. Res. Lett. 33, 2–5.

Vidal Jr, J.D., de Souza, A.P., Koch, I., 2019. Impacts of landscape composition, marginality of distribution, soil fertility and climatic stability on the patterns of woody plant endemism in the Cerrado. Glob. Ecol. Biogeogr. 28, 904–916.

Violle, C., Enquist, B.J., McGill, B.J., Jiang, L., Albert, C.H., Hulshof, C., et al., 2012. The return of the variance: intraspecific variability in community ecology. Trends Ecol. Evol. 27, 244–252.

Walck, J.L., Hidayati, S.N., Dixon, K.W., Thompson, K., Poschlod, P., 2011. Climate change and plant regeneration from seed. Glob. Change Biol. 17, 2145–2161.

Wang, X., Alvarez, M., Donohue, K., Ge, W., Cao, Y., Liu, K., et al., 2021. Elevation filters seed traits and germination strategies in the eastern Tibetan Plateau. Ecography 44, 242–254.

Wiggers, M.S., Hiers, J.K., Barnett, A., Boyd, R.S., Kirkman, L.K., 2017. Seed heat tolerance and germination of six legume species native to a fire-prone longleaf pine forest. Plant Ecol. 218, 151–171.

Willis, C.G., Baskin, C.C., Baskin, J.M., Auld, J.R., Venable, D.L., Cavender-Bares, J., et al., 2014. The evolution of seed dormancy: environmental cues, evolutionary hubs, and diversification of the seed plants. New Phytol. 203, 300–309.

Wyse, S.V., Dickie, J.B., 2017. Predicting the global incidence of seed desiccation sensitivity. J. Ecol. 105, 1082–1093.

Zirondi, H.L., Silveira, F.A.O., Fidelis, A., 2019. Fire effects on seed germination: heat shock and smoke on permeable vs impermeable seed coats. Flora 253, 98–106.

Zupo, T., Daibes, L.F., Pausas, J.G., Fidelis, A., 2021. Post-fire regeneration strategies in a frequently burned Cerrado community. J. Veg. Sci. 32, e12968. Available from: https://doi.org/10.1111/jvs.12968.

Chapter 15

Plant regeneration from seeds in savanna woodlands of Southern Africa

Emmanuel N. Chidumayo[1] and Gudeta W. Sileshi[2,3]

[1]Makeni Savanna Research Project, Lusaka, Zambia, [2]College of Natural and Computational Sciences, Addis Ababa University, Addis Ababa, Ethiopia, [3]School of Agricultural, Earth and Environmental Sciences, University of KwaZulu-Natal, Pietermaritzburg, South Africa

Introduction

Savanna woodlands in Sub-Saharan Africa are broadly categorized into the Sudanian and Zambezian woodlands (Assédé et al., 2020; White, 1983). With over 8500 known plant species (54% endemic), the Zambezian woodlands are three times more speciose than the Sudanian woodlands, which contain no more than 2750 plant species (Assédé et al., 2020). However, 46% of the species in Sudanian woodlands also occur in the Zambezian woodlands, while only 24% of the species in the latter occur in the Sudanian woodlands (Assédé et al., 2020).

The savanna woodlands of southern Africa extend over two main phytoregions (*sensu* White, 1983): the Zambezian Regional Centre of Endemism (ZRCE) and the Kalahari-Highveld Regional Transition Zone (KHRTZ). The KHRTZ is a transition zone between the ZRCE and the Karoo-Namib Regional Centre of Endemism and extends from the northwest in southern Angola and Namibia southeastwards to the Eastern Cape of South Africa over an area of about 122.3 million ha (Fig. 15.1). The ZRCE is the most extensive savanna woodland biome in southern Africa and covers about 377 million ha in the southern Democratic Republic of Congo, Angola, Botswana, Malawi, Mozambique, Namibia, Tanzania, Zambia, Zimbabwe, and the northeastern portion of South Africa (Fig. 15.1). The total flora in the ZRCE and KHRTZ is estimated to be 8500 and 3000 species, respectively (White, 1983).

The savanna woodlands of southern Africa experience a single rainy season (November-April), and annual rainfall ranges from about 950 mm in ZRCE to 440 mm in KHRTZ. The ZRCE therefore represents mesic woodlands, while KHRTZ represents semiarid woodlands. Average annual monthly temperature in ZRCE ranges from 17°C to 24°C with a mean of 21°C. In KHRTZ, the temperature range is 12°C–22°C with a mean of 18°C. The relatively humid period when precipitation is above average monthly temperature is 6 months in mesic woodlands and 4.5 months in semiarid woodlands.

Wyse and Dickie (2017) estimate that 1.9% of the plant species in savanna woodlands have desiccation-sensitive (recalcitrant) seeds. Thus, the majority of seed plants in southern Africa woodlands have orthodox seeds, which tolerate desiccation and germinate easily when water and temperature are favorable. The number of species with seed dormancy is not known, but no serotinous species have been reported in these woodlands. In spite of a large pool of plant species (>11,000) in southern Africa woodlands (White, 1983), germination has been studied in only a low proportion of them. Regeneration studies that consider climate change effects are even fewer.

Climate change refers to a significant change in the trend of a climate variable, such as temperature or rainfall, over time, although it can also refer to a change in the frequency of climatic events, such as floods and droughts. Another aspect of climate change is variability, which refers to the degree of departure of a climate variable from its mean value, and this can occur even in the absence of a significant trend. Climate change therefore is analyzed on the basis of long-term data spanning decades. In contrast, the majority of regeneration studies in African savanna woodlands is based on very short timescales of one to a few years and therefore is often inadequate for testing the linkage between climate change and regeneration. Not only can climate be described by a large number of variables, but it is important in regeneration studies to specify whether the effect of the variable is annual or occurring during a specific period of the year (Krebs and Berteaux, 2006) and which aspect of the regeneration process is affected. Unless the effect of a climate

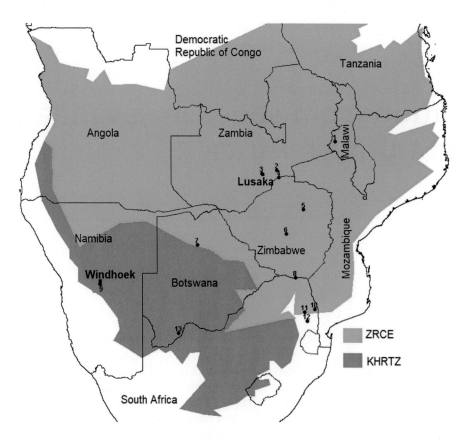

FIGURE 15.1 Location of 13 study sites (filled circles) at which plant regeneration from seeds and climate factors were investigated in the two savanna woodland biomes of southern Africa: the Zambezian Regional Centre of Endemism (ZRCE) and the Kalahari-Highveld Regional Transition Zone (KHRTZ). Brief details of the investigations conducted at the sites are given in Table 15.1. *Adapted from White, F., 1983. The Vegetation of Africa. UNESCO, Paris.*

variable is specified with regards to time within a phenological calendar, it is difficult to understand the mechanism by which it affects regeneration from seeds.

Many of the regeneration phases relate to the parental genotype and environment, while seed germination and establishment may be determined by the seed microenvironment. Indeed, seed traits, including their quality, are also influenced by the parental genetic constitution and environment. Therefore, climatic factors affecting the parent plant may manifest themselves in the seed germination behavior. Understanding these climatic effects on regeneration therefore requires appropriate study designs and implementation that most of the available studies in southern Africa often fail to meet. Furthermore, since climate cannot be manipulated, experiments on the effects of climate factors on regeneration from seeds have to be observational and long enough to identify a clear climate signal in the regeneration process.

This chapter represents the first step in identifying linkages between plant regeneration from seeds and climate change in southern African woodlands. We considered regeneration from seeds at three levels: (1) fruit production, (2) seed germination, and (3) seedling establishment and survival. Our focus ranges from the individual plant to population and community. We first review climate change patterns in the southern African region before considering plant regeneration studies and the potential responses to climate factors. We conclude the chapter with issues that need consideration in future studies to improve our mechanistic understanding of how climate change is likely to affect plant regeneration from seeds in a future warmer climate in southern Africa.

Data acquisition and analytical approach

This reassessment of plant regeneration from seeds is largely based on data from published and unpublished sources that were subjected to statistical analysis to determine simple linear relationships between a regeneration phase and specific climate factors. Data were obtained from tables in published papers. In cases where the data were in the form of graphs, these were scanned and the images digitized using the graph procedure in the Grapher software (Golden Software.com). Data presented as proportions, such as fruiting plants in a population, seed germination percentage, and seedling mortality, were first subjected to a normality test using the Shapiro–Wilk normality (W) test at a significance probability (P) of .05. If the data failed the normality test, they were arcsine-square root transformed to meet the requirements of a normal distribution (Sokal and Rohlf, 1995) prior to analysis. The data were then subjected to linear regression analysis to determine the significance of the relationship between a particular regeneration variable and a climate variable. If a significant relationship was found, we assessed if the climate factor(s) showed a trend over time in order to determine its potential to affect regeneration in a future climate regime. If the sample size was small (<5), we used the Spearman rank correlation to determine the relationship between the regeneration variable and the climate factor. All statistical analyses were done in Statistix 9 (Analytical Software, 1985–2008).

Regional climate trends in southern Africa

Although most precipitation indices in the ZRCE indicate a lack of consistent or statistically significant trends, average total precipitation has decreased over time (New et al., 2006). Parts of southern Africa are likely to experience a 5%–10% reduction in mean annual precipitation (Shongwe et al., 2009), while extreme events such as droughts and floods have occurred with increasing frequency, duration, and severity in the region (Adisa et al., 2020).

Extreme climatic events in southern Africa are driven by the El Niño–Southern Oscillation (ENSO). El Niño events are usually associated with below-normal rainfall over much of southern Africa (Manson, 2001). In the recent past, El Niño events in southern Africa occurred during the 1991/92, 1997/98, 2002/03, 2015/16, and 2019/20 seasons. La Niña, the opposite of El Niño, tends to be associated with above-average rainfall and occurs with lower frequency than El Niño, and in the recent past La Niña in southern Africa occurred in 1995/1996 and 1999/2000.

The savanna woodlands of southern Africa experienced cooler temperatures from the early 1900s up to the 1930s before stabilizing around the normal until the 1970s, after which temperatures rose rather steeply (Hulme et al., 2001). The scenarios discussed by Hulme et al. (2001) indicate that temperatures will increase by 1.34°C across African ecoregions by 2050. Unganai (1996) developed climate scenarios for southern Africa using two equilibrium Global Circulation Models (CCCM and GFDL3) and observed that both models gave simulated changes in mean surface air temperature of 2°C to 4°C across southern Africa under doubling of the CO_2 concentration. Other forecasts of mean annual temperature in southern Africa project an increase of 3.5°C to over 5.0°C by the end of the 21st century (Niang et al., 2014). Maximum and minimum temperatures in South Africa increased during the period 1960–2010, except for the central interior, where minimum temperatures decreased (MacKellar et al., 2014). This underscores the importance of recognizing that regional climate modeling studies may miss site-specific dynamics because local climatic processes may not always be adequately reflected in regional and global climate models (MacKellar et al., 2014). For example, it is not unusual for neighboring locations to experience opposite climatic trends (King'uyu et al., 2000). This contradiction may also occur at a regional scale as a study in the Limpopo Province of South Africa found that whereas much of the Province experienced an increase in mean annual temperature from 1950 to 1999 a few areas experienced a decrease (Tshiala et al., 2011).

Local climate trends in southern African woodlands

A plant species' niche includes the phenological niche, and the regeneration niche, which includes the production of viable seeds following flowering and pollination, as well as seed dispersal. Microsite or microhabitat conditions are critical for ensuring seed germination and establishment. It is for this reason that local, rather than regional or global, climate conditions become more important in studying regeneration from seeds in relation to climate change.

To demonstrate this, we describe the climate trends at Mount Makulu in Lusaka (Zambia) in the neighborhood of study site 3 (Makeni) and Windhoek (Namibia) in the neighborhood of study site 9 (Krumhuk) (Fig. 15.1). The climate trends at the two weather stations indicate that maximum and minimum temperatures have been increasing during 2008–20 at Windhoek ($R^2 \geq 0.59$, $P \leq .003$) but not at Lusaka ($R^2 < 0.11$, $P > .30$). The coefficient of variation (CV) in maximum temperatures is similar at the two stations with a value of about 3.4%, but with a CV in minimum temperatures of nearly 7% at Windhoek compared to 4% at Lusaka. The long-term trend in rainfall is not significant at either Lusaka ($R^2 = 0.001$, $P = .87$) or Windhoek ($R^2 = 0.04$, $P = .07$). However, the CV in rainfall increased from 39% during 1950–70 to 57% during 1990–2020 at Windhoek. Similarly, the CV in rainfall increased from 12% during 1950–70 to 25% during 1990–2020 at Lusaka. Therefore, although there appears to be no significant trend in rainfall at Lusaka or Windhoek during 1950–2020 annual rainfall variability has increased at both stations in the last 30 years.

Field observational studies on plant regeneration from seeds

Data on plant regeneration from seeds and climate were obtained from 11 studies conducted at 13 field sites (Fig. 15.1). Six of the studies were short term with a duration of less than 5 years, while the remaining five lasted for more than 5 years (Table 15.1). Two studies investigated the effect of desiccation (insolation and/or dry spells at the onset of the

TABLE 15.1 Description of 13 sites and investigations conducted on plant regeneration and climate factors in savanna woodlands of southern Africa.

Site code	Site name (Country)	Longitude (degree)	Latitude (degree)	Type of regeneration data	Study duration (years)	Climate factors investigated			Source
						Rainfall	Temperature	Desiccation	
1	Kasungu (Malawi)	33.208	−13.333	Seedling mortality	2			√	Robertson (1984)
2	Chakwenga (Zambia)	29.21	−15.26	Fruiting	10	√	√		Chidumayo (1997) and unpublished
3	Makeni (Zambia)	28.18	−15.47	Fruiting, seed germination, and seedling mortality	20	√	√		Chidumayo (2006, 2019) and unpublished
4	Mana Pools (Zimbabwe)	29.3667	−15.717	Fruiting	8	√	√		Dunham (1990)
5	Harare (Zimbabwe)	31.033	−17.833	Seed germination and seedling mortality	1			√	Strang (1966)
6	Gweru (Zimbabwe)	29.85	−19.45	Seed germination and seedling mortality	3	√			Mlambo and Nyathi (2004)
7	Maun (Botswana)	23.777	−20.16	Fruiting	3	√	√		Sekhwela and Yates (2007)
8	Venda (South Africa)	30.467	−22.317	Fruiting	3	√			Venter and Witkowski (2011)
9	Krumhuk (Namibia)	17.088	−22.734	Fruiting	9	√			Joubert et al. (2013)
10	Satara (South Africa)	31.778	−24.398	Seed germination and seedling mortality	2	√			Botha (2006)
11	Wits Rural Facility (South Africa)	31.103	−24.564	Fruiting	6	√	√		Helm et al. (2011)
12	Pretoriuskop (South Africa)	31.267	−25.167	Seed germination and seedling mortality	2	√			Botha (2006)
13	Tsabong (Botswana)	22.467	−25.944	Fruiting	3	√	√		Sekhwela and Yates (2007)

wet season) on seedling mortality and four considered only rainfall on fruiting phenology. The remaining five studies considered the effect of both rainfall and temperature on different aspects of regeneration. Six of the 11 studies involved a single species, while the remainder involved multiple species at each study site. Because of the small number of studies in the southern African savanna woodlands, we also included in this review, for comparison, two studies conducted in African tropical forests. One study, conducted at Kibale (30.425°E, 0.45°N) in Uganda (Chapman et al., 2005), included *Parinari excelsa* and *Dombeya kirkii*, tree species that also occur in ZRCE mesic woodlands of southern Africa. The Kibale study considered the effects of rainfall and temperature on reproductive phenology of tree species over a 13-year period. The other study was conducted at Lopé (11.583°E, −0.167°S) in Gabon considered rainfall and temperature effects on fruiting phenology of rainforest tree species over a period of 8 years (Tutin and Fernandez, 1993).

Fruit production and climate

In four woody plant species in which the proportion of plants in the population that produced fruits was studied for 10–12 years at Chakwenga, central Zambia, temperature was the common factor that significantly affected the proportion of fruiting plants (Table 15.2). Minimum temperature significantly reduced the proportion of fruiting plants. However, the time during which minimum temperature affected fruiting frequency varied from July in *Julbernardia globiflora* to November in *Lannea edulis* and December in *Uapaca kirkiana* (Table 15.2). In *Isoberlinia angolensis*, maximum temperatures in March explained one-third of the variance in the proportion of fruiting trees (Table 15.2).

Uapaca kirkiana flowers in December at the Chakwenga study site, and it is therefore probable that this is the reproductive phase that is affected and through which minimum temperatures in December determine the proportion of trees fruiting in the population. The trend in December minimum temperature at Lusaka, nearest weather station to the study site, shows a significant increase ($R^2 = 0.12$, $P = .02$) from the 1960s to date, which implies that the proportion of *U. kirkiana* trees fruiting may decrease in the future with potential negative effects on the species' regeneration from seeds (Fig. 15.2A). The mechanisms through which temperature affects fruiting frequency in the other three species are unclear and require further research. However, in the case of *I. angolensis* the increasing trend in maximum temperature in March ($R^2 = 0.12$, $P = .01$), which has a positive effect on the proportion of fruiting trees (Table 15.2), implies that fruiting in this species may potentially increase in the future, if a warming trend continues (Fig. 15.2B). The trend in minimum temperature in July and November, which affects the proportion of fruiting in *J. globiflora* and *L. edulis*, respectively, is not significant, implying that no significant changes are expected in the fruiting pattern in these species under future climate scenarios.

Chapman et al. (2005) studied the effects of rainfall and temperature on the reproductive phenology of trees over a 13-year period in Uganda that included *Parinari excelsa* and *Dombeya kirkii* that also occur in the ZRCE (White, 1962; Storrs, 1979). A reanalysis of the data in Chapman et al. (2005) confirmed that a significant percentage (43%, $P = .002$) of the variance in the proportion of fruiting *P. excelsa* trees was explained by annual maximum temperature, while 55% ($P = .006$) of the variance in the proportion of fruiting *D. kirkii* trees was explained by minimum temperature. Because the data used were on an annual basis, it is difficult to explain the mechanism through which temperature influenced the variance in the proportion of fruiting trees in these two species.

Fruit production has been studied in 11 plant species at eight sites in southern African woodlands (Table 15.2). No climate factor was found to significantly affect fruit production in *Senegalia (Acacia) mellifera*, *Vachellia (Acacia) luedertzii*, or *V. erioloba* at two sites in Botswana (Sekhwela and Yates, 2007). In contrast, a positive relationship between annual rainfall and fruit production was found in *S. mellifera* at Krumhuk in Namibia (based on data in Joubert et al., 2013). At this Namibian site, annual rainfall explained two-thirds of the variance in *S. mellifera* fruit production (Table 15.2). Botha (2006) reported that low rainfall at the onset of the rainy season resulted in low fruit production in *Sclerocarya birrea* in South Africa.

Although Venter and Witkowski (2011) found no relationship between rainfall and fruit production in *Adansonia digitata* over a 3-year period at Wits Rural Facility in South Africa, a reanalysis of their data using the Spearman rank correlation statistic revealed that high annual rainfall significantly reduced fruit production in *A. digitata* (Table 15.2). High annual rainfall was also among the factors that reduced fruit production in *A. digitata* in Benin (Assogbadjo et al., 2005), and Msalilwa et al. (2019) found a significant negative correlation between rainfall and *A. digitata* density in semiarid Tanzania, which may be linked to low fruit production and possibly low potential regeneration in high-rainfall areas of the species range.

At Mana Pools in Zimbabwe, rainfall from February to April explained 59% of the variance in mean fruits per *Faidherbia (Acacia) albida* tree (based on data in Dunham, 1990). Fruit development in *F. albida* occurs from February

TABLE 15.2 Climate impact on fruiting (i.e., proportion of plants with fruits) and fruit production in plant species in southern African woodlands.

Species	Data source	Woodland zone	Growth form	Site	Number of observation years	Significant climate factor	Explained variation (R^2)	Significance level (P)	Impact
A: Proportion of plants with fruits									
Isoberlinia angolensis	Chidumayo (1997) and unpublished	ZRCE	Tree	Chakwenga	12	MarT_{max}	0.33	.06	Positive
Julbernardia globiflora	Chidumayo (1997) and unpublished	ZRCE	Tree	Chakwenga	12	JulT_{min}	0.46	.02	Negative
Uapaca kirkiana	Chidumayo (unpublished)	ZRCE	Tree	Chakwenga	10	DecT_{min}	0.58	.02	Negative
Lannea edulis	Chidumayo (2006) and unpublished	ZRCE	Subshrub	Makeni	11	NovT_{min}	0.53	.02	Negative
B: Average fruits produced per plant									
Faidherbia albida	Dunham (1990)	ZRCE	Tree	Mana Pools	8	Feb–Apr rainfall	0.59	.04	Positive
Senegalia mellifera	Sekhwela and Yates (2007)	KHRTZ	Tree	Maun and Tsabong	3	None			
Vachellia luederitzii			Tree		3	None			
Vachellia erioloba			Tree		3	None			
Sclerocarya birrea	Helm et al. (2011)	KHRTZ	Tree	Wits Rural Facility	7	None			
Adansonia digitata	Venter and Witkowski (2011)	KHRTZ	Tree	Venda	3	Rainfall	−1.00^{Rs}		Negative
Senegalia mellifera	Joubert et al. (2013)	KHRTZ	Tree	Krumhuk	9	Rainfall	0.66	0.01	Positive
Isoberlinia angolensis	Chidumayo (1997) and unpublished	ZRCE	Tree	Chakwenga	12	OctT_{max}	0.43	0.02	Negative
Strychnos spinosa	Chidumayo (unpublished)	ZRCE	Tree	Makeni	16	MayT_{max}	0.53	0.002	Negative
Lannea edulis	Chidumayo (2006, 2019) and unpublished	ZRCE	Subshrub	Makeni	9	Nov Rainfall	0.70	0.009	Positive
Ledebouria sp.	Chidumayo (unpublished)	ZRCE	Bulbous herb	Makeni	11	SepT_{min}	0.63	0.004	Negative

Significant climate factors are abbreviated as follows: first three letters refer to month, T_{max} to maximum temperature, and T_{min} to minimum temperature. Superscript ^{Rs} for Adansonia digitata refers to the Spearman rank correlation statistic.

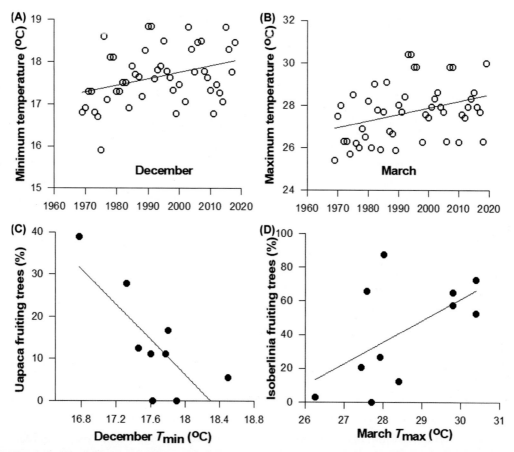

FIGURE 15.2 Trends in (A) minimum temperature in December and (B) maximum temperature in March at Lusaka, Zambia, and relationship between the proportion of fruiting trees in (C) *Uapaca kirkiana* and minimum temperature in December and (D) *Isoberlinia angolensis* and maximum temperature in March at Chakwenga study site in Zambia.

to April (Storrs, 1979). Therefore, it is probable that rainfall at the end of the wet season affects fruit production during the development phase, and low rainfall during this period may cause more fruit abortion, thereby resulting in a low number of mature fruits per tree. Since there is no significant trend in rainfall during February to April at Lusaka, 130 km to the northwest of Mana Pools, the current pattern in fruit production in *F. albida* is unlikely to change under a future climate.

In *Lannea edulis*, rainfall in November explained 70% of the variance in fruit production at the Makeni site in central Zambia (Table 15.2). Leaf flush and production in *L. edulis* occur in November (Chidumayo, 2006). Since this subshrub is often shaded by grasses and other herbs in the wet season (White, 1976), perhaps vegetative growth in November determines the quantity of food reserves produced and stored that subsequently support fruit production in the following year. There has been a significant decreasing trend in November rainfall at Lusaka ($R^2 = 0.10$, $P = .02$). Therefore, it is projected that fruit production in *L. edulis* at the study site will decrease under a future climate, and this will negatively impact regeneration from seeds in this species (Fig. 15.3).

Nearly 43% of the variance in fruit production per tree in *I. angolensis* at the Chakwenga site in central Zambia (Table 15.2) was explained by maximum temperature in October. This species flowers around October, and it is probable that maximum temperature in October negatively affects fruit production at the flowering phase, perhaps through flower abortion. There is an increasing trend in maximum temperature in October at Lusaka, and this is likely to reduce fruit production in *I. angolensis* under a future warmer climate (Fig. 15.3).

In *Strychnos spinosa*, 53% of the variance in fruit production is explained by maximum temperature in May at the Makeni site in central Zambia. Fruit maturation in *S. spinosa* occurs in winter (May–June). It is likely that maximum temperature in winter negatively affects fruit production through a high loss of fruit. However, there is no significant trend in maximum temperature in May at Lusaka, which implies that fruit production pattern in *S. spinosa* is unlikely to change under a warmer climate.

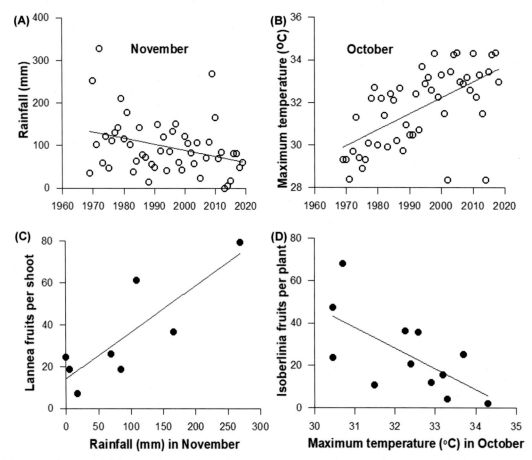

FIGURE 15.3 Trends in (A) rainfall in November and (B) maximum temperature in October at Lusaka, Zambia, and relationship between fruits produced in (C) *Lannea edulis* and rainfall in November and (D) *Isoberlinia angolensis* and maximum temperature in October.

In the bulbous perennial herb, *Ledebouria* sp., at the Makeni site in central Zambia, minimum temperature in September negatively affected fruit production. This species is dormant in the dry season, and therefore it is difficult to determine the mechanism through which minimum temperature affects fruit production. However, there is no significant trend in minimum temperature in September at Lusaka. Therefore, it is projected that the current pattern in fruit production in this species is unlikely to change under a future climate.

A study conducted in Lopé National Park in Gabon (Tutin and Fernandez, 1993) found that temperature significantly influenced fruit production in eight tropical wet forest trees. A reanalysis of their data in Tutin and Fernandez (1993) revealed that fruit production in *Dialium lopense* ($R^2 = 0.55$, $P = .03$), *Diospyros dendo* ($R^2 = 0.62$, $P = .02$), and *Ganophyllum giganteum* ($R^2 = 0.55$, $P = .03$) was significantly influenced by average minimum temperature during the dry period months of July and August, while average maximum temperature significantly influenced ($R^2 = 0.79$, $P = .02$) fruit production in *Cola lizae*. In *Heisteria parvifolia*, a combination of average maximum and minimum temperature significantly influenced fruit production ($R^2 = 0.92$, $P = .002$). In the remaining three species, the influence of minimum temperature on fruit production was marginally significant ($P = .07$) in *Diospyros zenkeri* and *D. polystemma*, but minimum and maximum temperature and rainfall during July–August failed to explain a significant proportion of the variance in fruit production in *Irvingia gabonensis*.

Seedling emergence and climate

Seed quality can be affected at several stages: flowering, pollination, development, and ripening. Identifying the specific climate factors affecting seed quality, and therefore seed germination, is critical in the search for mechanisms that explain changes in seed germination due to climate change.

In a 2-year study, Botha (2006) observed widespread seedling emergence in *Sclerocarya birrea* at Pretoriuskop and Satara in South Africa following a period of heavy rainfall. Generally, the effect of climate factors on seedling emergence has not been adequately investigated and published long-term data are scarce for species in southern African woodlands. We therefore relied on unpublished data collected by the first author (ENC) at the Makeni site in central Zambia (Table 15.3). For the eight woody species studied, rainfall was a significant factor influencing seedling mergence in half of the species, while in the other half temperature was a significant factor (Table 15.3). However, there was diversity in the timing of these significant factors.

Rainfall in January was a significant factor in *Dichrostachys cinerea* seedling emergence, while rainfall in February and April was a significant factor in influencing seedling emergence in *Securidaca longepedunculata* and *Senegalia polyacantha*, respectively (Table 15.3). In the subshrub, *Lannea edulis*, rainfall in October significantly affected seedling emergence. In *D. cinerea* fruit development occurs during the latter half of the wet season. Too much rainfall in January probably negatively affects this shallow-rooting shrub (Zhou et al., 2020), and its reproductive performance may be susceptible to too much rain in January that may result in production of seeds with lower seedling emergence. Currently, there is no significant trend in January rainfall at Lusaka; consequently, the pattern of seed germination in *D. cinerea* is unlikely to change under a future climate. Rainfall in February has a positive effect on seedling emergence in *S. longepedunculata*, and rainfall in April has a similar effect in *S. polyacantha*. Higher rainfall in February appears to facilitate fruit development in *S. longepedunculata* and probably the production of good seeds with high seedling emergence the following wet season. Similarly, more rainfall in April appears to facilitate flowering and probably the production of high-quality seeds in *S. polyacantha* that results in high germination percentages. The trend in rainfall in February and April at Lusaka is not significant; consequently, no change in seedling emergence is expected in *S. longepedunculata* and *S. polyacantha* under a future climate.

Fruit maturation in *L. edulis* occurs in October (Chidumayo, 2006), and more rain in this month may enhance fruit ripening that may increase seed germination. However, there is no significant trend in October rainfall at Lusaka, and therefore no change in seedling emergence is expected in *L. edulis* under a future climate.

TABLE 15.3 Climate impact on seed germination in plant species in southern African woodlands at the Makeni site (vegetation zone: ZRCE).

Species	Growth form	Number of observation years	Significant climate factor	Explained variation (R^2)	Significance level (P)	Impact
Strychnos spinosa	Tree	5	SepT_{max}	0.92	.01	Positive
Securidaca longepedunculata	Tree	5	Feb Rainfall	0.99	.007	Positive
Piliostigma thonningii	Tree	8	May T_{min}	0.82	.005	Negative
Senegalia polyacantha	Tree	6	Apr Rainfall	0.95	.006	Positive
Vachellia sieberiana	Tree	6	Oct T_{max}	0.81	.04	Positive
Dichrostachys cinerea	Tree	5	Jan Rainfall	0.83	.04	Negative
Ziziphus abyssinica	Tree	6	June T_{min}	0.81	.04	Negative
Lannea edulis	Subshrub	16	Oct Rainfall	0.34	.02	Positive
Ledebouria sp.	Bulbous herb	9	July T_{min}	0.83	.002	Positive
Lepeirousia rivularis	Cormous herb	10	Oct T_{max}	0.60	.01	Negative

Significant climate factors are abbreviated as follows: first three letters refer to month, T_{max} maximum temperature, and T_{min} minimum temperature. All analyses were conducted using linear regression on unpublished data collected by the first author (ENC).

Maximum temperature in September and October has positive effects on seedling emergence of *Strychnos spinosa* and *V. sieberiana*, respectively (Table 15.3). This linkage between maximum temperature and seed germination probably operates through the flowering phase, which occurs in October/November in both *S. spinosa* and *V. sieberiana*, which may result in the production of good quality seeds with high germination percentages. There is a significant increasing trend in both September ($R^2 = 0.38$, $P < .0001$) and October ($R^2 = 0.37$, $P < .0001$) maximum temperature at Lusaka, which implies that seedling emergence in *S. spinosa* and *V. sieberiana* may increase under a warmer climate.

Minimum temperature in May and June has significant negative effects on seedling emergence in *Piliostigma thonningii* and *Ziziphus abyssinica*, respectively (Table 15.3). In *P. thonningii*, fruit development occurs in winter, and high minimum temperatures in May may result in the production of low-quality seeds with low germination. A similar mechanism seems to be involved in *Z. abyssinica* (Fig. 15.4A and C). The significant decreasing trend in minimum temperatures in May ($R^2 = 0.45$, $P < .0001$) and June ($R^2 = 0.36$, $P < .0001$) at Lusaka therefore is likely to reduce seedling emergence in *P. thonningii* and *Z. abyssinica* under a future climate.

In the herbaceous *Ledebouria* sp., minimum temperature in July has a positive effect on seedling emergence, while in *Laperousia rivularis* maximum temperature in October has a significant negative effect on seedling emergence (Table 15.3). These species have belowground perennating organs, and the mechanism through which temperature affects seedling emergence is not clear because both species are dormant in the dry season. Nevertheless, there is a significant increasing trend in October maximum temperature (Fig. 15.4B) at Lusaka. Since a high maximum temperature in October reduces germination of *L. rivularis*, this species is likely to experience lower seedling emergence under a future climate (Fig. 15.4B and D). However, there is no significant trend in July minimum temperature at Lusaka, so no change is expected in the germination of *Ledebouria* sp. under a future climate.

Seedling mortality and climate

Most studies of seedling mortality in relation to climate have been conducted over periods of less than 3 years, with some conflicting results. High seedling mortality in *Julbernardia globiflora* at Harare in Zimbabwe was attributed to hot-dry spells at the onset of the rainy season (Strang, 1966). High mortality of newly emerged seedlings of

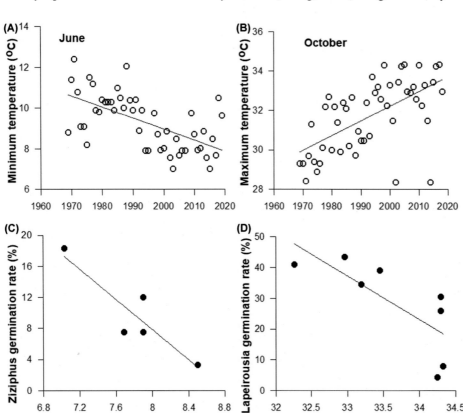

FIGURE 15.4 Trends in (A) minimum temperature in June and (B) maximum temperature in October at Lusaka, Zambia, and relationship between (C) *Ziziphus abyssinica* seed germination and minimum temperature in June and (D) *Lapeirousia rivularis* seed germination and maximum temperature in October.

Colophospermum mopane at Gweru in Zimbabwe was also attributed to water stress (Mlambo and Nyathi, 2004). However, a study at Kasungu in Malawi found that mortality of *Brachystegia spiciformis* seedlings was not linked to dry spells (Robertson, 1984). Botha (2006) also found that tree seedling mortality was highest immediately after germination at Pretoriuskop and Satara in South Africa and that this mortality occurred even under irrigation that minimized the effects of dry spells between rainfall events. Further research is required to understand the causes of high seedling mortality at the onset of the rainy season in southern African woodlands.

At the Makeni site in Zambia, the mortality of seedlings of *Laperousia rivularis* was significantly increased by maximum temperatures in November ($R^2 = 0.69$, $P = .04$). *L. rivularis* seedlings have very small corms located less than 4 cm below the soil surface (Chidumayo, 2002). Soil temperatures at 10 cm depth in savanna woodlands of southern Africa can reach 33°C–37°C in late afternoons during September to November (Jeffers and Boaler, 1966), and they are probably higher at 4 cm with the potential to severely desiccate the small corms of *L. rivularis*. Therefore, it is likely that high temperatures in November cause seedling death through desiccation of the tiny corms. November maximum temperatures have been increasing significantly at Lusaka since the 1960s ($R^2 = 0.38$, $P < .0001$), and it is likely that *L. rivularis* seedling mortality will increase under a future warmer climate and therefore negatively impact regeneration from seeds.

Conclusions and recommendations for future research

In this chapter, we have synthesized observational data, mostly collected over relatively short timescales (3–20 years), which show that rainfall and both maximum and minimum temperatures are significant determinants of regeneration from seeds in woodlands of southern Africa. Our findings show that although climate factors significantly affected regeneration in 10 species, the absence of any significant trend in the determinant climate factor leads us to conclude that changes in the regeneration patterns in these species are unlikely under a projected future climate. However, in another 10 species there was a significant trend in the determinant climate factor, which leads us to conclude that potential changes in the regeneration patterns are likely to occur in these species under a future climate regime.

Climate changes are usually analyzed at decadal periods, so that long-term monitoring of key variables that describe plant regeneration from seeds is necessary to facilitate assessments of the effects of climate factors on regeneration. The selection of the temporal scale of monitoring needs to be related to the duration of the regeneration phase. The use of annual climate data may not always be suitable in the search for mechanisms that are involved in the plant's responses to climate. Therefore, site-specific monthly climate data should be used in the analysis of the effects of climate factors on regeneration from seeds. Since climate cannot be easily manipulated, observational field experiments will play an important role in developing hypotheses that aim at testing mechanisms through which plants respond to climate factors. There is also a great need for research to be representative of sites, species, and stages of regeneration from seeds.

References

Adisa, O.M., Masinde, M., Botai, J.O., Botai, C.M., 2020. Bibliometric analysis of methods and tools for drought monitoring and prediction in Africa. Sustainability 12, 6516. Available from: https://doi.org/10.3390/su12166516.

Analytical Software, 1985–2008. Statistix 9.0. Analytical Software, Tallahassee, Florida.

Assédé, E.S.P., Azihou, A.F., Geldenhuys, C.J., Chirwa, P.W., Biaoua, S.S.H., 2020. Sudanian vs Zambezian woodlands of Africa: composition, ecology, biogeography and use. Acta Oecol. 107, 103599. Available from: https://doi.org/10.1016/j.actao.2020.103599.

Assogbadjo, A.E., Sinsin, B., Codjia, J.T.C., van Damme, P., 2005. Ecological diversity and pulp, seed and kernel production of baobab (*Adansonia digitata*) in Benini. Belgian J. Bot. 138, 47–56.

Botha, S., 2006. The Influence of Rainfall Variability on Savanna Tree Seedling Establishment (M.Sc. thesis). University of Cape Town.

Chapman, C.A., Chapman, L.J., Struhsaker, T.T., Zanne, A.E., Connie, J., Clark, C.J., et al., 2005. A long-term evaluation of fruiting phenology: importance of climate change. J. Trop. Ecol. 21, 1–14.

Chidumayo, E.N., 1997. Fruit production and seed predation in two miombo woodland trees in Zambia. Biotropica 29, 452–458.

Chidumayo, E.N., 2002. Population ecology of an afro-tropical savanna herb, *Lapeirousia rivularis*, in Zambia. Plant Ecol. 165, 275–286.

Chidumayo, E.N., 2006. Fitness implications of late bud break and time of burning in *Lannea edulis* (Sond.) Engl. (Anacardiaceae). Flora 201, 588–594.

Chidumayo, E.N., 2019. Biomass and population structure of a geoxyle, *Lannea edulis* (Sond.) Engl., at a savanna woodland site in Zambia. S. Afr. J. Bot. 125, 168–175.

Dunham, K.M., 1990. Fruit production by *Acacia albida* trees in Zambezi riverine woodlands. J. Trop. Ecol. 6, 445–457.

Helm, C.V., Scott, S.L., Witkowski, E.T.F., 2011. Reproductive potential and seed fate of *Sclerocarya birrea* subsp. *caffra* (marula) in the low altitude savannas of South Africa. S. Afr. J. Bot. 77, 650–664.

Hulme, M., Doherty, R.M., Ngara, T., New, M.G., Lister, D., 2001. African climate change: 1900–2100. Clim. Res. 17, 145–168.

Jeffers, J.N.R., Boaler, S.B., 1966. Ecology of a miombo site, Lupa North Forest Reserve, Tanzania: I. Weather and plant growth, 1962-64. J. Ecol. 54, 447–463.

Joubert, D.F., Smit, G.N., Hoffman, M.T., 2013. The influence of rainfall, competition and predation on seed production, germination and establishment of an encroaching *Acacia* in an arid Namibian savanna. J. Arid Environ. 9, 7–13.

King'uyu, S.M., Ogallo, L.A., Anyamba, E.K., 2000. Recent trends of minimum and maximum surface temperatures over Eastern Africa. J. Clim. 13, 2876–2886.

Krebs, C.J., Berteaux, D., 2006. Problems and pitfalls in relating climate variability to population dynamics. Clim. Res. 32, 143–149.

MacKellar, N., New, M., Jack, C., 2014. Observed and modelled trends in rainfall and temperature for South Africa: 1960–2010. S. Afr. J. Sci 110, 1–13.

Manson, S.J., 2001. El Niño, climate change, and southern African climate. Environmetrics 12, 327–345.

Mlambo, D., Nyathi, P., 2004. Seedling recruitment of *Colophospermum mopane* on the Highveld of Zimbabwe. S. Afr. For. J. 202, 45–53.

Msalilwa, U.L., Munishi, L.K., Makule, E.E., Ndakidemi, P.A., 2019. Pinpointing baobab (*Adansonia digitata* [Linn. 1759]) population hotspots in the semi-arid areas of Tanzania. Afr. J. Ecol. 58, 455–467.

New, M., Hewitson, B., Stephenson, D.B., Tsiga, A., Kruger, A., Manhique, A., et al., 2006. Evidence of trends in daily climate extremes over southern and west Africa. J. Geophys. Res. 11, D14102. Available from: https://doi.org/10.1029/2005JD006289.

Niang, I., Ruppel, O.C., Abdrabo, M.A., Essel, A., Lennard, C., Padgham, J., et al., 2014. Africa: Climate Change 2014: Impacts, Adaptation, and Vulnerability. Part B: Regional Aspects. Contribution of Working Group II to the Fifth Assessment Report of the Intergovernmental Panel on Climate Change. Cambridge University Press, Cambridge, pp. 1199–1265.

Robertson, E.F., 1984. Regrowth of Two African Woodland Types After Shifting Cultivation (Ph.D. thesis). University of Aberdeen.

Sekhwela, M.B.M., Yates, D.J., 2007. A phenological study of dominant acacia tree species in areas with different rainfall regimes in the Kalahari of Botswana. J. Arid Environ. 70, 1–17.

Shongwe, M.E., van Oldenborgh, G.J., van den Hurk, B.J.J.M., de Boer, B., Coelho, C.A.S., van Aalst, M.K., 2009. Projected changes in mean and extreme precipitation in Africa under global warming. Part I: Southern Africa. J. Clim. 22, 3819–3837.

Sokal, R.R., Rohlf, F.J., 1995. Biometry: The Principles and Practice of Statistics in Biological Research, Third ed. W.H. Freeman and Company, New York.

Storrs, A.E.G., 1979. Know Your Trees. Some of the Common Trees Found in Zambia. Forest Department, Ndola.

Strang, R.M., 1966. The spread and establishment of *Brachystegia spiciformis* (Benth.) and *Julbernardia globiflora* (Benth.) Troupin in the Rhodesia highveld. Commonw. For. Rev. 45, 253–256.

Tshiala, M.F., Olwoch, J.M., Engelbrecht, F.A., 2011. Analysis of temperature trends over Limpopo province, South Africa. J. Geogr. Geol. 3 (1), 1–13.

Tutin, C.E.G., Fernandez, M., 1993. Relationships between minimum temperature and fruit production in some tropical forest trees in Gabon. J. Trop. Ecol. 9, 241–248.

Unganai, L.S., 1996. Historic and future climatic change in Zimbabwe. Clim. Res. 6, 137–145.

Venter, S.M., Witkowski, E.T.F., 2011. Baobab (*Adansonia digitata* L.) fruit production in communal and conservation land-use types in Southern Africa. For. Ecol. Manage. 261, 630–639.

White, F., 1962. The forest flora of Northern Rhodesia. Oxford University Press, Oxford.

White, F., 1976. The underground forests of Africa: a preliminary review. Gardens' Bull. (Singap.) XXIX, 57–71.

White, F., 1983. The Vegetation of Africa. UNESCO, Paris.

Wyse, S.V., Dickie, J.B., 2017. Predicting the global incidence of seed desiccation Sensitivity. J. Ecol. 105, 1082–1093.

Zhou, Y., Wigley, B.J., Case, M.F., Coetsee, C., Staver, A.C., 2020. Rooting depth as a key woody functional trait in savannas. New Phytol. 227, 1350–1361.

Section II

Special Topics

Chapter 16

Effects of climate change on annual crops: the case of maize production in Africa

Carol C. Baskin[1,2]
[1]*Department of Biology, University of Kentucky, Lexington, KY, United States,* [2]*Department of Plant and Soil Sciences, University of Kentucky, Lexington, KY, United States*

Introduction

According to the Food and Agriculture Organization of the United Nations [FAO (Food and Agriculture Organization of the United Nations), 2019], about 820 million people in the world endured significant hunger in 2018, and over 2 billion people did not have regular access to nutritious, safe, and sufficient food. During 2018, moderate to severe food insecurity occurred for 29.5%—62.7%, 9.8%—34.3%, 31.5%, and 30.6% of the people living in Africa, Asia, Central American, and South America, respectively [FAO (Food and Agriculture Organization of the United Nations), 2019]. The causes of food insecurity include cultural/political/socioeconomic factors (Bohlken and Sergenti, 2010; Masambaya et al., 2018), shifts in market value of crops (Mihalache-O'Keef and Li, 2011; Tigchelaar et al., 2018), and low yield of crops (Sawe et al., 2018). Crop failure has been attributed to increased temperatures (Battisti and Naylor, 2009; Charles, 2014), variability/decrease in rainfall (Oseni and Masarirambi, 2011; Epule et al., 2017; Shumetie and Yismaw, 2017; Kim et al., 2019; Kinda and Badolo, 2019), or a combination of increased temperatures and drought (St. Clair and Lynch, 2010; Ochieng et al., 2016). Also, there may be reduced crop production due to low soil fertility and plant damage by pests and diseases (Tesso et al., 2012; Sawe et al., 2018; ten Berge et al., 2019). Increasingly, attention is being given to the relationship between the negative effects of climate change, such as increased temperatures and drought and food security (Gregory et al., 2005; Schlenker and Lobell, 2010; Di Falco et al., 2011; Thornton et al., 2011; Tesso et al., 2012; Asante and Amuakwa-Mensah, 2015; Ochieng et al., 2016; Tripathi et al., 2016; Ma et al., 2019; Chemura et al., 2020; Leisner, 2020).

The annual grasses (Poaceae, Poales) maize (*Zea mays* L.), wheat (*Triticum aestivum* L.), and rice (*Oryza sativa* L.) are the most important food crops for humans in the world, with a total world production for 2018—19 of 1123.3, 730.0, and 496.5 million metric tons, respectively [FAS-USDA (Foreign Agricultural Service USDA), 2020]. Zhao et al. (2017) compiled published research/modeling results and concluded that for each degree-Celsius increase in temperature total world yields of wheat, rice, maize, and soybean are reduced 6.0%, 3.2%, 7.5%, and 3.1%, respectively, with maize showing the strongest response. Studies in various parts of the world, including India (Tripathi et al., 2016), Africa (Table 16.1), the United States (Tigchelaar et al., 2018), Asia (Aryal et al., 2020), South America (Jones and Thornton, 2003), and Central America (Hannah et al., 2017), have shown that climate change may decrease the yield of maize.

Domestication of maize (corn) was a single event that occurred in the Mexican highlands about 9000 years ago (Matsuoka et al., 2002), and the wild ancestor was *Z. mays* subsp. *parviglumis* Iltis & Doebley (Matsuoka et al., 2002; Janzen and Hufford, 2016). By the time Christopher Columbus arrived in the New World in 1492, the use of maize for human food had spread to eastern Canada, northern Chile, the deserts of southwestern United States, Caribbean Islands, Amazon Basin, and Andes Mountains (Matsuoka et al., 2002). Maize is now cultivated in countries around the world, and in 2019 ≥ 1000 metric tons were produced in each of at least 113 countries, with the United States, China, Brazil, the European Union-27, Argentina, Ukraine, Mexico, India, Canada, the Russian Federation, South Africa, Indonesia,

TABLE 16.1 Effects of heat stress on gamete formation, pollination, and fertilization of maize.

Effect	References
Inhibition at tetrad stage of microsporogenesis	Begcy et al. (2019)
Decreased amount of pollen produced	Schoper et al. (1986)
Decreased anther emergence from tassel	Schoper et al. (1987)
Reduced time of pollen shedding	Wang et al. (2019b)
Decreased pollen viability	Schoper et al. (1986), Alam et al. (2017), Lizaso et al. (2018), Wang et al. (2020)
Decreased pollen germination	Herrero and Johnson (1980)
Inhibition of pollen germination and pollen tube growth	Singh et al. (2016)
Delay of flowering events (anthesis and silking)	Lonngquist and Jugenheimer (1943), Edreira et al. (2011), Cicchino et al. (2013)
Decreased floret differentiation	Edreira et al. (2011)
Decreased number of differentiated ovules	Shim et al. (2017)
Delayed stigma initiation	Alam et al. (2017)

Nigeria, Ethiopia, and the Philippines (in this order) being the top 15 producers [IndexMundi-corn, 2020; FAS-USDA (Foreign Agricultural Service USDA), 2020].

Maize was described as a "paramount staple crop" by Nuss and Tanumihardjo (2010), and these authors listed numerous traditional foods from many countries that are prepared with this cereal. The kernels, that is, grains/caryopses (diaspores), which are the dispersal/germination units of maize, are a good source of carbohydrates, fiber, protein, lipids, vitamins A and E, and minerals such as phosphorous, potassium, magnesium, and sulfur (Nuss and Tanumihardjo, 2010). Maize is a good source of carotenoids, phenolic compounds, and phytosterols, which are phytochemicals that promote good health of humans (Shah et al., 2016). However, maize is not a perfect food because the kernels (hereafter seeds) do not contain the essential amino acids lysine and tryptophan and have only low amounts of vitamins C and B-12, iron, and iodine (Nuss and Tanumihardjo, 2010). The vitamin niacin is found in maize seeds, but it is tightly bound and not available when humans eat them. However, niacin is released when seeds are soaked in calcium hydroxide (lime) and then ground (Caballero-Briones et al., 2000; Ranum et al., 2014).

Maize is the preferred cereal in Central America, Mexico, and Southern and Eastern Africa, with flour and meal being the two most popular ways to use it (Ranum et al., 2014). In African countries south of the Sahara, maize is a staple food, and consumption per person may be as high as 750 g per day (Ekpa et al., 2019). Many local dishes are prepared with maize flour and meal, and green maize on-the-cob is boiled, steamed, or roasted and then enjoyed (Ekpa et al., 2019). In addition to providing food for humans, maize is used for production of various products, including alcohol, fuel ethanol, glue, oil, starch, and sweeteners (Ranum et al., 2014). Also, maize is used as food for domesticated animals, and it is fed to them as seeds, fodder/silage (green plants including leaves, seeds, and cobs are harvested/ground), or stover (dried crop residue after the ears have been removed) (De Groote et al., 2013; Klopfenstein et al., 2013; Olaniyan, 2015).

The exact year and place of the introduction of maize to Africa are not known, but by the 16th-century travelers in western Africa were making notes about "miglio zaburro" (maize) in their diaries (Miracle, 1965). By the 19th century, maize had been spread over the African continent (Miracle, 1965; Olaniyan, 2015). Maize was quickly and widely adopted by farmers in Africa because the crop was not labor-intensive, had a short growing season (planting to maturity), and people liked to eat it. Maize is now an important staple food crop in many countries south of the Sahara. For example, in Lesotho, Malawi, and Zambia 50% of the total calories for humans come from maize, while in Zimbabwe and South Africa maize accounts for 43% and 31% of the total calories, respectively (Smale et al., 2011).

Given the important role of maize in food security of people in many African countries, as well as those in Latin America and parts of Asia (Shiferaw et al., 2011), this chapter will consider the potential effects of climate change on production of this paramount staple crop. In particular, the effects of increased temperature and drought on regeneration of maize from seeds will be evaluated. The various stages in the life cycle of maize will be discussed, including seed dormancy/germination, plant growth, gamete formation, pollination, fertilization, grain formation, and gain filling/yield. Also, the predicted changes in maize production in African countries due to climate change and possible strategies to help mitigate the negative effects of climate change on maize production will be considered. Many of the effects of climate change on production of maize apply to wheat and rice. Furthermore, the effects of climate change on various aspects of the life cycle maize provide insight with regard to the potential effects of global warming on production not only of annual food crops but also wild plant species. Thus, the broad implications of what has been learned about the effects of climate change on regeneration of maize from seeds in relation to effects of climate change on wild plant species will be considered briefly.

Effect of climate change on life cycle of maize

Seed dormancy/germination

In general, there is little or no dormancy in seeds of modern hybrids/lines/cultivars of maize (Simpson, 1990). However, seeds of the perennial species *Zea perennis* (Hitchc.) Reever & Mangelsd., one of the teosintes from Mexico, are dormant. Dormancy in seeds of *Z. perennis* is broken during dry storage (afterripening), and treatment with gibberellic acid (GA3) and mechanical scarification promotes germination (Mondrus-Engle, 1981). Germination data reported by López et al. (2011) also indicate that seeds of the perennial teosinte *Z. diploperennis* Iltis, Doebley & R. Guzmàn, as well as *Z. perennis*, from most of the several populations surveyed in Mexico had some dormancy; afterripening and mechanical scarification promoted germination. Since teosinte seeds become nondormant via afterripening, we can conclude that they have nondeep physiological dormancy, which is the only kind of dormancy reported in Poaceae (Baskin and Baskin, 1998, 2014).

In germination studies involving freshly matured seeds of *Z. mays*, a low amount of seed dormancy is sometimes detected. For example, only 57% of fresh seeds of maize (line/cultivar not given) germinated when sown 3 days after harvest, but 90% of them germinated when treated with 100 ppm GA3 (Al-Delaimy and Al-Mamoori, 2016). With a delay in planting from 3 to 5 days after harvest, germination increased from 57% to 64%, suggesting that afterripening was occurring in the seeds during dry storage.

The lack of dormancy in seeds of maize and other cereals can result in preharvest sprouting and thus loss of the grain for food or other purposes (Neill et al., 1987; Rodríguez et al., 2015). Preharvest sprouting (or lack thereof) is genetically controlled, and several viviparous genes have been found in maize (McCarty et al., 1991; Gale et al., 2002). Abscisic acid (ABA), which is required for normal seed maturation and drying in maize, will inhibit preharvest sprouting (Fong et al., 1983; Neill et al., 1987; White et al., 2000; White and Rivin, 2000). White et al. (2000), and White and Rivin (2000) concluded that the balance between gibberellins (GA) and ABA determines if seeds germinate while attached to the mother plant. Furthermore, GAs antagonize ABA signaling, but their effect varies with the stage of embryo development.

As numerous lines of maize have been developed (i.e., selected), much attention has been given to the minimum, optimum, and maximum temperatures for seed germination. In temperate regions, germination at early spring (low) temperatures is desirable, while in the tropics germination at high temperatures is desirable. Thus, lines/cultivars of maize have been developed for temperate and tropical regions. For example, lines 30, 64, 63, and 91 evaluated for germination at 15°C, 20°C, 25°C, 30°C, and 35°C in Lavras, Brasil, showed two patterns of temperature tolerance for germination (dos Santos et al., 2019). Lines 63 and 91 germinated to 6%–24% at 15°C and 20°C but to 58% and 38%, respectively, at 35°C. On the other hand, lines 30 and 64 germinated to 36%–58% at 15°C and 20°C but to 10% and 12%, respectively, at 35°C. All four lines germinated to 92%–96% at 25°C and to 64%–78% at 30°C.

Germination percentages declined for the temperate cultivar Maris Jade and the topical Pioneer hybrid X-105A at temperatures above 36°C–37°C, and they declined for the two Kenyan highland hybrids 632 and 614 at temperatures above 35°C (Riley, 1981). Riley found that rate of protein synthesis in the embryo declined significantly in Maris Jade seeds incubated at 41°C, and imbibed seeds lost viability. However, rate of protein synthesis in the embryo of Pioneer X-105A kernels imbibed at 41°C increased by 30% compared to that in seeds at 28°C.

The rate (speed) at which seeds of the four cultivars/hybrids of maize used in the studies by Riley (1981) imbibed water was not affected by temperatures between 23°C and 41°C. However, the amount of water imbibed has been shown to vary with the seed part. For *Z. mays* var. [cultivar] yellow dent, Stiles (1948) determined that the percentage increase in water (as a percentage of total water imbibed by kernel) for embryo, scutellum, endosperm, and pericarp was 57.0, 12.4, 26.3, and 4.3, respectively. The critical seed moisture content (MC) for germination of maize is *c.*30%

(Shaban, 2013). In a study of 12 maize seed lots, van der Venter and Hoffman (1988) found that germination percentage was significantly positively corrected with MC of seeds after 6–30 h of imbibition and with MC of embryos after 6–12 h of imbibition. Germination percentages were not related to embryo or seed mass. Seeds of *Z. mays* cv. Zhengtian 68 incubated over a range of water stresses (created with PEG 8000) from 0 to −1.2 MPa in light/dark at 25°C showed decreased water absorption and decreased germination with increased stress. At 0 and −1.2 MPa, water absorption by the seeds after 96 h had increased by *c.*150% and 50%, respectively, and *c.*95% and 0% of the seeds germinated, respectively (Li et al., 2017).

Plant growth from germination to tasseling

The growth of maize plants has been divided into different stages, the number of which can vary with the researcher. In general, however, there are three main stages between germination and tasseling (production of male inflorescence). During stage 1, the seed germinates, the coleoptile emerges above the soil surface, and the first foliage leaf emerges from the tip of the coleoptile. During stage 2, six to eight new leaves are produced, and prop roots grow from stem nodes above the soil surface. During stage 3, leaf production continues, the tassel reaches full length but its branches do not spread apart and ears reach a length of a few centimeters [AGDH (Australian Government Department of Health), 2008]. Ritchie and Hanway (1986) divided the growth of a maize plant into vegetative and reproductive stages with six or more subcategories in each of the two main stages.

Heat stress begins to occur when plants of maize are exposed to temperatures of ≥ 37.5°C (Crafts-Brandner and Salvucci, 2002). The lethal maximum temperature for leaf initiation, shoot growth, root growth, sowing to tassel initiation, and sowing to tassel initiation is 41.3°C, 38.9°C, 40.1°C, 40.2°C, and 39.2°C, respectively (Sánchez et al., 2014). Heat stress during the preanthesis growth period increases the rate of plant development, but plants are smaller than those developed under moderate temperatures (Harrison et al., 2011; Wang et al., 2011; Lizaso et al., 2018). The decreased size of heat-stressed plants is due to a reduction in leaf area index, photosynthetic rate per leaf area, and growth rate, which result in reduced accumulation of total biomass (Cicchino et al., 2010; Cairns et al., 2013a; Li et al., 2020).

Water stress (drought) during the preanthesis growth period reduces the number of leaves, plant leaf area, plant height, and total dry matter accumulation (NeSmith and Richie, 1992; Çakir, 2004; Efeoğlu et al., 2009; Rufino et al., 2018; Sah et al., 2020). Also, drought reduces root growth (Cutforth et al., 1986), height of first ear on plant, stem diameter, and number of rows of ovules per ear (Rufino et al., 2018), and it delays pollen dispersal and silk (stigma/style) emergence (NeSmith and Richie, 1992).

Gamete formation, pollination, and fertilization

Heat stress is generally more detrimental to the reproductive than to the vegetative stage of the life cycle of maize. However, the lethal maximum temperature for the sowing to tassel initiation stage is 39.2°C, it is 37.3°C and 36.0°C for the anthesis and grain-filling stages, respectively (Sánchez et al., 2014). Various aspects of gamete formation, pollination, and fertilization are negatively affected by heat stress (Table 16.1). During pollination, exposure to high temperatures for only short periods of time can significantly reduce the percentage of fertilized ovules. Exposure of freshly pollinated maize spikelets to 32°C, 36°C, and 40°C for 2 h resulted in only *c.* 57%, 55%, and 35% of the ovules being fertilized, and after a 4-h exposure to these temperatures 55%, 57%, and 10% of the ovules were fertilized. For the controls at 28°C, 65%–70% of the ovules were fertilized (Dupuis and Dumas, 1990).

Various studies have found that heat stress can rapidly reduce the percentage of viable pollen (Table 16.1). However, there are genotype differences in ability of pollen to survive exposure to high temperatures and then germinate. Fresh (control) pollen of genotypes A641, 0156rf, and MK386 germinated to 45.9%, 53.5%, and 89.1%, respectively, but after 10 min at 26°C germination was 45.6%, 14.3%, and 4.2%, respectively, and after 10 min at 35°C 13.4%, 1.3%, and 0%, respectively (Lyakh et al., 1991).

Drought can inhibit various events in the early male and female phases of the life cycle of maize (Table 16.2). In particular, water stress may slow silking more than tasseling/anther emergence, resulting in a lack of synchrony between pollen dispersal and stigma receptivity (Otegui et al., 1995; Sah et al., 2020). Synchrony of the male and female phases of the life cycle is essential for successful fertilization (Hedhly et al., 2008). Thus, when heat and/or water stress cause(s) a delay of flowering events, for example, stigma initiation (Alam et al., 2017), the "window of opportunity" for fertilization is reduced (Edreira et al., 2011).

TABLE 16.2 Effects of water stress on gamete production, pollination, and fertilization of maize.

Effect	References
Inhibition of meiosis of microsporocyte	Saini (1997)
Pollen killed at low moisture content of pollen (c. 30%)	Fonseca and Westgate (2004)
Ovary not developed	Mäkelä et al. (2005)
Increased anthesis to silking interval, which reduces percentage of fertilization	Sangoi and Salvador (1998)
Increased number of days to flowering and maturity and increased anthesis to silking interval	Sah et al. (2020)
At low water potential in silks (−0.78, −0.80 MPa), silks do not grow	Bassetti and Westgate (1993), Westgate and Boyer (1985)
Ovaries not exposing silks at time when pollen was shedding	Otegui et al. (1995)
Delay in silking	Moss and Downey (1971)
Inhibition of ear elongation	Wang et al. (2019a)
Embryo development prevented	Westgate and Boyer (1986)

Grain (seed) formation

The first step in seed formation of maize is fertilization of the egg in the ovule, after which embryo growth and endosperm development occur. However, heat and water stress can directly or indirectly lead to seed/embryo abortion. Seed abortion may occur shortly after fertilization, or it may be delayed until onset of the seed/grain stage of development.

Heat stress (35°C) applied 3 days after pollination caused 65% of the seeds to abort (Hanft and Jones, 1986). According to these authors, the aborting seeds did not accumulate dry matter, and there was little or no endosperm development. In another heat-stress study, young seeds were removed from the ear 3 days after pollination and cultured in vitro at 35°C for 0, 4, or 8 days and then allowed to develop at 25°C. After 30 days, 0%, 23%, and 97% of the seeds receiving 0, 4, and 8 days of heat stress, respectively, had aborted (Cheikh and Jones, 1994). Heat stress (c. 1°C above air temperature) applied to maize plants growing in the field in northeastern Spain during presilking significantly reduced the number of seeds produced (Ordóñez et al., 2015).

Westgate and Boyer (1986) found that embryos failed to grow when stigmas on plants experiencing water stress (leaf water potential −1.7 MPa) were pollinated with fresh pollen or if plants were rehydrated prior to being pollinated. Water stress, which delayed silking, reduced the number of seeds formed, although fresh pollen was applied to the late-emerging silks (Otegui et al., 1995). Zinselmeier et al. (1995) withheld water from maize plants, starting on the day the first silks appeared, and stigmas were pollinated 4 days later. At the time of pollination, silk and leaf water potentials were −1.0 and −1.8 MPa, respectively. Two days after pollination, plants were watered and kept well-watered until the seeds matured. The number of seeds produced in the water-stress treatment was 60% less than that of the control. In addition to inhibited ovary/seed growth, the level of reducing sugars in the ovary decreased, starch was depleted, sucrose concentration increased, and acid invertase activity was inhibited. Zinselmeier et al. (1995) attributed seed abortion to zygote abortion and suggested that embryo development stopped during water stress because the "normal flux of reduced carbon from the translocation stream to sites of metabolism within the ovaries was disrupted." Furthermore, they concluded that acid invertase activity was important for normal early growth of seeds.

Andersen et al. (2002) reported that water stress for 7 days after pollination repressed expression of the gene *Ivr2* for soluble invertase, which can cleave sucrose, thereby making hexose products (glucose and fructose) available for embryo growth. Starting on the first day of silking, McLaughlin and Boyer (2004a) withheld water from maize plants for 6 days. Stigmas were hand-pollinated on day 5 of the water-stress period, and 1 day later plants were watered. On day 4 of the water-stress period, sucrose was infused at one internode on a set of plants. The number of seeds matured per ear was c. 500, 330, and 10 for control, water-stress + sucrose, and water-stress plants, respectively. In the water-stress plants, there was early downregulation of genes for acid invertase and upregulation of genes for senescence. The infused sucrose apparently blocked the upregulation of the genes for senescence.

Oury et al. (2016) investigated seed abortion that occurred when maize plants were water-stressed while they were flowering. These authors measured (1) abundance of transcripts of genes involved in sugar metabolism and tissue expansion, (2) concentration of various sugars, and (3) activity of enzymes involved in carbon metabolism, and they concluded that changes in carbon metabolism were the consequences of seed abortion and not its cause.

Seed abortion that occurs during seed filling is related to availability of assimilates. That is, if assimilate are insufficient seed abortion occurs (Schussler and Westgate, 1991; Otegui et al., 1995; Bledsoe et al., 2017; Shen et al., 2018). Assimilates are the products of photosynthesis, and heat (Bird et al., 1977; Kim et al., 2007; Ben-Asher et al., 2008) and water (Westgate and Boyer, 1986; McLaughlin and Boyer, 2004b; Mäkelä et al., 2005) stress can reduce photosynthetic rates. Furthermore, transpiration is reduced due to stomate closure if plants are under water stress, which results in an increase in leaf temperatures on sunny days (Dalil and Ghassemi-Golezani, 2012).

Westgate and Boyer (1986) concluded that water stress in maize leaves interrupted photosynthesis, which prevented the formation of sufficient food reserves to supply the developing seeds. Mäkelä et al. (2005) reported that water stress and shading, which decreased photosynthesis, lead to seed abortion due to decreased carbohydrate delivery. The common lack of seed formation by the apical florets on the maize ear is attributed to low availability of assimilates for the apical seeds, resulting in their abortion (Shen et al., 2018, 2019; Yan et al., 2018). There is competition between basal and apical seeds for assimilates, but since basal stigmas are usually pollinated before the apical ones, basal seeds are less likely to abort than apical ones (Shen et al., 2019). That is, basal seeds outcompete apical seeds for assimilates (Ober et al., 1991). However, seed abortion can occur all over the ear—not just at the apical end.

There is a genetic component with regard to the proportion of seeds on an ear that is aborted (Gustin et al., 2018). Since there is a genetic aspect of seed abortion in maize, much research is being devoted to identifying the genes for seed abortion and exactly what they do. Notably, some of these genes are involved in supplying food to the embryo (Andersen et al., 2002; Ruan et al., 2012; Cagnola et al., 2018; Li and Lübberstedt, 2018). Furthermore, it is important to determine the action (or lack thereof) of these genes when plants have been subjected to heat and/or water stress (Cagnola et al., 2018).

Grain (seed) filling

Seeds of maize reach maximum dry matter accumulation (physiological maturity) 55−65 days after pollination [Egli and TeKrony, 1997; AGDH (Australian Government Department of Health), 2008], at which time their MC is 30%−55%, depending on maize line and differences in year to year environmental conditions (Brooking, 1990; Hunter et al., 1991; Sala et al., 2007). However, seeds need to be dried to c. 8%−13% MC for successful storage (Egli and TeKrony, 1997). Seed filling for maize has been divided into various stages to facilitate the reporting of results from research projects (Rench and Shaw, 1971). In general, these stages are: (1) seeds filled with clear liquid, that is, endosperm is a liquid; (2) seeds filled with a white liquid, *milk stage*; (3) seeds filled with a white paste, *dough stage*, during which starch is deposited in the endosperm; and (4) physiological maturity, at which time dry matter accumulation is completed, endosperm is solid, and a "black layer" has formed at the tip of the seed sealing it off from the water supply of the maize cob (Hunter et al., 1991).

Under optimal temperature and water conditions for growth of maize, there are three phases of seed growth and development (Egli, 2017). In phase I, which includes fertilization, rapid cell division occurs, and the various parts of the seed are formed. During phase II, seeds accumulate reserves, in particular starch, resulting in a rapid increase in dry mass. In phase III, dry mass plateaus and stops with physiological maturity. Seed water content is highest in phase I and declines during both phases II and III (Egli and TeKrony, 1997). After seeds reach physiological maturity, they continue to lose water (Martinez-Feria et al., 2019).

Reduction of seed development of maize occurs when temperature and water conditions are not favorable for dry mass accumulation. The optimum, lethal maximum, and lethal minimum temperatures for seed filling of maize are 26.4°C, 36.0°C, and 8.0°C, respectively (Sánchez et al., 2014). In general, as the temperature increases during seed filling rate of filling increases, but duration decreases (Muchow, 1990; Singletary et al., 1994; Wilhelm et al., 1999; Ghassemi-Golezani and Tajbakhsh, 2012; Prasad et al., 2018). Boehlein et al. (2019) evaluated the effects of elevated day and night temperatures [38 (day)/28 (night) °C] versus elevated day and normal night temperatures (38°C/17°C) on endosperm development and compared the results to normal day and night temperatures (28°C/17°C). The rate of endosperm development increased when both day and night temperatures were elevated but not when only the day temperature was elevated.

Various biochemical and molecular biology studies have been conducted on the responses of developing maize seeds to heat stress. A sample of the results is listed here: decreased starch synthesis (Singletary et al., 1994); loss of activity of soluble starch synthase (Keeling et al., 1994); reduced activity of ADP-glucose pyrophosphorylase (AGPase) (Wilhelm et al., 1999); disruption of cell division, sugar metabolism, and starch biosynthesis in endosperm and a reduction in protein and zein content of seeds (Monjardino et al., 2005); and reduced activity of starch synthase, AGPase, and pyruvate phosphate dikinase (Boehlein et al., 2019).

A major consequence of heat stress on maize seed development is that seed mass is reduced (Cheikh and Jones, 1994; Edreira and Otegui, 2013; Tao et al., 2016), primarily due to decreased availability of photosynthetic assimilates. Using a data set for more than 20,000 field trials for maize in Africa between 1999 and 2007, Lobell et al. (2011a) found that for each degree-day above 30°C yield was reduced by 1%; this was under optimal rain-fed conditions. Under drought conditions, the yield was decreased by 1.7% for each degree-day above 30°C.

In general, the early stages of reproductive development, that is, anthesis and silking, are more sensitive to plant water deficits than seeds in phase II of seed filling (Westgate and Grant, 1989; Ober et al., 1991; Schussler and Westgate, 1991; Westgate, 1994; Ouattar et al., 1987b). Westgate (1994) and Borrás et al. (2003) found that drought stress did not change the rate of dry matter accumulation in seeds, but it did decrease the duration of seed filling. Thus, water stress can cause seeds to reach physiological maturity earlier than they would under well-watered conditions (Ouattar et al., 1987a; Westgate, 1994). A reduced period of seed filling means a decrease in seed mass and ultimately decreased crop yield (Ober et al., 1991; Westgate 1994; Bruce et al., 2002; Wang et al., 2019b). In particular, drought stress can cause a reduction in number of endosperm nuclei and ultimately endosperm fresh mass (Ober et al., 1991).

Detailed studies of water relations during seed filling have revealed that seed growth can continue when plant water stress is severe enough to inhibit photosynthesis in the leaves (Ouattar et al., 1987b). Seed filling can continue because assimilates are remobilized from other plant parts, especially the stem, to the developing seeds (Jurgens et al., 1978; Ouattar et al., 1987b). However, we need to remember that water stress reduces plant photosynthetic rates, which ultimately reduces the amount of assimilates that can be translocated to the developing seeds (Çakir, 2004). With a 40% reduction in amount of water available for the growth of maize plants, Daryanto et al. (2016) concluded that the global reduction in maize yield would be 39.3%. Interestingly, however, in the absence of drought stress elevated levels of CO_2 do not have an effect on yield of field-grown maize. In an experiment conducted in the Mid-western (United States), maize grown under elevated CO_2 (550 vs 376 $\mu mol\ mol^{-1}$) in the absence of drought stress did not exhibit stimulation of photosynthesis or biomass accumulation of seeds or increase yield (Leakey et al., 2006).

In nature, heat and water stress of maize plants frequently occur at the same time. Thus the combination of heat and water stress is more prohibitive of growth and seed production than either heat or drought stress alone (Heyne and Brunson, 1940; Barnabás et al., 2008; Lamaouri et al., 2018; Hussain et al., 2019). Obata et al. (2015) concluded that the metabolic responses to the combination of heat and water stress in maize were not a new or novel response but "rather the sum of the effects of two individual stresses...."

There are genetic differences in the tolerance of maize lines to heat and water stress. Some lines are relatively tolerant of heat, others drought and still others both heat and drought (Cairns et al., 2013a). These authors found that tolerance of the combined stresses was genetically different from tolerance of heat or drought stress individually. Killi et al. (2017) concluded that maize genotype and a period of acclimation to heat stress prior to beginning of water stress are important in mitigating the effects of water stress on photosynthesis and stomatal conductance.

Predicted effect of climate change on maize production in Africa

Using CLIMEX, which is a statistical-mechanistic model that estimates the potential abundance and geographical distribution of organisms, Ramirez-Cabral et al. (2017) evaluated the future suitability of various regions on earth for maize cultivation. They predict that the area of land climatically suitable for maize production in 2100 will decrease in Africa (29%), South America (43%), and Australia/New Zealand (10%) and increase in Asia (11%), Europe (14%), and North America (20%). High latitudes, such as northern Europe, may have an increase in land area favorable for growth of maize due to increased temperatures.

In African countries, decreased yield of maize is attributed to increased temperatures (Ezeaku et al., 2014; Ochieng et al., 2016; Stevens and Madani, 2016; Luhunga, 2017), decreased precipitation (Östberg and Slegers, 2010; Ghebrezgabher et al., 2016; Msowoya et al., 2016; Shumetie and Yismaw, 2017), and the combination of increased temperatures and decreased precipitation (Makadho, 1996; Charles, 2014; Asante and Amuakwa-Mensah, 2015; Omoyo et al., 2015; Abera et al., 2018; Davenport et al., 2018; Mumo et al., 2018; Chemura et al., 2020).

In view of decreases in maize yield in various African countries in years with increased temperatures and/or drought, much modeling has been (is being) done to gain a perspective on how global warming will impact maize production in Africa (Lobell et al., 2011a,b; Cairns et al., 2013b). Using available meteorological and maize yield data,

TABLE 16.3 Examples of projected changes in yield of maize in Africa due to impact of climate change.

Place	Change (%)	References
Africa—Sub-Saharan (SS) region	−22	Schlenker and Lobell (2010)
Africa—SS region - Central	−13	Thornton et al. (2011)
Africa—SS region - East	−19	Thornton et al. (2011)
Africa—SS region - West	−23	Thornton et al. (2011)
Africa—SS region - South	−16	Thornton et al. (2011)
Africa—West	−7.4	Knox et al. (2012)
Africa—West	−10	Stuch et al. (2020)
Africa—Southern	+13	Stuch et al. (2020)
Benin	−30	Shi and Tao (2014)
Botswana	−30	Shi and Tao (2014)
Burundi	−30	Shi and Tao (2014)
Ethiopia—different locations	−24.5, −43, +51	Abera et al. (2018)
Ghana	−30	Shi and Tao (2014)
Guinea	−30	Shi and Tao (2014)
Guinea-Bissau	−30	Shi and Tao (2014)
Gambia	−30	Shi and Tao (2014)
Kenya	−20	Wandaka et al. (2016)
Kenya	−30	Shi and Tao (2014)
Kenya—East	−11	Davenport et al. (2018)
Kenya—West	−7	Davenport et al. (2018)
Lesotho	−30	Shi and Tao (2014)
Malawi	−14 to −33	Msowoya et al. (2016)
Malawi	−1.2 to +1.0	Stevens and Madani (2016)
Malawi	−30	Shi and Tao (2014)
Mauritania	−30	Shi and Tao (2014)
Mozambique	−30	Shi and Tao (2014)
Namibia	−30	Shi and Tao (2014)
Niger	−30	Shi and Tao (2014)
Nigeria	−13, −18	Ezeaku et al. (2014)
Nigeria	−30	Shi and Tao (2014)
Sahel realm	−12.6	Knox et al. (2012)
Sierra Leone	−30	Shi and Tao (2014)
Somalia	−30	Shi and Tao (2014)
South Africa	−11.4	Knox et al. (2012)
South Africa	−30	Shi and Tao (2014)
Swaziland	Decrease	Oseni and Masarirambi (2011)
Tanzania	−9.6	Luhunga (2017)
Tanzania	Decrease	Sawe et al. (2018)
Tanzania	−30	Shi and Tao (2014)
Uganda	−30	Shi and Tao (2014)
Zimbabwe	−11 to −17	Makadho (1996)
Zimbabwe	−13	Ruinda et al. (2015)
Zimbabwe	−30	Shi and Tao (2014)

various models, such as the Global Circulation (Makadho, 1996; Thornton et al., 2011; Knox et al., 2012; Ezeaku et al., 2014; Msuwoya et al., 2016; Stevens and Madani, 2016; Luhunga, 2017), Panel Dataset Analysis (Schlenker and Lobell, 2010), Agriculture Production Systems Simulator (Ruinda et al., 2015), Crop Simulation (Ogutu et al., 2018), Decision Support System for Agro-technology Transfer (Abera et al., 2018), and Spatial Production Allocation (Stuch et al., 2020), have been used to predict maize yield under rain-fed conditions for the next 20–30 years in various African countries.

In general, these models predict decreases in maize yield in African countries south of the Sahara Desert (Table 16.3). The projected decreases in maize yield range from 1.2% to 43% with a mean ($\pm$ s.e.) decrease of 22.7 $\pm$ 1.4% for the decreases shown in Table 16.3. However, depending on the methods used to evaluate the existing data predictions can vary for the same country, for example, predictions for Malawi range from -1.2% to -33%. Locations within the same country, for example, Ethiopia, with varying environmental conditions have received different predictions with regard to future maize yields. Depending on local environmental conditions in Ethiopia, changes in future yields of maize range from -43% to $+51\%$ (Abera et al., 2018).

Mitigation

The decreased yield by maize plants exposed to hot, dry weather has long been a problem for farmers, and many attempts to mitigate the effects of heat and drought have been (and are still being) made by researchers. Fortunately, there is much genetic diversity in maize, and selection for heat- and drought-tolerant cultivars/lines began many decades ago (e.g., Jenkins and Richey, 1931; Jenkins, 1932). In addition to development of heat- and drought-tolerant lines of maize (Table 16.4), selection for increased heat tolerance of pollen is being conducted (e.g., Mohapatra et al., 2020).

With regard to improving maize yield under hot, dry conditions, research is being conducted on the use of fertilizers in relation to heat/drought tolerance, improved use of water resources, and adjustment of time of seed sowing to avoid extreme heat especially during pollination (Table 16.4). With particular reference to countries in Africa south of the Sahara Desert, attention is being given to how readily farmers are willing to grow new cultivars of maize and how to safely store the seeds that are harvested (Table 16.3). Furthermore, to help alleviate food security problems and decrease dependence on maize, diversification of food crops is being promoted in Africa (Tadele, 2018).

Broad implications for wild plant species

The life cycle of maize from seed to seed has two stages, that is, fertilization and seed filling, which are especially sensitive to the negative effects of high temperatures and/or drought. The response of maize gametes and developing seeds to the predicted changes in climate due to global warming provides insight into specific questions that we need to be addressing about the regeneration from seeds of other food crops and in particular wild plant species.

TABLE 16.4 Examples of research efforts to develop strategies to help mitigate the negative effects of heat and drought on maize production.

Research efforts	References
Adjust time of crop sowing	Harrison et al. (2011), Omoyo et al. (2015), Muluneh et al. (2016)
Adoption/planting of drought-tolerant varieties	Fisher et al. (2015), Lunduka et al. (2019), Obunyali et al. (2019), Simtowe et al. (2019)
Improved storage of seeds on farms	Thamaga-Chitja et al. (2004), Tefra et al. (2011)
Improved use of water resources	Githui et al. (2009), Rockström and Falkenmark (2015), Muluneh et al. (2016)
Increased use of fertilizer	Aslam et al. (2013), ten Berge et al. (2019)
Seed priming with salt	Gebreegziabher and Qufa (2017)
Selection for drought tolerance	Chaves et al. (2003), Langner et al. (2019), Sah et al. (2020)
Selection for heat tolerance	Naveed et al. (2014), Gao et al. (2019)
Selection for pollen tolerance of heat stress	Mohapatra et al. (2020)

Some information is available for a small fraction of the seed plants on the effects of climate change on the early stages of seed formation. In a literature survey of the effects of low- and/or high-temperature stress on 33 species (not including *Zea mays*) in 19 families of angiosperms, Hedhly (2011) found that both male and female reproductive structures were sensitive to stress both in the pre- and postpollination stages. From the studies of Hedhly (2011), Zinn et al. (2010), and Rosbakh et al. (2018), we know that lack of successful seed production in plant species exposed to temperature stress may be due to many problems: low number of pollen grains formed, low pollen viability, low percentages of pollen germination, slow rate of pollen tube growth, abscission of flowers before they open, ovules do not develop, embryo sac either does not form or degenerates, early embryo abortion, slow rate of embryo growth, disruption of sugar metabolism, and decreased seed size and vigor.

Rosbakh and Poschlod (2016) determined the minimum, optimum, and maximum temperature for in vitro pollen tube growth of 25 herbaceous species with different elevational distributions in the Bavarian Alps (southeastern German). There was a significant positive relationship between (1) minimum temperature at which pollen tubes would grow and mean annual temperature at the collection site of the species in the Alps, and (2) maximum temperature for growth of pollen tubes and temperature when a species flowered in the field. Thus, the temperature requirements for pollen germination and pollen tube growth helped explain the distribution of species along the elevation/temperature gradient in the mountains.

Little information is available for wild plant species on the effects of climate change on the seed filling stage. As expected, much research attention has been given to food crops (e.g., Abdul-Baki, 1991; Begcy et al., 2018; Sehgal et al., 2018; Liu et al., 2019; Masouleh and Sassine, 2020). However, Marmagne et al. (2020) investigated seed filling by *Arabidopsis thaliana* plants exposed to heat, drought, and low nitrogen availability. These authors found that the ratio of harvest plant dry mass to nitrogen quantity in harvested plants was increased by drought and low nitrogen but decreased by heat stress.

For various wild plant species, it is known that germination/dormancy responses can vary when mother plants are grown under different moisture or temperature conditions. Furthermore, maternal environmental conditions can affect chemical composition and size of seeds, as well as thickness of the seed coat (Baskin and Baskin, 2014). However, we have little understanding of when these differences develop in seeds that are produced by plants exposed to heat and/or drought stress. Given that the seed filling stage of maize is sensitive to heat and drought stresses, it seems likely that much could be learned about the responses of wild plant species to climate change by studies that focused on the seed-filling stages. That is, studies on the effects of heat and drought stress on maize could be used as models for designing experiments on wild species.

References

Abdul-Baki, A.A., 1991. Tolerance of tomato cultivars and selected germplasm to heat stress. J. Am. Soc. Hort. Sci. 116, 1113–1116.
Abera, K., Crespo, O., Seid, J., Mequanent, F., 2018. Simulating the impact of climate change on maize production in Ethiopia, East Africa. Environ. Syst. Res. 7, 4. Available from: https://doi.org/10.1186/s40068-018-0107-z.
AGDH (Australian Government Department of Health), 2008. The biology of *Zea mays* L. ssp. *mays* (maize or corn). Office of the Gene Technology Regulator. <http://www.ogtr.gov.au/internet/ogtr/publishing.nsf/content/maize-3/$FILE/biologymaize08_2.pdf>.
Alam, M.A., Seetharam, K., Zaidi, P.H., Dinesh, A., Vinayan, M.T., Nath, U.K., 2017. Dissecting heat stress tolerance in tropical maize (*Zea mays* L.). Field Crop. Res. 204, 110–119.
Al-Delaimy, A.O.A., Al-Mamoori, A.H., 2016. Effect of gibberellic acid on breaking post harvest dormancy in seeds of *Zea mays* L. Res. J. Pharm. Biol. Chem. Sci. 7, 1898–1902.
Andersen, M.N., Asch, F., Wu, Y., Jensen, C.R., Næsted, H., Mogensen, V.O., et al., 2002. Soluble invertase expression is an early target of drought stress during the critical, abortion-sensitive phase of young ovary development in maize. Plant Physiol. 130, 591–604.
Aryal, J.P., Sapkota, T.B., Khurana, R., Khatri-Chhetri, A., Rahut, D.B., Jat, M.L., 2020. Climate change and agriculture in South Asia: adaptation options in smallholder production systems. Environ. Develop. Sustain. 22, 5045–5075.
Asante, F.A., Amuakwa-Mensah, F., 2015. Climate change and variability in Ghana: stocktaking. Climate 3, 78–99.
Aslam, M., Zamir, M.S.I., Afzal, I., Yaseen, M., Museen, M., Shoaib, A., 2013. Drought stress, its effect on maize production and development of drought tolerance through potassium application. Cercetari Agron. Moldova 46, 99–114.
Barnabás, B., Jäger, K., Fehér, A., 2008. The effect of drought and heat stress on reproductive processes in cereals. Plant Cell Environ. 31, 11–38.
Baskin, C.C., Baskin, J.M., 1998. Ecology of seed dormancy and germination in grasses. In: Cheplick, G.P. (Ed.), Population Biology of Grasses. Cambridge University Press, Cambridge, pp. 30–83.
Baskin, C.C., Baskin, J.M., 2014. Seeds: Ecology, Biogeography, and Evolution of Dormancy and Germination, Second ed. Academic Press/Elsevier, San Diego.
Bassetti, P., Westgate, M.E., 1993. Water deficit affects receptivity of maize silks. Crop Sci. 33, 279–282.
Battisti, D.S., Naylor, R.L., 2009. Historical warnings of future food insecurity with unprecedented seasonal heat. Science 323, 240–244.

Begcy, K., Sandhu, J., Walia, H., 2018. Transient heat stress during early seed development primes germination and seedling establishment in rice. Front. Plant Sci. 9, 1768. Available from: https://dx.doi.org/10.3389/pls.2018.01768.

Begcy, K., Nosenko, T., Zhou, L.-Z., Fragner, L., Weckwerth, W., Dresselhaus, T., 2019. Male sterility in maize after transient heat stress during the tetrad stage of pollen development. Plant Physiol. 181, 683–700.

Ben-Asher, J., Garcia, A.G.Y., Hoogenboom, G., 2008. Effect of high temperature on photosynthesis and transpiration of sweet corn (*Zea mays* L. var. *rugosa*). Photosynthetica 46, 595–603.

Bird, I.F., Cornelius, M.J., Keys, A.J., 1977. Effects of temperature on photosynthesis by maize and wheat. J. Exp. Bot. 28, 519–524.

Bledsoe, S.W., Henry, C., Griffiths, C.A., Paul, M.J., Feil, M.J., Lunn, J.E., et al., 2017. The role of Tre6P and SnRk1 in maize early kernel development and events leading to stress-induced kernel abortion. BMC Plant Biol. 17, 74. Available from: https://doi.org/10.1186/s12870-017-1018-2.

Boehlein, S.K., Webster, A., Ribeiro, C., Suzuki, M., Wu, S., Guan, J.-C., et al., 2019. Effects of long-term exposure to elevated temperature on *Zea mays* endosperm development during grain fill. Plant J. 99, 23–40.

Bohlken, A.T., Sergenti, E.J., 2010. Economic growth and ethnic violence: an empirical investigation of Hindu-Muslim riots in India. J. Peace Res. 47, 589–600.

Borrás, L., Westgate, M.E., Otegui, M.E., 2003. Control of kernel weight and kernel water relations by post-flowering source-sink ratio in maize. Ann. Bot. 91, 857–867.

Brooking, I.R., 1990. Maize ear moisture during grain-filling, and its relation to physiological maturity and grain-drying. Field Crops Res. 23, 55–68.

Bruce, W.B., Edmeades, G.O., Barker, T.C., 2002. Molecular and physiological approaches to maize improvement for drought tolerance. J. Exp. Bot. 53, 13–25.

Caballero-Briones, F., Iribarren, S., Peña, J.L., Castro-Rodríguez, R., Oliva, A.I., 2000. Recent advances on the understanding of the *nixtamalization* process. Superficies Vacio 10, 20–24.

Cagnola, J.I., Chassart, G.J.D., Ibarra, S.E., Chimenti, C., Ricardi, M.M., Delzer, B., et al., 2018. Reduced expression of selected *FASCICILIN-LIKE ARABINOGALACTAN PROTEIN* genes of *Zea mays* (maize) and of *Arabidopsis* seeds. Plant Cell Environ. 41, 661–674.

Cairns, J.E., Crossa, J., Zaidi, P.H., Grudloyma, P., Sanchez, C., Araus, J.L., et al., 2013a. Identification of drought, heat, and combined drought and heat tolerant donors in maize. Crop Sci. 53, 1335–1346.

Cairns, J.E., Hellin, J., Sonder, K., Araus, J.L., MacRobert, J.F., Thierfelder, C., et al., 2013b. Adapting maize production to climate change in sub-Saharan Africa. Food Sec. 5, 345–360.

Çakir, R., 2004. Effect of water stress at different development stages on vegetative and reproductive growth of corn. Field Crops Res. 89, 1–16.

Charles, N., 2014. Economic impacts of climate change on agriculture and implications for food security in Zimbabwe. Afr. J. Agric. Res. 9, 1001–1007.

Chaves, M.M., Maroco, J.P., Pereira, J.S., 2003. Understanding plant responses to drought – from genes to the whole plant. Funct. Plant Biol. 30, 239–264.

Cheikh, N., Jones, R.J., 1994. Disruption of maize kernel growth and development by heat stress. Plant Physiol. 106, 45–51.

Chemura, A., Schauberger, B., Gornott, C., 2020. Impacts of climate change on agro-climatic suitability of major food crops in Ghana. PLoS One 15, e0229881. Available from: https://doi.org/10.1371/journal.pone.0229881.

Cicchino, M.A., Edreira, J.I.R., Uribelarrea, M., Otegui, M.E., 2010. Heat stress in field-grown maize: response of physiological determinants of grain yield. Crop Sci. 50, 1438–1448.

Cicchino, M.A., Edreira, J.I.R., Otegui, M.E., 2013. Maize physiological responses to heat stress and hormonal plant growth regulators related to ethylene metabolism. Crop Sci. 53, 2135–2146.

Crafts-Brandner, S.J., Salvucci, M.E., 2002. Sensitivity of photosynthesis in a C_4 plant, maize, to heat stress. Plant Physiol. 129, 1773–1780.

Cutforth, H.W., Shaykewich, C.F., Cho, C.M., 1986. Effect of soil water and temperature on corn (*Zea mays* L.) root growth during emergence. Can. J. Soil Sci. 66, 51–58.

Dalil, B., Ghassemi-Golezani, K., 2012. Changes in leaf temperature and grain yield of maize under different levels of irrigation. Res. Crops 13, 481–485.

Daryanto, S., Wang, L., Jacinthe, P.-A., 2016. Global synthesis of drought effects on maize and wheat production. PLoS One 11, e0156362. Available from: https://doi.org/10.1371/journal.pone.0156362.

Davenport, F., Funk, C., Galu, G., 2018. How will East African maize yields respond to climate change and can agricultural development mitigate this response? Clim. Change 147, 491–506.

De Groote, H., Dema, G., Sonda, G.B., Gitonga, Z.M., 2013. Maize for food and feed in East Africa – the farmers' perspective. Field Crops Res. 153, 22–36.

Di Falco, S., Veronese, M., Yesuf, M., 2011. Does adaptation to climate change provide food security? A micro-perspective from Ethiopia. Am. J. Agric. Econ. 93, 829–846.

dos Santos, H.O., Vasconcellos, R.C.C., Pauli, B., Pires, R.M.O., Pereira, E.M., Tirelli, G.V., et al., 2019. Effect of soil temperature in the emergence of maize seeds. J. Agric. Sci. 11, 479–484.

Dupuis, I., Dumas, C., 1990. Influence of temperature stress on in vitro fertilization and heat shock protein synthesis in maize (*Zea mays* L.) reproductive tissues. Plant Physiol. 94, 665–670.

Edreira, J.I.R., Otegui, M.E., 2013. Heat stress in temperate and tropical maize hybrids: a novel approach for assessing sources of kernel loss in field conditions. Field Crops Res. 142, 58–67.

Edreira, J.I.R., Carpici, E.B., Smmarro, D., Otegui, M.E., 2011. Heat stress effects around flowering on kernel set of temperate and tropical maize hybrids. Field Crops Res. 123, 62–73.

Efeoğlu, B., Ekmekçi, Y., Çiçek, N., 2009. Physiological responses of three maize cultivars to drought stress and recovery. S. Afr. J. Bot. 75, 34–42.

Egli, D.B., 2017. Seed Biology and Yield of Grain Crops, Second ed. CAB International, Wallingford.

Egli, D.B., TeKrony, D.M., 1997. Species differences in seed water status during seed maturation and germination. Seed Sci. Res. 7, 3–11.

Ekpa, O., Palacios-Rojas, N., Kruseman, G., Fogliano, V., Linnemann, A.R., 2019. Sub-Saharan African maize-based foods – processing, practices, challenges and opportunities. Food Rev. Int. 35, 609–639.

Epule, T.E., Ford, J.D., Lwasa, S., Lepage, L., 2017. Vulnerability of maize yields to droughts in Uganda. Water 9, 181. Available from: https://doi.org/10.3390/w9030181.

Ezeaku, I.E., Okechukwu, E.C., Aba, C., 2014. Climate change effects on maize (*Zea mays*) production in Nigeria and strategies for mitigation. Asian J. Sci. Technol. 5, 862–871.

FAO (Food and Agriculture Organization of the United Nations), 2019. The state of food security and nutrition in the world 2019. Safeguarding against economic slowdowns and downturns. FAO, Rome.

FAS-USDA (Foreign Agricultural Service USDA), 2020. Global market analysis. <https://apps.fas.usda.gov/psdonline/app/index.html#/app/home> (Accessed 16.07.20).

Fisher, M., Abate, T., Lunduka, R.W., Asnake, W., Alemayehu, Y., Madulu, R.B., 2015. Drought tolerant maize for farmer adaptation to drought in sub-Saharan Africa: determinants of adoption in eastern and southern Africa. Clim. Change 133, 283–299.

Fong, F., Smith, J.D., Koehler, D.E., 1983. Early events in maize seed development. Plant Physiol. 73, 899–901.

Fonseca, A.E., Westgate, M.E., 2004. Relationship between desiccation and viability of maize pollen. Field Crops Res. 94, 114–125.

Gale, M.D., Flintham, J.E., Devos, K.M., 2002. Cereal comparative genetics and preharvest sprouting. Euphytica 126, 21–25.

Gao, J., Wang, S., Zhou, Z., Wang, S., Dong, C., Mu, C., 2019. Linkage mapping and genome-wide association reveal candidate genes conferring thermotolerance of seed-set in maize. J. Exp. Bot. 70, 4849–4863.

Gebreegziabher, B.G., Qufa, C.A., 2017. Plant physiological stimulation by seeds salt priming in maize (*Zea mays*): prospect for salt tolerance. Afr. J. Biotechnol. 16, 209–223.

Ghassemi-Golezani, K., Tajbakhsh, Z., 2012. Relationship of plant biomass and gain filling with grain yield of maize cultivars. Int. J. Agric. Crop Sci. 4, 1536–1539.

Ghebrezgabher, M.G., Yang, T., Yang, X., 2016. Long-term trend of climate change and drought assessment in the Horn of Africa. Adv. Meteorol. 2016, 8057641.

Githui, F., Gitau, W., Mutua, F., Bauwens, W., 2009. Climate change impact on SWAT simulated streamflow in western Kenya. Int. J. Climatol. 29, 1823–1834.

Gregory, P.J., Ingram, J.S.I., Brklacich, M., 2005. Climate change and food security. Philos. Trans. R. Soc. B 360, 2139–2148.

Gustin, J.L., Boehlein, S.K., Shaw, J.R., Junior, W., Settles, A.M., Webster, A., et al., 2018. Ovary abortion is prevalent in diverse maize inbred lines and is under genetic control. Sci. Rep. 8, 13032. Available from: https://doi.org/10.1038/s41598-018-31216-9.

Hanft, J.M., Jones, R.J., 1986. Kernel abortion in maize. I. Carbohydrate concentration patterns and acid invertase activity of maize kernel induced to abort *in vitro*. Plant Physiol. 81, 503–510.

Hannah, L., Donatti, C., Harvey, C.A., Alfaro, E., Rodriguez, D.A., Bouroncle, C., et al., 2017. Regional modeling of climate change impacts on smallholder agriculture and ecosystems in Central America. Clim. Change 141, 29–45.

Harrison, L., Michaelsen, J., Funk, C., Husak, G., 2011. Effects of temperature changes on maize production in Mozambique. Clim. Res. 46, 211–222.

Hedhly, A., 2011. Sensitivity of flowering plant gametophytes to temperature fluctuations. Environ. Exp. Bot. 74, 9–16.

Hedhly, A., Hormaza, J.L., Herrero, M., 2008. Global warming and sexual plant reproduction. Trends Plant Sci. 14, 30–36.

Herrero, M.P., Johnson, R.R., 1980. High temperature stress and pollen viability of maize. Crop Sci. 20, 796–800.

Heyne, E.G., Brunson, A.M., 1940. Genetic studies of heat and drought tolerance in maize. Agron. J. 32, 803–814.

Hunter, J.L., TeKrony, D.M., Miles, D.F., Egli, D.B., 1991. Corn seed maturity indicators and their relationship to uptake of carbon-14 assimilate. Crop Sci. 31, 1309–1313.

Hussain, H.A., Men, S., Hussain, S., Chen, Y., Ali, S., Zhang, S., et al., 2019. Interactive effects of drought and heat stresses on morphophysiological attributes, yield, nutrient uptake and oxidative status in maize hybrids. Sci. Rep. 9, 3890. Available from: https://doi.org/10.1038/s41598-019-40362-7.

IndexMundi-corn, 2020. <https://www.indexmundi.com/agriculture/?commodity = corn&graph = production>. (Accessed 01.08.20).

Janzen, G.M., Hufford, M.B., 2016. Crop domestication: a sneak-peek into the midpoint of maize evolution. Curr. Biol. 26, R1226–R1246.

Jenkins, M.T., 1932. Differential resistance of inbred and crossbred strains of corn to drought and heat injury. Agron. J. 24, 504–506.

Jenkins, M.T., Richey, F.D., 1931. Drought in 1930 showed some strains of corn to be drought resistant. U.S. Dept. Agric. Yearb. 1931, 198–200.

Jones, P.G., Thornton, P.K., 2003. The potential impacts of climate change on maize production in Africa and Latin America in 2055. Glob. Environ. Change 13, 51–59.

Jurgens, S.K., Johnson, R.R., Boyer, J.S., 1978. Dry matter production and translocation in maize subjected to drought during grain fill. Agron. J. 70, 678–682.

Keeling, P.L., Banisadr, R., Barone, L., Wasserman, B.P., Singletary, G.W., 1994. Effect of temperature on enzymes in the pathway of starch biosynthesis in developing wheat and maize grain. Aust. J. Plant Physiol. 21, 807–827.

Killi, D., Bussotti, F., Rasch, A., Haworth, M., 2017. Adaptation to high temperature mitigates the impact of water deficit during combined heat and drought stress in C3 sunflower and C4 maize varieties with contrasting drought tolerance. Physiol. Plant 159, 130–147.

Kim, S.-H., Gitz, D.C., Sicher, R.C., Baker, J.T., Timlin, D.J., Reddy, V.R., 2007. Temperature dependence of growth, development, and photosynthesis in maize under elevated CO_2. Environ. Exp. Bot. 61, 224–236.

Kim, W., Iizumi, T., Nishimori, M., 2019. Global patterns of crop production losses associated with droughts from 1983 to 2009. J. Appl. Meteorol. Climatol. 58, 1233–1244.

Kinda, S.R., Badolo, F., 2019. Does rainfall variability matter for food security in developing countries? Cogent Econ. Financ. 7, 1640098. Available from: https://doi.org/10.1080/23322039.2019.1640098.

Klopfenstein, T.J., Erickson, G.E., Berger, L.L., 2013. Maize is a critically important source of food, feed, energy and forage in the USA. Field Crops Res. 153, 5–11.

Knox, J., Hess, T., Daccache, A., Wheeler, T., 2012. Climate change impacts on crop productivity in Africa and South Asia. Environ. Res. Lett. 7, 034032. Available from: https://doi.org/10.1088/1748-9326/7/3/034032.

Lamaouri, M., Jemo, M., Datla, R., Bekkaoui, F., 2018. Heat and drought stresses in crops and approaches for their mitigation. Front. Chem. 6, 26. Available from: https://doi.org/10.3389/fchem.2018.00026.

Langner, J.A., Zanon, A.J., Streck, N.Q., Reiniger, L.R.S., Kaufmann, M.P., Alves, A.F., 2019. Maize: key agricultural crop in food security and sovereignty in a future with water scarcity. Rev. Brasil. Engenharia Agric. Amb. 23, 648–654.

Leakey, A.D.B., Uribelarrea, M., Ainsworth, E.A., Naidu, S.L., Rogers, A., Ort, D.R., et al., 2006. Photosynthesis, productivity, and yield of maize are not affected by open-air elevation of CO_2 concentration in the absence of drought. Plant Physiol. 140, 779–790.

Leisner, C.P., 2020. Review: climate change impacts on food security – focus on perennial cropping systems and nutritional value. Plant Sci. 293, 110412. Available from: https://doi.org/10.1016/j.plantsci.2020.110412.

Li, H.-Y., Lübberstedt, T., 2018. Molecular mechanisms controlling seed set in cereal crop species under stress and non-stress conditions. J. Integr. Agric. 17, 965–974.

Li, W., Zhang, X., Ashraf, U., Mo, Z., Suo, H., Li, G., 2017. Dynamics of seed germination, seedling growth and physiological responses of sweet corn under PEG-induced water stress. Pak. J. Bot. 49, 639–646.

Li, Y.-T., Xu, W.-W., Ren, B.-Z., Zhao, B., Zhang, J., Liu, P., et al., 2020. High temperature reduces photosynthesis in maize leaves by damaging chloroplast ultrastructure and photosystem II. J. Agron. Crop Sci. 206, 548–564.

Liu, Y., Zhu, Y., Jones, A., Rose, R.J., Song, Y., 2019. Heat stress in legume seed setting: effects, causes, and future prospects. Front. Plant Sci. 10, 938. Available from: https://doi.org/10.3389/fpls.2019.00938.

Lizaso, J.I., Ruiz-Ramos, M., Rodríguez, L., Gabaldon-Leal, C., Oliveira, J.A., Lorite, I.J., et al., 2018. Impact of high temperatures in maize: phenology and yield components. Field Crops Res. 216, 129–140.

Lobell, D.B., Bänziger, M., Magorokosho, C., Vivek, B., 2011a. Nonlinear heat effects on African maize as evidence by historical yield trials. Nat. Clim. Change 1, 42–45.

Lobell, D.B., Schlenker, W., Costa-Roberts, J., 2011b. Climate trends and global crop production since 1980. Science 333, 616–620.

Lonngquist, J.H., Jugenheimer, R.W., 1943. Factors affecting the success of pollination in corn. Agron. J. 35, 923–933.

López, A.N.A., González, J.J.S., Corral, J.A.R., Larios, L.D.L.C., Santacruz-Ruvalcaba, F., Hernández, C.V.S., et al., 2011. Seed dormancy in Mexican teosinte. Crop Sci. 51, 2056–2066.

Luhunga, P.M., 2017. Assessment of the impacts of climate change on maize production in the southern and western highlands sub-agro ecological zones of Tanzania. Front. Plant Sci. 5, 51. Available from: https://doi.org/10.3389/fenvs.2017.00051.

Lunduka, R.W., Mateva, K.I., Magorokosho, C., Manjeru, P., 2019. Impact of adoption of drought-tolerant maize varieties on total maize production in south Eastern Zimbabwe. Clim. Develop. 11, 35–46.

Lyakh, V.A., Kravchenko, A.N., Soroka, A.I., Dryuchina, E.N., 1991. Effects of high temperatures on mature pollen grains in wild and cultivated maize accessions. Euphytica 55, 203–207.

Ma, J., Weng, B., Bi, W., Xu, D., Xu, T., Yan, D., 2019. Impact of climate change on the growth of typical crops in karst areas: a case study of Guizhou Province. Adv. Meteorol. 2019, 1401402. Available from: https://doi.org/10.1155/2019/1401402.

Makadho, J.M., 1996. Potential effects of climate change on corn production in Zimbabwe. Clim. Res. 6, 147–151.

Mäkelä, P., McLaughlin, J.E., Boyer, J.S., 2005. Imaging and quantifying carbohydrate transport to the developing ovaries of maize. Ann. Bot. 96, 939–949.

Marmagne, A., Jasinski, S., Fagard, M., Bill, L., Guerche, P., Masclaus-Daubresse, C., et al., 2020. Post-flowering biotic and abiotic stresses impact nitrogen use efficiency and seed filling in *Arabidopsis thaliana*. J. Exp. Bot. 71, 4578–4590.

Martinez-Feria, R.A., Licht, M.A., Ordóñez, R.A., Hatfield, J.L., Coulter, J.A., Archontoulis, S.V., 2019. Evaluating maize and soybean grain drydown in the field with predictive algorithms and genotype-by-environment analysis. Sci. Rep. 9, 7167. Available from: https://doi.org/10.1038/s41598-019-43653-1.

Masambaya, F.N., Oludhe, C., Lukorito, C.B., Onwonga, R., 2018. Vulnerability of maize production to climate change in maize producing countries of Rift Valley Kenya: the indicator approach. Int. J. Sci. Res. Publ. 8, 8106. Available from: https://doi:10.29322/IJSRP.8.9.2018.p8106.

Masouleh, S.S.S., Sassine, Y.N., 2020. Molecular and biochemical responses of horticultural plants and crops to heat stress. Orn. Hort. 26, 148–158.

Matsuoka, Y., Vigouroux, Y., Goodman, M.M., Sanchez, G.J., Buckler, E., Doebley, J., 2002. A single domestication for maize shown by multilocus microsatellite genotyping. Proc. Natl. Acad. Sci. USA 99, 6080–6084.

McCarty, D.R., Hattori, T., Carson, C.B., Vassil, V., Lazar, M., Vasil, I.K., 1991. The *viviparous-1* developmental gene of maize encodes a novel transcriptional activator. Cell 66, 895–905.

McLaughlin, J.E., Boyer, J.S., 2004a. Sugar-responsive gene expression, invertase activity, and senescence in aborting maize ovaries at low water potentials. Ann. Bot. 94, 675–689.

McLaughlin, J.E., Boyer, J.S., 2004b. Glucose localization in maize ovaries when kernel number decreases at low water potential and sucrose is fed to the stems. Ann. Bot. 94, 75–86.

Mihalache-O'Keef, A., Li, Q., 2011. Modernization vs. dependency revisited: effects of foreign direct investment on food security in less developed countries. Int. Stud. Quart. 55, 71–93.

Miracle, M.P., 1965. The introduction and spread of maize in Africa. J. Afr. Hist. 6, 39–55.

Mohapatra, U., Singh, A., Ravikumar, R.L., 2020. Effect of gamete selection in improving of heat tolerance as demonstrated by shift in allele frequency in maize (*Zea mays* L.). Euphytica 216, 76. Available from: https://doi.org/10.1007/s10681-020-02603-z.

Mondrus-Engle, M., 1981. Tetraploid perennial teosinte seed dormancy and germination. J. Range Manage. 34, 59–61.

Monjardino, P., Smith, A.G., Jones, R.J., 2005. Heat stress effects on protein accumulation of maize endosperm. Crop Sci. 45, 1203–1210.

Moss, G.I., Downey, L.A., 1971. Influence of drought stress on female gametophyte development in corn (*Zea mays* L.) and subsequent grain yield. Crop Sci. 11, 368–372.

Msowoya, K., Madani, K., Davtalab, R., Mirchi, A., Lund, J.R., 2016. Climate change impacts on maize production in the Warm Heart of Africa. Water Resour. Manage. 30, 5299–5312.

Muchow, R.C., 1990. Effect of high temperature on grain-growth in field-grown maize. Field Crops Res. 23, 145–158.

Muluneh, A., Stroosnijder, L., Keesstra, S., Biazin, B., 2016. Adapting to climate change for food security in the Rift Valley dry lands of Ethiopia: supplemental irrigation, plant density and sowing date. J. Agric. Sci. 155, 703–724.

Mumo, L., Yu, J., Fang, K., 2018. Assessing impacts of seasonal climate variability on maize yield in Kenya. Int. J. Plant Prod. 12, 297–307.

Naveed, S., Aslam, M., Maqbool, M.A., Bano, S., Zaman, Q.U., Ahmad, R.M., 2014. Physiology of high temperature stress tolerance at reproductive stages in maize. J. Anim. Plant Sci. 24, 1141–1145.

Neill, S.J., Horgan, R., Rees, A.F., 1987. Seed development and vivipary in *Zea mays* L. Planta 171, 358–364.

NeSmith, D.S., Richie, J.T., 1992. Short- and long-term responses of corn to a pre-anthesis soil water deficit. Agron. J. 84, 107–113.

Nuss, E.T., Tanumihardjo, S.A., 2010. Maize: a paramount staple crop in the context of global nutrition. Comprehen. Rev. Food Sci. Food Safe. 9, 417–436.

Obata, T., Witt, S., Lisec, J., Palacios-Rojas, N., Florez-Sarasa, I., Yousfi, S., et al., 2015. Metabolite profiles of maize leaves in drought, heat, and combined stress field trials reveal the relationship between metabolism and grain yield. Plant Physiol. 169, 2665–2683.

Ober, E.S., Setter, T.L., Madison, J.T., Thompson, J.F., Shapiro, P.S., 1991. Influence of water deficit on maize endosperm development. Plant Physiol. 97, 154–164.

Obunyali, C.O., Karanja, J., Oikeh, S.O., Omanya, G.O., Mugo, S., Beyene, Y., et al., 2019. On-farm performance and farmers' perceptions of *DroughtTEGO*-Climate-Smart maize hybrids in Kenya. Agron. J. 111, 2754–2768.

Ochieng, J., Kirimi, L., Mathenge, M., 2016. Effects of climate variability and change on agricultural production: the case of small scale farmers in Kenya. NJAS-Wageningen. J. Life Sci. 77, 71–78.

Ogutu, G.E.O., Franssen, W.H.P., Supit, I., Omondi, P., Hutjes, R.W.A., 2018. Probabilistic maize yield prediction over East Africa using dynamic ensemble seasonal climate forecast. Agric. For. Meteorol. 250–251, 243–261.

Olaniyan, A.B., 2015. Maize: panacea for hunger in Nigeria, Afr. J. Plant Sci., 9. pp. 155–174.

Omoyo, N.N., Wakhungu, J., Oteng'i, S., 2015. Effects of climate variability on maize yield in the arid and semiarid lands of lower eastern Kenya. Agric. Food Sec. 4, 8. Available from: https://doi.org/10.1186/s40066-015-0028-2.

Ordóñez, R.A., Savin, R., Cossani, C.M., Slafter, G.A., 2015. Yield response to heat stress as affected by nitrogen availability in maize. Field Crops Res. 183, 184–203.

Oseni, T.O., Masarirambi, M.T., 2011. Effect of climate change on maize (*Zea mays*) production and food security in Swaziland. Amer.-Eurasian J. Agric. Environ. Sci. 11, 385–391.

Östberg, W., Slegers, M.F.W., 2010. Losing faith in the land: changing environmental perceptions in Burunge country, Tanzania. J. East Afr. Stud. 4, 247–265.

Otegui, M.E., Andrade, F.H., Suero, E.E., 1995. Growth, water use, and kernel abortion of maize subjected to drought at silking. Field Crops Res. 40, 87–94.

Oury, V., Caldeira, C.F., Prodhomme, D., Pichon, J.-P., Gibon, Y., Tardieu, F., et al., 2016. Is change in ovary carbon status a cause or a consequence of maize ovary abortion in water deficit during flowering? Plant Physiol. 171, 997–1008.

Ouattar, S., Jones, R.J., Crookston, R.K., 1987a. Effect of water deficit during grain filling on the pattern of maize kernel growth and development. Crop Sci. 27, 726–730.

Ouattar, S., Jones, R.J., Crookston, R.K., Kajeiou, M., 1987b. Effect of drought on water relations of developing maize kernels. Crop Sci. 27, 730–735.

Prasad, R., Gunn, S.K., Rotz, C.A., Karsten, H., Roth, G., Buda, A., et al., 2018. Projected climate and agronomic implications for corn production in the northeastern United States. PLoS One 13, e0198623. Available from: https://doi.org/10.1371/journal.pone.0198623.

Ramirez-Cabral, N.Y.Z., Kumar, L., Shabani, F., 2017. Global alternations in areas of suitability for maize production from climate change and using a mechanistic species distribution model (CLIMEX). Sci. Rep. 7, 5910. Available from: https://doi.org/10.1038/s41598-017-05804-0.

Ranum, P., Peña-Rosas, J.P., Garcia-Casal, M.N., 2014. Global maize production, utilization, and consumption. Ann. N. Y. Acad. Sci. 1312, 105–112.

Rench, W.E., Shaw, R.H., 1971. Black layer development in corn. Agron. J. 63, 303–305.
Riley, G.J.P., 1981. Effects of high temperature on the germination of maize (Zea mays L.). Planta 151, 68–74.
Ritchie, S.W., Hanway, J.J., 1986. How a corn plant develops. Iowa State University of Science and Technology Special Report No. 48.
Rockström, J., Falkenmark, M., 2015. Increase water harvesting in Africa. Nature 519, 283–286.
Rodríguez, M.V., Barrero, J.M., Corbineau, F., Gubler, F., Benech-Arnold, R.L., 2015. Dormancy in cereals (not too much, not so little): about the mechanisms behind this trait. Seed Sci. Res. 25, 99–119.
Rosbakh, S., Poschlod, P., 2016. Minimal temperature of pollen germination controls species distribution along a temperature gradient. Ann. Bot. 117, 1111–1120.
Rosbakh, S., Pacini, E., Nepi, M., Poschlod, P., 2018. An unexplored side of regeneration niche: seed quantity and quality are determined by the effect of temperature on pollen performance. Front. Plant Sci. 9, 1036. Available from: https://doi.org/10.3389/fpls.2018.01036.
Ruan, Y.-L., Patrick, J.W., Bouzayebn, M., Osorio, S., Fernie, A.R., 2012. Molecular regulation of seed and fruit set. Trends Plant Sci. 17, 1360–1385.
Rufino, C.A., Fernandes-Vieira, J., Martín-Gil, J., Júnior, J.S.A., Tavares, L.C., Fernandes-Correa, M., et al., 2018. Water stress influence on the vegetative period yield components of different maize genotypes. Agronomy 8, 151. Available from: https://doi.org/10.3390/agronomy8080151.
Ruinda, J., Van Wijk, M.T., Mapfumo, P., Descheemaeker, K., Supit, I., Giller, K.E., 2015. Climate change and maize yield in southern Africa: what can farm management do? Glob. Change Biol. 21, 4588–4601.
Sah, R.P., Chakraborty, M., Prasad, K., Pandit, M., Tudu, V.K., Chakravarty, M.K., et al., 2020. Impact of water deficit stress in maize: phenology and yield components. Sci. Rep. 10, 2944. Available from: https://doi.org/10.1038/s41598-020-59689-7.
Saini, H.S., 1997. Effects of water stress on male gametophyte development in plants. Sex Plant Reprod. 10, 67–73.
Sala, R.G., Andrade, F.H., Westgate, M.E., 2007. Maize kernel moisture at physiological maturity as affected by the source-sink relationship during grain filling. Crop Sci. 47, 711–716.
Sánchez, B., Rasmussen, A., Porter, J.R., 2014. Temperatures and the growth and development of maize and rice: a review. Glob. Change Biol. 20, 408–417.
Sangoi, L., Salvador, R.J., 1998. Maize susceptibility to drought at flowering: a new approach to overcome the problem. Cienc. Rural. 28, 699–706.
Sawe, J., Mung'ong'o, C.G., Kimaro, G.F., 2018. The impacts of climate change and variability on crop farming systems in semi-arid central Tanzania: the case of Manyoni District in Singida Region. Afr. J. Environ. Sci. Technol. 12, 323–334.
Schlenker, W., Lobell, D.B., 2010. Robust negative impacts of climate change on African agriculture. Environ. Res. Lett. 5, 014010. Available from: https://doi.org/10.1088/1748-9326/5/1/014010.
Schoper, J.B., Lambert, R.J., Vasilas, B.L., 1986. Maize pollen viability and ear receptivity under water and high temperature stress. Crop Sci. 26, 1029–1033.
Schoper, J.B., Lambert, R.J., Vasilas, B.L., 1987. Pollen viability, pollen shedding, and combining ability for tassel heat tolerance in maize. Crop Sci. 27, 27–31.
Schussler, J.R., Westgate, M.E., 1991. Maize kernel set at low water potential: I. Sensitivity to reduced assimilates during early kernel growth. Crop Sci. 31, 1189–1195.
Sehgal, A., Sita, J., Siddique, K.H.M., Kumar, R., Bhogireddy, S., Varshney, R.K., et al., 2018. Drought or/and heat-stress effects on seed filling in food crops: impacts on functional biochemistry, seed yields, and nutritional quality. Front. Plant Sci. 9, 1705. Available from: https://doi.org/10.3389/fpls.2018.01705.
Shaban, M., 2013. Effect of water and temperature on seed germination and emergence as a seed hydrothermal time model. Int. J. Adv. Biol. Biomed. Res. 1, 1686–1691.
Shah, T.R., Prasad, K., Kumar, P., 2016. Maize – a potential source of human nutrition and health: a review. Cogent Food Agric. 2, 1166995. Available from: https://doi.org/10.1080/23311932.2016.1166995.
Shen, S., Zhang, L., Liang, X.-G., Zhao, X., Lin, S., Qu, L.-H., et al., 2018. Delayed pollination and low availability of assimilates are major factors causing maize kernel abortion. J. Exp. Bot. 69, 1599–1613.
Shen, S., Liang, Z.-G., Zhang, L., Zhao, X., Liu, Y.-P., Lin, S., et al., 2019. Intervening in sibling competition for assimilates by controlled pollination prevents seed abortion under postpollination drought in maize. Plant Cell Environ. 43, 903–919.
Shi, W., Tao, F., 2014. Vulnerability of African maize yield to climate change and variability during 1961–2019. Food Sec. 6, 471–481.
Shiferaw, B., Prasanna, B.M., Hellin, J., Bänziger, M., 2011. Crops that feed the world 6. Past successes and future challenges to the role played by maize in global food security. Food Sec. 3, 307–327.
Shim, D., Lee, K.-J., Lee, B.-W., 2017. Response of phenology- and yield-related traits of maize to elevated temperature in a temperate region. Crops J. 5, 305–316.
Shumetie, A., Yismaw, M.A., 2017. Effect of climate variability on crop income and indigenous adaptation strategies of households. Int. J. Clim. Change Strat. Manage. 10, 580–595.
Simpson, G.M., 1990. Seed Dormancy in Grasses. Cambridge University Press, Cambridge.
Simtowe, F., Amondo, E., Marenya, P., Rahut, D., Sonder, K., Erenstein, O., 2019. Impacts of drought-tolerant maize varieties on productivity, risk, and resource use: evidence from Uganda. Land Use Policy 88, 104091. Available from: https://doi.org/10.1016/j.landusepol.2019.104091.
Singh, A., Ravikumar, R.L., Jingade, P., 2016. Genetic variability for gametophytic heat tolerance in maize inbreed lines. Sabrao J. Breed. Genet. 48, 41–49.
Singletary, G.W., Banisadr, R., Keeling, P.L., 1994. Heat stress during grain filling in maize: effects on carbohydrate storage and metabolism. Aust. J. Plant. Physiol. 21, 829–841.

Smale, M., Byerlee, D., Jayne, T., 2011. Maize Revolutions in Sub-Saharan Africa. The World Bank. Policy Research Working Paper No. 5659. <https://openknowledge.worldbank.org/handle/10986/3421>.

St.Clair, S.B., Lynch, J.P., 2010. The opening of Pandora's box: climate change impacts on soil fertility and crop nutrition in developing countries. Plant Soil 335, 101−115.

Stevens, T., Madani, K., 2016. Future climate impacts on maize farming and food security in Malawi. Sci. Rep. 6, 36241. Available from: https://doi.org/10.1038/srep36241.

Stiles, I.E., 1948. Relation of water to the germination of corn and cotton seeds. Plant Physiol. 23, 210−222.

Stuch, B., Alcamo, J., Schaldach, R., 2020. Projected climate change impacts on mean and year-to-year variability of yield of key smallholder crops in Sub-Saharan Africa. Clim. Develop. 13, 268−282. Available from: https://doi.org/10.1080/17565529.2020.1760771.

Tadele, Z., 2018. African orphan crops under abiotic stresses: challenges and opportunities. Scientifica 2018, 1451894. Available from: https://doi.org/10.1155/2018/1451894.

Tao, Z.-Q., Chen, Y.-Q., Li, C., Zou, J.-X., Peng, Y., Yuan, S.-F., et al., 2016. The causes and impacts for heat stress in spring maize during grain filling in the North China Plain − a review. J. Integr. Agric. 15, 2677−2687.

Tefra, T., Kanampiu, F., Groote, H.D., Hellin, J., Mugo, S., Kimenju, S., et al., 2011. The metal silo: an effective grain storage technology for reducing post-harvest insect and pathogen losses in maize while improving smallholder farmer's food security in developing countries. Crop Prot. 30, 240−245.

ten Berge, H.F.M., Hijbeek, R., van Loon, M.P., Rurinda, J., Tesfaye, K., Zingore, S., et al., 2019. Maize crop nutrient input requirements for food security in sub-Saharan Africa. Glob. Food Sec. 23, 9−21.

Tesso, G., Emana, B., Ketema, M., 2012. A time series analysis of climate variability and its impacts on food production in North Shewa Zone in Ethiopia. Afr. Crop. Sci. J. 20, 261−274.

Thamaga-Chitja, J.M., Hendriks, S.L., Ortmann, F.G., Green, M., 2004. Impact of maize storage on rural household food security in northern Kwazulu-Natal. Tyds. Gesinsekol. Verbruikerswet. 32, 8−15.

Thornton, P.K., Jones, P.G., Ericksen, P.J., Challinor, A.J., 2011. Agriculture and food systems in sub-Saharan Africa in a 4°C+ world. Philos. Trans. R. Soc. A 369, 117−136.

Tigchelaar, M., Battisti, D.S., Naylor, R.L., Ray, D.K., 2018. Future warming increases probability of globally synchronized maize production shocks. Proc. Natl. Acad. Sci. USA 115, 6644−6649.

Tripathi, A., Tripathi, D.K., Chauhan, D.K., Kumar, N., Singh, G.S., 2016. Paradigms of climate change impacts on some major food sources of the world: a review on current knowledge and future prospects. Agric. Ecosyst. Environ. 216, 356−373.

van der Venter, H.A., Hoffman, R., 1988. Germination rate of maize (*Zea mays* L.) kernels. I. Relationships with kernel properties and water uptake. S. Afr. J. Plant Soil 5, 189−192.

Wandaka, L.M., Kabubo-Mariara, J., Kimuyu, P., 2016. Economic impact of climate change on maize production in Kenya. Am. J. Agric. 1, 37−50.

Wang, M., Li, Y., Ye, W., Bornman, J.F., Yan, X., 2011. Effects of climate change on maize production, and potential adaptation measures: a case study in Jilin Province, China. Clim. Res. 46, 223−242.

Wang, B., Liu, C., Zhang, D., He, C., Zhang, J., Li, Z., 2019a. Effects of maize organ-specific drought stress response on yields from transcriptome analysis. BMC Plant Biol. 19, 335. Available from: https://doi.org/10.1186/s12870-019-1941-5.

Wang, Y., Tao, H., Tian, B., Sheng, D., Xu, C., Zhou, H., et al., 2019b. Flowering dynamics, pollen, and pistil contribution to grain yield in response to high temperature during maize flowering. Environ. Exp. Bot. 158, 80−88.

Wang, Y., Tao, H., Zhang, P., Hou, X., Sheng, D., Tian, B., et al., 2020. Reduction in seed set upon exposure to high night temperature during flowering in maize. Physiol. Plant 169, 73−82.

Westgate, M.E., 1994. Water stress and development of the maize endosperm and embryo during drought. Crop Sci. 34, 76−83.

Westgate, M.E., Boyer, J.S., 1985. Osmotic adjustment and the inhibition of leaf, root, stem and silk growth at low water potentials in maize. Planta 164, 540−549.

Westgate, M.E., Boyer, J.S., 1986. Reproduction at low silk and pollen water potentials in maize. Crop Sci. 26, 951−956.

Westgate, M.E., Grant, D.L.T., 1989. Response of the reproductive tissue to water deficits at anthesis and mid-grain fill. Plant Physiol. 91, 862−867.

White, C.M., Rivin, C.J., 2000. Gibberellins and seed development in maize. II. Gibberellin synthesis inhibition enhances abscisic acid signaling in cultured embryos. Plant Physiol. 122, 1089−1097.

White, C.N., Proebsting, W.M., Hedden, P., Rivin, C.J., 2000. Gibberellins and seed development in maize. I. Evidence that gibberellin/abscisic acid balance governs germination vs maturation pathways. Plant Physiol. 122, 1081−1088.

Wilhelm, E.P., Mullen, R.E., Keeling, P.L., Singletary, G.W., 1999. Heat stress during grain filling in maize: effects on kernel growth and metabolism. Crop Sci. 39, 1733−1741.

Yan, P., Chen, Y., Sui, P., Vogel, A., Zhang, X., 2018. Effect of maize plant morphology on the formation of apical kernels at different sowing dates and under different plant densities. Field Crops Res. 223, 83−92.

Zhao, C., Liu, B., Piao, S., Wang, X., Lobell, D.B., Huang, Y., et al., 2017. Temperature increase reduces global yields of major crops in four independent estimates. Proc. Natl. Acad. Sci. USA 114, 9326−9331.

Zinn, K.E., Tunc-Ozdemir, M., Harper, J.F., 2010. Temperature stress and plant sexual reproduction: uncovering the weakest links. J. Exp. Bot. 61, 1959−1968.

Zinselmeier, C., Westgate, M.E., Schussler, J.R., Jones, R.J., 1995. Low water potential disrupts carbohydrate metabolism in maize (*Zea mays* L.) ovaries. Plant Physiol. 107, 385−391.

Chapter 17

Fire and regeneration from seeds in a warming world, with emphasis on Australia

Mark K.J. Ooi[1], Ryan Tangney[1,2], and Tony D. Auld[1,3]

[1]*Centre for Ecosystem Science, School of Biological Earth and Environmental Sciences, University of New South Wales, Kensington, NSW, Australia,*
[2]*Kings Park Science, Biodiversity, and Conservation Science, Department of Biodiversity, Conservation, and Attractions, Kings Park, WA, Australia,*
[3]*School of Earth, Atmospheric and Life Sciences, University of Wollongong, Wollongong, NSW, Australia*

Introduction

Fire emerged as a critical element of the Earth's ecosystems 420 million years ago, increasing in prevalence as oxygen accumulated in our atmosphere and terrestrial plants became dominant (Scott and Glasspool, 2006; Bowman et al., 2009). Fire provides a natural disturbance that drives ecological processes and plays a key role in maintaining species diversity in forests, shrublands, and grasslands (Keith, 2012; Archibald et al., 2018; He et al., 2019a). Fire has been a part of the evolutionary history of many plant communities and biomes around the globe (Pausas and Rebeiro, 2017; Pausas and Keeley, 2019), selecting for traits that allow plant species to respond and persist under a particular fire regime (Keeley et al., 2011; Keith, 2012; He et al., 2019a; Lamont et al., 2019). Therefore, numerous traits and functional responses related to seeds and plant regeneration are driven by fire cues, including resprouting, flowering and seed production, seed release from canopies (pyriscence), dormancy-break, and germination.

Seeds, as a means for reproducing rapidly in the postfire environment, allow population regeneration and persistence since they replace individuals that have been killed during the fire and provide a mechanism for generational turnover. More broadly, components of the regeneration niche are strongly dependent on environmental filters, contributing to determining the distribution of a species and the assembly of plant communities (Jiménez-Alfaro et al., 2016; Smith et al., 2016). Any changes to these environmental filters, including fire and its occurrence, potentially will cause a shift in the regeneration niche (Staver et al., 2011; Smith et al., 2016). Quantifying such impacts is essential for understanding changes to fire-prone ecosystems under climate change (Bowman et al., 2014).

The importance of fire cues on the population dynamics of fire-prone species is reflected in the abundance of studies that found direct fire-related effects on seeds including (1) heat shock on breaking dormancy of physically dormant (PY) species (Keeley and Bond, 1997, Moreira et al., 2010; Zuloaga-Aguilar et al., 2010, Ooi et al., 2014; Luna et al., 2019), (2) smoke and smoke-derived chemicals on germination stimulation, primarily of physiologically (PD) and morphophysiologically (MPD) dormant species (Keeley and Fotheringham, 1998; Downes et al., 2014; Çatav et al., 2018, Carthey et al., 2018; King and Menges, 2018), (3) interactions between smoke, heat, or other factors (e.g., light) on germination (Tieu et al., 2001; Thomas et al., 2010; Mackenzie et al., 2016; Ramos et al., 2019), (4) the ability of seeds to tolerate high temperatures (Habrouk et al., 1999; Ramos et al., 2017; Daibes et al., 2019; Tangney et al., 2019), and (5) heat-induced release of serotinous [canopy-stored, nondormant (ND)] seeds (Enright and Lamont, 2006; Clarke et al., 2010; Milich et al., 2012; Lamont et al., 2020). Indirect mechanisms by which fires promote regeneration include fire-stimulated flowering (also known as postfire flowering) (Le Maitre and Brown, 1992; Brewer et al., 2009; Lamont and Downes, 2011; Zirondi et al., 2021) and canopy removal leading to subsequent changes in the soil seed bank environmental conditions (Auld and Bradstock, 1996; Santana et al., 2010, 2013; Hill and Auld, 2020). Fire also changes the conditions experienced by seedlings, reducing competition and increasing resource availability (Keith, 2012), and providing smoke, which can influence seedling growth (Nelson et al., 2012; Ramos et al., 2019).

230 SECTION | II Special Topics

A fire regime is defined as the frequency, season, intensity, type, and area covered that is most commonly experienced within an ecosystem (Bradstock, 2010). It consists of event-related and temporal elements of recurrent patterns of fire that are influenced by a broad range of factors, including climate, weather events, vegetation type and timing, and source of ignition events. All of these factors can influence the response of plants and animals in fire-prone landscapes via interacting with critical life-history traits (Whelan et al., 2002). Increasingly, the fire regime is changing due to the impacts of human activities, predominately clearing and fragmentation of habitat, increased human-caused ignitions, and a changing climate (Balch et al., 2017; Bowman et al., 2020). Climate change is arguably now the biggest driver of changes in fire regimes, and it is projected to increase the frequency, intensity, and size of fire events due to more frequent high fire-danger weather, longer and more sustained droughts, and heatwaves (Flannigan et al., 2013; Hoegh-Guldberg et al., 2018; Joseph et al., 2019).

In recent years, the effects of climate change have been manifested in record-breaking fires, which exceeded previous ones in extent, intensity, and frequency. In 2017, coined by some as the "year of the mega-fire," records for size and/or number of fires were broken in Brazil, Chile, Siberia, Portugal, Spain, Canada, United States, and other countries (Fidelis et al., 2018; Kirchmeier-Young et al., 2019; Turco et al., 2019, McCarty et al., 2020). In California, records were broken again in 2018 and 2020 (Williams et al., 2019; Higuera and Abatzoglou, 2021). The largest fires ever recorded in eastern Australia occurred during the 2019/2020 fire season and burnt 5.8 million hectares, which is 21% of the temperate forest biome in eastern Australia (Boer et al., 2020; Nolan et al., 2021). Much of the underlying cause of these multiple mega-fire events has been attributed to climate change (Nolan et al., 2020).

The potential for a warming world to impact the regeneration of plants from seeds in fire-prone regions is due to (1) changes to the conditions experienced by parent plants, seeds, and seedlings, and (2) changes to the fire regime itself (Davis et al., 2018) (Fig. 17.1). The interactions between these two drivers, as well as between these and other anthropogenic factors such as increasing disturbance, undoubtedly will impact regeneration from seeds.

Fire regime elements that are event-related, such as intensity (amount of vegetation burnt or organic matter consumed, i.e., "fire severity" sensu Keeley, 2009), seasonality, type and size, or frequency, that is, fire recurrence intervals and time since last fire, can disrupt the life cycles and population dynamics of fire-adapted species. One of the clearest

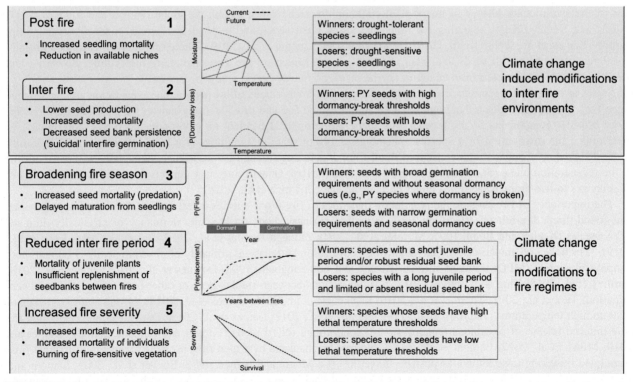

FIGURE 17.1 The potential mechanisms by which plants can be impacted under climate change scenarios in fire-prone ecosystems. Species can be impacted by the effects of changes to climate (Panels 1 and 2) and/or changes to the fire regime (Panels 3–5). Examples of characteristics are provided—species maintaining these characteristics may be subject to neutral or positive effects ("Winners") or negative effects ("Losers"). *PY*, Physical dormancy.

projections of the impacts of climate change is the switching of fire regimes away from historic patterns (Balch et al., 2017; Bowman et al., 2020; Boer et al., 2020), which is likely to cause changes in species abundance and persistence and associated state-shifts in community composition (Davis et al., 2018). In the worst-case scenario, shifts in fire regime will increase the risk of local extinction (Keith, 2012).

In this chapter, we highlight the impacts of climate change-induced shifts in three key elements of the fire regime, namely, severity, season, and frequency, on regeneration of plants from seeds and persistence of plant populations. In addition, we consider how such shifts will interact with other climate-induced changes, including longer and warmer summers and higher incidence of drought and heatwaves, to affect plant persistence.

Climate-related shifts in the fire regime and plant regeneration

In fire-prone systems, recruitment from seeds following fire is prolific and provides a pathway for population persistence and expansion in many but not all species. A number of species are "obligate seeders" (Ooi et al., 2006). Plants in this functional group do not survive 100% canopy scorch, and recruitment from seeds is the only means of population recovery following fire events. Seed survival of obligate seeders during passage of fire through the community is vital for postfire recruitment, which is achieved via long-lived persistent soil- or canopy-stored seed banks. Therefore, understanding how recruitment from seeds following fire is affected by changes in the fire regime, especially those associated with changes in climate, is required for predicting ecosystem shifts and informing conservation management.

Fire intensity and severity—what can happen when fires burn hotter?

Increasing drought and temperatures under climate change contribute to increasing the amount of fuel available to burn and the intensity of subsequent fires (Abatzoglou and Williams, 2016; Ruthrof et al., 2016; Nolan et al., 2020). Plant species exhibit two major responses to heat from a fire. Obligate seeders are killed and thus rely on seeds to recruit back into burnt habitats, while resprouters have dormant buds that survive the heat from a fire, allowing the plant to regrow (Keith, 2012). Seeds may be present in the plant canopy in serotinous (aerial) seed banks or in the soil (propagule persisters sensu Pausas et al., 2004), or they may need to be dispersed into the burnt site (propagule nonpersisters sensu Pausas et al., 2004). The degree of heating will affect plant survival. Thus, hotter fires can lead directly to increased plant mortality and decreased resprouting ability (Bowman et al., 2019), while interactions with prefire drought stress also increase mortality (Edwards and Krockenberger, 2006; van Mantgem et al., 2013). This means that as the climate gets warmer, there will be increased dependence on recruitment from seeds to replace plants that are killed. For example, in a study in Western Australia mortality of dominant tree species in karri forest was higher after an extreme-severity fire than after a low- or high-severity fire; however, recruitment after fire was not higher in low- or high-severity fire areas than in those exposed to extreme-severity fire (Etchells et al., 2020).

Survival of seeds and changes in fire severity

All predictions concerning recent mega-fire events point to future fires of increased intensity (energy output) and severity (Davis et al., 2018), which would increase the temperatures experienced by seeds stored in the plant canopy. The relationship between fire intensity and soil heating is variable, especially for prescribed burns (e.g., Bradstock and Auld, 1995). However, evidence indicates that extreme fire events also will result in higher soil temperatures. The soil layer near the surface contains the bulk of soil-stored seed banks, and it is exposed to the most extreme temperatures (Odion and Davis, 2000; Auld and Denham, 2006; Tangney et al., 2018). Hotter fires on the scale witnessed recently in eastern Australia, and that are forecast to increase under climate change, could impact seed survival and effectively reduce, or even eliminate, seed reserves.

For seeds stored in serotinous woody fruits or cones in plant canopies, the structures covering the seeds insulate them from lethal temperatures (Hanley and Lamont, 2000; Huss et al., 2019; Lamont, 2020), which can exceed 450°C (Bradstock and Myerscough, 1981). However, modeling of the impacts of high severity crown fires in jack pine (*Pinus banksiana*) and lodgepole pine (*P. contorta*) found that the seed bank decreased with extreme fires (Alexander and Cruz, 2012). For some taxa, interspecific variation in heat-induced mortality is determined by the thickness of the structures covering the seeds, which can determine how much heat a seed will experience (Bradstock et al., 1994). As such, an increase in the number of extremely hot fires occurring will not necessarily impact species equally. Canopy seed bank species with thin structures covering the seeds and thus with little insulating material will be most sensitive and at risk to seed bank loss.

Seeds of the majority of species in fire-prone systems are buried in the soil (Auld and Ooi, 2017), which insulates them from the heat of a fire (Céspedes et al., 2012; Ooi et al., 2012). Independent of fire intensity, seeds on the soil surface are likely to be exposed to lethal temperatures, unless there is significant spatial patchiness in the fire (Daibes et al., 2017). For seeds stored in the soil seed bank, successful postfire seed regeneration is contingent upon a trade-off between burial depth and fire survival. Temperatures experienced by seeds decrease with increasing soil depth (Bradstock and Auld, 1995), and during a fire temperatures lethal to dry seeds generally are found only on or near the soil surface (about 120°C–150°C for many species, Ooi et al., 2014; Tangney et al., 2020a). However, small-seeded species have less capacity and fewer resources with which to germinate (emerge) successfully from greater soil depths than large-seeded species (Bond et al., 1999; Hanley et al., 2003; Liyanage and Ooi, 2018; Tangney et al., 2019). For example, the lethal temperature threshold of the small seeds of the grass *Austrostipa elegantissima* is 117°C, and due to their size they can only emerge from a soil depth of 2 cm (Tangney et al., 2019, 2020a). Consequently, for *A. elegantissima* and other small-seeded species a future regime of extreme fires and greater soil heating means that more seeds on or near the soil surface will be killed during fire. This will favor large-seeded species and possibly reduce the available seed bank to depths beyond which successful germination is possible for some small-seeded species.

The impact of an increase in extreme fire events will depend on the thresholds of tolerance for individual species. Perhaps the simplest example of how impact may vary is found in PY species, where heat causes dormancy loss and germination follows when rainfall and ambient temperatures permit (Merritt et al., 2007). In these cases, there are lower temperature thresholds that affect seed dormancy-break and upper temperature thresholds that affect survival, with both mediated by duration of heating (Merritt et al., 2007; Ooi et al., 2014; Hill and Auld, 2020). Below the lower thresholds, seed dormancy remains intact, while above the upper thresholds seeds are killed. At temperatures between the two thresholds, PY is broken and germination becomes possible. Different species have different thresholds and ranges between the lower and upper thresholds (Auld and O'Connell, 1991), and there is often variation between coexisting species.

Fire season—what happens when fire season shifts from historic norms?

Climate change is influencing the timing and length of fire seasons in many ecosystems around the world, directly impacting postfire recruitment from seeds. Changes in the fire season are directly influenced by climate change by two interrelated but distinct pathways. First, fire seasons are getting longer, and the frequency of adverse fire weather is causing wildfires to occur outside the historic fire season (Jolly et al., 2015). Second, active fire mitigation practices (e.g., prescribed burns, hazard reduction burns) are used to combat increased fire risk (Fernandes and Botelho, 2003; Boer et al., 2009) since fuels are drying out quicker (Kelley et al., 2019) and temperatures are hotter for longer periods of time. Thus, increased temperatures and drying are causing fires to occur outside historic fire seasons, which along with increasing fire severity and coverage of larger areas, allow fires to burn areas historically considered fire-sensitive. For example, temperate rainforests were burned severely during the Black Summer fires in Australia during 2019/20 (Bradstock et al., 2020).

Plant regeneration from seeds is influenced by changes in fire season through multiple pathways (Miller et al., 2019; Cao et al., 2020; Tangney et al., 2020a), including germination timing, seed survival, flowering, seed production, and postfire seed predation.

Germination timing

Plant regeneration from seeds is directly influenced by fire season, since the seasonal timing of fire influences the timing of dormancy release cues for some seeds (notably those with PY), while controlling the initiation of germination for many smoke-responsive seeds, for example, those with PD and MPD (Tangney et al., 2020a). Changes in timing of fire alter the timing of seedling emergence, number of successful seedlings emerging, and subsequently delay flowering and maturation postfire (Ooi, 2019). For example, germination and seedling emergence of seeds with PD from aseasonal rainfall environments are delayed significantly following fires in spring. Notable examples of delayed germination following fires are available for southeastern Australia (Ooi, 2010; Collette and Ooi, 2020). Ooi (2010) found that seeds with PD that require a seasonal dormancy breaking cue (i.e., warm stratification) can delay germination by up to 12 months as a result of lack of overlap between seasonal dormancy release cues and germination cues generated by the fire. In association with annual dormancy cycling (Collette and Ooi, 2020), this mechanistic interaction (Tangney et al., 2020a) delays germination relative to PY species in the same climate. If seeds with PD (or MPD) that exhibit dormancy cycling receive germination cues during the dormant phase of the cycle, germination will not begin until the following year, after seeds have been exposed to dormancy-breaking seasonal temperatures. This potential delay of germination is evident only in aseasonal rainfall environments where germination is limited by temperature and not by moisture. In climates with strongly seasonal rainfall (e.g., hot, dry summer Mediterranean climates), we expect germination of all species to be delayed following fires that occur immediately after closure of the seasonal germination window, since seeds have germination requirements that align them with temperature and moisture conditions of the ecosystem.

Survival of seeds and changes to fire season

In ecosystems in which controlled fires are used to reduce fire risk, or where fires occur during periods when internal seed moisture is elevated, seeds with permeable seeds coats also will be at risk from changes in the fire season (Tangney et al., 2019). Tangney et al. (2019) showed that when seeds were exposed to high temperatures after being equilibrated to >95% RH (wet seeds) they died at a temperature about 30°C lower than they did when exposed to high temperatures after equilibration at ~15% RH (dry seeds). In some cases, wet seeds were killed at temperatures as low as 60°C (lethal temperature for wet seeds of 18 species ranged from 60°C to 110°C; Tangney et al., 2019).

Postfire flowering species and fire seasonality

A common mechanism driving postfire regeneration of plants is fire-stimulated or postfire flowering (PFF) (Lamont and Downes, 2011). PFF species experience enhanced flowering as a result of fire. In some species, flowering only occurs following fire (obligate PFF), while other PFF species experience enhanced flowering (facultative PFF) following fire. For both groups, there are limited opportunities to recruit in the absence of fire (Pyke, 2017). Changes in fire season can alter the intensity and synchrony of flowering and number of seeds produced by PFF species (Miller et al., 2019; Paroissien and Ooi, 2021). Deleterious effects on PFF species as a result of changes in fire season are most pronounced when fires occur during or shortly before the peak flowering season. Changes in the fire season potentially can result in delayed, reduced, or even failure of PFF, causing a reduction in seed production and a subsequent reduction in propagules available for seedling recruitment (Miller et al., 2019).

Perhaps the clearest evidence for shifts in fire season affecting PFF is that provided by Bowen and Pate (2004). These authors found that fires implemented in different seasons at sites of the facultative postfire flowering species *Stirlingia latifolia* from southwestern Western Australia influenced the number of inflorescences produced. Approximately 50% more inflorescences were produced following an autumn (April–June) fire compared to a spring (September–November) fire. In another example from southwestern Australia, Lamont et al. (2000) documented reduced flower and fruit production in the facultative PFF species *Xanthorrhoea preissii* following spring fires compared to both autumn and summer fires.

In the longleaf pine sandhill ecosystem of northern Florida (USA), Brewer et al. (2009) found that plants of the facultative PFF *Pinus palustris* exhibited increased fecundity and reduced herbivory of seed-producing cones following growing season burns (May and August) compared to dormant season burns (January), resulting in fewer seeds produced following dormant season burns. Furthermore, plant regeneration from seeds was highest following growing season burns, with approximately 90% greater seedling emergence following burns in May compared to those in January. This resulted in greater establishment, since only seedlings that emerged following May burns survived the 2 years postfire (Brewer et al., 2009).

For PFF species, regeneration from seeds is facilitated if it occurs soon after fire. Therefore, broadening the fire season under climate change may inhibit PFF species from recruiting until after the next fire event, which in some cases may be decades. The effects of changes in fire season on PFF species may be most pronounced in regions with a higher proportion of obligate seeders or facultative PFF species than of non-PFF species, as is common among herbaceous species (particularly geophytes) in temperate, subtropical, and tropical savanna ecosystems (e.g., Lamont and Downes, 2011; Massi et al., 2017; Zirondi et al., 2021).

Postfire seed predation and mortality

Plant regeneration from seeds requires that seeds survive the rigors of the fire and for long enough following fire to be exposed to suitable germination conditions. For some serotinous species, fires that release seeds from woody fruits or cones occur after the season when conditions are favorable for germination has passed. Thus, the released seeds are subjected to prolonged exposure to postfire conditions before experiencing suitable germination conditions. This period of exposure to postfire conditions renders seeds vulnerable to seed predators (Bond, 1984) and exposes them to physiological stresses that lead to reductions in seed vigor (Kranner et al., 2010).

For serotinous Proteaceae species of the southern Cape fynbos vegetation of South Africa, Bond (1984) estimated that between 67% and 72% of seeds released following a spring fire were lost to predation, in some cases resulting in nine times fewer seeds available for recruitment. Seeds released following spring fires were exposed to seed predators for 8–10 months postfire before the onset of suitable germination conditions. However, Heelemann et al. (2008) found that within the more aseasonal rainfall ecosystems of the Eastern Cape serotinous Proteaceae species in similar fynbos vegetation did not experience significant differences in postfire emergence between seasons. Compared to the Southern Cape, the Eastern Cape has more regular occurrences of suitable germination and recruitment conditions, thus reducing postfire exposure to seed predators to less than a month in most cases.

Evidence of increased predation following out-of-season fires has been found in Mediterranean climates, and in all cases it results in fewer viable seeds being present when conditions become suitable for germination (Bond, 1984; Cowling and Lamont, 1987; Enright and Lamont, 1989; Knox and Clarke, 2006). When this effect has been tested in aseasonal rainfall environments, there has been no effect on postfire predation between seasons (Heelemann et al., 2008; Kraaij et al., 2017). Thus, climate-related shifts in fire season will not impact different fire-prone regions equally, and they will interact with broader climate and weather patterns.

Fire frequency—what happens when fires occur at shorter intervals?

Increased fire frequency is a key prediction under climate change, and it can impact the capacity of plants to regenerate, from both resprouting and seeds. Fire frequency, specifically the frequency at which fire occurs at a particular site, is the most-studied aspect of the fire regime, and it is often the main focus when considering maintenance of biodiversity in fire-prone ecosystems (Miller and Murphy, 2017). For some vegetation types, such as grasslands and savannas, short-fire intervals are part of the historic fire regime, and they have selected for life-history traits that are adaptive under these conditions, including resprouting from protected storage organs, thick bark, and PFF (Lunt and Morgan, 2002; Bond and Keeley, 2005, Staver et al., 2011; Hoffman et al., 2012). However, increased fire frequency and thus shorter fire intervals than occurred historically can interrupt plant life cycles, reduce regenerative capacity, transform habitat structure, and increase susceptibility to invasion by alien species (Keith, 2012; Swab et al., 2012; Enright et al., 2014; Fairman et al., 2016).

Species from different postfire response functional groups may be vulnerable to short fire intervals. Individuals of species that resprout in response to fire can suffer from resprouting exhaustion from too-frequent fires, while hotter fires can increase mortality by physically damaging resprouting stems (Nolan et al., in review), placing greater reliance on postfire recruitment from seeds. Resprouters with slow regrowth would be particularly susceptible (Keith, 2012; Denham and Auld, 2012). Obligate seeders with a long primary juvenile period and a limited seed bank are another key group at risk. High-severity fires are likely to increase the number and size of areas in landscapes subject to high-fire frequency, in part because when weather conditions for fire are extreme fires can burn recently burnt areas or fire-sensitive communities usually considered fire refugia (Tolhurst and McCarthy, 2016; Collins et al., 2019). The loss of species diversity and ecosystem function associated with frequent fires is exacerbated by "conditions such as prolonged drought and heatwaves that increase the time required for plants to recover postfire, while simultaneously providing conditions that reduce the time between fire events" (Le Breton et al., 2022).

Seed bank depletion

High fire frequency can lead to seed bank depletion by promoting germination but not allowing enough time for plants to mature and replenish the seed bank before the next fire event (Bradstock et al., 1996; Keith, 2012). Obligate-seeding species with either soil- or canopy-stored seed banks are at risk, with serotinous species being more prone to total seed bank loss than soil seed banks. Spatially explicit life-history models found that the serotinous genus *Banksia* (Proteaceae) in Australia was susceptible to high fire frequency, irrespective of fire patch size (Bradstock et al., 1996). Similar outcomes have been found for obligate seeders with soil-stored seed banks. Using a spatially explicit stochastic matrix model linked with a dynamic niche model, Swab et al. (2012) found that increased fire frequency reduced expected minimum abundance of the obligate seeder *Leucopogon setiger* and that fire frequency was a greater threat to loss of local populations than climate-induced range shifts.

Fire-sensitive or fire-intolerant vegetation types, such as tropical and temperate rainforests or communities restricted to fire refugia, also are predicted to be burned over a significant area of their distribution as a result of climate change and its interactions with extreme drought and human activity (Nolan et al., in review). Recently, increased size of burned areas has occurred in tropical rainforests of the Amazon region of Brazil and southeastern Asia and in temperate rainforests of Australia (Langner and Siegert, 2009; Collins et al., 2019; Barlow et al., 2020; Boer et al., 2020). Depending on the component species, fires may put such communities at high risk of adult plant mortality and of reduction in both seed rain and seed banks. For example, in southeastern Amazon forests Dos Santos Cury et al. (2020) found that both annual and triennial fires reduced seed rain and seed banks in addition to negatively impacting resprouting.

Seed dispersal and persistence of postfire obligate colonizers

Beyond the expected impacts of reducing seed bank resilience and capacity, increased fire frequency as a result of climate change also may impact seed dispersal. The obligate seeder strategy appears to be one that could be negatively impacted by increasing fire area, with megafires delaying the rate of recovery of such colonizers in areas far from the fire edge, although the negative impact could be ameliorated by mixed fire severity within a burned area (Donato et al., 2009). However, Gill et al. (2020) found that increased fire frequency interacting with high fire severity also limited the recovery of propagule nonpersisters (obligate seeders) in subalpine forest in Wyoming (USA). They reported that the seed rain from conifers into recently burnt areas increased as fire interval increased in the surrounding forest. Younger and smaller trees in sites previously burnt at short intervals had lower seed input and dispersal distances up to eight times lower than those in the less frequently burnt older sites (Gill et al., 2020).

Interactions of multiple factors on plant regeneration from seeds under climate change

Components of fire regimes that shift under climate change will occur at the same time as other climate-driven changes, producing complex interactions that can impact the capacity for plant species to regenerate. One concept, outlined by Enright et al. (2015) and expressed as a series of interacting impacts, is collectively called "interval squeeze." The three core components—shifts in demographic rates, postfire recruitment rates, and fire interval—combine to reduce the window of time in which plant species can regenerate after fire. In particular, they are predicted to impact perennial obligate-seeding species and recruits of woody resprouters. Demographic shifts related to delayed and reduced seed production can be caused by greater stress from drought and heatwaves (e.g., Redmond et al., 2012), while similar climatic changes will reduce postfire seedling survival and growth (Enright et al., 2014). As such, a longer interfire period is required for populations to produce sufficient propagules for stand replacement in obligate seeders or for juveniles of resprouters to achieve fire resistance; however, these same climatic conditions will increase fire frequency.

Additional impacts of climate change also can contribute to, and potentially exacerbate, elements of interval squeeze. For example, shrub species in eastern Australian forests with soil-stored seeds with PD that have a seasonal component to their emergence postfire had longer primary juvenile periods, depending on the season when burning occurred (Ooi, 2010, 2019). Additionally, a number of studies found increased loss of seeds from soil seed banks during the interfire period as a result of hotter summers or heat waves, leading to reduced postfire seed bank capacity when regeneration from seeds was needed (Ooi et al., 2012; Cochrane, 2017). Thus, climate-driven changes in multiple elements of the fire regime (season and frequency), as well as concurrent changes in the weather, can produce complex interactions that limit ability of species to regenerate after fire under climate change.

Fire-prone regions and the potential for "winners" and "losers" under climate change

Plants and the ecosystems in which they occur have become adapted to the presence of evolutionarily stable fire regimes (Keeley et al., 2011; Lamont and He, 2017), and modifications to inherent characteristics of these regimes can shift plant composition, structure, and function (Bradstock, 2010; Fairman et al., 2017). However, changes to the fire regime and concurrent changes in climate and weather events will not impact all species equally, even those within the same plant community. Subsequently, while there may be reductions in the abundance of some species, and even local extinctions, others may persist or potentially thrive. For example, changes in the abundance of dominant species could lead to shifts in community structure and composition and shifts in vegetation types.

Several outcomes could occur as a result of changes in elements of the fire regime. For example, with hotter fires species whose seeds have a relatively low threshold for heat tolerance may experience increased seed mortality and thus a reduction in number of viable seeds in the seed bank after a fire (Auld and Denham, 2006), thereby reducing their ability to persist under more extreme and/or frequent fires. More broadly, species most at risk from extremely severe fires, independent of dormancy class, include small-seeded obligate seeders with low upper thermal thresholds and a reliance on seeds buried close to the soil surface for postfire germination. This group has an increased risk of seed mortality and depletion of the seed bank with increased soil heating combined with low ability of seedlings from small seeds to successfully emerge from deeper in the soil profile.

Changes in fire season also may impact the success of plant regeneration from seeds differently, depending on the species, through multiple mechanisms affecting the broad range of functional seed groups (Miller et al., 2019; Tangney et al., 2020b). For example, many species whose seeds have PD or MPD require seasonal dormancy release cues and therefore may be more at risk to changes in fire season than those with PY, which have minimal seasonal requirements. PY species also will be at an advantage since changes in fire season will not delay germination, emergence, and subsequent maturation of species identified in aseasonal rainfall regions (Ooi, 2010; Miller et al., 2019). Thus, changes in fire season in aseasonal rainfall regions may drive ecological communities toward dominance by species with PY, especially those PY species with the relatively high thermal thresholds.

For species endemic to ecosystems with strongly seasonal rainfall (e.g., Mediterranean regions), there may be a risk from changes in fire season across a broad suite of species in different functional groups (Miller et al., 2019). Regeneration requires that seeds be ND and in the right place at the right time to ensure germination and seedling recruitment. In seasonal rainfall environments, suitable conditions for regeneration occur only once a year, and timing of fire relative to those suitable conditions is one of the biggest drivers of regeneration success. That is, fire occurring before suitable conditions for germination increases regeneration compared to that to fires occurring after suitable conditions for germination.

While species with traits such as low tolerance to heat or seasonal emergence may be at greater risk of population decline from climate-induced fire regime shift, other species may flourish. Depending on their ability to withstand drought or other climate-related mortality drivers, species with broad germination and establishment requirements, long-lived seed banks, and short juvenile or secondary juvenile periods could persist under fire regimes influenced by climate change. For species with rapid recruitment, high-intensity fires occurring outside the normal (historical) fire season may have a limited impact on population persistence. For example, species with broad thresholds and relatively high upper thresholds of heat tolerance could be favored under fire regimes of increased intensity, while species with PY seeds may be favored over those with PD and MPD seeds under a shifting fire season in aseasonal regions. However, our understanding of the mechanisms important in determining a species' ability to respond to changes in fire-prone regions is still limited.

Adaptive capacity to ameliorate impacts of climate change in fire-prone regions

Sources of genetic variation that occur throughout natural plant populations may provide an avenue for species persistence under changes in fire regimes. Within-population variation (Gómez-González et al., 2020), maternal environmental effect during seed development (Penfield and MacGregor, 2016), and phenotypic plasticity (Richter et al., 2012) may provide sufficient adaptive capacity within some plant populations to ensure resilience to climate-change-induced shifts to fire regimes (He et al., 2019b).

As climate change impacts fire regimes, evolutionary pressure may act by selecting genotypes that provide adaptive advantage under future fire regimes. In Chile, the presence of frequent fires led to less round, more pubescent, and thicker-coated seeds of *Helenium aromaticum*, which is adaptive for this annual herb (Gómez-González et al., 2011). This and other examples of the relationship between fire and seed traits suggest that historical fire frequencies can have differential selective pressures across landscapes (Gómez-González et al., 2020). Knox and Morrison (2005) suggested that increase fire frequency may select for faster-maturing individuals with the potential to ameliorate population decline for some species.

The rate at which modifications to fire regimes occur under the influence of climate change may outpace the adaptive capacity of some long-lived species (Parmesan and Hanley, 2015). This is likely to be the case for obligate seeders in ecosystems where interfire recruitment is not possible (Enright et al., 1998), for example, *Banksia hookeriana* (Enright et al., 1996) and particularly species with narrow requirements for dormancy-break and germination (Rühl et al., 2015). Similarly, obligate-seeding fire-following ephemerals rely on long-lived soil seed banks to maintain populations over space and time. Often plants that are short-lived and rapidly produce seeds (Pate et al., 1985; Thanos and Rundel, 1995; Baker et al., 2005), such fire ephemerals, quickly accumulate seeds in the soil seed bank before dying (Keith et al., 2007). Because generational turnover is tightly linked with the occurrence of fire, which in some cases can be decades apart, obligate seeding fire ephemeral populations that lack sufficient adaptive capacity may be limited in germination, emergence, and establishment if microsite conditions significantly shift to hotter and drier during interfire periods.

Maternal environmental effects on seeds during development also may mitigate the effects of fire regime shift under climate change by influencing the temperature and environmental conditions that release dormancy and stimulate germination (Penfield and MacGregor, 2016). Rainfall and temperatures experienced by the maternal plant directly affect degree of dormancy and germination success of the seeds produced (Fenner, 1991). Furthermore, the environment of the mother plant had a strong influence on response of the PD seeds of *Brassica tournefortii* to the germination stimulant karrikinolide (Gorecki et al., 2012), a butenolide identified in smoke from burning plant material (Flematti et al., 2004). However, how maternal conditions translate into fire response cues like heat thresholds for PY seeds is yet to be fully resolved. Early evidence suggests that there are minimal effects of warmer maternal environments on PY for fire-related temperature thresholds but strong negative effects on the vigor of subsequent seedlings (M. Ooi, unpubl. data).

Conclusions and future research needs

Changes to natural fire-prone ecosystems under climate change are inevitable. While we have highlighted how impacts of climate change may be minimized within different plant functional groups, there are limitations to the capacity of a species to adapt, depending on the magnitude of change to fire regimes or climate conditions. For example, irrespective of the breadth of regeneration requirements that plant species have, either with narrow or generalist requirements, all of them will be equally disadvantaged if severe drought conditions are too frequent (Savage et al., 2013). Similarly, regular megafire events, such as those over the last few years, will negatively impact fire-sensitive plant communities previously considered fire refugia.

Solutions for minimizing global effects of climate change, such as increased likelihood of drought, require an urgent and concerted global effort. However, an understanding of the ecology of fire, plants, climate, and their interactions can help ameliorate impacts at the species, or even landscape, scale. Reducing the risk that many species face under higher fire frequency is arguably one of the biggest problems in fire-prone areas, and our considerable understanding of these impacts means we can target sensitive species and habitats for conservation. For other elements of climate-induced changes to the fire regime, our ecological knowledge needs to increase for us to better inform conservation efforts. As a result of targeting research to understand how fire season change can impact species (Miller et al., 2019), we know that (1) permanent changes in the fire season can reduce reproductive capacity of postfire flowering species, (2) the length of time seeds are exposed to predators (primarily serotinous species) may be changed, and (3) timing of germination and emergence of PD species in aseasonal rainfall regions may be altered. In managed ecosystems, this knowledge can be used to incorporate variation in the season of the year when areas are burned and in the severity of fires over time, as one way to ameliorate potential impacts. Additionally, we know that maternal environmental conditions and among-population trait variation can impart effects on the breadth of suitable germination conditions (Cochrane et al., 2015), potentially influencing the distribution of a species. Therefore, conservation actions aimed at maintenance of among-population variation may provide the safety net required for species to persist under shifts in fire regimes driven by climate change. Furthermore, study of the response to climate-driven shifts of species, particularly in the context of interval squeeze (see above), will help us predict the winners and losers under climate change and provide guidance for the actions required for conservation in fire-prone ecosystems.

References

Abatzoglou, J.T., Williams, A.P., 2016. Impact of anthropogenic climate change on wildfire across western US forests. Proc. Natl. Acad. Sci. USA 113, 11770–11775.

Alexander, M.E., Cruz, M.G., 2012. Modelling the effects of surface and crown fire behaviour on serotinous cone opening in jack pine and lodgepole pine forests. Int. J. Wildl. Fire 21, 709–721.

Archibald, S., Lehmann, C.E.R., Belcher, C.M., Bond, W.J., Bradstock, R.A., Daniau, A.-L., et al., 2018. Biological and geophysical feedbacks with fire in the Earth system. Environ. Res. Lett. 13, 033003. Available from: https://doi.org/10.1088/1748-9326/aa9ead.

Auld, T.D., O'Connell, M.A., 1991. Predicting patterns of post-fire germination in 35 eastern Australian Fabaceae. Aust. J. Ecol. 16, 53–70.

Auld, T.D., Bradstock, R.A., 1996. Do post-fire soil temperatures influence seed germination? Aust. J. Ecol. 21, 106–109.

Auld, T.D., Denham, A.D., 2006. How much seed remains in the soil after a fire? Plant Ecol. 187, 15–24.

Auld, T.D., Ooi, M.K.J., 2017. Plant life cycles above and below ground. In: Keith, D.A. (Ed.), Australian Vegetation, Third ed. Cambridge University Press, Cambridge, pp. 230–253.

Baker, K.S., Steadman, K.J., Plummer, J.A., Dixon, K.W., 2005. Seed dormancy and germination responses of nine Australian fire ephemerals. Plant Soil 277, 345–358.

Balch, J.K., Bradley, B.A., Abatzoglou, J.T., Nagy, R.C., Fusco, E.J., Mahood, A.L., 2017. Human-started wildfires expand the fire niche across the United States. Proc. Natl. Acad. Sci. USA 114, 2946–2951.

Barlow, J., Berenguer, E., Carmenta, R., Franca, F., 2020. Clarifying Amazonia's burning crisis. Glob. Change Biol. 26, 319–321.
Boer, M.M., Sadler, R.J., Wittkuhn, R.S., McCaw, L., Grierson, P.F., 2009. Long-term impacts of prescribed burning on regional extent and incidence of wildfires—evidence from 50 years of active fire management in SW Australian forests. For. Ecol. Manage. 259, 132–142.
Boer, M.M., Resco de Dios, V., Bradstock, R., 2020. Unprecedented burn area of Australian mega forest fires. Nat. Clim. Change 10, 171–172.
Bond, W.J., 1984. Fire survival of Cape Proteaceae - influence of fire season and seed predators. Vegetatio 56, 65–74.
Bond, W.J., Keeley, J.E., 2005. Fire as a global 'herbivore': the ecology and evolution of flammable ecosystems. Trends Ecol. Evol. 20, 387–394.
Bond, W.J., Honig, M., Maze, K.E., 1999. Seed size and seedling emergence: an allometric relationship and some ecological implications. Oecologia 120, 132–136.
Bowen, B.J., Pate, J.S., 2004. Effect of season of burn on shoot recovery and post-fire flowering performance in the resprouter *Stirlingia latifolia* R. Br. (Proteaceae). Austral Ecol. 29, 145–155.
Bowman, D.M.J.S., Balch, J.K., Artaxo, P., Bond, W.J., Carlson, J.M., Cochrane, M.A., et al., 2009. Fire in the earth system. Science 324, 481–484.
Bowman, D.M.J.S., Murphy, B.P., Williamsons, G.J., Cochrane, M.A., 2014. Pyrogeographic models, feedbacks and the future of global fire regimes. Glob. Ecol. Biogeogr. 23, 821–824.
Bowman, D.M.J.S., Bliss, A., Bowman, C.J.W., Prior, L.D., 2019. Fire caused demographic attrition of the Tasmanian palaeoendemic conifer *Athrotaxis cupressoides*. Austral Ecol. 44, 1322–1339.
Bowman, D.M., Kolden, C.A., Abatzaglou, J.T., Johnston, F.H., van der Werf, G.R., Flannigan, M.D., 2020. Vegetation fires in the Anthropocene. Nat. Rev. Earth Environ. 1, 500–515.
Bradstock, R.A., 2010. A biogeographic model of fire regimes in Australia: current and future implications. Glob. Ecol. Biogeogr. 19, 145–158.
Bradstock, R.A., Auld, T.D., 1995. Soil temperatures during experimental bushfires in relation to fire intensity: consequences for legume germination and fire management in south-eastern Australia. J. Appl. Ecol. 32, 76–84.
Bradstock, R.A., Myerscough, P.J., 1981. Fire effects on seed release and the emergence and establishment of seedlings in *Banksia ericifolia*. L.f. Aust. J. Bot. 29, 521–531.
Bradstock, R.A., Gill, A.M., Hastings, S.M., Moore, P.H.R., 1994. Survival of serotinous seedbanks during bushfires - comparative studies of *Hakea* species from southeastern Australia. Aust. J. Ecol. 19, 276–282.
Bradstock, R.A., Bedward, M., Scott, J., Keith, D.A., 1996. Simulation of the effect of spatial and temporal variation in fire regimes on the population viability of a *Banksia* species. Conserv. Biol. 10, 776–784.
Bradstock, R.A., Nolan, R.H., Collins, L., Resco de Dios, V., Clarke, H., Jenkins, M., et al., 2020. A broader perspective on the causes and consequences of eastern Australia's 2019–20 season of mega-fires: a response to Adams et al. Glob. Change Biol. 26, e8–e9. Available from: https://doi.org/10.1111/gcb.15111.
Brewer, J.S., Cunningham, A.L., Moore, T.P., Brooks, R.M., Waldrup, J.L., 2009. A six-year study of fire-related flowering cues and coexistence of two perennial grasses in a wet longleaf pine (*Pinus palustris*) savanna. Plant Ecol. 200, 141–154.
Carthey, A.J.R., Tims, A., Geedicke, I., Leishman, M.R., 2018. Broad-scale patterns in smoke-responsive germination from the south-eastern Australian flora. J. Veg. Sci. 29, 737–745.
Çatav, Ş.S., Küçükakyüz, K., Tavşanoğlu, Ç., Pausas, J.G., 2018. Effect of fire-derived chemicals on germination and seedling growth in Mediterranean plant species. Basic Appl. Ecol. 30, 65–75.
Cao, D., Baskin, C.C., Baskin, J.M., 2020. Dormancy class: another fire seasonality effect on plants. Trends Ecol. Evol. 35, 1055–1057.
Céspedes, B., Torres, I., Luna, B., Pérez, B., Moreno, J.M., 2012. Soil seed bank, fire season, and temporal patterns of germination in a seeder-dominated Mediterranean shrubland. Plant Ecol. 213, 383–393.
Clarke, P.J., Knox, K.J.E., Butler, D., 2010. Fire intensity, serotiny and seed release in 19 woody species: evidence for risk spreading among wind-dispersed and resprouting syndromes. Aust. J. Bot. 58, 629–636.
Cochrane, A., 2017. Are we underestimating the impact of rising summer temperatures on dormancy loss in hard-seeded species? Aust. J. Bot. 65, 248–256.
Cochrane, A., Yates, C.J., Hoyle, G.L., Nicotra, A.B., 2015. Will among-population variation in seed traits improve the chance of species persistence under climate change? Glob. Ecol. Biogeogr. 24, 12–24.
Collette, J.C., Ooi, M.K.J., 2020. Evidence for physiological seed dormancy cycling in the woody shrub *Asterolasia buxifolia* and its ecological significance in fire-prone systems. Plant Biol. 22, 745–749.
Collins, L., Bennet, A.F., Leonard, S.W.J., Penman, T., 2019. Wildfire refugia in forests: severe fire weather and drought mute the influence of topography and fuel age. Glob. Change Biol. 25, 3829–3843.
Cowling, R.M., Lamont, B.B., 1987. Post-fire recruitment of four co-occurring *Banksia* species. J. Appl. Ecol. 24, 645–658.
Daibes, L.F., Gorgone-Barbosa, E., Silveira, F.A.O., Fidelis, A., 2017. Gaps critical for the survival of exposed seeds during Cerrado fires. Aust. J. Bot. 66, 116–123.
Daibes, L.F., Pausas, J.G., Bonani, N., Nunes, J., Silveira, F.A.O., Fidelis, A., 2019. Fire and legume germination in a tropical savanna: ecological and historical factors. Ann. Bot. 123, 1219–1229.
Davis, K.T., Higuera, P.E., Sala, A., 2018. Anticipating fire-mediated impacts of climate change using a demographic framework. Funct. Ecol. 32, 1729–1745.
Denham, A.J., Auld, T.D., 2012. Population ecology of Waratahs, *Telopea speciosissima* (Proteaceae): implications for management of fire-prone habitats. Proc. Linn. Soc. NSW 134, B101–B111.
Donato, D.C., Fontaine, J.B., Campbell, J.L., Robinson, W.D., Kauffman, J.B., Law, B.E., 2009. Conifer regeneration in stand-replacement portions of a large mixed-severity wildfire in the Klamath-Siskiyou Mountains. Can. J. For. Res. 39, 823–838.

Dos Santos Cury, R.T., Montibeller-Santos, C., Balch, J.K., Brando, P.M., Torezan, J.M.D., 2020. Effects of fire frequency on seed sources and regeneration in southeastern Amazonia. Front. For. Glob. Change 3, 82. Available from: https://doi.org/10.3389/ffgc.2020.00082.

Downes, K.S., Light, M.E., Pošta, M., Kohout, L., van Staden, J., 2014. Do fire-related cues, including smoke-water, karrikinolide, glyceronitrile and nitrate, stimulate the germination of 17 *Anigozanthos* taxa and *Blancoa canescens* (Haemodoraceae)? Aust. J. Bot. 62, 347–358.

Edwards, W., Krockenberger, A., 2006. Seedling mortality due to drought and fire associated with the 2002 El Niño event in a tropical rain forest in north-east Queensland, Australia. Biotropica 38, 16–26.

Enright, N.J., Lamont, B.B., 1989. Seed banks, fire season, safe sites and seedling recruitment in five co-occurring *Banksia* species. J. Ecol. 77, 1111–1122.

Enright, N.J., Lamont, B.B., 2006. Fire temperature and follicle-opening requirements in 10 *Banksia* species. Austral Ecol. 14, 107–113.

Enright, N.J., Lamont, B.B., Marsula, R., 1996. Canopy seed bank dynamics and optimum fire regime for the highly serotinous shrub, *Banksia hookeriana*. J. Ecol. 84, 9–17.

Enright, N.J., Marsula, R., Lamont, B.B., Wissel, C., 1998. The ecological significance of canopy seed storage in fire-prone environments: a model for non-sprouting shrubs. J. Ecol. 86, 946–959.

Enright, N.J., Fontaine, J.B., Lamont, B.B., Miller, B.P., Westcott, V.C., 2014. Resistance and resilience to changing climate and fire regime depend on plant functional traits. J. Ecol. 102, 1572–1581.

Enright, N.J., Fontaine, J.B., Bowman, D.M.J.S., Bradstock, R.A., Williams, R.J., 2015. Interval squeeze: altered fire regimes and demographic responses interact to threaten woody species persistence as climate changes. Front. Ecol. Environ. 13, 265–272.

Etchells, H., O'Donnell, A.J., McCaw, W.L., Grierson, P.F., 2020. Fire severity impacts on tree mortality and post-fire recruitment in tall eucalypt forests of southwest Australia. For. Ecol. Manage. 459, 9. Available from: https://doi.org/10.1016/j.foreco.2019.117850.

Fairman, T.A., Nitschke, C.R., Bennett, L.T., 2016. Too much, too soon? A review of the effects of increasing wildfire frequency on tree mortality and regeneration in temperate eucalypt forests. Int. J. Wildl. Fire 25, 831–848.

Fairman, T.A., Bennett, L.T., Tupper, S., Nitschke, C.R., 2017. Frequent wildfires erode tree persistence and alter stand structure and initial composition of a fire-tolerant sub-alpine forest. J. Veg. Sci. 28, 1151–1165.

Fenner, M., 1991. The effects of the parent environment on seed germinability. Seed Sci. Res. 1, 75–84.

Fernandes, P.M., Botelho, H.S., 2003. A review of prescribed burning effectiveness in fire hazard reduction. Int. J. Wildl. Fire 12, 117–128.

Fidelis, A., Alvarado, S., Barradas, A., Pivello, V., 2018. The year 2017: megafires and management in the Cerrado. Fire 1, 49. Available from: https://doi.org/10.3390/fire1030049.

Flannigan, M.D., Cantin, A.S., de Groot, W.J., Wotton, M., Newbery, A., Gowman, L.M., 2013. Global wildland fire season severity in the 21st century. For. Ecol. Manage. 294, 54–61.

Flematti, G.R., Ghisalberti, E.L., Dixon, K.W., Trengove, R.D., 2004. A compound from smoke that promotes seed germination. Science 305, 977.

Gill, N.S., Hoecker, T.J., Turner, M.G., 2020. The propagule doesn't fall far from the tree, especially after short-interval, high-severity fire. Ecology 102, e03194. Available from: https://doi.org/10.1002/ecy.3194.

Gómez-González, S., Torres-Díaz, C., Bustos-Schindler, C., Gianoli, E., 2011. Anthropogenic fire drives the evolution of seed traits. Proc. Natl. Acad. Sci. USA 108, 18743–18747.

Gómez-González, S., Paniw, M., Durán, M., Picó, S., Martín-Rodríguez, I., Ojeda, F., 2020. Mediterranean heathland as a key habitat for fire adaptations: evidence from an experimental approach. Forests 11, 748. Available from: https://doi.org/10.3390/f11070748.

Gorecki, M.J., Long, R.L., Flematti, G.R., Stevens, J.C., 2012. Parental environment changes the dormancy state and karrikinolide response of *Brassica tournefortii* seeds. Ann. Bot. 109, 1369–1378.

Habrouk, A., Retana, J., Espelta, J., 1999. Role of heat tolerance and cone protection of seeds in the response of three pine species to wildfires. Plant Ecol. 145, 91–99.

Hanley, M.E., Lamont, B.B., 2000. Heat-shock and the germination of Western Australian plant species: effects on seeds of soil- and canopy-stored species. Acta Oecol. 21, 315–322.

Hanley, M., Unna, J., Darvill, B., 2003. Seed size and germination response: a relationship for fire-following plant species exposed to thermal shock. Oecologia 134, 18–22.

He, T., Lamont, B.B., Pausas, J.G., 2019a. Fire as a key driver of Earth's biodiversity. Biol. Rev. 94, 1983–2010.

He, T., Lamont, B.B., Enright, N.J., D'Agui, H.M., Stock, W., 2019b. Environmental drivers and genomic architecture of trait differentiation in fire-adapted *Banksia attenuata* ecotypes. J. Integr. Plant Biol. 61, 417–432.

Heelemann, S., Proches, Ş., Rebelo, A.G., van Wilgen, B., Porembski, S., Cowling, R.M., 2008. Fire season effects on the recruitment of non-sprouting serotinous Proteaceae in the eastern (bimodal rainfall) fynbos biome, South Africa. Austral Ecol. 33, 119–127.

Higuera, P.E., Abatzoglou, J.T., 2021. Record-setting climate enabled the extraordinary 2020 fire season in the western United States. Glob. Change Biol. 27, 1–2.

Hill, S.J., Auld, T.D., 2020. Seed size an important factor for the germination response of legume seeds subjected to simulated post-fire soil temperatures. Int. J. Wildl. Fire 29, 618–627.

Hoegh-Guldberg, O., Jacob, D., Taylor, M., Bindi, M., Brown, S., Camilloni, I., et al., 2018. Impacts of 1.5°C Global Warming on Natural and Human Systems. In: Masson-Delmotte, V., Zhai, P., Pörtner, H.-O., Roberts, D., Skea, J., Shukla, P.R., et al., (Eds.), Global Warming of 1.5° C. An IPCC Special Report on the impacts of global warming of 1.5° C above pre-industrial levels and related global greenhouse gas emission pathways, in the context of strengthening the global response to the threat of climate change, sustainable development, and efforts to eradicate poverty.

Hoffman, W.A., Geiger, E.L., Gotsch, S.G., Rossatto, D.R., Silva, L.C.R., Lau, O.L., et al., 2012. Ecological thresholds at the savanna-forest boundary: how plant traits, resources and fire govern the distribution of tropical biomes. Ecol. Lett. 15, 759–768.

Huss, J.C., Fratzl, P., Dunlop, J.W.C., Merritt, D.J., Miller, B.P., Eder, M., 2019. Protecting offspring against fire: lessons from *Banksia* seed pods. Front. Plant Sci. 10, 283. Available from: https://doi.org/10.3389/fpls.2019.00283.

Jiménez-Alfaro, B., Silveira, F.A.O., Fidelis, A., Poschlod, P., Commander, L.E., 2016. Seed germination traits can contribute better to plant community ecology. J. Veg. Sci. 27, 637–645.

Jolly, W.M., Cochrane, M.A., Freeborn, P.H., Holden, Z.A., Brown, T.J., Williamson, G.J., et al., 2015. Climate-induced variations in global wildfire danger from 1979 to 2013. Nat. Commun. 6, 75537. Available from: https://doi.org/10.1038/ncomms8537.

Joseph, M.B., Rossi, M.W., Mietkiewicz, N.P., Mahood, A.L., Cattau, M.E., St. Denis, L.A., et al., 2019. Spatiotemporal prediction of wildfire size extremes with Bayesian finite sample maxima. Ecol. Appl. 29, e01898. Available from: https://doi.org/10.1002/eap.1898.

Keeley, J.E., 2009. Fire intensity, fire severity and burn severity: a brief review and suggested usage. Int. J. Wildl. Fire 18, 116–126.

Keeley, J.E., Bond, W.J., 1997. Convergent seed germination in South African fynbos and Californian chaparral. Plant Ecol. 133, 153–167.

Keeley, J.E., Fotheringham, C., 1998. Smoke-induced seed germination in California chaparral. Ecology 79, 2320–2336.

Keeley, J.E., Pausas, J.G., Rundel, P.W., Bond, W.J., Bradstock, R.A., 2011. Fire as an evolutionary pressure shaping plant traits. Trends Plant Sci. 16, 1360–1385.

Keith, D.A., 2012. Functional traits: their roles in understanding and predicting biotic responses to fire regimes from individuals to landscapes. In: Williams, R.J., Gill, A.M., Bradstock, R.A. (Eds.), Flammable Australia: Fire Regimes, Biodiversity and Ecosystems in a Changing World. CSIRO Publishing, Clayton, pp. 97–126.

Keith, D.A., Holman, L., Rodoreda, S., Lemmon, J., Bedward, M., 2007. Plant functional types can predict decade-scale changes in fire-prone vegetation. J. Ecol. 95, 1324–1337.

Kelley, D.I., Bistinas, I., Whitley, R., Burton, C., Marthews, T.R., Dong, N., 2019. How contemporary bioclimatic and human controls change global fire regimes. Nat. Clim. Change 9, 690–696.

King, R.A., Menges, E.S., 2018. Effects of heat and smoke on the germination of six Florida scrub species. S. Afr. J. Bot. 115, 223–230.

Kirchmeier-Young, M.C., Gillett, N.P., Zwiers, F.W., Cannon, A.J., Anslow, F.S., 2019. Attribution of the influence of human-induced climate change on an extreme fire season. Earth's Future 7, 2–10.

Knox, K.J.E., Clarke, P.J., 2006. Fire season and intensity affect shrub recruitment in temperate sclerophyllous woodlands. Oecologia 149, 730–739.

Knox, K.J.E., Morrison, D.A., 2005. Effects of inter-fire intervals on the reproductive output of resprouters and obligate seeders in the Proteaceae. Austral Ecol. 30, 407–413.

Kraaij, T., Cowling, R.M., van Wilgen, B.W., Rikhotso, D.R., Difford, M., 2017. Vegetation responses to season of fire in an aseasonal, fire-prone fynbos shrubland. PeerJ 5, e3591. Available from: https://doi.org/10.7717/peerj.3591.

Kranner, I., Minibayeva, F.V., Beckett, R.P., Seal, C.E., 2010. What is stress? Concepts, definitions and applications in seed science. New Phytol. 188, 655–673.

Lamont, B.B., 2020. Evaluation of seven indices of on-plant seed storage (serotiny) shows that the linear slope is best. J. Ecol. 109, 13436. Available from: https://doi.org/10.1111/1365-2745.13436.

Lamont, B.B., Downes, K.S., 2011. Fire-stimulated flowering among resprouters and geophytes in Australia and South Africa. Plant Ecol. 212, 2111–2125.

Lamont, B.B., He, T., 2017. Fire-proneness as a prerequisite for the evolution of fire-adapted traits. Trends Plant Sci. 22, 278–288.

Lamont, B.B., He, T., Yan, Z., 2019. Evolutionary history of fire-stimulated resprouting, flowering, seed release and germination. Biol. Rev. 94, 903–928.

Lamont, B.B., Swanborough, P.W., Ward, D., 2000. Plant size and season of burn affect flowering and fruiting of the grasstree *Xanthorrhoea preissii*. Austral Ecol. 25, 268–272.

Lamont, B.B., Pausas, J.G., He, T., Witkowski, E.T.F., Hanley, M.E., 2020. Fire as a selective agent for both serotiny and nonserotiny over space and time. Crit. Rev. Plant Sci. 39, 140–172.

Langner, A., Siegert, F., 2009. Spatiotemporal fire occurrence in Borneo over a period of 10 years. Glob. Change Biol. 15, 48–62.

Le Breton, T.D., Lyons, M.B., Nolan, R.H., Penman, T., Williamson, G.J., Ooi, M.K.J, 2022. Megafire-induced interval squeeze threatens vegetation at landscape scale. Front. Ecol. Environ. In press.

Le Maitre, D.C., Brown, P.J., 1992. Life cycles and fire-stimulated flowering in geophytes. In: Van Wilgen, B.W., Richardson, D.M., Kruger, F.J., Van Hesenbergen, H.J. (Eds.), Fire in South African Mountain Fynbos. Springer, Berlin, pp. 145–160.

Liyanage, G.S., Ooi, M.K.J., 2018. Seed size-mediated dormancy thresholds: a case for the selective pressure of fire on physically dormant species. Biol. J. Linn. Soc. 123, 135–143.

Luna, B., Chamorro, D., Pérez, B., 2019. Effect of heat on seed germination and viability in species of Cistaceae. Plant Ecol. Divers. 12, 151–158.

Lunt, I.D., Morgan, J.W., 2002. The role of fire regimes in temperate lowland grasslands of southeastern Australia. In: Bradstock, R.A., Williams, J.E., Gill, A.M. (Eds.), Flammable Australia: The fire Regimes and Biodiversity of a Continent. Cambridge University Press, Cambridge, pp. 177–196.

Mackenzie, B.D.E., Auld, T.D., Keith, D.A., Hui, F.K.C., Ooi, M.K.J., 2016. The effect of seasonal ambient temperatures on fire stimulated germination of species with physiological dormancy: a case study using *Boronia* (Rutaceae). PLoS One 11, e0156142. Available from: https://doi.org/10.1371/journal.pone.0156142.

Massi, K.G., Eugênio, C.U.O., Franco, A.C., 2017. Post-fire reproduction of herbs at a savanna-gallery forest boundary in Distrito Federal, Brazil. Braz. J. Biol. 77, 876–886.

McCarty, J.L., Smith, T.E.L., Turetsky, M.R., 2020. Arctic fires re-emerging. Nat. Geosci. 13, 658–660.

Merritt, D.J., Turner, S.R., Clarke, S., Dixon, K.W., 2007. Seed dormancy and germination stimulation syndromes for Australian temperate species. Aust. J. Bot. 55, 336–344.

Milich, K.L., Stuart, J.D., Varner III, J.M., Merriam, K.E., 2012. Seed viability and fire-related temperature treatments in serotinous California native *Hesperocyparis* species. Fire Ecol. 8, 107–124.

Miller, B.P., Murphy, B., 2017. Fire and Australian vegetation. In: Keith, D.A. (Ed.), Australian Vegetation, Third ed. Cambridge University Press, Cambridge, pp. 113–134.

Miller, R.G., Tangney, R., Enright, N.J., Fontaine, J.B., Merritt, D.J., Ooi, M.K.J., et al., 2019. Mechanisms of fire seasonality effects on plant populations. Trends Ecol. Evol. 34, 1104–1117.

Moreira, B., Tormo, J., Estrelles, E., Pausas, J.G., 2010. Disentangling the role of heat and smoke as germination cues in Mediterranean Basin flora. Ann. Bot. 105, 627–635.

Nelson, D.C., Flematti, G.R., Ghisalberti, E.L., Dixon, K.W., Smith, S.M., 2012. Regulation of seed germination and seedling growth by chemical signals from burning vegetation. Annu. Rev. Plant Biol. 63, 107–130.

Nolan, R.H., Boer, M.M., Collins, L., Resco de Dios, V., Clarke, H., Jenkins, M., et al., 2020. Causes and consequences of eastern Australia's 2019–20 season of mega-fires. Glob. Change Biol. 26, 1039–1041.

Nolan, R.H., Collins, L., Leigh, A., Ooi, M.K.J., Curran, T.J., Fairman, T.A., et al., 2021. Limits to post-fire vegetation recovery under climate change. Plant Cell Environ. 44, 3471–3489.

Odion, D.C., Davis, F.W., 2000. Fire, soil heating, and the formation of vegetation patterns in chaparral. Ecol. Monogr. 70, 149–169.

Ooi, M.K.J., 2010. Delayed emergence and post-fire recruitment success: effects of seasonal germination, fire season and dormancy type. Aust. J. Bot. 58, 248–256.

Ooi, M.K.J., 2019. The importance of fire season when managing threatened plant species: a long-term case-study of a rare *Leucopogon* species (Ericaceae). J. Environ. Manage. 236, 17–24.

Ooi, M.K.J., Whelan, R.J., Auld, T.D., 2006. Persistence of obligate-seeding species at the population scale: effects of fire intensity, fire patchiness and long fire-free intervals. Int. J. Wildl. Fire 15, 261–269.

Ooi, M.K.J., Auld, T.D., Denham, A.J., 2012. Projected soil temperature increase and seed dormancy response along an altitudinal gradient: implications for seed bank persistence under climate change. Plant Soil 353, 289–303.

Ooi, M.K.J., Denham, A.J., Santana, V.M., Auld, T.D., 2014. Temperature thresholds of physically dormant seeds and plant functional response to fire: variation among species and relative impact of climate change. Ecol. Evol. 4, 656–671.

Parmesan, C., Hanley, M.E., 2015. Plants and climate change: complexities and surprises. Ann. Bot. 116, 849–864.

Paroissien, R., Ooi, M.K.J., 2021. Effects of fire season on the reproductive success of the post-fire flowerer *Doryanthes excelsa*. Environ. Exp. Bot. 192, 104634. Available from: https://doi.org/10.1016/j.envexpbot.2021.104634.

Pate, J., Casson, N., Rullo, J., Kuo, J., 1985. Biology of fire ephemerals of the sandplains of the kwongan of south-western Australia. Funct. Plant Biol. 12, 641–655.

Pausas, J.G., Keeley, J.E., 2019. Wildfire as an ecosystem service. Front. Ecol. Environ. 17, 289–295.

Pausas, J.G., Rebeiro, E., 2017. Fire and plant diversity at the global scale. Glob. Ecol. Biogeogr. 26, 889–897.

Pausas, J.G., Bradstock, R.A., Keith, D.A., Keeley, J.E., 2004. Plant functional traits in relation to fire in crown-fire ecosystems. Ecology 85, 1085–1100.

Penfield, S., MacGregor, D.R., 2016. Effects of environmental variation during seed production on seed dormancy and germination. J. Exp. Bot. 68, 819–825.

Pyke, G.H., 2017. Fire-stimulated flowering: a review and look to the future. Crit. Rev. Plant Sci. 36, 179–189.

Ramos, D.M., Liaffa, A.B.S., Diniz, P., Munhoz, C.B.R., Ooi, M.K.J., Borghetti, F., et al., 2017. Seed tolerance to heating is better predicted by seed dormancy than by habitat type in Neotropical savanna grasses. Int. J. Wildl. Fire 25, 1273–1280.

Ramos, D.M., Valls, J.F.M., Borghetti, F., Ooi, M.K.J., 2019. Fire cues trigger germination and stimulate seedling growth of grass species from Brazilian savannas. Am. J. Bot. 106, 1190–1201.

Redmond, M., Forcella, F., Barger, N., 2012. Declines in pinyon pine cone production associated with regional warming. Ecosphere 3, 1–14.

Richter, S., Kipfer, T., Wohlgemuth, T., Guerrero, C.C., Ghazoul, J., Moser, B., 2012. Phenotypic plasticity facilitates resistance to climate change in a highly variable environment. Oecologia 169, 269–279.

Rühl, A.T., Eckstein, R.L., Otte, A., Donath, T.W., 2015. Future challenge for endangered arable weed species facing global warming: low temperature optima and narrow moisture requirements. Biol. Conserv. 182, 262–269.

Ruthrof, K.X., Fontaine, J.B., Matusick, G., Breshears, D.D., Law, D.J., Powell, S., et al., 2016. How drought-induced forest die-off alters microclimate and increases fuel loadings and fire potentials. Int. J. Wildl. Fire 25, 819–830.

Santana, V.M., Bradstock, R.A., Ooi, M.K.J., Denham, A.J., Auld, T.D., Baeza, M.J., 2010. Effects of soil temperature regimes after fire on seed dormancy and germination in six Australian Fabaceae species. Aust. J. Bot. 58, 539–545.

Santana, V.M., Baeza, M.J., Blanes, M.C., 2013. Clarifying the role of fire heat and daily temperature fluctuations as germination cues for Mediterranean Basin obligate seeders. Ann. Bot. 111, 127–134.

Savage, M., Mast, J.N., Feddema, J.J., 2013. Double whammy: high-severity fire and drought in ponderosa pine forests of the Southwest. Can. J. For. Res. 43, 570–583.

Scott, A.C., Glasspool, I.J., 2006. The diversification of Paleozoic fire systems and fluctuations in atmospheric oxygen concentrations. Proc. Natl. Acad. Sci. USA 103, 10861–10865.

Smith, A.L., Blanchard, W., Blair, D.P., McBurney, L., Banks, S.C., Driscoll, D.A., et al., 2016. The dynamic regeneration niche of a forest fire following a rare disturbance event. Divers. Distrib. 22, 457–467.

Staver, A.C., Archibald, S., Levin, S.A., 2011. The global extent of savanna and forest as alternative biome states. Science 334, 230–232.

Swab, R.M., Regan, H.M., Keith, D.A., Regan, T.J., Ooi, M.K.J., 2012. Niche models tell half the story: spatial context and life-history traits influence species responses to global change. J. Biogeogr. 39, 1266–1277.

Tangney, R., Issa, N.A., Merritt, D.J., Callow, J.N., Miller, B.P., 2018. A method for extensive spatiotemporal assessment of soil temperatures during an experimental fire using distributed temperature sensing in optical fibre. Int. J. Wildl. Fire 27, 135–140.

Tangney, R., Merritt, D.J., Fontaine, J.B., Miller, B.P., 2019. Seed moisture content as a primary trait regulating the lethal temperature thresholds of seeds. J. Ecol. 107, 1093–1105.

Tangney, R., Merritt, D.J., Callow, N., Fontaine, J.B., Miller, B.P., 2020a. Seed traits determine species' response to fire under varying soil heating scenarios. Funct. Ecol. 34, 1967–1978.

Tangney, R., Miller, R.G., Enright, N.J., Fontaine, J.B., Merritt, D.J., Ooi, M.K.J., et al., 2020b. Seed dormancy interacts with fire seasonality mechanisms. Trends Ecol. Evol. 35, 1057–1059.

Thanos, C.A., Rundel, P.W., 1995. Fire-followers in chaparral: nitrogenous compounds trigger seed germination. J. Ecol. 83, 207–216.

Thomas, P.B., Morris, E.C., Auld, T.D., 2010. The interaction of fire-cues, temperature and water availability regulates seed germination in a fire-prone landscape. Oecologia 162, 293–302.

Tieu, A., Dixon, K.W., Meney, K.A., Sivasithamparam, K., 2001. The interaction of heat and smoke in the release of seed dormancy in seven species from southwestern Western Australia. Ann. Bot. 88, 259–265.

Tolhurst, K.G., McCarthy, G., 2016. Effect of prescribed burning on wildfire severity: a landscape-scale case study from the 2003 fires in Victoria. Aust. For. 79, 1–14.

Turco, M., Jerez, S., Augusto, S., Tarín-Carrasco, P., Ratola, N., Jiménez-Guerrero, P., et al., 2019. Climate drivers of the 2017 devastating fires in Portugal. Sci. Rep. 9, 13886. Available from: https://doi.org/10.1038/s41598-019-50281-2.

van Mantgem, P.J., Nesmith, J.C.B., Keifer, M.B., Knapp, E.E., Flint, A., Flint, L., 2013. Climatic stress increases forest fire severity across the western United States. Ecol. Lett. 16, 1151–1156.

Whelan, R.J., Rodgerson, L., Dickson, C.R., Sutherland, E.F., 2002. In: Bradstock, R.A., Williams, J.E., Gill, A.M. (Eds.), Flammable Australia: The Fire Regimes and Biodiversity of a Continent. Cambridge University Press, Cambridge, pp. 94–124.

Williams, A.P., Abatzoglou, J.T., Gershunov, A., Guzman-Morales, J., Bishop, D.A., Balch, J.K., et al., 2019. Observed impacts of anthropogenic climate change on wildfire in California. Earth's Future 7, 892–910.

Zirondi, H.L., Ooi, M.K.J., Fidelis, A., 2021. Fire-triggered flowering is the dominant post-fire strategy in a tropical savanna. J. Veg. Sci. 32, e12995. Available from: https://doi.org/10.1111/jvs.12995.

Zuloaga-Aguilar, S., Briones, O., Orozco-Segovia, A., 2010. Effect of heat shock on germination of 23 plant species in pine–oak and montane cloud forests in western Mexico. Int. J. Wildl. Fire 19, 759–773.

Chapter 18

Effects of global climate change on regeneration of invasive plant species from seeds

Cynthia D. Huebner
Northern Research Station, USDA Forest Service, Morgantown, WV, United States

Introduction and background

Plant species introduced to an area in response to intentional or accidental anthropogenic events are called nonnatives. If these species spread rapidly from introduction sites (Richardson et al., 2000) and have harmful effects on the economy, environment, or health (IUCN (International Union for Conservation of Nature), 2000), they are invasive nonnative species. There are approximately 14,000 nonnative plant species established globally (van Kleunen et al., 2015, 2019) and between 2500 (Pyšek et al., 2020) and 4375 (Pagad et al., 2018) of them are considered to be invasive. Invasive species occur in many plant families, but the Asteraceae, Fabaceae, Poaceae, and Rubiaceae have some of the highest numbers of species (Humair et al., 2015). Invasive plants have a range of reproductive and life-history traits (Perrins et al., 1992; Moles et al., 2008), and documenting the common characteristics among them improves our ability to predict how invasive plants are likely to respond to climate change.

Although some invasive plant species rely entirely on asexual reproduction and others benefit from both sexual and asexual reproduction once established, seeds are important for the introduction of many invasive plants into new environments (Barrett et al., 2008; Beckmann et al., 2011). Invasive plants, like native plants, have variable reproductive strategies under both stable and changing environmental conditions, including those predicted under a changing climate (Aronson et al., 2007; Walck et al., 2011). Ten characteristics of many invasive plants that may give them an advantage over natives in response to climate change are (1) rapid growth rates, (2) tolerance of a wide range of climates and environments, (3) short generation time, (4) prolific and reliable reproduction, (5) small seeds, (6) effective seed dispersal, (7) ability to reproduce with just one parent (self-compatible), (8) nonspecialized germination requirements, (9) effective competitive ability, and (10) effective defenses and/or lack of enemies (Baker, 1974; Whitney and Gabler, 2008; van Kleunen et al., 2010; Clements and DiTommaso, 2011). Plants with a horticultural history may share many traits that may be associated with successful landscaping but also with invasiveness (Nicotra et al., 2010). Many of these characteristics develop through human-mediated selection (Nicotra et al., 2010; Chrobock et al., 2011), but some of them are products of natural selection (Clements and DiTommaso, 2011, 2012). These traits, especially those associated with regeneration from seed (Walck et al., 2011), may give invasive plants an advantage when responding to climate change and make their responses to a changing environment predictable.

Many invasive plant species are predicted to expand their range to higher elevations or latitudes in response to global warming (Cunze et al., 2013; Allen and Bradley, 2016; Panda et al., 2018). For example, seed production of *Pueraria lobata* is limited in colder climates, but its range in the USA north from Kentucky and West Virginia into Ohio, Illinois, and Indiana expanded between 1971 and 2006 (Ziska et al., 2011). Phenotypic plasticity can give plants great flexibility in a changing environment, and thus it is a potential adaptation to new environments (Clements and DiTommaso, 2011; Bhowmik, 2014; Geng et al., 2016; Liao et al., 2016). However, phenotypic plasticity does not require a change in the genetic make-up of a species and may allow some plant populations to persist in situ with low genetic diversity (Benito Garzón et al., 2019). Nonetheless, if phenotypic plasticity increases plant fitness at a low cost to the plant, it may help maintain genetic diversity (Grenier et al., 2016).

Adaptation of invasive nonnative plants to novel habitats via natural selection is related to high genetic diversity resulting from multiple introductions, mutations, or hybridizations (Dlugosch and Parker, 2008; Zalapa et al., 2010; Ellegren and Galtier, 2016). The lag phase between the introduction and population expansion of many invasive plants may be the time needed for the species to adapt via natural selection (Clements and DiTommaso, 2011, 2012). However, phenotypic plasticity may be advantageous to a species when the environment changes more rapidly than a species can adapt via natural selection. The climate variability hypothesis predicts that species experiencing greater seasonality (e.g., at high latitudes) also will exhibit a greater range of tolerances to changes in environmental conditions, such as temperature, that is, be more phenotypically plastic (Mumladze et al., 2017). *Taraxacum officinalis*, which is invading along a latitudinal gradient in South America, shows increasing phenotypic plasticity with increasing southern latitudes (Molina-Montenegro and Naya, 2012).

Some invasive plant species have lower genetic diversity than their native counterparts, but instead of, or in addition to, phenotypic plasticity, they are epigenetically modified (e.g., by DNA methylation), which leads to rapid adaptation (Richards et al., 2012; Banerjee et al., 2019). Polyploidy is common in plants and having multiple copies of genes fosters significant genomic and epigenetic changes leading to rapid, reversible adaptation triggered by changes in the environment (Pikaard and Mittelsten Scheid, 2014). Such responses also may be reflected in shorter lag phases of population development (Pérez et al., 2006). For example, *Ambrosia artemisiifolia* has developed similar latitudinal clinal patterns in leaf surface area, plant size, growth, phenology, sex allocation, reproductive investment, and dichogamy in its two nonnative ranges (Europe and Australia) as exist in its native North American range. These patterns evolved repeatedly in both introduced ranges over only 100–150 years and under limited genetic variation in the Australian range, providing evidence for rapid adaptation (van Boheemen et al., 2019). A *meta*-analysis of studies on 56 invasive plants comparing phenotypic plasticity with local genetic adaptation concluded that changes in size, fecundity, and biomass allocation were due to phenotypic plasticity and changes in phenology due to local genetic adaptation (Liao et al., 2016).

Models predict that not all invasive species will perform well across their current nonnative range in response to climate change, although they typically do not account for plasticity or rapid adaptation (Bradley and Wilcove, 2009; Benito Garzón et al., 2019). Geographic ranges of many invasive plants are expected to contract along the southern latitudes in the Northern Hemisphere (Allen and Bradley, 2016). A *meta*-analysis of 204 native and 157 nonnative species by Sorte et al. (2013) found no difference in native versus nonnative terrestrial species (mostly plants) in response to changes in CO_2 and precipitation. However, nonnative species tended to perform better with increased precipitation and CO_2 and native species better with increasing temperatures and decreasing precipitation. In aquatic ecosystems (mostly animals), increased CO_2 and temperature were more inhibitory for native than nonnative species. The authors concluded that the risk of invasion increased if climate change increased the favorability of a site for plant growth, while risk of invasion decreased if sites became less favorable (Sorte et al., 2013).

Combining demography and phenology with climate models helps account for natural selection and rapid evolution and consequently should result in more accurate models about how native and nonnative species are likely to respond to climate change (Chapman et al., 2014; Merow et al., 2017). Optimistically, there may be an unrealized potential of some native plants that have not yet experienced a novel environment to adapt rapidly to climate change (Sow et al., 2018; Thiebaut et al., 2019). Indeed, shifts in geographic range of native species may be shifting their status from native to nonnative, possibly requiring a new definition of "native."

Restoration of ecosystems to a resilient state composed of native plants present prior to any invasion of nonnative species is considered an important initial step in mitigating the effects of global climate change and potential subsequent spread of invasive plants (Bradley et al., 2010; Allara et al., 2012; Chambers et al., 2014). Paradoxically, invaded systems are among the most resilient ones in terms of being able to sustain their altered stable state (Côté and Darling, 2010). Such resilience could be due in part to genetic diversity and subsequent selection or to epigenetic modifications within nonnative populations lacking the genetic diversity to respond to a novel environment. Resilience also likely is related to changes in species composition due to changes in fire frequency and soil nutrient cycling that favor invasives over natives (Gaertner et al., 2014). Moreover, climate-change mitigation efforts may include human-mediated introductions of even more species (including "natives") thought to be adapted to the predicted new climate.

The goal of this chapter is to summarize the literature on invasive plants in relation to climate change and includes (1) mating systems and phenology, (2) sexual reproductive capacity and seed dispersal, (3) seed dormancy, (4) seed germination and viability, (5) soil seed banks, and (6) biotic interactions. Consideration will be given to how each trait and biotic interaction may help ensure survival and spread of invasive species under a changing climate. This information will be linked to restoration efforts with the objective of defining a more informed climate-change mitigation strategy.

Mating systems and phenology

Changes in plant mating systems can impact gene flow, genetic diversity, gene recombination, and effective population size of plants, which in turn may affect the ability of an invasive species to respond to a changing climate (Eckert et al., 2010; Hargreaves and Eckert, 2014). Self-compatible plants are likely to become established outside their native range because they can reproduce from a single individual (Razanajatovo et al., 2016). Pollen development and pistil–pollen interactions are limited by moderate increases in temperature predicted to occur with global climate change, which, in turn, may impact regeneration by seeds (Snider and Oosterhuis, 2011). Disturbance and changes in climate also are associated with pollen limitation either by a reduction in pollinators or mates due to decreased population size; both may increase reliance on self-compatible mating systems and asexual reproduction (Barrett et al., 2008; Eckert et al., 2010). The risk of pollen limitation is higher for dioecious plants, and dioecy is rare in native and nonnative angiosperms (Käfer et al., 2017).

The gynodioecious species *Fallopia japonica* was introduced to the USA with only male sterile (female) plants, but it spread via rhizomes. However, this species has since hybridized with a less invasive nonnative knotweed (*F. sachalinensis*) to form *F.* x *bohemica*, which produces viable seeds, thereby promoting dispersal of the taxon. Although *F. japonica* has low genetic diversity, it is epigenetically diverse. However, *F.* x *bohemica* is 10 × more epigenetically diverse than *F. japonica*. Not only has the mating system of this species complex changed from mainly asexual to both asexual and sexual via hybridization, but the ability to respond epigenetically to stress has also increased its capacity to adapt rapidly (Richards et al., 2012; Gillies et al., 2016).

Global warming also impacts the phenology of invasive plant species. In recent decades, earlier flowering has been correlated with increasing temperatures (Ellwood et al., 2013). With increased temperatures, the growing season for plants starts earlier and/or ends later, and if both, is extended in duration. Likewise, longer flowering periods, which may occur with extended growing seasons (Dorji et al., 2020), allow plants to allocate more resources to reproduction and increase time for interaction of flowers with pollinators (Feng et al., 2016). Earlier flowering and phenological changes (e.g., first flowering) due to climate change have been documented. The historical dataset from Concord, Massachusetts, USA, ("Thoreau country") collected between 1851 and 2006 (Primack et al., 2009) shows that the invasive species *Alliaria petiolata*, *Cynanchium louiseae*, *Frangula alnus*, and *Lonicera morrowii* flowered 11 days earlier than native plants (Willis et al., 2010). Although 87% of these species follow predictions based on climate change (e.g., earlier and/or later flowering/budburst), some of them show no change or an opposite trend (Parmesan and Yohe, 2003). In response to a warmer and wetter environment in the lower latitudes of Europe, *Ambrosia artemisiifolia* flowers later, benefitting from larger late-season plants, and the time between pollen maturation and stigma receptivity is shorter than that for plants in its native North American range (van Boheemen et al., 2019).

Phenology of plants also is likely to be impacted by decreased precipitation due to climate change as documented by the response of the invasive annual grasses *Avena sterilis* and *Hordeum spontaneum* to an aridity gradient in their native Israel. With an increase in aridity, length of growing season was shortened and time to flowering decreased, but seed production increased as seed size decreased (Volis, 2007). Furthermore, *Bidens pilosa*, native to temperate and tropical America and a noxious weed in other regions worldwide, has a phenotype that flowers in February or March in the Southern Hemisphere and a phenotype that flowers 1–2 months earlier, both of which occur in its native range. Seeds of both phenotypes give rise to the typical phenotype under favorable growth conditions. The early flowering phenotype produces larger but fewer seeds than the typical-flowering phenotype. Thus, if climate change results in stress during the life cycle, we can expect an increase in abundance of the early phenotype (Gurvich et al., 2004). In contrast, *Lythrum salicaria* is predicted to have an extended growing season even with climate change because it occupies wetland habitats (Colautti et al., 2017).

Not all invasive plant species are expected to benefit from a change in growing season due to climate change. For instance, in the western USA the range of the annual invasive grass *Bromus tectorum* is predicted to decrease due to increases in drought conditions that result in a growing season too short for plant survival and/or reproduction (Bradley and Wilcove, 2009).

Sexual reproductive capacity and seed dispersal

High fecundity and small seed size are traits of many invasive plants (Radford and Cousens, 2000; Goergen and Daehler, 2001; Morris et al., 2002; Whitney and Gabler, 2008). However, plants from large seeds have higher survival (Moles and Westoby, 2004), greater competitive ability, and higher adult fecundity (Moravcová et al., 2007; Germain and Gilbert, 2014) than those from small seeds. The tradeoff between seed size and number is a response to stress such as drought, and species with large seeds are predicted to colonize more stressful habitats than those with small seeds (Muller-Landau, 2010). Variation in seed size of individuals of invasive plant species also occurs across environmental gradients (Cochrane et al., 2015).

Changes in seed mass of plants growing in different levels of stress provide insight into how climate change may affect seed mass. Seed mass increases in environments with consistently low amounts of precipitation, but it can vary if low levels of precipitation are temporally unpredictable (Volis and Bohrer, 2013). In contrast, seed mass of *Ambrosia artemisiifolia* is larger in the warmer, wetter climate of its nonnative European range than in its native North American range (van Boheemen et al., 2019). *Avena fatua, Festuca arundinacea,* and *Lolium multiflorum* tend to produce fewer and smaller seeds when exposed to higher temperatures (Wiesner and Grabe, 1972; Boyce et al., 1976; Peters, 1982). *Microstegium vimineum* tends to produce smaller seeds in relatively dry than in mesic environments within its invasive range (Huebner, 2011). However, there was no effect of increased temperature (1.5/3.0 degrees day/night) and CO_2 (600 ppmv) (combined) over a 6-year period on seed mass of the invasive *Centaurea diffusa* or *Linaria dalmatica* and two associated native grasses (*Bouteloua gracilis* and *Koeleria macrantha*) in Wyoming, USA (Li et al., 2018).

Seed size also may vary across generations, indicating adaptive transgenerational plasticity (Herman and Sultan, 2011). When *Microstegium vimineum* plants from seeds collected from two regions differing primarily in annual rainfall were grown under drought conditions in a greenhouse, plants derived from the drier region produced seeds with higher mass than those derived from the mesic region (Huebner and Waterland, unpublished data). As the maternal environment changes across generations, seed characteristics and responses to environmental cues also may change. Larger seed may confer greater drought resistance (Cochrane et al., 2015), which could support the differential transgenerational response noted for *M. vimineum*.

Efficient seed dispersal is a trait of many invasive plants (Honig et al., 1992; Vilà and D'Antonio, 1998), and range expansion is likely to be accompanied by long-distance dispersal. The more seeds produced the more likely a few of them will be dispersed to safe sites at a greater distance from the homesite in a changing climate (Clark et al., 2001; Hampe, 2011). Plant migration rates should be $3000-5000$ m yr^{-1} to track estimates of climate change rates (Petit et al., 2008). However, most plants, are estimated to expand their ranges <100 m yr^{-1} (Neilson et al., 2005; Petit et al., 2008; Nogués-Bravo et al., 2018), with a few approaching 610 m yr^{-1} (Parmesan and Yohe, 2003). Long-distance dispersal to new sites tends to favor self-compatibility, especially if marginal populations are small and the likelihood of being able to outcross is low (Hargreaves and Eckert, 2014). Thus, the ability to adapt rapidly may be a more important means of responding to climate change than having both high fecundity and the ability to disperse over long distances.

Since increased CO_2 increases plant height of some invasive species, including *Centaurea diffusa* (Reeves et al., 2015), *Cirsium arvense, Euphorbia esula, Sonchus arvensis* (Ziska, 2003; Ziska et al., 2011), and several crop weeds (Ramesh et al., 2017), it is expected that seed dispersal distance via wind will increase. Differences in plant height and capitulum drying time resulting from climate change conditions increased dispersal distance of *Carduus nutans* via wind by 38% (Teller et al., 2016). An increase in stormy weather, with increases in wind duration and speed, also may increase dispersal distance of seeds of both invasive and native species (Ziska et al., 2011). However, Jablonski et al. (2002) found that seed mass but not seed production of agricultural crop weeds increased in response to increasing CO_2.

Seed dormancy

Seed dormancy spreads germination of a cohort of seeds over time and thus the opportunity for seedling establishment over multiple seasons in environmentally unpredictable habitats (Venable and Brown, 1988). Dormancy of a species may differ between native and nonnative ranges, making responses to global change less predictable as species enter novel environments. For example, *Phragmites australis* seeds appear to have acquired physiological dormancy (PD) in its nonnative range, where 50% of seeds are dormant, compared to no dormancy in the native range (Kettenring and Whigham, 2009). Furthermore, seeds of *Cardamine hirsuta* in Japan (where it is invasive) have stronger PD than those from native European strains of the species under warmer temperatures, making it a strict winter annual in its nonnative range (Kudoh et al., 2007; Donohue et al., 2010).

Seeds of many invasive species have either no dormancy or nondeep PD. In nature, nondeep PD is broken either during exposure to summer temperatures or to low (moist) winter temperatures, depending on the species (Baskin and Baskin, 2014). Global warming may negatively impact the PD breaking requirements of species whose seeds require low temperatures and moist conditions in winter for dormancy-break, but impacts on species whose dormancy is broken in summer may be minimal (see Chapter 10). Invasive nonnative plants that require a relatively long period of cold stratification to break dormancy include *Alliaria petiolata* (Merow et al., 2017; Footitt et al., 2018), *Frangula alnus*

(Dukes et al., 2009), *Heracleum mantegazzianum* (Moravcová et al., 2006), and *Prunus serotina* (Phartyal et al., 2009). Temperature increases due to climate change may negatively impact dormancy-break and germination of these species especially in the southernmost part of their range. Effect of climate change on dormancy-break and germination of *A. petiolata* was investigated using a polyethylene tunnel in which temperature was increased by 0 to +4°C (above outside air temperature), depending on position in the tunnel. Warming in winter decreased seedling emergence, but a few seedlings emerged at the warmest end of the tunnel, suggesting adaptation to a warming climate via selection against dormancy is possible for *A. petiolata* (Footitt et al., 2018).

Seeds of *Cardiocrinum giganteum* var. *giganteum* (native of Japan and invading New Zealand) have an underdeveloped embryo that is physiologically dormant, that is, morphophysiological dormancy. To break dormancy of the embryo, a period of warm moist conditions of summer followed by cool moist conditions of autumn and cold moist conditions of winter are required, resulting in an 18–19-month period between dispersal and germination (Phartyal et al., 2012). Any negative effects of climate change on regeneration of this species from seeds likely would be related to decreased soil moisture.

Seeds of several invasive species, especially those in the Fabaceae, have physical dormancy (PY, water-impermeable seed coat). Acquisition of PY is related to seed drying to a certain moisture content, depending on the species (Baskin and Baskin, 2014). It is predicted that decreased precipitation will increase the proportion of seeds in species such as *Acacia saligna* with PY (vs. nondormancy) in temperate regions (Tozer and Ooi, 2014). In contrast, decreased precipitation may decrease seed dormancy breakage for *Mimosa pigra* and *Parkinsonia aculeata*, invaders of Australian wetlands, both of which require a wet-warm period to break PY (van Klinken and Goulier, 2013). Likewise, increased fire frequency and magnitude are possible consequences of climate change (IPCC, 2013, Ooi et al., Chapter 17), and fire can break PY and stimulate germination (Riveiro et al., 2020; Ooi et al., Chapter 17). High summer temperatures can promote dormancy break in some species with PY. Exposure of seeds of *Acacia dealbata*, an invasive shrub in southeastern Australia, to 60°C that mimics soil temperatures associated with climate change broke PY (Passos et al., 2017). Increased temperatures due to climate change could impact these species via subsequent germination and depletion of soil seed banks (Ooi et al., 2014).

Seed dormancy is a heritable trait under strong selection pressure (Baskin and Baskin, 2014), and in the invasive *Avena fatua* with dormant and nondormant genotypes it is controlled by multiple genes (Foley and Fennimore, 1998). The dormant genotypes are sensitive to temperature and drought experienced in the maternal environment, with higher germination percentages associated with higher temperatures and more severe drought conditions (Sawhney and Naylor, 1979; Naylor, 1983; Jana and Thai, 1987). The maternal plant environment can influence seed dormancy with higher temperatures and drought often reducing dormancy (Fenner, 1991). Seeds of *Parthenium hysterophorus*, an invasive herb in Australia (and other countries) originating in the New World tropics and subtropics, exhibit greater dormancy when maternal plants are grown under warm conditions, with even greater dormancy associated with seeds produced from plants grown in warm and dry conditions. In response to increasing atmospheric CO_2, warming temperatures, and decreasing moisture, *P. hysterophorus* grows to a larger size at a faster rate, has a shorter life span, produces more seeds, and has more dormant seeds than under normal conditions for growth. These results suggest that this species will perform well in parts of its invasive range where temperatures and drought are predicted to increase (Nguyen et al., 2017).

Seed germination and viability

After dormancy is broken, seeds of many invasive plants, including *Ailanthus altissima, Alliaria petiolata*, (Huebner et al., 2018), *Echium plantagineum* (Forcella et al., 1986), *Microstegium vimineum* (Huebner et al., 2018), *Physalis angulatus, P. philadelphicus* (Ozaslan et al., 2017), *Vulpia bromoides*, and *V. muryos* (Dillon and Forcella, 1984) germinate over a wider range of conditions in their invasive range than associated native species. One of the strongest shared patterns among invasive plants, including *Amaranthus retroflexus* (Ruprecht et al., 2014), *Ambrosia artimisiifolia* (Leiblein-Wild et al., 2014), *Echium plantagineum* (Forcella et al., 1986), *Eragrostis plana* (Guido et al., 2017), *Galinsoga ciliata* (Ruprecht et al., 2014), *Gunnera tinctoria* (Gioria et al., 2018), *Impatiens glandulifera* (Skálová et al., 2011; Ruprecht et al., 2014), *Plantago virginica* (Xu et al., 2019), *Rhododendron ponticum* (Erfmeier and Bruelheide, 2005), *Rudbeckia laciniata* (Ruprecht et al., 2014), *Senecio inaequidens* (Sans et al., 2004; Gioria and Pyšek, 2017), and *Ulmus pumila* (Hirsch et al., 2012) is earlier germination than associated native species, which may enable the invasive plants to grow under reduced competition (Wainwright and Cleland, 2013; Gioria and Pyšek, 2017). In contrast, while nonnative *Taraxacum officinale* had a higher germination percentage than the associated native *T. laevigatum*, this was only true under ideal environmental conditions that ensure seedling survival. *Taraxacum laevigatum* germinated to a higher percentage than *T. officinale* under stressful conditions, whereas a new invader *Taraxacum brevicorniculatum* had the highest germination among the three species at all test conditions (Luo and Cardina, 2012). These findings suggest that long-established nonnative plants may lose their ability to adapt quickly to novel environments such as those caused by climate change.

Seeds of *Achillea millefolium*, *Hieracium pilosella*, *Hypericum perforatum* (Beckmann et al., 2011), *Ludwigia peploides* (Gillard et al., 2017), and *Ulex europaeus* (Udo et al., 2017) germinate faster and at higher temperatures in their nonnative-warmer environment than in their native-colder environment, suggesting adaptation to warmer temperatures. Other invasive plant species that appear to be expanding their range into warmer, mesic environments include *Ambrosia artemisiifilia* (Leiblein-Wild et al., 2014), *Berberis thunbergii* (Merow et al., 2017), *Gunnera tinctoria* (Gioria et al., 2018), and *Leucaena leucocephala* (Marques et al., 2020). Rapid germination of species that are shifting their range north (Northern Hemisphere) may be somewhat risky for species intolerant of cold. Risk may be abated for some species by an increase in cold tolerance of seedlings as found for *Ambrosia artemisiifolia* in its nonnative European range (Leiblein-Wild et al., 2014).

Seeds of some invasive species are tolerant of high temperatures and water stress, suggesting that they would not be negatively affected by climate change (Hou et al., 2014). *Tithonia diversifolia* seeds germinated after a 30-day heat treatment at 80°C and about 20% of them germinated at -0.6 MPa (Wen, 2015). Seeds of *Ageratum conyzoides*, *Conyza canadensis*, and *Crassocephalum crepidioides* germinate over a broad range of temperatures (15°C–30°C), and those of *A. conyzoides* (the most tolerant species) germinated to about 25% at 35°C and 95% at 40/25°C (high temperature for 7 h per day). Seeds of *A. conyzoides* also germinated to 65% at -0.8 MPa, suggesting that germination and seedling establishment were possible under the temperature and water stress conditions of the introduced range in southern China (Yuan and Wen, 2018). Increased temperature enhances germination of seeds of *Oenothera biennis*, *Petiveria alliacea*, and *Syncarpia glomulifera* (Sershen et al., 2017). However, at 0.0 MPa seeds of *Cenchrus ciliaris* germinate to $\geq 60\%$ at 20°C–40°C but to only about 45% at -0.06 MPa (Tinoco-Ojanguren et al., 2016). Similarly, the combined effects of increased temperature and CO_2 had no impact on seed viability or overall germination percentages but increased germination rates of the invasive species *Centaurea diffusa* and *Linaria dalmatica* (Li et al., 2018).

A negative response of seeds of some invasive plant species to increasing temperature and/or water stress suggests that plant regeneration via seeds may be negatively affected by climate change. The palm *Archontophoenix alexandrae* (widely planted in tropical parts of China) can germinate only at temperatures between 20°C and 30°C and is highly sensitive to desiccation, with seed viability decreasing at temperatures above 60°C (Wen, 2019). Seedling emergence and survival of *Oenothera biennis* decreased in response to increasing temperatures (Sershen et al., 2017). *Piper aduncum* seeds did not germinate at constant temperatures above 35°C but germinated at an alternating temperature regime of 40/25°C. Germination of this species was inhibited by water potentials more negative than -0.06 MPa (Wen et al., 2015). Variation in winter precipitation decreased germination of the nonnative *Centaurea solstitialis* (Hierro et al., 2009).

Maternal plant environmental temperatures may impact seed germination, timing, and viability. Responses to elevated temperatures provide insight into how plants might respond to increased temperatures due to climate change. Seeds of *Carduus nutans* plants grown at temperatures moderately higher than those in current field conditions had higher germination percentages and rates than those from plants grown under ambient field conditions (Zhang et al., 2012). *Alliaria petiolata* seeds collected from populations across a latitudinal gradient in North America and sown in a common garden located at a northern latitude exhibited population differences after 13 years of monitoring. Seeds from southern populations had lower germination percentages than those of northern populations. However, after 6 years germination percentages had become more similar, with the southern-population germination percentages increasing, revealing local adaptation. Annual seedling emergence was correlated with spring temperatures, thus phenotypic plasticity also may play a role (Blossey et al., 2017). In a reciprocal seed transplant experiment with *Ludwigia peploides*, an invasive in the Mediterranean region of California (USA) and temperate climates of France, seed viability was higher in the Mediterranean climate seeds exposed to Mediterranean climate temperatures (average 24°C) than the temperate climate seeds exposed to temperate climate temperatures (average 18°C), However, seeds from both provenances germinated faster at 24 than at 18°C (Gillard et al., 2017).

Soil seed banks

Seed banks are a bet-hedging strategy in unpredictable environments (Venable and Brown, 1988). In addition, they may provide a genetic history of invasive plant species evolution. For *Gunnera tinctoria*, a long-established invasive plant in Ireland, the number of alleles, percentage of polymorphic loci, and genetic diversity decreased in seeds found at increasing soil depths. Furthermore, the greatest change (increase) in genetic diversity in the seed bank of this species occurred after the lag time of establishment, when range expansion began (Fennell et al., 2014).

The impact of climate change on seed banks is predicted to decrease the number of seeds in the soil, especially for species whose seeds have PY. For example, a 4°C increase in air temperature, as predicted by climate change for southeastern Australia, may increase soil temperature by about 10°C. Reduced seed viability or increased germination of buried seeds due to soil warming may deplete the soil seed bank (Ooi, 2012). With a 2°C increase in habitat temperature, 75% of the seed bank of *Leucaena leucocephala* was lost (Marques et al., 2020). For *Parkinsonia aculeata*, increases in soil temperature and soil moisture led to a decreased seed bank size via PY-break (van Klinken and Goulier, 2013), while most seeds of *Acacia saligna* in its invasive range lost viability (Cohen et al., 2019).

Invasive species that form persistent soil seed banks may be difficult to control (Marchante et al., 2011). Thus, much attention has been given to seed banks of invasive species (e.g., van Clef and Stiles, 2001; Gioria and Pyšek, 2016; Gioria et al., 2019). However, not all invasive species form large persistent soil seed banks, for example, *Ailanthus altissima* (Kowarik, 1995; Kostel-Hughes and Young, 1998), *Berberis thunbergii* (D'Appollonio, 2006), and *Lonicera maackii* (Luken and Mattimiro, 1991; Luken and Goessling, 1995; Hartman and McCarthy, 2008). Seeds of *Ambrosia artemisiifolia* (Fumanal et al., 2008) and *Ailanthus altissima* (Rebbeck and Jolliff, 2018) buried deeper than 5 cm in their invasive range had increased longevity. Seeds may be buried by disturbance, and they serve as a future seed source. Climate change is predicted to cause habitat disturbances and soil turn-over, which may increase the likelihood of increased seed burial of invasive species and formation of new seed banks (Fumanal et al., 2008).

Biotic interactions of invasive plant species

Evaluating the impacts of climate change on species assemblages in addition to individual species may ensure greater success of mitigation efforts. For example, although the invasive grass *Eragrostis plana* germinated more rapidly and to higher percentages than many of its associated native species when each species was tested separately in Petri dishes, its germination was delayed compared to that of native associated species in mixed-species cultures (Guido et al., 2017). Also, mismatches in timing between flowering and pollinators and between seed maturity and dispersers are likely to increase with climate change (Thomson et al., 2010). Some invasive shrubs may be able to resynchronize interactions more rapidly than native species because of high photosynthetic rates in response to increased length of the growing season (Fridley, 2012). Extended fruiting periods could delay departure of migratory birds or increase the number of broods of potential seed dispersers (Gallinat et al., 2015). Changes in pollinators and seed predators/diseases due to climate change also impact plant regeneration from seeds. A seed predator bruchid beetle (*Acanthoscelides macrophthalmus*) of *Leucaena leucocephala*, an invasive tree in Brazil and other tropical countries, does not injure the embryo but promotes germination by scarifying the water-impermeable seed coat (da Silva and Rossi, 2019). Increases in size of beetle populations due to climate change could increase germination of this species. Similarly, the invasive tree *Triadica sebifera* has higher germination percentages and rates and greater seed longevity after seeds have passed through the gut of birds (Renne et al., 2001).

Some pollinators, seed dispersers, and seed predators of invasive plants are invasive themselves. Thus an important consequence of climate change may be the presence of new animals in a plant community. For example, the red-whiskered bulbul (*Pycnonotus jocosus*), native to southeast Asia and introduced to Mauritius, consumes fruits of *Ligustrum robustum* and *Clidemia hirta*, native to southeast Asia and central and south America, respectively. Gut passage of seeds of these species increased germination success in their introduced habitat in Mauritius (Linnebjerg et al., 2009).

Birds that consume fruits of invasive plants without damaging the seeds serve as long-distance dispersal agents in addition to enhancing germination (Jordaan et al., 2011). Long-distance seed dispersal may be even more direct and easily predicted by the movement of cattle, other livestock, and deer that consume and then defecate seeds of invasives, such as *Acacia nilotica* (Kriticos et al., 2003), *Elaeagnus umbellata*, *Ligustrum vulgare* (Averill et al., 2016), *Lonicera maackii* (Williams and Ward, 2006; Guiden et al., 2015), *Lonicera morrowii* (Williams and Ward, 2006; Averill et al., 2016), *Rosa multiflora* (Williams and Ward, 2006), *Rubus phoenicolasius* (Williams and Ward, 2006), and *Stellaria media* (Myers et al., 2004). Northward migration in response to global warming in temperate regions may occur relatively rapidly for invasive species whose seeds are consumed by deer, whereas other invasives that deer avoid (e.g., *Alliaria petiolata*, *Berberis thunbergii*, and *Microstegium vimineum*) are less likely to migrate as fast (Averill et al., 2016).

Increases in extreme climatic events such as drought or flooding as well as increases in winter and nighttime temperatures are predicted to lead to range expansion of several diseases, either alone or in association with range expansion of invasive plants (Anderson et al., 2004). Pathogens associated with mature leaves and leaf litter of *Ageratina adenophora* increase time to germination and decrease germination percentages in China, where the species was established 20–80 years ago, suggesting that it may be accumulating pathogens with time (Fang et al., 2019). These results suggest that a changing environment may reset or reduce this accumulation of pathogens, thus removing enemies acquired by invasives established beyond their typical range.

Fungal effects on seed longevity also could be impacted by a changing climate. *Microstegium vimineum* has relatively high seed mortality due to fungal infections, resulting in longevity of only 2–4 years of seeds buried in the field. Increased precipitation and temperature may accelerate loss of seed viability of this species, potentially leading to no viable seed bank after 2–4 years (Redwood et al., 2018). *Prunus serotina*, an invasive tree in Europe, produces more viable seeds in well-drained, nutrient-poor soils than in moist, rich soils due to combined (negative) effects of fungal pathogens in the wetter soils and its inefficient nitrogen assimilation in nutrient-rich soils (Closset-Kopp et al., 2011). In contrast, seeds of *Ailanthus altissima* (Redwood et al., 2019) and *Alliaria petiolata* (Redwood et al., 2018) do not appear to be impacted by fungal infections, possibly due in part to allelopathic compounds they produce, and the seeds had low mortality when buried in the field. In tropical regions, *Ulex europaeus* occurs only above 1000 m in its invasive range due to high fungal infection in the warmer lower altitudes (Udo et al., 2017). Climate change may increase temperatures at higher elevations, which may increase fungal infections and reduce viability of *U. europaeus* seeds.

Interactions of invasive species with biocontrol agents are likely to change in response to climate change. Currently, the invasive species *Centaurea solstitialis* is controlled by two weevils and two picture-winged flies but only in the northernmost part of its range in Oregon (USA), where a shorter growing season impairs its ability to compensate for the damage done by the agents (Gutierrez et al., 2008). For this species, decreased precipitation due to climate change would decrease plant growth and enhance the effectiveness of control organisms. In the case of *C. diffusa* and its biocontrol agent (a weevil), weevil efficacy increases with elevated CO_2 and temperature. However, elevated CO_2 and temperature promote early flowering of *C. diffusa* resulting in more and larger seeds than in controls under normal conditions. Despite the weevil being able to infest more seeds in response to the earlier reproductive phenology of *C. diffusa*, increased seed predation did not eliminate the positive effects of CO_2 on seed production (Reeves et al., 2015).

Linking regeneration by seeds with climate change mitigation

Estimated rates of climate change may exceed the limits of adaptation and migration by both nonnative and native species, and thus assisted migration of plants is proposed as a potential solution (Vitt et al., 2010). Mitigation in the form of preemptively transplanting native species to their predicted future habitats will depend on knowledge about native as well as nonnative species mating systems, seed production, seed dormancy, germination requirements, soil seed banks, and avoidance of mismatches among positively or negatively interacting organisms (Seglias et al., 2018). Furthermore, manipulation of growth conditions of maternal plants during seed development by temperature, nutrients, and photoperiod such that dormancy is increased may help ensure long-term establishment of transplanted native species (Sharif-Zadeh and Murdoch, 2000).

It should be noted that range-shifting native species may be acting like invasive plants, colonizing novel environments with potential negative effects (Wallingford et al., 2020). Interactions between native and nonnative species may disrupt once-established native species assemblages that could survive major climatic events such as a drought. This has been demonstrated experimentally in field mesocosms of one, three, and six native species by adding one of two nonnative invasives, *Lupinus polyphyllus* or *Senecio inaequidens*. *Senecio inaequidens*, which is drought tolerant, outcompeted the natives even without drought, but the negative effect of *L. polyphyllus* on the native community depended on drought stress (Vetter et al., 2020).

Future research needs

Of the thousands of plants recognized as invasive globally, we have information about the effects of climate change on regeneration by seeds on less than 10% of them, and the studies typically are focused on only a few reproductive traits. There also may be geographic biases, with more research on invasive plants in North America and Oceania than elsewhere. Nonetheless, some patterns are evident, that is, more invasive species responding to climate change with decreased seed dormancy, earlier germination, and increased germination percentages (Fig. 18.1).

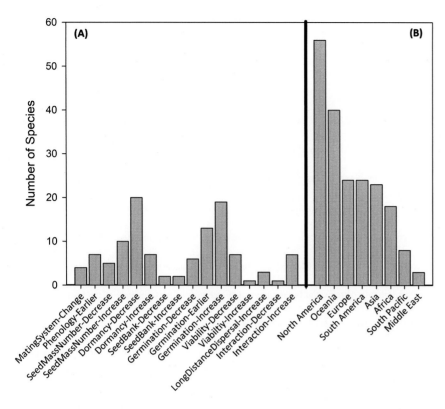

FIGURE 18.1 (A) Summary of number of invasive nonnative species that increase, decrease, or change (earlier or later) in mating system, phenology, seed mass/number, dormancy, seed bank, germination, viability, and interactions with other species. (B) Number of invasive species in each region for which response to climate change has been studied. Oceania includes Australia, New Zealand, Melanesia, Micronesia, and Polynesia.

Based on information summarized in this chapter, there is a need for research on the following aspects of the biology of invasive plant species:

- response of more nonnative invasives to climate change over a broad array of reproductive traits;
- impact of climate change on reproductive traits of assemblages of nonnative and native plant species;
- potential for success of applying climate change mitigation efforts; and
- potential for both native and nonnative plants to adapt rapidly to a changing climate.

References

Allara, M., Kugbei, S., Dusunceli, F., Gbehounou, G., 2012. Coping with changes in cropping systems: plant pests and seeds. In: Meybeck, A., Lankoski, J., Redfern, S., Azzu, N., Gitz, V. (Eds.), Building resilience for adaptation to climate change in the agriculture sector, Proceedings of a joint FAO/OECD Workshop, 23–24 April 2012, Rome, Italy. Rome: FAO, pp. 91–102.

Allen, J.M., Bradley, B.M., 2016. Out of the weeds? Reduced plant invasion risk with climate change in the continental United States. Biol. Conserv. 203, 306–312.

Anderson, P.K., Cunningham, A.A., Patel, N.G., Morales, F.J., Epstein, P.R., Daszak, P., 2004. Emerging infectious diseases of plants: pathogen pollution, climate change and agrotechnology drivers. Trends Ecol. Evol. 19, 535–544.

Aronson, M.F., Handel, S.N., Clemants, S.E., 2007. Fruit type, life form and origin determine the success of woody plant invaders in an urban landscape. Biol. Inv 9, 465–475.

Averill, K.M., Mortensen, D.A., Smithwick, E.A., Post, E., 2016. Deer feeding selectivity for invasive plants. Biol. Inv. 18, 1247–1263.

Baker, H.G., 1974. The evolution of weeds. Annu. Rev. Ecol. Syst. 5, 1–24.

Banerjee, A.K., Guo, W., Huang, Y., 2019. Genetic and epigenetic regulation of phenotypic variation in invasive plants — linking research trends towards a unified framework. NeoBiota 49, 77–103.

Barrett, S.C., Colautti, R.I., Eckert, C.G., 2008. Plant reproductive systems and evolution during biological invasion. Mol. Ecol. 17, 373–383.

Baskin, C.C., Baskin, J.M., 2014. Seeds: Ecology, Biogeography, and Evolution of Dormancy and Germination, Second ed. Academic Press/Elsevier, San Diego.

Beckmann, M., Bruelheide, H., Erfmeier, A., 2011. Germination responses of three grassland species differ between native and invasive origins. Ecol. Res. 26, 763–771.

Benito Garzón, M., Robson, T.M., Hampe, A., 2019. ΔTrait SDMs: species distribution models that account for local adaptation and phenotypic plasticity. New Phytol. 222, 1757–1765.

Bhowmik, P.C., 2014. Invasive weeds and climate change: past, present and future. J. Crop Weeds 10, 345–349.

Blossey, B., Nuzzo, V., Dávalos, A., 2017. Climate and rapid local adaptation as drivers of germination and seed bank dynamics of *Alliaria petiolata* (garlic mustard) in North America. J. Ecol. 105, 1485–1495.

Boyce, K.G., Cole, D.F., Chilcote, D.O., 1976. Effect of temperature and dormancy on germination of tall fescue. Crop Sci. 16, 15–18.

Bradley, B.A., Wilcove, D.S., 2009. When invasive plants disappear: transformative restoration possibilities in the western United States resulting from climate change. Restor. Ecol. 17, 715–721.

Bradley, B.A., Wilcove, D.S., Oppenheimer, M., 2010. Climate change increases risk of plant invasion in the eastern United States. Biol. Inv. 12, 1855–1872.

Chambers, F.M., Brain, S.A., Mauquoy, D., McCarroll, J., Daley, T., 2014. The 'Little Ice Age' in the Southern Hemisphere in the context of the last 3000 years: peat-based proxy-climate data from Tierra del Fuego. Holocene 24, 1649–1656.

Chapman, D.S., Haynes, T., Beal, S., Essl, F., Bullock, J.M., 2014. Phenology predicts the native and invasive range limits of common ragweed. Glob. Change Biol. 20, 192–202.

Chrobock, T., Kempel, A., Fischer, M., van Kleunen, M., 2011. Introduction bias: cultivated alien plant species germinate faster and more abundantly than native species in Switzerland. Basic Appl. Ecol. 12, 244–250.

Clark, J.S., Lewis, M., Horvath, L., 2001. Invasion by extremes: population spread with variation in dispersal and reproduction. Am. Nat. 157, 537–554.

Clements, D.R., DiTommaso, A., 2011. Climate change and weed adaptation: can evolution of invasive plants lead to greater range expansion than forecasted? Weed Res. 51, 227–240.

Clements, D.R., DiTommaso, A., 2012. Predicting weed invasion in Canada under climate change: evaluating evolutionary potential. Can. J. Plant Sci. 92, 1013–1020.

Closset-Kopp, D., Saguez, R., Decocq, G., 2011. Differential growth patterns and fitness may explain contrasted performances of the invasive *Prunus serotina* in its exotic range. Biol. Inv. 13, 1341–1355.

Cochrane, A., Yates, C.J., Hoyle, G.L., Nicotra, A.B., 2015. Will among-population variation in seed traits improve the chance of species persistence under climate change? Glob. Ecol. Biogeogr. 24, 12–24.

Cohen, O., Bar, P., Gamliel, A., Katan, J., Kurzbaum, E., Weber, G., et al., 2019. Rain-based soil solarization for reducing the persistent seed banks of invasive plants in natural ecosystems – *Acacia saligna* as a model. Pest Manage. Sci. 75, 1933–1941.

Colautti, R.I., Ågren, J., Anderson, J.T., 2017. Phenological shifts of native and invasive species under climate change: insights from the *Boechera–Lythrum* model. Philos. Trans. R. Soc. B 372, 20160032. Available from: https://doi.org/10.1098/rstb.2016.0032.

Cunze, S., Leiblein, M.C., Tackenberg, O., 2013. Range expansion of *Ambrosia artemisiifolia* in Europe is promoted by climate change. Int. Sch. Res. 2013, 610126. Available from: https://doi.org/10.1155/2013/610126.

Côté, I.M., Darling, E.S., 2010. Rethinking ecosystem resilience in the face of climate change. PLoS Biol. 8, e1000438. Available from: https://doi.10.1371/journal.pbio.1000438.

da Silva, A.V., Rossi, M.N., 2019. When a seed-feeding beetle is a predator and also increases the speed of seed germination: an intriguing interaction with an invasive plant. Evol. Ecol. 33, 211–232.

Dillon, S.P., Forcella, F., 1984. Germination, emergence, vegetative growth and flowering of two silvergrasses, *Vulpia bromoides* (L.) S.F. Gray and *V. myuros* (L.) C.C. Gmel. Aust. J. Bot. 32, 165–175.

Dlugosch, K.M., Parker, I.M., 2008. Founding events in species invasions: genetic variation, adaptive evolution, and the role of multiple introductions. Mol. Ecol. 17, 431–449.

Donohue, K., Rubio de Casas, R., Burghardt, L., Kovach, K., Willis, C.G., 2010. Germination, postgermination adaptation, and species ecological ranges. Annu. Rev. Ecol. Evol. Syst. 41, 293–319.

Dorji, T., Hopping, K.A., Meng, F., Wang, S., Jiang, L., Klein, J.A., 2020. Impacts of climate change on flowering phenology and production in alpine plants: the importance of end of flowering. Agric. Ecosyst. Environ. 291, 106795. Available from: https://doi.org/10.1016/j.agee.2019.106795.

Dukes, J.S., Pontius, J., Orwig, D., Garnas, J.R., Rodgers, V.L., Brazee, N., et al., 2009. Responses of insect pests, pathogens, and invasive plant species to climate change in the forests of northeastern North America: what can we predict? Can. J. For. Res. 39, 231–248.

D'Appollonio, J., 2006. Regeneration Strategies of Japanese Barberry (*Berberis thunbergii* DC) in Coastal Forests of Maine (M.Sc. University of Maine, Orono.

Eckert, C.G., Kalisz, S., Geber, M.A., Sargent, R., Elle, E., Cheptou, P.O., et al., 2010. Plant mating systems in a changing world. Trends Ecol. Evol. 25, 35–43.

Ellegren, H., Galtier, N., 2016. Determinants of genetic diversity. Nat. Rev. Genet. 17, 422–433.

Ellwood, E.R., Temple, S.A., Primack, R.B., Bradley, N.L., Davis, C.C., 2013. Record-breaking early flowering in the eastern United States. PLoS One 8, e53788. Available from: https://doi.org/10.1371/journal.pone.0053788.

Erfmeier, A., Bruelheide, H., 2005. Invasive and native *Rhododendron ponticum* populations: is there evidence for genotypic differences in germination and growth? Ecography 28, 417–428.

Fang, K., Chen, L., Zhou, J., Yang, Z.P., Dong, X.F., Zhang, H.B., 2019. Plant–soil–foliage feedbacks on seed germination and seedling growth of the invasive plant *Ageratina adenophora*. Proc. R. Soc. B 286, 20191520. Available from: https://10.1098/rspb.2019.1520.

Feng, Y., Maurel, N., Wang, Z., Ning, L., Yu, F.H., van Kleunen, M., 2016. Introduction history, climatic suitability, native range size, species traits and their interactions explain establishment of Chinese woody species in Europe. Glob. Ecol. Biogeogr. 25, 1356–1366.

Fennell, M., Gallagher, T., Vintro, L.L., Osborne, B., 2014. Using soil seed banks to assess temporal patterns of genetic variation in invasive plant populations. Ecol. Evol. 4, 1648–1658.

Fenner, M., 1991. The effects of the parent environment on seed germinability. Seed Sci. Res. 1, 75–84.
Foley, M.E., Fennimore, S.A., 1998. Genetic basis for seed dormancy. Seed Sci. Res. 8, 173–182.
Footitt, S., Huang, Z., Olcer-Footitt, H., Clay, H., Finch-Savage, W.E., 2018. The impact of global warming on germination and seedling emergence in *Alliaria petiolata*, a woodland species with dormancy loss dependent on low temperature. Plant Biol. 20, 682–690.
Forcella, F., Wood, J.T., Dillon, S.P., 1986. Characteristics distinguishing invasive weeds within *Echium* (Bugloss). Weed Res. 26, 351–364.
Fridley, J.D., 2012. Extended leaf phenology and the autumn niche in deciduous forest invasions. Nature 485, 359–362.
Fumanal, B., Gaudot, I., Bretagnolle, F., 2008. Seed-bank dynamics in the invasive plant, *Ambrosia artemisiifolia* L. Seed Sci. Res. 18, 101–114.
Gaertner, M., Biggs, R., Te Beest, M., Hui, C., Molofsky, J., Richardson, D.M., 2014. Invasive plants as drivers of regime shifts: identifying high-priority invaders that alter feedback relationships. Divers. Distrib. 20, 733–744.
Gallinat, A.S., Primack, R.B., Wagner, D.L., 2015. Autumn, the neglected season in climate change research. Trends Ecol. Evol. 30, 169–176.
Geng, Y., van Klinken, R.D., Sosa, A., Li, B., Chen, J., Xu, C.Y., 2016. The relative importance of genetic diversity and phenotypic plasticity in determining invasion success of a clonal weed in the USA and China. Front. Plant. Sci 7, 213. Available from: https://doi.org/10.3389/fpls.2016.00213.
Germain, R.M., Gilbert, B., 2014. Hidden responses to environmental variation: maternal effects reveal species niche dimensions. Ecol. Lett. 17, 662–669.
Gillard, M., Grewell, B.J., Futrell, C.J., Deleu, C., Thiébaut, G., 2017. Germination and seedling growth of water primroses: a cross experiment between two invaded ranges with contrasting climates. Front. Plant Sci. 8, 1677. Available from: https://doi.org/10.3389/fpls.2017.01677.
Gillies, S., Clements, D.R., Grenz, J., 2016. Knotweed (*Fallopia* spp.) invasion of North America utilizes hybridization, epigenetics, seed dispersal (unexpectedly), and an arsenal physiological tactics. Inv. Plant Sci. Manage. 9, 71–80.
Gioria, M., Pyšek, P., 2016. The legacy of plant invasions: changes in the soil seed bank of invaded plant communities. BioScience 66, 40–53.
Gioria, M., Pyšek, P., 2017. Early bird catches the worm: germination as a critical step in plant invasion. Biol. Inv. 19, 1055–1080.
Gioria, M., Pyšek, P., Osborne, B.A., 2018. Timing is everything: does early and late germination favor invasions by herbaceous alien plants? J. Plant Ecol. 11, 4–16.
Gioria, M., Le Roux, J.J., Hirsch, H., Moravcová, L., Pyšek, P., 2019. Characteristics of the soil seed bank of invasive and non-invasive plants in their native alien distribution range. Biol. Inv. 21, 2313–2332.
Goergen, E., Daehler, C.C., 2001. Reproductive ecology of a native Hawaiian grass (*Heteropogon contortus*; Poaceae) vs its invasive alien competitor (*Pennisetum setaceum*; Poaceae). Int. J. Plant Sci. 162, 317–326.
Grenier, S., Barre, P., Litrico, I., 2016. Phenotypic plasticity and selection: nonexclusive mechanisms of adaptation. Scientifica 2016, 7021701. Available from: https://doi.org/10.1155/2016/7021701.
Guiden, P., Gorchov, D.L., Nielsen, C., Schauber, E., 2015. Seed dispersal of an invasive shrub, Amur honeysuckle (*Lonicera maackii*), by white-tailed deer in a fragmented agricultural-forest matrix. Plant Ecol. 216, 939–950.
Guido, A., Hoss, D., Pillar, V.D., 2017. Exploring seed to seed effects for understanding invasive species success. Persp. Ecol. Conserv. 15, 234–238.
Gurvich, D.E., Enrico, L., Funes, G., Zak, M.R., 2004. Seed mass, seed production, germination and seedling traits in two phenological types of *Bidens pilosa* (Asteraceae). Aust. J. Bot. 52, 647–652.
Gutierrez, A.P., Ponti, L., d'Oultremont, T., Ellis, C.K., 2008. Climate change effects on poikilotherm tritrophic interactions. Clim. Change 87, 167–192.
Hampe, A., 2011. Plants on the move: the role of seed dispersal and initial population establishment for climate-driven range expansions. Acta Oecol. 37, 666–673.
Hargreaves, A.L., Eckert, C.G., 2014. Evolution of dispersal and mating systems along geographic gradients: implications for shifting ranges. Funct. Ecol. 28, 5–21.
Hartman, K.M., McCarthy, B.C., 2008. Changes in forest structure and species composition following invasion by a non-indigenous shrub, Amur honeysuckle (*Lonicera maackii*). J. Torrey Bot. Soc. 133, 245–259.
Herman, J.J., Sultan, S.E., 2011. Adaptive transgenerational plasticity in plants: case studies, mechanisms, and implications for natural populations. Front. Plant Sci. 2, 102. Available from: https://doi.org/10.3389/fpls.2011.00102. https://doi.org/10.3389/fpls.2011.00102.
Hierro, J.L., Eren, Ö., Khetsuriani, L., Diaconu, A., Török, K., Montesinos, D., et al., 2009. Germination responses of an invasive species in native and non-native ranges. Oikos 118, 529–538.
Hirsch, H., Wypior, C., von Wehrden, H., Wesche, K., Renison, D., Hensen, I., 2012. Germination performance of native and non-native *Ulmus pumila* populations. NeoBiota 15, 53–68.
Honig, M.A., Cowling, R.M., Richardson, D.M., 1992. The invasive potential of Australian banksias in South African fynbos: a comparison of the reproductive potential of *Banksia ericifolia* and *Leucadendron laureolum*. Aust. J. Ecol. 17, 305–314.
Hou, Q.Q., Chen, B.M., Peng, S.L., Chen, L.Y., 2014. Effects of extreme temperature on seedling establishment of nonnative invasive plants. Biol. Inv. 16, 2049–2061.
Huebner, C.D., 2011. Seed mass, viability, and germination of Japanese stiltgrass (*Microstegium vimineum*) under variable light and moisture conditions. Inv. Plant Sci. Manage. 4, 274–283.
Huebner, C.D., Regula, A.E., McGill, D.W., 2018. Germination, survival, and early growth of three invasive plants in response to five forest management regimes common to US northeastern deciduous forests. For. Ecol. Manage. 425, 100–118.
Humair, F., Humair, L., Kuhn, F., Kueffer, C., 2015. E-commerce trade in invasive plants. Conserv. Biol. 29, 1658–1665.
IPCC, 2013. Climate Change 2013. The Physical Science Basis. Working Group I Contribution to the Fifth Assessment Report of the Intergovernmental Panel on Climate Change. Cambridge University Press, Cambridge.

IUCN (International Union for Conservation of Nature), 2000. Guidelines for the Prevention of Biodiversity Loss Caused by Alien Invasive Species. 51st meeting IUCN Council, Gland, Switzerland, p. 25. Available from: https://portals.iucn.org/library/efiles/documents/Rep-2000-052.pdf.

Jablonski, L.M., Wang, X., Curtis, P.S., 2002. Plant reproduction under elevated CO_2 conditions: a *meta*-analysis of reports on 79 crop and wild species. New Phytol. 156, 9–26.

Jana, S., Thai, K.M., 1987. Patterns of changes of dormant genotypes in *Avena fatua* populations under different agricultural conditions. Can. J. Bot. 65, 1741–1745.

Jordaan, L.A., Johnson, S.D., Downs, C.T., 2011. The role of avian frugivores in germination of seeds of fleshy-fruited invasive alien plants. Biol. Inv 13, 1917–1930.

Kettenring, K.M., Whigham, D.F., 2009. Seed viability and seed dormancy of non-native *Phragmites australis* in suburbanized and forested watersheds of the Chesapeake Bay, USA. Aquat. Bot. 91, 199–204.

Kostel-Hughes, F., Young, T.P., 1998. The soil seed bank and its relationship to the aboveground vegetation in deciduous forests in New York City. Urban Ecosyst. 2, 43–59.

Kowarik, I., 1995. Clonal growth in *Ailanthus altissima* on a natural site in West Virginia. J. Veg. Sci. 6, 853–856.

Kriticos, D.J., Brown, J.R., Maywald, G.F., Radford, I.D., Nicholas, D.M., Sutherst, R.W., et al., 2003. SPAnDX: a process-based population dynamics model to explore management and climate change impacts on an invasive alien plant, *Acacia nilotica*. Ecol. Model. 163, 187–208.

Kudoh, H., Nakayama, M., Lihová, J., Marhold, K., 2007. Does invasion involve alternation of germination requirements? A comparative study between native and introduced strains of an annual Brassicaceae, *Cardamine hirsuta*. Ecol. Res. 22, 869–875.

Käfer, J., Marais, G.A.B., Pannell, J.R., 2017. On the rarity of dioecy in flowering plants. Mol. Ecol. 26, 1225–1241.

Leiblein-Wild, M.C., Kaviani, R., Tackenberg, O., 2014. Germination and seedling frost tolerance differ between the native and invasive range in common ragweed. Oecologia 174, 739–750.

Liao, H., D'Antonio, C.M., Chen, B., Huang, Q., Peng, S., 2016. How much do phenotypic plasticity and local genetic variation contribute to phenotypic divergences along environmental gradients in widespread invasive plants? A *meta*-analysis. Oikos 125, 905–917.

Linnebjerg, J.F., Hansen, D.M., Olesen, J.M., 2009. Gut passage effect of the introduced red-whiskered bulbul (*Pycnonotus jocosus*) on germination of invasive plant species in Mauritius. Austral Ecol. 34, 272–277.

Li, J., Ren, L., Bai, Y., Lecain, D., Blumenthal, D., Morgan, J., 2018. Seed traits and germination of native grasses and invasive forbs are largely insensitive to parental temperature and CO_2 concentration. Seed Sci. Res. 28, 303–311.

Luken, J.O., Goessling, N., 1995. Seedling distribution and potential persistence of the exotic shrub *Lonicera maackii* in fragmented forests. Am. Midl. Nat. 133, 124–130.

Luken, J.O., Mattimiro, D.T., 1991. Habitat-specific resilience of the invasive shrub Amur honeysuckle (*Lonicera maackii*) during repeated clipping. Ecol. Appl 1, 104–109.

Luo, J., Cardina, J., 2012. Germination patterns and implications for invasiveness in three *Taraxacum* (Asteraceae) species. Weed Res. 52, 112–121.

Marchante, H., Freitas, H., Hoffmann, J.H., 2011. The potential role of seed banks in the recovery of dune ecosystems after removal of invasive plant species. Appl. Veg. Sci. 14, 107–119.

Marques, A.R., Lima, L.L., Garcia, Q.S., Atman, A.P., 2020. A novel cellular automata approach: seed input/output of the alien species *Leucaena leucocephala* in the soil and the effects of climate changes. Plant Ecol. 221, 141–154.

Merow, C., Bois, S.T., Allen, J.M., Xie, Y., Silander Jr., J.A., 2017. Climate change both facilitates and inhibits invasive plant ranges in New England. Proc. Natl. Acad. Sci. USA 114, E3276–E3284.

Moles, A.T., Gruber, M.A., Bonser, S.P., 2008. A new framework for predicting invasive plant species. J. Ecol. 96, 13–17.

Moles, A.T., Westoby, M., 2004. Seedling survival and seed size: a synthesis of the literature. J. Ecol. 92, 372–383.

Molina-Montenegro, M.A., Naya, D.E., 2012. Latitudinal patterns in phenotypic plasticity and fitness-related traits: assessing the climatic variability hypothesis (CVH) with an invasive plant species. PLoS One 7, e47620. Available from: https://doi.org/10.1371/journal.pone.0047620.

Moravcová, L., Pyšek, P., Krinke, L., Pergl, J., Perglová, I., Thompson, K., 2007. Seed germination, dispersal and seed bank in *Heracleum mantegazzianum*. In: Pyšek, P., Cock, M.J.W., Nentwig, W., Ravn, H.P. (Eds.), Ecology and Management of Giant Hogweed (*Heracleum mantegazzianum*). CAB International, Wallingford, pp. 74–91.

Moravcová, L., Pyšek, P., Pergl, J., Perglová, I., Jarošík, V., 2006. Seasonal pattern of germination and seed longevity in the invasive species *Heracleum mantegazzianum*. Preslia 78, 287–301.

Morris, L.L., Walck, J.L., Hidayati, S.N., 2002. Growth and reproduction of the invasive *Ligustrum sinense* and native *Forestiera ligustrina* (Oleaceae): implications for the invasion and persistence of a nonnative shrub. Int. J. Plant Sci. 163, 1001–1010.

Muller-Landau, H.C., 2010. The tolerance–fecundity trade-off and the maintenance of diversity in seed size. Proc. Natl. Acad. Sci. USA 107, 4242–4247.

Mumladze, L., Asanidze, Z., Walther, F., Hausdorf, B., 2017. Beyond elevation: testing the climatic variability hypothesis vs. Rapoport's rule in vascular plant and snail species in the Caucasus. Biol. J. Linn. Soc. 121, 753–763.

Myers, J.A., Vellend, M., Gardescu, S., Marks, P.L., 2004. Seed dispersal by white-tailed deer: implications for long-distance dispersal, invasion, and migration of plants in eastern North America. Oecologia 139, 35–44.

Naylor, J.M., 1983. Studies on the genetic control of some physiological processes in seeds. Can. J. Bot. 61, 3561–3567.

Neilson, R.P., Pitelka, L.F., Solomon, A.M., Nathan, R., Midgley, G.F., Fragoso, J.M., et al., 2005. Forecasting regional to global plant migration in response to climate change. BioScience 55, 749–759.

Nguyen, T., Bajwa, A.A., Navie, S., O'Donnell, C., Adkins, S., 2017. Parthenium weed (*Parthenium hysterophorus* L.) and climate change: the effect of CO_2 concentration, temperature, and water deficit on growth and reproduction of two biotypes. Environ. Sci. Pollut. Res. 24, 10727–10739.

Nicotra, A.B., Atkin, O.K., Bonser, S.P., Davidson, A.M., Finnegan, E.J., Mathesius, U., et al., 2010. Plant phenotypic plasticity in a changing climate. Trends Plant Sci. 15, 684–692.

Nogués-Bravo, D., Rodríguez-Sánchez, F., Orsini, L., De Boer, E., Jansson, R., Morlon, H., et al., 2018. Cracking the code of biodiversity responses to past climate change. Trends Ecol. Evol. 33, 765–776.

Ooi, M.K.J., 2012. Seed bank persistence and climate change. Seed Sci. Res. 22, S53–S60.

Ooi, M.K., Denham, A.J., Santana, V.M., Auld, T.D., 2014. Temperature thresholds of physically dormant seeds and plant functional response to fire: variation among species and relative impact of climate change. Ecol. Evol. 4, 656–671.

Ozaslan, C., Farooq, S., Onen, H., Ozcan, S., Bukun, B., Gunal, H., 2017. Germination biology of two invasive *Physalis* species and implications for their management in arid and semi-arid regions. Sci. Rep. 7, 16960. Available from: https://doi.org/10.1038/s41598-017-17169-5.

Pagad, S., Genovesi, P., Carnevali, L., Schigel, D., McGeoch, M.A., 2018. Introducing the global register of introduced and invasive species. Sci. Data 5, 1–12.

Panda, R.M., Behera, M.D., Roy, P.S., 2018. Assessing distributions of two invasive species of contrasting habits in future climate. J. Environ. Manage. 213, 478–488.

Parmesan, C., Yohe, G., 2003. A globally coherent fingerprint of climate change impacts across natural systems. Nature 421, 37–42.

Passos, I., Marchante, H., Pinho, R., Marchante, E., 2017. What we don't seed: the role of long-lived seed banks as hidden legacies of invasive plants. Plant Ecol. 218, 1313–1324.

Perrins, J., Williamson, M., Fitter, A., 1992. A survey of differing views of weed classification: implications for regulation of introductions. Biol. Conserv. 60, 47–56.

Peters, N.C.B., 1982. Production and dormancy of wild oat (*Avena fatua*) seed from plants grown under soil waterstress. Ann. Appl. Biol. 100, 189–196.

Petit, R.J., Hu, F.S., Dick, C.W., 2008. Forests of the past: a window to future changes. Science 320, 1450–1452.

Phartyal, S.S., Godefroid, S., Koedam, N., 2009. Seed development and germination ecophysiology of the invasive tree *Prunus serotina* (Rosaceae) in a temperate forest in western Europe. Plant Ecol. 204, 285–294.

Phartyal, S.S., Kondo, T., Baskin, C.C., Baskin, J.M., 2012. Seed dormancy and germination in the giant Himalayan lily (*Cardiocrinum giganteum* var. *giganteum*): an assessment of its potential for naturalization in northern Japan. Ecol. Res. 27, 677–690.

Pikaard, C.S., Mittelsten Scheid, O., 2014. Epigenetic regulation in plants. CSH Perspect. Biol. 6, a019315. Available from: https://doi.org/10.1101/cshperspect.a019315.

Primack, R.B., Miller-Rushing, A.J., Dharaneeswaran, K., 2009. Changes in the flora of Thoreau's Concord. Biol. Conserv. 142, 500–508.

Pyšek, P., Hulme, P.E., Simberloff, D., Bacher, S., Blackburn, T.M., Carlton, J.T., et al., 2020. Scientists' warning on invasive alien species. Biol. Rev. 95, 1511–1534.

Pérez, J.E., Nirchio, M., Alfonsi, C., Muñoz, C., 2006. The biology of invasions: the genetic adaptation paradox. Biol. Inv. 8, 1115–1121.

Radford, I.J., Cousens, R.D., 2000. Invasiveness and comparative life-history traits of exotic and indigenous *Senecio* species in Australia. Oecologia 125, 531–542.

Ramesh, K., Matloob, A., Aslam, F., Florentine, S.K., Chauhan, B.S., 2017. Weeds in a changing climate: vulnerabilities, consequences, and implications for future weed management. Front. Plant Sci. 8, 95. Available from: https://doi.org/10.3389/fpls.2017.00095.

Razanajatovo, M., Maurel, N., Dawson, W., Essl, F., Kreft, H., Pergl, J., et al., 2016. Plants capable of selfing are more likely to become naturalized. Nat. Commun. 7, 13313. Available from: https://doi.org/10.1038/ncomms13313.

Rebbeck, J., Jolliff, J., 2018. How long do seeds of the invasive tree, *Ailanthus altissima* remain viable? For. Ecol. Manage. 429, 175–179.

Redwood, M.E., Matlack, G.R., Huebner, C.D., 2018. Seed longevity and dormancy state suggest management strategies for garlic mustard (*Alliaria petiolata*) and Japanese stiltgrass (*Microstegium vimineum*) in deciduous forest sites. Weed Sci. 66, 190–198.

Redwood, M.E., Matlack, G.R., Huebner, C.D., 2019. Seed longevity and dormancy state in an invasive tree species: *Ailanthus altissima* (Simaroubaceae). J. Torrey Bot. Soc. 146, 79–86.

Reeves, J.L., Blumenthal, D.M., Kray, J.A., Derner, J.D., 2015. Increased seed consumption by biological control weevil tempers positive CO_2 effect on invasive plant (*Centaurea diffusa*) fitness. Biol. Cont. 84, 36–43.

Renne, I.J., Spira, T.P., Bridges Jr., W.C., 2001. Effects of habitat, burial, age and passage through birds on germination and establishment of Chinese tallow tree in coastal South Carolina. J. Torrey Bot. Soc. 128, 109–119.

Richardson, D.M., Pyšek, P., Rejmánek, M., Barbour, M.G., Panetta, F.D., West, C.J., 2000. Naturalization and invasion of alien plants: concepts and definitions. Divers. Distrib. 6, 93–107.

Richards, C.L., Schrey, A.W., Pigliucci, M., 2012. Invasion of diverse habitats by few Japanese knotweed genotypes is correlated with epigenetic differentiation. Ecol. Lett. 15, 1016–1025.

Riveiro, S.F., Cruz, Ó., Casal, M., Reyes, O., 2020. Fire and seed maturity drive the viability, dormancy, and germination of two invasive species: *Acacia longifolia* (Andrews) Willd. and *Acacia mearnsii* De Wild. Ann. For. Sci. 77, 60. Available from: https://doi.org/10.1007/s13595-020-00965-x.

Ruprecht, E., Fenesi, A., Nijs, I., 2014. Are plasticity in functional traits and constancy in performance traits linked with invasiveness? An experimental test comparing invasive and naturalized plant species. Biol. Inv. 16, 1359–1372.

Sans, F.X., Garcia-Serrano, H., Afán, I., 2004. Life-history traits of alien and native *Senecio* species in the Mediterranean region. Acta Oecol. 26, 167–178.

Sawhney, R., Naylor, J.M., 1979. Dormancy studies in seed of *Avena fatua*. 9. Demonstration of genetic variability affecting the response to temperature during seed development. Can. J. Bot. 57, 59–63.

Seglias, A.E., Williams, E., Bilge, A., Kramer, A.T., 2018. Phylogeny and source climate impact seed dormancy and germination of restoration-relevant forb species. PLoS One 13, e0191931. Available from: https://doi.org/10.1371/journal.pone.0191931.

Sershen, Mdamba, B., Ramdhani, S., 2017. Propagule and seedling responses of three species naturalised in subtropical South Africa to elevated temperatures. Flora 229, 80–91.

Sharif-Zadeh, F., Murdoch, A.J., 2000. The effects of different maturation conditions on seed dormancy and germination of *Cenchrus ciliaris*. Seed Sci. Res. 10, 447–457.

Skálová, H., Moravcová, L., Pyšek, P., 2011. Germination dynamics and seedling frost resistance of invasive and native *Impatiens* species reflect local climatic conditions. Persp. Plant Ecol. Evol. Syst. 13, 173–180.

Snider, J.L., Oosterhuis, D.M., 2011. How does timing, duration, and severity of heat stress influence pollen-pistil interactions in angiosperms? Plant Signal. Behav. 6, 930–933.

Sorte, C.J.B., Ibáñez, I., Blumenthal, D.M., Molinari, N.A., Miller, L.P., Grosholz, E.D., et al., 2013. Poised to prosper? A cross-system comparison of climate change effects on native and non-native species performance. Ecol. Lett. 16, 261–270.

Sow, M.D., Allona, I., Ambroise, C., Conde, D., Fichot, R., Gribkova, S., et al., 2018. Epigenetics in forest trees: state of the art and potential implications for breeding and management in a context of climate change. Adv. Bot. Res. 88, 387–453.

Teller, B.J., Zhang, R., Shea, K., 2016. Seed release in a changing climate: initiation of movement increases spread of an invasive species under simulated climate warming. Divers. Distrib. 22, 708–716.

Thiebaut, F., Hemerly, A.S., Ferreira, P.C.G., 2019. A role for epigenetic regulation in the adaptation and stress responses of non-model plants. Front. Plant Sci. 10, 246. Available from: https://doi.org/10.3389/fpls.2019.00246.

Thomson, L.J., Macfadyen, S., Hoffmann, A.A., 2010. Predicting the effects of climate change on natural enemies of agricultural pests. Biol. Cont. 52, 296–306.

Tinoco-Ojanguren, C., Reyes-Ortega, I., Sánchez-Coronado, M.E., Molina-Freaner, F., Orozco-Segovia, A., 2016. Germination of an invasive *Cenchrus ciliaris* L. (buffel grass) population of the Sonoran Desert under various environmental conditions. S. Afr. J. Bot. 104, 112–117.

Tozer, M.G., Ooi, M.K.J., 2014. Humidity-regulated dormancy onset in the Fabaceae: a conceptual model and its ecological implications for the Australian wattle *Acacia saligna*. Ann. Bot. 114, 579–590.

Udo, N., Tarayre, M., Atlan, A., 2017. Evolution of germination strategy in the invasive species *Ulex europaeus*. J. Plant Ecol. 10, 375–385.

van Boheemen, L.A., Atwater, D.Z., Hodgins, K.A., 2019. Rapid and repeated local adaptation to climate in an invasive plant. N. Phytol. 222, 614–627.

van Clef, M., Stiles, E.W., 2001. Seed longevity in three pairs of native and non-native congeners: assessing invasive potential. Northe. Nat. 8, 301–310.

van Kleunen, M., Dawson, W., Essl, F., Pergl, J., Winter, M., Weber, et al., 2015. Global exchange and accumulation of non-native plants. Nature 525, 100–103.

van Kleunen, M., Pysek, P., Dawson, W., Essl, F., Kreft, H., Pergl, J., et al., 2019. The global naturalized alien Flora (GloNAF) database. Ecology 100, e02542. Available from: https://doi.org/10.1002/ecy.2542.

van Kleunen, M., Weber, E., Fischer, M., 2010. A *meta*-analysis of trait differences between invasive and non-invasive plant species. Ecol. Lett. 13, 235–245.

van Klinken, R.D., Goulier, J.B., 2013. Habitat-specific seed dormancy-release mechanisms in four legume species. Seed Sci. Res. 23, 181–188.

Venable, D.L., Brown, J.S., 1988. The selective interactions of dispersal, dormancy, and seed size as adaptations for reducing risk in variable environments. Am. Nat. 131, 360–384.

Vetter, V.M., Kreyling, J., Dengler, J., Apostolova, I., Arfin-Khan, M.A., Berauer, B.J., et al., 2020. Invader presence disrupts the stabilizing effect of species richness in plant community recovery after drought. Glob. Change Biol. 26, 3539–3551.

Vilà, M., D'Antonio, C.M., 1998. Fruit choice and seed dispersal of invasive vs. noninvasive *Carpobrotus* (Aizoaceae) in coastal California. Ecology 79, 1053–1060.

Vitt, P., Havens, K., Kramer, A.T., Sollenberger, D., Yates, E., 2010. Assisted migration of plants: changes in latitudes, changes in attitudes. Biol. Conserv. 143, 18–27.

Volis, S., 2007. Correlated patterns of variation in phenology and seed production in populations of two annual grasses along an aridity gradient. Evol. Ecol. 21, 381–393.

Volis, S., Bohrer, G., 2013. Joint evolution of seed traits along an aridity gradient: seed size and dormancy are not two substitutable evolutionary traits in temporally heterogeneous environment. New Phytol. 197, 655–667.

Wainwright, C.E., Cleland, E.E., 2013. Exotic species display greater germination plasticity and higher germination rates than native species across multiple cues. Biol. Inv. 15, 2253–2264.

Walck, J.L., Hidayati, S.N., Dixon, K.W., Thompson, K., Poschlod, P., 2011. Climate change and plant regeneration from seed. Glob. Change Biol. 17, 2145–2161.

Wallingford, P.D., Morelli, T.L., Allen, J.M., Beaury, E.M., Blumenthal, D.M., Bradley, B.A., et al., 2020. Adjusting the lens of invasion biology to focus on the impacts of climate-driven range shifts. Nat. Clim. Change 10, 398–405.

Wen, B., 2015. Effects of high temperature and water stress on seed germination of the invasive species Mexican sunflower. PLoS One 10, e0141567. Available from: https://doi.org/10.1371/journal.pone.0141567.

Wen, B., 2019. Seed germination ecology of Alexandra palm (*Archontophoenix alexandrae*) and its implication on invasiveness. Sci. Rep. 9, 4057. Available from: https://doi.org/10.1038/s41598-019-40733-0.

Wen, B., Xue, P., Zhang, N., Yan, Q., Ji, M., 2015. Seed germination of the invasive species *Piper aduncum* as affected by high temperature and water stress. Weed Res. 55, 155–162.

Whitney, K.D., Gabler, C.A., 2008. Rapid evolution in introduced species, 'invasive traits' and recipient communities: challenges for predicting invasive potential. Divers. Distrib. 14, 569–580.

Wiesner, L.E., Grabe, D.F., 1972. Effect of temperature preconditioning and cultivar on ryegrass (*Lolium* sp.) seed dormancy 1. Crop Sci. 12, 760–764.

Williams, S.C., Ward, J.S., 2006. Exotic seed dispersal by white-tailed deer in southern Connecticut. Nat. Area J. 26, 383–390.

Willis, C.G., Ruhfel, B.R., Primack, R.B., Miller-Rushing, A.J., Losos, J.B., Davis, C.C., 2010. Favorable climate change response explains non-native species' success in Thoreau's woods. PLoS One 5, e8878. Available from: https://doi.org/10.1371/journal.pone.0008878.

Xu, X., Wolfe, L., Diez, J., Zheng, Y., Guo, H., Hu, S., 2019. Differential germination strategies of native and introduced populations of the invasive species *Plantago virginica*. NeoBiota 43, 101–118.

Yuan, X., Wen, B., 2018. Seed germination response to high temperature and water stress in three invasive Asteraceae weeds from Xishuangbanna, SW China. PLoS One 13, e0191710. Available from: https://doi.org/10.1371/journal.pone.0191710.

Zalapa, J.E., Brunet, J., Guries, R.P., 2010. The extent of hybridization and its impact on the genetic diversity and population structure of an invasive tree, *Ulmus pumila* (Ulmaceae). Evol. Appl. 3, 157–168.

Zhang, R., Gallagher, R.S., Shea, K., 2012. Maternal warming affects early life stages of an invasive thistle. Plant Biol. 14, 783–788.

Ziska, L.H., 2003. Evaluation of the growth response of six invasive species to past, present and future atmospheric carbon dioxide. J. Exp. Bot. 54, 395–404.

Ziska, L.H., Blumenthal, D.M., Runion, G.B., Hunt, E.R., Diaz-Soltero, H., 2011. Invasive species and climate change: an agronomic perspective. Clim. Change 105, 13–42.

Chapter 19

Regeneration in recalcitrant-seeded species and risks from climate change

Hugh W. Pritchard[1,2], Sershen[3,4], Fui Ying Tsan[5], Bin Wen[6], Ganesh K. Jaganathan[7], Geângelo Calvi[8], Valerie C. Pence[9], Efisio Mattana[1], Isolde D.K. Ferraz[8] and Charlotte E. Seal[1]

[1]Royal Botanic Gardens, Kew, Wakehurst, West Sussex, United Kingdom, [2]Chinese Academy of Sciences, Kunming Institute of Botany, Kunming, Yunnan, P.R. China, [3]Institute of Natural Resources, Pietermaritzburg, South Africa, [4]University of the Western Cape, Cape Town, South Africa, [5]Universiti Teknologi MARA, Shah Alam, Selangor, Malaysia, [6]Xishuangbanna Tropical Botanical Garden, Chinese Academy of Sciences, Mengla, Yunnan, P.R. China, [7]Institute of Biothermal Science and Technology, University of Shanghai for Science and Technology, Shanghai, P.R. China, [8]Instituto Nacional de Pesquisas da Amazônia (INPA), Coordenação de Biodiversidade, Manaus, Amazonas, Brazil, [9]Center for Conservation and Research of Endangered Wildlife, Cincinnati Zoo and Botanical Garden, Cincinnati, OH, United States

Introduction

Conservationists helping to protect the world's plant diversity ex situ exploit a key seed trait, which is the ability to tolerate drying (Li and Pritchard, 2009). Conventional seed bank procedures include the equilibration of seeds to around 15% RH at about 15°C–20°C, followed by packaging in sealed containers to maintain low moisture content (MC) during subsequent storage at approximately −20°C (FAO, 2014). However, not all plants produce such orthodox seeds (Berjak et al., 1989; Black and Pritchard, 2002). Instead, so-called recalcitrant seeds are sensitive to drying below a high MC, often around 40%, and the seeds become completely nonviable when dried below about 25% MC, at which point there is no bulk cytoplasmic water present (Finch-Savage, 1992b; Pritchard, 2004; Walters et al., 2013). Since this physiological response precludes their ex situ seed banking, understanding the capacity of recalcitrant seeds for in situ survival and regeneration is important, particularly in relation to projected risks from a changing climate.

Recalcitrant-seeded species are estimated to constitute 8%–20% of the world's flora (Wyse and Dickie, 2018). Thus, potentially up to 70,000 of the c.350,000 seed plant species may produce recalcitrant seeds. There is considerable interest in the evolution of the seed desiccation tolerance trait with regard to whether it is derived or ancestral. Systematic analyses, mainly based on seed structural features regarded as primitive, have led to the proposal that desiccation sensitivity is the ancestral state. Accepting the difficulty in disentangling various co-correlated traits, such as life cycle, plant stress tolerance, and seed mass, it is thought that seed desiccation tolerance is at least as likely to be ancestral as is desiccation sensitivity (Dickie and Pritchard, 2002). The most parsimonious explanation for the wide distribution of recalcitrant seeds is by convergent loss of tolerance from tolerant ancestors in relation to one or more ecological trade-offs (Farnsworth, 2000; Dickie and Pritchard, 2002). In support of this perspective, an analysis of the evolutionary history of seed recalcitrance based on 721 species, 297 genera, and 84 families concluded that the trait was apomorphic (a specialized trait not present in an ancestral form) in numerous extant lineages, with evidence of a radiation of recalcitrant taxa from the Miocene (approximately 23–5 Ma) onwards (Subbiah et al., 2019).

According to recent estimates, 33% of the tree species worldwide are likely to produce desiccation-sensitive (recalcitrant) seeds (Wyse and Dickie, 2018), that is, about 20,000 + species. The highest percentage of such species is recorded in tropical and subtropical habitats (Tweddle et al., 2003; Daws et al., 2005, 2006; Pritchard et al., 2014; Wyse and Dickie, 2017). An analysis of 79 species from 30 families from two biodiversity hotspots in Brazil (Atlantic Forest and Cerrado) estimated that c. 47% of species produced orthodox seeds and c. 33% recalcitrant seeds, including all Lauraceae; the remaining c. 19% of the species produced seeds with intermediate storage behavior (Mayrinck et al., 2019). Of 101 woody species from the seasonal tropical rainforest of southern China, ∼50% are likely to have desiccation-sensitive seeds (Lan et al., 2014). The eastern Australian rainforest species are estimated to have about 49% nonorthodox-seeded species, based on an analysis of seeds of 100 species (Hamilton et al., 2013).

For 40 species of Caribbean woody species assessed, the mean mass of desiccation-intolerant seeds was much larger than that of desiccation-tolerant seeds (6.1 g vs 88 mg) (Mattana et al., 2020). Collated information on seed storage behavior of 67 tree species native to the Amazon rainforest of Brazil revealed that 38 (57%), 23 (34%), and 6 (9%) had orthodox, recalcitrant, and intermediate responses, respectively (Lima et al., 2014). Similarly, of the 20 Amazon liana species screened, the eight species with the largest seeds were desiccation sensitive (Roeder et al., 2013). Thus, in various biogeographical regions the tropical woody forest flora with larger seeds tends to have recalcitrant seeds.

Due to emission of greenhouse gasses, temperature is increasing in all vegetation zones on earth and is predicted to continue to increase through at least the 21st century. Furthermore, changes are predicted in amounts and timing of precipitation, resulting in increased drought and flooding, depending on location (IPCC, 2018). In particular, tropical wet forests are expected to receive increased intensity and duration of heatwaves and increased frequency of droughts (see Chapter 12, this book). Tropical forests also are under risk from microclimatic changes due to deforestation leading to fragmentation. Taking the Amazon as a case in point, species distribution models indicate that the combined effect of changes in climate (+2°C warming; RCP 8.5) and deforestation could cause a decrease of up to 58% in tree species richness by 2050, of which up to 37% could be due to climate change (Gomes et al., 2019). Ultimately, the Amazonian lowland rainforest may be cut into two blocks: one continuous block with 53% of the original forest area and a second one that is severely fragmented (Gomes et al., 2019). Based on studies near Manaus, Brazil, forest fragmentation is particularly stressful for big trees (>60 cm DBH), such that their mortality rate near fragment edges is nearly double that in the forest interior (Laurance et al., 2000). To this double jeopardy of climate change and forest fragmentation, we can add the risk of regeneration failure for tree species with desiccation-sensitive seeds, as the local environment becomes relatively warmer and drier. An associated change in solar irradiance might also impact germination success.

The relative light requirement for seed germination is highly positively correlated with seed mass, and thus many large recalcitrant seeds are light insensitive and germinate well in both light and dark. Desiccation-sensitive seeds of species in a tropical seasonal moist forest in Panama were significantly more likely to germinate quickly in light levels typical of intact forest understory (2.3% full sunlight) than in those of medium size gaps (18.5% full sunlight) (Daws et al., 2005). Recalcitrant seeds of *Euterpe edulis*, a key species of the Atlantic forest of Brazil, germinated optimally at 25°C in the dark, with little difference when light (78 μmol m^{-2} s^{-1} PAR) was applied continuously (Roberto and Habermann, 2010). Similarly, *Euterpe precatoria* seeds had the highest germination at 20°C in both darkness and light (Costa et al., 2018). Recalcitrant seeds of *Shorea wantianshuea* had a slightly higher germination percentage in the dark compared to continuous photon irradiance (PAR) of up to 168 μmol m^{-2} s^{-1} (Yan and Cao, 2006). There is evidence of a difference in light requirements for seed germination between species in Fagaceae (Pritchard and Manger, 1990; Finch-Savage and Clay, 1994). In both *Quercus robur* and *Castanea sativa*, germination was the same when seeds were exposed to nominal darkness (i.e., photon fluence rate of 2 W m^{-2} for 1 min every 3–4 d) or safe green light conditions (Pritchard and Manger, 1990). However, germination of *Q. robur* at 21°C exhibited the high irradiance reaction at photon doses above 30 mmol m^{-2} d^{-1} via inhibition of radicle extension. Germination was reduced further as photon dose increased, and photoinhibition was exacerbated at supraoptimal temperatures. In contrast, germination of *C. sativa* seeds at 26°C was little influenced by a photon dose of 752 mmol m^{-2} d^{-1} compared to darkness (Pritchard and Manger, 1990). It appears then that germination of recalcitrant seeds can be negatively affected by light, particularly when combined with elevated temperatures, as would be the case with continuing climate change and forest fragmentation. An illustration of the species-specific nature of light sensitivity during germination amongst desiccation-intolerant seeds is given in Case Study 1.

Case study 1: Regeneration and light

The ecology of seed germination of nonpioneer tree species can vary considerably in relation to niche and competition. Time of dispersal, seed size, relative desiccation tolerance, and optimal germination conditions differed amongst eight species (*Antiaris toxicaria, Baccaurea ramiflora, Castanopsis hystrix, Horsfieldia tetratepala, Horsfieldia pandurifolia, Litsea dilleniifolia, Litsea pierrei* var *szemaois,* and *Pometia tomentosa*) from the tropical seasonal rain forest of Xishuangbanna, SW China (Yu et al., 2008). After five days of desiccation in the presence of silica gel, seeds of seven species had lost the ability to germinate, but 30% of *Castanopsis hystrix* seeds germinated. The eight species exhibited a wide range of desiccation sensitivities, with 50% loss of seed viability estimated to be between 21% and 43% MC for seven of the species and much lower for *Horsfieldia pandurifolia* (12% MC). Compared to the other species, seeds of

A. toxicaria, *C. hystrix*, and *L. pierrei* were the slowest to lose water, perhaps indicating an adaptation to delay drying stress if dispersed into more open sites.

Most seeds of all eight species germinated in <30 days at constant 30°C in light. However, interspecies differences in light sensitivity for germination were noted under shade-house conditions. Light levels were controlled to 3.5%, 10%, and 30% of full sunlight, simulating solar irradiance in the forest understory, at the gap edge, and in the center of a gap, respectively. Four species did not differ in final germination percentages at the three irradiances. However, germination percentages of *Horsfieldia pandurifolia* and *Litsea pierrei* var. *szemaois* were significantly lower in 30% than in 3.5% or 10% light. In contrast, seeds of *Antiaris toxicaria* and *Castanopsis hystrix* germinated better in 30% and 10% than in 3.5% light. Therefore, when considering the germination response of recalcitrant seeds to climate change scenarios it is important to explore the potential sensitivity to light conditions, including a possible negative interaction with elevated temperature.

Evidence that recalcitrant seeds are most likely to be found in environments with very predictable seasons and abundant and constant water regimes suggests low selective pressure for seed desiccation tolerance (Marques et al., 2018; Wyse and Dickie, 2018). Such loss of function is often co-correlated with the diminution of other traits associated with "persistence" such as longevity and dormancy (and the possible acquisition of the capacity for fast germination). Consequently, any consideration of the potential impact of climate change on the regeneration of recalcitrant-seeded species demands an analysis of multiple traits and how they relate to the natural environment during seed development and postdispersal. Thus, in this review we consider seed development and dormancy, temperature requirements for germination, stopping germination during seed storage, and seed and seedling banks in relation to climate change. Such insights have potential practical value for the design of efficient propagation programs for habitat restoration.

Seed development

In orthodox seeds, desiccation tolerance is acquired during the three phases of development on the parent plant (Berjak et al., 1989; Vertucci and Farrant, 1995; Pammenter and Berjak, 1999), viz., histodifferentiation, reserve deposition, and maturation drying (Kermode and Finch-Savage, 2002; Bewley et al., 2013;). During the first phase, undifferentiated cells divide and develop into function-specific tissues. The second phase is when reserves in various forms are accumulated in the endosperm, or ultimately in the cotyledons, providing nutrients that sustain seedling development (Bewley et al., 2013). Dry and fresh mass increase during the first two phases, and while the duration of dry matter accumulation varies across species it always terminates with the vascular connection between the parent plant and seed being severed at physiological maturity (Kermode and Finch-Savage, 2002). This stage coincides with a metabolic shutdown accompanied by the start of water loss (and hence reduction in fresh mass) as part of the maturation drying (third) phase (Vertucci and Farrant, 1995; Kermode and Finch-Savage, 2002). Desiccation tolerance is acquired prior to maturation drying, and seeds are shed from the parent plant, usually upon completion of all three phases, and remain quiescent until water becomes available for germination (Farrant et al., 1993; Vertucci and Farrant, 1995; Kermode and Finch-Savage, 2002). Metabolic events associated with germination are not triggered in the dry state, and initially nondormant orthodox seeds (or those in which primary dormancy has been broken) will germinate only upon imbibition, if environmental conditions are favorable for them to do so.

Recalcitrant seeds, per contra, do not undergo maturation drying and are shed wet, often about 1 g H_2O g^{-1} DW or c. 50% moisture FW (Chin and Roberts, 1980; Tompsett and Pritchard, 1993; Berjak and Pammenter, 2004b) and are metabolically active (Berjak et al., 1989; Farnsworth, 2000; Kermode and Finch-Savage, 2002). The general pattern of development following initial histodifferentiation is similar across recalcitrant-seeded species and is akin to that of orthodox seeds before they attain maximum dry mass (Finch-Savage, 1992a; Farrant et al., 1997). Thus, the decrease in MC during the early stages of development is similar in orthodox and recalcitrant seeds such as *Aesculus hippocastanum* (Tompsett and Pritchard, 1993), *Coffea canephora* (Hong and Ellis, 1995), and *Quercus serrata* (Xia and Zhou, 2021). Most recalcitrant seeds appear to be least desiccation sensitive just before natural shedding, but this is only relative (Tompsett and Pritchard, 1993, Pammenter and Berjak, 1999; Daws et al., 2004b; Xia and Zhou, 2021). Co-correlated with the increase in desiccation tolerance during development in *Aesculus hippocastanum* seeds sampled across Europe was the ability to germinate over a wide range of temperatures (Daws et al., 2004b). These changes in the patterns of physiological traits in this species are related to developmental heat sum and provide a quantitative explanation for intraspecific variability in recalcitrant seed traits. Similarly, desiccation sensitivity of *Inga vera* embryos varies with environmental source conditions (Lamarca and Barbedo, 2015), and recalcitrant seeds are thought to have indeterminate development (Finch-Savage and Blake, 1994).

The high responsiveness of recalcitrant-seed developmental processes to the maternal environment makes this phase in the life cycle potentially extremely sensitive to climate change. Case Study 2 explores these potential risks within species of *Coffea*, which are distributed along a tropical environmental gradient and amongst varying forest types. In brief, the length of the rainy season corresponds to the period of seed development, and the length of the dry season postdispersal correlates with interspecies differences in desiccation tolerance (Dussert et al., 2000).

Case study 2: Regeneration in *Coffea* species

Within certain woody/shrubby genera, such as *Citrus* and *Coffea*, different degrees of seed desiccation sensitivity exist between species. Within nine African coffee species (*Coffea arabica, C. brevipes, C. canephora, C. eugenioides, C. humilis, C. liberica, C. pocsii, C. pseudozanguebariae*, and *C. stenophylla*), flowering occurs soon after the beginning of the main period of rainfall, marking the end of the dry season (Dussert et al., 2000). The species differ considerably in forest type (highland, rainforest, semideciduous, gallery, patches, and coastal dry), altitude (57−2120 m a.s.l.), and annual rainfall (739−2412 mm) where they occur. Simulations based on extensive rainfall patterns revealed a highly significant correlation between the length of seed development and the rainy season, from 2 (e.g., *C. pocsii*) to just over 10 (e.g., *C. liberica*) months.

Counterintuitively, mature seeds are dispersed at the beginning of dry season, and, intriguingly, the mean number of dry months after seed shedding is significantly correlated with the level of seed desiccation sensitivity. For example, *C. pseudozanguebariae* has a mid-point for viability loss on drying to 5.3% MC (FW basis) and the longest dry season of around 5 months. In contrast, *C. humilis* has comparable values of 27.6% MC and c. 2 months. Seeds of different *Coffea* species span the continuum of recalcitrant to orthodox desiccation tolerance. Moreover, an increased level of tolerance to seed drying amongst the species corresponds to an adaptation to drought.

The authors note that the simulations performed did not account for conditions in the natural environment that could impact the desiccation of the seeds, such as soil characteristics, temperature, and light conditions. This is important, since general habitat conditions such as mean annual rainfall can be a poor indicator of the level of seed desiccation sensitivity. Nonetheless, it seems likely that in coffee species any rapid change in the duration of the dry season as a result of climate change could have profound implications for both seed development on plants and cumulative desiccation stress following dispersal. Importantly, any reduction in the natural regeneration potential in these species will add significantly to the known extinction risk. Of 124 wild coffee species, at least 60% are threatened with extinction, and existing conservation measures are inadequate (Davis et al., 2019).

Is dormancy present in recalcitrant seeds?

After harvest or natural shedding, ongoing metabolism facilitates the completion of germination of recalcitrant seeds, and there is a progressive increase in desiccation sensitivity, for example, *Landolphia kirkii* (Berjak et al., 1992) and *Camellia sinensis* (Berjak et al., 1993). Germination may be achieved without the addition of water, and progression towards germination can be so rapid that some species are described as viviparous (King and Roberts, 1980; Farrant et al., 1988, 1989; Kioko et al., 1993, Sershen et al., 2008; Wen, 2009; Ismail et al., 2010). In *Avicennia marina*, such a response is similar to preharvest sprouting in nondormant orthodox seeds whilst attached to the parent plant. These studies suggest that recalcitrant seeds might transit seamlessly from the developmental to the germination stage of growth.

Whilst most recalcitrant seeds are nondormant, dormancy and recalcitrance are not necessarily mutually exclusive physiological traits. Using data available on 886 trees and shrubs, Tweddle et al. (2003) reported that 10% of the dormant seeds are desiccation sensitive, with physiological dormancy (PD) (sensu Baskin and Baskin, 2004) being the most common class, present in 14% of all the desiccation-sensitive species.

If freshly matured seeds are dormant and recalcitrant, there is a possibility that they will dry to a critically low MC and die before dormancy is broken. Thus, if the amount of precipitation in the habitat decreases or time of onset of the wet season is delayed due to climate change, recalcitrant seeds, especially dormant ones, may die. Despite their evolutionary and ecological significance, a combined assessment of dormancy and recalcitrance is relatively rare in comparative studies. In Lauraceae, such studies are complicated by the presence of a hard (tough) seed coat and its effect on germination, including after exposure to drying (Jaganathan et al., 2019). Similarly, seed/fruit morphology can affect the assessment of desiccation sensitivity in *Quercus* species since the total area of the pericarp scar affects the rate and route of water flow in and out of the acorn (Xia et al., 2012). Due to limited water flow through most of the pericarp, germination may be prevented during isolated showers, and the embryo axis (generally located opposite the scar) can maintain a high water content, thereby preventing mortality. A different morphological challenge exists in the

Arecaceae because of the presence in some species of a tough coat (endocarp) and a low embryo to endosperm ratio. Both traits impact germination. The evidence for coexistence of dormancy and desiccation sensitivity in some Arecaceae is compelling, with some kind of morphological or morphophysiological dormancy evident (Jaganathan, 2020) and 34 species known to be recalcitrant (Jaganathan, 2021).

Detailed studies on tree species with dormant, recalcitrant seeds are rare. In the case of *Quercus nigra* and *Quercus phellos* acorns, ≥18 weeks cold stratification is needed to break dormancy, suggesting the presence of deep PD (Hawkins, 2020). Possibly the most thoroughly investigated example of PD in a recalcitrant-seeded species is *Aesculus hippocastanum* (Pritchard et al., 1996; Tompsett and Pritchard, 1998). In this species, cold temperatures induce a systematic lowering of the minimum temperature for germination (Pritchard et al., 1999, Steadman and Pritchard, 2004). This also may be the case in *Quercus pagoda* acorns, since 20% of them germinated by 16 weeks during the cold stratification treatment, suggesting Type 2 nondeep PD (Hawkins, 2019). *Aesculus hippocastanum* seeds have the same kind of dormancy, which is under the control of abscisic acid (ABA; Obroucheva et al., 2016). A greater understanding of the role of plant growth regulators in recalcitrant seed germination may help to better elucidate the level(s) of PD in seeds. The balance between levels of plant growth regulators and the spatio-temporal expression of gene networks appears to be broadly similar in recalcitrant and orthodox seed germination (Barendse and Peeters, 1995). However, species differences can occur in the regulation of ABA levels (Pieruzzi et al., 2011), including recalcitrant-seeded species. For example, a comparative study of two recalcitrant-seeded species, *Araucaria angustifolia* (gymnosperm) and *Ocotea odorifera* (angiosperm), revealed that embryo IAA and polyamine levels increased in both species during germination. However, embryo ABA levels declined in *O. odorifera* and increased in *A. angustifolia*. ABA, which is produced by the maternal tissue, modulates embryo growth and development, plant and seed dormancy, and vivipary (Cheng et al., 2002; Frey et al., 2004; Kucera et al., 2005), suggesting the need for further studies in this area.

Interestingly, recalcitrant seeds of a few tropical rainforest species, all of which are trees, also have been shown to exhibit epicotyl dormancy, that is, the shoot emerges long after the radicle does. Examples include morphophysiological epicotyl dormancy in *Strychnos benthamii* of the Loganiaceae (Asanga et al., 2020) and physiological epicotyl dormancy in *Humboldtia laurifolia* of the Fabaceae (Jayasuriya et al., 2010). In two other Fabaceae species, *Cynometra cauliflora* and *Brownea coccinea*, with physiological epicotyl dormancy, shoot emergence lags behind root emergence by 1 and 2 months, respectively (Jayasuriya et al., 2012). Seeds of temperate *Quercus* species with epicotyl PD may require cold stratification for shoot emergence (Farmer, 1977), whereas tropical species require warm stratification for shoot emergence (Jayasuriya et al., 2010). The timing of germination has ecological significance. Rapid germination (root emergence) reduces the risk of high seed predation (Pritchard et al., 2004b), although the presence of toxins in seeds (a predation resistance trait) and the high mass of *Aesculus californica* seeds (a predation tolerance trait that allows heavily damaged seeds to germinate) seem to be an effective antipredation strategy (Mendoza and Dirzo, 2009). Delayed epicotyl emergence in *Quercus* species may mean the young leaves avoid exposure to freezing temperatures during winter.

If recalcitrant seeds are shed at a relatively immature stage, continued development is required before germination is initiated (Berjak et al., 1989). This is potentially a problem in endospermic seeds with a small embryo. In assessments of desiccation tolerance in palm seeds, it was noted that the "nondried" (i.e., MC as received) seeds held at 25°C as a control during the drying-treatment experiment (usually 1–2 months) attained a higher germination percentage after 1–2 months at 25°C than the fresh seeds, possibly indicating further maturation of the seed (Pritchard et al., 2004a; Wood et al., 2006). The embryo of palm seeds is underdeveloped, and morphoPD is common in the palm family (Baskin and Baskin, 2014). The presence of both desiccation sensitivity and dormancy in the palms makes it an interesting model family with which to further explore these physiological traits in an evolutionary and ecological context (Jaganathan, 2021). Also, palms are a good model with which to explore the effects of climate change on seed dormancy-break and germination and seedling survival.

Germination in relation to precipitation and temperature

The distribution of desiccation-sensitive species is highly concentrated in tropical wet forests, where water availability presumably is not limited (Pritchard et al., 2004b; Daws et al., 2005). In more seasonal environments, dispersal of recalcitrant seeds may (Pritchard et al., 2004b) or may not (Hill and Edwards, 2010) coincide tightly with the period of highest rainfall, depending on the extent of seasonality and total amount of rainfall. In a comparison of seed traits of 32 species with desiccation-sensitive and 183 species with desiccation-tolerant seeds from a tropical semideciduous forest in Panamá, desiccation-sensitive seeds were typically shed during the wet (as opposed to the dry) season (Daws et al., 2005). An exception is the recalcitrant seeds of *Virola sebifera*, which are dispersed at the beginning of the dry season.

Similarly, seeds of eight nonpioneer tree species in a tropical seasonal rain forest in southwest China, with varying levels of desiccation sensitivity, were dispersed from March to November, that is, from the late dry—early rainy season to the late rainy—early dry season (Yu et al., 2008).

In a drying environment, germination success of recalcitrant seeds depends on various factors. Maintaining a high seed internal water potential is important, and even exposure to a water stress as mild as −0.5 MPa greatly inhibited germination of *Baccaurea ramiflora* seeds (Wen et al., 2016). The rate of water loss is reduced by a low surface area: internal volume ratio of large seeds. Morphological adaptations that limit water flow can be important, such as the area of the pericarp scar in *Quercus* (Xia et al., 2012). Also, some level of desiccation resistance, as seen in three nonpioneer species of SW China seasonal rain forest, *Litsea pierrei, Castanopsis hystrix*, and *Antiaris toxicaria* (Yu et al., 2008), can help ensure germination success. Alternatively, seed MC might be buffered by the surrounding fruit tissues, as in *Swartzia langsdorffii* (Vaz et al., 2016, 2018). The seed internal morphology also may facilitate retention of water by the critical location of the embryonic axis, which in recalcitrant seeds of *Carapa surinamensis* and *Carapa guianensis* is between the fused cotyledons, with the cotyledons potentially acting as a water reserve (Amoedo and Ferraz, 2017). The potential for redistribution of water from the cotyledons to the embryonic axis during post-drying equilibration is known in *Aesculus hippocastanum* (Tompsett and Pritchard, 1998).

Whatever the adaptation to withstand environmental drying after seed dispersal, increased temperatures due to climate change likely will increase moisture loss. Furthermore, increased temperatures also might increase the level of desiccation sensitivity, as they did in *Hopea millosan* seeds dried between 10°C and 40°C (Wen, 2011). Overall, this range of adaptations might mean that even in seasonally dry tropical forest there might be enough water to enable recalcitrant seeds to complete the germination process. However, as explored in Case Study 3 the situation might be different in tropical ecosystems with a long dry season.

Case study 3: Regeneration timing in relation to precipitation in seasonally dry tropical vegetation

Successful regeneration of recalcitrant-seeded species requires the alignment of numerous phenological events. Preferably, the completion of seed development and release of propagules coincides with sufficient rainfall to sustain rapid germination and seedling establishment. Where such an ecological strategy might be most precarious is in seasonally dry vegetation such as savannahs with short periods of rainfall. However, one of the most important oilseed species, *Vitellaria paradoxa* (Sapotaceae), with recalcitrant seeds (Daws et al., 2004a) is found in such habitats across Africa. This species is the source of shea butter, which is used in cooking and in cosmetics.

In a study on the desiccation tolerance and germination characteristics of 10 dryland species, predominantly from Burkina Faso, Kenya, and Tanzania, a set of ecological correlates was assessed for their predictive value regarding the recalcitrant seed trait (Pritchard et al., 2004b). Seed storage data also were collected for 70 African dryland tree species to explore the relationship between seed mass, seed desiccation tolerance, and rainfall at the time of seed shedding. All species with desiccation-sensitive seeds had a seed mass >500 mg. Seed dispersal in *Vitellaria paradoxa* and in two other recalcitrant-seeded species (*Syzygium cumini* and *Trichilia emetica*) coincided with the mid-point of the rainy season, when precipitation was c. 200 mm per month. For *V. paradoxa* from Ouagadougou, Burkina Faso, the 7-month dry season had a cumulative rainfall of only c. 60 mm. In comparison, seed dispersal in the desiccation-tolerant species was not necessarily coincident with high rainfall, but it could be, such as with *Kigelia africana*. Thus, dispersal during peak rainfall is not sufficient to predict desiccation sensitivity. Nonetheless, seed desiccation sensitivity is comparatively infrequent in dryland environments.

For most recalcitrant-seeded species, germination starts soon after dispersal, assuming temperature and moisture are appropriate. For cool-temperate recalcitrant seeds, for example, *Quercus* and *Castanea* species, temperatures around 15°C−20°C are conducive for germination (Pritchard and Manger, 1990, Xia et al., 2015, Amimi et al., 2020), and also an average temperature close to that range; for example, 25/20°C (16 light/8 h dark) is appropriate for germination of Chinese chestnut (*Castanea mollissima*) (Du et al., 2020). Lower temperatures also are efficacious (Pritchard and Manger, 1990, Xia et al., 2015, Amimi et al., 2020). For example, the base temperature for germination of five temperate or Mediterranean, including mountain, *Quercus* species (*Q. robur, Q canariensis, Q coccifera, Q. ilex*, and *Q. suber*) and *Castanea sativa* varies from −1°C to 4°C (Pritchard and Manger, 1990; Amimi et al., 2020). Some differences can occur between *Quercus robur* provenances, with base temperatures of 0.8°C and 2.4°C for acorns from The Netherlands and England, respectively (Pritchard and Manger, 1990), and −3.3°C and −1.5°C for those from Italy and England, respectively (McCartan et al., 2015). The Mediterranean oaks *Quercus ilex* and *Q. canariensis*, which occur at high elevations, where frost events are frequent, also showed the lowest freezing sensitivity (Amimi et al., 2020).

In contrast, *Cyclobalanopsis* (Fagaceae) species from subtropical and tropical habitats germinate poorly at 15°C and reach high germination percentages at 25°C (Xia et al., 2015). Similar temperature responses have been observed in a wide range of tropical recalcitrant-seeded species, as reported in the "Database of tropical tree seed research, with special reference to the Dipterocarpaceae, Meliaceae, and Araucariaceae: user manual—DABATTS" (Tompsett and Kemp, 1996). Of 120 species covered in DABATTS, the germination characteristics are reported for 17 recalcitrant-seeded species in eight genera of Dipterocarpaceae (*Anisoptera, Cotylelobium, Dipterocarpus, Dryobalanopsis, Hopea, Parashorea, Shorea*, and *Vatica*). Seeds of all genera germinated to high percentages at 26°C–31°C. Within this thermal range, increasing the germination temperature from ambient (24°C/14°C, day/night) to 29°C/20°C did not compromise seed germination or subsequent seedling production in *Trichilia emetica* (Sershen et al., 2014).

Within the Dipterocarpaceae, estimates for recalcitrant seed germination base temperature range from 7.1°C (*Hopea odorata*) to 16.4°C (*Shorea argentifolia*), and thermal times ($G_{50}\%$) range from 38°Cd (*Shorea argentifolia*) to 624°Cd (*Dipterocarpus costatus*). For comparison, *Quercus robur* has a thermal time for germination of 145°Cd (Pritchard and Manger, 1990). Projected maximum (ceiling) temperatures for germination in temperate *Quercus robur* and *Castanea sativa* are around 40°C (Pritchard and Manger, 1990), but it seems unlikely that seedlings would survive if seeds germinated at such a high temperature. In *Carapa surinamensis*, the upper temperature limit for radicle emergence was >40°C, whereas for epicotyl growth it was between 35°C and 40°C (Amoedo and Ferraz, 2019). For *Oenocarpus bataua* seeds, the germination button (emergence of a mass of tissue to the outside of the seed from which the root and shoot emerge) formed over a large range of constant temperatures, but development of the second cataphyll (i.e., seedling development) was limited to 25°C–30°C (Bastos et al., 2017). Similarly, tropical recalcitrant seeds can lose viability after incubation at 40°C for a few days to 2 weeks (B. Wen, pers. observ.). In surface-dried recalcitrant seeds of *Baccaurea ramiflora* (Wen et al., 2016) and *Archontophoenix alexandrae* (Arecaceae) (Wen, 2019), heating for half an hour at 50°C–60°C and at ≥60°C, respectively, caused loss of viability. In comparison, some orthodox seeds such as spiny amaranth (Ye and Wen, 2017) and Mexican sunflower (Wen, 2015) can survive brief heating at 70°C.

Germination rate (speed) has been suggested as a distinguishing feature of recalcitrant versus orthodox seeds. One aspect of this argument has been that recalcitrant seeds have a lower seed coat ratio (SCR) and use relatively less energy for seed defense than orthodox seeds, which would facilitate the germination process (Daws et al., 2005; 2006). For the Mediterranean oaks *Q canariensis, Q coccifera, Q. ilex*, and *Q suber*, the time to 50% acorn germination at 13°C was 14 d, 11 d, 19 d, and 15 d, respectively (Amimi et al., 2020). The ratio of pericarp mass to acorn mass for these species is negatively correlated with germination rate (Amimi et al., 2020). Often such comparisons are made at a single temperature, which gives a rather limited perspective on germination rate or vigor. When vigor is considered as thermal time above a base temperature for the progression of germination, this perspective changes. We compared published thermal characteristics of germination of >70 tree species with either orthodox or recalcitrant seeds (Table 19.1). Whilst the optimum temperature for germination of recalcitrant seeds was significantly higher than that of the orthodox seeds, the other

TABLE 19.1 Suboptimal thermal time (θ_{50}) and cardinal temperatures of tree species.

Trait	Orthodox	Recalcitrant
θ_{50} (°Cd)	87.45 ± 117.08 (n = 37)[a]	148.71 ± 146.40 (n = 25)[a]
T_b (°C)	7.98 ± 4.16 (n = 58)[b]	9.70 ± 5.48 (n = 25)[b]
T_o (°C)	25.51 ± 4.54 (n = 23)[c]	27.86 ± 3.00 (n = 22)[c]
T_c (°C)	41.71 ± 6.10 (n = 23)[d]	40.00 ± 0.00 (n = 1)[d]

Values are means ± standard deviation, n = number of data records. Analysis by a two-sample t-test (in R version 4.0.3) showed that there were significant differences only between orthodox and recalcitrant for T_o ($P < .05$).
[a]t (43.774) = −1.7482, P = .08744.
[b]t (36.452) = −1.4049, P = .1685.
[c]t (38.274) = −2.0574, P = .04651.
[d]Insufficient data.
Source: The data are from Durr, C., Dickie, J.B., Yang, X-Y., Pritchard, H.W., 2015. Ranges of critical temperature and water potential values for the germination of species worldwide: contribution to a seed trait database. Agric. For. Meteorol. 200, 222–232; and from more recently published data as listed in Fig. 19.1.

parameters (base temperature [T_b], ceiling temperature [T_c], and thermal time [θ_{50}, °Cd]) were not significantly different. Moreover, recalcitrant seeds tended to have longer thermal times for germination than orthodox seeds (149°Cd and 88°Cd on average, respectively) (Durr et al., 2015). Thus, when the process of germination is quantified using comparable units above a threshold, recalcitrant seeds of trees do not necessarily germinate faster than orthodox seeds of trees. This may be due partly to the recalcitrant seeds studied being generally much larger, around 1–10 g (Fig. 19.1), than orthodox seeds. These results are broadly in support of the finding that small seeds germinate faster (shorter MTG) than large seeds amongst 1037 tree species across five tropical forests (Norden et al., 2009).

Stopping germination during seed storage

One of the challenges in using species with recalcitrant seeds in vegetation restoration projects is that the seeds may germinate and/or die during the time between collection and sowing them in the field or nursery. Thus much research has been conducted to better understand what happens to recalcitrant seeds during storage. The results of these studies provide insight into what may happen to seeds, particularly if nondormant, if conditions are not favorable for germination in the field immediately after dispersal. Indirectly, the results from storage experiments provide some understanding of how recalcitrant seeds may respond to climate change.

To facilitate short-term storage, seeds can be maintained at an MC close to that of newly dispersed seeds and at ambient or slightly reduced temperatures, that is, so-called "hydrated-storage" (Berjak et al., 1989; Kioko et al., 1993; Eggers et al., 2007). However, under these conditions recalcitrant seeds might initiate germination-associated events (Berjak et al., 1989; Bonner, 1990; Kioko et al., 1993; Pammenter et al., 1994; Motete et al., 1997; Berjak and Pammenter, 2004a, b; Sershen et al., 2008), culminating in cellular vacuolation and the start of cell division (Berjak et al., 1989). At this stage, additional water is required. If water is not supplied, the seeds are exposed to initially mild, but increasingly severe, water stress (Pammenter et al., 1994, 1997). The deleterious events associated with this stress, which include unbalanced metabolism and the consequential generation of damaging free radicals, culminate in seed death (Dussert et al., 2006; Ratajczak and Pukacka, 2006). However, ROS has a dual role in seeds (Bailly, 2004), including recalcitrant seeds, as evidenced by the high transcript levels for certain antioxidants in mature recalcitrant seeds of *Quercus ilex* (Romero-Rodríguez et al., 2018). ROS is both a trigger for germination and an oxidative stress molecule produced during embryo excision and/or drying, for example, *Castanea sativa* (Roach et al., 2010), *Trichilia dregeana* (Whitaker et al., 2010), and *Camellia sinensis* (Chen et al., 2012). Interestingly, pretreatment of recalcitrant seeds of *Madhuca latifolia* with ROS inhibitors reduced the rate of in situ germination during storage, without compromising viability over two months of hydrated storage at 25°C (Chandra et al., 2019).

Alternatively, the germination parameters can be used to identify a storage temperature that significantly slows germination progress. This is most effective for seeds that are not chilling-sensitive. Seeds of *Araucaria hunsteinii* (Pritchard et al., 1995b) and *Quercus* spp. (Connor and Sowa, 2002) can be stored fully hydrated at refrigerator temperatures (c. 5°C) for many months. In these cases, the storage temperature is probably close to the minimum temperature for germination progress. In contrast, seeds of *Aesculus hippocastanum* survived more than 1 year at 16°C, which is below the minimum temperature for fresh seed germination of about 20°C (Pritchard et al., 1996). For more

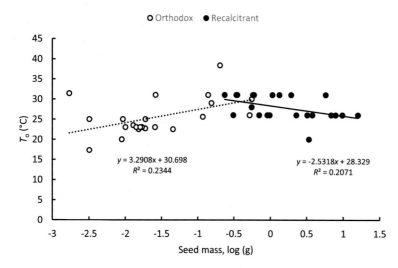

FIGURE 19.1 Relations between optimum temperature for germination rate (speed) (T_o) and seed mass for 45 tree species with orthodox (open circle) and recalcitrant (closed circle) seeds. *The data are mainly from Durr, C., Dickie, J.B., Yang, X-Y., Pritchard, H.W., 2015. Ranges of critical temperature and water potential values for the germination of species worldwide: contribution to a seed trait database. Agric. For. Meteorol. 200, 222–232; and from other sources (Arana et al., 2016; Brüning et al., 2011; Daibes et al., 2019; Daibes and Cardoso, 2018; Dantas et al., 2020; Duncan et al., 2019; Gomes et al., 2019; Gomes et al., 2019; Macera et al., 2017; Martínez-Villegas et al., 2018; Midmore et al., 2015; Ordoñez-Salanueva et al., 2021; Porceddu et al., 2013; Rampart, 2019; Ranno et al., 2020; Soltani, 2011; de Souza et al., 2015).*

chilling-sensitive species, germination might be delayed by only a few weeks to three months, for example, for *Scadoxus membranaceus* and *Landolphia kirkii* (Farrant et al., 1989), *Symphonia globulifera* (Corbineau and Côme, 1986, 1988), *Azadirachta indica* (Nayal et al., 2000; Neya et al., 2004), and various amaryllid species (Sershen et al., 2008). In the case of tropical *Inga* spp. (Pritchard et al., 1995a; Pritchard 2004) and *Vitellaria paradoxa* (Gaméné et al., 2005), storage at 15°C, which is close to the germination temperature minima for these tropical recalcitrant seeds, is recommended. Clearly, normal habitat temperatures in tropical wet forests will be detrimental to maintaining the viability of recalcitrant seeds. Whilst increased temperatures are likely to promote germination of nondormant recalcitrant seeds if they are hydrated, more advanced germination stages also tend to have higher desiccation sensitivity, such as in *Oenocarpus bacaba* (Bastos et al., 2021). Thus, we predict that increased temperatures due to climate change will enhance the speed at which seeds will germinate and also the speed at which they will lose viability.

Seed and seedling banks

After dispersal, recalcitrant seeds often complete germination quickly, and then they may form a seedling bank. Energy reserves in large recalcitrant seeds are far greater than those that the seedlings need, for example, *Quercus mongolica* (Yi et al. 2015). However, there is evidence that some species can form a transient seed bank, for example, *Quercus ilex* (Joët et al., 2016) and *Podocarpus angustifolius* (Ferrandis et al., 2011). Recalcitrant seeds of *Cryptocarya aschersoniana* in the Brazilian Atlantic Forest are dispersed at the end of the rainy season but do not germinate immediately, with a germination peak starting about 200 days after burial and coinciding with the onset of the rainy season. In this period, seed mortality did not exceed 28% in the natural environment compared to a viability loss of 90% in a disturbed environment. Also, germinated seeds in the disturbed environment did not establish seedlings (Tonetti et al., 2016). The recalcitrant seeds of *Swartzia langsdorffii* are dispersed during the dry season and remain viable within the dispersed fruits for up to 7 months, until the rainy season begins, because of multiple diaspore features that slow desiccation (Vaz et al., 2016, 2018). Seeds of *Garcinia cowa* also are dispersed during the dry season in seasonal rainforests and are protected by a hard tough coat (Liu et al., 2005).

When placed in the forest, recalcitrant seeds of *Araucaria angustifolia* from the Atlantic Forest (Brazil) maintained high viability, and seedlings emerged at 270 d. However, when located at the forest edge, seeds died by 120 days due to water loss and high predation (Gasparin et al., 2017, 2020). Similarly, a decrease in germination of *Quercus rubra* seeds was associated with a reduction in soil water content (García et al., 2002), and mortality of *Quercus ilex* seeds in winter coincided with them reaching a critically low water content during periods of drought (Joët et al., 2016). *Podocarpus angustifolius* seeds extracted from soil samples also have low viability because of aging (Ferrandis et al., 2011). Not only drought but also temperature conditions, resulting from habitat fragmentation might have a negative effect on recalcitrant seed germination. Germination of *Baccaurea ramiflora* was delayed in nonrainforest conditions compared with the tropical seasonal rainforest understory, and only one-third of the seeds placed to an open site germinated, with the seeds being particularly sensitive to warming above 35°C (Wen and Cai, 2014).

Natural regeneration of late-successional trees in fragmented and degraded landscapes can be strongly limited by a lack of seed dispersal into successional habitats and is a particular problem for larger seeded animal-dispersed trees (Wijdeven and Kuzee, 2000; Rodrigues da Silva, Matos, 2006). Movement of large seeds beyond the edges of forest fragments is rare in the tropics (Dosch et al., 2007; del Castillo and Rios, 2008; Cole, 2009), and establishment of these species by natural regeneration is limited even after many decades of succession (Aide et al., 2000). Even when seed availability is not limited, such as by intentional "seeding," seedling establishment may be very low. For *Quercus schottkyana*, only 0.6% of seedlings survived because of the combined effects of desiccation and predation by weevils and bark beetles (Xia et al., 2018). Thus, for many reasons the future survival and germination of recalcitrant-seeded species are seriously threatened by deforestation and climate change (Marques et al., 2018), whilst the natural regeneration of orthodox-seeded pioneer species will be favored.

Future research needs

This chapter has revealed that there are few concerted, integrated studies on the impact of climate change on regeneration of recalcitrant-seeded species from seeds. Unlike orthodox seeds, most recalcitrant seeds have no quiescent phase, little dormancy and no long-term persistence in the soil seed bank. Clearly, much is yet to be learned, and progress on the five research areas described below could provide insights into the scale and scope of the problem of long-term persistence of recalcitrant-seeded species from a global warming perspective. Furthermore, new insights ultimately would lead to improved strategies for risk mitigation and intervention in response to the impacts of climate change.

1. Estimates of the frequency of recalcitrant seeded tree species and their associated seed ecological correlates (SCR, mass, dispersal time) have been generated for aseasonal and seasonal tropical rainforests of Brazil (Atlantic Forest and Amazon, respectively), SW China, Panama, the Caribbean region, and eastern Australia. However, such studies are needed from other biodiversity hotspots in MesoAmerica, West Africa, Madagascar, Sundaland, and IndoBurma. Greater coverage of plant families also will have the added benefit of increasing the opportunity to consider evolutionary patterns associated with the gain/loss of the desiccation tolerance trait.
2. The ability of recalcitrant-seeded species to regenerate via seeds depends on a high level of synchrony between the timing of seed dispersal and the availability of adequate water to avoid cumulative dehydration stress and support the completion of the germination process. Understanding the variation in phenological plasticity within and amongst genetic limits to the environmental conditions within particular habitats in recalcitrant-seeded species could inform predictive models for the likely risks to regeneration from seed under future climate change scenarios.
3. The varied morphological adaptations of recalcitrant seeds from tropical regions are poorly explored in relation to desiccation resistance, desiccation avoidance, and the capacity to redistribute water to critical tissues. Yet, such morphological analyses of a wider range of families with a high number of recalcitrant-seeded species (e.g., Annonaceae, Arecaceae, Clusiaceae, Lauraceae, Moraceae, Sapindaceae, and Sapotaceae) also will provide valuable insights on the dormancy trait and its kind and frequency and on relative embryo maturity at dispersal. This information will facilitate the mapping of the multidimensional continuum of traits in recalcitrant seeds, not just for desiccation sensitivity.
4. The temperature parameters of seed germination in relation to threshold models have been determined for hundreds of species with orthodox seeds, but for only c. 25 species with recalcitrant seeds. There is a particular dearth of data on the ceiling temperature for recalcitrant seed germination rate (speed) and on base water potential for germination. Such information and how thresholds might shift at the seed-seedling transition will provide a better understanding of the risks from heat stress under future climates, particularly when combined with "drought" conditions.
5. Recalcitrant seed germination generally is light insensitive, as is often the case for large seeds in general. However, an appropriate light level is important for establishment of a seedling bank and subsequent growth of the seedlings. This is because gaps and edges created in the forest (potentially as a result of tree extraction or land conversion) reduce the buffering effect of shade against high temperatures, resulting in recalcitrant seedlings potentially being exposed to multiple risks at once (thermal gain, high irradiance, and desiccating conditions). Whilst seedling emergence and survival of orthodox-seeded species from the soil seed bank have been explored, more attention should be given to the environmental resilience of seedlings from recalcitrant seeds in order to gain a better understanding of the potential for regeneration success at this stage of the life cycle.

Finally, since in situ survival of recalcitrant-seeded species in some environments appears to be precarious, efforts to provide improved ex situ conservation options should be increased. This will require a better understanding of the variation in seed tissue physiology, cell structure, cryobiotechnology potential (Pritchard, 2018), sensitivity to chemical protectants, and in vitro culture conditions. Since the greatest concern about recalcitrant-seeded species is in the tropics, where most of them occur, this is where a greater investment is needed in training seed biologists and cryobiologists.

Acknowledgments

The Royal Botanic Gardens, Kew receives grant-in-aid from Defra. HWP acknowledges the support of the Garfield Weston Foundation Tree Seed Bank Project.

References

Aide, T.M., Zimmerman, J.K., Pascarella, J.B., Rivera, L., Marcano-Vega, H., 2000. Forest regeneration in a chronosequence of tropical abandoned pastures: implications for restoration ecology. Restor. Ecol. 8, 328–338.

Amimi, N., Dussert, S., Vaissayre, V., Ghouil, H., Doulbeau, S., Costantini, C., et al., 2020. Variation in seed traits among Mediterranean oaks in Tunisia and their ecological significance. Ann. Bot. 125, 891–904.

Amoedo, S.C., Ferraz, I.D.K., 2017. Seed quality evaluation by tetrazolium staining during a desiccation study of the recalcitrant seeds of *Carapa guianesis* Aubl. and *Carapa surinamensis* Miq.- Meliaceae. Afr. J. Agric. Res. 12, 1005–1013.

Amoedo, S.C., Ferraz, I.D.K., 2019. A comparative study of the thermal ranges of three germination criteria of a tropical tree with bioeconomic interest: *Carapa surinamensis* Miq. - Meliaceae. Brazil. J. Biol. 79, 213–219.

Arana, M.V., Gonzalez-Polo, M., Martinez-Meier, A., Gallo, L.A., Benech-Arnold, R.L., Sánchez, Rodolfo A, et al., 2016. Seed dormancy responses to temperature relate to *Nothofagus* species distribution and determine temporal patterns of germination across altitudes in Patagonia. New Phytol. 209, 507–520.

Asanga, W.M.G., Wijetunga, S.T.B., Jayasuriya, K.M.G.G., 2020. Epicotyl morphophysiological dormancy and storage behaviour of seeds of *Strychnos nux-vomica*, *Strychnos potatorum* and *Strychnos benthamii* (Loganiaceae). Seed Sci. Res. 30, 284–292.

Bailly, C., 2004. Active oxygen species and antioxidants in seed biology. Seed Sci. Res. 14, 93–107.

Barendse, G.W.M., Peeters, T.J.M., 1995. Multiple hormonal control in plants. Acta Bot. Neerl. 44, 3–17.

Baskin, J.M., Baskin, C.C., 2004. A classification system for seed dormancy. Seed Sci. Res. 14, 1–16.

Baskin, J.M., Baskin, C.C., 2014. What kind of seed dormancy might palms have? Seed Sci. Res. 24, 17–22.

Bastos, L.L.S., Calvi, G.P., Lima Junior, M.J.V., Ferraz, I.D.K., 2021. Degree of desiccation sensitivity of the Amazonian palm *Oenocarpus bacaba* depends on the criterion for germination. Acta Amazon. 51, 85–90.

Bastos, L.L.S., Ferraz, I.D.K., Lima Junior, M.J.V., Pritchard, H.W., 2017. Variation in limits to germination temperature and rates across the seed-seedling transition in the palm *Oenocarpus bataua* from the Brazilian Amazon. Seed Sci. Technol. 45, 1–13.

Berjak, P., Farrant, J.M., Pammenter, N.W., 1989. The basis of recalcitrant seed behaviour. In: Taylorson, R.B. (Ed.), Recent Advances in the Development and Germination of Seeds. Plenum Press, New York, pp. 89–105.

Berjak, P., Pammenter, N.W., 2004a. Biotechnological aspects of non-orthodox seeds: an African perspective. S. Afr. J. Bot. 70, 102–108.

Berjak, P., Pammenter, N.W., 2004b. Recalcitrant seeds. In: Benech-Arnold, R.L., Sánchez, R.A. (Eds.), Handbook of Seed Physiology: Applications to Agriculture. Haworth Reference Press, New York, pp. 305–345.

Berjak, P., Pammenter, N.W., Vertucci, C., 1992. Homoiohydrous (recalcitrant) seeds: developmental status, desiccation sensitivity and the state of water in axes of *Landolphia kirkii* Dyer. Planta 186, 249–261.

Berjak, P., Vertucci, C.W., Pammenter, N.W., 1993. Effects of developmental status, and dehydration rate on characteristics of water and desiccation-sensitivity in recalcitrant seeds of *Camellia sinensis*. Seed Sci. Res. 3, 155–166.

Bewley, D.J., Bradford, K.J., Hilhorst, H.W.M., Nonogaki, H., 2013. Seeds: Physiology of Development and Germination, Third ed. Springer Science & Business Media, Berlin.

Black, M., Pritchard, H.W. (Eds.), 2002. Desiccation and Survival in Plants: Drying without Dying. CABI Publishing, Wallingford.

Bonner, F.T., 1990. Storage of seeds: potential and limitations for germplasm conservation. For. Ecol. Manage. 35, 35–43.

Brüning, F.D.O., Lúcio, A.D.C., Muniz, M.F.B., 2011. Limits for germination, purity, humidity and 1000-seed weight in seed analysis of native tree species in Rio Grande do Sul state. Ciência Florestal 21, 193–202.

Chandra, J., Sershen, Varghese, B., Keshavkant, S., 2019. The potential of ROS inhibitors and hydrated storage in improving the storability of recalcitrant *Madhuca latifolia* seeds. Seed Sci. Technol. 47, 33–45.

Chen, H., Pritchard, H.W., Seal, C.E., Nadarajan, J., Li, W., Yang, S., et al., 2012. Post desiccation germination of mature seeds of tea (*Camellia sinensis* L.) can be enhanced by pro-oxidant treatment, but partial desiccation tolerance does not ensure survival at −20°C. Plant Sci. 184, 36–44.

Cheng, W.H., Endo, A., Zhou, L., Penney, J., Chen, H.-C., Arroyo, A., et al., 2002. A unique short-chain dehydrogenase/reductase in *Arabidopsis* glucose signaling and abscisic acid biosynthesis and functions. Plant Cell 14, 2723–2743.

Chin, H.F., Roberts, E.H., 1980. Recalcitrant Crop Seeds. Tropical Press, Kuala Lumpur.

Cole, R.J., 2009. Post-dispersal seed fate of tropical montane trees in an agricultural landscape, southern Costa Rica. Biotropica 41, 319–327.

Connor, K.F., Sowa, S., 2002. Recalcitrant behaviour of temperate forest tree seeds: storage, biochemistry, and physiology. In: Kenneth, W. (Ed.), Proceedings of the 11th Biennial Southern Silvicultural Research Conference. SRS, North Carolina, pp. 47–50.

Corbineau, F., Côme, D., 1986. Experiments on the storage of seeds and seedlings of *Symphonia globulifera* L.f. (Guttiferae). Seed Sci. Technol. 14, 585–591.

Corbineau, F., Côme, D., 1988. Storage of recalcitrant seeds of four tropical species. Seed Sci. Technol. 16, 97–103.

Costa, C.R.X., Pivetta, K.F.L., de Souza, G.R.B., Mazzini-Guedes, R.B., Pereira, S.T.S., Nogueira, M.R., 2018. Effects of temperature, light and seed moisture content on germination of *Euterpe precatoria* palm. Am. J. Plant Sci. 9, 98–106.

Daibes, L.F., Amoêdo, S.C., do Nascimento Moraes, J., Fenelon, N., da Silva, D.R., Vargas, L.A., et al., 2019. Thermal requirements of seed germination of ten tree species occurring in the western Brazilian Amazon. Seed Sci. Res. 29, 115–123.

Daibes, L.F., Cardoso, V.J., 2018. Seed germination of a South American forest tree described by linear thermal time models. J. Thermal Biol. 76, 156–164.

Dantas, B.F., Moura, M.S., Pelacani, C.R., Angelotti, F., Taura, T.A., Oliveira, G.M., et al., 2020. Rainfall, not soil temperature, will limit the seed germination of dry forest species with climate change. Oecologia 192, 529–541.

Davis P., A., Chadburn, H., Moat, J., O'Sullivan, R., Hargreaves, S., Nic Lughadha, E., 2019. High extinction risk for wild coffee species and implications for coffee sector sustainability. Sci. Adv. 5, aav3473. Available from: https://doi.org/10.1126/sciadv.aav3473.

Daws, M.I., Gaméné, C.S., Glidewell, S.M., Pritchard, H.W., 2004a. Seed mass variation potentially masks a single critical water content in recalcitrant seeds. Seed Sci. Res. 14, 185–195.

Daws, M.I., Garwood, N.C., Pritchard, H.W., 2005. Traits of recalcitrant seeds in a semi-deciduous tropical forest in Panamá: some ecological implications. Funct. Ecol. 19, 874–885.

Daws, M.I., Garwood, N.C., Pritchard, H.W., 2006. Prediction of desiccation sensitivity in seeds of woody species: a probabilistic model based on two seed traits and 104 species. Ann. Bot. 97, 667–674.

Daws, M.I., Lydall, E., Chmielarz, P., Leprince, O., Matthews, S., Thanos, C.A., et al., 2004b. Developmental heat sum influences recalcitrant seed traits in *Aesculus hippocastanum* across Europe. New Phytol. 162, 157–166.

de Souza, T.V., Torres, I.C., Steiner, N., Paulilo, M.T.S., 2015. Seed dormancy in tree species of the Tropical Brazilian Atlantic Forest and its relationships with seed traits and environmental conditions. Braz. J. Bot. 38, 243–264.

del Castillo, R.F., Rios, M.A.P., 2008. Changes in seed rain during secondary succession in a tropical montane cloud forest region in Oaxaca, Mexico. J. Trop. Ecol. 24, 433–444.

Dickie, J.B., Pritchard, H.W., 2002. Systematic and evolutionary aspects of desiccation tolerance in seeds. In: Black, M., Pritchard, H.W. (Eds.), Desiccation and Survival in Plants: Drying without Dying. CABI Publishing, Wallingford, pp. 239–259.

Dosch, J.J., Peterson, C.J., Haines, B.L., 2007. Seed rain during initial colonization of abandoned pastures in the premontane wet forest zone of southern Costa Rica. J. Trop. Ecol. 23, 151–159.

Du, C., Chen, W., Wu, Y., Wang, G., Zhao, J., Sun, J., et al., 2020. Effects of GABA and vigabatrin on the germination of Chinese chestnut recalcitrant seeds and its implications for seed dormancy and storage. Plants 9, 449. Available from: https://dx.doi.org/10.3390/plants9040449.

Duncan, C., Schultz, N.L., Good, M.K., Lewandrowski, W., Cook, S., 2019. The risk-takers and-avoiders: germination sensitivity to water stress in an arid zone with unpredictable rainfall. AoB Plants 11, plz066. Available from: https://doi.org/10.1093/aobpla/plz066.

Durr, C., Dickie, J.B., Yang, X.-Y., Pritchard, H.W., 2015. Ranges of critical temperature and water potential values for the germination of species worldwide: contribution to a seed trait database. Agric. For. Meteorol. 200, 222–232.

Dussert, S., Chabrillange, N., Engelmann, R., Anthony, R., Louarn, J., Hamon, S., 2000. Relationship between seed desiccation sensitivity, seed water content at maturity and climatic characteristics of native environments of nine *Coffea* L. species. Seed Sci. Res. 10, 293–300.

Dussert, S., Davey, M.W., Laffargue, A., Doulbeau, S., Swennen, R., Etienne, H., 2006. Oxidative stress, phospholipid loss and lipid hydrolysis during drying and storage of intermediate seeds. Physiol. Plant 127, 192–204.

Eggers, S., Erdey, D., Pammenter, N.W., Berjak, P., 2007. Storage and germination responses of recalcitrant seeds subjected to mild dehydration. In: Adkins, S., Ashmore, S., Navie, S.C. (Eds.), Seeds: Biology, Development and Ecology. CABI Publishing, Wallingford, pp. 85–92.

FAO, 2014. Genebank Standards for Plant Genetic Resources for Food and Agriculture, Revised edition. Food and Agriculture Organization, Rome.

Farmer Jr., R.E., 1977. Epicotyl dormancy in white and chestnut oaks. For. Sci. 23, 329–332.

Farnsworth, E., 2000. The ecology and physiology of viviparous and recalcitrant seeds. Annu. Rev. Ecol. Syst. 31, 107–138.

Farrant, J.M., Pammenter, N.W., Berjak, P., 1988. Recalcitrance - a current assessment. Seed Sci. Technol. 16, 155–166.

Farrant, J.M., Pammenter, N.W., Berjak, P., 1989. Germination-associated events and the desiccation sensitivity of recalcitrant seeds - a case study in three unrelated species. Planta 178, 189–198.

Farrant, J.M., Pammenter, N.W., Berjak, P., 1993. Seed development in relation to desiccation tolerance: a comparison between desiccation-sensitive (recalcitrant) seeds of *Avicennia* and desiccation-tolerant types. Seed Sci. Res. 3, 29–41.

Farrant, J.M., Pammenter, N.W., Berjak, P., Walters, C., 1997. Subcellular organization and metabolic activity during the development of seeds that attain different levels of desiccation tolerance. Seed Sci. Res. 7, 135–144.

Ferrandis, P., Bonilla, M., del Carmen Osorio, L., 2011. Germination and soil seed bank traits of *Podocarpus angustifolius* (Podocarpaceae): an endemic tree species from Cuban rain forests. Rev. Biol. Trop. 59, 1061–1069.

Finch-Savage, W., 1992a. Seed development in the recalcitrant species *Quercus robur* L.: germinability and desiccation tolerance. Seed Sci. Res. 2, 17–22.

Finch-Savage, W.E., 1992b. Embryo water status and survival in the recalcitrant species *Quercus robur* L.: evidence for a critical moisture content. J. Exp. Bot. 43, 663–669.

Finch-Savage, W.E., Blake, P., 1994. Indeterminate development in desiccation-sensitive seeds of *Quercus robur* L. Seed Sci. Res. 4, 127–133.

Finch-Savage, W.E., Clay, H.A., 1994. Evidence that ethylene, light and abscisic acid interact to inhibit germination in recalcitrant seeds of *Quercus robur* L. J. Exp. Bot. 45, 1295–1299.

Frey, A., Godin, B., Bonnet, M., Sotta, B., Marion-Poll, A., 2004. Maternal synthesis of abscisic acid controls seed development and yield in *Nicotiana plumbaginifolia*. Planta 218, 958–964.

Gaméné, C.S., Pritchard, H.W., Daws, M.I., 2005. Effects of desiccation and storage on *Vitellaria paradoxa* seed viability. In: Sacandé, M., Jøker, D., Dulloo, M.E., Thomsen, K.A. (Eds.), Comparative Storage Biology of Tropical Tree Seeds. International Plant Genetic Resources Institute, Rome, pp. 57–66.

García, D., Bañuelos, M.-J., Houle, G., 2002. Differential effects of acorn burial and litter cover on *Quercus rubra* recruitment at the limit of its range in eastern North America. Can. J. Bot. 80, 1115–1120.

Gasparin, E., Faria, J.M.R., José, A.C., Hilhorst, H.W.M., 2017. Physiological and ultrastructural responses during drying of recalcitrant seeds of *Araucaria angustifolia*. Seed Sci. Technol. 45, 112–129.

Gasparin, E., Faria, J.M.R., José, A.C., Tonetti, O.A.O., de Melo, R.A., Hilhorst, H.W.M., 2020. Viability of recalcitrant *Araucaria angustifolia* seeds in storage and in a soil seed bank. J. For. Res. 31, 2413–2422.

Gomes, V.H.F., Vieira, I.C.G., Salomão, R.P., ter Steege, H., 2019. Amazonian tree species threatened by deforestation and climate change. Nat. Clim. Change 9, 547–553.

Gomes, S.E.V., de Oliveira, G.M., do Nascimento Araujo, M., Seal, C.E., Dantas, B.F., 2019. Influence of current and future climate on the seed germination of *Cenostigma microphyllum* (Mart. ex G. Don) E. Gagnon & GP Lewis. Folia Geobot. 54, 19-28.

Hamilton, K.N., Offord, C.A., Cuneo, P., Deseo, M.A., 2013. A comparative study of seed morphology in relation to desiccation tolerance and other physiological responses in 71 Eastern Australian rainforest species. Plant Species Biol. 28, 51–62.

Hawkins, T.S., 2019. Regulating acorn germination and seedling emergence in *Quercus pagoda* (Raf.) as it relates to natural and artificial regeneration. New For. 50, 425–436.

Hawkins, T.S., 2020. Dormancy break and germination requirements in acorns of two bottomland *Quercus* species (Sect. Lobatae) of the eastern United States with references to ecology and phylogeny. Seed Sci. Res. 30, 199–205.

Hill, J.P., Edwards, W., 2010. Dispersal of desiccation-sensitive seeds is not coincident with high rainfall in a seasonal tropical forest in Australia. Biotropica 42, 271–275.

Hong, T.D., Ellis, R.H., 1995. Interspecific variation in seed storage behavior within two genera - *Coffea* and *Citrus*. Seed Sci. Technol. 23, 65–81.

IPCC (Intergovernmental Panel on Climate Change), 2018. Summary for policymakers. In: Masson-Delmotte, V., Zhai, P., Pőrtner, H.-O., Skea, J., Shukla, P.R., Pirani, A., et al.,Global warming of 1.5°C. Special Report on the Impacts of Global Warming of 1.5°C above Pre-industrial Levels and Related Global Greenhouse Gas Emission Pathways, in the Context of Strengthening the Global Response to theTreat of Climate Change, Sustainable Development, and Efforts to Eradicate Poverty.

Ismail, F.A., Nitsch, L.M., Wolters-Arts, M.M., Mariani, C., Derksen, J.W., 2010. Semi-viviparous embryo development and dehydrin expression in the mangrove *Rhizophora mucronata* Lam. Sex. Plant Reprod. 23, 95–103.

Jaganathan, G.K., 2020. Defining correct dormancy class matters: morphological and morphophysiological dormancy in Arecaceae. Ann. For. Sci. 77, 100. Available from: https://doi.org/10.1007/s13595-020-01010-7.

Jaganathan, G.K., 2021. Ecological insights into the coexistence of dormancy and desiccation-sensitivity in Arecaceae species. Ann. For. Sci. 78, 10. Available from: https://doi.org/10.1007/s13595-021-01032-9.

Jaganathan, G.K., Li, J., Yang, Y., Han, Y., Liu, B., 2019. Complexities in identifying seed storage behavior of hard seed-coated species: a special focus on Lauraceae. Bot. Lett. 166, 70–79.

Jayasuriya, K.M.G.G., Wijetunga, A.S.T.B., Baskin, J.M., Baskin, C.C., 2010. Recalcitrancy and a new kind of epicotyl dormancy in seeds of the understory tropical rainforest tree *Humboldtia laurifolia* (Fabaceae, Caesalpinioideae). Amer. J. Bot. 97, 15–26.

Jayasuriya, K.M.G., Wijetunga, A.S., Baskin, J.M., Baskin, C.C., 2012. Physiological epicotyl dormancy and recalcitrant storage behaviour in seeds of two tropical Fabaceae (subfamily Caesalpinioideae) species. AoB Plants 2012, pls044. Available from: https://doi.org/10.1093/aobpla/pls044.

Joët, T., Ourcival, J.-M., Capelli, M., Dussert, S., Morin, X., 2016. Explanatory ecological factors for the persistence of desiccation-sensitive seeds in transient soil seed banks: *Quercus ilex* as a case study. Ann. Bot. 117, 165–176.

Kermode, A.R., Finch-Savage, B.E., 2002. Desiccation sensitivity in orthodox and recalcitrant seeds in relation to development. In: Black, M., Pritchard, H.W. (Eds.), Desiccation and Survival in Plants: Drying without Dying. CABI Publishing, Wallingford, pp. 149–184.

King, M.W., Roberts, E.H., 1980. Maintenance of recalcitrant seeds in storage. In: Chin, H.F., Roberts, E.H. (Eds.), Recalcitrant Crop Seeds. Tropical Press, Kuala Lumpur, pp. 53–89.

Kioko, J., Albrecht, J., Uncovsky, S., 1993. Seed collection and handling. In: Albrecht, J. (Ed.), Tree Seed Handbook of Kenya. GTZ Forestry Centre, Nairobi, pp. 34–54.

Kucera, B., Cohn, A., Leubner-Metzger, G., 2005. Plant hormone interactions during seed dormancy release and germination. Seed Sci. Res. 15, 281–307.

Lamarca, E.V., Barbedo, C.J., 2015. Desiccation sensitivity of embryos of *Inga vera* Willd. Obtained from different environmental conditions, 39. Rev. Árvore, pp. 1083–1092.

Lan, Q., Xia, K., Wang, X., Liu, J., Zhao, J., Tan, Y., 2014. Seed storage behaviour of 101 woody species from the tropical rainforest of southern China: a test of the seed-coat ratio–seed mass (SCR–SM) model for determination of desiccation sensitivity. Aust. J. Bot. 62, 305–311.

Laurance, W., Delamônica, P., Laurance, S., Vasconcelos, H.L., Lovejoy, T.E., 2000. Rainforest fragmentation kills big trees. Nature 404, 836.

Li, D.-Z., Pritchard, H.W., 2009. The science and economics of ex situ plant conservation. Trends Plant Sci. 14, 614–621.

Lima Jr., M., Hong, T.D., Arruda, Y.M.B.C., Mendes, A.M.S., Ellis, R.H., 2014. Classification of seed storage behaviour of 67 Amazonian tree species. Seed Sci. Technol. 42, 363–392.

Liu, Y., Qui, Y.-P., Zhang, L., Chen, J., 2005. Dormancy breaking and storage behaviour of *Garcinia cowa* Roxb. (Guttiferae) seeds: implications for ecological function and germplasm conservation. J. Integr. Plant Biol. 47, 38–49.

Macera, L.G., Pereira, S.R., Souza, A.L.T.D., 2017. Survival and growth of tree seedlings as a function of seed size in a gallery forest under restoration. Acta Bot. Brasil. 31, 539–545.

Marques, A., Buijs, G., Ligterink, W., Hilhorst, H., 2018. Evolutionary ecophysiology of seed desiccation sensitivity. Funct. Plant Biol. 45, 1083–1095.

Martínez-Villegas, J.A., Castillo-Argüero, S., Márquez-Guzmán, J., Orozco-Segovia, A., 2018. Plant attributes and their relationship to the germination response to different temperatures of 18 species from central Mexico. Plant Biol 20, 1042–1052.

Mattana, E., Peguero, B., Di Sacco, A., Agramonte, W., Encarnación Castillo, W.R., Jiménez, F., et al., 2020. Assessing seed desiccation responses of native trees in the Caribbean. New For. 51, 705–721.

Mayrinck, R.C., Vilela, L.C., Pereira, T.M., Rodrigues-Junior, A.G., Davide, A.C., Vaz, T.A.A., 2019. Seed desiccation tolerance/sensitivity of tree species from Brazilian biodiversity hotspots: considerations for conservation. Trees 33, 777–785.

McCartan, S.A., Jinks, R.L., Barsoum, N., 2015. Using thermal time models to predict the impact of assisted migration on the synchronization of germination and shoot emergence of oak (*Quercus robur* L.). Ann. For. Sci. 72, 479–487.

Mendoza, E., Dirzo, R., 2009. Seed tolerance to predation: evidence from the toxic seeds of the buckeye tree (*Aesculus californica*; Sapindaceae). Am. J. Bot. 96, 1255–1261.

Midmore, E.K., McCartan, S.A., Jinks, R.L., Cahalan, C.M., 2015. Using thermal time models to predict germination of five provenances of silver birch (*Betula pendula* Roth) in southern England. Silva Fennica 49, 1266.

Motete, N., Pammenter, N., Berjak, P., Frédéric, J., 1997. Response of the recalcitrant seeds of *Avicennia marina* to hydrated storage: events occurring at the root primordia. Seed Sci. Res. 7, 169–178.

Nayal, J.S., Thapliyal, R.C., Rawat, M.M.S., Phartyal, S.S., 2000. Desiccation tolerance and storage behaviour of neem (*Azadirachta indica* A. Juss.) seeds. Seed Sci. Technol. 28, 761–767.

Neya, O., Golovina, E.A., Nijsse, J., Hoekstra, F.A., 2004. Ageing increases the sensitivity of neem (*Azadirachta indica*) seeds to imbibitional stress. Seed Sci. Res. 14, 205–217.

Norden, N., Daws, M.I., Antoine, C., Gonzalez, M.A., Garwood, N.C., Chave, J., 2009. The relationship between seed mass and mean time to germination for 1037 tree species across five tropical forests. Funct. Ecol. 23, 203–210.

Obroucheva, N., Sinkevich, I., Lityagina, S., 2016. Physiological aspects of seed recalcitrance: a case study on the tree *Aesculus hippocastanum*. Tree Physiol. 36, 1127–1150.

Ordoñez-Salanueva, C.A., Orozco-Segovia, A., Mattana, E., Castillo-Lorenzo, E., Davila-Aranda, P., Pritchard, H.W., et al., 2021. Thermal niche for germination and early seedling establishment at the leading edge of two pine species, under a changing climate. Environ. Exp. Bot. 181, 104288. Available from: https://doi.org/10.1016/j.envexpbot.2020.104288.

Pammenter, N.W., Berjak, P., 1999. A review of recalcitrant seed physiology in relation to desiccation-tolerance mechanisms. Seed Sci. Res. 9, 13–37.

Pammenter, N.W., Berjak, P., Farrant, J.M., Smith, M.T., Ross, G., 1994. Why do stored hydrated recalcitrant seeds die? Seed Sci. Res. 4, 187–191.

Pammenter, N.W., Motete, N., Berjak, P., 1997. The responses of hydrated recalcitrant seeds to long-term storage. In: Ellis, R.H., Black, M., Murdoch, A.J., Hong, T.D. (Eds.), Basic and Applied Aspects of Seed Biology. Kluwer Academic Publishers, Dordrecht, pp. 671–687.

Pieruzzi, F.P., Dias, L.L., Balbuena, T.S., Santa-Catarina, C., Santos, A.L.D., Floh, E.I., 2011. Polyamines, IAA and ABA during germination in two recalcitrant seeds: *Araucaria angustifolia* (Gymnosperm) and *Ocotea odorifera* (Angiosperm). Ann. Bot. 108, 337–345.

Porceddu, M., Mattana, E., Pritchard, H.W., Bacchetta, G., 2013. Thermal niche for in situ seed germination by Mediterranean mountain streams: model prediction and validation for *Rhamnus persicifolia* seeds. Ann. Bot. 112, 1887–1897.

Pritchard, H.W., 2004. Classification of seed storage 'types' for ex situ conservation in relation to temperature and moisture. In: Guerrant, E.O., Havens, K., Maunder, M. (Eds.), Ex Situ Plant Conservation: Supporting Species Survival in the Wild. Island Press, Washington, DC, pp. 139–161.

Pritchard, H.W., Wood, C.B., Hodges, S., Vautier, H.J., 2004a. 100-seed test for desiccation tolerance and germination: a case study on eight tropical palm species. Seed Sci. Technol. 32, 393–403.

Pritchard, H.W., 2018. The rise of plant cryobiotechnology and demise of plant cryopreservation? Cryobiology 85, 160–161.

Pritchard, H.W., Daws, M.I., Fletcher, B.J., Gamene, C.S., Msanga, H.P., Omondi, W., 2004b. Ecological correlates of seed desiccation tolerance in tropical African dryland trees. Am. J. Bot. 91, 863–870.

Pritchard, H.W., Haye, A.J., Wright, W.J., Steadman, K.J., 1995a. A comparative study of seed viability in *Inga* species: desiccation tolerance in relation to the physical characteristics and chemical composition of the embryo. Seed Sci. Technol. 23, 85–100.

Pritchard, H.W., Manger, K.R., 1990. Quantal response of fruit and seed germination rate in *Quercus robur* L. and *Castanea sativa* Mill, to constant temperatures and photon dose. J. Exp. Bot. 41, 1549–1557.

Pritchard, H.W., Moat, J.F., Ferraz, J.B.S., Marks, T.R., Camargo, J.L.C., Nadarajan, J., et al., 2014. Innovative approaches to the preservation of forest trees. For. Ecol. Manage. 333, 88–98.

Pritchard, H.W., Steadman, K.J., Nash, J.V., Jones, C., 1999. Kinetics of dormancy release and the high temperature germination response in *Aesculus hippocastanum* seeds. J. Exp. Bot. 50, 1507–1514.

Pritchard, H., Tompsett, P., Manger, K., 1996. Development of a thermal time model for the quantification of dormancy loss in *Aesculus hippocastanum* seeds. Seed Sci. Res. 6, 127–135.

Pritchard, H.W., Tompsett, P.B., Manger, K., Smidt, W.J., 1995b. The effect of moisture content on the low temperature responses of *Araucaria hunsteinii* seed and embryos. Ann. Bot. 76, 79–88.

Rampart, M., 2019. Seed germination variation among crop years from a *Pinus sylvestris* clonal seed orchard. Asian J. Res. Agric. Forest. 3, 1–14.

Ranno, V., Blandino, C., Giusso del Galdo, G., 2020. A comparative study on temperature and water potential thresholds for the germination of *Betula pendula* and two Mediterranean endemic birches, *Betula aetnensis* and *Betula fontqueri*. Seed Sci. Res. 30, 249–261.

Ratajczak, E., Pukacka, S., 2006. Changes in the ascorbate-glutathione system during storage of recalcitrant seeds. Acta Soc. Bot. Poloniae 75, 23–27.

Roach, T., Beckett, R.P., Minibayeva, F.V., Colville, L., Whitaker, C., Chen, H., et al., 2010. Extracellular superoxide production, viability and redox poise in response to desiccation in recalcitrant *Castanea sativa* seeds. Plant Cell Environ. 33, 59–75.

Roberto, G.G., Habermann, G., 2010. Morphological and physiological responses of the recalcitrant *Euterpe edulis* seeds to light, temperature and gibberellins. Seed Sci. Technol. 38, 367–378.

Rodrigues da Silva, U.D.S., Matos, D., 2006. The invasion of *Pteridium aquilinum* and the impoverishment of the seed bank in fire prone areas of Brazilian Atlantic Forest. Biodiv. Conserv. 15, 3035–3043.

Roeder, M., Ferraz, I.D.K., Hölscher, D., 2013. Seed and germination characteristics of 20 Amazonian liana species. Plants 2, 1–15.

Romero-Rodríguez, M.C., Archidona-Yuste, A., Abril, N., Gil-Serrano, A.M., Meijón, M., Jorrín-Novo, J.V., 2018. Germination and early seedling development in Q*uercus ilex* recalcitrant and non-dormant seeds: targeted transcriptional, hormonal, and sugar analysis. Front. Plant Sci. 9, 1508. Available from: https://doi.org/10.3389/fpls.2018.01508.

Sershen, Pammenter, N.W., Berjak, P., 2008. Post-harvest behaviour and short- to medium-term storage of recalcitrant seeds and encapsulated embryonic axes of selected amaryllid species. Seed Sci. Technol. 36, 133–147.

Sershen, Perumal, A., Varghese, B., Govender, P., Ramdhani, S., Berjak, P., 2014. Effects of elevated temperatures on germination and subsequent seedling vigour in recalcitrant *Trichilia emetica* seeds. S. Afr. J. Bot. 90, 153–162.

Soltani, A., 2011. Seed germination response of *Haloxylon persicum* (Chenopodiaceae) to different hydrothermal conditions and sand burial depths. Caspian J. Environ. Sci. 9, 211−221.

Steadman, K.J., Pritchard, H.W., 2004. Germination of *Aesculus hippocastanum* seeds following cold-induced dormancy loss can be described in relation to a temperature-dependent reduction in base temperature (T_b) and thermal time. New Phytol. 161, 415−425.

Subbiah, A., Ramdhani, S., Pammenter, N.W., Macdonald, A.H.H., Sershen, 2019. Towards understanding the incidence and evolutionary history of seed recalcitrance: an analytical review. Perspect. Plant Ecol. Evol. Syst. 37, 11−19.

Tompsett, P.B., Kemp, R., 1996. DABATTS - Database of Tropical Tree Seed Research, with Special Reference to the Dipterocarpaceae, Meliaceae and Araucariaceae: User Manual. Royal Botanic Gardens, Kew.

Tompsett, P.B., Pritchard, H.W., 1993. Water status changes during development in relation to the germination and desiccation tolerance of *Aesculus hippocastanum* L. seeds. Ann. Bot. 71, 107−116.

Tompsett, P., Pritchard, H.W., 1998. The effect of chilling and moisture status on the germination, desiccation tolerance and longevity of *Aesculus hippocastanum* L. seed. Ann. Bot. 82, 249−261.

Tonetti, O.A.O., Faria, J.M.R., José, A.C., Oliveira, T.G.S., Martins, J.C., 2016. Seed survival of the tropical tree *Cryptocarya aschersoniana* (Lauraceae): consequences of habitat disturbance. Austral Ecol. 41, 248−254.

Tweddle, J.C., Dickie, J.B., Baskin, C.C., Baskin, J.M., 2003. Ecological aspects of seed desiccation sensitivity. J. Ecol. 91, 294−304.

Vaz, T., Davide, A., Rodrigues-Junior, A.G., Nakamura, A., Tonetti, O., Da Silva, E., 2016. *Swartzia langsdorffii* Raddi: morphophysiological traits of a recalcitrant seed dispersed during the dry season. Seed Sci. Res. 26, 47−56.

Vaz, T.A.A., Rodrigues-Junior, A.G., Davide, A.C., Nakamura, A.T., Toorop, P.E., 2018. A role for fruit structure in seed survival and germination of *Swartzia langsdorffii* Raddi beyond dispersal. Plant Biol. 20, 263−270.

Vertucci, C.W., Farrant, J.M., 1995. Acquisition and loss of desiccation tolerance. In: Kigel, J., Galili, G. (Eds.), Seed Development and Germination. Marcel Dekker Inc, New York, pp. 237−271.

Walters, C., Berjak, P., Pammenter, N., Kennedy, K., Raven, P., 2013. Preservation of recalcitrant seeds. Science 339, 915−916.

Wen, B., 2009. Storage of recalcitrant seeds: a case study of the Chinese fan palm, *Livistona chinensis*. Seed Sci. Technol. 37, 167−179.

Wen, B., 2011. Changes in the moisture and germination of recalcitrant *Hopea mollissima* seeds (Dipterocarpaceae) in different desiccation regimes. Seed Sci. Technol. 39, 214−218.

Wen, B., 2015. Effects of high temperature and water stress on seed germination of the invasive species Mexican sunflower. PLoS One 10, e0141567. Available from: https://doi.org/10.1371/journal.pone.0141567.

Wen, B., 2019. Seed germination ecology of Alexandra palm (*Archontophoenix alexandrae*) and its implication on invasiveness. Sci. Rep. 9, 4057. Available from: https://doi.org/10.1038/s41598-019-40733-0.

Wen, B., Cai, Y., 2014. Seed viability as a function of moisture and temperature in the recalcitrant rainforest species *Baccaurea ramiflora* (Euphorbiaceae). Ann. For. Sci. 71, 853−861.

Wen, B., Liu, M., Tan, Y., Liu, Q., 2016. Sensitivity to high temperature and water stress in recalcitrant *Baccaurea ramiflora* seeds. J. Plant Res. 129, 637−645.

Whitaker, C., Beckett, R.P., Minibayeva, F.V., Kranner, I., 2010. Production of reactive oxygen species in excised, desiccated and cryopreserved explants of *Trichilia dregeana* Sond. S. Afr. J. Bot. 76, 112−118.

Wijdeven, S.M.J., Kuzee, M.E., 2000. Seed availability as a limiting factor in forest recovery processes in Costa Rica. Restor. Ecol. 8, 414−424.

Wood, C.B., Vautier, H.J., Bin, W., Rakotondranony, L.G., Pritchard, H.W., 2006. Conservation biology for seven palm species from diverse genera. Aliso 22, 278−284.

Wyse, S.V., Dickie, J.B., 2017. Predicting the global incidence of seed desiccation sensitivity. J. Ecol. 105, 1082−1093.

Wyse, S.V., Dickie, J.B., 2018. Taxonomic affinity, habitat and seed mass strongly predict seed desiccation response: a boosted regression trees analysis based on 17 539 species. Ann. Bot. 121, 71−83.

Xia, K., Daws, M.I., Stuppy, W., Zhou, Z.-K., Pritchard, H.W., 2012. Rates of water loss and uptake in recalcitrant fruits of *Quercus* species are determined by pericarp anatomy. PLoS One 7, e47368. Available from: https://doi.org/10.1371/journal.pone.0047368.

Xia, K., Daws, M.I., Zhou, Z.-K., Pritchard, H.W., 2015. Habitat-linked temperature requirements for fruit germination in *Quercus* species: a comparative study of *Quercus* subgenus *Cyclobalanopsis* (Asian evergreen oaks) and *Quercus* subgenus *Quercus*. S. Afr. J. Bot. 100, 108−113.

Xia, K., Turkington, R., Tan, H.-Y., Fan, L., 2018. Factors limiting the recruitment of *Quercus schottkyana*, a dominant evergreen oak in SW China. Plant Divers. 40, 277−283.

Xia, K., Zhou, Z.-Q., 2021. Characterization of physiological traits during development of the recalcitrant seeds of *Quercus serrata*. Plant Biol. 23 (6), 1000−1005.

Yan, X., Cao, M., 2006. Influence of light and temperature on the germination of *Shorea wantianshuea* (Dipterocarpaceae) seeds. Chin. Bull. Bot. 23, 642−650.

Ye, J., Wen, B., 2017. Seed germination in relation to the invasiveness in spiny amaranth and edible amaranth in Xishuangbanna, SW China. PLoS One 12, e0175948. Available from: https://doi.org/10.1371/journal.pone.0175948.

Yi, X., Wang, Z., Liu, C., Liu, G., Zhang, M., 2015. Acorn cotyledons are larger than their seedlings' need: evidence from artificial cutting experiments. Sci. Rep. 5, 8112. Available from: https://doi.org/10.1038/srep08112.

Yu, Y., Baskin, J.M., Baskin, C.C., Tang, Y., Cao, M., 2008. Ecology of seed germination of eight non-pioneer tree species from a tropical seasonal rain forest in southwest China. Plant Ecol. 197, 1−16.

Chapter 20

Effect of climate change on regeneration of seagrasses from seeds

Gary A. Kendrick[1], Robert J. Orth[2], Elizabeth A. Sinclair[1] and John Statton[1]
[1]School of Biological Sciences and Oceans Institute, The University of Western Australia, Crawley, WA, Australia, [2]Virginia Institute of Marine Science, College of William and Mary, Gloucester Point, VA, United States

Introduction

Seagrasses are an ancient group of marine flowering plants that live in shallow marine waters (Larkum et al., 2018) and provide important ecosystem services to humanity (Cullen-Unsworth and Unsworth, 2018). Although they are highly clonal, seed production and dissemination of seeds and seedlings are important in both ecological and evolutionary connectivity among seagrass populations (Kendrick et al., 2012, 2017; McMahon et al., 2014; Sinclair et al., 2018). Global climate change threatens to disrupt seagrass growth, survival, and geographical distribution through sea-level rise and increases in temperature, pH, extreme heating events, storminess, and changing patterns of ocean climates and currents.

Historically, the most significant threat to seagrasses has been increasing nutrients and sediments in runoff from coastal development, leading to decreased light penetration and loss of shallow-water dwelling meadows (Orth et al., 2006a). Coastal development also led to large-scale dredging for harbors and marinas and to the deepening of channels for large sea-going vessels. Dredging increases turbidity of the water and thereby reduces light levels, leading to losses of seagrass meadows (Fraser et al., 2017). Thus, there is a need to assess risk to seagrasses of extended dredging programs (Wu et al., 2018). Additional threats from humans are related to recreational and commercial vessels operating in shallow water, creating large-scale removal of plants in seagrass meadows by boat propellers (Orth et al., 2017) and swinging boat moorings (Unsworth et al., 2017; Glasby and West, 2018). Consequently, seagrasses are under threat from the combination of global climate change and other anthropogenic impacts (Orth et al., 2006a; Waycott et al., 2009). Furthermore, hurricanes, typhoons, and cyclones have resulted in significant losses of seagrasses in regions where these natural events are most prevalent, but seagrasses have recovered naturally from extreme weather events at the landscape scale (Wilson et al., 2020).

Gradual changes in ocean warming due to anthropogenically caused climate change increases metabolic (respiratory) load in seagrasses (e.g., Pedersen et al., 2016) that may be offset by increases in pCO_2 in seawater associated with ocean acidification since seagrasses generally are pCO_2 limited (Borum et al., 2016; Olsen et al., 2018). Similarly, building breakwaters and seawalls along coastlines to reduce impacts from sea level rise restricts seagrasses and other coastal and marine organisms such as algae from colonizing shallow waters. Thus, these activities prevent seagrasses from naturally tracking sea-level change, and they may be extirpated from many coastal margins of the world. Unlike research on terrestrial plants, understanding climate change effects on seagrass growth and reproduction is in its infancy.

The biggest immediate threat to seagrasses is extreme climate events, especially marine heatwaves, which already have resulted in loss of ecosystem resilience and evidence of collapse of seagrass communities in the Mediterranean Sea (Diaz-Almela et al., 2007; Marba and Duarte, 2010), Chesapeake Bay (Johnson et al., 2021) in the United States, and Shark Bay in Australia (Kendrick et al., 2019a; Strydom et al., 2020). These events provide insight into how seagrasses will respond to further gradual climate change and to the timing of recovery following heatwaves, thereby informing us on the resilience of seagrass ecosystems (e.g., Kendrick et al., 2019a).

Restoration and assisted ecosystem recovery of seagrass-dominated ecosystems has developed from initial experiments done in the 1970s and 1980s (van Katwijk et al., 2016) to relatively large-scale restoration successes (e.g., Orth et al., 2020; Sinclair et al., 2021). Seed settlement, recruitment, and seedling establishment have proven to be major bottlenecks to seagrass colonization and recovery in many environments (Kendrick et al., 2017). Seeds are vitally important in the recovery dynamics of seagrasses (Orth et al., 2006b; Kendrick et al., 2017), and thus a thorough understanding of how seeds behave in marine environments (e.g., Guerrero-Meseguer et al., 2018; Kendrick et al., 2019b) is a crucial element in the success of seed-based restoration strategies. However, a critical issue for restoration scientists is the effects of climate change on regeneration of seagrasses from seeds.

This chapter reviews our knowledge and predictions about the effects of climate change on regeneration of seagrasses from seeds. We need to be able to predict the effects of climate change on seeds in order to successfully restore damaged seagrass meadows. We briefly summarize aspects of the evolution of seagrasses and discuss their sexual reproduction, seed germination, seedling establishment, and recruitment; outline the bottlenecks in the regeneration of seagrasses from seeds in a fully submerged marine environment; and consider the predicted effects of climate change on them. The chapter concludes with a summary of gaps in our knowledge of seagrasses in relation to climate change and restoration.

Evolution of seagrasses

Seagrasses are a polyphyletic group that evolved from the Alismatales, an ancient lineage of monocotyledons (Les et al., 1997). They appeared in the fossil record in the Late Cretaceous (approx. 80 Ma), when the marine environment was warmer and more enriched in gaseous CO_2 than it is at present (Fig. 20.1). Subsequently, seagrasses adapted to live in contemporary marine environments with lower mean temperature and lower availability of CO_2 (Beer and Koch, 1996). They have persisted through the transition from a warmer (peak c.55 Ma, i.e., Paleocene–Eocene Thermal Maximum) to a cooler (Eocene–Oligocene transition, about 34 Ma) global ocean (Fig. 20.1) under limitations of carbon, nitrogen, and phosphorus, and major rapid changes in sea level (Fig. 20.1; Miller et al., 2011). All genera and species of seagrasses have broad biogeographical ranges (Short et al., 2011). Thus, it is predicted that seagrasses will do well under increased temperature and pCO_2 due to climate change.

Physiologically, seagrasses may be better adapted to climate change than other marine organisms such as hard corals and seaweeds. However, the interacting effects of grazing, disease, and rate of climate-driven environmental change on their biology are of concern (Hyndes et al., 2016; Larkum et al., 2017). Seagrasses have higher pCO_2 requirements than

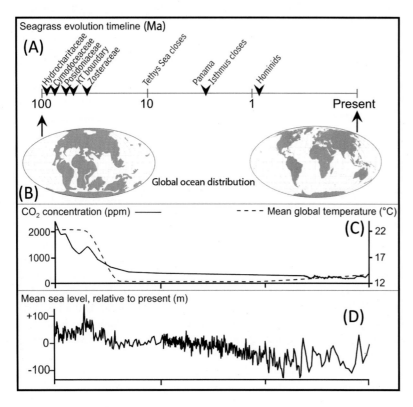

FIGURE 20.1 Geological timeline of seagrass evolution over the past 80 million years (A) in relation to changes in global ocean distribution (B), atmospheric CO_2 concentration and mean global temperature (C), and mean sea level (D). KT, Cretaceous–Tertiary boundary [approximately 65 million years ago (Ma)]. *From Orth, R.J., Carruthers, T.J.B., Dennison, W.C., Duarte, C.M., Fourqurean, J.W., Heck, K.L., Jr., et al., 2006a. A global crisis for seagrass ecosystems. BioScience 56, 987–996, published with permission.*

other photosynthesizing marine organisms (e.g., Borum et al., 2016; Larkum et al., 2017; Zimmerman et al., 2017), and the multiple anatomical, physiological, and biochemical mechanisms by which they concentrate CO_2 are being investigated (Larkum et al., 2017).

Comparative studies of whole genomes across seagrass lineages have given insight into how they solved the main physiological and reproductive challenges in adapting to life under water (e.g., Lee et al., 2018). Whole genomes have been sequenced for several seagrass species with small genomes. These include representatives from two of the three independent seagrass-return-to-the-sea events: Zosteraceae, *Zostera marina* (Olsen et al., 2016) and *Z. muelleri* (Golicz et al., 2015; Lee et al., 2016); and Hydrocharitaceae, *Halophila ovalis* (Lee et al., 2018) and *H. stipulacea* (Tsakogiannis et al., 2020).

Independent evolutionary history of seagrass lineages facilitates the identification of unique and/or shared adaptive traits that may provide resilience to a changing marine environment under global climate change (Wissler et al., 2011; Lee et al., 2018). Several key gene differences associated with adaptation of seagrasses to the marine environment include loss of genes associated with (1) stomatal development (Golicz et al., 2015; Lee et al., 2018), (2) ethylene biosynthesis and signaling, along with other gas-mediation genes (Golicz et al., 2015; Lee et al., 2016, 2018), (3) sepal and petal development (Golicz et al., 2015), and (4) loss of the NADH dehydrogenase-like complexes involved in nitrogen assimilation in *H. ovalis* (Lee et al., 2018). Furthermore, there is variation within and among genera and species of seagrasses in the northern and southern hemispheres, suggesting that evolutionary solutions have evolved independently as regional climate and geography of the oceans have changed across the biogeographical ranges of species and genera.

Seagrass mating systems

Seagrass species are strongly clonal and reproduce vegetatively via rhizomes and also sexually by seeds. They have unique underwater pollination systems that involve both biotic (zoobenthophilous, van Tussenbroek et al., 2016) and abiotic (water buoyant/floating, McMahon et al., 2014) movement of pollen. Flowering, pollination and seed production, release, and dispersal are tightly linked to sea-water temperature. Increases in sea-water temperature resulting from global warming have been shown to influence reproductive events and can impact flowering and seed production.

Approximately 56% of seagrasses are dioecious (Les, 1988); therefore, successful pollination must occur through outcrossing, for example, the genera *Amphibolis*, *Enhalus*, *Halodule*, *Phyllospadix*, *Thalassia*, and *Thalassodendron*. Dioecy can limit seed production. For example, in *Phyllospadix* species pollen limitation has been reported in populations in which the male:female ratio was skewed in favor of females (Shelton, 2008). The other 44% of seagrasses are hermaphroditic (*Posidonia* species) or monecious (some species of *Halophila* and *Heterozostera* and *Zostera* species) (Orth et al., 2000). In hermaphroditic and monecious species, there is a greater chance of self-pollination and inbreeding. To reduce inbreeding, mating strategies have evolved within the hermaphroditic genus *Posidonia*: Northern Hemisphere species *P. oceanica* (protogyny; Entrambasaguas et al., 2017) and Southern Hemisphere species *P. australis* (protandry; McConchie and Knox, 1989). Similarly, in the monecious *Zostera marina* outcrossing rates were positively associated with temporal separation of male and female flowering (protogyny) (Ruckelshaus, 1995).

Plant populations at their geographic range edges can provide a perspective on how species may respond to climate change. Typically, they have a smaller effective population size (Ne), reduced sexual reproduction, and lower levels of genetic connectivity than those in the center of the range (Eckert et al., 2008). It is often populations at the low latitude parts of the range of seagrasses that are most impacted by changing climates. For example, seagrasses in the low latitude warming range-edge populations in Shark Bay (Australia) experienced significant declines during and after an extreme marine heatwave in 2011; 1300 km^2 of 4600 km^2 (28%) of seagrass meadows disappeared (Kendrick et al., 2019a; Strydom et al., 2020). Seagrass plants were defoliated when sea temperatures reached 4°C warmer than in previous summers (Kendrick et al., 2019a). Flowering and seed production were impacted for 3 years after the heatwave, slowing meadow recovery (Kendrick et al., 2019a).

Clonal growth enables seagrass genotypes to persist almost indefinitely if undisturbed. Some examples of extensive clones that may be 100s to 1000s of years old include those of *Zostera marina* in the Baltic Sea at the northern edge of its range (Reusch et al., 1999), *Cymodocea nodosa* at the limit of its northern Atlantic Ocean range (Alberto et al., 2001), *Posidonia oceanica* in the Mediterranean Sea (Arnaud-Haond et al., 2012), and *Thalassia testudinum* at the limit of its northern Atlantic Ocean range in the Indian River Lagoon, Florida (United States) (Bricker et al., 2018). Importantly, individual clones of these species have persisted under extreme fluctuations in sea level, salinity, and temperature. This ability of clones to survive these changing environmental conditions suggests a remarkable level of phenotypic plasticity within seagrass genomes and may reflect a trade-off that has resulted in their present-day low species diversity.

Flowering, seed production, and seed germination

The greatest climate change impact on flowering, seed production, and seed germination will be an increase in seawater temperatures since flowering phenology in seagrasses is controlled primarily by temperature. For example, along the east coast of the United States, where *Z. marina* is found at 35°N to 45°N, anthesis to fruit maturity occurs in the southernmost populations from January to April as water temperatures increase from 3°C to 20°C. Dates for flowering to fruit maturity occur later in more northern populations. At 45°N, these same reproductive events occur from April to July (Silberhorn et al., 1983). Seed release from fruits of *Z. marina* follows a similar pattern, occurring earlier in southern populations than in northern populations (Orth et al., 2000). Flowering shoot densities and seed production in *Z. marina* will depend on whether the population is annual or perennial. Seed banks in this species are transient, with most seeds germinating within a year after release from the mother plant. Seed germination in *Z. marina* is promoted by low oxygen and a decrease in temperature to 15°C (Moore et al., 1993). Large declines in size of populations of this species have occurred in southern Virginia (United States) in response to a high-temperature event and to a high precipitation event, but in each case the population recovered (Johnson et al., 2021).

A global analysis of the influence of increased water temperature on flowering phenology of *Z. marina* across a wide latitudinal range (26.8°N–56.8°N) found that the time of flowering was advanced significantly with increased capacity for sexual reproduction in northern populations with ocean warming (Blok et al., 2018). Overall, this analysis indicated that an increase of 1°C advanced the time of flowering and maturation of seeds by 12 and 10.8 days, respectively. Seed germination and recruitment were advanced by 9.7 days, with warm temperate populations germinating in autumn and cold temperate populations in spring and early summer. These regional and global studies do not take into account adaptation of *Z. marina* to variation in local temperature, in which case survival of local populations may not be easy to predict. For example, in Korea higher sea temperatures resulted in faster seed maturation, whereas greater seed production of *Z. marina* occurred in cooler locations (Qin et al., 2020).

Different seagrass species within a genus respond to temperature in different ways. Marine heatwaves induce flowering in *Posidonia oceanica* (Diaz-Almela et al., 2007) in the northwestern part of the Mediterranean Sea (sea-water temperature range 12°C–25°C). However, marine heatwaves had the opposite effect on *P. australis* in Shark Bay (Australia) (sea temperature range 18°C–28°C), where seed abortion and pseudovivipary occur during and after extreme marine heatwaves (sea temperature 30°C) (Sinclair et al., 2016; Kendrick et al., 2019a). In temperate Australia, *Posidonia* species flower in mid-winter (16°C–18°C), and a single large seed develops within each fruit. Mature fruits float after they are released from the mother plants in early summer (19°C–22°C) (Sinclair et al., 2021). Under climate change, we expect higher winter sea temperatures and a shift in timing of flowering in these *Posidonia* species similar to that described above for *Z. marina*. With higher and seasonally shifting water temperatures, we predict an increase in flowering events for *P. oceanica* at its northern distribution in the Mediterranean Sea but reduced flowering and seed set and increased selfing and abortion in warmer North African populations, like those observed for *P. australis* in Shark Bay, Australia. Predictions based solely on temperature tolerance of temperate adult plants suggest that *P. australis* will become locally extinct in Shark Bay and that its geographic range will decrease southward by 100s of kilometers (Hyndes et al., 2016). Temperatures in Shark Bay waters have increased gradually and with extreme marine heatwaves, but *P. australis* has not become locally extinct. The species has a mixed mating strategy (selfing and outcrossing) with high seed abortion, low recruitment from seeds, high clonality, and pseudovivipary (Sinclair et al., 2016, 2020).

In terms of restoration, there is a small window of opportunity to collect seeds for regeneration from genera with recalcitrant seeds that are direct-developing (viviparous), that is, the single seed germinates producing a radicle and plumule while it is still attached to the mother plant (e.g., *Enhalus, Posidonia*, and *Thalassia*). For direct seeding (Sinclair et al., 2021) or for growing seedlings in nurseries for out-planting (Statton et al., 2013, 2014), seeds need to be collected either from fruits still attached to the mother plant or from those recently released and floating on the sea surface. Restoration efforts to use seeds and maintain genetic diversity in local populations through seedling recruitment will be impacted significantly by a warming climate. For example, the high rates of outcrossing in temperate *P. australis* meadows may shift to a mixed mating system and an increase in self-pollination, as observed in Shark Bay. Climate-adjusted seed provenancing is an obvious strategy that could be applied to seed-based seagrass restoration in preparation for further global warming. Seed provenancing has been recommended for terrestrial plant restoration (Prober et al., 2015) and remnant flowering meadows in tropical and warm temperate locations (e.g., Shark Bay, Australia, and the Mediterranean Sea in North Africa) (Sinclair et al., 2020; Pazzaglia et al., 2021), and it may hold the key for increasing resilience in temperate seagrass ecosystems in the next 50–100 years.

The ephemeral seagrass *Halophila decipiens* grows in tropical environments and has strong annual patterns of growth and reproduction that are associated with seasonal monsoons. Growth of plants in meadows occurs during the dry season and flowering and seed production at the beginning of the wet season. The species survives the wet season as seeds in a sediment seed bank, and seeds germinate at the beginning of the following dry season (Hovey et al., 2015; Fortes et al., 2018). There is no simple relationship between phases of the life cycle and water temperature, and other factors such as turbidity, storm frequency, and freshwater runoff also play a role in timing of different phases of the life cycle. Climate-driven changes in seasonal rains and storms will influence both seasonality of the life cycle and germination of seeds in the sediment seed bank, but their influence is difficult to predict and may be driven by regional shifts in climate. Similarly, in more temperate environments strong seasonality of the life cycle of *H. ovalis* is connected to multiple environmental cues: warming waters (15°C–20°C), increase in light availability, and shifts in light quality toward red wavelengths. These environmental factors are associated with spring warming, longer daylengths, and increased freshwater runoff that serve as cues for germination, which is followed by rapid plant growth and reproduction from summer into autumn, with dieback occurring at the onset of winter (Statton et al., 2017; Strydom et al., 2017; Waite et al., 2021). With the trend of climates drying across temperate regions of the globe, we expect reduced levels of seed germination, with clonal persistence and vegetative recruitment becoming more prevalent in maintaining populations. Marine populations of *H. ovalis* already persist with relatively low levels of recruitment from seeds, and some marine populations are highly clonal (McMahon et al., 2017).

Seed dispersal

Seed dispersal will be influenced by a reduction in flowering and seed production under climate change and thus by availability of seeds and changes in ocean conditions and grazing pressure during seed release. There are both biotically and abiotically driven patterns of dispersal in seagrasses. These include sea-surface dispersal of buoyant fruits and floating plant parts; sediment surface movement of negatively buoyant seeds; and ingestion and transport of reproductive fragments by birds, fishes, dugongs, and turtles.

The genera *Enhalus*, *Posidonia*, and *Thalassia* have floating fruits that contain one to many negatively buoyant seeds (Kendrick et al., 2012; McMahon et al., 2014). Fruits floating on the sea surface are influenced by wind and currents and can be moved 10s–100s of kilometers before they dehisce and release the seeds (e.g., Ruiz-Montoya et al., 2012), resulting in frequent recruitment into and connectivity among distant populations over evolutionary time scales (Kendrick et al., 2012; Sinclair et al., 2018). The genera *Cymodocea*, *Halodule*, *Halophila*, and *Syringodium* primarily disperse seeds locally (1s–100s m), but secondary resuspension during sediment movement due to wave action during storms can increase dispersal distances. Also, consumption and transport of reproductive fragments by birds and marine grazers (fishes, dugongs, and turtles) can result in long-distance dispersal and increased regional connectivity among meadows (Sumoski and Orth, 2012; McMahon et al., 2014; Tol et al., 2017). Genera such as *Zostera* have floating reproductive shoots (rhipidia or spathes) that also increase dispersal distances to 10s–100s of kilometers. Seeds of *Z. marina* generally are dispersed only a few meters from the parent plant, but those released with an attached air bubble can be dispersed 10s of meters (Churchill et al., 1985). Dispersal of floating rhipidia with viable seeds accounts for long-distance dispersal of 100s of meters to 100s of kilometers (Harwell and Orth, 2002; Kendrick et al., 2012).

How climate change will influence seed dispersal in seagrasses is difficult to predict given the number of modes (floating fruits, rhipidia, and seeds) and vectors (sea surface wind, waves and currents, seafloor currents and swells, and birds, fishes, turtles, and dugongs). Climate-driven changes in seed dispersal are likely to be restricted to regions where (1) low numbers of seeds are produced, for example, temperature-sensitive populations at the edge of the species' range (Sinclair et al., 2020); (2) climate-driven increases in disturbance from storms influence clonal diversity, thus impacting reproductive success and seed production (e.g., McMahon et al., 2017); or (3) variability in ocean conditions and storminess impact fertilization of flowers and/or seed dispersal (e.g., Kendrick et al., 2017).

Seed settlement and early seedling survival

Seed settlement and early seedling survival are influenced by changes in the timing, frequency, and duration of storms. The major limitation on seeds is that they settle into hydrodynamically active environments, where drag and lift forces are up to three times greater than they are in air (Kendrick et al., 2019b). Seed settlement to seedling establishment is the most critical bottleneck in the seagrass life cycle (Statton et al., 2017); however, several traits assist in overcoming these challenges. Seeds of *P. oceanica* in the Mediterranean Sea develop many microscopic adhesive hairs on the seeds that promote attachment to sand grains while the root develops (Guerrero-Meseguer et al., 2018). Similarly, in *Z. muelleri* adhesive hypocotyl hairs develop as seeds germinate (Stafford-Bell et al., 2016). Several genera have seeds (*Phyllospadix*) or viviparous seedlings (*Amphibolis* and *Thalassodendron*) with morphological "anchor-like" structures that allow them to "hook" onto either algae or other firm surfaces (Rivers et al., 2011; Kendrick et al., 2012; McMahon et al., 2014). Seeds on the sediment surface do not move far from where they settle, and they can be buried rapidly (Orth et al., 1994; Delefosse and Kristensen, 2012; Blackburn and Orth, 2013; Kendrick et al., 2019b). Any climate-driven increases in populations of bioturbators and seed predators also will have a major effect on seed and seedling survival and recruitment. Sediment reworking by bioturbators such as polychaetes move seeds below the sediment surface, where conditions are conducive for germination, and loss of seeds due to mobile surface-dwelling seed predators is minimized (Fishman and Orth, 1996). Seedlings from seeds buried too deeply, for example, as a result of storm-driven wave action, may be unable to grow enough to reach the sediment surface (Marion and Orth, 2012; Marion et al., 2021).

Climate change and seagrass regeneration from seeds—the future

Climate change has the potential to significantly influence natural populations of seagrasses and the practice and outcomes of seagrass conservation and seed-based restoration. Importantly, more detailed climate studies in terms of seed resilience are needed across a range of genera, not just the three to five major seagrass genera utilized for restoration. This would enhance our ability to predict more accurately the impact of climate change and therefore conservation and restoration responses, especially in seagrass biodiversity hotspots.

For genera that have seeds with a hard (but water-permeable, i.e., not physical dormancy) pericarp or seed coat (*Cymodocea, Halodule, Halophila,* and *Zostera*), viability, dormancy, and germination are important areas for understanding the direct and indirect impacts of climate change. Changes in the environment due to global warming have the potential to alter timing of seed availability or germination and thus the life cycle of species, for example, annual and perennial plants of *Zostera marina* that occur near the low latitude edge of the species' range (Johnson et al., 2021). Temporal changes in fecundity, seed quality and settlement, seed banks, and seedling emergence also will impact the size of populations and their demographic structures (e.g., out-of-season triggers for germination result in cohort death; Johnson et al., 2021).

For genera with no seed dormancy (*Enhalus, Posidonia,* and *Thalassia*), we need to understand the effect of shifting seasonality, increased storminess, and changing temperatures on dispersal of seeds/seedlings to the sediment and stability of seedlings during attachment and initial root development. The first year of life for these species is critical, and it is a significant bottleneck to population recovery or restoration (Statton et al., 2017) due to their limited ability to become reestablished after secondary dispersal.

Climate-driven changes in the distribution of seagrass-associated biota could profoundly impact seagrass communities and the outcome of restoration. Increases in abundance of animals or changes in species distributions of bioturbators or seed-predators could drastically alter community composition by increasing sediment disturbance regimes (exposure or burial of the seed bank and/or burial of emerging seedlings) (Johnson et al., 2018) or by overgrazing (Orth et al., 2006b).

We need to determine temperature thresholds for seeds, seedlings, and adult plants for all seagrass species in order to generate the data needed to improve our understanding of the capacity of seeds to germinate in a warming climate, especially for restoration projects. Climate simulations based on temperature thresholds for adult plants already have established a framework for understanding the general effects of climate change on distribution of seagrasses (Hyndes et al., 2016; Chefaoui et al., 2018). Our predictive capability would be improved greatly if we knew how temperature and other climate-driven influences affect seed viability, dormancy-break and germination, seedling settlement and attachment, and plant growth and seed production for a large number of seagrass species, especially those along climatic and latitudinal gradients. We presently have this kind of information for only a few species (e.g., *Z. marina*; Blok et al., 2018).

Restoration of seagrasses is becoming an accepted and necessary way to maintain resilience in seagrass ecosystems (e.g., Tan et al., 2020). Seed-based restoration in seagrasses is still in its infancy, but success can be extraordinary with whole ecosystems being restored over decades (Orth et al., 2020; Sinclair et al., 2021). We need to add climate-adjusted provenance seed sourcing (e.g., Prober et al., 2015) to our restoration protocols in order to build resilience in seagrasses to a warming ocean.

Acknowledgments

Gary A. Kendrick, Robert J. Orth, John Statton and Elizabeth A. Sinclair were funded by the Australian Research Council (ARC Discovery DP180100668, DP210101932, and ARC Linkage LP160101011). This is contribution 4030 from the Virginia Institute of Marine Science, College of William & Mary.

References

Alberto, F., Mata, L., Santos, R., 2001. Genetic homogeneity in the seagrass *Cymodocea nodosa* at its northern Atlantic limit revealed through RAPD. Mar. Ecol. Prog. Ser. 221, 299–301.

Arnaud-Haond, S., Duarte, C.M., Diaz-Almela, E., Marbà, N., Sintes, T., Serrão, E.A., 2012. Implications of extreme life span in clonal organisms: millenary clones in meadows of the threatened seagrass *Posidonia oceanica*. PLoS One 7, e30454. Available from: https://doi.org/10.1371/journal.pone.0030454.

Beer, S., Koch, E., 1996. Photosynthesis of marine macroalgae and seagrasses in globally changing CO_2 environments. Mar. Ecol. Prog. Ser. 141, 199–204.

Blackburn, N.J., Orth, R.J., 2013. Seed burial in *Zostera marina* (eelgrass): the role of infauna. Mar. Ecol. Prog. Ser. 474, 135–145.

Blok, S.E., Olesen, B., Krause-Jensen, D., 2018. Life history events of eelgrass *Zostera marina* L. populations across gradients of latitude and temperature. Mar. Ecol. Prog. Ser. 590, 79–93.

Borum, J., Pedersen, O., Kotula, L., Fraser, M.W., Statton, J., Colmer, T.D., et al., 2016. Photosynthetic response to globally increasing CO_2 of co-occurring temperate seagrass species. Plant Cell Environ. 39, 1240–1250.

Bricker, E., Calladine, A., Virnstein, R., Waycott, M., 2018. Mega clonality in an aquatic plant - a potential survival strategy in a changing environment. Front. Plant Sci. 9, 435. Available from: https://doi.org/10.3389/fpls.2018.00435.

Chefaoui, R.M., Duarte, C.M., Serrão, E.A., 2018. Dramatic loss of seagrass habitat under projected climate change in the Mediterranean Sea. Glob. Change Biol. 24, 4919–4928.

Churchill, A.C., Nieves, G., Brenowitz, A.H., 1985. Flotation and dispersal of eelgrass seeds by gas bubbles. Estuaries 8, 352–354.

Cullen-Unsworth, L.C., Unsworth, R., 2018. A call for seagrass protection. Science 361, 446–448.

Delefosse, M., Kristensen, E., 2012. Burial of *Zostera marina* seeds in sediment inhabited by three polychaetes: laboratory and field studies. J. Sea Res. 71, 41–49.

Diaz-Almela, E., Marba, N., Duarte, C.M., 2007. Consequences of Mediterranean warming events in seagrass (*Posidonia oceanica*) flowering records. Glob. Change Biol. 13, 224–235.

Eckert, C.G., Samis, K.E., Lougheed, S.C., 2008. Genetic variation across species' geographical ranges: the central-marginal hypothesis and beyond. Mol. Ecol. 17, 1170–1188.

Entrambasaguas, L., Jahnke, M., Biffali, E., Borra, M., Sanges, R., Marín-Guirao, L., et al., 2017. Tissue-specific transcriptomic profiling provides new insights into the reproductive ecology and biology of the iconic seagrass species *Posidonia oceanica*. Mar. Genom. 35, 51–61.

Fishman, J.R., Orth, R.J., 1996. Effects of predation on *Zostera marina* seed abundance. J. Exp. Mar. Biol. Ecol. 198, 11–26.

Fortes, M., Ooi, J.L.S., Tan, Y.M., Prathep, A., Bujang, J.S., Yaakub, S.M., 2018. Seagrass habitats in Southeast Asia: a review of status and knowledge gaps, and a roadmap for conservation. Bot. Marina (Spec. Issue) 61 (3), 269–288.

Fraser, M.W., Short, J., Kendrick, G.A., McLean, D., Keesing, J., Byrne, M., et al., 2017. Effects of dredging on critical ecological processes for marine invertebrates, seagrasses and macroalgae, and the potential for management with environmental windows using Western Australia as a case study. Ecol. Indicat. 78, 229–242.

Glasby, T.M., West, G., 2018. Dragging the chain: quantifying continued losses of seagrasses from boat moorings. Aquat. Conserv. Mar. Freshw. Ecosyst. 28, 383–394.

Golicz, A.A., Schliep, M., Lee, H.T., Larkum, A.W.D., Dolferus, R., Batley, J., et al., 2015. Genome-wide survey of the seagrass *Zostera muelleri* suggests modification of the ethylene signalling network. J. Exp. Bot. 66, 1489–1498.

Guerrero-Meseguer, L., Sanz-Lázaro, C., Marín, A., 2018. Understanding the sexual recruitment of one of the oldest and largest organisms on Earth, the seagrass *Posidonia oceanica*. PLoS One 13, e0207345. Available from: https://doi.org/10.1371/journal.pone.0207345.

Harwell, M.C., Orth, R.J., 2002. Long distance dispersal potential in a marine macrophyte. Ecology 83, 3319–3330.

Hovey, R.K., Statton, J., Fraser, M.W., Kendrick, G.A., 2015. Strategy for assessing impacts in ephemeral tropical seagrasses. Mar. Poll. Bull. 101, 594–599.

Hyndes, G.A., Heck Jr., K.L., Harvey, E.S., Kendrick, G.A., Lavery, P.S., McMahon, K., et al., 2016. Accelerating tropicalization and the transformation of temperate seagrass meadows. Biosciences 16, 938–948.

Johnson, A.J., Statton, R., Orth, R.J., Kendrick, G.A., 2018. A sediment bioturbator bottleneck to seedling recruitment for the seagrass *Posidonia australis*. Mar. Ecol. Prog. Ser. 595, 89–103.

Johnson, A.J., Shields, E.C., Kendrick, G.A., Orth, R.J., 2021. Recovery dynamics of the seagrass *Zostera marina* following mass mortalities from two extreme climatic events. Estuar. Coasts 44, 535–544.

Kendrick, G.A., Waycott, M., Carruthers, T.J.B., Cambridge, M.L., Hovey, R.K., Krauss, S.L., et al., 2012. The central role of dispersal in the maintenance and persistence of seagrass populations. BioScience 62, 56–65.

Kendrick, G.A., Orth, R.J., Statton, J., Hovey, R.K., Ruiz Montoya, L., Lowe, R.J., et al., 2017. Demographic and genetic connectivity: the role and consequences of reproduction, dispersal and recruitment in seagrasses. Biol. Rev. 92, 921–938.

Kendrick, G.A., Nowicki, R.J., Olsen, Y.S., Strydom, S., Fraser, M.W., Sinclair, E.A., et al., 2019a. A systematic review of how multiple stressors from an extreme event drove ecosystem-wide loss of resilience in an iconic seagrass community. Front. Mar. Sci. 6, 455. Available from: https://doi.org/10.3389/fmars.2019.00455.

Kendrick, G.A., Pomeroy, A.W., Orth, R.J., Cambridge, M.L., Shaw, J., Kotula, L., et al., 2019b. A novel adaptation facilitates seed establishment under marine turbulent flows. Sci. Rep. 9, 19693. Available from: https://doi.org/10.1038/s41598-019-56202-7.

Larkum, A., Davey, P.A., Kuo, J., Ralph, P.J., Raven, J.A., 2017. Carbon-concentrating mechanisms in seagrasses. J. Exp. Bot. 68, 3773–3784.

Larkum, A., Kendrick, G.A., Ralph, P. (Eds.), 2018. Seagrasses of Australia - Structure, Ecology and Conservation. Springer, Dordrecht.

Lee, H.T., Golicz, A.A., Bayer, P.E., Jiao, Y., Tang, H., Paterson, A.H., et al., 2016. The genome of a southern hemisphere seagrass species (*Zostera muelleri*). Plant Physiol. 172, 272–283.

Lee, H.T., Golicz, A.A., Bayer, P.E., Chan, C.-K.K., Batley, J., Kendrick, G.A., et al., 2018. Genomic comparison of two independent seagrass return-to-the-sea events. J. Exp. Bot. 69, 3689–3702.

Les, D.H., 1988. Breeding systems, population structure and evolution in hydrophilous angiosperms. Ann. Mo. Bot. Gard. 75, 819–835.

Les, D.H., Cleland, M.A., Waycott, M., 1997. Phylogenetic studies in Alismatidae, II: evolution of marine angiosperms (seagrasses) and hydrophily. Syst. Bot. 22, 443–463.

Marba, N., Duarte, C.M., 2010. Mediterranean warming triggers seagrass (*Posidonia oceanica*) shoot mortality. Glob. Change Biol. 16, 2366–2375.

Marion, S.R., Orth, R.J., 2012. Seedling establishment in eelgrass: seed burial effects on winter losses of developing seedlings. Mar. Ecol. Prog. Ser. 448, 197–207.

Marion, S.R., Orth, R.J., Fonseca, M., Malhotra, A., 2021. Seed burial alleviates wave energy constraints on *Zostera marina* (eelgrass) seedling establishment at restoration-relevant scales. Estuar. Coasts 44, 352–366.

McConchie, C.A., Knox, R.B., 1989. Pollen-stigma interaction in the seagrass *Posidonia australis*. Ann. Bot. 63, 235–248.

McMahon, K., van Dijk, K.-J., Ruiz-Montoya, L., Kendrick, G.A., Krauss, S.L., Waycott, M., et al., 2014. The movement ecology of seagrasses. Proc. R. Soc. B Biol. Sci. 281, 20140878. Available from: https://doi.org/10.1098/rspb.2014.0878.

McMahon, K.M., Evans, R.D., van Dijk, K., Hernawan, U., Kendrick, G.A., Lavery, P.S., et al., 2017. Disturbance is an important driver of clonal richness in tropical seagrasses. Front. Plant Sci. 8, 2026. Available from: https://doi.org/10.3389/fpls.2017.02026.

Miller, K.G., Mountain, G.S., Wright, J.D., Browning, J.V., 2011. A 180-million-year record of sea level and ice volume variations from continental margin and deep-sea isotopic records. Oceanography 24, 40–53.

Moore, K.A., Orth, R.J., Nowak, J.F., 1993. Environmental regulation of seed germination in *Zostera marina* L. (eelgrass) in Chesapeake Bay: effects of light, oxygen, and sediment burial depth. Aquat. Bot. 45, 79–91.

Olsen, J.L., Rouzé, P., Verhelst, B., Lin, Y.-C., Bayer, T., Collen, J., et al., 2016. The genome of the seagrass *Zostera marina* reveals angiosperm adaptation to the sea. Nature 530, 331–335.

Olsen, Y.S., Collier, C., Ow, Y.X., Kendrick, G.A., 2018. Global warming and ocean acidification: effects on Australian seagrass ecosystems. In: Larkum, A.W., Kendrick, G.A., Ralph, P. (Eds.), Seagrasses of Australia - Structure, Ecology and Conservation. Springer, Dordrecht, pp. 705–742.

Orth, R.J., Luckenbach, M.L., Moore, K.A., 1994. Seed dispersal in a marine macrophyte: implications for colonization and restoration. Ecology 75, 1927–1939.

Orth, R.J., Harwell, M.C., Bailey, E.M., Bartholomew, A., Jawad, J.T., Lombana, A.V., et al., 2000. A review of issues in seagrass seed dormancy and germination: implications for conservation and restoration. Mar. Ecol. Prog. Ser. 200, 277–288.

Orth, R.J., Carruthers, T.J.B., Dennison, W.C., Duarte, C.M., Fourqurean, J.W., Heck Jr., K.L., et al., 2006a. A global crisis for seagrass ecosystems. BioScience 56, 987–996.

Orth, R.J., Kendrick, G.A., Marion, S.R., 2006b. *Posidonia australis* seed predation in seagrass habitats of Rottnest Island, Western Australia: patterns and predators. Mar. Ecol. Prog. Ser. 313, 105–114.

Orth, R.J., Lefcheck, J.S., Wilcox, D.J., 2017. Boat propeller scarring of seagrass beds in lower Chesapeake Bay, USA: patterns, causes, recovery, and management. Estuar. Coasts 40, 1666–1676.

Orth, R.J., Lefcheck, J.S., McGlathery, K.S., Aoki, L., Luckenback, M.L., Moore, K.A., et al., 2020. Restoration of seagrass habitat leads to rapid recovery of coastal ecosystem services. Sci. Adv. 6, eabc6434. Available from: https://doi.org/10.1126/sciadv.abc6434.

Pazzaglia, J., Nguyen, H.M., Santillan-Sarmiento, A., Ruocco, M., Dattolo, E., Marin-Guirao, L., et al., 2021. The genetic component of seagrass restoration: what we know and the way forwards. Water 13, 829. Available from: https://doi.org/10.3390/w13060829.

Pedersen, O., Colmer, T.D., Borum, J., Zavala-Perez, A., Kendrick, G.A., 2016. Heat stress of two tropical seagrass species during low tides – impact on underwater net photosynthesis, dark respiration and diel in situ internal aeration. New Phytol. 210, 1207–1218.

Prober, S.M., Byrne, M., McLean, E.H., Steane, D.A., Potts, B.M., Vaillancourt, R.E., et al., 2015. Climate-adjusted provenancing: a strategy for climate-resilient ecological restoration. Front. Ecol. Evol. 3, 65. Available from: https://doi.org/10.3389/fevo.2015.00065.

Qin, L.Z., Kim, S.H., Song, H.J., Suonan, Z., Kim, H., Kwon, O., et al., 2020. Influence of regional water temperature variability on the flowering phenology and sexual reproduction of the seagrass *Zostera marina* in Korean coastal waters. Estuar. Coasts 43, 449–462.

Reusch, T.B.H., Bostrom, C., Stam, W.T., Olsen, J.L., 1999. An ancient eelgrass clone in the Baltic Sea. Mar. Ecol. Prog. Ser. 183, 301–304.

Rivers, D.O., Kendrick, G.A., Walker, D.I., 2011. Microsites play an important role for seedling survival in the seagrass *Amphibolis antarctica*. J. Exp. Mar. Biol. Ecol. 401, 29–35.

Ruckelshaus, M.H., 1995. Estimates of outcrossing rates and of inbreeding depression in a population of the marine angiosperm *Zostera marina*. Mar. Biol. 123, 583–593.

Ruiz-Montoya, L., Lowe, R.J., Van Niel, K.P., Kendrick, G.A., 2012. The role of hydrodynamics on seed dispersal in seagrasses. Limnol. Oceanogr. 57, 1257–1265.

Shelton, A.O., 2008. Skewed sex ratios, pollen limitation, and reproductive failure in the dioecious seagrass *Phyllospadix*. Ecology 89, 3020–3029.

Short, F.T., Polidoro, B., Livingstone, S.R., Carpenter, K.E., Bandeira, S., Bujang, J.S., et al., 2011. Extinction risk assessment of the world's seagrass species. Biol. Conserv. 144, 1961–1971.

Silberhorn, G., Orth, R.J., Moore, K.A., 1983. Anthesis and seed production in *Zostera marina* L. (eelgrass) from the Chesapeake Bay. Aquat. Bot. 15, 133–144.

Sinclair, E.A., Statton, J., Hovey, R.K., Anthony, J.M., Dixon, K.W., Kendrick, G.A., 2016. Reproduction at the extremes: pseudovivipary and genetic mosaicism in *Posidonia australis* Hooker (Posidoniaceae). Ann. Bot. 117, 237−247.

Sinclair, E.A., Ruiz-Montoya, L., Krauss, S.L., Anthony, J.M., Hovey, R.K., Lowe, R.J., et al., 2018. Seeds in motion: genetic assignment and hydrodynamic models demonstrate concordant patterns of seagrass dispersal. Mol. Ecol. 27, 5019−5034.

Sinclair, E.A., Edgeloe, J.M., Anthony, J.M., Statton, J., Breed, M.F., Kendrick, G.A., 2020. Variation in reproductive effort, genetic diversity and mating systems across *Posidonia australis* seagrass meadows in Western Australia. AoB Plants 12, plaa038. Available from: https://doi.org/10.1093/aobpla/plaa038.

Sinclair, E.A., Sherman, C.D.H., Statton, J., Copeland, C., Matthews, A., Waycott, M., et al., 2021. Advances in approaches to seagrass restoration in Australia. Ecol. Manage. Restor. 22, 10−21.

Stafford-Bell, R.E., Chariton, A.A., Robinson, R.W., 2016. Germination and early-stage development in the seagrass, *Zostera muelleri* Irmisch ex Asch. in response to multiple stressors. Aquat. Bot. 128, 18−25.

Statton, J., Cambridge, M.L., Dixon, K.W., Kendrick, G.A., 2013. Aquaculture of *Posidonia australis* seedlings for seagrass restoration programs: effect of sediment type and organic enrichment on growth. Restor. Ecol. 21, 250−259.

Statton, J., Kendrick, G.A., Dixon, K.W., Cambridge, M.L., 2014. Inorganic nutrient supplements constrain restoration potential of seedlings of the seagrass, *Posidonia australis*. Restor. Ecol. 22, 196−203.

Statton, J., Ruiz-Montoya, L., Orth, R.J., Dixon, K.W., Kendrick, G.A., 2017. Identifying critical recruitment bottlenecks limiting seedling establishment in a degraded seagrass ecosystem. Sci. Rep. 7, 14786. Available from: https://doi.org/10.1038/s41598-017-13833-y.

Strydom, S., McMahon, K.M., Kendrick, G.A., Lavery, P.S., Statton, J., 2017. Seagrass *Halophila ovalis* is affected by light quality across different life history stages. Mar. Ecol. Prog. Ser. 572, 103−116.

Strydom, S., Murray, K., Wilson, S., Huntley, B., Rule, M., Heithaus, M., et al., 2020. Too hot to handle: unprecedented seagrass death driven by marine heatwave in a World Heritage Area. Glob. Change Biol. 3525−3538.

Sumoski, S., Orth, R.J., 2012. Biotic dispersal in eelgrass *Zostera marina*. Mar. Ecol. Prog. Ser. 471, 1−10.

Tan, Y.M., Dalby, O., Kendrick, G.A., Statton, J., Sinclair, E.A., Fraser, M.W., et al., 2020. Seagrass restoration is possible: insights and lessons from Australia and New Zealand. Front. Mar. Sci. 7, 617. Available from: https://doi.org/10.3389/fmars.2020.00617.

Tol, S., Jarvis, J., York, P., Grech, A., Congdon, B., Coles, R., 2017. Long distance biotic dispersal of tropical seagrass seeds by marine megaherbivores. Sci. Rep. 7, 4458. Available from: https://doi.org/10.1038/s41598-017-04421-1.

Tsakogiannis, A., Manousaki, T., Anagnostopoulou, V., Stavroulaki, M., Apostolaki, E.T., 2020. The importance of genomics for deciphering the invasion success of the seagrass *Halophila stipulacea* in the changing Mediterranean Sea. Diversity 12, 263. Available from: https://doi.org/10.3390/d12070263.

Unsworth, R.K.F., Williams, B., Jones, B.L., Cullen-Unsworth, L.C., 2017. Rocking the boat: damage to eelgrass by swinging boat moorings. Front. Plant Sci. 8, 1309. Available from: https://doi.org/10.3389/fpls.2017.01309.

van Katwijk, M.M., Thorhaug, A., Marbà, N., Orth, R.J., Duarte, C.M., Kendrick, G.A., et al., 2016. Global review of seagrass restoration and the importance of large-scale planting. J. Appl. Ecol. 53, 567−578.

van Tussenbroek, B., Villamil, N., Márquez-Guzmán, J., Wong, R., Monroy-Velázquez, L.V., Solis-Weiss, V., 2016. Experimental evidence of pollination in marine flowers by invertebrate fauna. Nat. Commun. 7, 12980. Available from: https://doi.org/10.1038/ncomms12980.

Waite, B., Statton, J., Kendrick, G.A., 2021. Temperature stratification and monochromatic light break dormancy and facilitate on-demand *in situ* germination in the seagrass *Halophila ovalis*, with seed viability determined by a novel x-ray analysis. Estuar. Coasts 44, 412−421.

Waycott, M., Duarte, C.M., Carruthers, T.J.B., Orth, R.J., Dennison, W.C., Olyarnik, S., et al., 2009. Accelerating loss of seagrasses across the globe threatens coastal ecosystems. Proc. Natl. Acad. Sci. USA 106, 12377−12381.

Wilson, S.S., Furman, B.T., Hall, M.O., Fourqurean, J.W., 2020. Assessment of Hurricane Irma impacts on south Florida seagrass communities using long-term monitoring programs. Estuar. Coasts 43, 1119−1132.

Wissler, L., Codoner, F.M., Gu, J., Reusch, T.B.H., Olsen, J.L., Procaccini, G., et al., 2011. Back to the sea twice: identifying candidate plant genes for molecular evolution to marine life. BMC Evol. Biol. 11, 8. Available from: https://doi.org/10.1186/1471-2148-11-8.

Wu, P., McMahon, K.M., Rasheed, M.A., Kendrick, G.A., York, P.H., Chartrand, K., et al., 2018. Managing seagrass resilience under cumulative dredging affecting light: predicting risk using dynamic Bayesian networks. J. Appl. Ecol. 55, 1339−1350.

Zimmerman, R.C., Hill, V.J., Jinuntuya, M., Celebi, B., Ruble, D., Smith, M., et al., 2017. Experimental impacts of climate warming and ocean carbonation on eelgrass *Zostera marina*. Mar. Ecol. Prog. Ser. 566, 1−15.

Chapter 21

Soil seed banks under a warming climate

Margherita Gioria[1], Bruce A. Osborne[2,3] and Petr Pyšek[1,4]

[1]*Department of Invasion Ecology, Institute of Botany, Czech Academy of Sciences, Průhonice, Czech Republic,* [2]*School of Biology and Environmental Science, University College Dublin, Dublin, Ireland,* [3]*Earth Institute, University College Dublin, Dublin, Ireland,* [4]*Department of Ecology, Faculty of Science, Charles University, Prague, Czech Republic*

Introduction

Global warming is an important driver of recent and projected future changes in the distribution of species (Urban, 2015; Kerr, 2020), and its effects have become evident in various ecosystems and taxonomic groups worldwide (Parmesan and Yohe, 2003; Parmesan, 2006; Baldwin et al., 2014; IPCC, 2014). Along with global warming, increases in the frequency and magnitude of extreme climatic events (IPCC, 2013, 2014) are being proposed as one of the causes of present-day extinction of species (Root et al., 2003; Urban, 2015). A warming climate can have cascading effects on plant populations and community dynamics and ultimately on plant distribution. Among the most evident effects of climate change are alterations in community composition (Parmesan and Yohe, 2003), phenology (Miller-Rushing et al., 2010; Ovaskainen et al., 2013; Orsenigo et al., 2015), selective adaptation (Hoffmann and Sgrò, 2011; Merritt et al., 2014), and species distribution. The distribution of a species is associated with conditions that have become unsuitable, or for some species more suitable, for survival, growth, and/or recruitment. Distributional changes include the loss of populations from previously occupied areas and the poleward and upward/elevational expansion in the distribution of many plants (Grabherr et al., 1994; Lenoir et al., 2008; Doak and Morris, 2010; Urban, 2015; Manish et al., 2016; Kerr, 2020), supporting earlier predictions of a widespread redistribution of species, as well as accelerated extinction rates (Peters and Darling, 1985).

In some cases, shifts in the distribution of individual species associated with a warming climate have resulted in the formation of novel communities (Parmesan and Yohe, 2003; Manish et al., 2016; Giménez-Benavides et al., 2018; Løkken et al., 2019), while in cases where individual species have similar responses, community composition has remained largely intact (Beckage et al., 2008; Lenoir et al., 2008; Shevtsova et al., 2009; Pucko et al., 2011). However, changes in the distribution of individual species will likely become increasingly divergent as they approach warming-related environmental thresholds (Pucko et al., 2011). Moreover, while climate change is expected to have the greatest impact at the margins of a species' distribution (Manish et al., 2016), there is evidence that it also can affect the core of their distribution (Lenoir et al., 2008).

Since many distributional shifts are occurring more rapidly than originally anticipated (Beckage et al., 2008), predicting the magnitude, direction, and speed of response to range modifications and the resilience of plant communities to any changes is a critical conservation issue (Corlett and Westcott, 2013; Van Looy et al., 2016; Løkken et al., 2019). Ultimately, the ability of a species to disperse to new sites and to track the altered climatic conditions will be critical in shaping its future distribution (Peters and Darling, 1985; Bertrand et al., 2011; Baldwin et al., 2014). It is also considered unlikely that most plant species can disperse fast enough to keep pace with the rate of rapid climate change (Bertrand et al., 2011; Corlett and Westcott, 2013). The question then is whether these species can adapt to the new conditions (Skelly et al., 2007) via selection of individuals with increased fitness or whether they possess the phenotypic plasticity required to survive in the changed environment (Parmesan, 2006; Hoffmann and Sgrò, 2011; Manish et al., 2016; Colautti et al., 2017).

For terrestrial plants, limitations in short-distance seed dispersal (Bertrand et al., 2011) and/or abiotic constraints on germination, growth, and establishment (Lloret et al., 2005; Shevtsova et al., 2009) can limit their ability to keep pace with the changing climatic conditions (Baldwin et al., 2014). Alternatively, the ability to persist in a community as latent propagules/seeds that allow a species to spread not only in space but also through time (Harper, 1977, Fenner and Thompson, 2005) could play a key role in determining future plant community dynamics (Ooi et al., 2009; Walck et al., 2011). A key distinction for understanding the potential contribution of soil seed banks (hereafter seed banks) to the future responses of individual plants and communities to a warming climate is the difference between persistent and transient seed banks. Persistent seed banks are composed of seeds that retain their viability in the soil for >1 year or to the second germination season, while transient seed banks are those whose seeds lose viability and/or germinate in ≤1 year or do not persist until the second germination season (Thompson et al., 1997; Walck et al., 2005).

Relying on seeds that may persist in the soil over multiple germination seasons has long been regarded as a bet-hedging strategy against the risks of reproductive failure associated with unpredictable environmental conditions (Cohen, 1966; Venable and Brown, 1988; Venable, 2007; Tielbörger et al., 2012; Larson and Funk, 2016). As reserves of genetic variability (Templeton and Levin, 1979; Honnay et al., 2008), persistent seed banks also might play an important role in determining the evolutionary response of seed plants to environmental unpredictability (Venable and Brown, 1988; Baskin et al., 1998; Donohue et al., 2005, 2010), although there is little evidence for this. However, the ability to form a persistent seed bank has been recognized as a major component of ecosystem resilience (Hopfensperger, 2007; González-Alday et al., 2009; Walck et al., 2011; Plue et al., 2013, 2021; Blossey et al., 2017; Ma et al., 2019).

Since temperature is a critical factor influencing the persistence of seeds in the soil (Baskin and Baskin, 2014), a warming climate can have profound effects on the composition and structure of the seed bank and, in turn, on the persistence of individual species and communities (Walck et al., 2011; Hoyle et al., 2013; Long et al., 2015; Bernareggi et al., 2016; Giménez-Benavides et al., 2018). However, assessments of the effects of a warming climate on the seed bank have received less attention than the standing vegetation (Grime et al., 2000, 2008; Ooi et al., 2009; Briceño et al., 2015; Basto et al., 2018). Here, we review recent evidence for climate and climate-related changes in the persistence and structure (richness, size, and composition) of seed banks across different ecosystems and biomes. While excellent studies have described the mechanisms by which a warming climate may affect seed bank persistence and regeneration from seeds (Walck et al., 2011; Ooi, 2012; Jaganathan et al., 2015), we focus on the observed and predicted changes in the seed bank and discuss their role in promoting ecosystem resilience by preventing species extinctions and contributing to the migration of species.

Effects of a warming climate on seed bank persistence and density

A warming climate may have a cascading effect on the persistence and size of the seed bank of individual species. Seed persistence in the soil is a function of a range of seed traits and pre- and postdispersal biotic and abiotic conditions (Thompson et al., 1993, 2003; Bekker et al., 1998; Long et al., 2008, 2015), which define the seed ecological spectrum (Saatkamp et al., 2019). Seed bank size is defined here as the number of seeds in/on the soil per surface area of the soil and represents the balance between seed inputs and seed outputs.

There are concerns that global warming may accelerate losses of species from the seed bank through adverse effects on seed aging, viability, and longevity (Fig. 21.1). Several experimental studies provide evidence for accelerated seed aging under high temperatures (Bekker et al., 1998; Leishman et al., 2000; Murdoch and Ellis, 2000; Parmesan, 2006; Long et al., 2008, 2015; Kochanek et al., 2011; Bernareggi et al., 2015; Panetta et al., 2018; Luna, 2020), although the response to artificial aging conditions may differ from that of seeds aging naturally in the seed bank (Roach et al., 2018). Also, the imposition of short-term increases in temperature in laboratory experiments may not reflect the generally slower longer-term temperature increases expected in the field. Field evidence suggests that cold and wet postdispersal environments can reduce seed deterioration (Cavieres and Arroyo, 2000; Bewley et al., 2013), thereby promoting seed persistence and the accumulation of seeds in the soil (Pakeman et al., 1999; Cummins and Miller, 2002). In contrast, decreased persistence in the soil due to accelerated aging and reduced viability could deplete the seed bank and its regeneration potential (Ooi, 2012; Bewley et al., 2013). However, shorter-term environmental variations can result in marked transgenerational changes in seed longevity (Kochanek et al., 2011). Increases in seed longevity resulting from warming-related population shifts that select phenotypes with increased resistance to warming could play an important role in promoting survival of plants and long-term adaptation to a rapidly changing environment (Nicotra et al., 2010; Mondoni et al., 2014; Bernareggi et al., 2015). Selection of phenotypes resistant to warming might explain why species from warmer and drier regions generally produce longer-lived seeds than species from cooler regions (Probert et al., 2009; Mondoni et al., 2011; Merritt et al., 2014). Clearly, the effects of a warmer parental environment on seed longevity may interact with other environmental conditions, including the generally drier, but possibly wetter, conditions associated with warming (Probert et al., 2009; Kochanek et al., 2011).

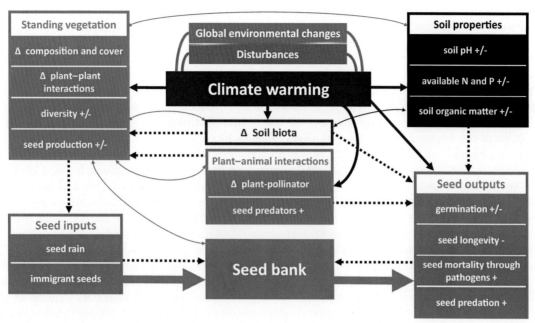

FIGURE 21.1 A summary of the main direct and indirect effects of climate warming on the soil seed bank, including interactions with the standing vegetation, soil physiochemical properties, and soil biota. Solid *blue arrows* indicate the direct effects of climate warming and dashed *blue arrows* indirect effects on seed bank inputs and outputs, with inflows and outflows from the seed bank shown by thick *orange lines*. Interactions between climate warming and other global environmental changes and disturbances are shown by solid lines, and thin arrows represent double-loop feedback processes. The main positive (+) or negative (−) effects and changes (Δ) associated with climate warming identified in the literature are highlighted.

Another important way for loss of seeds from the seed bank is germination (Fig. 21.1). The documented impacts of increased pre- and postdispersal temperatures on germination include dormancy alleviation for increasing proportions of physically dormant seeds and of physiologically dormant seeds of many species (e.g., Auld and O'Connell, 1991; Baskin and Baskin, 2014; Bernareggi et al., 2016; Aragón-Gastélum et al., 2018; Footitt et al., 2018). Also, soil warming may increase or reduce germination and shift the timing of germination (Petrů and Tielbörger, 2008; Milbau et al., 2009; Ooi et al., 2009; Shevtsova et al., 2009; Walck et al., 2011; Jaganathan et al., 2015; Orsenigo et al., 2015; Footitt et al., 2018), although soil type has been shown to buffer some of these effects (Petrů and Tielbörger, 2008). Germination responses to warming may be species-specific, depending on the temperature requirements for dormancy-break and germination, the class of dormancy, and seasonal environmental variability in the extent of dormancy (Walck et al., 2011; Baskin and Baskin, 2014; Blossey et al., 2017; Rubio de Casas et al., 2017; Footitt et al., 2018).

The longer-term effects of accelerated germination on the seed bank depend on the postgermination environmental risks experienced by seedlings and their effects on the persistence of species in a community. In polar and subpolar zones and alpine ecosystems characterized by a short growing season, higher, earlier, or faster germination might improve the probability of seedling survival and thus successful regeneration from the seed bank, allowing for the exploitation of the longer growing season (Milbau et al., 2009). However, in other climate zones increased germination may lead to higher seedling mortality by exposing young plants to environmental constraints unfavorable for growth (Shevtsova et al., 2009; Ooi et al., 2012; Porceddu et al., 2020) as well as increasing predation or reducing the formation of mutualisms (Graee et al., 2008; Connolly and Orrock, 2015; Gómez-Ruiz and Lacher, 2019). The effects of warming will be greater for species with narrow temperature windows for germination, although this may depend on interactions between temperature and moisture availability (Cochrane, 2016). Moreover, parental environmental effects on the germination response of seeds to the environment might mediate the effects of increased soil temperatures (Long et al., 2008; Kochanek et al., 2011; Ooi et al., 2012; Meineri et al., 2013; Mondoni et al., 2014, 2015; Bernareggi et al., 2016).

One of the main direct effects of a warming climate on the seed bank is through changes in the seed rain associated with alterations in the structure and composition of the standing vegetation and/or changes in seed production (Fig. 21.1; Cleland et al., 2007; Springer et al., 2008; Briceño et al., 2015; Ma et al., 2019, 2020; An et al., 2020; Prevéy, 2020). Evidence from wetlands and grasslands suggests that climatic warming might alter both the standing vegetation and the seed bank through negative effects on species coexistence and competitive interactions associated with changes in the timing of germination of individual species. Changes in the standing vegetation include the displacement of subordinate species by dominant species (Brock, 2011; Baldwin et al., 2014; Basto et al., 2018) and shifts from perennial to annual communities (Ma et al., 2010) and from perennial- to woody-dominated communities (Fridley and Wright, 2018). Positive feedbacks between the standing vegetation and seed bank are to be expected (del Cacho et al., 2012; Panetta et al., 2018), with changes in the seed bank likely to be exacerbated by any potential negative effects of warming on the standing vegetation.

Demographic compensation mechanisms, such as increased seed production (Akinola et al., 1998; García-Camacho et al., 2012; Ibáñez et al., 2017), could mitigate any adverse effects of climate warming on the seed bank (Doak and Morris, 2010; Sheth and Angert, 2018). Mitigation would explain why only minor changes in the size, composition, and diversity of the seed bank have been observed under warmer conditions, at least in the short term (Akinola et al., 1998). In alpine systems, increased seed production and seed quality associated with earlier snowmelt and a longer growing season have been reported (Arft et al., 1999; Springer et al., 2008). Shifts in flowering time and an increase in the length of the flowering period can also increase the probability and frequency of seed set, thereby increasing the number of flowers and ultimately the number of seeds produced (Thórhallsdóttir, 1998; Teller et al., 2016). However, increased seed production might not be sufficient to compensate for increased losses from the seed bank if warmer conditions reduce seedling survival and establishment. Potential reduction in the size of the seed bank is more likely to occur in ecosystems characterized by low seed production, such as arctic and alpine ecosystems (Onipchenko et al., 1998) and other communities dominated by perennial species. Seed production may also be decreased if a warmer climate results in asynchronous phenology between the flowering of plants and their pollinators (Gilman et al., 2010; Gómez-Ruiz and Lacher, 2019) or in the disruption of other multitrophic relationships (Miller-Rushing et al., 2010). Thus, the need to include biotic interactions across trophic levels to understand and predict the response of species to climate change has been recommended (Van der Putten et al., 2010).

Experiments examining the effects of projected temperature increases on seed aging, longevity, and germination responses provide indirect information on the potential impact of warming on the seed bank. However, the complexity of the effects of warming on ecosystems makes it difficult to extrapolate many results to the field and to estimate seed viability and longevity in the soil (Cooper et al., 2004; Walck et al., 2011; Blossey et al., 2017). Moreover, warmer temperatures are only one component of a warming climate (Jaganathan and Dalrymple, 2016). Our inability to make large-scale conclusions on the response of individual plant species and communities to climate warming is also confounded by differences in seed collection methods, dormancy-breaking treatments, and germination test conditions used in various studies (Jaganathan and Dalrymple, 2016). Long-term, field studies can however provide important information on the effects of a warming climate on the species composition of seed banks as well as the standing vegetation. For this reason, studies of the potential effects of a warming climate on the seed bank have increased in recent years using a variety of approaches that consider individual species, groups of species, and entire communities.

Changes in the composition and structure of seed banks under a warming climate

Mountain ecosystems and elevation gradients

Studies on the potential effects of climate change on the seed bank have revealed the difficulty of untangling direct and indirect effects (Fig. 21.1). These difficulties include indirect effects related to modifications in soil properties (Ma et al., 2019, 2020; An et al., 2020), plant-soil feedbacks (Pugnaire et al., 2019), and the impact of pathogens (Sharma et al., 2006; Pucko et al., 2011; Ma et al., 2020). A warming climate also has been shown to increase pre- and postdispersal seed predations that have negative effects on the seed bank (McKone et al., 1998; Arroyo et al., 2006; Pucko et al., 2011; Del Cacho et al., 2012; Noroozi et al., 2016; Naoe et al., 2019). Evidence of increased postdispersal seed predation in transient compared to persistent seed banks (Hulme, 1998) suggests that the indirect effects of climate change on seed predation will be stronger for species that form only transient seed banks. In contrast, dispersal by animals can help plants avoid climate warming (Naoe et al., 2016, 2019; González-Varo et al., 2017), albeit temporarily. However, it is still unclear whether this will be sufficient to allow some plants to escape current global warming (Naoe et al., 2019).

Despite these difficulties, an increasing number of studies recognize the critical role of seed banks for plant regeneration in many types of communities. To date, most information is from studies examining seed bank communities along elevational or latitudinal gradients with natural climatic gradients, although differences in climate might not be the main or only driver of seed bank variation. Moreover, the temperature gradient in these studies might not be broad enough to detect significant patterns, since above- and/or belowground effects of temperature might become evident only when a certain temperature threshold is exceeded (An et al., 2020). These studies have reported increases (Funes et al., 2003; Espinosa et al., 2013), decreases (Ortega et al., 1997; Cummins and Miller, 2002; An et al., 2020), hump-shaped (Hegazy et al., 2009), or no (Lippok et al., 2013) changes in seed densities and species richness along elevational gradients. The overall effects of warming in mountain regions depend on habitat-specific characteristics, with evidence from mountain systems that the correlation between seed bank properties and elevation varies with the type of vegetation (Erfanzadeh et al., 2013; Ma et al., 2020), disturbance regime (Cooper et al., 2004; Espinosa et al., 2013; Hoyle et al., 2013), and interactions between climatic variables and elevational gradients (Espinosa et al., 2013).

Mountain ecosystems are among those that have been examined most extensively for the potential effects of climate warming on seed banks, given their recognized vulnerability to even small temperature increases (Grabherr et al., 1994; Hughes et al., 2003; Cramer et al., 2014). Increasing evidence indicates that the formation of persistent seed banks is a key survival strategy in these systems. Seed banks buffer the systems against the effects of environmental variability and climate change, although the importance of vegetative propagation relative to regeneration from seeds tends to increase with elevation (Onipchenko et al., 1998). However, the presence of large, species-rich, long-term persistent seed banks in alpine ecosystems supports experimental evidence that seed bank persistence is a life-history trait that has been selected for in these environments (Arroyo et al., 1999; Cavieres and Arroyo, 2001). The same environmental conditions that constrain biomass and seed production can, in fact, promote the formation of persistent seed banks by reducing seed deterioration (Cavieres and Arroyo, 2000; Murdoch and Ellis, 2000; Walck et al., 2005, 2011; Ma et al., 2010).

An interesting example of how climate warming might affect natural seed banks in alpine regions is from recent studies on the Tibetan Plateau. These seed banks have a large number of species that are absent from the vegetation, and the seeds have high longevity (Ma et al., 2010). Several studies have examined variations in seed bank communities on the Tibetan Plateau along elevational gradients (or levels) and the role of seed banks as drivers of ecosystem resilience in different types of meadows (e.g., Ma et al., 2010, 2013, 2017, 2019, 2020; An et al., 2020). Mean annual temperature and precipitation appear to have an important effect on the seed bank by directly affecting the richness and abundance of the standing vegetation as well as influencing soil properties (mainly pH but also total N and P), which in turn affect the alpine vegetation. Patterns in the seed bank on the Tibetan Plateau tend to be driven by increases in the proportion of perennial versus annual species with increasing elevation (Ma et al., 2010; An et al., 2020). These results support earlier suggestions that clonal propagation is a better survival strategy in cold and unstable systems, where biomass and seed production are constrained by a short growing season (Onipchenko et al., 1998). However, the seed banks of these alpine meadows tend to be dominated by perennial species, indicating that seed bank formation is an important strategy not only for annuals (Venable, 2007) but also for perennial herbs that can rely on other strategies for persistence in a community (Grime, 2001; Honnay and Bossuyt, 2005; Clarke et al., 2013). This finding is further supported by evidence that the formation of a persistent (as opposed to a transient) seed bank is the most frequent strategy at high elevations in these ecosystems (Ma et al., 2010).

That a persistent seed bank is a key survival strategy for many species at high elevations (Funes et al., 2003) and contributes to an increased resilience is supported by field observations globally. Persistent seed banks play an important role in maintaining species diversity in the Australian Alps, including the diversity of obligate alpine species, supporting species range shifts, and moderating dominance along elevational gradients (Venn and Morgan, 2010; Hoyle et al., 2013). Increases in the proportion of species forming persistent seed banks as elevation increases have been documented in tall tussock grasslands in the Córdoba mountains, central Argentina (Funes et al., 2003) and in Mediterranean pasture communities in central Spain (Ortega et al., 1997), with certain species forming persistent seed banks only at high but not at low elevations. Espinosa et al. (2013) found differences in seed bank richness and density along elevational gradients in dry mountain scrub communities, but such differences were not significant when only the persistent component of the seed bank was accounted for, providing further evidence that persistent, but not transient, seed banks can increase ecosystem resilience.

Seed banks confer increased resilience in extreme environments

Evidence for the importance of persistent seed banks as a survival strategy in response to a warming climate also comes from Arctic and Antarctic ecosystems, where large and persistent seed banks have been recorded (e.g., McGraw and Vavrek, 1989; Lévesque and Svoboda, 1995; McGraw and Day, 1997; Arroyo et al., 2004; Cooper et al., 2004; Jónsdóttir, 2011; Williams et al., 2016). Based on the available knowledge of plant survival in the high Arctic since pre-Holocene times and examination of contemporary populations in these regions, Crawford and Abbott (1994) concluded that some Arctic species may form long-term persistent seed banks that might confer increased resilience against climate change. Field evidence of this resilience was provided by Cooper et al. (2004), who examined seedling emergence in six dry-mesic habitats on Svalbard. These authors found that 50 of 72 species were present in the standing vegetation as mature plants emerged from the seed bank. However, thermophilic species failed to germinate under natural conditions. Moreover, some species present at/in several sites/habitats germinated only from the thermophilic heath seed bank, suggesting that current climatic conditions constrain any recruitment from seeds, while warmer conditions could deplete the seed bank by promoting germination.

Concerns that a warming climate might compromise the bet-hedging role of soil seed banks also comes from studies on arid and semiarid ecosystems, where formation of a persistent seed bank is a critical survival strategy for species with short life cycles (Venable and Brown, 1988; Arroyo et al., 2006; Venable, 2007). In these systems, warming is expected to exacerbate the effects of extreme climatic events (Alpert et al., 2008; Kafle and Bruins, 2009; Ooi et al., 2009, 2012; del Cacho and Lloret, 2012; Basto et al., 2018) through reduced seed viability, increased soil temperatures and subsequent increases in seedling mortality due to temperature-related reductions in water availability (Ooi et al., 2012). Increased germination under warmer conditions may also deplete the seed bank, even when projected increases in temperature fall within the thermal germination range of a species (Aragón-Gastélum et al., 2018). Increased lightning strikes in arid regions (Veblen et al., 2011) may increase fire frequency and intensity, further compromising the persistence of plant populations dependent on long-term persistent seed banks (Ooi et al., 2012, 2014). Manipulation studies in these systems have shown reductions in seed bank richness and density with increased temperatures, especially for short-lived species, potentially resulting in positive feedbacks that exacerbate the loss of vegetation cover (del Cacho et al., 2012) and a reduction in species with facultative pyrogenic dormancy (Ooi et al., 2014).

Buffering the effects of climate warming: temporary resilience?

Although there is evidence that persistent seed banks can play a critical role in buffering the effects of climate warming, this might be only temporary (Plue et al., 2021). Calcareous grasslands that support high vascular plant species richness, including many rare and threatened species (Hutchings and Stewart, 2002; Van Looy et al., 2016) are resistant to climatic changes in the short term (Akinola et al., 1998) but not in the longer term (Basto et al., 2018). In a manipulation study in a species-rich calcareous grassland, Basto et al. (2018) found significant changes in the composition of the seed banks after 14 years, with decreases in both seed bank richness and density. Changes in the seed bank were also larger than those in the vegetation and did not reflect only aboveground changes, suggesting that changed climatic conditions altered seed viability and longevity and/or seed production. Since perennial species often dominate calcareous grasslands, a modified climate might reduce the importance of seed production for regeneration relative to vegetative propagation, partly explaining why only minor changes in productivity and/or composition have been observed in these systems (Grime et al., 2008).

The well-known importance of persistent seed banks in promoting species persistence and ecosystem resilience in wetlands (Leck et al., 1989) might also be only temporary. In ephemeral wetlands, persistent seed banks contribute to inter- and intraannual environmental resilience (Deil, 2005) and act as a reservoir for protected and rare annual species that are absent in the standing vegetation (Aponte et al., 2010). These seed banks tend to be dominated by seeds of annual species (Deil, 2005; Leck et al., 1989; Aponte et al., 2010). A warming climate might affect these communities by increasing the duration of dry versus wet phases. For example, Brock (2011) found that five Australian temporary wetlands supported species-rich, long-term persistent seed banks that were not depleted by successive germination events. However, seed bank composition changed with increasing duration of the dry phases and the number of successive germination events, indicating that the resilience role of the seed bank might be only temporary. Brock (2011) suggested that the "most resilient" species pool in these systems consisted of species that survive the longest dry periods and several wetting and drying events with little depletion of the seed bank, while the "least resilient" species are characterized by shorter survival times and their rapid decline in the seed bank.

Seed banks and plant migration potential under a warming climate

Knowledge of changes in the structure of the soil seed bank is a key factor in understanding the long-term implications of a warming climate on plant communities. Formation of a persistent seed bank may facilitate the ability of a species to colonize new areas and is often regarded as a potential indicator of community trajectories (González-Alday et al., 2009; Kottler and Gedan, 2020). The seed bank shows which species can disperse into the area and can persist and subsequently germinate under suitable environmental conditions (Wang et al., 2013). In fact, a warming climate might create spatial or temporal niches that promote the germination and establishment of seeds of species dispersed from distant localities, assisting the migration of species and contributing to a reduced risk of extinction. Large-scale evidence based on the distribution of European plant species indicates that those forming persistent seed banks and those with a high dispersal capacity have the smallest climate-related range limitations (Estrada et al., 2015). Persistent seed banks can assist migration of species under warmer conditions in different ecosystems (Erfanzadeh et al., 2013; Hoyle et al., 2013; Estrada et al., 2015; Kottler and Gedan, 2020). Clearly, migration of species might facilitate survival of individual species, but at the community level it could promote successful establishment of alien species or species generally regarded to have low conservation value, potentially exacerbating the negative effects of climatic warming on native and endemic species (Hoyle et al., 2013).

Challenges and future research directions

Our understanding of the long-term implications of a warming climate on plant regeneration from seeds through its effects on the seed bank is limited by the lack of studies examining the impact of global warming on the various sources and sinks of seeds in the soil and the early recruitment processes, including any demographic changes. More information is needed on the impact of increased temperatures on seed production. Although increasing evidence indicates that warming might deplete soil seed banks, increased seed production could mitigate or counteract any negative effects (Akinola et al., 1998; Doak and Morris, 2010; Ibáñez et al., 2017; Sheth and Angert, 2018).

Our review strongly points to the need for improved protocols in studies estimating both the direct and indirect effects of climate warming on seed banks and the magnitude and direction of feedbacks on the seed bank and standing vegetation. Only a combination of long-term observations and manipulation studies that examine the response and adaptive capacity of seed banks to climate change can improve our ability to predict the future risk of extinction vis-a-vis modifications in the distribution of species and whole communities, especially where they are subjected to temporally stochastic disturbances (Parmesan, 2006; Keith et al., 2008; Walck et al., 2011; Ooi et al., 2012; Jaganathan and Dalrymple, 2016). Temperature has a significant impact on plants at all stages of their life cycle (Trudgill et al., 2005) and is a major factor controlling plant distribution. However, a complete understanding of the future effect of global warming on soil seed banks needs to recognize the interacting effects of warming with other climate-related changes, including alterations in precipitation patterns and water availability (Ooi et al., 2009, 2012; Basto et al., 2015, 2018; An et al., 2020; Ma et al., 2020) and increases in atmospheric CO_2 (Seibert et al., 2019). Changes in land use (Ortega et al., 1997; Espinosa et al., 2013), atmospheric nitrogen deposition (Grime et al., 2000; Basto et al., 2015; Ma et al., 2020), changes in fire frequency and intensity (Ooi et al., 2009; Enright et al., 2014; Camac et al., 2017), and the introduction of alien species (Hoyle et al., 2013; Gioria and Osborne, 2014; Hou et al., 2014) are also expected to have major impacts on soil seed banks. The effects of antagonistic or beneficial biotic interactions among species also need to be considered (Pucko et al., 2011; Ash et al., 2017; Ma et al., 2017, 2020; Gómez-Ruiz and Lacher, 2019; An et al., 2020; Giejsztowt et al., 2020). The magnitude and frequency of extreme climatic events also require further consideration since they may have disproportionate effects on plant communities and ecosystems and thus could undermine predictions based on short-term field experiments.

A better understanding of how seed banks may affect the evolutionary responses of plants to environmental change is much needed. The adaptive ability of species relying on long-term persistent seed banks for survival will depend on several factors, including generation time, time between recruitment events, and the level of change required to adapt to new conditions. Genetic adaptation of seed and seedling traits may be more rapid in annuals due to their shorter life cycles than in perennial herbaceous or woody species (Smith and Beaulieu, 2009). However, whether genetic adaptations can track the speed of climate change remains largely unknown (Walck et al., 2011; Parmesan and Hanley, 2015). Rapid evolutionary changes that optimize the timing of germination or the ability of seeds to survive in the seed bank have been observed for some invasive alien plants (Blossey et al., 2017). In this respect, insights into the rate of adaptation can be obtained from studies examining alien species that have recently expanded their geographical ranges into new areas.

Knowledge of how seed bank properties are distributed globally across latitudes and habitats can help identify the species most likely to persist and contribute to the improved resilience of communities exposed to a warming climate and other environmental changes. Seed bank data based on field collections are becoming increasingly available from most regions and biomes (Jaganathan et al., 2015; Gioria et al., 2020). Analyses of global seed bank data from the native range of 2350 species of flowering plants show that climate and latitude have relatively smaller effects on local seed bank persistence and densities than habitat-related variables (Gioria et al., 2020). These results are consistent with field evidence that disturbances mediate the effects of a warming climate on soil seed banks and regeneration from seeds (del Cacho et al., 2012; Espinosa et al., 2013; Ma et al., 2013, 2018), although this needs to be explored more broadly.

It is possible that microclimatic conditions might mask the effects of broad-scale climatic patterns. For example, facilitative plant—plant interactions are thought to ameliorate the severe microenvironmental conditions in alpine plant communities (Cavieres and Sierra-Almeida, 2012). Although future climate change scenarios will likely include increased temperatures and altered precipitation patterns (IPCC, 2013, 2014), specific microenvironmental conditions could create favorable microhabitats and preserve the bet-hedging role of seed banks, buffering local populations against rapid climatic changes (Denney et al., 2020). However, more evidence for the preservation of the bet-hedging role of seed banks under climate change is needed.

Concluding remarks

Studies of natural seed banks are a key factor for an improved understanding of the long-term implications of a warming climate on plant distribution and diversity. Increasing evidence shows that persistent seed banks provide resilience to a warming climate, especially in the most vulnerable ecosystems. Seed banks reduce the likelihood of ecosystem extinction by supporting a high diversity of native and endemic species, while allowing the survival of species no longer present in the standing vegetation. This resilience has been observed not only for annual species, but also for perennial species.

There is growing concern that climate warming might rapidly and negatively impact the bet-hedging role of persistent seed banks, through both direct and indirect effects on seed persistence in the soil and increased seedling mortality. Furthermore, the role of persistent seed banks in buffering the effects of climatic changes may be only temporary. Little is known about the extent to which compensation via genetic adaptation or phenotypic plasticity might determine the long-term responses of plants to a warming climate. More information is needed on the role of seed dispersers in facilitating plant distributional shifts and in preventing their extinction.

Ultimately, the long-term implications of a warming climate on plant communities via the seed bank will depend on its effect on seed longevity and on the net balance between seed input and losses from the seed bank, as well as on the risks associated with postgermination environmental conditions. Accelerated germination will be important only if it ultimately balances any seed losses. The ability to form long-term persistent seed banks could become more important in the future, and it has been suggested that plant species may respond to a changing environment by relying less on dispersal through space and more through time (Johnson et al., 2019). The ability to form a persistent seed bank is expected to be especially critical for the survival of species with limited dispersal ability, such as alpine species (Morgan and Venn, 2017), species unable to adapt rapidly to a warming climate, and those at the limits of their distributional range (Hughes et al., 1996; Parmesan and Yohe, 2003). In contrast, transient or short-term persistent seed banks are less likely to prevent species extinctions. There is also evidence for lower resilience in communities composed of short- than of long-term persistent seed banks (Van Looy et al., 2016). Expanding on the concept of Brock (2011) on the resilience of plants in ephemeral wetlands, the species less likely to be impacted by climate warming will be those whose seeds can survive in persistent seed banks that are not rapidly depleted by temperature increases or other related environmental changes.

References

Akinola, M.O., Thompson, K., Buckland, S.M., 1998. Soil seed bank of an upland calcareous grassland after 6 years of climate and management manipulations. J. Appl. Ecol. 35, 544—552.

Alpert, P., Krichak, S.O., Shafir, H., Osetinsky, I., 2008. Climatic trends to extremes employing regional modeling and statistical interpretation over the E. Mediterranean. Glob. Planet. Change 63, 163—170.

An, H., Zhao, Y., Ma, M., 2020. Precipitation controls seed bank size and its role in alpine meadow community regeneration with increasing altitude. Glob. Change Biol. 26, 5767—5777.

Aponte, C., Kazakis, G., Ghosn, D., Papanastasis, V.P., 2010. Characteristics of the soil seed bank in Mediterranean temporary ponds and its role in ecosystem dynamics. Wetl. Ecol. Manage. 18, 243—253.

Aragón-Gastélum, J.L., Flores, J., Jurado, E., Ramírez-Tobías, H.M., Robles-Díaz, E., Rodas-Ortiz, J.P., et al., 2018. Potential impact of global warming on seed bank, dormancy and germination of three succulent species from the Chihuahuan Desert. Seed Sci. Res. 28, 312—318.

Arft, A.M., Walker, M.D., Gurevitch, J.E.A., Alatalo, J.M., Bret-Harte, M.S., Dale, M., et al., 1999. Responses of tundra plants to experimental warming: *meta*-analysis of the international tundra experiment. Ecol. Monogr. 69, 491—511.

Arroyo, M.T.K., Cavieres, L.A., Castor, C., Humaña, A.M., 1999. Persistent seed bank and standing vegetation in a high alpine site in the central Chilean Andes. Oecologia 119, 126—132.

Arroyo, M.T.K., Cavieres, L.A., Humaña, A.M., 2004. Experimental evidence of potential for persistent seed bank formation at a subantarctic alpine site in Tierra del Fuego, Chile. Ann. Missouri Bot. Gard. 91, 357–365.

Arroyo, M.T.K., Chacon, P., Cavieres, L.A., 2006. Relationship between seed bank expression, adult longevity and aridity in species of *Chaetanthera* (Asteraceae) in central Chile. Ann. Bot. 98, 591–600.

Ash, J.D., Givinsh, T.J., Waller, D.M., 2017. Tracking lags in historical plant species' shifts in relation to regional climate change. Glob. Change Biol. 23, 1305–1315.

Auld, T.D., O'Connell, M.A., 1991. Predicting patterns of post-fire seed germination in 35 eastern Australian Fabaceae. Aust. J. Ecol. 16, 53–70.

Baldwin, A.H., Jensen, K., Schönfeldt, M., 2014. Warming increases plant biomass and reduces diversity across continents, latitudes, and species migration scenarios in experimental wetland communities. Glob. Change Biol. 20, 835–850.

Baskin, C.C., Baskin, J.M., 2014. Seeds: Ecology, Biogeography, and Evolution of Dormancy and Germination, Second ed. Academic Press/Elsevier, San Diego.

Baskin, J.M., Nan, X.-Y., Baskin, C.C., 1998. A comparative study of seed dormancy and germination in an annual and a perennial species of *Senna* (Fabaceae). Seed Sci. Res. 8, 501–512.

Basto, S., Thompson, K., Grime, P.J., Fridley, J.D., Calhim, S., Askew, A.P., et al., 2018. Severe effects of long-term drought on calcareous grassland seed banks. Clim. Atmos. Sci. 1, 1. Available from: https://doi.org/10.1038/s41612-017-0007-3.

Basto, S., Thompson, K., Phoenix, G., Sloan, V., Leake, J., Rees, M., 2015. Long-term nitrogen deposition depletes grassland seed banks. Nat. Commun. 6, 6185. Available from: https://doi.org/10.1038/ncomms7185.

Beckage, B., Osborne, B., Gavin, D.G., Pucko, C., Siccama, T., Perkins, T., 2008. A rapid upward shift of forest ecotone during 40 years of warming in the Green Mountains of Vermont. Proc. Natl. Acad. Sci. USA 105, 4197–4202.

Bekker, R.M., Bakker, J.P., Grandin, U., Kalamees, R., Milberg, P., Poschlod, P., et al., 1998. Seed size, shape and vertical distribution in the soil: indicators of seed longevity. Funct. Ecol. 12, 834–842.

Bernareggi, G., Carbognani, M., Mondoni, A., Petraglia, A., 2016. Seed dormancy and germination changes of snowbed species under climate warming: the role of pre- and post-dispersal temperatures. Ann. Bot. 118, 529–539.

Bernareggi, G., Carbognani, M., Petraglia, A., Mondoni, A., 2015. Climate warming could increase seed longevity of alpine snowbed plants. Alp. Bot. 125, 69–78.

Bertrand, R., Lenoir, J., Piedallu, C., Riofrío-Dillon, G., de Ruffray, P., Vidal, C., et al., 2011. Changes in plant community composition lag behind climate warming in lowland forests. Nature 479, 517–520.

Bewley, J.D., Bradford, K.J., Hilhorst, H.W.M., Nonogaki, N., 2013. Seeds: Physiology of Development, Germination and Dormancy, Third ed. Springer, New York.

Blossey, B., Nuzzo, V., Dávalos, A., 2017. Annual emergence of *Alliaria petiolata* in North America (alien range) was positively correlated with spring temperature and inversely correlated with number of spring days with minimum temperature below freezing. J. Ecol. 105, 1485–1495.

Briceño, V.F., Hoyle, G.L., Nicotra, A.B., 2015. Seeds at risk: how will a changing alpine climate affect regeneration from seeds in alpine areas? Alp. Bot. 125, 59–68.

Brock, M.A., 2011. Persistence of seed banks in Australian temporary wetlands. Freshw. Biol. 56, 1312–1327.

Camac, J.S., Williams, R.J., Wahren, C.-H., Hoffmann, A.A., Vesk, P.A., 2017. Climatic warming strengthens a positive feedback between alpine shrubs and fire. Glob. Change Biol. 23, 3249–3258.

Cavieres, L.A., Arroyo, M.T.K., 2000. Seed germination response to cold stratification period and thermal regime in *Phacelia secunda* (Hydrophyllaceae): altitudinal variation in the Mediterranean Andes of Central Chile. Plant Ecol. 149, 1–8.

Cavieres, L.A., Arroyo, M.T.K., 2001. Persistent soil seed banks in *Phacelia secunda* (Hydrophyllaceae): experimental detection of variation along an altitudinal gradient in the Andes of central Chile (33°S). J. Ecol. 89, 31–39.

Cavieres, L.A., Sierra-Almeida, A., 2012. Facilitative interactions do not wane with warming at high elevations in the Andes. Oecologia 170, 575–584.

Clarke, P.J., Lawes, M.J., Midgley, J.J., Lamont, B.B., Ojeda, F., Burrows, G.E., et al., 2013. Resprouting as a key functional trait: how buds, protection and resources drive persistence after fire. New Phytol. 197, 19–35.

Cleland, E.E., Chuine, I., Menzel, A., Mooney, H.A., Schwartz, M.D., 2007. Shifting plant phenology in response to global change. Trends Ecol. Evol. 22, 357–365.

Cochrane, A., 2016. Can sensitivity to temperature during germination help predict global warming vulnerability? Seed Sci. Res. 26, 14–29.

Cohen, D., 1966. Optimizing reproduction in a randomly varying environment. J. Theoret. Biol. 12, 119–129.

Colautti, R.I., Ågren, J., Anderson, J.T., 2017. Phenological shifts of native and invasive species under climate change: insights from the *Boechera–Lythrum* model. Philos. Trans. R. Soc. B 372, 20160032. Available from: https://doi.org/10.1098/rstb.2016.0032.

Connolly, B.M., Orrock, J.L., 2015. Climatic variation and seed persistence: freeze–thaw cycles lower survival via the joint action of abiotic stress and fungal pathogens. Oecologia 179, 609–616.

Cooper, E.J., Alsos, I.G., Hagen, D., Smith, M., Coulson, S.J., Hodkinson, I.D., 2004. Plant recruitment in the High Arctic: seed bank and seedling emergence on Svalbard. J. Veg. Sci. 15, 115–124.

Corlett, R.T., Westcott, D.A., 2013. Will plant movements keep up with climate change? Trends Ecol. Evol. 28, 482–488.

Cramer, W., Yohe, G.W., Auffhammer, M., Huggel, C., Molau, U., da Silva Dias, M.A.F., et al. (Eds.), 2014. Climate Change 2014: Impacts, Adaptation, and Vulnerability. Part A: Global and Sectoral Aspects. Contribution of Working Group II to the Fifth Assessment Report of the Intergovernmental Panel of Climate Change. Cambridge University Press, Cambridge, pp. 979–1037.

Crawford, R.M.M., Abbott, R.J., 1994. Pre-adaptation of Arctic plants to climate change. Plant Biol. 107, 271–278.
Cummins, R.P., Miller, G.R., 2002. Altitudinal gradients in seed dynamics of *Calluna vulgaris* in eastern Scotland. J. Veg. Sci. 13, 859–866.
Deil, U., 2005. A review on habitats, plant traits and vegetation of ephemeral wetlands: a global perspective. Phytocoenologia 35, 533–705.
del Cacho, M., Lloret, F., 2012. Resilience of Mediterranean shrubland to a severe drought episode: the role of seed bank and seedling emergence. Plant Biol. 14, 458–466.
del Cacho, M., Saura-Mas, S., Estiarte, M., Peñuelas, J., Lloret, F., 2012. Effect of experimentally induced climate change on the seed bank of a Mediterranean shrubland. J. Veg. Sci. 23, 280–291.
Denney, D.A., Jameel, M.I., Bemmels, J.B., Rochford, M.E., Anderson, J.T., 2020. Small spaces, big impacts: contributions of micro-environmental variation to population persistence under climate change. AoB Plants 12, plaa005. Available from: https://doi.org/10.1093/aobpla/plaa005.
Doak, D.F., Morris, W.F., 2010. Demographic compensation and tipping points in climate-induced range shifts. Nature 467, 959–962.
Donohue, K., Dorn, L., Griffith, C., Kim, E., Aguilera, A., Polisetty, C.R., et al., 2005. The evolutionary ecology of seed germination of *Arabidopsis thaliana*: variable natural selection on germination timing. Evolution 59, 758–770.
Donohue, K., Rubio de Casas, R., Burghardt, L., Kovach, K., Willis, C.G., 2010. Germination, post-germination adaptation, and species ecological ranges. Annu. Rev. Ecol. Evol. Syst. 41, 293–319.
Enright, N.J., Fontaine, J.B., Lamont, B.B., Miller, B.P., Westcott, V.C., Cornelissen, H., 2014. Resistance and resilience to changing climate and fire regime depend on plant functional traits. J. Ecol. 102, 1572–1581.
Erfanzadeh, R., Kahnuj, S.H., Azarnivand, H., Pétillon, J., 2013. Comparison of soil seed banks of habitats distributed along an altitudinal gradient in northern Iran. Flora 208, 312–320.
Espinosa, C.I., Luzuriaga, A.L., de la Cruz, M., Montero, M., Escudero, A., 2013. Co-occurring grazing and climate stressors have different effects on the total seed bank when compared to the persistent seed bank. J. Veg. Sci. 24, 1098–1107.
Estrada, A., Meireles, C., Morales-Castilla, I., Poschlod, P., Vieites, D., Araújo, M.B., et al., 2015. Species' intrinsic traits inform their range limitations and vulnerability under environmental change. Glob. Ecol. Biogeogr. 24, 849–858.
Fenner, M., Thompson, K., 2005. The Ecology of Seeds, Second ed. Cambridge University Press, Cambridge.
Footitt, S., Huang, Z., Ölcer-Footitt, H., Clay, H.A., Finch-Savage, W.E., Kranner, I., 2018. The impact of global warming on germination and seedling emergence in *Alliaria petiolata*, a woodland species with dormancy loss dependent on low temperature. Plant Biol. 20, 682–690.
Fridley, J.D., Wright, J.P., 2018. Temperature accelerates the rate fields become forests. Proc. Natl. Acad. Sci. USA 115, 4702–4706.
Funes, G., Basconcelo, S., Diáz, S., Cabido, M., 2003. Seed bank dynamics in tall-tussock grasslands along an altitudinal gradient. J. Veg. Sci. 14, 253–258.
García-Camacho, R., Albert, M.J., Escudero, A., 2012. Small-scale demographic compensation in a high-mountain endemic: the low edge stands still. Plant Ecol. Divers. 5, 37–44.
Giejsztowt, J., Classen, A.T., Deslippe, J.R., 2020. Climate change and invasion may synergistically affect native plant reproduction. Ecology 101, e02913. Available from: https://doi.org/10.1002/ecy.2913.
Gilman, S.E., Urban, M.C., Tewksbury, J., Gilchrist, G.W., Holt, R.D., 2010. A framework for community interactions under climate change. Trends Ecol. Evol. 25, 325–331.
Giménez-Benavides, J., Escudero, A., García-Camacho, R., García-Fernández, A., Iriondo, J.M., Lara-Romero, C., et al., 2018. How does climate change affect regeneration of Mediterranean high-mountain plants? An integration and synthesis of current knowledge. Plant Biol. 50 (Suppl. 1), 50–62.
Gioria, M., Osborne, B.A., 2014. Resource competition in plant invasions: emerging patterns and research needs. Front. Plant Sci. 5, 501. Available from: https://doi.org/10.3389/fpls.2014.00501.
Gioria, M., Pyšek, P., Baskin, C.C., Carta, A., 2020. Phylogenetic relatedness mediates persistence and density of soil seed banks. J. Ecol. 108, 2121–2131.
González-Alday, J., Marrs, R.H., Martínez-Ruiza, C., 2009. Soil seed bank formation during early revegetation after hydroseeding in reclaimed coal wastes. Ecol. Eng. 35, 1062–1069.
González-Varo, J.P., López-Bao, J.V., Guitián, J., 2017. Seed dispersers help plants to escape global warming. Oikos 126, 1600–1606.
Grabherr, G., Gottfried, M., Pauli, H., 1994. Climate effects on mountain plants. Nature 369, 448.
Graee, B.J., Alsos, I.G., Ejrnaes, R., 2008. The impact of temperature regimes on development, dormancy breaking and germination of dwarf shrub seeds from arctic, alpine and boreal sites. Plant Ecol. 198, 275–284.
Grime, J.P., 2001. Plant Strategies, Vegetation Processes, and Ecosystem Properties, Second ed. John Wiley and Sons, Oxford.
Grime, J.P., Brown, V.K., Thompson, K., Masters, G.J., Hillier, S.H., Clarke, I.P., et al., 2000. The response of two contrasted grasslands to simulated climate change. Science 289, 762–765.
Grime, J.P., Fridley, J.D., Askew, A.P., Thompson, K., Hodgson, J.G., Bennett, C.R., 2008. Long-term resistance to simulated climate change in an infertile grassland. Proc. Natl. Acad. Sci. USA 105, 10028–10032.
Gómez-Ruiz, E.P., Lacher Jr., T.E., 2019. Climate change, range shifts, and the disruption of a pollinator-plant complex. Sci. Rep. 9, 14048. Available from: https://doi.org/10.1038/s41598-019-50059-6.
Harper, J., 1997. The population biology of plants. Academic Press, London.
Hegazy, A.K., Hammouda, O., Lovett-Doust, J., Gomaa, N.H., 2009. Variations of the germinable soil seed bank along the altitudinal gradient in the northwestern Red Sea region. Acta Ecol. Sin. 29, 20–29.
Hoffmann, A.A., Sgrò, C.M., 2011. Climate change and evolutionary adaptation. Nature 470, 479–485.

Honnay, O., Bossuyt, B., 2005. Prolonged clonal growth: escape route or route to extinction? Oikos 108, 427–432.

Honnay, O., Bossuyt, B., Jacquemyn, H., Shimono, A., Uchiyama, K., 2008. Can a seed bank maintain the genetic variation in the above ground plant population? Oikos 117, 1–5.

Hopfensperger, K.N., 2007. A review of similarity between seed bank and standing vegetation across ecosystems. Oikos 116, 1438–1448.

Hou, Q.-Q., Chen, B.-M., Peng, S.-L., Chen, L.-Y., 2014. Effects of extreme temperature on seedling establishment of nonnative invasive plants. Biol. Inv. 16, 2049–2061.

Hoyle, G.L., Venn, S.E., Steadman, K.J., Good, R.B., McAuliffe, E.J., Williams, E.R., et al., 2013. Soil warming increases plant species richness but decreases germination from the alpine soil seed bank. Glob. Change Biol. 19, 1549–1561.

Hughes, T.P., Baird, A.H., Bellwood, D.R., Card, M., Connolly, S.R., Folke, C., et al., 2003. Climate change, human impacts, and the resilience of coral reefs. Science 301, 929–933.

Hughes, L., Cawsey, E.M., Westoby, M., 1996. Climatic range sizes of Eucalyptus species in relation to future climate change. Glob. Ecol. Biogeogr. Lett. 5, 23–29.

Hulme, P.E., 1998. Post-dispersal seed predation: consequences for plant demography and evolution. Persp. Plant Ecol. Evol. Syst. 1, 32–46.

Hutchings, M.J., Stewart, A.J.A., 2002. Calcareous grasslands. In: Perrow, M.R., Davy, A.J. (Eds.), Handbook of Ecological Restoration. Restoration in Practice, vol. 2. Cambridge University Press, Cambridge, pp. 419–444.

Ibáñez, I., Katz, D.S.W., Lee, B.R., 2017. The contrasting effects of short-term climate change on the early recruitment of tree species. Oecologia 184, 701–713.

IPCC (Intergovernmental Panel on Climate Change), 2013. Climate change 2013: the physical science basis. Contribution of Working Group I to the Fifth Assessment Report of the Intergovernmental Panel on Climate Change. Cambridge University Press, Cambridge.

IPCC (Intergovernmental Panel on Climate Change), 2014. Climate change 2014: impacts, adaptation, and vulnerability. Contributions of Working Group II to the Fifth Assessment Report. Cambridge University Press, Cambridge.

Jaganathan, G.K., Dalrymple, S.E., 2016. Inconclusive predictions and contradictions: a lack of consensus on seed germination response to climate change at high altitude and high latitude. J. Bot. 2016, 6973808. Available from: https://doi.org/10.1155/2016/6973808.

Jaganathan, G.K., Dalrymple, S.E., Liu, B., 2015. Towards an understanding of factors controlling seed bank composition and longevity in the alpine environment. Bot. Rev. 81, 70–103.

Johnson, J.S., Cantrell, R.S., Cosner, C., Hartig, F., Hastings, A., Rogers, H.S., et al., 2019. Rapid changes in seed dispersal traits may modify plant responses to global change. AoB Plants 11, plz020. Available from: https://doi.org/10.1093/aobpla/plz020.

Jónsdóttir, I.S., 2011. Diversity of plant life histories in the Arctic. Preslia 83, 281–300.

Kafle, H.K., Bruins, H.J., 2009. Climatic trends in Israel 1970–2002: warmer and increasing aridity inland. Clim. Change 96, 63–77.

Keith, D.A., Akçakaya, H.R., Thuiller, W., Midgley, G.F., Pearson, R.G., Phillips, S.J., et al., 2008. Predicting extinction risk under climate change: coupling stochastic population models with dynamic bioclimatic habitat models. Biol. Lett. 4, 560–563.

Kerr, J.T., 2020. Racing against change: understanding dispersal and persistence to improve species' conservation prospects. Proc. R. Soc. B 287, 20202061. Available from: https://doi.org/10.1098/rspb.2020.2061.

Kochanek, J., Steadman, K.J., Probert, R.J., Adkins, S.W., 2011. Parental effects modulate seed longevity: exploring parental and offspring phenotypes to elucidate pre-zygotic environmental influences. New Phytol. 191, 223–233.

Kottler, E.J., Gedan, K., 2020. Seeds of change: characterizing the soil seed bank of a migrating salt marsh. Ann. Bot. 125, 335–344.

Larson, J.E., Funk, J.L., 2016. Regeneration: an overlooked aspect of trait-based plant community assembly models. J. Ecol. 104, 1284–1298.

Leck, M.A., Parker, V.T., Simpson, R.L., 1989. Ecology of Soil Seed Banks. Academic Press, San Diego.

Leishman, M.R., Masters, G.J., Clarke, I.P., Brown, V.K., 2000. Seed bank dynamics: the role of fungal pathogens and climate change. Funct. Ecol. 14, 293–299.

Lenoir, J., Gégout, J.C., Marquet, P.A., de Ruffray, P., Brisse, H., 2008. A significant upward shift in plant species optimum elevation during the 20th century. Science 320, 1768–1771.

Lippok, D., Walter, F., Hensen, I., Beck, S.G., Schleuning, M., 2013. Effects of disturbance and altitude on soil seed banks of tropical montane forests. J. Trop. Ecol. 29, 523–529.

Lloret, F., Peñuelas, J., Estiarte, M., 2005. Effects of vegetation canopy and climate on seedling establishment in Mediterranean shrubland. J. Veg. Sci. 16, 67–76.

Long, R.L., Gorecki, M.J., Renton, M., Scott, J.K., Colville, L., Goggin, D.E., et al., 2015. The ecophysiology of seed persistence: a mechanistic view of the journey to germination or demise. Biol. Rev. 90, 31–59.

Long, R.L., Panetta, F.D., Steadman, K.J., Probert, R., Bekker, R.M., Brooks, S., et al., 2008. Seed persistence in the field may be predicted by laboratory-controlled aging. Weed Sci. 56, 523–528.

Luna, B., 2020. Fire and summer temperatures work together breaking physical seed dormancy. Sci. Rep. 10, 6031. Available from: https://doi.org/10.1038/s41598-020-62909-9.

Lévesque, E., Svoboda, J., 1995. Germinable seed bank from polar desert stands, Central Ellesmere Island, Canada. In: Callaghan, T.V., Molau, U., Holten, J. (Eds.), Global Change and Arctic Terrestrial Ecosystems. European Commission, Brussels, pp. 98–107. , Ecosystems Research Report 10.

Løkken, J.O., Hofgaard, A., Dalen, L., Hytteborn, H., 2019. Grazing and warming effects on shrub growth and plant species composition in subalpine dry tundra: an experimental approach. J. Veg. Sci. 30, 698–708.

Ma, M., Baskin, C.C., Li, W., Zhao, Y., Zhao, Y., Zhao, L., et al., 2019. Seed banks trigger ecological resilience in subalpine meadows abandoned after arable farming on the Tibetan Plateau. Ecol. Appl. 29, e01959. Available from: https://doi.org/10.1002/eap.1959.

Ma, M., Collins, S.L., Du, G., 2020. Direct and indirect effects of temperature and precipitation on alpine seed banks in the Tibetan Plateau. Ecol. Appl. 30, e02096. Available from: https://doi.org/10.1002/eap.2096.

Ma, M., Dalling, J.W., Ma, Z., Zhou, X., 2017. Soil environmental factors drive seed density across vegetation types on the Tibetan Plateau. Plant Soil 419, 349–361.

Ma, M., Walck, J.L., Ma, Z., Wang, L., Du, G., 2018. Grazing disturbance increases transient but decreases persistent soil seed bank. Ecol. Appl. 28, 1020–1031.

Ma, M., Zhou, X., Du, G., 2013. Effects of disturbance intensity on seasonal dynamics of alpine meadow soil seed banks on the Tibetan Plateau. Plant Soil 369, 283–295.

Ma, M., Zhou, X., Wang, G., Ma, Z., Du, G., 2010. Seasonal dynamics in alpine meadow seed banks along an altitudinal gradient on the Tibetan Plateau. Plant Soil 336, 291–302.

Manish, K., Telwala, Y., Nautiyal, D.C., Pandit, M.K., 2016. Modelling the impacts of future climate change on plant communities in the Himalaya: a case study from Eastern Himalaya, India. Model. Earth Syst. Environ. 2, 92. Available from: https://doi.org/10.1007/s40808-016-0163-1.

McGraw, J.B., Day, T.A., 1997. Size and characteristics of a natural seed bank in Antarctica. Arct. Antarct. Alp. Res. 29, 213–216.

McGraw, J.B., Vavrek, M.C., 1989. The role of buried viable seeds in arctic and alpine plant communities. In: Leek, M.A., Parker, V.T., Simpson, R. L. (Eds.), Ecology of Soil Seed Banks. Academic Press, San Diego, pp. 91–106.

McKone, M.J., Kelly, D., Lee, W.G., 1998. Effect of climate change on mast-seeding species: frequency of mass flowering and escape from specialist insect seed predators. Glob. Change Biol. 4, 591–596.

Meineri, E., Spindelböck, J., Vandvik, V., 2013. Seedling emergence responds to both seed source and recruitment site climates: a climate change experiment combining transplant and gradient approaches. Plant Ecol. 214, 607–619.

Merritt, D.J., Martyn, A.J., Ainsley, P., Young, R.E., Seed, L.U., Thorpe, M., et al., 2014. A continental-scale study of seed lifespan in experimental storage examining seed, plant, and environmental traits associated with longevity. Biodivers. Conserv. 23, 1081–1104.

Milbau, A., Graae, B.J., Shevtsova, A., Nijs, I., 2009. Effects of a warmer climate on seed germination in the subarctic. Ann. Bot. 104, 287–296.

Miller-Rushing, A.J., Høye, T.T., Inouye, D.W., Post, E., 2010. The effects of phenological mismatches on demography. Philos. Trans. R. Soc. B 365, 3177–3186.

Mondoni, A., Orsenigo, S., Donà, M., Balestrazzi, A., Probert, R.J., Hay, F.R., et al., 2014. Environmentally induced transgenerational changes in seed longevity: maternal and genetic influence. Ann. Bot. 113, 1257–1263.

Mondoni, A., Pedrini, S., Bernareggi, G., Rossi, G., Abeli, T., Probert, R.J., et al., 2015. Climate warming could increase recruitment success in glacier foreland plants. Ann. Bot. 116, 907–916.

Mondoni, A., Probert, R.J., Rossi, G., Vegini, E., Hay, F.R., 2011. Seeds of alpine plants are short lived: implications for long-term conservation. Ann. Bot. 107, 171–179.

Morgan, J.W., Venn, S.E., 2017. Alpine plant species have limited capacity for long-distance seed dispersal. Plant Ecol. 218, 813–819.

Murdoch, A.J., Ellis, R.H., 2000. Dormancy, viability and longevity. In: Fenner, M. (Ed.), Seeds: The Ecology of Regeneration in Plant Communities, Second ed. CABI Publishing, Wallingford, pp. 183–214.

Naoe, S., Tayasu, I., Sakai, Y., Masaki, T., Kobayashi, K., Nakajima, A., et al., 2016. Mountain-climbing bears protect cherry species from global warming through vertical seed dispersal. Curr. Biol. 26, R315–R316.

Naoe, S., Tayasu, I., Sakai, Y., Masaki, T., Kobayashi, K., Nakajima, A., et al., 2019. Downhill seed dispersal by temperate mammals: a potential threat to plant escape from global warming. Sci. Rep. 9, 14932. Available from: https://doi.org/10.1038/s41598-019-51376-6.

Nicotra, A.B., Atkin, O.K., Bonser, S.P., Davidson, A.M., Finnegan, E.J., Mathesius, U., et al., 2010. Plant phenotypic plasticity in a changing climate. Trends Plant Sci. 15, 684–692.

Noroozi, S., Alizadeh, H., Mashhadi, H.R., 2016. Temperature influences postdispersal predation of weed seeds. Weed Biol. Manage. 161, 24–33.

Onipchenko, V.G., Semenova, G.V., van der Maarel, E., 1998. Population strategies in severe environments: alpine plants in the northwestern Caucasus. J. Veg. Sci. 9, 27–40.

Ooi, M.K.J., 2012. Seed bank persistence and climate change. Seed Sci. Res. 22 (Suppl. S1), S53–S60.

Ooi, M.K.J., Auld, T.D., Denham, A.J., 2009. Climate change and bet-hedging: interactions between increased soil temperatures and seed bank persistence. Glob. Change Biol. 15, 2375–2386.

Ooi, M.K.J., Auld, T.D., Denham, A.J., 2012. Projected soil temperature increase and seed dormancy response along an altitudinal gradient: implications for seed bank persistence under climate change. Plant Soil 353, 289–303.

Ooi, M.K.J., Denham, A.J., Santana, V.M., Auld, T.D., 2014. Temperature thresholds of physically dormant seeds and plant functional response to fire: variation among species and relative impact of climate change. Ecol. Evol. 4, 656–671.

Orsenigo, S., Abeli, T., Rossi, G., Bonasoni, P., Pasquaretta, C., Gandini, M., et al., 2015. Effects of autumn and spring heat waves on seed germination of high mountain plants. PLoS One 10, e0133626. Available from: https://doi.org/10.1371/journal.pone.0133626.

Ortega, M., Levassor, C., Peco, B., 1997. Seasonal dynamics of Mediterranean pasture seed banks along environmental gradients. J. Biogeogr. 24, 177–195.

Ovaskainen, O., Skorokhodov, S., Yakovlev, M., Sukhov, A., Kutenkov, A., Kutenkov, N., et al., 2013. Community-level phenological response to climate change. Proc. Natl. Acad. Sci. USA 110, 13434–13439.

Pakeman, R.J., Cummins, R.P., Miller, G.R., Roy, D.R., 1999. Potential climatic control of seedbank density. Seed Sci. Res. 9, 101–110.

Panetta, A.M., Stanton, M.L., Harte, J., 2018. Climate warming drives local extinction: evidence from observation and experimentation. Sci. Adv. 4, eaaq1819. Available from: https://doi.org/10.1126/sciadv.aaq1819.

Parmesan, C., 2006. Ecological and evolutionary responses to recent climate change. Annu. Rev. Ecol. Evol. Syst. 37, 637–669.
Parmesan, C., Hanley, M.E., 2015. Plants and climate change: complexities and surprises. Ann. Bot. 116, 849–864.
Parmesan, C., Yohe, G., 2003. A globally coherent fingerprint of climate change impacts across natural systems. Nature 421, 37–42.
Peters, R.L., Darling, J.D.S., 1985. The greenhouse effect and nature reserves. Global warming would diminish biological diversity by causing extinctions among reserve species. BioScience 35, 707–717.
Petrů, M., Tielbörger, K., 2008. Germination behaviour of annual plants under changing climatic conditions: separating local and regional environmental effects. Oecologia 155, 717–728.
Plue, J., De Frenne, P., Acharya, K., Brunet, J., Chabrerie, O., Decocq, G., et al., 2013. Climatic control of forest herb seed banks along a latitudinal gradient. Glob. Ecol. Biogeogr. 22, 1106–1117.
Plue, J., Van Calster, H., Auestad, I., Basto, S., Bekker, R.M., Bruun, H.H., et al., 2021. Buffering effects of soil seed banks on plant community composition in response to land use and climate. Glob. Ecol. Biogeogr. 30, 128–139.
Porceddu, M., Pritchard, H.W., Mattana, E., Bacchetta, G., 2020. Differential interpretation of mountain temperatures by endospermic seeds of three endemic species impacts the timing of in situ germination. Plants 9, 1382. Available from: https://doi.org/10.3390/plants9101382.
Prevéy, J.S., 2020. Climate change: flowering time may be shifting in surprising ways. Curr. Biol. 30, R112–R133.
Probert, R.J., Daws, M.I., Hay, F.R., 2009. Ecological correlates of ex situ seed longevity: a comparative study on 195 species. Ann. Bot. 104, 57–69.
Pucko, C., Beckage, B., Perkins, T., Keeton, W.S., 2011. Species shifts in response to climate change: individual or shared responses? J. Torrey Bot. Soc. 138, 156–176.
Pugnaire, F.I., Morillo, J.A., Peñuelas, J., Reich, P.B., Bardgett, R.D., Gaxiola, A., et al., 2019. Climate change effects on plant-soil feedbacks and consequences for biodiversity and functioning of terrestrial ecosystems. Sci. Adv. 5, eaaz1834. Available from: https://doi.org/10.1126/sciadv.aaz1834.
Roach, T., Nagel, M., Börner, A., Eberle, C., Kranner, I., 2018. Changes in tocochromanols and glutathione reveal differences in the mechanisms of seed ageing under seedbank conditions and controlled deterioration in barley. Environ. Exp. Bot. 156, 8–15.
Root, T.L., Price, J.T., Hall, K.R., Schneider, S.H., Rosenzweig, C., Pounds, J.A., 2003. Fingerprints of global warming on wild animals and plants. Nature 421, 57–60.
Rubio de Casas, R., Willis, C.G., Pearse, W.D., Baskin, C.C., Baskin, J.M., Cavender-Bares, J., 2017. Global biogeography of seed dormancy is determined by seasonality and seed size: a case study in the legumes. New Phytol. 214, 1527–1536.
Saatkamp, A., Cochrane, A., Commander, L., Guja, L.K., Jimenez-Alfaro, B., Larson, J., et al., 2019. A research agenda for seed-trait functional ecology. New Phytol. 221, 1764–1775.
Seibert, R., Grünhage, L., Müller, C., Otte, A., Donath, T.W., 2019. Raised atmospheric CO_2 levels affect soil seed bank composition of temperate grasslands. J. Veg. Sci. 30, 86–97.
Sharma, S., Szele, Z., Schilling, R., Munch, J.C., Schloter, M., 2006. Influence of freeze–thaw stress on the structure and function of microbial communities and denitrifying populations in soil. Appl. Environ. Microbiol. 72, 2148–2154.
Sheth, S.N., Angert, A.L., 2018. Demographic compensation does not rescue populations at a trailing range edge. Proc. Natl. Acad. Sci. USA 115, 2413–2418.
Shevtsova, A., Graae, B.J., Jochum, T., Milbau, A., Kockelbergh, F., Beyens, l, et al., 2009. Critical periods for impact of climate warming on early seedling establishment in subarctic tundra. Glob. Change Biol. 15, 2662–2680.
Skelly, D.K., Joseph, L.N., Possingham, H.P., Freidenburg, L.K., Farrugia, T.J., Kinnison, M.T., et al., 2007. Evolutionary responses to climate change. Conserv. Biol. 21, 1353–1355.
Smith, S.A., Beaulieu, J.M., 2009. Life history influences rates of climatic niche evolution in flowering plants. Proc. R. Soc. B 276, 4345–4352.
Springer, C.J., Orozco, R.A., Kelly, J.K., Ward, J.K., 2008. Elevated CO_2 influences the expression of floral-initiation genes in *Arabidopsis thaliana*. New Phytol. 178, 63–67.
Teller, B.J., Zhang, R., Shea, K., 2016. Seed release in a changing climate: initiation of movement increases spread of an invasive species under simulated climate warming. Divers. Distrib. 22, 708–716.
Templeton, A., Levin, D., 1979. Evolutionary consequences of seed pools. Am. Nat. 114, 232–249.
Thompson, K., Bakker, J.P., Bekker, R.M., 1997. Soil Seed Banks of NW Europe: Methodology, Density and Longevity. Cambridge University Press, Cambridge.
Thompson, K., Band, S., Hodgson, J., 1993. Seed size and shape predict persistence in soil. Funct. Ecol. 7, 236–241.
Thompson, K., Ceriani, R.M., Bakker, J.P., Bekker, R.M., 2003. Are seed dormancy and persistence in soil related? Seed Sci. Res. 13, 97–100.
Thórhallsdóttir, T.E., 1998. Flowering phenology in the central highland of Iceland and implications for climatic warming in the Arctic. Oecologia 114, 43–49.
Tielbörger, K., Petrů, M., Lampei, C., 2012. Bet-hedging germination in annual plants: a sound empirical test of the theoretical foundations. Oikos 121, 1860–1868.
Trudgill, D.L., Honek, A., Li, D., Van Straalen, N.M., 2005. Thermal time – concepts and utility. Ann. Appl. Biol. 146, 1–14.
Urban, M.C., 2015. Accelerating extinction risk from climate change. Science 348, 571–573.
Van der Putten, W.H., Macel, M., Visser, M.E., 2010. Predicting species distribution and abundance responses to climate change: why it is essential to include biotic interactions across trophic levels. Philos. Trans. R. Soc. B 365, 2025–2034.
Van Looy, K., Lejeune, M., Verbeke, W., 2016. Indicators and mechanisms of stability and resilience to climatic and landscape changes in a remnant calcareous grassland. Ecol. Indic. 70, 498–506.

Veblen, T.T., Holz, A., Paritsis, J., Raffaele, E., Kitzberger, T., Blackhall, M., 2011. Adapting to global environmental change in Patagonia: what role for disturbance ecology? Austral Ecol. 36, 891–903.

Venable, D.L., 2007. Bet hedging in a guild of desert annuals. Ecology 88, 1086–1090.

Venable, D.L., Brown, J.S., 1988. The selective interactions of dispersal, dormancy, and seed size as adaptations for reducing risk in variable environments. Am. Nat. 131, 360–384.

Venn, S.E., Morgan, J.W., 2010. Soil seedbank composition and dynamics across alpine summits in southeastern Australia. Aust. J. Bot. 58, 349–362.

Walck, J.L., Baskin, J.M., Baskin, C.C., Hidayati, S.N., 2005. Defining transient and persistent seed banks in species with pronounced seasonal dormancy and germination patterns. Seed Sci. Res. 15, 189–196.

Walck, J.L., Hidayati, S.N., Dixon, K.W., Thompson, K., Poschlod, P., 2011. Climate change and plant regeneration from seed. Glob. Change Biol. 17, 2145–2161.

Wang, Y., Jiang, D., Toshio, O., Zhou, Q., 2013. Recent advances in soil seed bank research. Contemp. Probl. Ecol. 6, 520–524.

Williams, L.K., Kristiansen, P., Sindel, B.M., Wilson, S.C., Shaw, J.D., 2016. Quantifying the seed bank of an invasive grass in the sub-Antarctic: seed density, depth, persistence and viability. Biol. Inv. 18, 2093–2106.

Section III

Conclusion

Section III

Conclusion

Chapter 22

Summary and general conclusions

Carol C. Baskin[1,2] and Jerry M. Baskin[1]
[1]Department of Biology, University of Kentucky, Lexington, KY, United States, [2]Department of Plant and Soil Sciences, University of Kentucky, Lexington, KY, United States

Seed production

In general, increased temperatures and well-watered soil promote plant growth and thus flowering and seed production (a component of fitness), but drought stress can override the benefits of increased temperatures on seed production and may cause a decrease in seed quality. However, a short rainless period is an environmental cue for onset of flowering of some trees in tropical wet forests, especially if it is accompanied by a slight decrease in temperature. Heatwaves at the time of pollination can have detrimental effects on pollen viability, thereby causing low seed set, for example, in *Zea mays*. Notably, in hot and cold deserts and in a Mediterranean climate the timing of adequate soil moisture for seed germination (autumn-winter vs spring), especially for annuals, has significant effects not only on seedling survival but also on plant growth and seed production. Increased atmospheric nitrogen (N) deposition has the potential to increase the growth and seed production of plants if soil moisture is not limiting. In general, CO_2 enrichment increases seed production, but the response is highly species-specific.

At high elevations and latitudes, warming in early spring increases the length of the snow-free growing season, but a late spring frost and/or summer drought stress can lead to species differences in flower and seed production. Increased habitat temperatures decrease seed production and seed mass of some cold-climate species, but they can increase both seed production and seed mass of other cold-climate species. In tropical savannas, increased rainfall during fruit development may cause fruit abortion. Changes in time of onset of the growing season due to increased temperatures and/or changes in patterns and amount of precipitation can modify the phenology of flowering and flower—pollinator interactions. Decreased precipitation results in increased soil salinity in arid and semiarid regions, which negatively impacts plant growth and seed production. Fires can destroy seeds either before or after dispersal and potentially the plants that produce them. Increased temperatures promote seed production of some seagrasses, but they may cause seed abortion and pseudovivipary in others. Drought stress may be one of the factors selecting for early flowering phenotypes and early seed set of some invasive species.

Seed dormancy and germination

In temperate and cold climates, seeds of numerous species have physiological dormancy (PD) that is broken by cold stratification ($c.0°C–10°C$) during winter. Seeds of other species have morphophysiological dormancy (MPD), and the PD component of MPD may be broken by cold stratification. The concern about these species is: will increased temperatures during winter interfere with dormancy break? For species such as *Gentiana lutea* subsp. *lutea* and *Vitis vinifera* subsp. *sylvestris*, it is predicted that in the future the length of the cold stratification period at low elevations will not be long enough to break seed dormancy. For seeds of many temperate-zone species with PD or MPD that become nondormant during winter in the present climate, a decrease in number of hours of cold stratification means that seeds will germinate at a higher temperature in spring than they would if they had received the full period of cold stratification (see Fig. 10.1). That is, with shortening of the cold stratification period the seeds will not become fully nondormant during winter and therefore cannot germinate at the lowest temperatures possible for the taxon. An adaptation of many invasive species (seeds of many of which have PD) is that seeds require only a short period of cold stratification to break dormancy, and thus they germinate earlier in spring than associated native species. It seems unlikely that increased temperatures due to climate change will have negative effects on dormancy-break of seeds with PD or MPD that become nondormant during the warm season of the year. In fact, higher temperatures may promote dormancy-break in some species. Increased habitat temperatures and heat from fires may break dormancy of seeds with water-impermeable seed/fruit coats (i.e., physical dormancy).

Increased temperatures and decreased precipitation can result in seeds with lower (less intense) PD than if they developed in a habitat with moderate temperatures and moist soil. That is, the maternal environment during seed development has an influence on dormancy of the offspring seeds. However, variation in requirements for dormancy break may be due to genetic effects, and for a species that occurs in a diversity of habitats adaptation to local conditions results in ecotypic differentiation, which gives the species flexibility with regard to its response to the effects of climate change.

Increased temperatures potentially will promote germination of nondormant seeds in cold climates. Climate warming in arctic/alpine and boreal/subalpine regions may stimulate germination due to early snowmelt, but the resulting seedlings may be vulnerable to frost damage. One effect of decreased snow depth in winter is that temperatures of the soil surface and at shallow soil depths, where seeds are located, may be <0°C and thus too cold for the breaking of PD by cold stratification. Although warming may inhibit germination of nondormant seeds that germinate only at low temperatures, it may not have much, if any, direct effect on germination of a wide range of species because their nondormant seeds can germinate over a wide range of temperatures. In some cases, increased temperatures, if accompanied by moist soil, could promote germination of seeds in the soil and thus deplete the soil seed bank.

Decreases in precipitation are expected to decrease germination percentages of wild and crop species in all vegetation zones and possibly change the timing of germination. If rainfall is delayed until after the "temperature window" for germination of a species has "closed," then seeds will not germinate. If changes occur in the temperature in deserts during rainfall events, there may be changes in the species whose seeds germinate at a given time, resulting in potential shifts in species composition of the community. Species with recalcitrant seeds may be at high risk for extirpation from a community if drought consistently follows seed dispersal for a number of years or if length of the dry season increases. In general, decreased precipitation is predicted to be a more serious threat to germination than increased temperatures. CO_2 enrichment has species-dependent effects on seed mass and on germination percentages and rates. Modification in the fire regime due to global warming can cause changes in habitat conditions that promote seed germination and seedling survival and growth.

Seedling survival and growth

In cold climates, the short snow-free growing season is a major factor limiting the survival and growth of seedlings. Thus, warming, especially an increase in the night temperature, due to climate change may increase seedling survival, but the benefits of increased temperature could be overridden by increased drought stress and damage from frost, pathogens, diseases, and herbivores that potentially could kill seedlings. Another aspect of warming in cold climates is that temperatures in autumn may be high enough to promote germination of seeds that under the present climate do not germinate until spring. The question is: can the seedlings survive the winter? It is predicted that seedlings can survive winter under snow, but this prediction needs to be investigated in view of decreasing and increasing, depending on location, depth of the snow layer during winter in response to climate change.

In arid and semiarid (including seasonally dry) climates, timing and amount of precipitation not only determine when seeds germinate but also seedling survival. A short period of drying after radicle emergence from the seeds is known to be tolerated by seedlings of a few species, but in general desiccation of the radicle results in death of the seedling. Delays in time of onset of the wet season and/or decrease in amount of precipitation decrease(s) the length of the favorable period for plant growth and may decrease seedling survival. Another consequence of drought stress is that it can increase the susceptibility of a plant community to invasion by exotic species. Precipitation is predicted to increase in spring in some locations (e.g., cold deserts of Central Asia), which could enhance the establishment and growth of seedlings from autumn- and spring-germinated seeds, and especially in the case of annuals lead to increased seed production.

Seedlings of species in tropical wet forests differ in their ability to tolerate both drought and shade. As expected, seedlings of early-successional species are more tolerant of drought than of shading, while the reverse is true for late-successional species. It is predicted that erratic precipitation and a shortened wet season in the tropics will cause many seedlings to die due to drought stress. Furthermore, a warmer and drier climate will increase the vapor pressure gradient between the leaf and surrounding air, thereby increasing the effects of climate change on seedling survival. In tropical savannas, decreased precipitation will increase fire frequency, which will kill seedlings and/or juvenile plants before they grow large enough to produce seeds. Increased fire frequency will favor survival of resprouting species over that of nonresprouting species.

Shifts in species composition of plant communities

Changes in seed production, dormancy-break and germination, and seedling survival and growth in response to climate change could promote regeneration from seeds of some species in a community over that of others. Thus, the number/biomass of individuals of species in a community could change, thereby shifting the relative importance of species. The species most negatively affected by climate change could stop producing new individuals and eventually become extirpated from the community unless they can reproduce vegetatively. On the other hand, the changed environment may become suitable for establishment of immigrant species, which could be natives undergoing range shifts due to the impact of climate change or invasive species.

Increased temperatures, if soil moisture is adequate for plant growth, will increase biomass production in a community, which could provide fuel for increased fire frequency. Fire may select for species with rapid recruitment, and it can facilitate the invasion of exotics into grasslands, savannas, and Mediterranean woodlands. On the Tibet Plateau, warming decreases species richness in herbaceous plant communities, due in part to increased vegetative growth of some species that inhibits the growth of others. Warming is predicted to increase graminoids over forbs in temperate grasslands and cause an increase in the C4:C3 species ratio. The invasion of shrub species into the steppe grasslands on the Mongolian Plateau is attributed to increased temperatures.

Changes in amount of annual precipitation are expected to lead to changes in plant communities. Increasing precipitation in summer is correlated with an increase in shrubs in cold desert communities of Central Asia dominated by herbaceous species. On the other hand, a decrease in amount of annual precipitation is related to decreased survival of nurse shrubs in hot deserts and contraction of the distribution range of species in Mediterranean and tropical plant communities away from the most drought-stressed part of the range. Increased deposition of atmospheric N is decreasing species richness of communities on the Tibet Plateau. Elevated CO_2 is predicted to have a strong positive response on N-fixing legume trees in the wet tropics, but low P and Mo levels in the soil could limit N-fixation as CO_2 increases, potentially influencing abundance of legumes.

Much research attention is being given to migration of species in response to the effects of climate change. Plants growing at the warmer/drier edge of the distribution of a species at low latitudes and elevations become especially vulnerable as temperatures and/or drought increase(s). Thus, the direction of migration of a species generally is toward increased latitudes and elevations. Migration may result in a species disappearing from one plant community and appearing in another one. The most drastic predictions for shifts in species composition of plant communities are that some tropical dry forests in Central and South America will become savannas and that some rainforests in Brazil will become tropical dry forests or savannas. Formation of a seed bank helps to ensure that a species remains in a community, although seeds may not be produced every year. However, unless the species with a seed bank can at least occasionally produce seeds that give rise to new plants the seed bank is only a temporary means of preventing the species from being extirpated from the community.

It is well known that migration of many plant species has occurred in the past, for example, in the Holocene following the retreat of glaciers in the Northern Hemisphere. However, in today's world migration of plant species from nonfavorable to favorable regions of the earth cannot be viewed as a solution to preventing extinction of species. The problem with migration being a way to help prevent species extinction is that the natural biomes in many parts of the world are now highly fragmented by cities, farms, and other human-created obstacles that greatly reduce the possibility of a species migrating.

Future research needs

A general conclusion from all the chapters in this book is that much more research remains to be done before we have a good understanding of the full impact of climate change on regeneration of plants from seeds. Furthermore, we need more accurate predictive models of the effects of climate change on plant species distributions in the future. However, enough information is available to help identify specific research questions that need to be addressed.

One important question about the effect of global warming relates to changes in pattern and amount of precipitation in relation to timing of seed germination and survival of seedlings. It is predicted that the germination temperature niche for seeds of a high proportion of species in various parts of the world is wide enough for seeds to germinate in a warming world, if sufficient water is available in the soil. However, increased temperatures could decrease reproduction of seagrasses from seeds and thus impact the future distribution range of at least some species. For terrestrial species, the big question is will enough water be present in the soil to ensure the survival and growth of seedlings to mature plants. This question is being asked about plant species in all vegetation regions on earth. The effect of global warming on water availability for seedling survival and growth includes the impact of early snow melt, which may be followed by summer drought. Although seedlings probably are more vulnerable to environmental stress than seeds, except for recalcitrant seeds, which are most common in tropical wet forests, more research is needed to determine the tolerance of seeds and seedlings to increased temperatures and drought. It is proposed that (1) the combined as well as individual effects of heat and drought stress on seeds and seedling be investigated, and (2) more attention be given to whole communities, or at least to major functional groups of plants within communities. Furthermore, in planning experiments weather data from sites where the study species are growing naturally should be considered/used. It is agreed that studies conducted in the field would enhance development of testable hypotheses about the effects of heat and drought stress on regeneration of plants from seeds.

There is an ongoing need for research on the effects of deposition of atmospheric N and elevated CO_2 concentrations on seedling survival and establishment, plant growth, and community species composition in all vegetation regions on earth. Thus, long-term field studies that evaluate the interaction of N and elevated CO_2 in relation to plant regeneration from seeds and species persistence in communities are needed. In tropical wet forests, the effects of N deposition and elevated CO_2 need to be evaluated in relation to low P availability.

Attention is being given to restoration and conservation in the context of increasing heat and drought stress, and two important points are made in this book about climate change and restoration/conservation. One, variation among populations of a species may ensure persistence of the species in at least some population sites as the climate changes due to global warming. Two, in selecting plant materials to use in restoration projects climate-adjusted provenance seed sourcing needs to be considered. This strategy differs from traditional seed sourcing for restoration, in which it is assumed that seeds from local populations yield plants that are the best adapted to the local sites. In the climate-adjusted provenance seed sourcing strategy, the restored population consists of plants from seeds collected from the warmer/drier edge of the species' range of distribution.

Authors of about half the chapters in the book suggest that future research on plant responses to climate change should involve various kinds of field studies and that temperature and soil moisture need to be monitored. The proposed field research includes the use of reciprocal transplants, common gardens at different elevations and latitudes, open-top warming chambers, and free air CO_2 enrichment techniques. These field studies need to consider seed germination and postgermination traits such as seedling survival, plant growth, and seed production. Long-term monitoring of plant communities would identify shifts in species abundance in the community and document species extirpations and immigrations. Furthermore, long-term monitoring of natural soil seed banks over environmental gradients is recommended.

One goal of studies on responses of plant species to climate change is to collect data that can be incorporated into climate change models to increase their accuracy. To fully accomplish this goal, much research must be done. For example, there are parts of the earth for which only a little or no information is available on responses of plants to climate change, for example, alpine zone in the tropics. There is a need to integrate research across all trophic levels to understand how climate change will affect the whole community, including pollinators, seed dispersers and predators, herbivores, pathogens, and mutualists such as mycorrhizae and symbiont N-fixers. An investigation of the reproductive traits of nonnative invasive species in response to increased temperatures and drought will help determine if these species have the potential for rapid adaptation to the changing climate. In fact, exotic plant species can provide insight into the rate of adaptation when species are exposed to changes in environmental conditions. Climate change is predicted to increase fire severity and frequency and modify time of the fire season. Thus, in fire-prone habitats more information is needed on the effects of postfire substrate conditions, temperature, and soil water availability on seed germination and seedling survival. In tropical wet forests, the hypothesis that there is a trade-off between seedling drought and shade tolerance needs additional testing. Furthermore, in these tropical forests how will global warming impact synchronized mass flowering and the survival and germination of recalcitrant seeds?

Much remains to be learned about the response of most plant species to the environmental stress caused by climate change. Do species have enough genetic variation to adapt to (track) climate change or enough phenotypic plasticity to "tolerate" the changed environment? What is the role of epigenetic modification in the response of species to climate change? Furthermore, do species exhibit transgenerational plasticity in traits related to regeneration from seeds? A better understanding of how persistent seed banks will influence the evolutionary response of plant species to environmental change is needed.

In various species, differences have been found in the degree of seed dormancy, especially PD for seeds collected from plants growing at different elevations and latitudes or in a variety of habitats. Genetic variation has been documented in dormancy-breaking and germination requirements of a few species, suggesting that the germination phase of the life cycle may have the potential to adapt to climate change. However, for most species known to exhibit variation in degree of seed dormancy it is not known if the differences are due to genetics (G), environment (E) of the mother plant during seed development, or to G × E interactions. If genetic variation occurs in a species, there is a possibility that increased temperatures and/or drought could select for ecotypes that could grow and reproduce under the stressful conditions. In the case of *Zea mays*, geneticists have selected cultivars/lines with increased heat and drought tolerance, thereby helping to ensure food security for humans.

Conclusions

The research reviewed in this book indicates that all types of vegetation on earth likely will be (are being) affected by climate change. Furthermore, all phases in the regeneration of plants from seeds, including seed production, dormancy-break and germination; seedling survival and establishment; and plant growth to maturity, may be affected by climate change. In general, it is predicted that temperature, fire frequency, and drought severity will increase, the results of which potentially are detrimental to the regeneration of at least some plant species from seeds in all types of vegetation on earth. However, in a few situations, for example, cold deserts of Central Asia, climate change is predicted to result in increased precipitation in spring, which watering experiments show would increase seedling survival and seed production.

The greatest threat from climate change to a high percentage of species, including terrestrial species with recalcitrant seeds and seagrasses, is that populations on the warmest and/or driest edge of the range may fail to regenerate from seeds due to increased temperature, drought severity, or other factors associated with climate change. Unless a species is clonal, failure to reproduce from seeds means that it eventually will be extirpated from the site. However, a persistent seed bank could delay extirpation of the species. Presumably, species that are being extirpated from the warm-dry part of their range can/will migrate to a relatively cool-moist region. However, for most species we do not know if migration can/will occur, especially since the landscape in many regions of the world is highly fragmented due to human activities. Furthermore, we do not know the degree of phenotypic plasticity and/or ability of most species to adapt genetically to changing environmental conditions.

Due to species extirpations and immigrations, one of the major effects of climate change on the regeneration of plants from seeds is predicted to be shifts in species composition of plant communities in all vegetation regions on earth. An unanswered question is: how will climate change affect species composition of communities qualitatively and quantitatively? Furthermore, what proportion of the immigrants into communities will be natives from nearby relatively warmer and drier communities, and what proportion will be invasive exotic weeds? Another consequence of climate change on the regeneration of plants from seeds is that the regions of the earth suitable for the production of annual food crops such as *Zea mays* are predicted to shift, resulting in some areas becoming less favorable and others more favorable for crop production than at present. One way to mitigate the problem of some land areas becoming unfavorable for crop production is development of heat- and drought-tolerant lines of crop species. Clearly, there remain many questions to answer and challenges to meet with regard to the effects of climate change on regeneration of plants from seeds.

Index

Note: Page numbers followed by "*b*," "*f*," and "*t*" refer to boxes, figures, and tables, respectively.

A

ABA. *See* Abscisic acid (ABA)
Above-ground net primary production (ANPP), 75–76
Abscisic acid (ABA), 215, 263
Acacia awestoniana, 119
Acacia erioloba, 175
Acacia raddiana, 54
Acanthoscelides macrophthalmus, 249
Acer saccharum, 138
Actinotus leucocephalus, 119
Adansonia digitata, 203
Adenostoma sparsifolium, 106
ADP-glucose pyrophosphorylase (AGPase), 218–219
Aesculus californica, 263
Aesculus hippocastanum, 261, 263
Africa
  crop failure, 213
  heat stress, effects of, 213, 214*t*
  maize, 213–214
    life cycle, climate change on, 215–219
    maize production, climate change on, 219–221
    miglio zaburro, 214
    mitigation, 221
    role of, 214–215
    statistical-mechanistic model, 219
    wild plant species, broad implications for, 221–222
AG. *See* Autumn-germinating (AG)
AGPase. *See* ADP-glucose pyrophosphorylase (AGPase)
Agriophyllum squarrosum, 79–81
Agropyron mongolicum, 79
Agrostis capillaris, 89
Ailanthus altissima, 249–250
Alliaria petiolata, 248, 250
Ambrosia artemisiifolia, 244–246, 248–249
Ammopiptanthus mongolicus, 78
Amygdalus mongolica, 79
Andersonia echinocephala, 118
Andropogon gerardii, 91, 93–94
Annual primary production (ANPP), 42
ANPP. *See* Above-ground net primary production (ANPP); Annual primary production (ANPP)
Antelope bitterbrush (*Purshia tridentata*), 67–68
Anticipated snowmelt (ASM), 6–7

Arabidopsis thaliana, 136, 222
Araucaria angustifolia, 263, 267
Archontophoenix alexandrae, 248, 265
Arctic-alpine plants, 11–12
  climate warming and plant regeneration from seeds in, 5–10
    ecological framework, 5
    seed dormancy and germination, 8–9
    seedling emergence and establishment, 9–10
    seed production and seed mass, 6–7
    time and space, seed dispersal in, 7–8
    Web of Science, 5
  climate warming in, 4
    effects of, 4–5
    model projections, 5
    physiological and phenological changes, 5
    summer warming, 4
  habitats, 3
  warming, rate of, 3
Arctostaphylos pungens, 108
Asclepias incarnata, 134, 136
Asclepias syriaca, 134
Asclepias verticillata, 134
Asian Water Tower, 145
ASM. *See* Anticipated snowmelt (ASM)
Aspalathus linearis, 122
Astragalus lehmannianus, 36
Atmospheric nitrogen (N) deposition, 81
Austrodanthonia caespitosa, 93–94
Austrostipa elegantissima, 232
Autumn-germinating (AG), 37
  Diptychocarpus strictus, 40
  Eremopyrum distans, 40
  Erodium oxyrhinchum, 40–41
Avena fatua, 247
Avena sterilis, 245
Avicennia marina, 262
Azorella madreporica, 121

B

Baccaurea ramiflora, 265
Banksia hookeriana, 236
Barro Colorado Island (BCI), 157–158
Basin wildrye (*Leymus cinereus*), 68
BCI. *See* Barro Colorado Island (BCI)
Beilschmiedia miersii, 121
Bidens pilosa, 245

Big sagebrush (*Artemisia tridentata*), 62–63, 65–66
Blackbrush (*Coleogyne ramosissima*), 62–63, 67
Boreal zone, 27–28
  large- and small-scale disturbances, 20–21
  literature search, 22, 22*b*
  predicted climate change and impact, 21–22
  seed dormancy and germination, 24–26
    Anemone nemorosa, 24
    cold stratification, 24
    Empetrum nigrum, 24
    germination-cueing mechanisms, 25
    interspecific and intraspecific levels, 25
    natural fire regimes, 26
    subarctic and subalpine Scandinavia, 25
    warming macroclimate, 25–26
  seedling survival and growth, 26–27
  seed production, 23
  seeds, recruitment from
    long-term consequences of climate, 27
    soil seed banks, 21
    tree layer and/or ground layer vegetation, 21
Bottlebrush squirreltail (*Elymus elymoides*), 68
Bouteloua gracilis, 93
Brachypodium hybridum, 50
Brachypodium pinnatum, 138
Brassica tournefortii, 237
Bromus erectus, 138
Bromus tectorum, 69, 161
Brownea coccinea, 263

C

Calluna vulgaris, 21, 25
Carapa surinamensis, 265
Carduus nutans, 246, 248
Castanea sativa, 260
Castanopsis hystrix, 260–261
Ceanothus cuneatus, 108
Celaenodendron mexicanum, 174
Cenostigma microphyllum, 173
Centaurea diffusa, 250
Centaurea solstitialis, 248, 250
Central Asia, cold deserts of, 43
  annual and seasonal precipitation, 34
  global warming, 34
  plant life history traits
    biomass accumulation/allocation, 37

Central Asia, cold deserts of (Continued)
  dry mass accumulation and allocation, 40–41
  heterodiaspory, importance of, 41–42
  phenology, 37–40
  plant growth and development, 37
  plant morphological characters, 40
  seed dormancy/germination, 34–36
  seedling survival and growth, 36–37
  seed production, 37
  seeds produced, number of, 41
  vegetation and community dynamics, 42–43
Cephalotus follicularis, 118–119
Cheatgrass or downy brome, 69–70
Cirsium pitcheri, 138–139
Cistus libanotis, 107
Clidemia hirta, 249
Climate change on plant regeneration from seeds, 61, 70
  arctic-alpine plants. See Arctic-alpine plants
  boreal zone. See Boreal zone
  Central Asia, cold deserts of. See Central Asia, cold deserts of
  current and future climates, 61–62
    mean annual precipitation, 62
    seasonal precipitation distribution, 61
  grasslands. See Grasslands, plants regeneration from seeds in
  hot deserts. See Hot deserts, plant regeneration by seeds in
  Mediterranean regions of, Northern Hemisphere. See Mediterranean regions of, Northern Hemisphere
  northern China steppes and semideserts of. See Northern China steppes and semideserts of
  seagrasses. See Seagrasses
  seed regeneration strategies and climate change, 62–70
    among-population, role of, 63
    antelope bitterbrush (Purshia tridentata), 67–68
    big sagebrush (Artemisia tridentata), 62–63, 65–66
    blackbrush (Coleogyne ramosissima), 62–63, 67
    genus Penstemon, 63–65
    invasive annual grass, 69–70
    native herbaceous dicots, 68–69
    native perennial grasses, 68
    population matrix models, 63
    process-based model, 62–63
    rubber rabbitbrush (Ericameria nauseosa), 66–67
    seed germination syndrome, 68
    shadscale (Atriplex confertifolia), 67
    within-population genetic variation, 63
  TDF. See Tropical dry forests (TDF)
  tropical wet forests. See Tropical wet forests
Coffea humilis, 262
Corydalis ambigua, 136
Crescentia alata, 175
Cynometra cauliflora, 263

D
Delphinium nelsonii, 23
Dichrostachys cinerea, 207
Diospyros melanoxylon, 173
Dipteryx alata, 189–190
Diptychocarpus strictus, 37, 41
Dombeya kirkii, 203

E
El Niño Southern Oscillation (ENSO), 158–159, 201
Empetrum nigrum, 24
ENSO. See El Niño Southern Oscillation (ENSO)
Eremosparton songoricum, 36–37
Erica multiflora, 107
Erodium oxyrhinchum, 36–37, 41
  AG and SG plants of, 37
Eucalyptus microcarpa, 120
Euterpe edulis, 260
Euterpe precatoria, 259

F
Fagus sylvatica, 106–107
Faidherbia (Acacia) albida, 203–205
Fallopia japonica, 245
Fire regime
  climate change, minimizing global effects of, 237
  direct fire-related effects, 229
  environmental filters, 229
  fire frequency
    obligate seeders, 234
    postfire obligate colonizers, seed dispersal and persistence of, 235
    seed bank depletion, 234–235
    shorter fire intervals, 234
  fire-prone regions, 230–231
    ameliorate impacts, 236–237
    fire season, 236
    genetic variation, sources of, 236
    hotter fires species, 235–236
    maternal environmental effects, 237
    seasonal rainfall environments, 236
  fire severity, 230–231
  indirect mechanisms, 229
  intensity and severity
    fire season, 232
    germination timing, 232–233
    hotter fires, 231
    obligate seeders, 231
    PFF species and fire seasonality, 233
    postfire seed predation and mortality, 233–234
    seeds and changes, survival of, 231–232
    seeds and changes to fire season, survival of, 233
  interacting impacts, 235
  interval squeeze, 235
  natural fire-prone ecosystems, 237
  obligate seeders, 231
  record-breaking fires, 230
Flowering phenology, 6

Fraxinus nigra, 138
Free AirCO$_2$ Enrichment (FACE), 87
Fumana thymifolia, 108

G
Galanthus nivalis, 132
General flowering (GF), 159
Germination syndromes, 70
  big sagebrush, 65–66
  Penstemon, 64
  rubber rabbit brush, 66–67
GF. See General flowering (GF)
Globularia alypum, 107
Grain (seed) filling, 218–219
  heat stress, 219
  seed development, reduction of, 218
  seed growth and development, phases of, 218
  stages, 218
  water stress, 218–219
Grasslands, plants regeneration from seeds in, 94–95
  climate change in, 87–88
  FACE experiments, 87
  natural or seminatural ecosystems, 87
  reproductive phenology and seed production
    advanced flowering and increased seed production, 91–92
    artificial warming, 90
    climate change impacts, 91
    CO$_2$ enrichment, effects of, 91
    comprehensive observational study, 90
    meta-analysis, 91
    phenological responses, 90–91
    Tibetan Plateau grasslands, 90
  seed germination
    Andropogon gerardii, 93
    annual herbaceous species, 92–93
    climate change, direct effects of, 92
    germination, meta-analysis of, 92
    invasive species, 92
    parental climate change, effects of, 92–93
    parental warming and CO$_2$ enrichment, combination of, 93, 94t
    Poa annua, plants of, 93
    seedling recruitment, success of, 93
  seed physiology, 93–94
  species composition and population dynamics, 88–89
    biomass production and density of forbs, 89
    bunchgrasses, 89
    C3 and C4 species, 88–89
    sod-forming graminoids, 89
    vegetative reproduction, 89
    warming and CO$_2$
      effects of, 88
  types, 87
Gunnera tinctoria, 248–249

H
Halophila decipiens, 278–279
HDi-LDo. See High dispersal–low dormancy (HDi-LDo)

Heat stress, 216
Heisteria parvifolia, 206
Helenium aromaticum, 236
High dispersal—low dormancy (HDi-LDo), 41—42
Hordeum comosum, 121
Hordeum spontaneum, 245
Horsfieldia pandurifolia, 261
Hot deserts, plant regeneration by seeds in, 48—49, 55—56
   decreasing precipitation, effects on
      bimodal precipitation, 55
      dormancy and seed banks, 54
      germination, 54
      interact with increasing temperatures, 54—55
      precipitation variability, 54—55
   increasing temperature, effects on, 49—51
      dormancy and seed banks, 50
      germination, 51
      seed survival, 49—50
   nitrogen deposition, 55
Hydrophyllum capitatum, 90
Hyparrhenia rufa, 92

I

Ilex aquifolium, 107
Intergovernmental Panel on Climate Change (IPCC), 75
Invasive plant species, 250—251
   adaptation of, 244
   aquatic ecosystems, 244
   biology, aspects of, 251
   biotic interactions of, 249—250
      Ailanthus altissima, 250
      Alliaria petiolata, 250
      Centaurea diffusa, 250
      Centaurea solstitialis, 250
      Eragrostis plana, 249
      extreme climatic events, 249—250
      long-distance seed dispersal, 249
      Microstegium vimineum, 250
      Prunus serotina, 250
   characteristics, 243
   climate change mitigation, 250
   invasive nonnative species, 243
   mating systems and phenology, 245
      Bidens pilosa, 245
      Fallopia japonica, 245
      global warming impacts, 245
      Lythrum salicaria, 245
      pollen limitation, risk of, 245
   meta-analysis, 244
   phenotypic plasticity, 243
   polyploidy, 244
   resilience, 244
   seed dormancy
      Alliaria petiolata, 246—248
      Ambrosia artemisiifolia, 248
      Archontophoenix alexandrae, 248
      Cardamine hirsuta, seeds of, 246
      cool moist conditions, 247
      high summer temperatures, 247
      maternal plant environment, 247—248
      nondeep PD, 246—247
      nonnative-warmer environment, 248
      Oenothera biennis, 248
      Parthenium hysterophorus, 247
      Phragmites australis seeds, 246
      Piper aduncum, 248
      seed germination and viability, 247—248
      Taraxacum brevicorniculatum, 247—248
      Taraxacum laevigatum, 247—248
      Taraxacum officinale, 247—248
      Tithonia diversifolia, 248
   sexual and asexual reproduction, 243
   sexual reproductive capacity and seed dispersal, 245—246
      Microstegium vimineum, 246
      plant migration rates, 246
      seed mass, 246
      seed size, 245—246
   soil seed banks, 248—249
IPCC. *See* Intergovernmental Panel on Climate Change (IPCC)
Ipomopsis tenuituba, 23
Isoberlinia angolensis, 203, 205

J

Julbernardia globiflora, 203, 208—209

K

Kalahari-Highveld Regional Transition Zone (KHRTZ), 199
KHRTZ. *See* Kalahari-Highveld Regional Transition Zone (KHRTZ)
Kobresia humilis, 147
Koeleria cristata, 76—78
Koenigia islandica, 147

L

Lannea edulis, 203, 205, 207
Laperousia rivularis, 208—209
LDi-HDo. *See* Low dispersal-high dormancy (LDi-HDo)
Leucaena leucocephala, 249
Leucocoryne dimorphopetala, 121
Leymus chinensis, 93
Ligustrum robustum, 249
Limonium girardianum, 105
Limonium narbonense, 105
Limonium santapolense, 105
Limonium virgatum, 105
Linum perenne, 68—69
Litsea pierrei var. *szemaois*, 261
Lonicera maackii, 134
Lotus corniculatus, 94
Low dispersal—high dormancy (LDi-HDo), 41—42
Lupinus polyphyllus, 250
Lythrum salicaria, 245

M

Maize, Africa
   life cycle, climate change on, 215—219
      gamete formation, 216
      germination to tasseling, plant growth from, 216
      grain (seed) filling, 218—219
      grain (seed) formation, 217—218
      pollination and fertilization, 216
      seed dormancy/germination, 215—216
   maize production, climate change on, 219—221
   mitigation, 221
   statistical-mechanistic model, 219
   wild plant species, broad implications for, 221—222
Medicago lupulina, 94
Mediterranean regions of, Northern Hemisphere
   altered fire regimes, 105—106
   altered precipitation regimes and germination, 105
   amount of precipitation, 101—102
   Anthropocene, 109
   climate change, temporal and spatial effects of, 101—102, 102f
   cold-cued germination, 109
   cold-wet stratification, 104b
   decreased cold stratification, 103
   facilitation and drought stress, 107—108
   global warming clashes, 109
   harsher summers, 109—110
   increased germination temperatures, 104—105
   local adaptation and phenotypic plasticity, 108
   Mediterranean germination syndrome, 104b
   mild-wet winters and hot-dry summers, 101—103
   mild winters, 103
   sea level rise and salinity stress, 105
   seed dormancy-breaking and germination requirements, 103
   seedling survival, 106—107
      cold-adapted tree species, 107
      hotter and drier summers, 106
      local edaphic conditions, 106
      Quercus ilex, survival of, 106
      seedling establishment, 106—107
   seed production, 103
   warmer winters, 109—110
   warm-wet stratification, 104b
   wildfire intensity and frequency, 102—103
Mediterranean temporary ponds (MTPs), 109b
Mediterranean-type ecosystems (MTEs), 115—116
   climate change impacts, 123—124
   Southern Mediterranean regions. *See* Southern Mediterranean regions
Megathyrsus maximus, 92
Melinis minutiflora, 92
Mesua ferrea, 173
Microstegium vimineum, 250

Milium effusum, 138
Mimosa leiocephala, 188
Mimosa pigra, 247
Morphophysiological dormancy (MPD), 24, 36
MPD. *See* Morphophysiological dormancy (MPD)
MTEs. *See* Mediterranean-type ecosystems (MTEs)
MTPs. *See* Mediterranean temporary ponds (MTPs)
Myracrodruon urundeuva, 173

N

Narcissus pseudonarcissus, 132
Native herbaceous dicots, 68−69
  Lewis flax, 68−69
  seed germination syndrome, 69
Native perennial grasses
  basin wildrye (*Leymus cinereus*), 68
  bluebunch wheatgrass (*Pseudoroegneria spicata*), 68
  bottlebrush squirreltail (*Elymus elymoides*), 68
Natural experiments, 23
NDVI. *See* Normalized difference vegetation index (NDVI)
Neurachne alopecuroidea, 118
Nitraria tangutorum, 79−81
Normalized difference vegetation index (NDVI), 42
Northern China steppes and semideserts of
  ANPP, 75−76
  atmospheric nitrogen deposition, effect of, 81
  climate warming, effect of, 76−79
  dynamic land ecosystem model, 75−76
  germination requirements, 76
  laboratory tests or green house experiments, 83
  mean annual temperature, changes in, 75
  model-based analysis on, 76
  precipitation, change in, 75−76
  precipitation, projected climate warming and changes in, 83
  salt stress, effect of, 81−83
  temperature and precipitation, 76
N resorption efficiency (NRE), 151

O

Ocotea odorifera, 263
Oenocarpus bataua, 265
Oenothera biennis, 248
Open-top chambers (OTCs), 135, 150
Ormosia macrocalyx, 163
OTCs. *See* Open-top chambers (OTCs)

P

Parinari excelsa, 203
Parkinsonia aculeata, 247, 249
Parthenium hysterophorus, 247
Parthenocissus quinquefolia, 134
PD. *See* Physiological dormancy (PD)

Pedicularis fletcheri, 148
Penstemon, 63−65
Penstemon digitalis, 134
Phragmites australis, 246
Physalis longifolia, 134
Physiological dormancy (PD), 24, 246, 301−302
Picea mongolica, 76−78
Piliostigma thonningii, 208
Pinus jeffreyi, 107
Pinus nigra, 107−108
Pinus sylvestris, 78, 107−108
Piper aduncum, 248
Plants regeneration from seeds, 305
  Central Asia, cold deserts of, 305
  Central Chile
    anthropogenic fires, 120
    indirect impacts, multitude of, 121
    lightning-ignited (natural) fires, 120
    Placea (Amaryllidaceae), 120
    seed densities, 121
  drought stress, 301
  fire regime. *See* Fire regime
  genetic variation, 304−305
  in grasslands. *See* Grasslands, plants regeneration from seeds in
  heatwaves, 301
  Mediterranean regions of, Northern Hemisphere. *See* Mediterranean regions of, Northern Hemisphere
  N and elevated CO_2 interaction of, 304
  northern China, steppes and semideserts of. *See* Northern China steppes and semideserts of
  plant communities, species composition of, 303
  restoration, 304
  seed dormancy and germination, 301−302
  seedling survival and growth, 302−303
  snow-free growing season, 301
  South Africa Cape Region, 121−123
    demographic distribution models, 122
    Erica, 122
    extreme weather events, 123
    heat and drought impacts, 123
    low-lying areas, 122
    MTE, 123
    Oxalis, seeds of, 123
    Proteaceae species, 121−122
    Proteoid species, 121−122
    seedling recruitment, reductions in, 121
    soil-stored seeds, 121
    succulent karoo community, 122−123
  South American savannas, Brazilian Cerrado. *See* South American savannas, Brazilian Cerrado
  southwestern and southern Australia, 117−120
    Acacia awestoniana, 119
    Andersonia echinocephala, 118
    Banksia woodlands, 119−120
    Cephalotus follicularis, 118−119
    Eucalyptus species, 120

    native and nonnative woodland species, 119
    Neurachne alopecuroidea, 118
    seed age and air temperature, 118−119
    serotinous species, 119
    South Australian *Brachyscome* species, 118−119
    Sphenotoma drummondii, 118
    water-stressed native species, 120
  TDF. *See* Tropical dry forests (TDF)
  temperate deciduous forest zone. *See* temperate deciduous forest zone
  terrestrial species, 303−304
  Tibet Plateau in China. *See* Tibet Plateau in China
  tropical forests, 304
  tropical wet forests. *See* Tropical wet forests
  Zea mays, 304−305
Podocarpus angustifolius, 267
Posidonia oceanica, 278
Postfire flowering (PFF), 233
Primula alpicola, 148
Prosopis juliflora, 51
Prosopis laevigata, 55
Prunus serotina, 250
Pterocarpus macrocarpus, 173
Puccinia psidii, 120
Pueraria lobata, 243
Pycnonotus jocosus, 249

Q

Quercus douglasii, 106
Quercus ilex, 107−108, 267
Quercus kelloggii, 106
Quercus mongolica, 267
Quercus nigra, 263
Quercus petraea, 136
Quercus robur, 132, 260
Quercus schottkyana, 267

R

Racinaea aerisincola, 189
Ranunculus parviflorus, 134−135
Recalcitrant-seeded species, regeneration of, 267−268
  Caribbean woody species, 260
  Castanea sativa, 260
  conventional seed bank procedures, 259
  desiccation-sensitive seeds, 260
  dormancy present, 262−263
  Euterpe edulis, 260
  Euterpe precatoria, 259
  greenhouse gasses, emission of, 260
  precipitation and temperature, germination in relation to, 263−264
  Quercus robur, 260
  regeneration and light, 260−261
  regeneration in *Coffea* species, 262
  seasonally dry tropical vegetation, precipitation in, 264−266
  Carapa surinamensis, 265
  Castanea species, 264−265

Dipterocarpaceae, 265
Oenocarpus bataua seeds, 265
Quercus species, 264–265
Vitellaria paradoxa, 264
seed and seedling banks, 267
seed development, 261–262
Shorea wantianshuea, 259
stopping germination during seed storage, 266–267
hydrated-storage, 266
ROS, 266
storage temperature, 266–267
systematic analyses, 259
Relative growth rate (RGR), 174
RGR. *See* Relative growth rate (RGR)
Rhinanthus minor, 24
Rubber rabbitbrush (*Ericameria nauseosa*), 66–67

S

Sanguisorba minor, 93
Sarcopoterium spinosum, 107
Sclerocarya birrea, 207
Seagrasses
   climate change, 280
      environment, changes in, 280
      seed-based restoration, 280
   evolution of, 276–277
   extreme climate events, 275
   flowering, 278–279
   losses of, 275
   marine flowering plants, 275
   marine heatwaves, 275
   ocean warming, 275
   seagrass mating systems, 277
   seed-based restoration strategies, 275–276
   seed dispersal, 279
   seed germination, 278–279
   seed production, 278–279
   seed settlement and early seedling survival, 279–280
Securidaca longepedunculata, 207
Seed coat ratio (SCR), 265–266
Seed dispersal, 7–8
Seed ecological spectrum of arctic-alpine plants
   ecological framework, 5
   seed dormancy and germination, 8–9
   seedling emergence and establishment, 9–10
   seed production and seed mass, 6–7
   time and space, seed dispersal in, 7–8
   Web of Science, 5
Seed germination syndrome, 8, 68
Senecio inaequidens, 250
Senegalia polyacantha, 207
Senna multijuga, 188
SG. *See* Spring-germinating (SG)
Shadscale (*Atriplex confertifolia*), 67
Shorea leprosula, 159–160
Shorea macroptera, 159–160
Shorea wantianshuea, 259
Soil (sand) water content (SWC), 37
South American savannas, Brazilian Cerrado
   annual precipitation, changes in, 192
   anthropic factors, 192–193
   Cerrado vegetation, 183–184
   climate change effects in, 184–185
   germination ecology, 185–187
      closed vegetation, 186–187
      regeneration stages, 185b
      soil seed banks, 185
   hydrothermal time models, 190–191
   increased fire frequency, 187–188
      fire-sensitive vegetation, 188
      heat shock treatments, 188
      heat tolerance, 187–188
      smoke-stimulated effects, 188
   intraspecific variation, 191–192
   mean annual rainfall, decrease in, 188–189
   mean temperature, increase in, 188
   plant populations and communities, climate change on, 189–190
      species abundance and community composition, changes in, 190
      species distribution, changes in, 189–190
   seed germination and dormancy, 183
Southern Africa, savanna woodlands of, 209
   climate variable, 199–200
   data acquisition and analytical approach, 200–201
   field observational studies, 201–203
   fruit production and climate, 203–206
      Adansonia digitata, 203
      Dombeya kirkii, 203
      Faidherbia (*Acacia*) *albida* tree, 203–205
      Isoberlinia angolensis, 203, 205
      Lannea edulis, 205
      Ledebouria sp., 206
      Parinari excelsa, 203
      Senegalia (*Acacia*) *mellifera*, 203
      Strychnos spinosa, 205
      Uapaca kirkiana flowers, 203, 205f
   KHRTZ, 199, 200f
   local climate trends in, 201
   regional climate trends in, 201
   seedling emergence and climate
      Dichrostachys cinerea, 207
      Lannea edulis, 207
      Laperousia rivularis, 208
      Piliostigma thonningii, 208
      Sclerocarya birrea, 207
      Securidaca longepedunculata, 207
      Senegalia polyacantha, 207
      Strychnos spinosa, 208
      Ziziphus abyssinica, 208
   seedling mortality and climate, 208–209
   ZRCE, 199, 200f
Southern Mediterranean regions, 123–124
   global warming
      annual rainfall ranges, 116
      Cape Region of South Africa, 117
      Central Chile, 117
      southwestern and southern Australia, 116–117
      winter growing conditions, 116
   plant regeneration from seeds
      Central Chile, 120–121
      South Africa Cape Region, 121–123
      southwestern and southern Australia, 117–120
   Southern Hemisphere of, MTEs, 115–116
      fire role, 116
      plant migration, 115
      seed germination, 116
      shrublands and heathlands, 115
Sphenotoma drummondii, 118
Spondias mombin, 174
Spring-germinating (SG), 37
   Diptychocarpus strictus, 40–41
   Erodium oxyrhinchum, 40
Spring not-watered (SNW), 37
Stachys sylvatica, 138
Stipa tenacissima, 104–105
Stirlingia latifolia, 233
Streptanthus tortuosus, 108
Strychnos benthamii, 263
Strychnos spinosa, 208
Swartzia langsdorffii, 189, 267
SWC. *See* Soil (sand) water content (SWC)

T

Taraxacum brevicorniculatum, 247–248
Taraxacum laevigatum, 247–248
Taraxacum officinale, 247–248
TDF. *See* Tropical dry forests (TDF)
Temperate deciduous Forest zone, 138–139
   CO_2
   concentration of, 131–132
   elevated CO_2, 136
   geographical range shifts, 137–138
   physical dormancy, 131
   precipitation, changes in, 134–135
   seed mass, 138–139
   seed production, 136–137
   snow cover, 131–132
   snow cover reduction, 135–136
   soil seed banks, 137
   temperature, changes in, 132–134
      artificial warming, 132
      cold stratification period, 134
      dormancy breaking period, 134
      earlier spring germination, 133
      germination phenology, 132–133
      Type 2 nondeep physiological dormancy, 132–133, 133f
   warming and changes in moisture with biotic factors, interaction of, 135
   winter deciduous trees, 131
   winter temperatures, 131–132
Tephrosia sphaerospora, 49–50
Terminalia chebula, 173
Tibet Plateau in China, 151
   broad-leaved forest, 145
   climate change, 146–147
   climate warming, 146
   N deposition, 147
   community structure and composition, change in, 150–151
      alpine ecosystems, 150–151
      N addition, 151
      NRE, 151
      OTCs, 150
      seeds, dormancy breaking of, 151
      warming, effect of, 151
   coniferous forest, 145

Tibet Plateau in China (*Continued*)
  seed dormancy and germination, 148–149
    cold-wet conditions, 148–149
    dry-warm conditions, 149
    soil moisture, 149
  seedling emergence and establishment, 149–150
  seed production, 147–148
    alpine meadow, 147
    elevated CO_2, 147
    environmental factors, variation in, 148
    extreme climatic events, effect of, 148
    grazed meadow, 147
    increased precipitation, 147–148
    soil seed bank, 148
    vegetation types, 145, 146f
Tithonia diversifolia, 248
Trachypogon plumosus, 92
Tradescantia ohiensis, 136
Triadica sebifera, 249
Trifolium pratense, 93
Tropical dry forests (TDF), 176
  flowering, 171
  predicted climate changes, 170–171
  seed dormancy and germination, 171–173
    orthodox seeds, 172
    physical and physiological dormancy, 172–173
    recalcitrant seeds, 172
    seed attributes, 171
    temperature, effect of, 173
  seedling growth and survival, 174
  seed production and dispersal, 171
  soil seed bank, 174–175
    Acacia erioloba, 175
    increased air temperature, 175
    open- *vs.* closed-canopy sites, 175
    precipitation events, 175
    role of, 174
    species distribution, community composition and shifts in, 176
  structural and functional characteristics, 169
  β diversity values, 169
Tropical wet forests, 163–164
  elevated CO_2 and nutrient limitation, 163
  elevated temperature and drought, seed responses, 160–161
    desiccation-sensitive seeds, 161
    fungal infections, 161
    maternal plant, environmental conditions of, 160
    physical dormancy, 160–161
    short droughts, 161
    soybean plants, 160
    subtropical and temperate distributions, 160
    tropical rain forests, 160
    warmer conditions, 160
  global carbon cycle, 157
  global warming, impacts of, 157
  nutrient availability, climate change on, 163
  observed and predicted climate change in, 158–159
  reproductive phenology, climate change on, 159–160
    environmental factors, temporal fluctuation of, 159
    GF events, 159
    SE Asia, warming in, 160
    Shorea macroptera, 159–160
    supraannual flowering events, 159
  seedling responses to increased temperatures, 161–162
    carbon loss, 162
    VPD, 161–162
  seedlings, drought tolerance and shade tolerance trade-offs for, 162
  seeds, critical regeneration stages of, 158
  species composition in, 157–158
  wet tropical forests, features of, 163–164

U

Uapaca kirkiana, 203
Urochloa decumbens, 92

V

Vaccinium myrtillus, 23
Vapor pressure deficit (VPD), 161
Vitellaria paradoxa, 264
Vochysia alata, 187–188
VPD. *See* Vapor pressure deficit (VPD)

W

Warming climate
  global warming, 285
  long-term implications, 291–292
  microclimatic conditions, 292
  microenvironmental conditions, 292
  plant distribution, 285
  seed bank persistence and density, 286–288, 291–292
    Arctic and Antarctic ecosystems, 289–290
    demographic compensation mechanisms, 288
    direct and indirect effects of, 286, 287f, 291–292
    mountain ecosystems and elevation gradients, 288–289
    parental environmental effects, 287
    plant migration, 290–291
    short-term environmental variations, 286
    temporary resilience, 290
    wetlands and grasslands, 287–288
  terrestrial plants, 285–286
Water stress, 216, 217t, 218

X

Xanthorrhoea preissii, 233

Z

Zambezian Regional Centre of Endemism (ZRCE), 199
Zea mays, 215–216, 304–305
Zea perennis, 215
Ziziphus abyssinica, 208
Zostera marina, 277–278, 280
ZRCE. *See* Zambezian Regional Centre of Endemism (ZRCE)

Printed in the United States
by Baker & Taylor Publisher Services